SALSA PICANTE (HOT SAUCE)

2 SMALL TOMATOES, PEELED (½ lb)
¼ C. CHOPPED ONIONS
1 GARLIC CLOVE
½ TO 1 CANNED PICKLED JALAPEÑO CHILE
1 t VEGETABLE OIL
½ t LIQUID FROM CHILES
¼ t DRIED LEAF OREGANO, CRUSHED
¼ t SALT

COMBINE TOMATOE, ONION, GARLIC + CHILE IN BLENDER
OR FOOD PROCESSOR. BLEND UNTIL PUREED. HEAT OIL IN
SMALL SAUCEPAN, ADD TOMATO MIXTURE, CHILE LIQUID,
OREGANO + SALT. BRING TO BOIL. COOK 10 MIN. LET SAUCE
STAND FOR 2 HRS. MAKES ABOUT ¾ CUP.

MECHANICAL METALLURGY
principles and applications

MARC ANDRÉ MEYERS

Department of Metallurgical and Materials Engineering
New Mexico Institute of Mining and Technology
Socorro, New Mexico

KRISHAN KUMAR CHAWLA

Center for Materials Research
Military Institute of Engineering
Rio de Janeiro, Brazil

Prentice-Hall, Inc., Englewood Cliffs, New Jersey 07632

Library of Congress Cataloging in Publication Data

MEYERS, MARC A. (date)
  Mechanical metallurgy.

  Includes bibliographies and index.
    1. Metals—Mechanical properties.   2. Physical
metallurgy.   I. Chawla, Krishan Kumar, (date)
  II. Title.
TA460.M466 1984   620.1′6   83–552
ISBN–0–13–569863–4

Editorial /production supervision and
  interior design: *Bette Kurtz /Theresa Soler /Natalie Krivanek*
Cover design: *Edsal Enterprises*
Manufacturing buyer: *Anthony Caruso*

Cover photograph courtesy of the General Electric Research and Development Center,
Michael A. Hemberger, photographer.

Printed in the United States of America

10   9   8   7   6   5   4   3   2

ISBN 0-13-569863-4

Prentice-Hall International, Inc., *London*
Prentice-Hall of Australia Pty. Limited, *Sydney*
Editora Prentice-Hall do Brasil, Ltda., *Rio de Janeiro*
Prentice-Hall Canada Inc., *Toronto*
Prentice-Hall of India Private Limited, *New Delhi*
Prentice-Hall of Japan, Inc., *Tokyo*
Prentice-Hall of Southeast Asia Pte. Ltd., *Singapore*
Whitehall Books Limited, *Wellington, New Zealand*

TO OUR CHILDREN

Marc Henri Patrick Meyers
Maria Cristina Alexandra Meyers
Nikhilesh Chawla
Kanika Chawla

CONTENTS

* Sections marked with a ‡ indicate recommended senior-level one-semester course material.

Contents   **ix**

FOREWORD

No characteristic of a metal is more important than its response to applied stresses. The use of metals in every phase of modern civilization relies on their ability to withstand the stresses encountered in service without breaking, plastically deforming, or weakening. We are constantly aware that machinery, engines, tools, structures both large and small, and instruments of all kinds must be designed with a knowledge in mind of the mechanical properties of the metals they contain and property changes that might occur in service. Even before entering service, the mechanical characteristics of the constituent metals are basic to the fabrication and assembly of parts from castings or plates, sheets, bars, rods, and wires, which, in turn, have already been shaped by suitable mechanical processes.

The tremendous variety of applications in which mechanical properties are critical and the vast numbers of different alloys, old and new, used for these applications have stimulated countless mechanical metallurgy investigations and accumulated mountains of specific data. To correlate these data and deduce the underlying general principles there have been continuing developments on the theoretical side, with input from the basic sciences of physics, chemistry, and mathematics. There has also been a determined effort to discover the basic mechanisms that operate when a metal responds to an applied stress and to devise theories and equations that can quantitatively account for this response and predict it under all conditions that will be encountered in service.

The aim has been to develop knowledge and quantitative understanding not only in the range of size of objects of concern to engineers, manufacturers, and users, but also in the microscopic and atomic size range, where the basic events take place that govern the engineering properties. The extent to which this aim has been achieved in the twentieth century, especially in the last few decades, has been truly remarkable.

The magnitude of the task, with its background in the basic sciences, applied sciences, and engineering, and its reliance on both experimental and theoretical research, has brought about a vital need for all kinds of reviews, summaries, data tabulations, handbooks, and textbooks.

Any textbook that is written by authors familiar with the complexities of mechanical metallurgy on both the empirical and the theoretical sides is to be welcomed. Especially welcome is a textbook such as this, written by two authors who have a firm grasp not only of the older concepts and facts, but also of the important newer concepts, and who have had experience in teaching the subject in the classroom and are aware of the difficulties students often have in understanding some of the ideas.

Charles S. Barrett

PREFACE

Rather than a technical or scientific discipline, mechanical metallurgy is an interface. It uses the principles of mechanics and of metallurgy with the objective of rationalizing, predicting, modifying, and describing the response of metals to loads. The volume of knowledge and the number of methods encompassed by mechanics and metallurgy are immense. Hence great care has to be taken in choosing the material that is to be included in a mechanical metallurgy text. Cohen et al.* have proposed a schematic framework that explains the complex relationships that form the field of materials. The central theme is the structure—properties—performance triangle. These three entities are biuniquely related and changes in one are inseparably related to changes in the others; these changes are introduced by processing. Both scientific and technical developments contribute to the pool of knowledge.

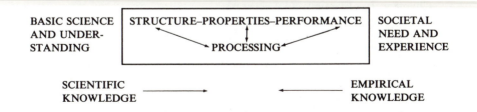

* M. Cohen, B. H. Kear, and R. Mehrabian, "Rapid Solidification Processing—An Outlook," *Rapid Solidification Processing, Principles, and Technologies,"* ed. M. Cohen, B. H. Kear, and R. Mehrabian, Claitor's Publ. Division, Baton Rouge, La., 1980, p. 1.

In this framework, mechanical metallurgy represents the portion of the linkage in which mechanical properties and performance are involved.

This book was written to serve as a textbook for a mechanical metallurgy course. Virtually all metallurgy and materials programs have added to their curriculum a course on mechanical metallurgy, usually taught at the senior level. Aware of the diversity of topics covered under this heading, we have incorporated into this book much more material than can possibly be covered in one semester. Additionally, we have included a considerable amount of background material, which should prove useful to nonmetallurgy students. We have opted for a quadrangular structure, dividing the book into:

Part I. Mechanics of Deformation (Chapters 1, 2, 3). Deformation is treated from a primarily mechanical point of view, in the sequence in which it takes place: elastic deformation, plastic deformation, and fracture.

Part II. Metallurgy of Deformation (Chapters 4, 5, 6, 7, 8). The structural imperfections responsible for the mechanical properties of metals are defined, discussed, and analyzed.

Part III. Strengthening Mechanisms (Chapters 9, 10, 11, 12, 13, 14, 15). Metallurgists have used their ingenuity to an incredible extent to obtain a wide range of mechanical responses of metals. The different strengthening mechanisms are treated in separate chapters.

Part IV. Mechanical Testing (Chapters 16, 17, 18, 19, 20, 21). The vital linkage between mechanical properties and performance is provided by mechanical testing. The various mechanical properties and testing procedures are studied.

The choice of material for a mechanical metallurgy course rests, of course, on the instructor. We have written this textbook at the Military Institute of Engineering, South Dakota School of Mines and Technology, and New Mexico Institute of Mining and Technology and have found the following sequence of subjects very satisfactory (the sections are marked with a ‡ in the text and index):

UNDERGRADUATE MECHANICAL METALLURGY COURSE
(TOTAL: 42 LECTURES)

Chapter 1: Elasticity (4 Lectures)

  A. Elementary Treatment (Part A)

Chapter 2: Plasticity (5 Lectures)

    Introduction (Section 2.1)
    Flow Criteria (Section 2.2)
    Deformation Processing (Sections 2.5.1, 2.5.2, 2.5.3)

Chapter 3: Fracture and Fracture Toughness (5 Lectures)

    Introduction (Section 3.1)
    Stress Concentration and Griffith Criterion of Fracture (Sections 3.2.1, 3.2.2, 3.2.3)

Chapter 17: Hardness Testing (2 Lectures)

Introduction (Section 17.1)
Macroindentation Tests (Section 17.2.1)
Microindentation Tests (Section 17.2.2)

Chapter 18: Formability Testing (1 Lecture)

Important Parameters (Section 18.1)
Punch–Stretch Tests and Forming-Limit Curves
(Section 18.5)

Chapter 19: Fracture Testing (2 Lectures)

Why Fracture Testing (Section 19.1)
Impact Tester (Section 19.2)
Charpy Impact Test (Section 19.3)

Chapter 20: Creep and Creep Testing (3 Lectures)

Introduction (Section 20.1)
Correlation and Extrapolation Methods (Section 20.2)
Deformation-Mechanism Maps (Section 20.4)
Heat-Resistant Alloys (Section 19.5)

Chapter 21: Fatigue and Fatigue Testing (2 Lectures)

Conventional Fatigue Tests (Section 21.2.1)
Non-Conventional Fatigue Testing (Section 21.2.2)

For a graduate course in mechanical metallurgy there is much more free-dom of material; nevertheless, we recommend the following sequence.

GRADUATE MECHANICAL METALLURGY COURSE

Chapter 1: Elasticity

B. Advanced Treatment (Sections 1.2 through 1.7)

Chapter 2: Plasticity (Sections 2.1 through 2.5)

Chapter 3: Fracture and Fracture Toughness (If no course on fracture mechanics is offered, should be completely covered; otherwise, should be left out)

Chapter 6: Line Defects (Should be covered completely if no course on dislocation theory is offered; otherwise, should be left out)

Chapter 8: Additional Dislocation Effects and Geometry of Deformation (Sections 8.1 through 8.7)

Chapter 20: Creep and Creep Testing (In case Chapters 3 and/or 6 are not covered)

Chapter 21: Fatigue and Fatigue Testing (In case Chapters 3 and/or 6 are not covered)

Although not as common as mechanical metallurgy, courses on strengthen-ing mechanisms are part of the offerings of a number of metallurgy/materials departments. This book would serve well as a text for such a course; indeed, it has been class-tested, and the following sequence is recommended.

STRENGTHENING MECHANISMS COURSE

Chapter 6:   Line Defects

               Dislocation Sources (Section 6.8)

Chapter 7:   Planar Defects

               Grain Boundaries (Section 7.2)

Chapter 8:   Additional Dislocation Effects and Geometry of Deformation

               Stress Required for Slip (Section 8.6.2)

The three chapters above provide background material needed to treat mathematically and fully understand the strengthening mechanisms. This introduction should be followed by the full and detailed coverage of Chapters 9 through 15.

This book is based on the class notes that we have developed during the last 9 years. We have enriched these notes with materials from many sources. These sources are cited throughout the text, and each chapter contains a list of references. Additionally, there are suggested readings at the end of the chapters; readers should consult the references if they need to expand a specific point and the suggested reading if they want to broaden knowledge in an area. Full acknowledgment is given in the text to all sources of tables and illustrations. To save space, the figure captions and table headings reprinted or adapted from other sources have a reference number and page number, the source being given in the reference list. We might have inadvertently forgotten to cite some of the sources in the final text; we sincerely apologize if we have failed to do so.

Our students provided, by their intelligent questions and valuable criticisms, the most important input to the book; we are very grateful for their contributions. We would like to thank our colleagues and fellow scientists who have, through painstaking effort and unselfish devotion, proposed the concepts, performed the critical experiments, and developed the theories that form the framework of an emerging quantitative understanding of the mechanical behavior of metals. The patient and competent typing of the manuscript by Elizabeth Fraissinet and drafting by Robert Colpitts is gratefully acknowledged. Krishan Chawla would also like to thank his wife, Nivi, for her patience. Marc Meyers acknowledges the continued support of the National Science Foundation through the grants: DMR-7728278, DMR-7927102, and DMR-8115127. The inspiration provided by his grandfather, Jean-Pierre Meyers, and father, Henri Meyers, both metallurgists who devoted their lives to the profession, has given Marc Meyers the strength to complete this work.

Marc André Meyers
Krishan Kumar Chawla

Part I

MECHANICS
OF DEFORMATION

Chapters 1, 2, and 3, which compose Part I, treat the deformation of metals in the sequence in which it takes place, when subjected to an external load: elastic (recoverable) deformation, plastic (nonrecoverable) deformation, and fracture. The phenomena involved in these three different regimes are very different. Chapter 1 introduces the theory of elasticity at both an elementary and an advanced level; the two treatments are intended for undergraduate and graduate students, respectively. The advanced treatment uses the tensorial approach to describe stresses and strains; it is very helpful in more complex problems, such as stress fields around cracks and dislocations. Chapter 2 starts with flow criteria, which establish the stress level at which one goes from an elastic to a plastic regime. The theoretical treatment of plasticity is kept at a minimum. The Levy–von Mises theory, which is very helpful in a number of analyses of plasticity problems, is briefly presented. Deformation processing is discussed at length, starting with a general description and classification of the processes. Stress analyses (using the equilibrium-of-forces method) are conducted for specific processes: rolling, wire drawing, forging. Chapter 3 explores the mechanical and microstructural aspects of fracture. The Griffith criterion, applicable to brittle solids, is derived. Linear elastic fracture mechanics is introduced; the stress fields at the vicinity of the crack tip are calculated; and the concepts of fracture toughness, crack extension force, crack opening displacement, J integral, and R curve are presented.

ELASTICITY

Chapter 1

‡1.1 INTRODUCTION

Elasticity deals with elastic stresses and strains, their relationship, and with the external forces that cause them. An elastic strain is defined as a strain that disappears instantaneously once the forces that cause it are removed. The theory of elasticity for Hookean solids—that are characterized by the proportionality between stress and strain—is rather complex in its more rigorous treatment. However, it is essential to the understanding of micro- and macromechanical problems. Examples of the former are stress fields around dislocations, incompatibility of stresses at the interface between grains, and dislocation interactions in work hardening; examples of the latter are the stresses developed in wire drawing, rolling, and the analysis of specimen–machine interactions in tensile testing. This chapter is structured in such a way as to satisfy the needs of both the undergraduate and graduate student. Part A of this chapter provides a simplified treatment of elasticity, perfectly adjusted to treat the subsequent problems in an undergraduate course. Stresses and strains are presented for a few simplified cases; the tridimensional treatment is kept at a minimum. A graphic method for the solution of two-dimensional stress problems (the Mohr circle) is described. On the other hand, the graduate student needs more powerful tools to handle problems that are somewhat more involved. The more rigorous treatment of elasticity, given in Part B of this chapter, uses the tensor approach. Stresses and strains are treated as second-rank tensors, and the mathematical framework of tensors can be readily utilized. This allows the consideration of anisotropy effects. In most cases, the stress and strain systems in tridimensional bodies can be better treated as tensors, with the indicial notation. Once this tensor approach is understood, the student will have acquired a very helpful visualization of stresses and strains as tridimensional entities. Important problems whose solutions require this treatment are the analyses of stresses around dislocations, interactions between dislocations and solute atoms, fracture mechanics, plastic waves in solids, stress concentrations caused by precipitates, anisotropy of individual grains, and stress state in a composite material. The simplified treatment (Part A) is recommended to the graduate student as preliminary reading before delving into the tensor approach.

‡A. ELASTICITY—ELEMENTARY TREATMENT

‡1.2 LONGITUDINAL STRESS AND STRAIN

Figure 1.1 shows a cylindrical specimen being stressed in a tensile testing machine. The upper part is screwed to the cross-head of the machine. The coupled rotation of the two lateral screws produces the movement of the cross-head. The load cell is a transducer that measures the load and sends it to a recorder; the increase in length of the specimen can be read by strain gages, extensometers, or indirectly, from the velocity of motion of the cross-head. Chapter 16 discusses tensile testing in greater detail; another type of machine, called a servohydraulic machine, is also used. Assuming that at a certain moment the force applied on the specimen by the machine is F, there will be a tendency to "stretch" the specimen, breaking the internal bonds. This breaking tendency is opposed by internal reactions, called *stresses*. The best way of visualizing them is by means of the method of analysis used in mechanics of materials: the specimen is "sectioned" and the missing part is replaced by the forces that it exerts on the other. This procedure is indicated in Fig. 1.1. This "resistance" is, in the present case, uniformly distributed over the normal section; it is represented by three modest arrows. The normal stress σ is defined as this "resistance" per unit area. Applying the equilibrium-of-forces equation from mechanics of materials to the lower portion of the specimen, we have

$$\Sigma F = 0$$

$$F - \sigma A = 0 \tag{1.1}$$

$$\sigma = \frac{F}{A}$$

This is the internal resisting stress opposing the externally applied load, avoiding the breaking of the specimen. The following stress convention is used: tensile stresses are positive and compressive stresses are negative. In geology and rock mechanics, on the other hand, the opposite sign convention is used because compressive stresses are much more common.

As the applied force F increases, so does the length of the specimen.

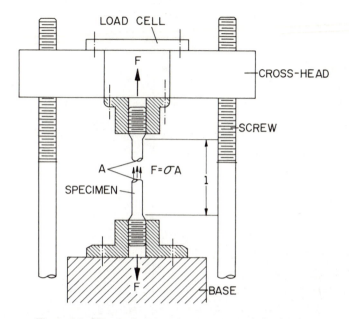

Figure 1.1 Sketch of screw-driven tensile testing machine.

For an increase of dF, the length l increases by dl. The normalized (per unit length) increase in length is equal to

$$d\epsilon = \frac{dl}{l} \qquad \text{or, integrating} \qquad \epsilon = \int_{l_0}^{l_1} \frac{dl}{l} = ln\frac{l_1}{l_0} \qquad (1.2)$$

This parameter is known as the *longitudinal strain*. The usual sign convention is the same as that for stresses: tensile strains are positive, compressive strains are negative. In Fig. 1.2 two stress–strain curves (in tension) are shown; both specimens exhibit elastic behavior. The full lines describe the loading trajectory and the dashed lines describe the unloading. For perfectly elastic solids they should coincide if thermal effects are neglected. The curve of Fig. 1.2(a) is characteristic of metals and ceramics; the elastic regime can be satisfactorily described by a straight line. The curve of Fig. 1.2(b) is characteristic of rubber; σ and ϵ are not proportional. Nevertheless, the strain returns to zero once the stress is removed. The reader can verify this by stretching a rubber band. First, you will notice that the resistance to stretching increases slightly with extension. After considerable deformation the rubber band "stiffens up," and further deformation will eventually lead to rupture. The whole process (except failure) is elastic. A conceptual error often made is to assume that elastic behavior is *always* linear; the latter example shows very clearly that there are notable exceptions. However, for metals the stress and strain can be assumed to be proportional in the elastic regime; these solids are known as Hookean solids. For polymers anelastic effects are very important. Anelasticity results in different trajectories for loading and unloading, with the formation of a hysteresis loop. The area of the hysteresis loop is the energy lost in the process per unit volume.

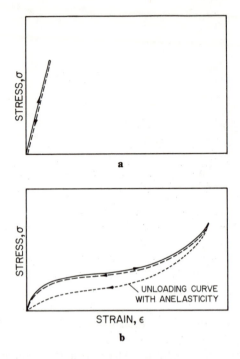

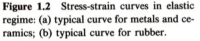

Figure 1.2 Stress-strain curves in elastic regime: (a) typical curve for metals and ceramics; (b) typical curve for rubber.

Metals also exhibit some anelasticity, but it is most often neglected. Anelasticity is due to time-dependent microscopic processes accompanying deformation. An analogy that applies well is the attachment of a spring and dashpot in parallel. The spring represents the elastic portion; the dashpot represents the anelastic portion.

In 1678, Robert Hooke performed the experiments that demonstrated the proportionality between stress and strain. As was customary at that time, he proposed his law as an anagram: "ceiiinossssttuv," which in Latin is "ut tensio sic vis." The meaning is: "As the tension goes so does the stretch." In its most simplified form, we express it as

$$E = \frac{\sigma}{\epsilon} \tag{1.3}$$

where E is Young's modulus. For metals and ceramics it has a very high value. A typical value is 210 GPa for iron. Chapter 4 devotes some effort to the derivation of E for materials from first principles. E depends mainly on the composition, crystallographic structure, and nature of bonding of elements. Heat and mechanical treatments have little effect on E as long as they do not affect the former parameters. Hence, annealed and cold-rolled steel should have the same Young's modulus; there are, of course, small differences due to the formation of the cold-rolling texture. E decreases slightly with temperature increase, as shown in Chapter 4. In monocrystals, E shows different values for different crystallographic orientations. In polycrystalline aggregates that do not exhibit any texture, E is isotropic: it has the same value in all directions. The values of E given in tables (e.g., Table 1.5) are usually obtained by dynamic methods involving elastic-wave propagation, not in conventional stress–strain tests. An elastic wave is passed through a sample; the velocity of the wave is related to the elastic constants by means of mathematical expressions, given in Section 1.13.

‡1.3 SHEAR STRESS AND STRAIN

Imagine the loading arrangement of Fig. 1.3(a). The specimen is placed between a punch and a base having a cylindrical orifice; the punch compresses the specimen. The internal resistance to the external forces has now a shear nature. The small cube in Fig. 1.3(b) was removed from the region being sheared (between punch and base). It is distorted in such a way that the perpendicularity of the faces is lost. The shear stresses and strains are defined as

$$\tau = \frac{F}{A} \qquad \gamma = \frac{dl}{l} = \tan \theta \simeq \theta \tag{1.4}$$

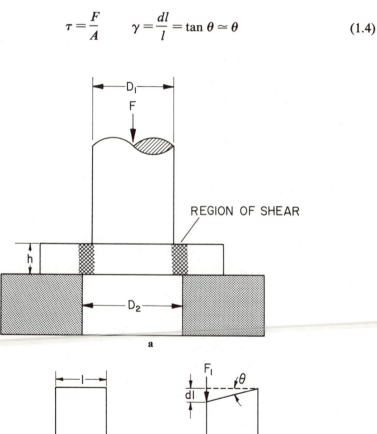

Figure 1.3 (a) Specimen subjected to shear force; (b) strain undergone by small cube in shear region.

The sign convention for shear stresses is given in Section 1.6. The area A is the area of the surface that undergoes shear. It is approximately equal to

$$A \simeq \pi \left(\frac{D_1 + D_2}{2} \right) h$$

The average of the two diameters was taken since D_2 is slightly larger than D_1.

A mechanical test commonly used to find the shear stresses and strains is the torsion test. Dieter[1]† provides a detailed description. The equations that give the shear stresses and strains in terms of the torque are also given in mechanics of materials texts. For metals, ceramics, and certain polymers (the Hookean solids) the proportionality between τ and γ is observed in the elastic regime. In analogy with Young's modulus, a transverse elasticity, called rigidity or shear modulus, is defined as

$$G = \frac{\tau}{\gamma} \tag{1.5}$$

G is numerically inferior to E. It is related to E by Poisson's ratio, which is discussed in the next section. Values of G for different materials are given in Table 1.5; it can be seen that it varies between one-third and one-half of E.

‡1.4 POISSON'S RATIO

A body, upon being pulled in tension, tends to contract laterally. The cube shown in Fig. 1.4(b) shows this behavior. The stresses are now defined in a tridimensional body, and they have two indices. The first indicates the plane (or normal to) in which they are acting; the second subscript indicates the direction in which they are pointing. These stresses are schematically shown acting on three faces of a unit cube in Fig. 1.4(a). The normal stresses have two identical subscripts: σ_{11}, σ_{22}, σ_{33}. The shear stresses have two different subscripts: σ_{12}, σ_{13}, σ_{23}. These subscripts refer to the reference system $Ox_1x_2x_3$. If this notation is used, both normal and shear stresses are designated by the same letter, lowercase sigma. On the other hand, in more simplified cases where we are dealing with only one normal and one shear stress component, σ and τ will be used, respectively; this notation will be maintained throughout the text. In Fig. 1.4, the stress σ_{33} generates strains ϵ_{11}, ϵ_{22}, ϵ_{33} (the same convention is used for stresses and strains). Since the initial dimensions of the cube are equal to 1, the length changes are equal to the strains. Poisson's ratio is defined as the ratio between the lateral and the longitudinal strains. Both ϵ_{11} and ϵ_{22} are negative (decrease in length) and ϵ_{33} is positive. In order for Poisson's ratio to be positive, the negative sign is used. Hence,

$$\nu = - \frac{\epsilon_{11}}{\epsilon_{33}} = - \frac{\epsilon_{22}}{\epsilon_{33}} \tag{1.6}$$

† Bracketed numbers (typeset as superscripts within the text) refer the reader to the Reference list at the end of each chapter.

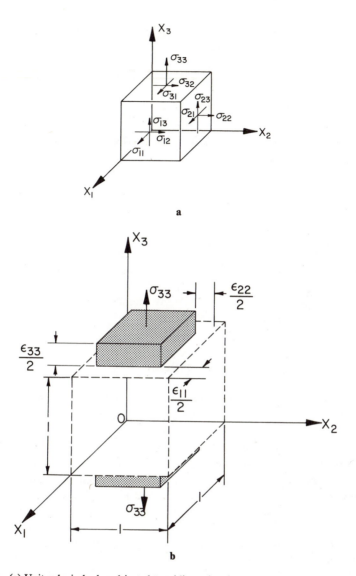

Figure 1.4 (a) Unit cube in body subjected to tridimensional stress; only stresses on three exposed faces of cube are shown; (b) unit cube being extended in direction Ox_3.

ϵ_{11} is equal to ϵ_{22} for an isotropic material. We can calculate the value of V for two extreme cases: (1) when the volume remains constant and (2) when there is no lateral contraction. When the volume is constant, the initial and final volumes, V_0 and V, respectively, are equal:

$$V_0 = 1$$

$$V = (1 + \epsilon_{11})(1 + \epsilon_{22})(1 + \epsilon_{33})$$

Neglecting the cross-products of the strains, because they are orders of magnitude smaller than the strains themselves,

$$V = 1 + \epsilon_{11} + \epsilon_{22} + \epsilon_{33}$$

Since $V = V_0$,

$$\epsilon_{11} + \epsilon_{22} + \epsilon_{33} = 0$$

For the isotropic case the two lateral contractions are the same ($\epsilon_{11} = \epsilon_{22}$). Hence,

$$2\epsilon_{11} = -\epsilon_{33} \tag{1.7}$$

Substituting Eq. 1.7 into Eq. 1.6, we arrive at

$$\nu = 0.5$$

For the case in which there is no lateral contraction, ν is equal to zero. Poisson's ratio for metals is usually around 0.3 (see Table 1.5). It should be emphasized that the values given in Table 1.5 apply to the elastic regime. In the plastic regime ν increases to 0.5, since the volume remains constant. Figure 2.2 shows the variation of Poisson's ratio with strain.

‡1.5 MORE COMPLEX STATES OF STRESS

The relationships between stress and strain described in Sections 1.2 and 1.3 do not apply to bidimensional and tridimensional states of stress; they are unidimensional or uniaxial stress states. The most general state of stress can be represented by the unit cube of Fig. 1.4(a). It is described in greater detail in Section 1.8. In spite of the complexity of the more rigorous tensor analysis it is possible to obtain the relations between stress and strain in a simplified manner by making some assumptions. The generalized Hooke's law (as the set of equations relating tridimensional stresses and strains is called) is derived next, for an isotropic solid. It is assumed that shear stresses can *only* generate shear strains. Thus, the longitudinal strains are exclusively produced by the normal stresses. σ_{11} generates the following strains:

$$\epsilon_{11} = \frac{\sigma_{11}}{E} \tag{1.8}$$

Since $\nu = -\epsilon_{22}/\epsilon_{11} = -\epsilon_{33}/\epsilon_{11}$ for stress σ_{11}, we also have

$$\epsilon_{22} = \epsilon_{33} = -\frac{\nu\sigma_{11}}{E}$$

The stress σ_{22}, in its turn, generates the following strains:

$$\epsilon_{22} = \frac{\sigma_{22}}{E} \quad \text{and} \quad \epsilon_{11} = \epsilon_{33} = -\frac{\nu\sigma_{22}}{E} \tag{1.9}$$

for σ_{33}:

$$\epsilon_{33} = \frac{\sigma_{33}}{E} \quad \text{and} \quad \epsilon_{11} = \epsilon_{22} = -\frac{\nu\sigma_{33}}{E} \tag{1.10}$$

The shear stresses only generate shear strains, in this treatment:

$$\gamma_{12} = \frac{\sigma_{12}}{G} \qquad \gamma_{13} = \frac{\sigma_{13}}{G} \qquad \gamma_{23} = \frac{\sigma_{23}}{G}$$

The second simplifying assumption is made now. It is called the "principle of superposition." The total strain in one direction is considered to be equal to the sum of the strains generated by the various stresses along that direction. Hence, the total ϵ_{11} is the sum of ϵ_{11} produced by σ_{11}, σ_{22}, and σ_{33}. Adding Eqs. 1.8, 1.9, and 1.10, we obtain

$$\epsilon_{11} = \frac{1}{E} [\sigma_{11} - \nu(\sigma_{22} + \sigma_{33})]$$

Similarly,

$$\epsilon_{22} = \frac{1}{E} [\sigma_{22} - \nu(\sigma_{11} + \sigma_{33})]$$

$$\epsilon_{33} = \frac{1}{E} [\sigma_{33} - \nu(\sigma_{11} + \sigma_{22})] \tag{1.11}$$

$$\gamma_{12} = \frac{\sigma_{12}}{G} \qquad \gamma_{13} = \frac{\sigma_{13}}{G} \qquad \gamma_{23} = \frac{\sigma_{23}}{G}$$

Equations 1.11 are identical to Eqs. 1.119, which will be derived using tensor analysis. However, no simplifying assumptions were needed in the derivation of Eqs. 1.119. Equations 1.11 are needed to determine the stress fields around dislocations (Chapter 6). Applying them to a hydrostatic stress situation ($\sigma_{11} = \sigma_{22} = \sigma_{33} = -p$), we can see perfectly that there are no distortions in the cube ($\gamma_{12} = \gamma_{13} = \gamma_{23} = 0$) and that $\epsilon_{11} = \epsilon_{22} = \epsilon_{33}$.

The triaxial state of stress is difficult to treat in elasticity (and even more difficult in plasticity). In the great majority of cases we try to assume a more simplified state of stress that resembles the tridimensional stress. This is often justified by the geometry of the body and by the loading configuration. The example discussed in Section 1.2 is the simplest state (uniaxial stress). It occurs when beams are axially loaded (in tension or compression). In sheets and plates (where one dimension can be neglected with respect to the two other ones) the state of stress can be assumed to be bidimensional. This state of stress is also known as plane stress. This is due to the fact that normal stresses (normal to the surface) are zero at the surface, as are shear stresses (parallel to the surface) at the surface. In Fig. 1.4(a) one would be left with σ_{11}, σ_{12}, σ_{22} if Ox_1x_2 were the plane of the sheet. Since the sheet is thin there is no space for buildup of the stresses that are zero at the surface. The solution to this problem is approached graphically in Section 1.6. The opposite case, in which one of the dimensions is infinite with respect to the other two, is treated under the assumption of plane strain. If one dimension is infinite, strain in it is constrained; hence, one has two dimensions left. This state is called bidimensional or, more commonly, plane strain. It also occurs when strain is constrained in one direction by some other means. A long dam is an example where deformation in the direction of the dam is constrained. There is also a state of uniaxial strain, which takes place when shock waves travel through a solid. They are treated in Chapter 15. Yet another state of stress is pure shear, when there are no normal stresses (Section 1.7).

‡1.6 GRAPHICAL SOLUTION OF A BIAXIAL STATE OF STRESS— THE MOHR CIRCLE

Figure 1.5(a) shows a biaxial (or bidimensional) state of stress. The graphical scheme developed by 0. Mohr allows the determination of the normal and shear stresses in any orientation in the plane. The reader should be warned, right at the onset, that a *change in sign convention* for the shear stresses has to be introduced here. The former sign convention—positive shear stresses pointing

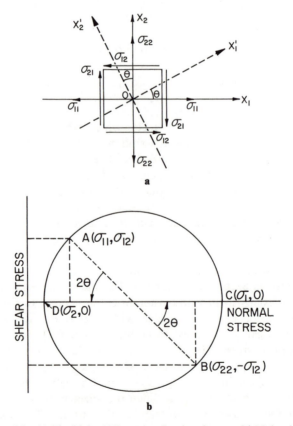

Figure 1.5 (a) Biaxial (or bidimensional) state of stress; (b) Mohr circle.

toward the positive direction of axes in faces shown in Fig. 1.4(a)—has to be *temporarily* abandoned and the following convention is adopted: positive shear stresses produce clockwise rotation of cube (or square), and negative shear stresses produce counterclockwise rotation. The sign convention for normal stresses remains the same. Figure 1.5(b) shows Mohr's construction. The normal stresses are plotted in the abcissa while the shear stresses are plotted in the ordinate axis. Point A in the diagram corresponds to a state of stress on the face of the cube perpendicular to Ox_1; point B represents the state of stress on the face perpendicular to Ox_2. From A and B we construct a circle with center in the axis of the abcissa and passing through A and B. The center is the point where the segment AB intersects the abcissa. The stress states for all orientations of the square (in the same plane) correspond to points diametrically opposed in Mohr's circle. Hence, we can determine the state of stress for any orientation. The rotations in the square (real rotations) and in Mohr's circle have the same sense; however, a rotation of θ in the square corresponds to 2θ in Mohr's circle. For instance, a rotation of 2θ in the counterclockwise direction leads to a state of stress defined by C and D in Mohr's circle. The shear stresses are zero for this orientation and the normal stresses are called *principal* stresses. One subscript is sufficient to designate the stresses at these special orientations: σ_1, σ_2, σ_3. We use the convention $\sigma_1 > \sigma_2 > \sigma_3$. In Fig. 1.5(a) a rotation of only θ was done in the same counterclockwise sense, leading to the same principal stresses. The orientations Ox_1 and Ox_2 are called *principal axes* (or directions).

‡1.7 PURE SHEAR—RELATIONSHIP BETWEEN G AND E

There is a special case of bidimensional stress in which $\sigma_{22} = -\sigma_{11}$. This state of stress is represented in Fig. 1.6(a). It can be seen that $\sigma_{12} = 0$, implying that σ_{11} and σ_{22} are principal stresses. Hence, we can use the special subscripts for principal stresses, and write $\sigma_2 = -\sigma_1$. In Mohr's circle of Fig. 1.6(b) the center coincides with the origin of the axes. We can see that a rotation of 90° (on the circle) leads to a state of stress in which the normal stresses are zero. This rotation is equivalent to a 45° rotation in the body (real space). The magnitude of the shear stress at this orientation is equal to the radius of the circle. Hence, the square shown in Fig. 1.6(c) is deformed to a losange under the combined effect of the shear stresses. This state of stress is called *pure shear*.

It is possible for this particular case to obtain a relationship between G and E; furthermore, this relationship has a general nature. The strain ϵ_{11} is, for this case,

$$\epsilon_{11} = \frac{1}{E} (\sigma_1 - \nu\sigma_2) = \frac{\sigma_1}{E} (1 + \nu) \tag{1.12}$$

We have, for the shear stresses (using the normal and not the Mohr sign convention),

$$\tau = -\sigma_1 \tag{1.13}$$

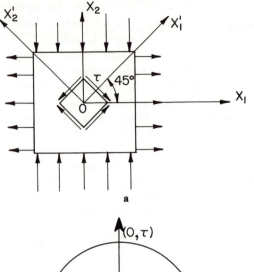

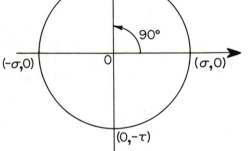

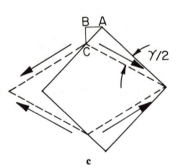

Figure 1.6 Pure shear.

But we also have (Eq. 1.5)

$$\tau = G\gamma \tag{1.14}$$

Substituting Eqs. 1.13 and 1.14 into Eq. 1.12 yields

$$\epsilon_{11} = -\frac{G\gamma}{E}\,(1+\nu)$$

It is possible, by means of geometrical considerations on the triangle ABC in Fig. 1.6(c), to show that

$$2\epsilon_{11} = -\gamma$$

The reader should do this, as an exercise. Hence,

$$G = \frac{E}{2(1 + \nu)} \tag{1.15}$$

Consequently, G is related to E by means of Poisson's ratio. This theoretical relationship between E and G is in good agreement with experimental results. Table 1.5 presents values of ν. For a typical metal having $\nu = 0.3$, we have $G = E/2.6$. The maximum value of G is $E/2$.

B. ELASTICITY—ADVANCED TREATMENT

1.8 CONCEPT OF STRESS

If external forces are acting on a body, or if one part of the body exerts a force on its neighboring parts, this body is said to be under stress. Stress is the internal resistance of a body under the effect of forces. Figure 1.7 shows a body upon which external forces F_i are acting. Considering the body as divided into parts I and II (and this is the well-known method of sections from mechanics of materials), we can say that upon the element of area dA a force per unit area equal to P is acting. This force per unit area is the stress; it may be decomposed into a normal stress σ_1 and a shear stress σ_2. We say that a stress system (σ_1, σ_2) is acting on the element of area dA.

We usually define the system of stresses acting in a tridimensional body by specifying the stresses acting on the faces of a cube having small dimensions and embedded in this body. Two types of forces can act on a body. First, the body forces, which are proportional to the volume of the body (if the density and composition are homogeneous). Examples are the gravitational, electric, and magnetic forces. They act directly on the individual atoms that compose the body. Second, the surface forces, that act on the surface; the internal atoms feel their presence by means of the intermediary entity, the stress.

First, the state of stress acting on a body will be studied for a very simplified case:

1. The stress is assumed to be homogeneous; that is, the stresses acting on an element of area having constant orientation and shape are the same, independent of the position of this element of area in the body.
2. All parts of the body are in static equilibrium.
3. No body forces or moments are acting.

The state of stress in an elemental cube (a cube with unit dimensions, also called a unit cube) is shown in Fig. 1.8. The stresses are assumed to be

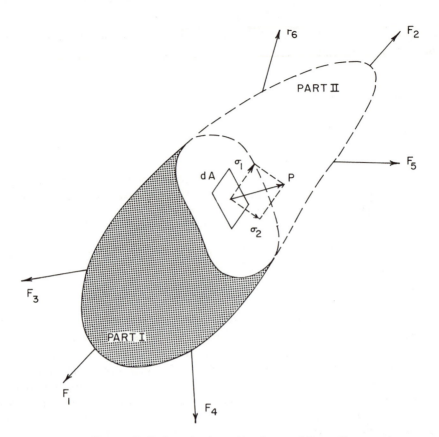

Figure 1.7 Body under the action of external forces F_i.

positive in the directions indicated. Hence, normal tensile stresses are positive, and normal compressive stresses are negative; shear stresses are positive if they point in the positive direction of the axis while acting on the faces seen in Fig. 1.8; if they are acting on the back faces, they are positive when pointing to the negative direction of the axes. The state of stress of Fig. 1.8 is the same for any position of the cube in the body as long as the orientation remains constant (homogeneous stresses). The sign convention used here will be maintained throughout the text. There is only one situation in which it has to be changed—in the graphical solution for biaxial stresses (Section 1.6). Two subscripts are used to designate the stress components: the first indicates the direction of the force and the second indicates the plane on which it is acting (actually, the normal to the plane). Hence, σ_{23} has the direction Ox_2 and is acting on the plane perpendicular to Ox_3. When the two subscripts are identical (e.g., σ_{11}, σ_{22}) we have normal stresses; when they are different (e.g., σ_{12}, σ_{23}) we have shear stresses. Different texts use different notations, but this one is the best because it facilitates the tensorial treatment. During his or her career, the reader will come across all kinds of symbols: σ_x, X_x, $\widehat{xx}$, τ_{11} (for σ_{11}), τ_{xy}, X_y, $\widehat{xy}$, τ_{12} (for σ_{12}). This should not be discouraging, as one can perform the mental transformations very quickly. Throughout this text, wherever the treatment is very simplified and only one normal and one shear stress are used, the foregoing notation is replaced by the simple (σ, τ) without subscript. The treatment used for stresses and strains in this chapter is based on Nye's excellent book[2]. Whenever the reader encounters some difficulty, this source should be consulted.

In Fig. 1.7, two components of the stress are acting on the area. If a system of references were defined in such a way that Ox_3 would be normal to the plane and Ox_1x_2 would be parallel to it, the shear stress σ_2 would have to be decomposed into two components, parallel to Ox_1 and Ox_2, respectively. Hence, there are three components of stress—one normal and two shear—acting on a plane. This is exactly the situation in Fig. 1.8, in which there are three stresses acting on each face of the cube. The number of stress components acting on the three faces is nine; there are nine more, equal to the first ones, acting on the back faces. Hence, the stress is, in its most general form, an entity that has nine components. It is the result of the application of a force (a vector, with three components) on an area (whose orientation and magnitude can also be specified by a vector). The force $\mathbf{P}$ can be decomposed into ($\mathbf{p}_1$, $\mathbf{p}_2$, $\mathbf{p}_3$) and the area $d\mathbf{A}$ can be decomposed into ($d\mathbf{A}_1$, $d\mathbf{A}_2$, $d\mathbf{A}_3$). Another way of expressing that a vector has three components is by saying that it has a magnitude, a direction, and a sense. Vector analysis provides the mathematical basis for the treatment of these entities. Scalars and vectors are also known as zero- and first-rank (or order) tensors; and this leads right into stresses, which are second-rank tensors. Tensor analysis, a natural extension of vector analysis,

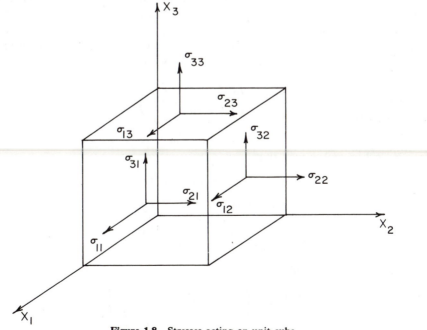

Figure 1.8 Stresses acting on unit cube.

is used to treat these entities. Other physical properties that are second-rank tensors (require nine components to be completely satisfied) are strain, electrical conductivity, thermal conductivity, magnetic susceptibility, and permeability. There are also higher-order tensors. For example, the relationship between two second-rank tensors (such as stress and strain) provides a fourth-rank tensor (the elastic constants). It can be seen that the mathematically rigorous treatment of stresses, strains, and elasticity requires tensor analysis. Section 1.9 provides the rudiments of tensors needed to treat the subsequent problems.

1.9 TENSORS

Tensors are specified in the following manner:

1. A zero-rank tensor is specified by a sole component, independent of the system of reference (e.g., mass, density).
2. A first-rank tensor is specified by three components, each associated with one reference axis (e.g., force).
3. A second-rank tensor is specified by nine components, each associated simultaneously with two reference axes (e.g., stress, strain).
4. A fourth-rank tensor is specified by 81 components, each associated simultaneously with four reference axes (e.g., elastic stiffness, compliance).

The indicial (also called dummy suffix) notation will be used here. The number of indices (subscripts) associated with a tensor is equal to its rank. Hence, density does not have any subscript (ρ), force has one (F_1, F_2, etc.), and stress has two (σ_{12}, σ_{22}, etc). The easiest way of representing the components of a second-rank tensor is as a matrix. We have, for the tensor T,

$$T = \begin{pmatrix} T_{11} & T_{12} & T_{13} \\ T_{21} & T_{22} & T_{23} \\ T_{31} & T_{32} & T_{33} \end{pmatrix}$$

In general, a property T that relates two vectors $\mathbf{p} = [p_1, p_2, p_3]$ and $\mathbf{q} = [q_1, q_2, q_3]$ in such a way that

$$p_1 = T_{11}q_1 + T_{12}q_2 + T_{13}q_3 \qquad (1.16)$$

$$p_2 = T_{21}q_1 + T_{22}q_2 + T_{23}q_3 \qquad (1.17)$$

$$p_3 = T_{31}q_1 + T_{32}q_2 + T_{33}q_3 \qquad (1.18)$$

and where T_{11}, T_{12}, . . . , T_{33} are constants is a second-rank tensor. The system of equations 1.16, 1.17, and 1.18 can be expressed matricially as

$$\begin{bmatrix} p_1 \\ p_2 \\ p_3 \end{bmatrix} = \begin{bmatrix} T_{11} & T_{12} & T_{13} \\ T_{21} & T_{22} & T_{23} \\ T_{31} & T_{32} & T_{33} \end{bmatrix} \begin{bmatrix} q_1 \\ q_2 \\ q_3 \end{bmatrix}$$

1.9.1 Indicial Notation

The indicial notation, which simplifies the treatment tremendously, is introduced here. Equations 1.16, 1.17, and 1.18 can be expressed as

$$p_1 = \sum_{j=1}^{3} T_{1j} q_j$$

$$p_2 = \sum_{j=1}^{3} T_{2j} q_j$$

$$p_3 = \sum_{j=1}^{3} T_{3j} q_j$$

We can simplify it further to

$$p_i = \sum_{i=1}^{3} \sum_{j=1}^{3} T_{ij} q_j$$

The symbol Σ is usually omitted. We now introduce Einstein's summation rule: When a subscript appears twice in the same term it is implied that one is summing with respect to this subscript in this term. We have

$$p_i = T_{ij} q_j \qquad (i, j = 1, 2, 3) \tag{1.19}$$

The subscript i is called a free subscript; j (which appears twice) is called a "dummy" suffix, for some unknown reason.

1.9.2 Transformations

It is important to know how the components of a tensor $[T_{ij}]$ vary with the system of reference. It is known that a vector **p** has components $[p_1, p_2, p_3]$ in reference system $Ox_1x_2x_3$. First we will see how the $[p_i']$ can be expressed in terms of $[p_i]$. Then, second-rank tensors will be transformed. Figure 1.9 shows two Cartesian systems of reference: the old system $Ox_1x_2x_3$ and the new system $Ox_1'x_2'x_3'$, rotated with respect to the old one. The orientation of the new system can be established with respect to the old one by specifying the angles between the axes. The symbols l_{ij} are used to specify the cosines of the angles between Ox_i' and Ox_j. Hence,

$$\cos x_1' x_1 = l_{11}$$

$$\cos x_2' x_3 = l_{23}$$

The nine angles that the two systems form are as follows:

		old system		
		x_1	x_2	x_3
new	x_1'	l_{11}	l_{12}	l_{13}
system	x_2'	l_{21}	l_{22}	l_{23}
	x_3'	l_{31}	l_{32}	l_{33}

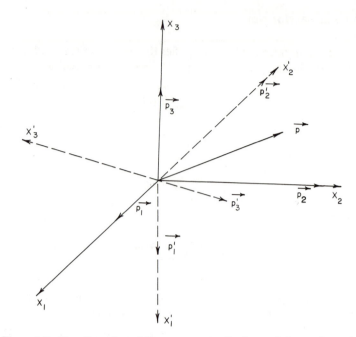

Figure 1.9 Transformation of the components of a first-rank tensor (vector).

It is recommended that the reader make the construction above whenever a transformation of axes is being made. The orientation of axis Ox_1' is defined by l_{11}, l_{12}, and l_{13}. The first subscript always indicates the "new" system; the second, the "old" system. The nine l_{ij} are not independent; it should be noted that, in general, $l_{ij} \neq l_{ji}$. We have (and the student can prove this)

$$l_{ij}^2 = 1$$

The components $[p_i]$ and $[p_i']$ of vector **p** in the two systems are also shown in Fig. 1.9. We have, for p_1',

$$p_1' = p_1 \cos x_1' x_1 + p_2 \cos x_1' x_2 + p_3 \cos x_1' x_3$$

or

$$p_1' = p_1 l_{11} + p_2 l_{12} + p_3 l_{13}$$

The same applies to p_2 and p_3, and we have, in indicial notation,

$$p_i' = p_j l_{ij} \tag{1.20}$$

Another rule for the indicial notation is introduced at this point; whenever the "new" components are expressed in terms of the "old" ones, the dummy suffixes are placed as close as possible. The correct form of Eq. 1.20 is, therefore,

$$p_i' = l_{ij} p_j \tag{1.21}$$

On the other hand, if the opposite operation is being performed (old in terms of new), we put them as far apart as possible:

$$p_i = l_{ji}p'_j \qquad (1.22)$$

It is worth noting that the coordinates of a point transform in the same manner as the components of a vector. Point P, which has coordinates $[x_1, x_2, x_3]$ in the old system, will have coordinates $[x'_1, x'_2, x'_3]$ in the new system given by

$$x'_i = l_{ij}x_j$$

In order to transform the components of a second-rank tensor T_{ij} it is first required to transform the components of the first-rank tensors related to them by Eq. 1.19. The following sequence of equations is used:

$$p' \xrightarrow{\;1.23\;} p \xrightarrow{\;1.24\;} q \xrightarrow{\;1.25\;} q'$$

So we determine q as a function of q', p as a function of q, and p' as a function of p. Combining them, we have the relationship between p' and q'. Equation 1.23 in the above sequence can be expressed as

$$p'_i = l_{ik}p_k \qquad (1.23)$$

Using the dummy suffix of Eq. 1.23 as the free suffix of Eq. 1.24, we get

$$p_k = T_{kl}q_l \qquad (1.24)$$

For Eq. 1.25 we have

$$q_l = l_{jl}q'_j \qquad (1.25)$$

Combining Eqs. 1.23, 1.24, and 1.25, we get

$$p'_i = l_{ik}T_{kl}q_l = l_{ik}T_{kl}l_{jl}q'_j \qquad (1.26)$$

$$p'_i = l_{ik}l_{jl}T_{kl}q'_j \qquad (1.27)$$

It is known that $[T_{ij}]$ is a physical entity that relates two vectors according to Eq. 1.19. The physical integrity of the vectors $[p_i]$ and $[q_i]$ is not changed when the system of reference is changed; only the components are transformed and the vectors remain the same. By the same token, $[T_{ij}]$ represents the same physical entity independent of the system of reference. Consequently, the relationship between $[p]$ and $[q]$ in the new system has to be the same tensor $[T_{ij}]$. Therefore, we have to have

$$p'_i = T'_{ij}q'_j \qquad (1.28)$$

Comparing Eqs. 1.27 and 1.28, we get

$$T'_{ij} = l_{ik}l_{jl}T_{kl} \qquad (1.29)$$

Equation 1.29 is the transformation law for tensors. It can be seen that the dummy suffixes k and l are as close as possible. If we go from the new system to the old one, we have the corresponding relationship (with the dummy suffixes as far apart as possible):

$$T_{ij} = l_{ki}l_{lj}T'_{kl} \qquad (1.30)$$

The simplification involved in the indicial notation can now be fully realized. Equation 1.30 represents nine equations, each with nine terms. As an example:

$$T'_{11} = l_{11}l_{11}T_{11} + l_{11}l_{12}T_{12} + l_{11}l_{13}T_{13} + l_{12}l_{11}T_{21}$$
$$+ l_{12}l_{12}T_{22} + l_{12}l_{13}T_{23} + l_{13}l_{11}T_{31} + l_{13}l_{12}T_{32} + l_{13}l_{13}T_{33}$$

The transformation law for tensors is so important that it is used to define them. Hence, a second-rank tensor can be defined as a quantity that has nine components with respect to a system of reference (Ox_i) that transform themselves to a new system according to the transformation law of second-rank tensors. This definition is actually used to prove that both the stress and the strain are second-rank tensors.

It should be emphasized that (l_{ij}) and $[T_{ij}]$ are completely different in nature, although both have nine components that can be put in matrix form. (l_{ij}) is the relationship between two systems of reference, having no other physical significance; $[T_{ij}]$ is a physical entity related to a specific system of reference.

1.9.3 Symmetric and Antisymmetric Tensors

Symmetric tensors are defined by

$$T_{ij} = T_{ji} \tag{1.31}$$

Antisymmetric tensors are defined by

$$T_{ij} = -T_{ji} \tag{1.32}$$

The matricial form is, for symmetric and antisymmetric tensors, respectively:

$$\begin{bmatrix} T_{11} & T_{12} & T_{13} \\ T_{12} & T_{22} & T_{23} \\ T_{13} & T_{23} & T_{33} \end{bmatrix} \qquad \begin{bmatrix} 0 & T_{12} & T_{13} \\ -T_{12} & 0 & T_{23} \\ -T_{13} & -T_{23} & 0 \end{bmatrix}$$

1.9.4 Quadratic Representation of Tensors and Principal Axes

Tensors can be graphically represented by means of a very convenient analogy. This graphic analogy is very helpful in the solution of problems and explains why we have principal directions. From analytical geometry, the reader might remember that the quadratic equation has, in its most general form, the following configuration:

$$Ax_1^2 + Bx_2^2 + Cx_3^2 + Dx_1x_2 + Ex_1x_3 + Fx_2x_3 = 1$$

Replacing coefficients A, B, etc., by others that are more descriptive of their position in the equation, we have

$$S_{11}x_1^2 + S_{22}x_2^2 + S_{33}x_3^2 + 2S_{12}x_1x_2 + 2S_{13}x_1x_3 + 2S_{23}x_2x_3 = 1$$

In indicial notation,

$$S_{ij}x_ix_j = 1 \tag{1.33}$$

To transform the quadratic equation to the new system of axes, we have to transform the coordinates of (x_i) and (x_j):

$$x_i = l_{ki}x'_k \tag{1.34}$$

$$x_j = l_{lj}x'_l \tag{1.35}$$

Substituting Eqs. 1.34 and 1.35 into 1.33 yields

$$S_{ij}l_{ki}l_{lj}x'_k x'_l = 1 \tag{1.36}$$

Since the quadratic curve should not change with the system of reference, we have

$$S'_{kl}x'_k x'_l = 1 \tag{1.37}$$

Hence, from Eqs. 1.36 and 1.37 we get

$$S'_{kl} = S_{ij}l_{ki}l_{lj}$$

Rearranging the terms yields

$$S'_{kl} = l_{ki}l_{lj}S_{ij} \tag{1.38}$$

Equation 1.38 shows that the coefficients of the quadratic equation transform in the same way as the components of a symmetric second-rank tensor. This leads to the important analogy: a symmetric second-rank tensor can be represented as a quadratic equation.

All quadratic equations have a peculiar property; they can be referred to a special system of axes, called *principal axes*, for which the equation is simplified to

$$S_{11}x_1^2 + S_{22}x_2^2 + S_{33}x_3^2 = 1 \tag{1.39}$$

The principal axes are the symmetry axes of these quadratic curves; Fig. 1.10 shows an ellipsoid referred to its principal axes. The principal axes coincide with the axes of the ellipsoid. The intercepts are equal to $S_{11}^{-1/2}$, $S_{22}^{-1/2}$, and $S_{33}^{-1/2}$. By the same token, a symmetric second-rank tensor can be referred to its principal axes. When referred to them, $[T_{ij}]$ is expressed by the following components:

$$\begin{bmatrix} T_{11} & T_{12} & T_{13} \\ T_{12} & T_{22} & T_{23} \\ T_{13} & T_{23} & T_{33} \end{bmatrix} \Rightarrow \begin{bmatrix} T'_{11} & 0 & 0 \\ 0 & T'_{22} & 0 \\ 0 & 0 & T'_{33} \end{bmatrix}$$

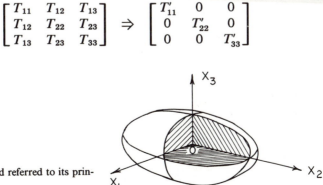

Figure 1.10 Ellipsoid referred to its principal axes.

The same mathematical procedure used in analytical geometry is used to determine the transformation of axes that leads to the principal components of $[T_{ij}]$. For simplicity, one sole subscript is used when $[T_{ij}]$ is referred to its principal directions:

$$\begin{bmatrix} T_1 & 0 & 0 \\ 0 & T_2 & 0 \\ 0 & 0 & T_3 \end{bmatrix}$$

1.10 STRESS

1.10.1 Stress as a Symmetric Tensor

The only requirement for stress to be a symmetrical tensor is that no body moments be present. Figure 1.11 shows a parallelepiped with sides δx_1, δx_2, and δx_3, embedded in a body. Since the state of stress is not homogeneous, the forces acting on the parallelepiped depend on its position in the body. Figure 1.11(a) shows a section perpendicular to the axis Ox_3. The axes are taken as orthogonal for simplicity. The stresses acting on the origin are σ_{ij}. Consequently, the stresses acting on the parallelepiped are changed, due to the drift in stress throughout the body. Applying Newton's second law in the direction Ox_1, we get

$$\Sigma Fx_1 = ma_{x_1} = m\ddot{x}_1$$

$$\left(\sigma_{11} + \frac{\partial \sigma_{11}}{\partial x_1} \times \frac{1}{2} \delta x_1 - \sigma_{11} + \frac{\partial \sigma_{11}}{\partial x_1} \times \frac{1}{2} \delta x_1 \right) \delta x_2 \, \delta x_3$$

$$+ \left(\sigma_{12} + \frac{\partial \sigma_{12}}{\partial x_2} \times \frac{1}{2} \delta x_2 - \sigma_{12} + \frac{\partial \sigma_{12}}{\partial x_2} \times \frac{1}{2} \delta x_2 \right) \delta x_1 \, \delta x_3$$

$$+ \left(\sigma_{13} + \frac{\partial \sigma_{13}}{\partial x_3} + \frac{1}{2} \delta x_3 - \sigma_{13} + \frac{\partial \sigma_{13}}{\partial x_3} \times \frac{1}{2} \delta x_3 \right) \delta x_1 \, \delta x_2 = m\ddot{x}_1$$

Introducing the simplifications gives

$$\left(\frac{\partial \sigma_{11}}{\partial x_1} + \frac{\partial \sigma_{12}}{\partial x_2} + \frac{\partial \sigma_{13}}{\partial x_3} \right) \delta x_1 \, \delta x_2 \, \delta x_3 = m\ddot{x}_1$$

But $m = V \times \rho$, where V and ρ are the volume and density of the parallelepiped. The volume, in turn, is

$$V = \delta x_1 \, \delta x_2 \, \delta x_3$$

Hence,

$$\frac{\partial \sigma_{11}}{\partial x_1} + \frac{\partial \sigma_{12}}{\partial x_2} + \frac{\partial \sigma_{13}}{\partial x_3} = \rho \ddot{x}_1$$

Mechanics of Deformation   Part I

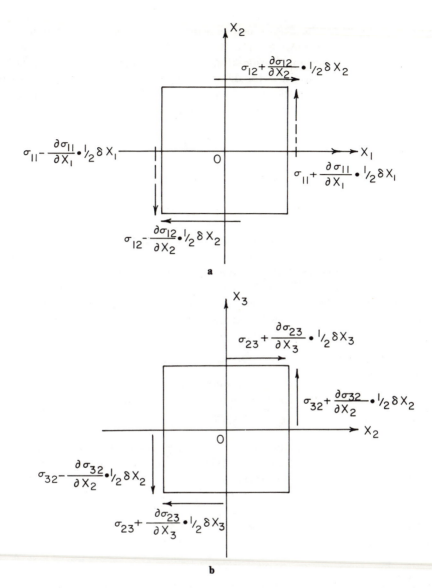

Figure 1.11 Inhomogeneous state of stress; (a) section perpendicular to Ox_3; (b) section perpendicular to Ox_1.

Similar equations are obtained for the other directions, Ox_2 and Ox_3. The general equation, in indicial notation, is

$$\frac{\partial \sigma_{ij}}{\partial x_j} = \rho \ddot{x}_i \qquad (1.40)$$

Equation 1.40 correlates the variation of stresses through a body to the acceleration of the particles ($\ddot{x}_i$). In spite of its academic appearance, it is eminently practical, since it is the basis for the study of elastic waves. From it the velocity of elastic waves can be determined. Elastic waves, in their turn, are important because they are the basis for the study of plastic waves. These, in turn, are one of the most important considerations in dynamic deformation. An additional application of elastic waves is the determination of the elastic constants. Section 1.13 shows how the elastic constants are related to the propagation velocity of elastic waves. For static equilibrium, we have

$$\frac{\partial \sigma_{ij}}{\partial x_j} = 0 \tag{1.41}$$

In the presence of body forces such as gravity,

$$\frac{\partial \sigma_{ij}}{\partial x_j} + \rho g_i = 0 \tag{1.42}$$

These are the equations of equilibrium of the theory of elasticity.

Equations 1.40 and 1.41 express the summation-of-forces conditions of mechanics of materials. For a complete analysis, the summation of moments has to be considered. Figure 1.11(b) shows the section of the parallelepiped perpendicular to Ox_1. The normal components of stress whose lines of action pass through the origin do not produce moments. We have the general equation from mechanics,

$$\Sigma M_{x_1} = I_1 \ddot{\theta}_1 \tag{1.43}$$

where I_1 is the moment of inertia and $\ddot{\theta}_1$ is the angular acceleration. This is Newton's law for moments. Equation 1.43 corresponds to

$$(\sigma_{32} - \sigma_{23}) \, \delta x_1 \, \delta x_2 \, \delta x_3 = I_1 \ddot{\theta}_1 \tag{1.44}$$

In Eq. 1.44, $\delta x_2 \, \delta x_3$ is the area of the faces upon which σ_{32} and σ_{23} are acting. When we have static equilibrium, there are no internal rotations or internal vortices and $\ddot{\theta}_i = 0$. Hence,

$$\sigma_{32} = \sigma_{23}$$

Applying the equations equivalent to Eq. 1.43 with respect to the axes Ox_2 and Ox_3, we obtain the equality between the other two shear stresses:

$$\sigma_{13} = \sigma_{31} \qquad \text{and} \qquad \sigma_{12} = \sigma_{21} \tag{1.45}$$

In indicial notation, $\sigma_{ij} = \sigma_{ji}$. The condition for Eq. 1.45 is that there are no body moments. However, it should be noticed that we do not need static equilibrium ($\ddot{x}_i \neq 0$), and that the state of stress is not homogeneous. Hence, σ_{ij} is a symmetric tensor and it can be referred to a principal system of axes.

1.10.2 Stress Acting in a General Direction

Figure 1.12 shows a force $\mathbf{p} \, ds$ acting on a surface element ds. The area ds is defined by the unit vector (normal to it) $\boldsymbol{\mu}$ that passes through R. If we want to know how $\mathbf{p} \, ds$ changes as the orientation of the area changes, we have to consider a system of orthogonal axes and take the summation of forces.

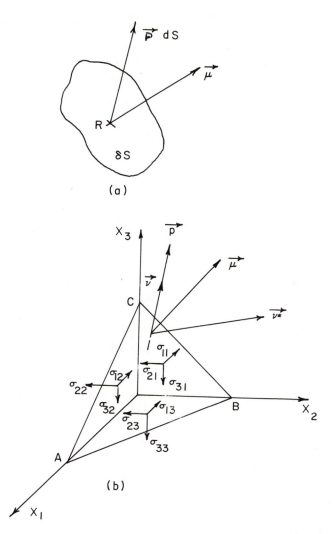

Figure 1.12 (a) Force **p** *ds* acting on surface element *ds*; (b) same situation referred to a system of orthogonal axes.

In static equilibrium, when the stresses are homogeneous and there are no body moments, we have

$$p_1(ABC) = \sigma_{11}(OBC) + \sigma_{12}(AOC) + \sigma_{13}(AOB)$$

where (ABC), (OBC), (AOC), and (AOB) are areas.

$$p_1 = \sigma_{11}\frac{(OBC)}{(ABC)} + \sigma_{12}\frac{(AOC)}{(ABC)} + \sigma_{13}\frac{(AOB)}{(ABC)}$$

But

$$\cos \mu x_1 = \frac{(OBC)}{(ABC)}$$

$$\cos \mu x_2 = \frac{(OAC)}{(ABC)}$$

$$\cos \mu x_3 = \frac{(AOB)}{(ABC)}$$

These direction cosines can be called $l_{\mu 1}$, $l_{\mu 2}$, and $l_{\mu 3}$.

$$p_1 = \sigma_{11} l_{\mu 1} + \sigma_{12} l_{\mu 2} + \sigma_{13} l_{\mu 3}$$

$$p_2 = \sigma_{21} l_{\mu 1} + \sigma_{22} l_{\mu 2} + \sigma_{23} l_{\mu 3}$$

$$p_3 = \sigma_{31} l_{\mu 1} + \sigma_{32} l_{\mu 2} + \sigma_{33} l_{\mu 3}$$

In indicial notation,

$$p_i = \sigma_{ij} l_{\mu j} \tag{1.46}$$

Incidentally, Eq. 1.46 serves to define σ_{ij} as a tensor, because it relates to vectors $\mathbf{p}$ and $\boldsymbol{\mu}$ according to the relationship 1.19. Now, two new directions are defined in area ABC (by their unit vectors): ν^* is a general direction and ν is the direction passing through force $\mathbf{p}$, the resultant force acting on the area. The stress along a general direction ν^* is given by

$$\sigma_{\nu^* \mu}(ABC) = \mathbf{p}(ABC) l_{\nu^* \nu}$$

$\sigma_{\nu^* \mu}$ is the component of the stress acting in surface ABC along direction μ, and $l_{\nu^* \nu}$ is the cosine of the angle $\nu^* \nu$. Decomposing $\sigma_{\nu^* \mu}$ yields

$$\sigma_{\nu^* \mu} = p_1 l_{\nu^* 1} + p_2 l_{\nu^* 2} + p_3 l_{\nu^* 3}$$

Applying Eq. 1.46 yields

$$\sigma_{\nu^* \mu} = \sigma_{1j} l_{\mu j} l_{\nu^* 1} + \sigma_{2j} l_{\mu j} l_{\nu^* 2} + \sigma_{3j} l_{\mu j} l_{\nu^* 3}$$

Hence, we have, in indicial notation,

$$\sigma_{\nu^* \mu} = l_{\nu^* i} l_{\mu j} \sigma_{ij} \tag{1.47}$$

To determine the normal stress acting on surface ABC, we have to make ν^* coincide with μ:

$$\sigma_N = \sigma_{\mu \mu} = l_{\mu i} l_{\mu j} \sigma_{ij} \tag{1.48}$$

Expanding Eq. 1.48, we have

$$\sigma_N = l_{\mu 1}^2 \sigma_{11} + l_{\mu 2}^2 \sigma_{22} + l_{\mu 3}^2 \sigma_{33} + 2 l_{\mu 1} l_{\mu 2} \sigma_{12} + 2 l_{\mu 2} l_{\mu 3} \sigma_{23} + 2 l_{\mu 3} l_{\mu 1} \sigma_{13} \tag{1.49}$$

1.10.3 Determination of Principal Stresses

The shear stresses acting on the faces of a cube referred to its principal axes are zero; in other words, the total stress is equal to the normal stress. The total stress is given by Eq. 1.46 and the normal stress is given by Eq. 1.48. If p_i and σ_N coincide:

$$p_i = \sigma_N l_{\mu i} \qquad (1.50)$$

Applying Eqs. 1.46 and 1.50 to p_1, we have

$$p_1 = l_{\mu 1}\sigma_{11} + l_{\mu 2}\sigma_{12} + l_{\mu 3}\sigma_{13} = \sigma_N l_{\mu 1} \qquad (1.51)$$

or

$$(\sigma_{11} - \sigma_N)l_{\mu 1} + \sigma_{12}l_{\mu 2} + \sigma_{13}l_{\mu 3} = 0 \qquad (1.52)$$

Similarly, for p_2 and p_3,

$$\sigma_{12}l_{\mu 1} + (\sigma_{22} - \sigma_N)l_{\mu 2} + \sigma_{23}l_{\mu 3} = 0 \qquad (1.53)$$

$$\sigma_{13}l_{\mu 1} + \sigma_{23}l_{\mu 2} + (\sigma_{33} - \sigma_N)l_{\mu 3} = 0 \qquad (1.54)$$

The solution of the system of Eqs. 1.52, 1.53, and 1.54 is given by the following determinant:

$$\begin{bmatrix} \sigma_{11} - \sigma_N & \sigma_{12} & \sigma_{13} \\ \sigma_{12} & \sigma_{22} - \sigma_N & \sigma_{23} \\ \sigma_{13} & \sigma_{23} & \sigma_{33} - \sigma_N \end{bmatrix} = 0 \qquad (1.55)$$

$$\sigma_N^3 - (\sigma_{11} + \sigma_{22} + \sigma_{33})\sigma_N^2$$
$$+ (\sigma_{11}\sigma_{22} + \sigma_{22}\sigma_{33} + \sigma_{33}\sigma_{11} - \sigma_{12}^2 - \sigma_{23}^2 - \sigma_{13}^2)\sigma_N \qquad (1.56)$$
$$- (\sigma_{11}\sigma_{22}\sigma_{33} + 2\sigma_{12}\sigma_{23}\sigma_{31} - \sigma_{11}\sigma_{23}^2 - \sigma_{22}\sigma_{13}^2 - \sigma_{33}\sigma_{12}^2) = 0$$

This cubic equation has three roots: the three principal stresses σ_1, σ_2, σ_3. If we want to determine the orientation of the principal stresses with the existing system of axes, we substitute σ_1, σ_2, and σ_3 successively back into Eqs. 1.52, 1.53, and 1.54 and determine the direction cosines $l_{\mu i}$. It should be remembered that we have an additional relationship between these direction cosines:

$$l_{\mu i}^2 = 1$$

1.10.4 Principal Shear Stresses

The principal shear stresses are defined in analogy with the principal normal stresses; they are called τ_1, τ_2, τ_3. The shear stresses in the planes that bisect the principal axes are the highest. They have the following numerical values:

$$\tau_1 = \frac{\sigma_2 - \sigma_3}{2}$$

$$\tau_2 = \frac{\sigma_1 - \sigma_3}{2} \qquad (1.57)$$

$$\tau_3 = \frac{\sigma_1 - \sigma_2}{2}$$

Since $\sigma_1 > \sigma_2 > \sigma_3$, τ_2 is the maximum shear stress acting on the body. This is an important parameter; in materials that fail by shear (as most metals

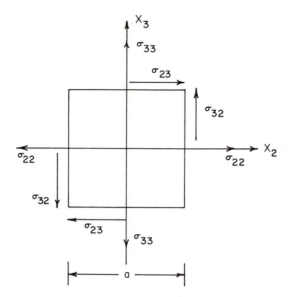

Figure 1.13 Principal shear stresses.

do) the orientation of the maximum shear stress is very important. Figure 1.13 shows the principal shear stresses and the planes on which they are acting.

1.10.5 Stress Invariants

Equation 1.55 has three solutions, which are the three principal stresses. For a given state of stress to which a body is subjected, there is one and only one set of principal stresses. Hence, the solution for Eq. 1.55 is always the same, independent of the orientation of the axes of which $[\sigma_{ij}]$ is taken. This is tantamount to saying that Eq. 1.56 is of the form

$$\sigma^3 - I_1\sigma^2 - I_2\sigma - I_3 = 0 \tag{1.58}$$

I_1, I_2, and I_3 are constants for a given state of stress; they are called the first, second, and third stress invariants, respectively. We have

$$I_1 = \sigma_{11} + \sigma_{22} + \sigma_{33}$$

$$I_2 = -(\sigma_{11}\sigma_{22} + \sigma_{22}\sigma_{33} + \sigma_{33}\sigma_{11}) + \sigma_{12}^2 + \sigma_{13}^2 + \sigma_{23}^2 \tag{1.59}$$

$$I_3 = \sigma_{11}\sigma_{22}\sigma_{33} + 2\sigma_{12}\sigma_{23}\sigma_{31} - (\sigma_{11}\sigma_{23}^2 + \sigma_{22}\sigma_{13}^2 + \sigma_{33}\sigma_{12}^2)$$

Since they are invariant, we can use the orientation for which the stresses are principal:

$$I_1 = \sigma_1 + \sigma_2 + \sigma_3$$

$$I_2 = -(\sigma_1\sigma_2 + \sigma_2\sigma_3 + \sigma_1\sigma_3) \tag{1.60}$$

$$I_3 = \sigma_1\sigma_2\sigma_3$$

I_1, I_2, and I_3 describe the state of stress as well as σ_1, σ_2, and σ_3.

1.10.6 Special Forms of the Stress Tensor

There are special states of stress that occur rather frequently or that are assumed in the simplifications made to render the problems more tractable. They are given below, referred to their principal stresses, and are sketched in Fig. 1.14.

1. Uniaxial stress: it is given by

$$\begin{bmatrix} \sigma & 0 & 0 \\ 0 & 0 & 0 \\ 0 & 0 & 0 \end{bmatrix}$$

2. Biaxial stress:

$$\begin{bmatrix} \sigma_1 & 0 & 0 \\ 0 & \sigma_2 & 0 \\ 0 & 0 & 0 \end{bmatrix}$$

3. Hydrostatic pressure:

$$\begin{bmatrix} -p & 0 & 0 \\ 0 & -p & 0 \\ 0 & 0 & -p \end{bmatrix}$$

It occurs in fluids; it is a special case of triaxial stress, when the three principal stresses are equal.

4. Pure shear:

$$\begin{bmatrix} 0 & \sigma & 0 \\ \sigma & 0 & 0 \\ 0 & 0 & 0 \end{bmatrix}$$

It is a special case of biaxial stress, as will be seen by performing a 45° rotation [Fig. 1.14(d)]. Applying the transformation law, we have

$$\text{old}$$

$$\text{new} \quad \begin{array}{c|cc} & l_{11} & l_{12} \\ \hline & l_{21} & l_{22} \end{array}$$

$$l_{11} = \frac{\sqrt{2}}{2} \qquad l_{12} = \frac{\sqrt{2}}{2}$$

$$l_{22} = \frac{\sqrt{2}}{2} \qquad l_{21} = -\frac{\sqrt{2}}{2}$$

From Eq. 1.29,

$$\sigma'_{11} = 2l_{11}l_{12}\sigma_{12} = \sigma_{12}$$

$$\sigma'_{22} = 2l_{21}l_{22}\sigma_{12} = -\sigma_{12}$$

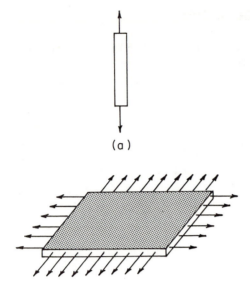

(a)

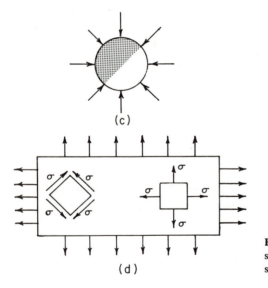

(b)

(c)

(d)

Figure 1.14 Examples of special states of stress: (a) uniaxial; (b) biaxial; (c) hydrostatic; (d) pure shear.

Hence, we have

$$\begin{bmatrix} 0 & \sigma \\ \sigma & 0 \end{bmatrix} = \begin{bmatrix} \sigma & 0 \\ 0 & -\sigma \end{bmatrix}$$

1.10.7 Important Stresses in Plasticity

The hydrostatic (or dilatational) and deviatoric (or shear) components of stress, as well as the octahedral and effective stresses, are often used in plasticity; they are described in this section.

It is often useful to decompose $[\sigma_{ij}]$ into two components:

1. Spherical, or hydrostatic, or dilatational stress tensor, designated by $[\delta_{ij}\sigma''_{ij}]$, where δ_{ij} is Kronecker's delta ($\delta_{ij} = 1$ when $i = j$, and $\delta_{ij} = 0$ when $i \neq j$)
2. Deviatoric, or deviator, or shear stress tensor, designated by $[\sigma'_{ij}]$

This decomposition is due to the fact that materials can undergo high hydrostatic stresses without plastic deformation; the deviatoric stresses, on the other hand, are responsible for the failure (under shear). An example that allows one to grasp this concept easily is the shrimp. Shrimp can live at great ocean depths. Consider a shrimp at a depth of 100 m. The hydrostatic pressure acting on it is equal to about 1.0 MPa. However, this tensor does not seem to bother the shrimp. On the other hand, if it is squeezed between our fingers, it will crush at an applied stress (uniaxial) of about 0.5 MPa. The difference between the two cases is that the deviatoric stresses are zero for hydrostatic pressure and equal to 0.25 MPa for the uniaxial compression case. We have the following equations:

$$\sigma_{ij} = \sigma'_{ij} + \delta_{ij}\sigma''_{ij}$$

$$[\sigma_{ij}] = \begin{bmatrix} \sigma_{11} - \sigma_m & \sigma_{12} & \sigma_{13} \\ \sigma_{12} & \sigma_{22} - \sigma_m & \sigma_{23} \\ \sigma_{13} & \sigma_{23} & \sigma_{33} - \sigma_m \end{bmatrix} + \begin{bmatrix} \sigma_m & 0 & 0 \\ 0 & \sigma_m & 0 \\ 0 & 0 & \sigma_m \end{bmatrix} \quad (1.61)$$

It can be found that σ_m has to be equal to

$$\sigma_m = \frac{\sigma_{11} + \sigma_{22} + \sigma_{33}}{3} = \frac{I_1}{3} \quad (1.62)$$

$[\sigma'_{ij}]$ is responsible for the distortion, while $[\sigma''_{ij}]$ produces the change in volume. Referring $[\sigma'_{ij}]$ to its principal axes, we have

$$\sigma'_1 = \sigma_1 - \sigma_m = \frac{2\sigma_1 - \sigma_2 - \sigma_3}{3}$$

$$\sigma'_2 = \sigma_2 - \sigma_m = \frac{2\sigma_2 - \sigma_1 - \sigma_3}{3} \quad (1.63)$$

$$\sigma'_3 = \sigma_3 - \sigma_m = \frac{2\sigma_3 - \sigma_2 - \sigma_1}{3}$$

It should be noted that

$$\sigma'_1 + \sigma'_2 + \sigma'_3 = 0$$

The invariants of the deviatoric stress are found in the same way as the invariants of stress (Section 1.10.5). They are designated by J_1, J_2, and J_3.

$$[\sigma'_{ij}] = \begin{bmatrix} \sigma'_{11} & \sigma'_{12} & \sigma'_{13} \\ \sigma'_{21} & \sigma'_{22} & \sigma'_{23} \\ \sigma'_{31} & \sigma'_{32} & \sigma'_{33} \end{bmatrix}$$

The solution of the equation for deviatoric stresses is

$$\sigma'^3 - J_1\sigma'^2 - J_2\sigma' - J_3 = 0$$

Hence,

$$J_1 = \sigma'_{11} + \sigma'_{22} + \sigma'_{33} = \sigma'_1 + \sigma'_2 + \sigma'_3 = 0$$

$$J_2 = -(\sigma'_1\sigma'_2 + \sigma'_2\sigma'_3 + \sigma'_1\sigma'_3) \qquad (1.64)$$

$$J_3 = \sigma'_1\sigma'_2\sigma'_3$$

The *octahedral* stresses are the stresses acting on the faces of a regular octahedron composed of the eight (111) planes. Figure 1.15 shows these faces. The angles of the face with the three axes are the same. From Section 1.10.2 we have

$$l_{\mu 1}^2 + l_{\mu 2}^2 + l_{\mu 3}^2 = 1$$

$$3l_{\mu 1}^2 = 1$$

$$l_{\mu 1} = \cos 54°44' = \frac{1}{\sqrt{3}}$$

Using Eq. 1.48, we obtain the normal octahedral stress:

$$\sigma_{\text{oct}} = \tfrac{1}{3}(\sigma_1 + \sigma_2 + \sigma_3) \qquad (1.65)$$

The shear component is found by finding the total stress and decomposing it into a normal component (Eq. 1.65) and into a shear component.

$$\tau_{\text{oct}} = \tfrac{1}{3}[(\sigma_1 - \sigma_2)^2 + (\sigma_1 - \sigma_3)^2 + (\sigma_3 - \sigma_2)^2]^{1/2} \qquad (1.66)$$

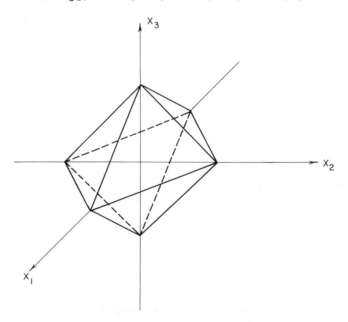

Figure 1.15 Faces on which octahedral stresses act.

It should be noticed that the normal and shear octahedral stresses are equal to the hydrostatic and deviatoric stresses, respectively. Hence, σ_{oct} does not introduce distortions; these are produced by τ_{oct}.

The *effective* (or significant) stress is also used often. It is defined as

$$\sigma_{eff} = \frac{\sqrt{2}}{2}[(\sigma_1 - \sigma_2)^2 + (\sigma_1 - \sigma_3)^2 + (\sigma_3 - \sigma_2)^2]^{1/2} \qquad (1.67)$$

It represents the conversion of a triaxial state of stress into an "effective" uniaxial stress. In other words, σ_{eff} is the uniaxial stress that would produce (approximately) the same effect as the triaxial stress state defined by σ_1, σ_2, and σ_3.

1.11 STRAIN

The state of deformation in a body will be treated analytically. Deformation and strain will be discussed for uni-, bi-, and tridimensional strains.

1.11.1 Unidimensional Deformation

Consider a linear body (e.g., a rubber band); mark an origin O and a general point P on this body. The distance between P and O is x [Fig. 1.16(a)]. Point P moves to P' (and then distance x increases to $x + u$) when the body is deformed. The plot of x versus u [Fig. 1.16(b)] shows that the body is undergoing homogeneous stretching. Taking a second point Q close to P [Fig. 1.16(c)], we have their distance Δx becoming $\Delta x + \Delta u$, where x and u represent position and displacement, respectively. The deformation of the segment PQ is defined as

$$\frac{\text{increase in length}}{\text{initial length}} = \frac{P'Q' - PQ}{PQ} = \frac{\Delta u}{\Delta x}$$

The deformation at P is defined as

$$e = \lim_{\Delta x \to 0} \frac{\Delta u}{\Delta x} = \frac{du}{dx} \qquad (1.68)$$

1.11.2 Bidimensional Deformation

Consider now a bidimensional body being stretched; a good example is a rubber sheet. Taking O as the origin (imagine that the rubber sheet is nailed down at this point) and defining two arbitrary points P and Q on the sheet, one has the configuration of Fig. 1.17(a). After a certain deformation, P and Q move to P' and Q', respectively. These points are given by the coordinates

$P(x_1, x_2)$ $\qquad\qquad\qquad$ $P'(x_1 + u_1, x_2 + u_2)$

$Q(x_1 + \Delta x_1, x_2 + \Delta x_2)$ $\qquad$ $Q'(x_1 + u_1 + \Delta x_1 + \Delta u_1, x_2 + u_2 + \Delta x_2 + \Delta u_2)$

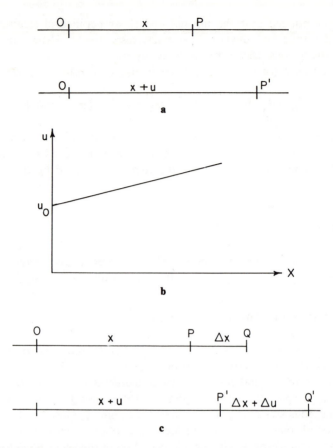

Figure 1.16 (a) Unidimensional body being deformed; (b) plot of displacement versus position; (c) strain in unidimensional body.

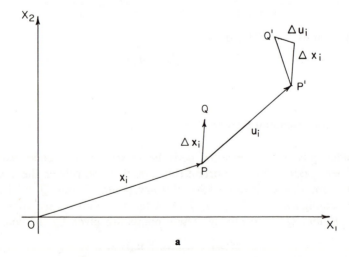

Figure 1.17 (a) Bidimensional body being deformed. (Continued on next page.)

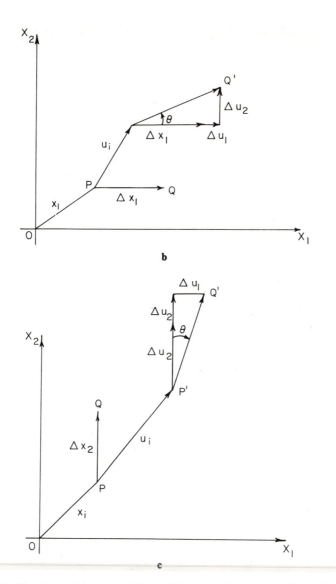

b

c

Figure 1.17 (Continued) (b) segment PQ parallel to Ox_1; (c) segment PQ parallel to Ox_2.

We need four parameters to describe deformation in the bidimensional case:

$$e_{11} = \frac{\partial u_1}{\partial x_1}; \qquad e_{22} = \frac{\partial u_2}{\partial x_2}; \qquad e_{12} = \frac{\partial u_1}{\partial x_2}; \qquad e_{21} = \frac{\partial u_2}{\partial x_1}$$

In indicial notation,

$$e_{ij} = \frac{\partial u_i}{\partial x_j} \qquad (i, j = 1, 2) \tag{1.69}$$

The physical significance of $[e_{ij}]$ is given next. To do this, we have to define the components of displacement $[u_i]$ as a function of the components of position $[x_i]$.

$$\Delta u_1 = \frac{\partial u_1}{\partial x_1}\Delta x_1 + \frac{\partial u_1}{\partial x_2}\Delta x_2$$

$$\Delta u_2 = \frac{\partial u_2}{\partial x_1}\Delta x_1 + \frac{\partial u_2}{\partial x_2}\Delta x_2$$

In indicial notation,

$$\Delta u_i = \frac{\partial u_i}{\partial x_j}\Delta x_j \tag{1.70}$$

Substituting Eq. 1.69 to 1.70 yields

$$\Delta u_i = e_{ij}\,\Delta x_j \tag{1.71}$$

Since both Δu_i and Δx_j are vectors, $[e_{ij}]$ is a second-rank tensor (see Eq. 1.19). There are two special positions of the segment PQ that help in understanding the physical significance of the various components of $[e_{ij}]$. The first position, parallel to Ox_1, is shown in Fig. 1.17(b). We have $\Delta x_2 = 0$. After deformation, we have

$$\Delta u_1 = \frac{\partial u_1}{\partial x_1}\Delta x_1 = e_{11}\,\Delta x_1$$

$$\Delta u_2 = \frac{\partial u_2}{\partial x_1}\Delta x_1 = e_{21}\,\Delta x_1$$

Hence, e_{11} is the measure of the extension per unit length along Ox_1. The angle θ [Fig. 1.17(b)] is given by

$$\tan \theta = \frac{\Delta u_2}{\Delta x_1 + \Delta u_1}$$

If small extensions are assumed, we have, approximately,

$$\tan \theta \simeq \frac{\Delta u_2}{\Delta x_1}$$

Since Δu_2 is small with respect to Δx_1,

$$\theta \simeq \frac{\Delta u_2}{\Delta x_1} = e_{21}$$

Hence, e_{21} is approximately equal to the angle of rotation of a segment initially parallel to Ox_1, toward Ox_2. To conceptualize e_{12} we use the special configuration of Fig. 1.17(c). In this case, PQ is parallel to Ox_2. Repeating the foregoing procedure, *mutatis mutandis*, we arrive at: e_{22} measures the extension, per unit length, of a segment along Ox_2; e_{12} measures the angle θ of rotation of a segment parallel to Ox_2, toward Ox_1.

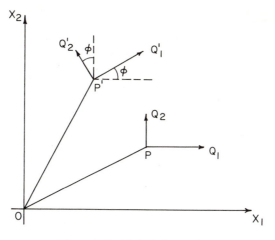

Figure 1.18 Rigid-body rotation.

The tensor e_{ij} thus provides a good representation of the deformation undergone by a two-dimensional body. It measures the extensions and rotations due to the deformation. There is, however, one problem. In case there is no deformation or extension, $[e_{ij}]$ should be zero. Figure 1.18 shows that such is not the case. A rigid body rotates by θ in Fig. 1.18. The components of $[e_{ij}]$ are

$$e_{11} = \frac{\partial u_1}{\partial x_1} = 0$$

$$e_{22} = \frac{\partial u_2}{\partial x_2} = 0$$

$$e_{21} = \phi \qquad \text{and} \qquad e_{12} = -\phi$$

Therefore, it can be concluded that $[e_{ij}]$ is not a perfect representation of deformation. To subtract the effect of rigid-body rotation, we decompose $[e_{ij}]$ into

$$e_{ij} = \epsilon_{ij} + \omega_{ij} \tag{1.72}$$

where

$$\epsilon_{ij} = \tfrac{1}{2}(e_{ij} + e_{ji}) \tag{1.73}$$

$$\omega_{ij} = \tfrac{1}{2}(e_{ij} - e_{ji}) \tag{1.74}$$

$[\epsilon_{ij}]$ is a symmetric tensor, since

$$\epsilon_{ij} = \epsilon_{ji} = \tfrac{1}{2}(e_{ij} + e_{ji}) = \tfrac{1}{2}(e_{ji} + e_{ij})$$

ω_{ij} is an antisymmetric tensor. In matricial notation, we have

$$\begin{bmatrix} e_{11} & e_{12} \\ e_{21} & e_{22} \end{bmatrix} = \begin{bmatrix} e_{11} & \tfrac{1}{2}(e_{12} + e_{21}) \\ \tfrac{1}{2}(e_{12} + e_{21}) & e_{22} \end{bmatrix} + \begin{bmatrix} 0 & \tfrac{1}{2}(e_{12} - e_{21}) \\ \tfrac{1}{2}(e_{21} - e_{12}) & 0 \end{bmatrix}$$

$[\epsilon_{ij}]$ represents deformation, while $[\omega_{ij}]$ represents the rotation of a rigid body. In Fig. 1.19 $[e_{ij}]$ is decomposed graphically.

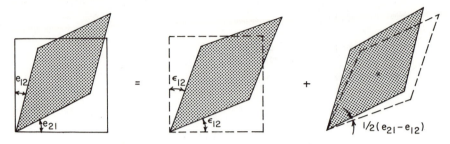

Figure 1.19 Decomposition of $[e_{ij}]$ into $[\epsilon_{ij}]$ and $[\omega_{ij}]$.

1.11.3 Tridimensional Deformation

The same procedure is followed for tridimensional strain:

$$e_{ij} = \frac{1}{2}\left(\frac{\partial u_i}{\partial x_j} + \frac{\partial u_j}{\partial x_i}\right) \qquad (i, j = 1, 2, 3) \qquad (1.75)$$

The significance of the components of $[e_{ij}]$ is the same as in the bidimensional case; e_{11}, e_{22}, e_{33} are extensions, while e_{12}, e_{21}, etc., are distortions. To eliminate the effect of rigid-body rotation—which is not, obviously, a deformation—we define $[\epsilon_{ij}]$ in such a way that

$$e_{ij} = \epsilon_{ij} + \omega_{ij}$$

Again, $[\epsilon_{ij}]$ is a symmetric and $[\omega_{ij}]$ an antisymmetric tensor. Since $[\epsilon_{ij}]$, the strain tensor, is symmetric, it can be treated as a quadratic curve. Hence, it has principal directions, and we have principal strains.

$$\begin{bmatrix} \epsilon_{11} & \epsilon_{12} & \epsilon_{13} \\ \epsilon_{12} & \epsilon_{22} & \epsilon_{23} \\ \epsilon_{13} & \epsilon_{23} & \epsilon_{33} \end{bmatrix} \Rightarrow \begin{bmatrix} \epsilon_1 & 0 & 0 \\ 0 & \epsilon_2 & 0 \\ 0 & 0 & \epsilon_3 \end{bmatrix}$$

The procedure to obtain the principal strains is identical to the one used for the principal stresses (Eq. 1.55). It should be noted that the shear strains are zero when the unit cube is oriented along the principal axes; this is shown in Fig. 1.20. The faces of the cube remain perpendicular on straining. The change in volume of the cube is called dilatation (or contraction) and is equal to

$$\Delta = (1 + \epsilon_1)(1 + \epsilon_2)(1 + \epsilon_3) - 1$$

For small deformations we can neglect the cross products of the strains.

$$\Delta \approx \epsilon_1 + \epsilon_2 + \epsilon_3$$

The shear strains γ_{ij} are often used instead of ϵ_{ij}. We have

$$\gamma_{ij} = 2\epsilon_{ij}$$

It should be emphasized that they are not true tensor quantities; they are called engineering shear strains. The strain matrix takes the configuration (notice the $\frac{1}{2}$ terms)

$$\begin{bmatrix} \epsilon_{11} & \frac{1}{2}\gamma_{12} & \frac{1}{2}\gamma_{13} \\ \frac{1}{2}\gamma_{12} & \epsilon_{22} & \frac{1}{2}\gamma_{23} \\ \frac{1}{2}\gamma_{13} & \frac{1}{2}\gamma_{23} & \epsilon_{33} \end{bmatrix} \tag{1.76}$$

A state of strain of particular importance is plane strain. When one of the principal strains is zero (deformation constrained in one direction) we have a plane-strain state:

$$\begin{bmatrix} \epsilon_1 & 0 & 0 \\ 0 & \epsilon_2 & 0 \\ 0 & 0 & 0 \end{bmatrix}$$

This type of strain occurs when one of the dimensions of the body is very large with respect to the other two or when flow of material in one direction is inhibited by some other means.

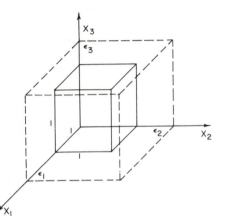

Figure 1.20 Principal strains.

1.11.4 Strain Invariants—
Hydrostatic, Deviatoric, and Octahedral Strains

In analogy with stresses, the strain tensor $[\epsilon_{ij}]$ can receive the same treatment. Hence, there are strain invariants I'_1, I'_2, and I'_3, just as in Section 1.10.5; octahedral strains, just as in Section 1.10.7; and hydrostatic (or dilatational) and deviatoric strains, just as in Section 1.10.7. We have

$$I'_1 = \epsilon_1 + \epsilon_2 + \epsilon_3 \tag{1.77}$$

$$I'_2 = -(\epsilon_1\epsilon_2 + \epsilon_2\epsilon_3 + \epsilon_1\epsilon_3) \tag{1.78}$$

$$I'_3 = \epsilon_1\epsilon_2\epsilon_3 \tag{1.79}$$

$$\epsilon_{\text{oct}} = \tfrac{1}{3}(\epsilon_1 + \epsilon_2 + \epsilon_3) \tag{1.80}$$

$$\epsilon_{\text{eff}} = \frac{\sqrt{2}}{3}[(\epsilon_1 - \epsilon_2)^2 + (\epsilon_1 - \epsilon_3)^2 + (\epsilon_2 - \epsilon_3)^2]^{1/2} \tag{1.81}$$

$$\epsilon_{ij} = \epsilon'_{ij} + \delta_{ij}\epsilon''_{ij} \tag{1.82}$$

$$J_1' = \epsilon_{11}' + \epsilon_{22}' + \epsilon_{33}' = \epsilon_1' + \epsilon_2' + \epsilon_3' \tag{1.83}$$

$$J_2' = -(\epsilon_1'\epsilon_2' + \epsilon_2'\epsilon_3' + \epsilon_1'\epsilon_3') \tag{1.84}$$

$$J_3' = \epsilon_1'\epsilon_2'\epsilon_3' \tag{1.85}$$

1.11.5 Strain Energy (or Deformation Energy)

When work is done on a body, the dimensions of this body change. This work (W) is converted into heat (q) and an increase in internal energy (u) according to the second law of thermodynamics:

$$du = \delta q - \delta W$$

For solids, elastic work produces amounts of heat that can be neglected without too much error. For gases, this is not the case. Hence, the work is converted into internal energy; the energy stored in the material by this process is commonly called strain energy. Figure 1.21(a) shows an elemental cube being extended in tension by σ_{11}. The work being done is the force versus the increase in length. Figure 1.21(b) shows the proportionality between force and extension characterized by elastic deformation. The work is equal to the area OAB:

$$dW_{11} = \tfrac{1}{2}\sigma_{11}(dx_2 \, dx_3)\epsilon_{11} \, dx_1$$

$$= \tfrac{1}{2}\sigma_{11}\epsilon_{11}(dx_1 \, dx_2 \, dx_3)$$

This work, per unit volume, is

$$W_{11} = \frac{1}{2}\,\sigma_{11}\epsilon_{11} = \frac{dW_{11}}{dx_1 \, dx_2 \, dx_3}$$

Considering the work performed by the other normal stresses, we obtain similar equations. For the shear stresses, we assume that they generate only shear strains. Figure 1.21(c) shows this configuration. σ_{31} generates the shear strain $\gamma_{31}(=2\epsilon_{31})$. Since it is applied on the face with dimensions $dx_2 \, dx_3$, and moves it by an amount equal to $\gamma_{31} \, dx_1$, the total work is

$$dW_{31} = \tfrac{1}{2}\sigma_{31}(dx_2 \, dx_3)\gamma_{31} \, dx_1$$

The work per unit volume is

$$W_{31} = \tfrac{1}{2}\sigma_{13}\gamma_{13}$$

The principle of superposition states that two or more stresses (or strains) may be combined by superposition. Applying it, we can calculate the total work:

$$W = W_{11} + W_{22} + W_{33} + W_{12} + W_{13} + W_{23}$$

$$= \tfrac{1}{2}(\sigma_{11}\epsilon_{11} + \sigma_{22}\epsilon_{22} + \sigma_{33}\epsilon_{33} + 2\sigma_{12}\epsilon_{12} + 2\sigma_{13}\epsilon_{13} + 2\sigma_{23}\epsilon_{23})$$

In indicial notation, we may write

$$U = W = \tfrac{1}{2}\sigma_{ij}\epsilon_{ij} \tag{1.86}$$

Equation 1.86 expresses the strain energy in indicial notation.

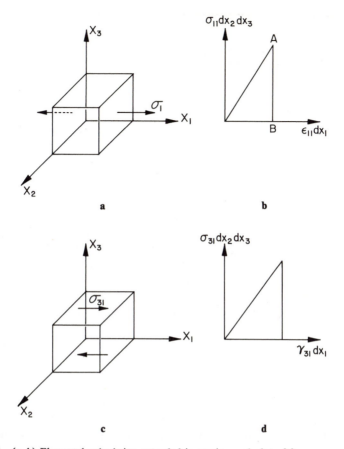

Figure 1.21 (a, b) Elemental cube being extended in tension and plot of force versus increase in length; (c, d) elemental cube being sheared by stress σ_{31} and plot of force versus displacement.

1.12 RELATIONSHIP BETWEEN STRESS AND STRAIN

1.12.1 Hooke's Law

A solid body undergoes deformation upon being subjected to a stress. When this strain is recoverable (i.e., it disappears once the stress has been removed), we are dealing with elastic behavior. One particular case of elastic behavior is when the stress and strain are proportional. For many solids—metals, ceramics, and certain plastics—the stress is observed to be proportional to strain. These bodies are called *Hookean* bodies. Actually, there are deviations from ideal Hookean behavior, even at low values of strain. Figure 1.22 illustrates this phenomenon. A tensile specimen is loaded to successively higher stress levels and unloaded.[3] At the loading step, step 6, a permanent strain of 10^{-6} is observed. This is, so to speak, the true elastic limit. However, this strain level cannot be detected in common tensile testing machines, and the elastic

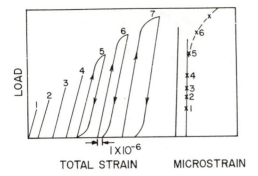

LOAD

TOTAL STRAIN   MICROSTRAIN

1×10^{-6}

Figure 1.22 Microstrain tests showing that the onset of plastic deformation occurs at $\epsilon_p = 10^{-6}$; this is only a fraction of the conventional yield stress. (Reprinted with permission from [3], p. 990.)

regime is considered to go up to much higher stress levels. For instance,[3] copper has a microyield stress of ~27 MPa($\epsilon_p = 2 \times 10^{-3}$); low-carbon steel (AISI 1010) has micro- and macroyield stresses of 75 MPa and 300 MPa, respectively. Hence, three-fourths of what is called elastic regime is, if we want to be rigorous, a microplastic regime.

Imagine a bar of initial length l_0 and cross-sectional area A_0 being subjected to a weight P. Its length increases to l, and the stress and strain along its length can be expressed as

$$\sigma = \frac{P}{A} \simeq \frac{P}{A_0} \qquad \epsilon = \int_{l_0}^{l} \frac{dl}{l} \simeq \frac{\Delta l}{l_0}$$

The equality between what is called true and engineering stress, and true strain and engineering strain, is due to the fact that the total elastic strain, for metals, rarely exceeds 0.005. The definitions of true and engineering stresses and strains are given in Section 1.2. If the bar exhibits a Hookean response, we have

$$\epsilon = S\sigma \tag{1.87}$$

S is a proportionality constant called *compliance*. Alternatively, Eq. 1.87 can be expressed as

$$\sigma = C\epsilon \tag{1.88}$$

C is called stiffness. It should be noted that, for some unknown reason, the letters describing these parameters are the initials of the names, but inverted: S for compliance, C for stiffness.

When we have a tridimensional state of stress and strain, we have second-rank tensors. Two second-rank tensors are related by a fourth-rank tensor. Each component of the strain is related to all stress components, and vice versa.

$$\epsilon_{11} = S_{1111}\sigma_{11} + S_{1112}\sigma_{12} + S_{1113}\sigma_{13}$$
$$+ S_{1112}\sigma_{21} + S_{1122}\sigma_{22} + S_{1123}\sigma_{23}$$
$$+ S_{1131}\sigma_{31} + S_{1132}\sigma_{32} + S_{1133}\sigma_{33}$$

In indicial notation,

$$\epsilon_{ij} = S_{ijkl} \sigma_{kl} \qquad (1.89)$$

and for the stiffness,

$$\sigma_{ij} = C_{ijkl} \epsilon_{kl} \qquad (1.90)$$

There are, in the most general case, 81 components of the stiffness (or compliance). But when there are no body moments, both the stress and strain are symmetric tensors. Hence,

$$S_{ijkl} = S_{jikl}$$

$$S_{ijkl} = S_{ijlk}$$

This reduces the number of compliance (and stiffness) components to 36.

A fourth-rank tensor is, as a second-rank tensor, defined by the transformation law for fourth-rank tensors. The transformation law for fourth-rank tensors has the following form:

$$T'_{ijkl} = l_{im} l_{jn} l_{ko} l_{lp} T_{mnop} \qquad (1.91)$$

Comparing Eqs. 1.29 and 1.91, we see that they are of an analogous nature. We will try to show that the elastic compliance matrix obeys this law. The following sequence of operations will be performed:

$$\epsilon' \xrightarrow{\ 1.92\ } \epsilon \xrightarrow{\ 1.93\ } \sigma \xrightarrow{\ 1.94\ } \sigma'$$

Equations 1.92, 1.93, and 1.94 are known:

$$\epsilon'_{ij} = l_{ik} l_{jl} \epsilon_{kl} \qquad (1.92)$$

$$\epsilon_{kl} = S_{klmn} \sigma_{mn} \qquad (1.93)$$

$$\sigma_{mn} = l_{om} l_{pn} \sigma'_{op} \qquad (1.94)$$

Substituting Eq. 1.94 into Eq. 1.93 and Eq. 1.93 into Eq. 1.92, we have

$$\epsilon'_{ij} = l_{ik} l_{jl} l_{om} l_{pn} S_{klmn} \sigma'_{op} \qquad (1.95)$$

But $[\epsilon'_{ij}]$ and $[\sigma'_{ij}]$ are second-rank tensors and they retain their physical identity with the change in coordinate system. Hence,

$$\epsilon'_{ij} = S'_{ijop} \sigma'_{op} \qquad (1.96)$$

Equations 1.95 and 1.96 are identical. This implies that

$$S'_{ijop} = l_{ik} l_{jl} l_{om} l_{pn} S_{klmn} \qquad (1.97)$$

This shows that S_{ijkl} is a fourth-rank tensor. If we go from the new system of axes to the old one, the position of the subscripts is inverted; the dummy suffixes are as far as possible,

$$S_{ijop} = l_{ki} l_{lj} l_{mo} l_{np} S'_{klmn} \qquad (1.98)$$

For the stiffness one has equations identical to Eqs. 1.97 and 1.98.

1.12.2 Matrix Notation

A different notation is introduced at this point. It is called matrix notation and it simplifies the treatment of elastic stiffness and compliance. Two subscripts are replaced by one, according to the rule

$$\begin{bmatrix} \sigma_{11} & \sigma_{12} & \sigma_{13} \\ \sigma_{21} & \sigma_{22} & \sigma_{23} \\ \sigma_{31} & \sigma_{32} & \sigma_{33} \end{bmatrix} \Rightarrow \begin{bmatrix} \sigma_1 & \sigma_6 & \sigma_5 \\ \sigma_6 & \sigma_2 & \sigma_4 \\ \sigma_5 & \sigma_4 & \sigma_3 \end{bmatrix}$$

$$\begin{bmatrix} \epsilon_{11} & \epsilon_{12} & \epsilon_{13} \\ \epsilon_{21} & \epsilon_{22} & \epsilon_{23} \\ \epsilon_{31} & \epsilon_{32} & \epsilon_{33} \end{bmatrix} \Rightarrow \begin{bmatrix} \epsilon_1 & \frac{1}{2}\epsilon_6 & \frac{1}{2}\epsilon_5 \\ \frac{1}{2}\epsilon_6 & \epsilon_2 & \frac{1}{2}\epsilon_4 \\ \frac{1}{2}\epsilon_5 & \frac{1}{2}\epsilon_4 & \epsilon_3 \end{bmatrix}$$

The arrows indicate the direction in which one proceeds from 1 to 6. It should be noted that

$$\epsilon_6 = 2\epsilon_{12}$$

$$\epsilon_5 = 2\epsilon_{13}$$

$$\epsilon_4 = 2\epsilon_{23}$$

while the stresses are numerically equal. For the compliances, it can be shown (by expanding the equations) that

$$S_{ijkl} = S_{mn} \qquad \text{when } m \text{ and } n \text{ are 1, 2, or 3}$$

$$2S_{ijkl} = S_{mn} \qquad \text{when } m \text{ or } n \text{ are 4, 5, or 6}$$

$$4S_{ijkl} = S_{mn} \qquad \text{when both } m \text{ and } n \text{ are 4, 5, or 6}$$

where i,j and k,l become m and n, respectively; the indices in matrix notation. For instance,

$$\epsilon_1 = S_{11}\sigma_1 + \tfrac{1}{2}S_{16}\sigma_6 + \tfrac{1}{2}S_{15}\sigma_5 + \tfrac{1}{2}S_{16}\sigma_6$$
$$+ S_{12}\sigma_2 + \tfrac{1}{2}S_{14}\sigma_4 + \tfrac{1}{2}S_{15}\sigma_5 + \tfrac{1}{2}S_{14}\sigma_4 + S_{13}\sigma_3$$

In general,

$$\epsilon_i = S_{ij}\sigma_j \qquad (i, j = 1, 2, 3, 4, 5, 6) \tag{1.99}$$

The factors of 2 and 4 were introduced in order to maintain, in matrix notation, the same type of equation (1.99) as in tensor notation (1.89).

For the stiffnesses we have the same transformation and the same equation as Eq. 1.99 except that no factors of 2 or 4 are required.

$$C_{ijkl} = C_{mn} \qquad \text{for any } m \text{ and } n$$

The matricial notation allows expressing the stiffnesses and compliances as 6×6 matrices:

$$\begin{bmatrix} S_{11} & S_{12} & S_{13} & S_{14} & S_{15} & S_{16} \\ S_{21} & S_{22} & S_{23} & S_{24} & S_{25} & S_{26} \\ S_{31} & S_{32} & S_{33} & S_{34} & S_{35} & S_{36} \\ S_{41} & S_{42} & S_{43} & S_{44} & S_{45} & S_{46} \\ S_{51} & S_{52} & S_{53} & S_{54} & S_{55} & S_{56} \\ S_{61} & S_{62} & S_{63} & S_{64} & S_{65} & S_{66} \end{bmatrix}$$

It should be observed that S_{ij} are *not* the components of a fourth-rank tensor and therefore do not obey the transformation law for fourth-rank tensors. If we want to perform the transformation, we have to return to the tensor notation.

1.12.3 Crystalline Symmetry

We will study how the 36 components of C and S behave for the different crystalline structures. For the most general—and anisotropic—of the cases, there are 36 (6 × 6) components. As the crystalline symmetries are introduced, the number of components is reduced, until we arrive at an isotropic solid; the latter has only two independent components. The first observation to be made is that elasticity is a centrosymmetric property. In other words, if the reference system is transformed by a center-of-symmetry operation, the components of C and S should remain unchanged. Figure 1.23 shows a center-of-symmetry operation. The direction cosines are

	x_1	x_2	x_3
x'_1	-1	0	0
x'_2	0	-1	0
x'_3	0	0	-1

Hence, we can use the symbol

$$l_{ij} = -\delta_{ij}$$

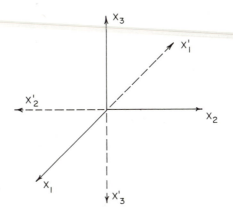

Figure 1.23 Center-of-symmetry operation.

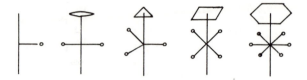

Figure 1.24 Symmetry operations and symbols.

where δ_{ij} is Kronecker's delta. We have, for the compliance,

$$S'_{ijkl} = l_{im}l_{jn}l_{ko}l_{lp}S_{mnop}$$

$$S'_{ijkl} = (-\delta_{im})(-\delta_{jn})(-\delta_{ko})(-\delta_{lp})S_{mnop}$$

$$S'_{ijkl} = \delta_{im}\delta_{jn}\delta_{ko}\delta_{lp}S_{mnop}$$

Hence, we have

$$S'_{1111} = S_{1111}$$

$$S'_{1212} = S_{1212} \qquad \text{etc.}$$

The different crystal systems can be characterized exclusively by their symmetries. The proof of this is beyond the objectives of the book. However, it is sufficient to say that the cubic system can be perfectly described by four threefold rotations. The seven crystalline systems can be perfectly described by their axes of rotation. The Hermann–Mauguin notation for the rotation axes shown in Fig. 1.24 is 1, 2, 3, 4, and 6 (for one-, two-, three-, four-, and sixfold rotations, respectively). Table 1.1 presents the different symmetry operations defining the seven crystal systems. For example, a threefold rotation is a rotation of 120° ($3 \times 120° = 360°$); after 120°, the crystal system comes to an identical position to the initial one. The hexagonal system exhibits a sixfold rotation around the c axis; after each 60°, the structure superimposes itself.

There are essentially two methods to determine the components of C and S for the different crystal systems. The approach in both methods is to effect a transformation of axes according to the symmetry operation. This transformation of axes should maintain the components of C and S unchanged, because

TABLE 1.1 **Minimum Number of Symmetry Operations in the Various Systems**[a]

System	Rotation
Triclinic	None (or center-of-symmetry)
Monoclinic	1 twofold rotation
Orthorhombic	2 perpendicular twofold rotations
Tetragonal	1 fourfold rotation
Rhombohedral	1 threefold rotation
Hexagonal	1 sixfold rotation
Cubic	4 threefold rotations (cube diagonals)

[a] Reprinted with permission from [2], p. 23.

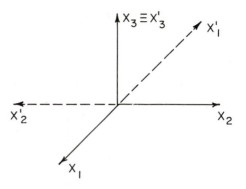

Figure 1.25 Second-order rotation.

the crystal structure returns to itself. The equality of components before and after transformation allows the determination of special relations.

The first method may be called *direct* inspection. Two- and fourfold rotations are studied by direct inspection. The axes have to change in such a way that they make angles that are multiples of $\pi/2$. Hence, we can say that, for instance, x_1 becomes x'_3. A specific example is shown in Fig. 1.25. We have

$$x_1 \longrightarrow -x'_1$$
$$x_2 \longrightarrow -x'_2$$
$$x_3 \longrightarrow x'_3$$

In simplified notation,

$$1 \longrightarrow -1$$
$$2 \longrightarrow -2$$
$$3 \longrightarrow 3$$

Combining these indices into groups of two, in such a way that the signs are identical to the sign of their product, we have

$$11 \longrightarrow 11$$
$$22 \longrightarrow 22$$
$$33 \longrightarrow 33$$
$$12 \longrightarrow 12 \; [(-1) \times (-2) = 1 \times 2]$$
$$21 \longrightarrow 21$$
$$13 \longrightarrow -13$$
$$31 \longrightarrow -31$$
$$23 \longrightarrow -23$$
$$32 \longrightarrow -32$$

Switching to matrix notation, we have

$$1 \longrightarrow 1 \qquad 4 \longrightarrow -4$$

$$2 \longrightarrow 2 \qquad 5 \longrightarrow -5$$

$$3 \longrightarrow 3 \qquad 6 \longrightarrow 6$$

Hence, we have the transformation (only the indices are indicated in the matrices below)

$$
\begin{bmatrix}
11 & 12 & 13 & 14 & 15 & 16 \\
. & 22 & 23 & 24 & 25 & 26 \\
. & . & 33 & 34 & 35 & 36 \\
. & . & . & 44 & 45 & 46 \\
. & . & . & . & 55 & 56 \\
. & . & . & . & . & 66
\end{bmatrix}
\Rightarrow
\begin{bmatrix}
11 & 12 & 13 & -14 & -15 & 16 \\
. & 22 & 23 & -24 & -25 & 26 \\
. & . & 33 & -34 & -35 & 36 \\
. & . & . & 44 & 45 & -46 \\
. & . & . & . & 55 & -56 \\
. & . & . & . & . & 66
\end{bmatrix}
$$

Since both matrices have to represent exactly the same situation (if the crystal exhibits this particular symmetry), we must have

$$C_{14} = -C_{14} = 0$$

$$C_{15} = -C_{15} = 0$$

$$C_{24} = -C_{24} = 0 \qquad \text{etc.}$$

The equalities above hold true only when the stiffnesses (or compliances) are zero. Hence, we have

$$
\begin{bmatrix}
11 & 12 & 13 & 0 & 0 & 16 \\
. & 22 & 23 & 0 & 0 & 26 \\
. & . & 33 & 0 & 0 & 36 \\
. & . & . & 44 & 45 & 0 \\
. & . & . & . & 55 & 0 \\
. & . & . & . & . & 66
\end{bmatrix}
$$

Table 1.1 shows that the twofold rotation executed corresponds to the symmetry that defines the monoclinic system. Hence, the matrix above is the one for the monoclinic system; there are 13 independent elastic constants. If we want to obtain the elastic constants for the orthorhombic and tetragonal system, we take this monoclinic system and apply one twofold (around either Ox_2 or Ox_1) and one fourfold rotation (around Ox_3) to the foregoing matrix, respectively. We obtain

Orthorhombic

$$
\begin{bmatrix}
11 & 12 & 13 & 0 & 0 & 0 \\
. & 22 & 23 & 0 & 0 & 0 \\
. & . & 33 & 0 & 0 & 0 \\
. & . & . & 44 & 0 & 0 \\
. & . & . & . & 55 & 0 \\
. & . & . & . & . & 66
\end{bmatrix}
$$

Tetragonal $(4, \bar{4}, 4/m)$

$$
\begin{bmatrix}
11 & 12 & 13 & 0 & 0 & 16 \\
. & 11 & 13 & 0 & 0 & -16 \\
. & . & 33 & 0 & 0 & 0 \\
. & . & . & 44 & 0 & 0 \\
. & . & . & . & 44 & 0 \\
. & . & . & . & . & 66
\end{bmatrix}
$$

For the rhombohedral, hexagonal, and cubic system, we have three- and sixfold rotations, and the direct inspection method cannot be applied. We must use the analytical method, and we proceed in the following manner.

1. We apply the symmetry operation, determining the direction cosines of the new axes with respect to the old ones.
2. The stresses and strains in the new system are expressed in terms of the old system:

$$\sigma'_{ij} = l_{ik}l_{jl}\sigma_{kl} \tag{1.100}$$

$$\epsilon'_{ij} = l_{ik}l_{jl}\epsilon_{kl} \tag{1.101}$$

3. The following equations are applied:

$$\sigma_{ij} = C_{ijkl}\epsilon_{kl} \tag{1.102}$$

and

$$\sigma'_{ij} = C'_{ijkl}\epsilon'_{kl} \tag{1.103}$$

4. Substituting the σ_{kl} of Eq. 1.100 by the values from Eq. 1.102, we obtain an equation of the form

$$\sigma'_{ij} = f_1(l_{ij}, C_{ijkl}, \epsilon_{kl}) \tag{1.104}$$

5. Taking Eq. 1.103 and substituting the ϵ_{kl} by the values from Eq. 1.101 yields

$$\sigma'_{ij} = f_2(l_{ij}, C'_{ijkl}, \epsilon_{kl}) \tag{1.105}$$

6. We equate Eqs. 1.104 and 1.105. From the initial condition that the symmetry operation does not change the components of the stiffness,

$$C_{ijkl} = C'_{ijkl}$$

Therefore, Eqs. 1.104 and 1.105 become

$$f_1(l_{ij}, C_{ijkl}, \epsilon_{kl}) = f_2(l_{ij}, C_{ijkl}, \epsilon_{kl})$$

Grouping the coefficients of ϵ_{ij}, we arrive at

$$g_i(l_{ij}, C_{ijkl})\epsilon_{ij} = 0$$

Since $\epsilon_{ij} \neq 0$,

$$g_i(l_{ij}, C_{ijkl}) = 0$$

This leads to the determination of C_{ijkl}.

For the cubic system, we have

$$
\begin{bmatrix}
11 & 12 & 12 & 0 & 0 & 0 \\
. & 11 & 12 & 0 & 0 & 0 \\
. & . & 11 & 0 & 0 & 0 \\
. & . & . & 44 & 0 & 0 \\
. & . & . & . & 44 & 0 \\
. & . & . & . & . & 44
\end{bmatrix}
\tag{1.106}
$$

The number of independent elastic constants has been reduced to three.

1.12.4 Relationships for Isotropic Materials

A great number of materials can be treated as isotropic, although they are not microscopically isotropic. The individual grains exhibit the crystalline anisotropy and symmetry; when they form a polycrystalline aggregate and are randomly oriented, the material is macroscopically isotropic (i.e., the elastic constants are the same in all directions). Often, we do not have complete isotropy; if the elastic modulus (E) is different along three perpendicular directions, we have a case of orthotropy.

The elastic modulus can be determined, in a cubic material, along any orientation, from the elastic constants, by application of the equation

$$
\frac{1}{E_{ijk}} = S_{11} - 2\left(S_{11} - S_{12} - \frac{1}{2} S_{44}\right)(l_{i1}^2 l_{j2}^2 + l_{j2}^2 l_{k3}^2 + l_{i1}^2 l_{k3}^2)
\tag{1.107}
$$

E_{ijk} is Young's modulus in the [ijk] direction; l_{i1}, l_{j2}, and l_{k3} are the direction cosines of the direction [ijk]. For a cubic material in which

$$
2(S_{11} - S_{12}) = S_{44}
\tag{1.108}
$$

we should have the same Young's modulus along all directions; hence, it is the isotropy case.

It will be shown below that Eq. 1.108 (or its equivalent, for stiffnesses) is obeyed for the isotropic system. The elastic matrix for an isotropic system is found by applying an arbitrary rotation to the cubic matrix and applying the condition that the elastic constants remain unchanged. This can be done starting from any other crystal system; however, the cubic is the simplest one, and it is easier to start from it. Figure 1.26 shows an arbitrary rotation applied to the Ox_3 axis; for simplicity, an angle of 45° was chosen. The direction cosine matrix is

old

		x_1	x_2	x_3
	x_1'	$\dfrac{\sqrt{2}}{2}$	$\dfrac{\sqrt{2}}{2}$	0
new	x_2'	$-\dfrac{\sqrt{2}}{2}$	$\dfrac{\sqrt{2}}{2}$	0
	x_3'	0	0	1

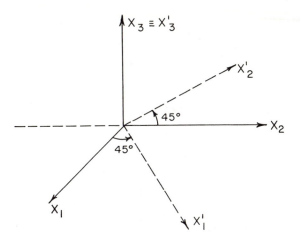

Figure 1.26 45° rotation around Ox_3.

Applying steps 1 through 6 delineated in Section 1.12.3 yields

$$\sigma'_{11} = \tfrac{1}{2}\sigma_{11} + \tfrac{1}{2}\sigma_{22} + \sigma_{12}$$

$$\sigma'_{22} = \tfrac{1}{2}\sigma_{11} + \tfrac{1}{2}\sigma_{22} - \sigma_{12}$$

$$\sigma'_{33} = \sigma_{33}$$

$$\sigma'_{12} = -\tfrac{1}{2}\sigma_{11} + \tfrac{1}{2}\sigma_{22}$$

$$\sigma'_{23} = \left(-\frac{\sqrt{2}}{2}\right)\sigma_{13} + \left(\frac{\sqrt{2}}{2}\right)\sigma_{23}$$

$$\sigma'_{13} = \left(\frac{\sqrt{2}}{2}\right)\sigma_{13} + \left(\frac{\sqrt{2}}{2}\right)\sigma_{23}$$

Changing the stiffnesses to matrix notation, we get

$$\sigma'_{11} = C'_{11}\epsilon'_{11} + C'_{12}\epsilon'_{22} + C'_{13}\epsilon'_{33} + 2C'_{16}\epsilon'_{12} + 2C'_{15}\epsilon'_{13}$$
$$+ 2C'_{14}\epsilon'_{23}$$

From Eq. 1.106,

$$\sigma'_{11} = C'_{11}\epsilon'_{11} + C'_{12}\epsilon'_{22} + C'_{12}\epsilon'_{33}$$

But $C'_{ij} = C_{ij}$ and

$$\sigma'_{11} = C_{11}\epsilon'_{11} + C_{12}\epsilon'_{22} + C_{12}\epsilon'_{33}$$

Substituting the values of ϵ'_{ij} (same equations as the ones above for stress) yields

$$\sigma'_{11} = C_{11}(\tfrac{1}{2}\epsilon_{11} + \tfrac{1}{2}\epsilon_{22} + \epsilon_{12}) + C_{12}(\tfrac{1}{2}\epsilon_{11} + \tfrac{1}{2}\epsilon_{22} - \epsilon_{12}) + C_{12}\epsilon_{33}$$

But we also have

$$\sigma'_{11} = \tfrac{1}{2}(C_{11}\epsilon_{11} + C_{12}\epsilon_{22}) + \tfrac{1}{2}(C_{12}\epsilon_{11} + C_{11}\epsilon_{22}) + 2C_{44}\epsilon_{12} + C_{12}\epsilon_{33}$$

Equating, we get

$$\epsilon_{11}(\tfrac{1}{2}C_{11} + \tfrac{1}{2}C_{12}) + \epsilon_{22}(\tfrac{1}{2}C_{11} + \tfrac{1}{2}C_{12}) + \epsilon_{12}(C_{11} - C_{12}) + C_{12}\epsilon_{33} =$$
$$\epsilon_{11}(\tfrac{1}{2}C_{11} + \tfrac{1}{2}C_{12}) + \epsilon_{22}(\tfrac{1}{2}C_{12} + \tfrac{1}{2}C_{11}) + 2C_{44}\epsilon_{12} + C_{12}\epsilon_{33}$$

We have

$$\tfrac{1}{2}C_{11} + \tfrac{1}{2}C_{12} = \tfrac{1}{2}C_{11} + \tfrac{1}{2}C_{12} \tag{1.109}$$

$$C_{44} = \frac{C_{11} - C_{12}}{2} \tag{1.110}$$

The matrix is

$$\begin{bmatrix} C_{11} & C_{12} & C_{12} & 0 & 0 & 0 \\ \cdot & C_{11} & C_{12} & 0 & 0 & 0 \\ \cdot & \cdot & C_{11} & 0 & 0 & 0 \\ \cdot & \cdot & \cdot & \dfrac{C_{11} - C_{12}}{2} & 0 & 0 \\ \cdot & \cdot & \cdot & \cdot & \dfrac{C_{11} - C_{12}}{2} & 0 \\ \cdot & \cdot & \cdot & \cdot & \cdot & \dfrac{C_{11} - C_{12}}{2} \end{bmatrix} \tag{1.111}$$

For cubic systems, 1.110 does not apply, and we define an anisotropy ratio:

$$A = \frac{2C_{44}}{C_{11} - C_{12}} \neq 0 \tag{1.112}$$

Some metals have high anisotropy ratios, whereas others, such as aluminum and tungsten, have values of A very close to 1. For the latter, even single crystals are almost isotropic.

For the elastic compliances we have

$$\begin{bmatrix} S_{11} & S_{12} & S_{12} & 0 & 0 & 0 \\ \cdot & S_{11} & S_{12} & 0 & 0 & 0 \\ \cdot & \cdot & S_{11} & 0 & 0 & 0 \\ \cdot & \cdot & \cdot & 2(S_{11} - S_{12}) & 0 & 0 \\ \cdot & \cdot & \cdot & \cdot & 2(S_{11} - S_{12}) & 0 \\ \cdot & \cdot & \cdot & \cdot & \cdot & 2(S_{11} - S_{12}) \end{bmatrix} \tag{1.113}$$

Hence, the 81 components of the elastic constants have been reduced to two independent ones, for the cubic system. However, it is not under this form that the elastic constants are usually known. Several parameters are used to describe the elastic properties of isotropic materials. They are related to the elastic stiffness and compliance constants by the equations
Young's modulus:

$$E = \frac{1}{S_{11}} \tag{1.114}$$

Rigidity or shear modulus:

$$G = \frac{1}{2(S_{11} - S_{12})} \tag{1.115}$$

Compressibility (K) and bulk modulus (B):

$$K = \frac{1}{B} = \frac{\epsilon_{11} + \epsilon_{22} + \epsilon_{33}}{-\frac{1}{3}(\sigma_{11} + \sigma_{22} + \sigma_{33})} \tag{1.116}$$

Poisson's ratio:

$$\nu = -\frac{S_{12}}{S_{11}} \tag{1.117}$$

Lamé's constants:

$$\mu = C_{44} = \tfrac{1}{2}(C_{11} - C_{12}) = \frac{1}{S_{44}} \tag{1.118}$$

$$\lambda = C_{12}$$

There are a number of equations interrelating the foregoing parameters. They are given in Table 1.2.

The relationships between stresses and strains for isotropic materials become

$$\epsilon_1 = S_{11}\sigma_1 + S_{12}\sigma_2 + S_{12}\sigma_3 = \frac{1}{E}\left[\sigma_1 - \nu(\sigma_2 + \sigma_3)\right]$$

$$\epsilon_2 = S_{12}\sigma_1 + S_{11}\sigma_2 + S_{12}\sigma_3 = \frac{1}{E}\left[\sigma_2 - \nu(\sigma_1 + \sigma_3)\right]$$

$$\epsilon_3 = S_{12}\sigma_1 + S_{12}\sigma_2 + S_{11}\sigma_3 = \frac{1}{E}\left[\sigma_3 - \nu(\sigma_1 + \sigma_2)\right]$$

$$\epsilon_4 = 2(S_{11} - S_{12})\sigma_4 \qquad = \frac{1}{G}\sigma_4 \tag{1.119}$$

$$\epsilon_5 = 2(S_{11} - S_{12})\sigma_5 \qquad = \frac{1}{G}\sigma_5$$

$$\epsilon_6 = 2(S_{11} - S_{12})\sigma_6 \qquad = \frac{1}{G}\sigma_6$$

Expressing the strains as function of the stresses, we have

$$\sigma_1 = C_{11}\epsilon_1 + C_{12}\epsilon_2 + C_{12}\epsilon_3 = (2\mu + \lambda)\epsilon_1 + \lambda\epsilon_2 + \lambda\epsilon_3$$

$$\sigma_2 = C_{12}\epsilon_1 + C_{11}\epsilon_2 + C_{12}\epsilon_3 = \lambda\epsilon_1 + (2\mu + \lambda)\epsilon_2 + \lambda\epsilon_3$$

$$\sigma_3 = C_{12}\epsilon_1 + C_{12}\epsilon_2 + C_{11}\epsilon_3 = \lambda\epsilon_1 + \lambda\epsilon_2 + (2\mu + \lambda)\epsilon_3$$

$$\sigma_4 = \tfrac{1}{2}(C_{11} - C_{12})\epsilon_4 \qquad = \mu\epsilon_4 \tag{1.120}$$

$$\sigma_5 = \tfrac{1}{2}(C_{11} - C_{12})\epsilon_5 \qquad = \mu\epsilon_5$$

$$\sigma_6 = \tfrac{1}{2}(C_{11} - C_{12})\epsilon_6 \qquad = \mu\epsilon_6$$

TABLE 1.2 Relations Between the Elastic Constants for Isotropic Materials[a]

Elastic Constants	In Terms of:				
	E, ν	E, G	B, ν	B, G	λ, μ
E	$= E$	$= E$	$= 3(1-2\nu)B$	$= \dfrac{9B}{1+3B/G}$	$= \dfrac{\mu(3+2\mu/\lambda)}{1+\mu/\lambda}$
ν	$= \nu$	$= -1 + \dfrac{E}{2G}$	$= \nu$	$= \dfrac{1-2G/3B}{2+2G/3B}$	$= \dfrac{1}{2(1+\mu/\lambda)}$
G	$= \dfrac{E}{2(1+\nu)}$	$= G$	$= \dfrac{3(1-2\nu)B}{2(1+\nu)}$	$= G$	$= \mu$
B	$= \dfrac{E}{3(1-2\nu)}$	$= \dfrac{E}{9-3E/G}$	$= B$	$= B$	$= \lambda + \dfrac{2\mu}{3}$
λ	$= \dfrac{E\nu}{(1+\nu)(1-2\nu)}$	$= \dfrac{E(1-2G/E)}{3-E/G}$	$= \dfrac{3B\nu}{1+\nu}$	$= B - \dfrac{2G}{3}$	$= \lambda$
μ	$= \dfrac{E}{2(1+\nu)}$	$= G$	$= \dfrac{3(1-2\nu)B}{2(1+\nu)}$	$= G$	$= \mu$

[a] Reprinted with permission from [4], p. 80.

TABLE 1.3   Elastic Stiffnesses of Monocrystals at Ambient Temperature
(1 Unit = 10 GPa)[a]

Element	Structure	C_{11}	C_{44}	C_{12}	C_{33}	C_{66}	C_{13}	C_{14}
Ag	C	12.40	4.61	9.34				
Al	C	10.82	2.85	6.13				
Au	C	18.60	4.20	15.7				
C	Diamond C	107.6	56.6	12.5				
Ni	C	24.65	12.47	14.73				
Pb	C	4.95	1.49	4.23				
Fe	C	22.8	11.65	13.2				
Mo	C	46	11.0	17.6				
Ta	C	26.7	8.25	16.1				
W	C	50.10	15.14	19.8				
Co	HCP	30.70	7.53	16.50	35.81		10.3	
Zn	HCP	16.10	3.83	3.42	6.10		5.01	
Sn	Tetr.	7.35	2.2	2.34	8.7	2.26	2.8	
In	Tetr.	4.45	0.66	3.95	4.44	1.22	4.05	
Hg	Rhomb. (−190°C)	3.60	1.29	2.89	5.05		3.03	0.5
Cu	C	16.84	7.54	12.14				
Ti	HCP	16.24	4.67	9.20	18.07	6.90		
Be	HCP	29.23	16.25	2.67	33.64	1.4		
Zr	HCP	14.34	3.20	7.28	16.48	6.53		
Mg	HCP	5.97	1.67	2.62	6.17	2.17		

[a] Reprinted with permission from [7], p. 163.

If the stresses and strains are defined in cylindrical coordinates, we have exactly the same equations, except for the indices, which become rr, $\theta\theta$, $\phi\phi$, $r\phi$, $r\theta$, $\phi\theta$.

Tables 1.3 and 1.4 show the elastic stiffnesses and compliances, respectively, of monocrystals. One of the most complete compilations of elastic constants for crystals is the one by Simmons and Wang.[5] The elastic constants for a number of polycrystalline metals are given in Table 1.5. We can also determine the polycrystalline (isotropic) elastic constants from the monocrystalline ones. McGregor Tegart[6] describes such a procedure, warning the reader of the complexity of the problem. In a polycrystalline aggregate the deformation of one grain is not independent of the deformation of its neighbor. The compatibility requirements are such that we have to apply either one of two possible simplifying assumptions:

1. The local strain is equal to the mean strain (all grains undergo the same strain); this is called the Voigt average.

$$E = \frac{(E - G + 3H)(F + 2G)}{2F + 3G + H}$$

Element	Structure	S_{11}	S_{44}	S_{12}	S_{33}	S_{13}
Ag	C	2.29	2.17	−0.983		
Al	C	1.57	3.51	−0.568		
Au	C	2.33	2.38	−1.065		
Cu	C	1.498	1.326	−0.629		
Ni	C	0.734	0.802	−0.274		
Pb	C	9.51	6.72	−4.38		
Fe	C	0.762	0.858	−0.279		
Mo	C	0.28	0.91	−0.078		
Nb	C	0.69	3.42	−0.249		
Ta	C	0.685	1.21	−0.258		
W	C	0.257	0.66	−0.073		
Bl	HCP	0.348	0.616	−0.030	0.298	−0.031
Mg	HCP	2.20	6.1	−0.785	1.97	−0.50
Ti	HCP	0.958	2.14	−0.462	0.698	−0.189
Zr	HCP	1.013	3.13	−0.404	0.799	−0.241

[a] Adapted with permission from [7], p. 163.

TABLE 1.5 Elastic and Shear Moduli and
Poisson Ratios for Polycrystalline Metals[a]

Metal (20°C)	E (GPa)	G (GPa)	ν
Aluminum	70.3	26.1	0.345
Cadmium	49.9	19.2	0.300
Chromium	279.1	115.4	0.210
Copper	129.8	48.3	0.343
Gold	78.0	27.0	0.440
Iron	211.4	81.6	0.293
Magnesium	44.7	17.3	0.291
Nickel	199.5	76.0	0.312
Niobium	104.9	37.5	0.397
Silver	82.7	30.3	0.367
Tantalum	185.7	69.2	0.342
Titanium	115.7	43.8	0.321
Tungsten	411.0	160.6	0.280
Vanadium	127.6	46.7	0.365

[a] Adapted with permission from [8], p. 8.

where

$$F = \tfrac{1}{2}(C_{11} + C_{22} + C_{33})$$

$$G = \tfrac{1}{3}(C_{12} + C_{23} + C_{13})$$

$$H = \tfrac{1}{3}(C_{44} + C_{55} + C_{66})$$

2. The local stress is equal to the mean stress (all grains are under the same stress); this is called the Reuss average.

$$\frac{1}{E} = \tfrac{1}{5}(3F' + 2G' + H')$$

$$F' = \tfrac{1}{3}(S_{11} + S_{22} + S_{32})$$

$$G' = \tfrac{1}{3}(S_{12} + S_{23} + S_{13})$$

$$H' = \tfrac{1}{3}(S_{44} + S_{55} + S_{66})$$

The actual stress and strain configuration is probably between the two.

1.12.5 Theory of Elasticity

1.12.5.1 Elasticity equations. The elasticity does not restrict itself to the relationships between the stresses and the strains (called constitutive equations). The forces that are applied to the body were not considered up to this point. There is also a need for the strains to be compatible, and for the stresses and forces to be consistent at the boundary. Hence, we must consider:

1. Equilibrium equations
2. Compatibility equations
3. Boundary conditions

In the absence of acceleration and body moments we have static equilibrium. This is expressed by:

$$\frac{\partial \sigma_{ij}}{\partial x_j} = 0$$

One of the requirements of great importance is *compatibility*: deformation on each element has to be such that the continuity of the body is maintained. In other words, no voids are formed and two bodies cannot simultaneously occupy the same volume of space. Starting from the general equation for strain (Eq. 1.75), we arrive at

$$\epsilon_{11} = \frac{\partial u_1}{\partial x_1} \tag{1.121}$$

$$\epsilon_{22} = \frac{\partial u_2}{\partial x_2} \tag{1.122}$$

$$\epsilon_{33} = \frac{\partial u_3}{\partial x_3} \tag{1.123}$$

$$\gamma_{12} = \frac{\partial u_1}{\partial x_2} + \frac{\partial u_2}{\partial x_1} \tag{1.124}$$

$$\gamma_{23} = \frac{\partial u_2}{\partial x_3} + \frac{\partial u_3}{\partial x_2} \tag{1.125}$$

$$\gamma_{13} = \frac{\partial u_1}{\partial x_3} + \frac{\partial u_3}{\partial x_1} \tag{1.126}$$

Combining Eqs. 1.21 through 1.26 into one equation, we make sure that all the strains are compatible. This is done by taking the following partial derivatives:

$$\frac{\partial^2(1.121)}{\partial x_2^2} \tag{1.127}$$

$$\frac{\partial^2(1.122)}{\partial x_1^2} \tag{1.128}$$

$$(1.127) + (1.128) \longrightarrow (1.129) \tag{1.129}$$

$$\frac{\partial(1.124)}{\partial x_1\, \partial x_2} \tag{1.130}$$

From Eqs. 1.129 and 1.130, we obtain the first compatibility equation:

$$\frac{\partial^2 \epsilon_{11}}{\partial x_2^2} + \frac{\partial^2 \epsilon_{22}}{\partial x_1^2} = \frac{\partial^2 \gamma_{12}}{\partial x_1\, \partial x_2} \tag{1.131}$$

Similarly, we obtain the other equations:

$$\frac{\partial^2 \epsilon_{22}}{\partial x_3^2} + \frac{\partial^2 \epsilon_{33}}{\partial x_2^2} = \frac{\partial^2 \gamma_{23}}{\partial x_2\, \partial x_3} \tag{1.132}$$

$$\frac{\partial^2 \epsilon_{33}}{\partial x_1^2} + \frac{\partial^2 \epsilon_{11}}{\partial x_3^2} = \frac{\partial^2 \gamma_{13}}{\partial x_1\, \partial x_3} \tag{1.133}$$

$$2\frac{\partial^2 \epsilon_{11}}{\partial x_2\, \partial x_3} = \frac{\partial}{\partial x_1}\left(-\frac{\partial \gamma_{23}}{\partial x_1} + \frac{\partial \gamma_{13}}{\partial x_2} + \frac{\partial \gamma_{12}}{\partial x_3}\right) \tag{1.134,}$$

$$2\frac{\partial^2 \epsilon_{22}}{\partial x_1\, \partial x_3} = \frac{\partial}{\partial x_2}\left(-\frac{\partial \gamma_{13}}{\partial x_2} + \frac{\partial \gamma_{12}}{\partial x_3} + \frac{\partial \gamma_{23}}{\partial x_1}\right) \tag{1.135}$$

$$2\frac{\partial^2 \epsilon_{33}}{\partial x_1\, \partial x_2} = \frac{\partial}{\partial x_3}\left(-\frac{\partial \gamma_{12}}{\partial x_3} + \frac{\partial \gamma_{13}}{\partial x_2} + \frac{\partial \gamma_{23}}{\partial x_1}\right) \tag{1.136}$$

Substituting in the values of the stresses (using Eqs. 1.119), we have, for Eq. 1.131:

$$\frac{\partial^2 \sigma_{11}}{\partial x_2^2} + \frac{\partial^2 \sigma_{22}}{\partial x_1^2} - \nu\left(\frac{\partial^2 \sigma_{11}}{\partial x_1^2} + \frac{\partial^2 \sigma_{22}}{\partial x_2^2} + \frac{\partial^2 \sigma_{33}}{\partial x_1^2} + \frac{\partial^2 \sigma_{33}}{\partial x_2^2}\right)$$

$$= 2(1+\nu)\frac{\partial^2 \sigma_{12}}{\partial x_1\, \partial x_2} \tag{1.137}$$

Following the same procedure for Eqs. 1.132 through 1.136, we obtain the other equations of compatibility.

1.12.5.2 Solution of the plane-strain problem.

The three conditions specified in Section 1.12.5.1—equilibrium, compatibility, and boundary—will be applied to a state of plane strain. The solution of the problem is required in Chapter 6 for the stress field around a dislocation. We have

$$\epsilon_{33} = \gamma_{13} = \gamma_{23} = 0$$

Equations 1.133 through 1.136 are identically zero, and we are left solely with Eq. 1.132. Substituting stresses for strains, we are left with Eq. 1.137.

The equilibria are

$$\frac{\partial \sigma_{ij}}{\partial x_j} = 0$$

or

$$\frac{\partial \sigma_{11}}{\partial x_1} + \frac{\partial \sigma_{12}}{\partial x_2} = 0 \tag{1.138}$$

$$\frac{\partial \sigma_{21}}{\partial x_1} + \frac{\partial \sigma_{22}}{\partial x_2} = 0 \tag{1.139}$$

$$\frac{\partial (1.138)}{\partial x_1}: \quad \frac{\partial^2 \sigma_{11}}{\partial x_1^2} + \frac{\partial^2 \sigma_{12}}{\partial x_1 \, \partial x_2} = 0 \tag{1.140}$$

$$\frac{\partial (1.139)}{\partial x_2}: \quad \frac{\partial^2 \sigma_{22}}{\partial x_2^2} + \frac{\partial^2 \sigma_{12}}{\partial x_1 \, \partial x_2} = 0 \tag{1.141}$$

Adding 1.140 to 1.141 yields

$$\frac{2 \partial \sigma_{12}}{\partial x_1 \, \partial x_2} = - \left(\frac{\partial^2 \sigma_{11}}{\partial x_1^2} + \frac{\partial^2 \sigma_{22}}{\partial x_2^2} \right) \tag{1.142}$$

We now take Eq. 1.119, which expresses ϵ_{33}:

$$\sigma_{33} = \nu(\sigma_{11} + \sigma_{22}) \tag{1.143}$$

Substituting Eq. 1.143 into 1.137, we have

$$\frac{\partial^2 \sigma_{11}}{\partial x_2^2} + \frac{\partial^2 \sigma_{22}}{\partial x_1^2} - \nu \left(\frac{\partial^2 \sigma_{11}}{\partial x_1^2} + \frac{\partial^2 \sigma_{22}}{\partial x_2^2} + \frac{\nu \partial^2 \sigma_{11}}{\partial x_1^2} + \nu \frac{\partial^2 \sigma_{22}}{\partial x_1^2} + \nu \frac{\partial^2 \sigma_{11}}{\partial x_2^2} + \nu \frac{\partial^2 \sigma_{22}}{\partial x_2^2} \right)$$

$$= 2(1 + \nu) \frac{\partial^2 \sigma_{12}}{\partial x_1 \, \partial x_2}$$

Now, using Eq. 1.142, we get

$$\frac{\partial^2 \sigma_{11}}{\partial x_2^2} + \frac{\partial^2 \sigma_{22}}{\partial x_1^2} - \nu \left[(1 + \nu) \frac{\partial^2 \sigma_{11}}{\partial x_1^2} + (1 + \nu) \frac{\partial^2 \sigma_{22}}{\partial x_2^2} \right] - \nu^2 \frac{\sigma_{22}}{\partial x_1^2} - \nu^2 \frac{\partial^2 \sigma_{11}}{\partial x_2^2}$$

$$= 2(1 + \nu) \frac{\partial^2 \sigma_{12}}{\partial x_1 \, \partial x_2} - (1 + \nu) \left[\frac{\partial^2 \sigma_{11}}{\partial x_1^2} + \frac{\partial^2 \sigma_{22}}{\partial x_2^2} \right]$$

$$= (1 - \nu^2) \frac{\partial^2 \sigma_{11}}{\partial x_2^2} + (1 - \nu^2) \frac{\partial^2 \sigma_{22}}{\partial x_1^2} - \nu(1 + \nu) \left(\frac{\partial^2 \sigma_{11}}{\partial x_1^2} + \frac{\partial^2 \sigma_{22}}{\partial x_2^2} \right)$$

$$= -(1 + \nu) \left(\frac{\partial^2 \sigma_{11}}{\partial x_1^2} + \frac{\partial^2 \sigma_{22}}{\partial x_2^2} \right)$$

$$(1 - \nu^2) \left(\frac{\partial^2 \sigma_{11}}{\partial x_2^2} + \frac{\partial^2 \sigma_{22}}{\partial x_1^2} \right) - \nu^2 \left(\frac{\partial^2 \sigma_{11}}{\partial x_1^2} + \frac{\partial^2 \sigma_{22}}{\partial x_2^2} \right) = - \left(\frac{\partial^2 \sigma_{11}}{\partial x_1^2} + \frac{\partial^2 \sigma_{22}}{\partial x_2^2} \right)$$

$$(1 - \nu^2) \left(\frac{\partial^2 \sigma_{11}}{\partial x_2^2} + \frac{\partial^2 \sigma_{22}}{\partial x_1^2} \right) + (1 - \nu^2) \left(\frac{\partial^2 \sigma_{11}}{\partial x_1^2} + \frac{\partial^2 \sigma_{22}}{\partial x_2^2} \right) = 0$$

$$(1 - \nu^2) \left(\frac{\partial^2 \sigma_{11}}{\partial x_2^2} + \frac{\partial^2 \sigma_{22}}{\partial x_1^2} + \frac{\partial^2 \sigma_{11}}{\partial x_1^2} + \frac{\partial^2 \sigma_{22}}{\partial x_2^2} \right) = 0$$

But $(1 - \nu^2) \neq 0$, so

$$\left(\frac{\partial^2}{\partial x_1^2} + \frac{\partial^2}{\partial x_2^2} \right) (\sigma_{11} + \sigma_{22}) = 0$$

or

$$\nabla^2(\sigma_{11} + \sigma_{22}) = 0 \tag{1.144}$$

The solution of this particular elasticity problem consists in determining σ_{11}, σ_{12}, and σ_{22} by means of Eqs. 1.138, 1.139, and 1.144, applying the boundary conditions (i.e., the external loads). This solution involves the simultaneous solution of three differential equations. Airy solved this problem introducing the stress function known as Airy stress function. It is equal to

$$\Phi = f(x_1, \ x_2, \text{ expresses stress in terms of strains})$$

Airy showed that there is one function such that, in the absence of gravity,

$$\sigma_{11} = \frac{\partial^2 \Phi}{\partial x_2^2}$$

$$\sigma_{22} = \frac{\partial^2 \Phi}{\partial x_1^2} \tag{1.145}$$

$$\sigma_{12} = -\frac{\partial^2 \Phi}{\partial x_1 \, \partial x_2}$$

These equations satisfy the equilibrium conditions. Substituting them into Eq. 1.144, we have

$$\nabla^2(\nabla^2 \Phi) = 0$$

or

$$\nabla^4 \Phi = 0 \tag{1.146}$$

Thus, if an Airy function is found that satisfies Eq. 1.146, we can easily obtain the stresses applying it to Eq. 1.145. These functions have been derived and are reported in the literature for a number of configurations.

The method of solution of Eq. 1.146 is described briefly below. Equation 1.146 is called a biharmonic equation. The solution is of the form

$$\Phi = a_0 + a_1\psi_1 + a_2\psi_2 + \cdots$$

However, the ψ_i also have to be harmonic:

$$\nabla^2\psi_i = 0$$

The solution of an harmonic equation is a function with complex variables:

$$w = f(z) = u_i + iu_2$$

where

$$z = x_1 + ix_2$$

The solution involves the Cauchy–Riemann conditions; the reader should consult a source on differential equations.

1.13 PROPAGATION OF ELASTIC WAVES IN SOLIDS

When the body is not in static equilibrium (i.e., when we have a situation such as the one depicted in Fig. 1.11), the stress (and strain) disturbance will travel through the body. From Eq. 1.40 we can derive the equations that provide the velocities at which these disturbances propagate. There are, in an unbounded medium that is isotropic, two types of waves: (1) irrotational (or dilatational, or longitudinal) waves, and (2) equivoluminal (or distortional, or shear) waves. The detailed derivation is given in the classic book by Kolsky[9] and in Meyers and Murr,[10] among others. We have, for an isotropic solid,

$$\vartheta_{\text{lon}} = \left(\frac{2\mu + \lambda}{\rho}\right)^{1/2} \tag{1.147}$$

$$\vartheta_{\text{shear}} = \left(\frac{\mu}{\rho}\right)^{1/2} \tag{1.148}$$

where ρ is the material density.

The velocity of elastic disturbances can be used in the determination of the elastic constants, as we can see from Eqs. 1.147 and 1.148. Indeed, this is a well-established procedure. These "dynamic" techniques can be classified into three categories:[11]

1. Transit-time measurements
2. Pulse-echo methods
3. Ultrasonic interferometry

They provide a high degree of precision.

1.14 OTHER METHODS FOR THE SOLUTION OF ELASTICITY PROBLEMS

The analytical methods to solve elasticity problems can be very complex, as one can see from Section 1.12.5. Even the plane-strain problems, for which the Airy stress function provides a ready solution route, is not without complications. This renders numerical methods using computers and analog methods using physical models very attractive. Among the numerical methods, the first to be introduced was the finite difference method, used in solving the biharmonic equation (Eq. 1.146). The exact differential is replaced by a difference equation and approximate solutions are found for "cells." The finite difference method consists of replacing the derivatives by differences.

The finite element method is, today, the most popular numerical method for treating elasticity problems. It is a very powerful technique and will have a lasting impact. The body is divided into a mesh of "finite elements" having in general triangular or quadrilateral shape (for the plane problem). These elements are interconnected at nodes (vertices). These elements form the basic structural unit to which a "stiffness" factor is applied at the nodes. The unknown forces and/or displacements at each node are calculated; the solution for a system of simultaneous equations is obtained by matrix analysis, which lends itself very well to computer treatment. Additional details are given in Section

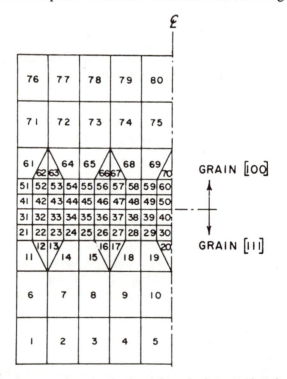

Figure 1.27   Finite element mesh used to simulate deformation in two cubic grains having a common interface and orientations [100] and [111]. (Reprinted with permission from [12], p. 754.)

Figure 1.28 Cross section of underground excavation with isoshear stress contours made evident by photoelastic technique. (Reprinted with permission from [19], p. 18.)

2.5.4.6. Figure 1.27 shows an illustration of the application of the FEM. Two grains, with orientations [111] and [100] and having a cubic shape, are deformed in tension, under the same stress. The effective Young's moduli being different, different strains will result, leading to additional incompatibility stresses (and strains) at the interface. The grains were divided into rectangles for the computer code used (described in detail by Desai and Abel[13]); the computer code automatically divides each rectangle into four triangles, by passing the diagonals through each of them. The external forces can be applied as one wishes (at the nodes). For this, the applied stress has to be converted into forces applied at the nodes. The elements were made smaller close to the interface to allow greater resolution. The number of elements depends on the capacity of the computer. These computer programs are burdensome to develop (especially the mesh mapping), but the potential user should know that software packages are available. An example of such a code is GIFTS-4.[14]

Additionally, there is a group of numerical methods called "integral methods."[15] In one of these methods, called the *boundary-element method*, the interior of the body is assumed to be an elastic continuum while the boundary is divided into finite elements.

Experimental stress analysis methods of different types have been used. Examples are the brittle coating method, the electrical resistance network, the membrane method, and the photoelastic method.[16] Photoelasticity is a widely used analog method; it can be applied to virtually any two-dimensional stress situation. A model material in which it is possible to see the stress pattern by special techniques is used; a scaled model is built, and the external forces (scaled) reproduce the ones applied on the real system. Glass, celluloid, certain grades of Bakelite, plexiglass, epoxy resin, and Columbia resin[17] have the desired bire-fringence properties. Stresses will change the optical properties, and the stress patterns are readily observable under plane-polarized light. A detailed treatment is given by Frocht.[18] Figure 1.28 shows an example; an underground excavation system is modeled, and the isochromatic (black and white) lines represent regions of constant maximum shear stress ($=\sigma_1 - \sigma_3$). The T's represent tension, and the numbers designate the amplitudes of the maximum shear stresses. The stress concentrations between different excavations are clearly shown, as are the regions of tension (T) on the excavation walls. Combination of the photoelastic methods with other methods allows determination of the principal stresses and their orientations at all positions within the body. It is the feeling of the authors that there is a great potential in numerical and experimental methods in the modeling of microelastic problems.

EXERCISES

1.1. Two specimens, having an initial length of 5 cm, are tested, one in compression and one in tension. If the engineering strains are -0.5 and $+0.5$, respectively, what will be the final lengths of the specimens? What are the true strains, and why are they numerically different?

1.2. An aluminum polycrystalline specimen is being elastically compressed in plane strain. If the true strain along the compression direction is -2×10^{-4}, what are the other two longitudinal strains?

1.3. Determine B, λ, and G for polycrystalline niobium, titanium, and iron.

1.4. A state of stress is given by

$$\sigma_{11} = 350 \text{ MPa}$$

$$\sigma_{12} = 70 \text{ MPa}$$

$$\sigma_{22} = 210 \text{ MPa}$$

Determine the principal stresses, the maximum shear stress, and their angle with the given direction.

1.5. Calculate the anisotropy ratio for the cubic metals in Table 1.3.

1.6. Prove that the transformation law for second-rank tensors is

$$T_{ij} = l_{ki}l_{lj}T'_{kl}$$

1.7. Prove that the transformation law for fourth-rank tensors is

$$T'_{ijkl} = l_{im}l_{jn}l_{ko}l_{lp}T_{mnop}$$

1.8. Prove, using the indicial notation, that the square of the length of a vector $[p_i]$, defined as $p_i p_i$, is not altered by a transformation of axes and is therefore a scalar.

1.9. If p_i and q_i are vectors, p_1/q_1, p_2/q_2, and p_3/q_3 are three numbers whose values are defined for any system of axes. Are they the components of a vector?

1.10. Show that a uniaxial hydrostatic compressive stress can be decomposed into a hydrostatic pressure and two states of pure shear. Use sketches if necessary.

1.11. Determine the principal stresses and the maximum shear stress, as well as their angles with the system of reference given by the following stresses:

$$[\sigma_{ij}] = \begin{pmatrix} 3 & 3 & 2 \\ 3 & 3 & 2 \\ 2 & 2 & 0 \end{pmatrix} \times 10^6 \text{ Pa}$$

1.12. For the state of deformation defined below, obtain the strain and rigid-body-rotation tensors.

$$[\epsilon_{ij}] = \begin{pmatrix} 5 & 3 & -2 \\ 1 & 4 & 2 \\ 2 & -1 & 0 \end{pmatrix} \times 10^{-6}$$

1.13. Determine the principal strains and their orientation for Exercise 1.12.

1.14. Find the effective and octahedral stresses and strains, as well as the stress and strain invariants, for the matrices given in Exercises 1.11 and 1.12.

1.15. A body is subjected to a state of pure shear by being stressed in tension along Ox_1 and in compression along Ox_2. The displacements are $u_1 = ex_1$ and $u_2 = ex_2$ along Ox_1 and Ox_2, respectively (e is small). Determine the strain tensor referred to a system of axes $Ox_1x_2x_3$ obtained by a counterclockwise rotation of $40°$ of Ox_1 and Ox_2 around Ox_3.

1.16. A body is subjected to an elastic tensile strain along Ox_1. The displacements are $u_1 = ex_1$, $u_2 = -vex_2$, and $u_3 = -vex_3$, along Ox_1, Ox_2, and Ox_3, respectively. Derive the expression for the strain tensor referred to the system of axes $Ox_1x_2x_3$ and $Ox_1x_2x_3$ defined in Exercise 1.15.

1.17. An elastic deformation is given by

$$[\epsilon_{ij}] = \begin{pmatrix} 8 & -1 & -1 \\ 1 & 6 & 0 \\ -5 & 0 & 2 \end{pmatrix} \times 10^{-5}$$

Determine $[\epsilon_{ij}]$, $[w_{ij}]$, the principal strains, and their orientation.

1.18. For an isotropic metal with Poisson ratio equal to 0.30 and $E = 210$ GPa, determine the strain tensor corresponding to following the stress state:

$$\sigma_{ij} = \begin{pmatrix} -4.2 & 7 & -14 \\ 7 & 3.5 & -21 \\ -14 & -21 & 0.7 \end{pmatrix} \times 10^6 \text{ Pa}$$

1.19. Prove the relationship on the third line and third column of Table 1.2.

1.20. Extensometers attached to the external surface of a steel pressure vessel indicate $\epsilon_l = 0.002$ and $\epsilon_t = 0.005$ along the longitudinal and transverse directions, respectively. Determine the corresponding stresses. What would be the error if Poisson's ratio were not considered?

1.21. Prove that for a monocrystal, Young's modulus along a certain direction is given by the equation below. When a tensile specimen is extended along a certain direction, the ratio between the longitudinal strain and the normal stress is given by

$$\frac{1}{E} = S_{11} - 2(S_{11} - S_{12} - \tfrac{1}{4} S_{44})(l_{11}^2 l_{12}^2 + l_{12}^2 l_{13}^2 + l_{11}^2 l_{13}^2)$$

(*Hint*: Transform the compliance S_{11} to the new axes with Ox_1 parallel to the desired direction.)

1.22. Calculate Young's and shear moduli for monocrystalline iron along [100], [110], and [111].

1.23. From the values obtained in Exercise 1.22, obtain a rough estimate of the Young's modulus of a polycrystalline aggregate, assuming that there are only three orientations for the grains ([100], [110], and [111]) and that they occur proportionally to their multiplicity factors. Compare this with the predictions of Voigt averages (isostrain) and Reuss averages (isostress).

1.24. A silver monocrystal is being extended along [100]. Obtain the values for the Young's and shear moduli, as well as Poisson's ratio, using the equation from Exercise 1.21 and data from Table 1.4.

1.25. Prove that a state of homogeneous shear $\sigma_{12} = g$ transforms a circle $x_1^2 + x_2^2 = 1$ into an ellipsis whose major axes makes an angle with Ox_1 such that $\tan 2\theta = 2/g$.

1.26. Determine the stiffness matrix for the cubic structure starting with the rhombohedral one and applying threefold rotations around the cube diagonals. The rhombohedral matrix is

$$\begin{bmatrix} C_{11} & C_{12} & C_{13} & C_{14} & C_{15} & 0 \\ \cdot & C_{11} & C_{13} & -C_{14} & -C_{15} & 0 \\ \cdot & \cdot & C_{33} & 0 & 0 & 0 \\ \cdot & \cdot & \cdot & C_{44} & 0 & -2C_{15} \\ \cdot & \cdot & \cdot & \cdot & C_{44} & 2C_{14} \\ \cdot & \cdot & \cdot & \cdot & \cdot & \tfrac{1}{2}(C_{11} - C_{12}) \end{bmatrix}$$

1.27. An isotropic specimen is being extended in tension. Are there any shear stresses applied on it? Along what plane are they greatest? Are there any strains other than in the extension direction? Illustrate using the compliance matrix of Eq. 1.113.

REFERENCES

[1] G.E. Dieter, *Mechanical Metallurgy*, McGraw-Hill, New York, 2nd. ed., 1976, p. 378.

[2] J.F. Nye, *Physical Properties of Crystals*, Oxford University Press, London, 1957.

[3] J. Carnahan, *J. Metals*, 16 (Dec. 1964) 990.

[4] F.A. McClintock and A.S. Argon (eds.), *Mechanical Behavior of Materials*, Addison-Wesley, Reading, Mass., 1966, p. 80.

[5] G. Simmons and H. Wang, *Single Crystal Elastic Constants*, MIT Press, Cambridge, Mass., 1971.

[6] W.J. McGregor Tegart, *Elements of Mechanical Metallurgy*, Macmillan, New York, 1966, p. 93.

[7] A. Kelly and G.W. Groves, *Crystallography and Crystal Defects*, Addison-Wesley, Reading, Mass., 1970.

[8] R.W. Hertzberg, *Deformation and Fracture Mechanics of Engineering Materials*, Wiley, New York, 1976.

[9] H. Kolsky, *Stress Waves in Solids*, Dover, New York, 1963, p. 10.

[10] M.A. Meyers and L.E. Murr, "Propagation of Stress Waves in Metals," in *Explosive, Welding, Forming and Compaction*, T.Z. Blazynski (ed.), Applied Science Publishers, Essex, England, 1983, Chap. 2, p. 17.

[11] E. Schreiber, O.L. Anderson, and N. Soga, *Elastic Constants and Their Measurement*, McGraw-Hill, New York, 1974, p. 360.

[12] M.A. Meyers and E. Ashworth, Philosophical Magazine A, 46 (1982) 737.

[13] C.S. Desai and J.F. Abel, *Introduction to the Finite Element Method*, Van Nostrand, Reinhold, New York, 1972, p. 439.

[14] "GIFTS 4: Graphics—Oriented Interactive Finite Element Time-Sharing System," University of Arizona, Tucson, Ariz.

[15] J.C. Jaeger and N.G.W. Cook, *Fundamentals of Rock Mechanics*, 3rd ed., Chapman & Hall, London, 1979, p. 305.

[16] M. Hetenyi, *Handbook of Experimental Stress Analysis*, Wiley, New York, 1950.

[17] E.P. Popov, *Mechanics of Materials*, 2nd ed., Prentice-Hall, Englewood Cliffs, N.J., 1976, p. 298.

[18] M.M. Frocht, *Photoelasticity*, Vols. I and II, Wiley, New York, 1941 and 1948.

[19] K.I. Oravecz, "An Investigation into the Stability of Workings of a Shallow Auriferous Deposit," M.Sc. thesis, University of Durham, Durham, England, 1961.

SUGGESTED READING

HUNTINGTON, H.B., *The Elastic Constants of Crystals*, Academic Press, New York, 1958.

KELLY, A., and G.W. GROVES, *Crystallography and Crystal Defects*, Addison-Wesley, Reading, Mass., 1970.

LOVE, A.E.H., *The Mathematical Theory of Elasticity*, Dover, New York, 1952.

MCCLINTOCK, F.A., and A.S. ARGON (eds.), *Mechanical Behavior of Materials*, Addison-Wesley, Reading, Mass., 1966.

NYE, J.F., *Physical Properties of Crystals*, Oxford University Press, London, 1957.

SIMMONS, G., and H. WANG, *Single Crystal Elastic Constants*, MIT Press, Cambridge, Mass., 1971.

SOKOLNIKOFF, I.S., *Mathematical Theory of Elasticity*, 2nd ed., McGraw-Hill, New York, 1956.

TIMOSHENKO, S., and J.N. GOODIER, *Theory of Elasticity*, McGraw-Hill, New York, 1951.

WESTERGAARD, H.M., *Theory of Elasticity and Plasticity*, Dover, New York. 1952.

Chapter 2

PLASTICITY

‡2.1 INTRODUCTION

A material, upon being mechanically stressed, will in general exhibit the following sequence of responses: elastic deformation, plastic deformation, and fracture. This chapter addresses the second response: plastic deformation. A sound knowledge of plasticity is of great importance because:

1. A greater and greater number of projects are executed in which small plastic deformations of the structure are accepted. The "theory of limit design" is used in applications where the weight factor is critical, such as space vehicles and rockets. The rationale for accepting a limited plastic deformation is that the material will work-harden at that region, and plastic deformation will cease once the flow stress (due to work hardening) reaches the applied stress.

2. It is very important to know the stresses and strains involved in deformation processing, such as rolling, forging, extrusion, drawing, and so on. All these processes involve substantial plastic deformation and the response of the material will depend on their plastic behavior. The application of plasticity theory to these processes is presented in Section 2.2 and following.

3. The mechanism of fracture involves plastic deformation at the crack tip. The way in which the high stresses developed at the crack can be accommodated by the surrounding material is of utmost importance in the propagation of the crack.

4. An overall knowledge of plasticity theory is essential to the understanding of the mechanical strength and deformability of materials.

This chapter is restricted to macroplasticity; the material is assumed to be a continuum. The microscopic deformation modes (dislocations, twinning, vacancies) will be studied in Part II and constitute the working tools of microplasticity.

The first contributions to the theory of plasticity were made around 1870 by Saint-Venant and von Mises for a state of plane strain. In the forties, after a period of little activity, advances were made by Hencky and Prandtl both in the theory of plasticity and in the solution methods for a state of plane strain. Systematic investigations to determine the laws of plasticity for more complex states of deformation have also been conducted. More recently, numerical methods that are very well suited to solution in the computer have been used with great success. The numerical methods involving computer use will probably, in the future, surpass the older analytical methods.

The methods of plasticity are the ones commonly used in the analysis of deformable bodies. The first step is to establish the basic laws for the plastic deformation based on experimental results and, if possible, using fundamental equations from physics. Next, a system of equations is set up; once solved, it will describe the plastic deformation of the body under various conditions. One of the great problems of plasticity resides in the nonlinearity of the main laws, and, consequently, of the equations. Hence, the mathematical difficulties are enormous and the classical methods of solution are not applicable. Therefore, it is necessary to develop new investigative methods and solution techniques as each new problem is tackled.

Several simplifying hypotheses are required to make the problem tractable. The most common simplifications are:

1. The material is assumed to be isotropic. It was seen in Chapter 1 that metals are elastically anisotropic. This anisotropy extends itself to the plastic range. As an example, Fig. 8.33 shows the stress–strain curves along two different crystallographic directions in a single crystal of aluminum; they are definitely not similar. On the other hand, if one considers a polycrystalline aggregate in which the individual grains are randomly oriented and in which they are all equiaxed in shape, one can assume that this aggregate is, macroscopically, isotropic. In this chapter, in the few instances in which the material will be treated as anisotropic, a clear statement will be made.

2. The deformations will be assumed to be independent of time. Hence, the strain rate does not have any effect on the final state of deformation. There are cases in which this assumption is not applicable. They are treated separately in viscoplasticity (an example is creep).

3. It is assumed that the material obeys Hooke's law up to yielding or that its elastic modulus is infinite up to that point. The latter assumption is reasonable for large strains (greater than 0.10) because the elastic strains usually do not exceed 0.005 and can therefore be neglected. Another effect that is ne-

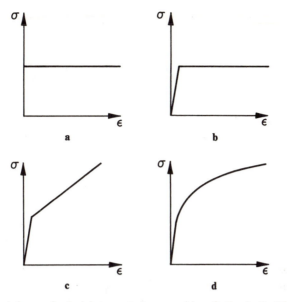

Figure 2.1 Idealized shapes of uniaxial stress–strain curve: (a) perfectly plastic; (b) ideal elastoplastic; (c) ideal elastoplastic with linear work hardening; (d) parabolic work hardening ($\sigma = \sigma_0 + k\epsilon^n$).

glected is anelasticity, which is manifested by a hysteresis in the elastic range (different paths in loading and unloading) when the material is loaded in the elastic range.

4. The plastic range of the uniaxial stress–strain curve is also assumed to have a simplified shape. One of the configurations displayed in Fig. 2.1 can be assumed. Configuration (a) is called perfectly plastic; the elastic strains are assumed zero. When the elastic deformation is assumed zero, the body is called rigid. When these assumptions cannot be made, configurations (b) and (c) are used. Configuration (b) is known as ideal elastoplastic. More complex plastic configurations [such as the one shown in Fig. 2.1(d)] are described separately in Section 9.7.1. The volume of the material is assumed to be constant in plastic deformation. It is known that such is not the case in elastic deformation. As shown in Section 1.4, the constancy in volume implies that

$$\epsilon_{11} + \epsilon_{22} + \epsilon_{33} = 0 \tag{2.1}$$

or

$$\epsilon_1 + \epsilon_2 + \epsilon_3 = 0 \tag{2.2}$$

and that Poisson's ratio is 0.5. Figure 2.2 shows that this assumption is reasonable and that ν rises from 0.3 to 0.5 as deformation goes from elastic to plastic.

However, prior to delving into the plasticity theories, we have to know, for a complex state of stress, the stress level at which the body starts to flow plastically. The methods developed to determine this are called flow criteria.

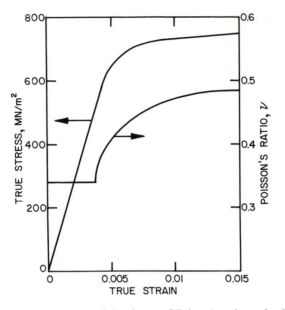

Figure 2.2 Schematic representation of the change of Poisson's ratio as the deformation regime changes from elastic to plastic.

‡2.2 FLOW (OR YIELD, OR FAILURE) CRITERIA

The term "flow criterion" will be used here rather than "yield criterion" or "failure criterion." The term "failure criterion" has its historical origin in applications where the onset of plastic deformation indicated failure. However, in deformation-processing operations this is obviously not the case, and plastic flow is desired. The term "yield criterion" applies only to materials that are in the annealed condition. It is known that, when a material is previously deformed by, for instance, rolling, its yield stress increases due to work hardening (see Chapter 9). The term "flow stress" is usually reserved for the onset of plastic flow of a previously deformed material. For the reasons stated above, it is felt that the term "flow criterion" is the most appropriate.

To be completely general, a flow criterion has to be valid for any stress state. In a uniaxial stress state, plastic flow starts when the stress–strain curve deviates from its initial linear range. Uniaxial stress–strain curves are very easily obtained experimentally (see Chapter 16) and the deformation response of metals is usually known for this situation. The main function of the flow criteria is to predict the onset of plastic deformation in a complex state of stress knowing the flow stress (under uniaxial tension) of the material. It should be noted that this value is strongly dependent on the state of stress, and if this effect is not considered, it can lead to potentially dangerous errors in design. Four flow criteria will be presented.

‡2.2.1 Maximum Stress Criterion (Rankine)

According to it, plastic flow takes place when the greatest principal stress in a complex state of stress reaches the flow stress in uniaxial stress in tension. Since $\sigma_1 > \sigma_2 > \sigma_3$, we have

$$\sigma_0(\text{tension}) < \sigma_1 < \sigma_0(\text{compression})$$

where σ_0 is the flow stress of the material. The great weakness of this criterion is that it predicts plastic flow of a material under a hydrostatic state of stress; however, this is impossible, as shown by the example below. It is well known that tiny shrimp can live at very great depths. The hydrostatic pressure due to water is equivalent to 1 atm (10^5 N/m²) for every 10 m; at 1000 m below the surface the shrimp would be subjected to a hydrostatic stress of 10^7 N/m². Hence,

$$-p = \sigma_1 = \sigma_2 = \sigma_3 = -10^7 \text{ N/m}^2$$

A quick experiment to determine the yield stress of the shrimp could be conducted by carefully holding it between two fingers and pressing it. By doing the test with a live shrimp the flow stress could be defined as the stress at which the amplitude of the tail wiggling would have become less than a critical value. This would certainly occur at a stress of about 10^5 N/m² (14.5 psi). Hence,

$$\sigma_0 = -10^5 \text{ N/m}^2$$

Rankine's criterion would produce shrimp failure at

$$p \equiv -\sigma_0 = 10^5 \text{ N/m}^2$$

This corresponds to a depth of only 10 m. Fortunately for all lovers of crustaceans, this is not the case, and hydrostatic stresses do not contribute to plastic flow.

‡2.2.2 Maximum Deformation-Energy Criterion

In a state of triaxial stress the accumulated energy is (see Section 1.11.5)

$$U = \tfrac{1}{2}\sigma_i \epsilon_i$$

According to this criterion, plastic flow occurs when the deformation energy reaches a critical value. Again, it can be seen that it predicts plastic flow under hydrostatic pressure.

‡2.2.3 Maximum Shear Stress Criterion (Tresca)[1]

Plastic flow starts when the maximum shear stress in a complex state of deformation reaches a value equal to the maximum shear stress at the onset of flow in uniaxial tension (or compression).

The maximum shear stress is given by (see Section 1.10.4)

$$\tau_{max} = \frac{\sigma_1 - \sigma_3}{2} \tag{2.3}$$

For uniaxial stress state we have, at the onset of plastic flow,

$$\sigma_1 = \sigma_0; \qquad \sigma_2 = \sigma_3 = 0$$

so

$$\tau_{max} = \frac{\sigma_0}{2}$$

Therefore,

$$\sigma_0 = \sigma_1 - \sigma_3 \tag{2.4}$$

This criterion corresponds to taking the difference between σ_1 and σ_3 and making it equal to the flow stress in uniaxial tension (or compression). It can be seen that it does not predict failure under hydrostatic stress, because we would have $\sigma_1 = \sigma_3 = p$ and no resulting shear stress.

‡2.2.4 Maximum Distortion-Energy Criterion (von Mises)[2]

This criterion was originally proposed by Huber as: "When the expression

$$\frac{\sqrt{2}}{2}[(\sigma_1 - \sigma_2)^2 + (\sigma_2 - \sigma_3)^2 + (\sigma_1 - \sigma_3)^2]^{1/2} > \sigma_0 \tag{2.5}$$

then the material will plastically flow." This stress is known as effective stress (Eq. 1.67, Section 1.10.7). This criterion was stated by von Mises without a physical interpretation. It is now accepted that the criterion expresses the critical value of the deviatoric component of the deformation energy of a body. Based on this interpretation, a body flows plastically in a complex state of stress when the distortional deformation energy is equal to the distortional deformation energy in uniaxial stress (tension or compression). This will be shown below.

The total deformation energy is the sum of the deviatoric (or distortional) and hydrostatic components:

$$U = U' + U''$$

The hydrostatic component is

$$U'' = \tfrac{1}{2} \sigma_i'' \epsilon_i''$$

Equation 1.61 of Section 1.10.7 gives

$$U'' = \tfrac{1}{2} \sigma_m (\epsilon_1'' + \epsilon_2'' + \epsilon_3'')$$

From Section 1.11.4,

$$\epsilon_1'' + \epsilon_2'' + \epsilon_3'' = \epsilon_1 + \epsilon_2 + \epsilon_3$$

Mechanics of Deformation   Part I

so

$$U'' = \tfrac{1}{2}\sigma_m(\epsilon_1 + \epsilon_2 + \epsilon_3)$$

The distortional component of deformation energy becomes

$$U' = \tfrac{1}{2}(\sigma_1\epsilon_1 + \sigma_2\epsilon_2 + \sigma_3\epsilon_3) - \tfrac{1}{2}\sigma_m(\epsilon_1 + \epsilon_2 + \epsilon_3)$$

From Eqs. 1.11, we have, for isotropic materials,

$$\epsilon_1 + \epsilon_2 + \epsilon_3 = \frac{1 - 2\nu}{E}(\sigma_1 + \sigma_2 + \sigma_3)$$

and

$$U' = \frac{1}{2}(\sigma_1\epsilon_1 + \sigma_2\epsilon_2 + \sigma_3\epsilon_3) - \frac{1 - 2\nu}{6E}(\sigma_1 + \sigma_2 + \sigma_3)^2$$

Substituting all the strains for stresses, according to Eq. 1.11 we have,

$$U' = \frac{1}{2E}(\sigma_1^2 + \sigma_2^2 + \sigma_3^2) - \frac{\nu}{E}(\sigma_1\sigma_2 + \sigma_2\sigma_3 + \sigma_1\sigma_3)$$

$$- \frac{1 - 2\nu}{6E}(\sigma_1 + \sigma_2 + \sigma_3)^2 \qquad (2.6)$$

It suffices now to apply Eq. 2.6 to a state of uniaxial stress, at the onset of plastic flow ($\sigma_1 = \sigma_0$, $\sigma_2 = \sigma_3 = 0$):

$$U_0' = \frac{\sigma_0^2}{2E} - \left(\frac{1 - 2\nu}{6E}\right)\sigma_0^2 = \left(\frac{1 + \nu}{3E}\right)\sigma_0^2 \qquad (2.7)$$

Setting Eq. 2.7 equal to Eq. 2.6, we should obtain Eq. 2.5; von Mises' criterion therefore expresses the distortion-energy criterion.

‡2.2.5 Graphical Representation and Experimental Verification of Tresca's and von Mises' Criteria

There is a convenient way to represent both criteria for a plane state of stress. For this, one makes $\sigma_3 = 0$ and has σ_1 and σ_2. It will be necessary to momentarily forget the convention of $\sigma_1 > \sigma_2 > \sigma_3$ because it would not be obeyed for $\sigma_2 < 0$; we have $\sigma_2 < \sigma_3 = 0$. Figure 2.3 shows a σ_1 versus σ_2 plot. According to Tresca's criterion, plastic flow starts when

$$\tau_{\max} = \frac{\sigma_0}{2}$$

The four quadrants have to be analyzed separately. In the first quadrant, there are two possible situations. For σ_1 greater than σ_2, $\tau_{\max} = (\sigma_1 - \sigma_3)/2$ and $\sigma_1 = \sigma_0$. This is a line passing through $\sigma_1 = \sigma_0$ and parallel to $O\sigma_1$.

In the second quadrant, $\sigma_2 < 0$ and $\sigma_1 > 0$. We have

$$\tau_{\max} = \frac{\sigma_1 - \sigma_2}{2} \qquad \text{and} \qquad \sigma_1 - \sigma_2 = \sigma_0$$

This equation represents a straight line intersecting the $O\sigma_1$ axis at σ_0 and the $O\sigma_2$ axis at $-\sigma_0$.

The flow criteria for quadrants III and IV are found in a similar way. For von Mises' criterion we have, from Eq. 2.5 and $\sigma_3 = 0$,

$$\sigma_0 = \frac{\sqrt{2}}{2}\,[(\sigma_1 - \sigma_2)^2 + \sigma_2^2 + \sigma_1^2]^{1/2}$$

$$\sigma_1^2 - \sigma_1\sigma_2 + \sigma_2^2 = \sigma_0^2$$

This is the equation for an ellipse whose major and minor axes were rotated 45° from the orthogonal axes $O\sigma_1$ and $O\sigma_2$, respectively. This can be easily shown by applying a rotation of axes to the equation of an ellipse referred to its axes:

$$\left(\frac{\sigma_1}{a}\right)^2 + \left(\frac{\sigma_2}{b}\right)^2 = k^2 \tag{2.8}$$

From the observation of Eq. 2.8 it can be seen that Tresca's criterion is more conservative than von Mises'. Tresca's criterion would predict plastic flow for the stress state defined by point P_1, whereas von Mises' would not. However, both criteria are fairly close. It can be seen from Fig. 2.3 that plastic flow may require a stress σ_1 greater than σ_0 for a combined state of stress (see point P_2). However, there are regions (when one stress is tensile and another is compressive) when plastic flow starts where both stresses are within the interval

$$\sigma_0 < \sigma_1,\ \sigma_2 < \sigma_0$$

This occurs in the second and fourth quadrants. Point P_2 shows this very clearly. This shows that the correct application of a yield criterion is very

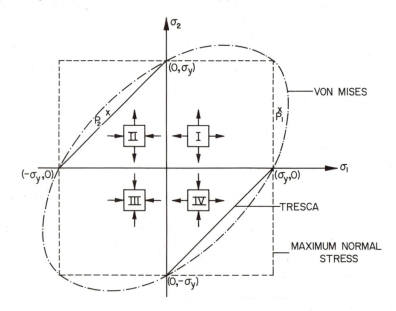

Figure 2.3 Comparison of the Rankine, von Mises, and Tresca criteria.

Mechanics of Deformation   Part I

important for design purposes. For comparison purposes, the maximum stress criterion is also drawn in Fig. 2.3. It is just a square with sides parallel to the $0\sigma_1$ and $0\sigma_2$ axes and intersecting them at $(\sigma_0, 0)$, $(-\sigma_0, 0)$ $(0, \sigma_0)$, and $(0, -\sigma_0)$. We see that there is a considerable difference between Rankine's criterion, on the one hand, and Tresca's and von Mises' criteria, on the other hand, for quadrants II and IV. This difference is readily explained by the fact that Rankine's criterion applies to brittle solids (including cast irons and steel below the ductile–brittle transition temperature) where failure (or fracture) is produced by tensile stresses. For this reason these criteria have traditionally been separated into failure (Rankine) and yield (von Mises and Tresca). Figure 2.4 shows the three criteria, together with experimental results for copper, aluminum, steel, and cast iron. While copper and aluminum tend to follow von Mises' criterion (and, in a more conservative way, Tresca's), cast iron clearly obeys Rankine's criterion. This is clearly in line with the low ductility exhibited by cast iron. The reader is warned that the ratio $\sigma/\sigma_{\text{ult}}$ and not σ/σ_0 is used in Fig. 2.4. Nevertheless, it serves to illustrate the difference in response between ductile

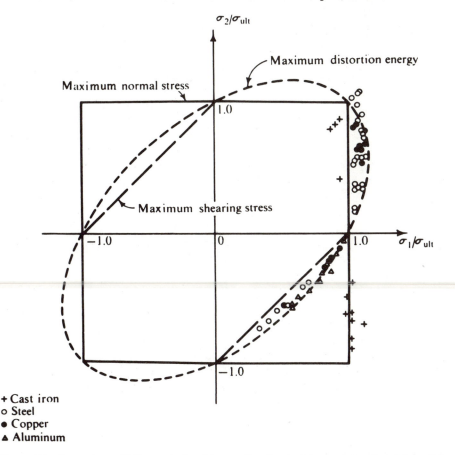

+ Cast iron
o Steel
• Copper
▲ Aluminum

Figure 2.4 Comparison of failure criteria with test. (Reprinted with permission from [3], p. 298, and [4], p. 83.)

and brittle materials. For brittle materials, the flow (or fracture) stress under compression is much higher than under tension. Therefore, the square defined by Rankine's criterion is not symmetrical with respect to the system of axes. σ_0 under compression is usually several times σ_0 under tension. Other criteria, such as the Babel–Sines and McLaughlin criteria, have been developed for brittle materials.

The determination of the flow locus is usually conducted in biaxial testing machines, which operate in a combined tension–torsion or tension–hydrostatic pressure mode. These two modes use tubular specimens and one has to determine the appropriate calculations to find the principal stresses. As the material is plastically deformed, we have an expansion of the flow locus. For von Mises' criterion, we can envision concentric ellipses having increasing major and minor axes. This is illustrated in Fig. 2.5.

Kuhn and Downey[5] extended von Mises' flow criterion to sintered powder compacts exhibiting porosity. We can readily see that, for these materials, some unique deviations from the accepted criteria of plasticity take place.

1. The volume is not constant during plastic deformation ($\nu < 0.5$).
2. A hydrostatic pressure produces plastic deformation.

Based on this, Kuhn and Downey[5] proposed the following flow criterion:

$$\sigma_0 = \frac{\sqrt{2}}{2} [(\sigma_1 - \sigma_2)^2 + (\sigma_1 - \sigma_3)^2 + (\sigma_2 - \sigma_3)^2 - 2(1 - 2\nu)J_2]^{1/2}$$

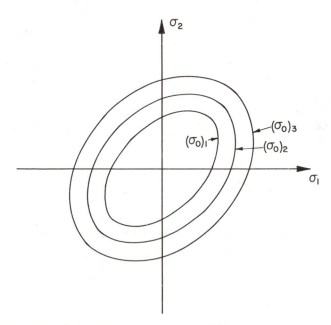

Figure 2.5 Expansion of the yield locus as the flow stress of the material increases due to work hardening.

Mechanics of Deformation   Part I

where J_2 is the second invariant of the stress deviator (Eq. 1.64). We see that, when the compact reaches its theoretical density, ν becomes equal to 0.5 and the Kuhn–Downey criterion reduces itself to the von Mises criterion. The analytical treatment of powder forging would require the criterion above.

2.3 THEORIES OF PLASTICITY

The process of plastic deformation is irreversible; the greatest part of the deformation work is transformed into heat by internal dissipative processes that will be described later. Careful calorimetric investigations[6] have been conducted and indicated that only 10% (and, in some cases, much less) of the deformation energy is stored in the material. When metals are very rapidly deformed, the process can be considered as adiabatic and the temperature increases can be very substantial. For instance, if the material is deformed at temperatures slightly below the melting point, the temperature rise might be enough to cause localized melting. High-velocity machining causes severe (and potentially damaging) temperature rises.

Because the process of plastic deformation is irreversible, the stresses and strains in the final state are path dependent. Consequently, the equations that describe plastic deformation cannot be finite relationships that relate the stress and strain components, such as the equations of linear elasticity theory; they are nonintegrable differential equations. In elasticity, a state of strain can be expressed as a function of a state of stress at a certain temperature, because the process is reversible. On the other hand, this direct relationship cannot be obtained in plasticity. The whole deformation history will affect the state of strain, as shown below. Consequently, one has to introduce strain increments (in plasticity, $d\epsilon_{ij}$ will replace ϵ_{ij}). The total deformation is obtained by summing up the deformation increments over the whole deformation path. An example that illustrates the basic difference between elasticity and plasticity is the following. Imagine a cylinder that is first compressed from l to $l - \Delta l$ and then extended from $l - \Delta l$ to l. If the process were elastic, the work applied in the compression would be regained in the extension and the total strain would be zero. In plasticity, work would have to be applied both in compression and extension, and the total strain would be counted as twice the compression strain.

Some authors divide the plasticity theories into two classes: flow theories and deformation theories. The former relate the stress with the variation in strain. Since the flow theories deal with instantaneous strain, they are better for large deformations. The deformation theories correlate stress and strain and use an approximation method to represent the deformation history; they relate the total plastic strains to the final stress. They can be applied to cases where the loading is proportional, that is, when we have

$$\frac{d\sigma_1}{\sigma_1} = \frac{d\sigma_2}{\sigma_2} = \frac{d\sigma_3}{\sigma_3}$$

(2.9)

The stresses increase in proportion to their instantaneous values. Another possible situation is proportional straining.

2.3.1 Theory of Levy–von Mises (Flow)

Around 1870, Saint-Venant[7] proposed that, for plane strain, the principal axes of the strain increments coincide with the principal axes of stress. Levy[8] in 1870, and later von Mises[2] in 1913, established a theory for ideal plastic ($\sigma_0 =$ constant) rigid ($E = \infty$) bodies. The equations can be put in the following form:

$$\sigma'_{ij} = 2\lambda\dot{\epsilon}'_{ij} \tag{2.10}$$

$$\epsilon''_{ij} = 0 \tag{2.11}$$

Equation 2.10 expresses the proportionality between the deviator and the time derivative of the strain deviator. This equation should not be confused with Newton's law for viscous flow. The similarity is apparent only because of the different nature of λ. Another form of the equation that emphasizes the incremental character of the strain is

$$\sigma'_{ij} = \frac{2\lambda}{dt}\, d\epsilon'_{ij}$$

$$\sigma'_{ij} = \lambda'\, d\epsilon'_{ij}$$

Equation 2.10 is a generalization of experimental results in complex loading situations. According to experiments, the increments of strain deviators have been found to be proportional to the respective stress deviators. The other equation, 2.11, corresponds to the condition of constancy in volume, because

$$\epsilon_{ij} = \epsilon'_{ij} + \delta_{ij}\epsilon''_{ij} = \epsilon'_{ij} + \frac{\epsilon_m}{3} = \epsilon'_{ij} + \frac{\epsilon_1 + \epsilon_2 + \epsilon_3}{3}$$

and since

$$\epsilon_1 + \epsilon_2 + \epsilon_3 = 0$$

$$\epsilon_{ij} = \epsilon'_{ij} \quad \text{or} \quad d\epsilon_{ij} = d\epsilon'_{ij}$$

and

$$\sigma'_{ij} = \lambda'\, d\epsilon_{ij} \tag{2.12}$$

Equation 2.12 requires that if the components of one of the tensors is taken with respect to the principal axes, the other will also have to be taken. In other words, the principal axes of the stress deviator have to be parallel to the ones of the strain increments. This condition is slightly different from the one proposed by Saint-Venant. To find the value for the parameter λ', we have to involve effective stresses and strains (see Eqs. 1.67 and 1.81). In a uniaxial tensile test, we would have, for Eq. 2.12:

$$\sigma'_{11} = \lambda'\, d\epsilon_{11} \tag{2.13}$$

Mechanics of Deformation   Part I

Since $\sigma_{22} = \sigma_{33} = 0$, the value of σ'_{11} is (from Eq. 1.63)

$$\sigma'_{11} = \frac{2\sigma_{11} - \sigma_{22} - \sigma_{33}}{3} = \frac{2\sigma_{11}}{3} \tag{2.14}$$

For a uniaxial stress the effective stress is

$$\sigma_{\text{eff}} = \frac{\sqrt{2}}{2}[(\sigma_1 - \sigma_2)^2 + (\sigma_1 - \sigma_3)^2 + (\sigma_2 - \sigma_3)^2]^{1/2} = \sigma_1 = \sigma_{11} \tag{2.15}$$

Substituting Eq. 2.15 into Eq. 2.14, we have

$$\sigma'_{11} = \tfrac{2}{3}\sigma_{\text{eff}} \tag{2.16}$$

For the strains

$$d\epsilon_{\text{eff}} = \frac{\sqrt{2}}{3}[(d\epsilon_1 - d\epsilon_2)^2 + (d\epsilon_1 - d\epsilon_3)^2 + (d\epsilon_2 - d\epsilon_3)^2]^{1/2}$$

By substituting the appropriate values of $d\epsilon_2$ and $d\epsilon_3$ (with $\nu = 0.5$), we obtain

$$d\epsilon_{\text{eff}} = d\epsilon_{11} = d\epsilon_1 \tag{2.17}$$

Substituting Eqs. 2.16 and 2.17 into 2.12, we get

$$\lambda' = \frac{2}{3}\frac{\sigma_{\text{eff}}}{d\epsilon_{\text{eff}}}$$

Substitution into Eq. 2.12 yields

$$d\epsilon_{ij} = \frac{3}{2}\frac{d\epsilon_{\text{eff}}}{\sigma_{\text{eff}}}\sigma'_{ij} \tag{2.18}$$

Application of Eqs. 2.15 and 2.16 into 2.18 gives us

$$d\epsilon_{11} = \frac{d\epsilon_{\text{eff}}}{\sigma_{\text{eff}}}\left[\sigma_{11} - \frac{1}{2}(\sigma_{22} + \sigma_{33})\right]$$

$$d\epsilon_{22} = \frac{d\epsilon_{\text{eff}}}{\sigma_{\text{eff}}}\left[\sigma_{22} - \frac{1}{2}(\sigma_{11} + \sigma_{33})\right]$$

$$d\epsilon_{33} = \frac{d\epsilon_{\text{eff}}}{\sigma_{\text{eff}}}\left[\sigma_{33} - \frac{1}{2}(\sigma_{11} + \sigma_{22})\right]$$

$$d\epsilon_{12} = \frac{3}{2}\frac{d\epsilon_{\text{eff}}}{\sigma_{\text{eff}}}\sigma_{12}$$

$$d\epsilon_{13} = \frac{3}{2}\frac{d\epsilon_{\text{eff}}}{\sigma_{\text{eff}}}\sigma_{13}$$

$$d\epsilon_{23} = \frac{3}{2}\frac{d\epsilon_{\text{eff}}}{\sigma_{\text{eff}}}\sigma_{23}$$

$$\tag{2.19}$$

This set of equations is of great importance in the analysis of deformation processing and will be used in Sections 2.5.5 and 2.5.6. The similarity between

Eqs. 2.19 and 1.119 (generalized Hooke's law for isotropic materials) should be carefully analyzed. Poisson's ratio is replaced by $\frac{1}{2}$, its value in the plastic range. $1/E$ is replaced by $d\epsilon_{\text{eff}}/\sigma_{\text{eff}}$. The latter parameter can be obtained directly from an effective stress–effective strain curve by taking the inverse of the slope and is the exact correspondent of E. There is also a correspondence between $\frac{1}{2}G$ (in Eq. 1.119) and $\frac{3}{2}$ ($\epsilon_{\text{eff}}/\sigma_{\text{eff}}$) (Eq. 2.19); they both represent the resistance of the material to distortion. The relationship between G and E is, in elasticity (Table 1.2),

$$G = \frac{E}{2(1 + \nu)}$$

For $\nu = 0.5$, we have $G = E/3$ and

$$\frac{1}{2G} = \frac{3}{2E}$$

Consequently, there is a total correspondence. An alternative way is to obtain the different values of ($d\epsilon_{\text{eff}}/\sigma_{\text{eff}}$) from a work-hardening relation, such as $\sigma_{\text{eff}} = k(\epsilon_{\text{eff}})^n$. One can conclude that the Levy–von Mises equations are the equivalent in plasticity to the constitutive equations for isotropic materials in elasticity.

The Levy–von Mises theory was extended by Prandtl[9] to elastoplastic bodies and by Reuss[10] for the general case.

In the flow theories described above, the strain increments are related to the stress deviators. On the other hand, in the deformation theories the total plastic strain is related to the stress deviator. Both Hencky's[11] and Nadai's[12] theories are identical to Levy–von Mises' theory if the principal stresses and strains are coaxial during the whole extent of deformation and if the loading is assumed to be proportional.

2.4 EFFECT OF HYDROSTATIC PRESSURE ON PLASTIC DEFORMATION

The application of a hydrostatic pressure to a component being formed has significant effects; the main advantage is a large increase in ductility. This enhancement of ductility is especially helpful in the case of hard-to-form metals. Superalloys (Udimet 700, Inconel 718, Stellite 21, TD nickel, among others), and AISI 4340, Marage 20, Marage 350 steels, aluminum, and titanium alloys have been successfully hydrostatically extruded.[13]

Figure 2.6 shows the increase in ductility of a number of metals with hydrostatic pressure. The most dramatic increases occur for the hexagonal close-packed (HCP) metals zinc and magnesium. The ductility is expressed as the logarithm of the ratio A_0/A_f, where A is the cross-sectional area. This is the true strain. The reason for this increase in ductility is the interaction of the hydrostatic pressure with the fracture propagation mechanism.[13] In brittle fracture and intergranular fracture, the normal tensile stresses are responsible for

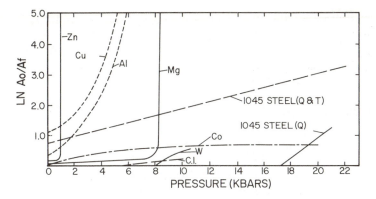

Figure 2.6 Effect of hydrostatic pressure on ductility of several metals. (Reprinted with permission from [13], p. 537.)

the growth of the crack (mode I fracture). A superimposed hydrostatic stress will close these cracks or make the sliding over the crack faces more difficult; consequently, they cannot propagate any longer. Shear cracks (mode III), on the other hand, are not affected by hydrostatic stresses. Therefore, if fracture takes place, at atmospheric pressure, by the propagation of cleavage or intergranular cracks, the hydrostatic pressure will inhibit these mechanisms; the fracture mechanism shifts toward those dependent on more and more shear strain.

Extrusion, drawing, and closed-die forging have been used successfully with hydrostatic pressure. In the case of extrusion, for instance, the billet is surrounded by a hydrostatic pressure, so that the pressure acts on the portion of the component that has been formed.

2.5 DEFORMATION PROCESSING

‡2.5.1 General

In the technologically very important area of deformation processing, the analysis of stresses, strains, strain rates, temperatures, and loads involved is very important. These analyses rest on the yield criteria and theory of plasticity. The use of computers in the analysis of deformation processing has increased dramatically in the seventies; examples of approaches that are made possible by the use of numerical methods are slip-line field theory, the upper bound method, and the finite element method. Deformation processing will be presented briefly in the next sections. For more in-depth treatment the reader is referred to "Suggested Reading." Section 2.5.2 will present a general description of the various processes; the classification of the wide range of deformation processes according to three points of view is presented in Section 2.5.3. Section 2.5.4 delineates the methods of analysis; four deformation processing systems are analyzed briefly in Sections 2.5.5 to 2.5.8: rolling, wire drawing, forging, and sheet-metal forming. For the first three, the equilibrium (or slab) method will be applied with the objective of determining the loads required to effect the deformation as well as critical conditions. For sheet-metal forming, two very important tests will be described. The forming-limit diagram, an important development in the change of this process from an art to a science, is introduced and discussed separately in Chapter 18.

‡2.5.2 Deformation Processing Systems

The plastic forming of metals—commonly referred to as deformation processing—is of utmost technological importance. There are essentially three ways by which a metal can be processed to its final shape: (1) by casting from the liquid state, (2) by the powder metallurgy route (fine metal powders are pressed and then sintered), and (3) by deformation processing. The latter class includes the well-known processes of rolling, forging, wire drawing, sheet-metal forming, spinning, and extrusion. Metal-removal processes such as boring, drilling, lathing, milling, and blancher grinding are usually not included in this classification. They constitute a separate group of operations.

In deformation processing two changes are accomplished: the shape of the workpiece is changed by plastic deformation and its mechanical properties are altered because of structural changes; these two changes are connected and it is up to the metallurgist's ingenuity to derive the greatest performance increase from the modifications in both shape and properties. The approach used in the development of deformation processing has been mostly an empirical one. As such, the rigorous scientific description of these processes in mathematical terms is only now reaching the stage where the understanding obtained can lead to further advancements. There are a great variety of special techniques and the detailed study of plastic stresses and strains in all of them would be overly lengthy and complex. Figure 2.7 shows a general "deformation processing system." The generalized approach to the study of deformation processing system

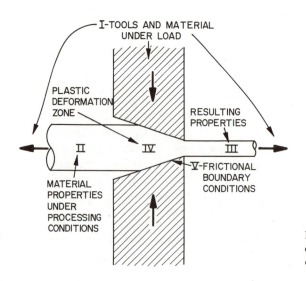

Figure 2.7 Generalized deformation processing system for a geometry of converging dies.

is e... igh there is a great diversity of deformation pr... spects that can be studied from a general po... Every deformation processing system has to... plastic deformation zone where plasticity th... ctional stresses between material and tools (V... roperties of the material (II) as well as its properties during and after deformation (III) are important parameters. However, this generalized and systematic approach will not be used in this book because of the extensive treatment required. Figures 2.8 through 2.15 show schematically the most common deformation processes. These processes will be described briefly. In *rolling* (Fig. 2.8) the material is passed between two cylinders of equal radius rotating in opposite senses. The most common application of rolling is in the production of plates and sheets. However, the cylinders are not necessarily smooth, and a wide variety of different structural shapes can be rolled: I beams, T beams, rails, and so on. The number of passes required to roll a certain type of rail is shown in Fig. 2.8(b). The design of the intermediate shapes is more of an art than a science, and great craftsmanship and experience are required. One starts with a billet and progressively arrives at the final shape; 10 passes are needed in the specific example. After the fourth pass the bar is rotated 90°.

Forging is shown in Fig. 2.9. Open-die forging is often used for preforming the workpiece; the dies are either flat or have very simple shapes. The upsetting of a cylindrical material is treated in greater detail in Section 2.5.7. In closed-die forging [Fig. 2.9(b)], on the other hand, one uses dies that have carefully machined shapes and the material is deformed to a specific shape that is either the final shape or very close to it. The dies are designed in such a way that the volume circumscribed by them is slightly lower than the volume of the workpiece, which was preshaped. Therefore, there is a need for a flash gutter, shown in Fig. 2.9(b). The excess material ("flash") is forced into the "gutter." After forging, the flash is removed in a separate operation. The objective of having a die-opening volume smaller than the volume of the workpiece is to assure that no open spaces are left in the workpiece. Additionally, the flash regulates the escape of material out of the die-opening region. Consequently, the width of this channel is important in establishing the pressure within the die-opening region; it has to be narrow enough to allow the pressure to build up to the values sufficiently high to direct the material to all the recesses of the die. *Wire drawing* is shown in Fig. 2.10. The wire is pulled in tension through a die that forces its diameter to be reduced because of its conical cross section. Other shapes, such as rounds, hexagons, squares, small angles, channels, and tubes, can be cold drawn. The part of the die that is in direct contact with the wire is called a nib; it is usually made of tungsten carbide, although diamond dies are used for very thin wires and high-precision work. Lubrication is used to decrease the high frictional forces.

Extrusion is similar to wire drawing, except that the material is under compression and not tension. Figure 2.11 shows three slightly distinct techniques. In direct extrusion the flow of material and the ram have the same sense; since

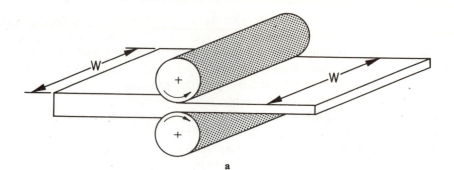

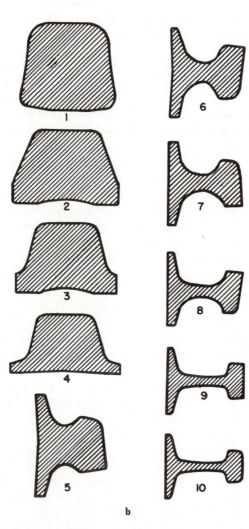

Figure 2.8 (a) Schematic representation of rolling of a plate. Notice the state of plane strain, whereby the width of the plate remains unchanged; (b) roll-pass contours for producing a 132-lb RE rail. (Reprinted with permission from [15], p. 758.)

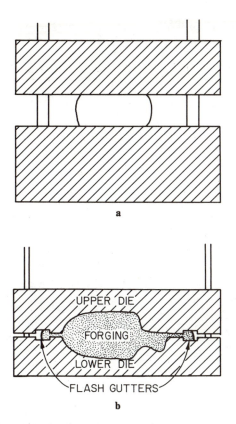

a

UPPER DIE

FORGING

LOWER DIE

FLASH GUTTERS

b

Figure 2.9 (a) Open-die forging; (b) closed-die forging.

the material within the chamber is flowing toward the die, the frictional forces between the chamber and the billet will have to be overcome by the ram. In indirect extrusion, on the other hand, the ram and extruded material move in oppposite senses; the billet is stationary inside the chamber. The power requirements for indirect extrusion are lower than for direct extrusion. In impact extrusion [Fig. 2.11(c)], advantage is taken of the quasiadiabatic temperature rise

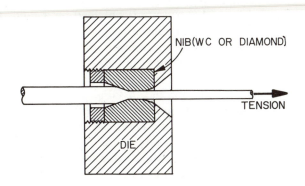

NIB(WC OR DIAMOND)

TENSION

DIE

Figure 2.10 Wire drawing.

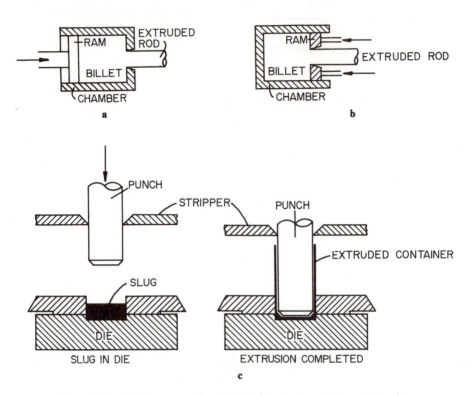

Figure 2.11 (a) Direct extrusion; (b) indirect extrusion; (c) impact extrusion.

(because there is no time for heat transfer from the metal being deformed to the tools) of the material; this results in a decrease in strength. The material is quickly extruded through the space between the die and the plunger. Thin-walled items such as aluminum beer cans are made by this process. The stress is applied to the part of the material that is "upstream" of the deformation zone. It is a very common deformation processing technique for very ductile metals such as aluminum.

In cases where the material is not so ductile, extrusion can be conducted in a high-pressure chamber. The pressure is obtained by a fluid. Beryllium is a typical example of a metal whose tensile properties are strongly dependent on pressure. This metal has an HCP structure and the limited slip systems cannot easily perform all the deformation required by the compatibility of strains. For example,[16] beryllium exhibits a total elongation of 0.8% at atmospheric pressure; this value goes up to 54% at 2.4 GPa. The reason for this strong pressure dependence of ductility is that the pressure will (1) prevent the formation of voids during plastic deformation; and (2) effect the closure of voids already existing in the material due to previous processing. These pores would otherwise grow into larger pores or cracks. Hydrostatic extrusion has also been successfully applied to other alloys exhibiting low ductility (such as tool steels).

Another industrially important deformation process is *sheet-metal forming*.

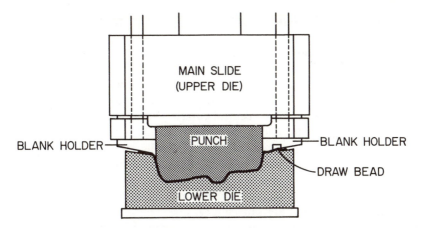

Figure 2.12 Sheet-metal stamping. Notice two actions. First, the blank holder secures the blank; then the upper die brings punch down.

Figure 2.12 shows schematically a stamping press where a blank is being formed, between a die and a punch, into a specific shape. Stamping is by no means the only method of forming parts from sheet metal, but it is economically the most important; other sheet-metal forming processes are deep drawing and spinning. The double-action mechanical press shown in Fig. 2.12 involves the motion of two rams. First, the blank holder comes down and clamps the blank; the next action pushes down the main slide, which has the punch attached to it, forming the part. The draw beads help to control the flow of the metal by making it bend and unbend, eliminating possible "wrinkles."

Tubes are classified, according to their manufacturing processes, into three classes: casting tubes, welded tubes, and seamless tubes. The Mannesmann process of *tube piercing* is illustrated in Fig. 2.13; it is a well-known and ingenious method for the production of seamless tubes. It was patented in 1885 and the principle of helical rolling is employed. The two rolls do not have parallel axes; they make between 6 and 12° with the horizontal centerline of the mill. The rolling surfaces are not cylindrical but contoured in such a way that they produce a net lateral advance of the billet; contrary to conventional rolling, both rolls have the same sense of rotation. The billet is forced to advance against

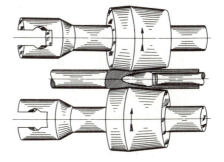

Figure 2.13 Mannesmann tube-piercing system. (Reprinted with permission from [15], p. 892.)

a stationary piercer. The billet rotates and advances in relation to the piercer, which consequently penetrates into it. It is interesting to notice that the action of the piercer is only to aid the formation of the center hole; a small but somewhat irregular hole would form even in the absence of the piercer. Indeed, a hole can be opened at the centerline of a cylindrical body by compressing it (with the direction of motion perpendicular to the cylinder axis). This causes the well-known forging defect called "central burst." The vertical load sets up tensile stresses that are horizontal and perpendicular to the axis of the cylinder. By rolling the cylinder, one creates the state of tensile stress successively along all radial orientations. The purpose of the nonparallelism between the axes is to make the billet advance as its center is pressed. After being pierced, the tubes receive further processing (plug rolling, reeling, sizing) until their diameter, surface, and wall thickness are satisfactory.

Operations that essentially involve *shearing* are blanking, notching, trimming, nibbling, and perforating; in these operations the material is actually cut by means of high internal shear stresses generated by appropriately applied and positioned external loads. Blanking consists of the preparation of blanks for subsequent sheet-metal forming. Nibbling consists of cutting the workpiece by continuously punching slots in it. Figure 2.14 shows a typical perforating operation. The workpiece is first elastically, then plastically, deformed by the action of the punch. The center is therefore depressed. Fracture starts at the regions of maximum stress concentration (sharp edges of punch and die) and propagates into the workpiece along regions of maximum shear stress. If the punch-die clearance is appropriate, the cracks intersect in the center of the workpiece [Fig. 2.14(a)]. If not, additional tearing is required [Fig. 2.14(b)]. Figure 2.14(a) shows the appearance of the fracture after shearing. One can see the region of plastic deformation that preceded burnishing, which in its turn preceded the shear fracture. On the other hand, if the clearance is allowed,

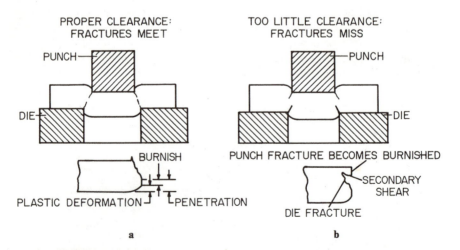

Figure 2.14 Cross sections of a typical punching operation showing the effect of clearance and the characteristics of the sheared slug. (Reprinted with permission from [17], p. 577.)

Mechanics of Deformation   Part I

the burnishing from the punch intersects the shear crack. Specific clearance and shearing loads are given elsewhere.[18]

There is a group of deformation processes that are known as high-energy rate fabrication (HERF).[19] They are unique in the sense that the workpiece is subjected to extremely high strain rates. In conventional forging, shearing, and forming operations the changes in dimension are obtained by the transfer of the kinetic energy of a large mass (hammer, punch, main slide) to a workpiece. The classical expression for the kinetic energy ($\frac{1}{2}mv^2$) shows clearly that the energy is proportional to the mass and to the *square* of the velocity. This provides a good idea of the great effectiveness of increasing the velocity of the die. If the velocity is doubled, the energy output is quadrupled. The high-energy rate (HER) processes are distinct from conventional high-velocity forming machines; the impact velocity in drop hammers varies between 0.3 and 7 m/s, for high-velocity forming machines it goes up to 20 m/s; on the other hand, the range of HERF machines is between approximately 20 and 300 m/s. The three principal types of HERF systems are electrohydraulic and electromagnetic machines in which electrical energy stored is suddenly released (by, for instance, the discharge of a capacitor) and the explosive process, in which the sudden release of chemical energy from the detonating explosive provides the driving means. The three systems have industrial applications. Figure 2.15 shows a common experimental setup for explosive forming. Water is used as a pressure medium. The blank is held between the die and a draw ring and the void in the die is evacuated. The explosive charge is placed at a distance s from the blank and detonated. The choice of explosives is dictated by the cost, availability, physical properties and safety. Typically, the following explosives (with their detonation velocities) are used: TNT (6900 m/s), RDX (8350 m/s) composition B (7800 m/s), and composition C-4 (8000 m/s). Two factors are responsible for the deformation of the workpiece. First, the shock wave generated by the detonation propagates through the water and transmits its kinetic energy to the blank. The shock wave carries about 60% of the detonation energy (chemical energy liberated by explosion). Then a gas bubble [shown in Fig. 2.15(b)] forms; it carries about 40% of the energy. The expansion of this bubble will force it against the already partially formed workpiece, completing the process. In the simplified picture described above, any important interactions were ignored. A detailed study, with design of actual systems, is given by Ezra.[20]

The following advantages are offered by HERF over more conventional metalworking processes:[21]

1. Large integral parts can be produced with little tooling and capital equipment expense when only one or a few are required.
2. Difficult-to-form metals can be formed more readily.
3. Instability strains and accordingly, draw depths, tend to be higher.
4. Springback is minimized.
5. Components can be fabricated which could not be formed by any other process.

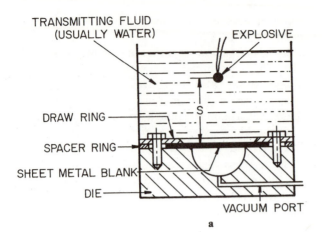

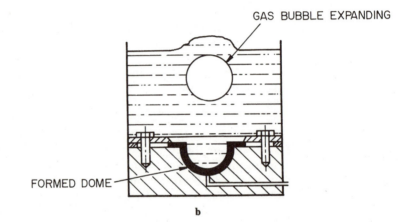

Figure 2.15 Explosive forming of metal dome: (a) prior to detonation; (b) after detonation. (Reprinted with permission from [19], p. 494.)

As a consequence, HERF processes have found some specific industrial applications. Other HERF processes are explosive welding, cladding, cutting, engraving, and hardening; the latter process is treated in greater detail in Chapter 15.

There are additional deformation processes not presented here. Bending, flanging, rubber forming, and spinning are some of these. Additionally, as technology develops, new processes are introduced. Examples are:

1. Powder rolling, whereby hot metal powder is forced between rolls, resulting in the production of sheet and bars. A variety of metals are being made by this process, including nickel sheet and strip for electronic applications coinage.[22]

2. Superplastic forming,[23] whereby advantage is taken of the extraordinary ductility exhibited by some alloys (e.g., Zn–22% Al) under certain conditions. This behavior is described in greater detail in Section 16.1.5. The

outstanding formability of these alloys lends them to forming techniques of the glass and plastics industries.

3. High-pressure processing,[24] the favorable effect of hydrostatic pressure superimposed on the stresses required for forming, was previously discussed in the context of extrusion. However, the application of high-pressure forming, deep drawing, tube expansion, hole punching, and wire extrusion are being or have been developed. The metallurgical causes for the enhancement of ductility by hydrostatic pressure are discussed in Section 2.4.

4. Hot isostatic pressing (HIP) is a technique by which cast components are heated and subjected to a pressure; the existing imperfections (voids, etc.) close by creep. HIP has been very successful in superalloys and shows promise in rapidly-solidified powders.

‡2.5.3 Classification of Deformation Processes

The classification of the processes depends on the point of view used. One may classify them according to the *continuity of the process*. In this case, they can be:

1. Steady-state processes, when the zone of plastic deformation remains fixed in shape and size and does not alter with time. Examples are rolling, wire drawing, extrusion, and tube piercing. The tool geometry is usually called "geometry of converging channels" because the material is forced through a region with a smaller and smaller cross section.

2. Non-steady-state processes, in which the region of plastic deformation changes in shape and extent with time. These processes are discontinuous and examples are forging, spinning, bending, and shearing.

On the other hand, deformation processes can also be classified according to the state of stress generated in the workpiece. If this criterion is used, one has to make certain approximations because large deformations usually involve tridimensional stresses and strains. However, for classification purposes one considers the state of stress as approximately fitting into one category. The classification scheme adopted by Blazynski[25] is satisfactory in this regard:

I. Tensile–compressive systems
    A. Biaxial tension–uniaxial compression
        1. Under a roll of a two-roll rotary piercer
        2. Under a roll of a two- and three-roll rotary elongator
        3. Under a roll of the helical rolling process
    B. Uniaxial tension–uniaxial compression
        1. Between the rolls in roll forming
        2. In the flange of a cup in deep drawing
    C. Uniaxial tension–biaxial compression
        1. In the die in the wire and bar drawing
        2. In the die in tube sinking and plug or mandrel tube drawing

II. Compressive stress systems
   A. Triaxial stresses
      1. In the oblique zone (i.e., in front of the plug) in three-roll rotary piercing
      2. Throughout the whole region in forging and upsetting in close dies
      3. In the region near the die throat in forward and backward extrusion of bar
      4. In the region under the punch in tube extrusion
   B. Biaxial stress
      1. Between the rolls of a rolling mill when operating without the front and/or back tensions
      2. Throughout the whole region when upsetting in open dies
III. Tensile stress systems
   A. Biaxial stress
      1. Stretch forming
      2. Bulging
IV. Shear stress systems
   A. Punch piercing
   B. Blanking

Yet another classification scheme is the temperature of the deformation process. According to this point of view, one has cold, warm, and hot working. Hot working requires a combination of temperature and strain rates so that essentially no strain hardening is produced by the deformation. The recovery of the original properties of the material may occur by either dynamic recovery or dynamic recrystallization. Cold working, on the other hand, is accompanied by strain hardening. Figure 2.16 shows the increase of hardness associated with rolling nickel at ambient temperature. The material was initially in the annealed

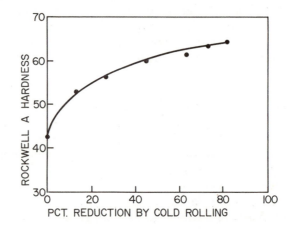

Figure 2.16 Increase in hardness for a nickel plate (99.5% pure, annealed, 25.4 mm thick) after successive passes by cold rolling; 2 to 3% reduction per pass. (Courtesy of C. Y. Hsu, South Dakota School of Mines and Technology.)

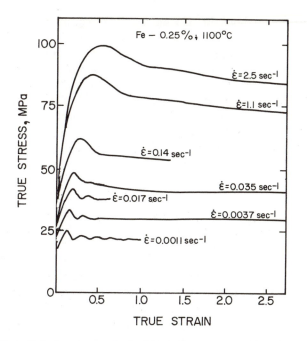

Figure 2.17 Effect of strain rate on plastic deformation response of Fe–0.25% C. [Reprinted with permission from [26], p. 222, Fig. 3(a).]

condition; after every few passes, the hardness was recorded. It can be seen that the rate of hardening decreases as more deformation is imparted. This parabolic behavior resembles the strain hardening of the material in tensile testing. Cold working is preferred over hot working when the added strength associated with the deformation is desired or when close dimensional control and good surface finish are desired. The name "warm working" is reserved for deformation processes imparted at intermediate temperatures. Modern deformation processes often use a combination of temperature ranges; this subject is treated in greater detail in Section 15.1.

The response of metals in hot working is typified by Fig. 2.17, which shows the compressive strength of Fe-0.25% C at 1100°C as a function of strain rate. At the lower strain rates the plastic range is reasonably flat, indicating that the rate of dynamic recovery (and/or recrystallization) is equal to the rate of work hardening. At the higher strain rates, the material actually exhibits flow softening after a certain critical strain. These effects have been studied in detail by Jonas and Luton.[26] A marked decrease in workability is characteristic of warm working for steels and titanium alloys.

2.5.4 Methods of Analysis

The analysis of loads and stresses involved in deformation processing systems has undergone great development in the past years; in particular, numerical methods using the computer have a great potential. There are a considerable number of possible methods, in varying degrees of complexity. The best known methods can be classified into five groups, which are described briefly in this section. All methods use the constancy of volume assumption, a specific yield criterion, and a deformation dependence of the flow stress.

2.5.4.1 The equilibrium (or slab) method.

This method consists essentially of separating an element (or slab) of the workpiece being deformed and of applying the equilibrium conditions to it; from there, one arrives at differential equations that, when solved, provide the internal stresses and loads applied by the tools on the workpiece. We impose the following conditions:

$$\Sigma\, F_{x_1} = \Sigma\, F_{x_2} = \Sigma\, F_{x_3} = 0 \qquad (2.20)$$

In its simplest form it assumes that the workpiece is being homogeneously deformed, as shown in Fig. 2.18(a); this means that there are no shear strains along the orientation indicated. Consequently, there are no *redundant* stresses and strains; that is, the perpendicularity between the lines in the imaginary grid drawn in Fig. 2.18 is maintained. This is the simplest way of deforming a metal. However, the frictional boundary conditions in real deformation processing systems are such that redundant strain is superimposed on homogeneous strain. In drawing and rolling they create distortions that are opposite in sense [Fig. 2.18(b) and (c)]. The redundant deformation requires additional work; therefore, the assumption of homogeneous deformation underestimates the work of deformation.

2.5.4.2 Slip-line field analysis.

In slip-line field analysis, as the name indicates, one tries to establish the directions along which the shear stresses are maximum; these are the directions along which deformation will take place. This method is used for plane-strain geometries. While the slab method only is involved in the determination of stresses and loads, the slip-line field theory links stresses with strains, and emphasizes the pattern of flow in the workpiece. The penetration of a punch into a metal slab with finite depth is illustrated in Fig. 2.19. A frictionless flat rectangular punch is used and the workpiece is aluminum. The slip lines can be clearly seen; there is a triangular region directly below the punch where no deformation is produced.[27] The fields in Fig. 2.19(a) were computer generated; the complexity of the problem can be perceived if one sees that one has an anisotropic material; on one side, the fan of slip fields has an angle of 70°; on the other side, it is equal to 62°. In recent years computational techniques have been successfully applied to these problems.

2.5.4.3 Upper- and lower-bound solutions.

This method of analysis is described in great detail by Avitzur (see "Suggested Reading"), who devotes two chapters to it. In these methods, one computes the power required to overcome the resistance of the workpiece to deformation and the frictional boundary conditions. The upper bound establishes the highest possible power, while the lower-bound analysis establishes the lower limit. By refining these techniques, one can bring the two solutions closer and closer together. The upper- and

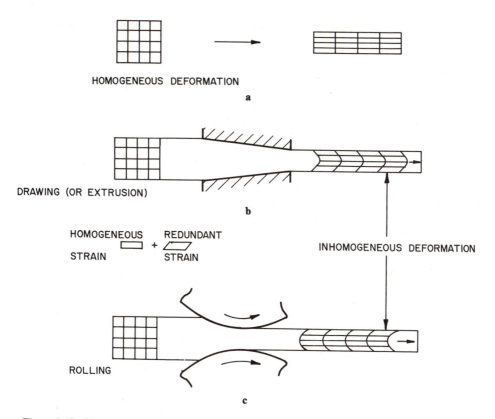

HOMOGENEOUS DEFORMATION

a

DRAWING (OR EXTRUSION)

b

HOMOGENEOUS STRAIN + REDUNDANT STRAIN

INHOMOGENEOUS DEFORMATION

ROLLING

c

Figure 2.18 Homogeneous (a) and inhomogeneous (b, c) deformation. Notice frictional boundary conditions producing opposite shear strains for (b) and (c).

lower-bound techniques are based on two theorems by Prager and Hodge.[28] The upper-bound theorem involves tangential velocity discontinuities in the workpiece. The deformation generates heat; consequently, a region in the workpiece that is more strained heats up more than the other and temperature jumps occur in a material that is being plastically deformed. Since both the flow stress and work hardening are temperature dependent, the flow pattern of material will depend on the temperature and internal velocity discontinuities that are generated. The higher the strain rate, the larger these velocity discontinuities will be, because no time for heat transfer is allowed. Prager and Hodge's upper-bound theorem is: "If surfaces of velocity discontinuities are to be included, among all mathematically admissible strain-rate fields the actual one minimizes the expression

$$J = \dot{W}_I + \dot{W}_S - \dot{W}_T"$$ (2.21)

where J is the upper bound on power, $\dot{W}_I$ the power required for deformation of the piece, $\dot{W}_S$ the shear power over surfaces of velocity discontinuities, including the tool–workpiece interfaces, and $\dot{W}_T$ the power supplied by predetermined body tractions (such as back and front tensions in rolling).

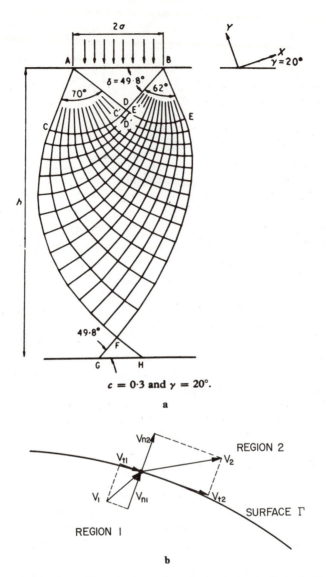

Figure 2.19 (a) Slip-line field for the indentation of an anisotropic, finite-depth medium by a punch. (Reprinted with permission from [27], p. 21.) (b) Surface of velocity discontinuity. Observe that $v_{n1} = v_{n2}$, but $v_{t1} \neq v_{t2}$.

A better understanding of these powers is obtained if we express them. Imagine a certain boundary between two fields, moving at velocities v_1 and v_2 [Fig. 2.19(b)]. The normal components of the velocities have to be the same; otherwise, the continuity at the interface could not be maintained. In a fluid analog, we can think of two streams moving at different velocities. At the interface, there has to be plastic deformation. If we imagine the state of stress to be of pure shear, as a first approximation, and applies von Mises' yield criterion, we have

$$\sigma_0 = \frac{\sqrt{2}}{2} [(\sigma_1 - (-\sigma_1))^2 + (\sigma_1 - 0)^2 + (0 + \sigma_1)^2]^{1/2} \qquad (2.22)$$

This is due to the fact that the pure shear is a particular case of biaxial stress where $\sigma_3 = -\sigma_1$ (see Section 1.10.6). Consequently,

$$\sigma_1 = \frac{\sigma_0}{\sqrt{3}}$$

and since

$$\tau_{max} = \frac{\sigma_1 - \sigma_3}{2} = \sigma_1 \qquad (2.23)$$

we have

$$\tau = \frac{\sigma_0}{\sqrt{3}} \qquad (2.24)$$

The power $\dot{W}_s$ can be expressed as

$$\dot{W}_s = \int_\Gamma (v_{t1} - v_{t2}) \frac{\sigma_0}{\sqrt{3}} ds \qquad (2.25)$$

where Γ is the surface of velocity discontinuity and ds is the space differential. $\dot{W}_I$ can also be easily expressed. From Eq. 1.86, we have

$$\dot{W} = \tfrac{1}{2} \sigma_{ij} \dot{\epsilon}_{ij} \qquad (2.26)$$

The total power is the integral over the volume:

$$\dot{W}_I = \int_V \tfrac{1}{2} \sigma_{ij} \dot{\epsilon}_{ij} \, dV \qquad (2.27)$$

The power due to the body tractions is also computed and Eq. 2.21 becomes

$$J = \tfrac{1}{2} \int_V \sigma_{ij} \dot{\epsilon}_{ij} \, dV + \frac{\sigma_0}{\sqrt{3}} \int_\Gamma (v_{t1} - v_{t2}) \, ds - \int_S T_i v_i \, ds \qquad (2.28)$$

where T_i are surface tractions.

The lower bound on power consists of Prager and Hodge's first theorem. The ideal power of deformation is calculated. The lower bound on power is by far less helpful than the upper bound. For this reason it is used much less frequently.

2.5.4.4 Visioplasticity.

As its name indicates, this method involves the actual observation of the deformation pattern in the workpiece. Once the flow pattern has been established experimentally, it is possible by computational techniques to establish the stress, strain, and strain rates in any section within

Figure 2.20 Flow of material in plasticine bar rolled between two wooden rolls. (Reprinted with permission from [30], p. 17.)

the deformation zone of a workpiece. Since no assumptions are made, the solution is correct as long as the experimental techniques are correct. This method actually serves as a "check" for the more involved mathematical treatments.

Several experimental techniques have been developed. For axisymmetric deformation processes—extrusion and wire drawing, for instance—we split the workpiece in two along the axis of symmetry and imprint, by either mechanical or photographic means, a grid in the inner face of the section; the two halves are rejoined together and deformed. The analysis of the deformation of the grid allows the determination of stresses, strains, and velocity fields; consequently, the strain rates can also be determined. The "laminate" technique lends itself to other processes[29]; a composite billet is produced by sandwiching at regular intervals a foil of a marker metal between the metal being actually deformed. We can also drill holes in the workpiece and insert in them wires or rods of a marker material. The matrix and marker have to have very close mechanical properties if reliable results are desired. Radioactive markers have also been used.

The use of metallographic techniques can be extremely helpful in establishing the flow pattern of the material. We can, on the one hand, use macrographs and follow the flow pattern of the material by applying an appropriate macroetch. Quantitative microscopy provides the same evidence on a much more reduced scale. For instance, we can determine the deformation undergone by grains that were initially equiaxed (characteristic of annealed material) by making sections along three orthogonal planes and establishing average quantitative relationships between the dimensions of the grains along different orientations. If we make the reasonable assumption that the grains undergo, on the average, the same deformation as the workpiece, we can obtain the flow.

Yet another approach is to use model materials in order to simulate the actual process. Materials that have been widely used are plasticine and lead; a whole range of waxes and clays, in addition to work-hardened aluminum, have been introduced. Figure 2.20 shows a composite plasticine bar rolled between wooden rolls. The inhomogeneity of deformation can be seen very clearly. After determining the various flow patterns in the model material, it is possible to optimize the process and "scale" it up to the real material. One problem that is encountered is that the strain-rate sensitivity and temperature dependence of flow stress may be very different in model and real material; this might have a drastic effect on the flow pattern, especially in deformation processes involving high strain rates.

2.5.4.5 Instrumented machines.

The use of instrumented rolling mills, drawing benches, and forging and stamping presses is becoming more and more common in laboratory experiments. Load cells, torque meters, and strain gages provide a full range of instruments that can establish the loads and power requirements in the various deformation processes.

2.5.4.6 The finite element method.

This powerful numerical method has developed simultaneously with the use of computers; essentially, it consists of dividing a continuous body into a specific number of discrete elements, called *finite elements*. These elements are considered interconnected at joints, called *nodes*. In each element, simple functions are chosen to describe the displacements or their variation throughout the element. In this way, the displacements at the nodes of all elements can be calculated.

In spite of its wide application since the early 1960s to elasticity problems, only since 1972 has finite element analysis been applied to plastic deformation processes; the reason for this is the greater complexity of the problem. Some modifications in the mathematical analysis were required. The method presents an almost unlimited potential. The international symposium on "Metal Forming Plasticity" organized by the International Union of Theoretical and Applied Mechanics"[31] devoted a great deal of attention to the finite element method. In order to assess the different numerical methods, 14 research groups from throughout the world were asked to analyze the problem of a cylindrical workpiece being upset between two plates to a reduction of height of 60%, assuming sticking friction (no sliding).[32] Figure 2.21(a) shows one possible way of discretizing the continuous body; only one-fourth of the body is shown. Forty-eight elements interconnected by 32 nodes were used by Yamada et al.[33] Figure 2.21(b) shows the distortion of a grid after 50% reduction. There was considerable variation between the results of the different research groups; however, this standard problem served to "tune up" their numerical methods; it is expected that in the future problems of much greater complexity will be successfully tackled by the finite element method.

2.5.5 Rolling

2.5.5.1 General.

The passage of a plate, sheet, strip, or structural shape between cylindrical rolls is accomplished by providing the power to any combination of uncoiler (back tension) and windup reels (front tension) and rolls. Figure 2.22 shows a schematic diagram. The windup reel and uncoiler are often replaced by other rolling stands in the same rolling mill; rolling stands are usually grouped in tandem in industrial operations. For the situation shown, we have the applied torque T_1, the back tension σ_b, and the front tension σ_f. If the initial velocity of the plate is v_i, it is changed to v_f through the passage between the rolls. The relationship between them is easily obtained from the mass action law:

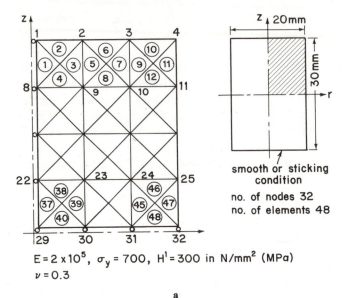

$$E = 2 \times 10^5, \quad \sigma_y = 700, \quad H^1 = 300 \text{ in N/mm}^2 \text{ (MPa)}$$
$$\nu = 0.3$$

a

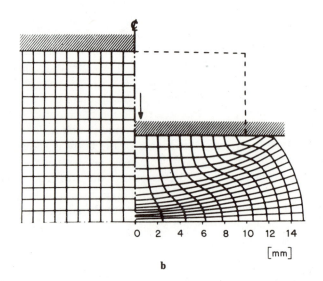

b

Figure 2.21 (a) Grid of finite elements on diametral section of cylindrical workpiece upset in compression; (b) distorted grid after 50% reduction in height and under sticking friction conditions. (Reprinted with permission from [32], p. 169, and [33], p. 395.)

input minus output equals accumulation. Since the accumulation of material between the rolls is zero, we have, per unit time,

$$\text{input} = \text{output}$$
$$v_i h_i b = v_f h_f b \tag{2.29}$$

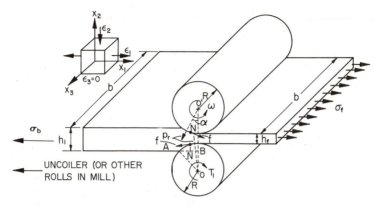

Figure 2.22 Forces involved in rolling.

This is so because the density of the material does not vary with deformation. The width b is constant and we have

$$v_i h_i = v_f h_f$$

and

$$v_f = v_i \frac{h_i}{h_f} \qquad (2.30)$$

The angular velocity of the rolls being ω, their tangential velocity will be

$$v = \omega R \qquad (2.31)$$

This velocity is related to v_i and v_f:

$$v_i < v < v_f$$

Consequently, at the entry side the rolls have a greater velocity than the plate. At the outcoming side, the opposite is the case. There is one point in the arc of contact in which we have

$$v = v_p \qquad (2.32)$$

where v_p is the plate velocity at that point. At that point there is no sliding between the rolls and plate; this is called the neutral point. It is indicated by N in Fig. 2.22. The rolls exert a friction on the plate which varies in sense depending on the relative velocities of the plate and rolls. It can be seen from Fig. 2.22 that f has the same sense as v at the left of the neutral point (between the entry section and the no-slip point); on the right of the neutral point (between the no-slip point and the exit section) f and v have opposite senses. The rolls exert a radial pressure p_r on the plate. The value of this pressure depends on the angle α; it will be calculated in the next few pages. It will be seen that it has a maximum somewhere along the neutral point. By appropriate integration of p_r (taking into account the change in orientation) one can calculate the total rolling load P, which is the force exerted by the rolls on the workpiece.

There are some geometrical relationships between the components that are worth deriving. The length of the projected arc of contact between the roll and plate, *AB*, is equal to

$$AB^2 = AO^2 - BO^2 \qquad (2.33)$$

$$= R^2 - \left(R - \frac{h_i - h_f}{2} \right)^2$$

$$= R(h_i - h_f) - \frac{(h_i - h_f)^2}{4} \qquad (2.34)$$

But, in general, $(h_i = h_f)$ (the reduction) is much smaller than the roll radius, and we can neglect the last term:

$$AB = \sqrt{R(h_i - h_f)} \qquad (2.35)$$

The angle α between the entrance plane and the axis passing through the center of the circles (line OO' in Fig. 2.22) is called the angle of contact or angle of bite. If a plate is brought into contact with the moving rolls, it might be "bitten" by the rolls or might just slide. The limiting condition for unaided entry of the plate into the rolls can be established by looking at the equilibrium of forces at the entry section. Taking the summation of forces along axis x_1, we have

$$f \cos \alpha = p_r \sin \alpha \qquad (2.36)$$

The area was not entered in the expression above because it is the same for both sides. Assuming that the frictional forces are coulombic, we have

$$\frac{f}{p_r} = \mu \qquad (2.37)$$

where μ is the Coulomb coefficient of friction. Equation (2.37) comes from mechanics. It simply states that the friction force is proportional to the pressure. It is a reasonable approximation for sliding friction situations. We have, substituting Eq. 2.36 into Eq. 2.37,

$$\mu = \tan \alpha \qquad (2.38)$$

But we can see from Fig. 2.22 that the maximum value of α is given by

$$\tan \alpha = \frac{AB}{OB} = \frac{[R(h_i - h_f)]^{1/2}}{R - (h_i - h_f)/2} \qquad (2.39)$$

But if the angle α is small, we can assume that the reduction per pass, $h_i - h_f$, is small in comparison to R. Consequently, the condition for the workpiece to enter into the rolls becomes

$$\mu = \left(\frac{h_i - h_f}{R} \right)^{1/2} \qquad (2.40)$$

and

$$h_i - h_f = \mu^2 R \qquad (2.41)$$

Equation 2.41 is helpful in determining the maximum reduction per pass, for a given initial thickness and roll diameters. The values of μ are given in Fig. 2.30.

2.5.5.2 Calculation of stresses.

The theory of rolling was developed originally by von Kármán. The differential equation at which we arrive is, however, very complex and several simplifications and assumptions have to be made in order to arrive at an acceptable solution. The easiest treatment is the equilibrium of forces (or slab) approach. The method of solution presented below is essentially the one by Hoffman and Sachs.[34] A small element of the metal being deformed is shown in Fig. 2.23.

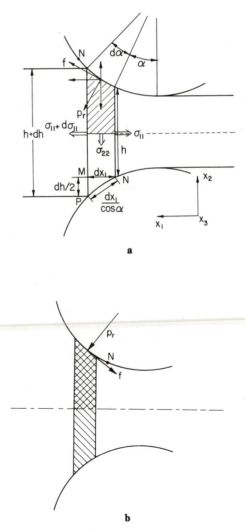

Figure 2.23 "Slab" or volume element upon which equilibrium of forces equations are applied.

Let us consider the equilibrium of a volume element (shown crosshatched); for symmetry reasons only half of the element is used.

$$\Sigma F_{x_1} = 0:$$

$$-(\sigma_{11} + d\sigma_{11})\left(\frac{h + dh}{2}\right) b + \frac{\sigma_{11}hb}{2} - bf \cos \alpha \frac{dx_1}{\cos \alpha}$$

$$-p_r b \sin \alpha \frac{dx_1}{\cos \alpha} = 0 \tag{2.42}$$

But $d\sigma_1 \, dh$ can be neglected and b can be eliminated.

$$-\sigma_{11} \, dh - h \, d\sigma_{11} + \sigma_{11}h - 2f \, dx_1 - 2p_r \tan \alpha \, dx_1 = 0$$

$$d(\sigma_{11}h) + 2f \, dx_1 + 2p_r \tan \alpha \, dx_1 = 0 \tag{2.43}$$

But

$$f = \mu p_r$$

$$d(\sigma_{11}h) = -2p_r(\mu + \tan \alpha) \, dx_1 \tag{2.44}$$

On the other hand, between the neutral point and the entry region the horizontal components of p_r and f have opposite senses [Fig. 2.23(b)] and we have

$$d(\sigma_{11}h) = -2p_r(\mu - \tan \alpha) \, dx_1 \tag{2.45}$$

These are called the von Kármán equations of rolling. An additional relationship between σ_1 and σ_2 is provided by the yield criterion. Assuming that the von Mises criterion holds, and since there is a state of plane strain, we have

$$2\sigma_0^2 = (\sigma_1 - \sigma_2)^2 + (\sigma_2 - \sigma_3)^2 + (\sigma_3 - \sigma_1)^2 \tag{2.46}$$

Using the Levy–von Mises equation of plasticity for $\epsilon_3 = 0$ yields

$$d\epsilon_3 = 0 = \frac{d\epsilon_{\text{eff}}}{\sigma_{\text{eff}}} [\sigma_{33} - 0.5(\sigma_{22} + \sigma_{11})] \tag{2.47}$$

$$\sigma_{33} = 0.5(\sigma_{22} + \sigma_{11}) \tag{2.48}$$

But we assume a state of homogeneous deformation, meaning that there are no shear strains acting along Ox_1, Ox_2, Ox_3. Therefore, the other Levy–von Mises equations indicate that $\sigma_{12} = \sigma_{13} = \sigma_{23} = 0$ and that $\sigma_1 = \sigma_{11}$, $\sigma_3 = \sigma_{22}$, $\sigma_2 = \sigma_{33}$. This latter inversion ($\sigma_3 = \sigma_{22}$ instead of $\sigma_3 = \sigma_{33}$) was done in order to maintain the $\sigma_1 > \sigma_2 > \sigma_3$ convention. We have

$$\sigma_2 = \frac{\sigma_3 + \sigma_1}{2} \tag{2.49}$$

and substituting Eq. 2.46 into Eq. 2.49, we have

$$\sigma_1 - \sigma_3 = \frac{2}{\sqrt{3}} \sigma_0 \tag{2.50}$$

For convenience purposes, we define

$$\frac{2}{\sqrt{3}} \sigma_0 = \sigma_0' \qquad (2.51)$$

In order to continue the derivation and to arrive at mathematically tractable equations, two assumptions have to be made. First, the angle α_{max} is assumed to be small. This results in

$$\sin \alpha \simeq \alpha$$
$$\cos \alpha \simeq 1 \qquad (2.52)$$
$$\tan \alpha \simeq \alpha$$

Second, it is assumed that $\sigma_3 = -p_r$. This, of course, introduces some error in the expressions. We have, substituting it in Eq. 2.50,

$$\sigma_1 + p_r = \sigma_0' \qquad (2.53)$$
$$d\sigma_1 = -dp_r \qquad (2.54)$$

Going back to von Kármán's equations and expressing them in terms of the two variables p_r and σ_1 yields

$$d(\sigma_0' - p_r)h = -2p_r(\tan \alpha \pm \mu) \, dx_1 \qquad (2.55)$$

or

$$-h \, dp_r + (\sigma_0' - p_r) \, dh = -2p_r(\tan \alpha \pm \mu) \, dx_1 \qquad (2.55a)$$

In triangle MNP,

$$dx_1 = \frac{dh}{2 \tan \alpha} \qquad (2.56)$$

A relationship between h and α has to be obtained to eliminate h in Eq. 2.55. To simplify the computations, the circle with radius R will be replaced by a second-order parabola; it has the same tangent and the same curvature as the circle at the exit section ($x_1 = 0$) and approximates it very well. We therefore have

$$h = h_f + \frac{x_1^2}{R} \qquad (2.57)$$

Differentiating Eq. 2.56 gives

$$dh = \frac{2x_1 \, dx_1}{R} \qquad (2.58)$$

Substituting Eq. 2.58 into Eq. 2.56, we have

$$x_1 = R \tan \alpha \qquad (2.59)$$

and Eq. 2.59 into Eq. 2.57 yields

$$h = h_f + R \tan^2 \alpha \qquad (2.60)$$

Differentiation of Eq. 2.59 provides

$$dx_1 = R \sec^2 \alpha \, d\alpha \qquad (2.61)$$

$$dh = 2R \tan \alpha \sec^2 \alpha \, d\alpha \qquad (2.62)$$

Substituting Eqs. 2.60, 2.61, and 2.62 into von Kármán's equation (Eq. 2.55a) gives us

$$(h_f + R \tan^2 \alpha) \, dp_r - (\sigma'_0 - p_r)2R \tan \alpha \sec^2 \alpha \, d\alpha$$
$$= 2p_r(\tan \alpha \pm \mu)R \sec^2 \alpha \, d\alpha \qquad (2.63)$$

Making the approximations indicated in Eqs. 2.52 and noting that, if α is small, $\sec \alpha \simeq 1$, $\tan^2 \alpha \simeq \alpha^2 \simeq 0$, then Eq. 2.63 reduces to

$$h_f \, dp_r = (2\sigma'_0 R \alpha \pm 2\mu p_r R) \, d\alpha$$

or

$$dp_r = (\sigma'_0 \alpha \pm \mu p_r) \frac{2R}{h_f} \, d\alpha \qquad (2.64)$$

The solution of this differential equation is obtained by the method of the integrating factor:

$$dp_r - A\sigma'_0 \alpha \, d\alpha \mp \mu A p_r \, d\alpha = 0 \qquad (2.65)$$

where

$$A = \frac{2R}{h_f}$$

Multiplying it by $e^{\mp \mu A\alpha}$ (− for the exit region, + for the entry region), we get

$$e^{\mp \mu A\alpha}(dp_r \mp \mu A p_r \, d\alpha) = A\sigma'_0 e^{\mp \mu A\alpha} \alpha \, d\alpha$$
$$d(p_r e^{\pm \mu A\alpha}) = A\sigma'_0 e^{\mp \mu A\alpha} \alpha \, d\alpha \qquad (2.66)$$

Integrating both sides yields

$$p_r e^{\mp \mu A\alpha} = A\sigma'_0 \int e^{\mp \mu A\alpha} \alpha \, d\alpha + C \qquad (2.67)$$

From the integration tables, we obtain

$$p_r e^{\mp \mu A\alpha} = A\sigma'_0 \frac{e^{\mu A\alpha}(-1 \mp \mu A\alpha)}{(\mu A)^2} + C \qquad (2.68)$$

The boundary conditions are now applied to obtain the value of the constant C. The reader should be reminded that the signs in front of μ were inverted. From Fig. 2.22, assuming both front and back tensions (the more general case), we have:

$$\begin{aligned} \text{exit:} \quad & \sigma_{11} = \sigma_f \quad && \text{at } x = 0(\alpha = 0) \\ \text{entry:} \quad & \sigma_{11} = \sigma_b \quad && \text{at } x = AB(\alpha = \alpha_{max}) \end{aligned} \qquad (2.69)$$

From Eq. 2.53 and for the entry region,

$$\frac{\sigma_0' - \sigma_b}{\sigma_0'} = C_1 + A e^{\mu A \alpha_{max}} \frac{\mu A \alpha_{max} - 1}{A^2 \mu^2} \tag{2.70}$$

and

$$C_1 = e^{\mu A \alpha_{max}} \left(1 - \frac{\sigma_b}{\sigma_0'} + \frac{-1 + \mu A \alpha_{max}}{A \mu^2}\right)$$

Substituting this value into Eq. 2.68 with a negative sign in the exponent, we get

$$\frac{p_r}{\sigma_0'} = -\frac{1 - \mu A \alpha}{\mu^2 A} + e^{+\mu A(\alpha_{max} - \alpha)} \left(1 - \frac{\sigma_b}{\sigma_0'} + \frac{1 - \mu A \alpha_{max}}{\mu^2 A}\right) \tag{2.71}$$

where

$$A = \frac{2R}{h}$$

For the neutral point-exit region, we apply the second set of boundary conditions:

$$\frac{\sigma_0' - \sigma_f}{\sigma_0'} = A e^{-\mu A(0)} \frac{-\mu A(0) - 1}{\mu^2 A^2} + C_2$$

$$C_2 = 1 - \frac{\sigma_f}{\sigma_0'} + \frac{1}{\mu^2 A}$$

Substituting it into Eq. 2.68 (with the negative sign of the exponent) gives us

$$\frac{p_r}{\sigma_0'} = \frac{-(\mu A \alpha + 1)}{\mu^2 A} + e^{\mu A \alpha} \left(\frac{\sigma_0' - \sigma_f}{\sigma_0'} + \frac{1}{\mu^2 A}\right) \tag{2.72}$$

Figure 2.24 shows a plot of Eqs. 2.71 and 2.72 for various back and front tensions. As the back and front tensions are decreased, the friction hill is lowered. The effects as shown can be substantial. Another very interesting effect, shown in Fig. 2.25, is the roll diameter. For a reduction of 15%, from an initial thickness of 10 mm, it can be seen that the higher friction hill corresponds to the 1.0-m roll radius.

We can also calculate, to a first approximation, the power required for rolling. This is shown next.

Assuming that $p = p_r$ (i.e., that the radial and vertical components of pressure are the same), we can compute the total rolling load by integrating over the whole arc of contact. This is provided by the area under the curve in Fig. 2.24 after appropriately adjusting the axes.

$$dP = p(bR \, d\alpha) \tag{2.73}$$

$$P = Rb \int_0^\alpha p \, d\alpha \tag{2.74}$$

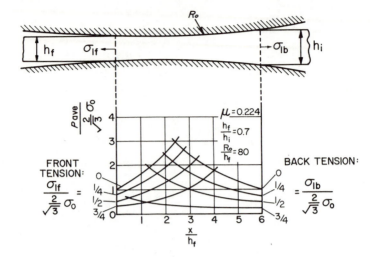

Figure 2.24 The friction hill in rolling. (Reprinted with permission from [34], p. 218.)

We can compute the work that will be made by the rolls in the deformation process therefrom. For this, we assume, for simplicity, that the load P is concentrated at a point P, as shown in Fig. 2.26. The work for one revolution is

$$W = 2(2\pi a)P \tag{2.75}$$

The radius a is established empirically. We have, as a first approximation,

$$a = 0.5(AB)$$

$$a = \tfrac{1}{2}[R(h_i - h_f)]^{1/2} \tag{2.76}$$

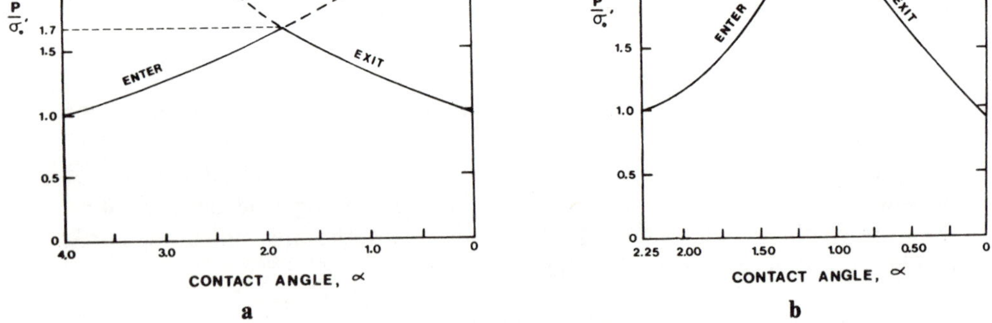

Figure 2.25 Effect of a roll diameter on configuration of friction hill. Both cases apply to a reduction of 15% per pass and to an initial thickness of 10 mm. (a) Roll diameter 0.6 m. (b) Roll diameter 2.0 m. (Courtesy of E. Krosche and C. T. Aimone, New Mexico Institute of Mining and Technology.)

Mechanics of Deformation Part I

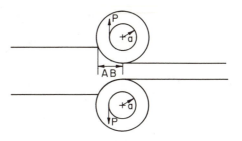

Figure 2.26 Torque applied by rolls on plate.

and

$$W = 2\pi \left[R(h_i - h_f) \right]^{1/2} Rb \int_0^\alpha P \, d\alpha \qquad (2.77)$$

The power can be computed from the work by taking the work per unit time. In real systems, however, there are many losses; for a typical case given below, only 60% of the power goes into the actual deformation of the metal. In a reversing primary mill (used to reduce a steel ingot to a slab or bloom), we have[35]

Power required for rolling	59.00%
Friction of pinions and mill	5.90%
Loss in reversing motor	10.84%
Loss in electrical connections	0.75%
Loss in generators	11.44%
Loss in flywheel	1.32%
Loss in slip regulator	2.69%
Loss in induction motor	6.40%
Loss in exciters, blowers, etc.	1.66%
Total	100.00%

We might recognize that the calculations leading to Eqs. 2.71, 2.72, and 2.77 are only approximate. More rigorous analyses will provide results that are closer to the real loads and torques required.

2.5.6 Wire Drawing

2.5.6.1 Calculation of stresses. The approach used here—equilibrium of forces or slab method—is essentially the one presented by Hoffman and Sachs[34] and Avitzur.[16] Apparently, Hill and Tupper[36] presented this solution for the first time. Figure 2.27 shows a wire being drawn through a die with die semiangle α. A slab of thickness dx, situated at a distance x from the angle apex, is separated and the summation of forces acting on it is determined. A simplifying assumption is made: deformation is considered homogeneous. This implies that the shear strains, in the system of axes shown in Fig. 2.27(a), are zero. From the Levy–von Mises equations we can conclude that the shear stresses are zero. Thus, the axes Ox_1, Ox_2, Ox_3 are the principal axes. The wire axis is an axis of symmetry due to the symmetry of the deforma-

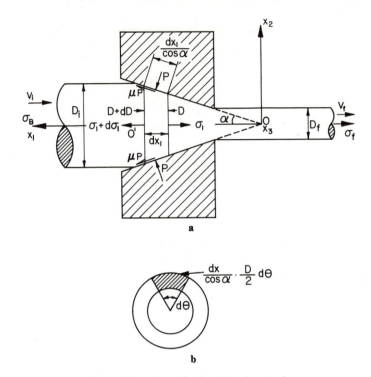

Figure 2.27 Forces involved in wire drawing.

tion. Therefore, the two principal stresses on planes Ox_1x_3 and Ox_1x_2 are equal. Since they are both compressive, we have $\sigma_2 = \sigma_3$ and σ_1 is the tensile stress along Ox_1. In order to calculate the components of pressure and friction along the Ox_1 direction, we have to determine the surface of the area they are acting upon. This is shown in Fig. 2.27(b).

$$A = \int_0^\theta \frac{dx}{\cos \alpha} \frac{D}{2} d\theta = \frac{\pi D \, dx}{\cos \alpha} \tag{2.78}$$

In Fig. 2.20(a),

$$\Sigma F_{x_1} = 0$$

$$(\sigma_1 + d\sigma_1)\frac{\pi}{4}(D + dD)^2 - \sigma_1 \pi \frac{D^2}{4} + \mu p \frac{\cos \alpha \pi D \, dx}{\cos \alpha}$$
$$+ p \frac{\sin \alpha \pi D \, dx}{\cos \alpha} = 0 \tag{2.79}$$

It should be noted that coulombic friction is assumed. At very high velocities or when special lubricants are used, this is not a very good approximation, and hydrodynamic lubrication dominates. In hydrodynamic lubrication a liquid film forms between the tool and workpiece, so that they are not in direct contact.

Mechanics of Deformation   Part I

However, this condition will not be studied here. Neglecting the products $d\sigma_1\, dD$ and dD^2 (they are very small) and since

$$dx_1 = \frac{dD}{2\tan\alpha} \tag{2.80}$$

we arrive at

$$\sigma_1\, dD + D\, d\sigma_1 + \frac{\mu p}{\tan\alpha}\, dD + p\, dD = 0 \tag{2.81}$$

$$d\,(\sigma_1 D) + p\, dD\left(\frac{\mu}{\tan\alpha} + 1\right) = 0 \tag{2.82}$$

By applying von Mises' yield criterion, we can express p in terms of σ_1. The solution is greatly simplified if we assume, as on rolling, that the direction of the pressure p is approximately normal to the direction of the axis Ox_1; this is a reasonable assumption since the angle α is usually small. We have

$$\sigma_0 = \frac{\sqrt{2}}{2}[(\sigma_1 - \sigma_2)^2 + (\sigma_1 - \sigma_3)^2 + (\sigma_2 - \sigma_3)^2]^{1/2}$$

$$= \frac{\sqrt{2}}{2}[(\sigma_1 - p)^2 + (\sigma_1 - p)^2]^{1/2} \tag{2.83}$$

$$p = \sigma_0 - \sigma_1$$

Substituting Eq. 2.63 into Eq. 2.62, we have

$$d\,(\sigma_1 D) + (\sigma_0 - \sigma_1)\, dD\left(\frac{\mu}{\tan\alpha} + 1\right) = 0$$

Making $\mu/\tan\alpha = B$ gives us

$$\frac{d\sigma_1}{\sigma_1 B - \sigma_0(1 + B)} = 2\frac{dD}{D} \tag{2.84}$$

Integrating both sides of Eq. 2.84 and applying the boundary conditions, we get

$$\frac{1}{B}\, ln\,[\sigma_1 B - \sigma_0(1 + B)] = 2\, ln\, D + C \tag{2.85}$$

At entry:

$$\sigma_1 = \sigma_{1b} \quad \text{(backpull)}$$

$$D = D_i$$

At exit:

$$\sigma_1 = \sigma_{1f} \quad \text{(front tension)}$$

$$D = D_f$$

Rearranging Eq. 2.85 yields

$$\sigma_1 = C' \frac{D^{2B} + \sigma_0(1 + B)}{B} \tag{2.86}$$

Applying the entry boundary conditions, we get

$$C' = \frac{\sigma_{1b}B}{D_i^{2B} + \sigma_0(1 + B)} \tag{2.87}$$

Substituting Eq. 2.87 into Eq. 2.86 and applying the boundary condition for the exit gives us

$$\frac{\sigma_{1f}}{\sigma_0} = \frac{1 + B}{B}\left[1 - \left(\frac{D_f}{D_i}\right)^{2B}\right] + \frac{\sigma_{1b}}{\sigma_0}\left(\frac{D_f}{D_i}\right)^{2B} \tag{2.88}$$

If there is no backpull, the front tension is given by

$$\frac{\sigma_{1f}}{\sigma_0} = \frac{1 + B}{B} - \left[1 - \left(\frac{D_f}{D_i}\right)^{2B}\right] \tag{2.89}$$

The maximum reduction takes place when the front tension is equal to the flow stress of the wire. Beyond this value the front tension will tend to deform the wire, and will result in a defective product. So, for $\sigma_{1f} = \sigma_0$,

$$\left(\frac{D_f}{D_i}\right)^{2\mu/\tan\alpha} = \frac{1}{1 + \mu/\tan\alpha} \tag{2.90}$$

We can eliminate σ_1 in Eq. 2.84 and solve the problem for p; in this case it is possible to find the pressure applied by the wire on the die as a function of position. In usual operations B is approximately equal to 1. This corresponds to a ratio of diameter D_f/D_i equal to 0.71. The slab method apparently[37] is accurate for small die angles, but underestimates the loads at higher angles. This is because it does not take into account the redundant deformation (deformation in addition to the one strictly needed to change shape; see Fig. 2.18). On the other hand, the energy method makes predictions within 20 to 50% from the actual loads for die angles in the range 5 to 31°.

2.5.6.2 Wire-drawing parameters.
The most important parameters in wire drawing are the amount of reduction, the die angle 2α, the lubrication, and the mechanical properties of the workpiece. The reduction and die angle can be determined experimentally and compared to theoretical results. For each reduction there is an optimum die angle that minimizes the front tension. Avitzur[16] presents a detailed analysis of the theoretical methods used in the optimization. Figure 2.28(a) shows the drawing stress as a function of the die semicone angle. The curve was not obtained from Eq. 2.90 but from more complex theories. The method described in the preceding pages assumes homogeneous deformation; it can be seen in Fig. 2.28(a) that only the sound flow situation does not present any redundant deformation. As the angle α is increased, a dead zone is formed. The material adjoining the die is immobilized.

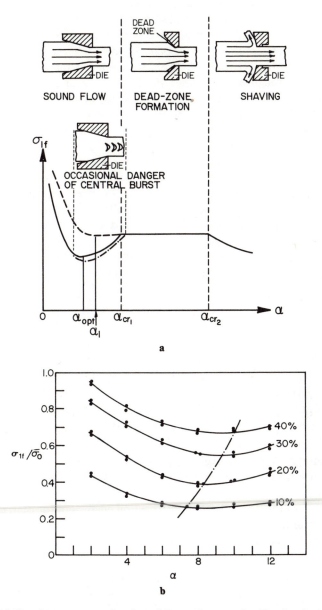

Figure 2.28 (a) Drawing stress as a function of die semicone angle. (Reprinted with permission from [16], p. 167.) (b) Effect of die angle α on $\sigma_{1f}/\overline{\sigma_0}$ (σ_{1f}: frontal stress; $\overline{\sigma_0}$: mean flow stress of wire) for AISI 1010 wire. (Adapted with permission from [38], p. 220.)

An internal region of velocity discontinuity is formed, and this corresponds to an increase in friction. The angle at which it starts is called the first critical angle. Eventually, a second critical angle will be reached [α_{cr2} in Fig. 2.28(a)] where the convergence of material flow will cease. This phenomenon is called shaving. The material is actually diverted from the die. Since there is no more convergence of the material, the velocity of the material is the same before and after passing through the die. Both dead-zone formation and shaving are usually avoided, because the former produces a large amount of redundant work and the latter results in losses of material; however, shaving is utilized in processes where the purpose is removal of undesired (descaling). Figure 2.28(a) also shows another region of material response; central bursting (or chevroning) might take place. These are internal cracks that are formed under a certain unique combination of parameters. It happens when the plastic deformation zone does not go up to the center of the section. On the entry region, the undeformed material travels with velocity v_1; on the exit region, the velocity is v_2. If the core of the wire does not undergo plastic deformation, strong tensile stresses will develop, since $v_2 > v_1$. This will produce the cracks at approximately 45° to the tensile axis. Because of the symmetry, these cracks will have a cone shape. Central bursting is by no means the only problem in wire drawing. Some of these problems are briefly described below.

1. Snakeskin or fishskin takes place when the central core flows more rapidly than the surface, putting the latter under tensile stresses; it tends to break with a "snakeskin" appearance. This phenomenon is the opposite of central bursting, where the core is under tension.

2. Cavity or piping effect. A cavity or "pipe" forms along the core of the wire, starting at the end. As reduction takes place, the cavity advances. This problem is eliminated by reducing the die angle.

3. Distortion. Some degree of distortion is always present, because deformation has some redundancy in it.

4. Pressure peak at the beginning of flow. The force required to pull the wire through the die is higher at the start of the process because of several reasons. One important reason is that the static coefficient of friction is higher than the dynamic Coulomb friction coefficient (or than the hydrodynamic friction coefficient). Another is that in the first part, the deformation zone has to be built up.

The optimum die semi-angle (i.e., the one requiring least stress) is dependent on the amount of reduction. Figure 2.28(b) shows the variation of the ratio $\sigma_{1f}/\bar{\sigma}_0$ ($\bar{\sigma}_0$ is the mean flow stress of the wire) as a function of α for an AISI 1010 wire (2 mm diameter) being drawn at 0.1 m/min. This angle increases from 6 to 10° when the reduction increases from 10 to 40%.

The metallurgical effects of wire drawing were the object of systematic investigations by Cohen and coworkers.[39-41]

2.5.6.3 Determination of strain rate. Wire drawing can involve very high strain rates because the wire velocities may be very high (over 30 m/s, in modern mills). The reductions in diameter per pass are of the order of 30%. It is instructive to determine the strain rate for a situation such as that. It should be noticed that the strain rate is not constant, but increases as the wire travels from the entry toward the exit. Figure 2.27 will be used for the derivation with a slight change. The origin of the axes will be changed from O to O', for convenience. The positive direction of the Ox_1 axis now points to the right. The constancy of volume relationship allows determination of the velocities.

$$\left(\frac{\pi D_i^2}{4}\right) L_i = \left(\frac{\pi D_f^2}{4}\right) L_f = \left(\frac{\pi D^2}{4}\right) L = \text{constant}$$

$$A_i L_i = A_f L_f = AL = \text{constant}$$

(2.91)

where A_i, A_f, and A are the initial, final, and instantaneous areas. L is an arbitrary section of material that was marked and whose length changes as it traverses the die. Dividing the expression above by the time required to displace L, we obtain

$$Av = \text{constant}$$

Thus, we have

$$AL = \text{constant}; \qquad \text{therefore, } A\,dL = -L\,dA \tag{2.92}$$

The strain at a certain point is

$$d\epsilon = \frac{dL}{L} = -\frac{dA}{A} \tag{2.93}$$

The strain rate is

$$\dot{\epsilon} = \frac{d\epsilon}{dt} = -\frac{1}{A}\frac{dA}{dt} = -\frac{1}{A}\frac{dA}{dx}\frac{dx}{dt} \tag{2.94}$$

The area A at a certain distance x_1 from the origin O in the deformation region can be expressed as

$$A = \frac{\pi D^2}{4} = \frac{\pi}{4}(D_i - 2x \tan \alpha)^2 \tag{2.95}$$

$$\frac{dA}{dx} = -\pi \tan \alpha\, (D_i - 2x \tan \alpha) \tag{2.96}$$

From Eq. 2.92,

$$\frac{dx}{dt} = v = v_i \frac{D_i^2}{D^2} = v_i \frac{D_i^2}{(D_i - 2x \tan \alpha)^2} \tag{2.97}$$

Substituting Eqs. 2.95, 2.96, and 2.97 into Eq. 2.94, we have

$$\dot{\epsilon} = \frac{4}{\pi(D_i - 2x \tan \alpha)} \, \pi \tan \alpha (D_i - 2x \tan \alpha) \frac{v_i D_i^2}{(D_i - 2x \tan \alpha)^2}$$

(2.98)

$$\dot{\epsilon} = \frac{4 \tan \alpha \, v_i D_i^2}{(D_i - 2x \tan \alpha)^3}$$

The strain rate at the entry is

$$\dot{\epsilon}_i = \frac{4 \tan \alpha \, v_i}{D_i}$$

(2.99)

The strain rate at the exit is

$$\dot{\epsilon}_f = \frac{4 \tan \alpha \, v_i D_i^2}{D_f^3}$$

(2.100)

For a hypothetical situation in which the wire diameter is being reduced from 10 mm to 7 mm, with a die semiangle of 20° and an initial velocity of 20 m/s, we would have entry and exit strain rates of 7.2×10 and 2×10^2 s^{-1}, respectively. On the other hand, if a wire is being reduced from 1 mm to 0.7 mm diameter under the same conditions, the entry and exit strain rates will be 7.2×10^3 and 2×10^4 s^{-1}. This increase in strain rate for a decrease in wire diameter is due to the fact that the deformation zone decreases.

2.5.7 Forging

The simplest situation occurring in forging will be treated here: that of a cylinder being upset (compressed) between two parallel dies. At first sight it would seem that the pressure required for upsetting the cylinder would be equal to the flow stress of the material. The equilibrium of forces approach will show that friction at the cylinder–die interfaces will significantly increase the pressure requirements. The situation analyzed here is similar to the one presented in Fig. 2.21; the approach is in essence the one used by Dieter.[42] Figure 2.29 shows the disk during deformation. Some simplifying assumptions are introduced. The state of deformation is assumed to be homogeneous; therefore, no barreling takes place. The material–die interface friction is assumed to obey a Coulomb behavior under sliding conditions. The friction force τ_{rz} is shown in the element drawn in Fig. 2.29(a). We then separate the element shown in Fig. 2.29(a) and look at the summation of forces acting on it, assuming equilibrium. The stresses acting on the element shown are referred to cylindrical coordinates. A cylindrical system of coordinates was chosen because the problem has an axis of symmetry. We have $d\epsilon_\theta = d\epsilon_{rr}$, because of axial symmetry. Applying the Levy–von Mises equations with the indices modified in order to fit the cylindrical coordinates gives us

$$\frac{d\epsilon_{\text{eff}}}{\sigma_{\text{eff}}} [\sigma_r - \tfrac{1}{2}(\sigma_\theta + \sigma_z)] = \frac{d\epsilon_{\text{eff}}}{\sigma_{\text{eff}}} [\sigma_\theta - \tfrac{1}{2}(\sigma_r + \sigma_z)]$$

(2.101)

$$\sigma_r = \sigma_\theta$$

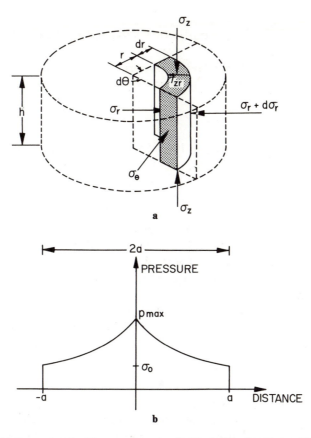

a

b

Figure 2.29 (a) Forces involved in upsetting of cylindrical disk; (b) friction hill, showing that pressure is maximum at center.

Applying the equilibrium of forces condition in the radial direction, we have

$$\sigma_r hr\, d\theta - (\sigma_r + d\sigma_r)(r + dr)h\, d\theta$$

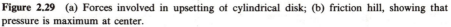

$$+ 2\sigma_\theta h\, dr \sin\frac{d\theta}{2} - 2\tau_{zr} r\, dr\, d\theta = 0 \tag{2.102}$$

But since $d\theta$ is small, $\sin d\theta/2 \simeq d\theta/2$. Applying Coulomb's friction law yields

$$\tau_{rz} = \mu\sigma_z \tag{2.103}$$

The product $d\sigma_r dr$ can be neglected because it is very small. We arrive at

$$\frac{d\sigma_r}{dr} + \frac{2\mu\sigma_z}{h} = 0 \tag{2.104}$$

Applying von Mises' flow criterion, we can obtain a relationship between σ_r and σ_z:

$$\sigma_0 = \frac{\sqrt{2}}{2} [(\sigma_1 - \sigma_2)^2 + (\sigma_1 - \sigma_3)^2 + (\sigma_2 - \sigma_3)^2]^{1/2}$$

Making the assumption that σ_r, σ_z, σ_θ are principal stresses (and this introduces some error, because there is a shear stress τ_{rz}), we have $\sigma_1 = \sigma_\theta = \sigma_2 = \sigma_r$, $\sigma_3 = \sigma_z$:

$$\sigma_0 = \sigma_1 - \sigma_3 = \sigma_r - \sigma_z \qquad (2.105)$$

Substituting Eq. 2.105 into Eq. 2.104 and solving the differential equation by integrating both sides yields

$$ln\ \sigma_z = -\frac{2\mu r}{h} + C \qquad (2.106)$$

A pressure p will now be defined in such a way that $p = -\sigma_z$. p is the pressure that the dies exert on the workpiece. Applying the boundary conditions to Eq. 2.106, we can obtain the value of C. For $r = a$, we will have $\sigma_r = 0$ (normal stresses at a free surface are zero). Therefore,

$$C = ln\ \sigma_0 + \frac{2\mu a}{h} \qquad (2.107)$$

Substituting Eq. 2.107 into Eq. 2.106 after σ_z has been replaced by $-p$, we obtain

$$p = \sigma_0\ e^{2\mu(a-r)/h} \qquad (2.108)$$

This is the equation for the "friction hill." It is schematically plotted in Fig. 2.29(b). The pressure rises exponentially toward the center of the cylinder. The greater the coefficient of friction, the greater p_{max}. The height of the friction hill can be found by integrating over the whole region and dividing the result by the area of the top (or bottom) surface of the disk. We have

$$\bar{p} = \frac{\int_0^a 2\pi r\ dr}{\pi a^2} = \frac{\sigma_0}{2} \left(\frac{h}{\mu a}\right)^2 \left(e^{2\mu a/h} - \frac{2\mu a}{h} - 1\right) \qquad (2.109)$$

The effect of the Coulomb friction coefficient on the mean pressure is shown in Fig. 2.30. It can be seen that the mean pressure is about eight times greater than the flow stress of the material, for $a/h = 25$ and $\mu = 0.15$. The general conclusion that can be obtained from Eq. 2.109 and Fig. 2.30 is that friction plays a very important role in the determination of the loads required for forging. Similar conclusions can be reached for forging under plane-strain or other conditions.

If no sliding were to occur between the workpiece and the die, the configuration, after reduction, would be as shown in Fig. 2.21. Under sticking conditions, the friction coefficient can be calculated by assuming that the friction stress is equal to the flow shear stress of the material. This is equivalent to saying that the shear strength of the boundary layer is the same as that of the bulk material. Under hot-working conditions lubrication is often difficult and we encounter

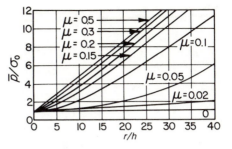

Figure 2.30 Average deformation pressure in compression of a disk as a function of the Coulomb friction coefficient μ and the ratio between radius r and height h. (Reprinted with permission from [42], p. 564.)

sticking friction situations. The friction coefficient can be calculated from the relationship

$$\tau_0 = \tau_{rz}$$

By applying the Coulomb equation to the outer part of the disk and considering the von Mises criterion under pure shear, we have

$$\tau_{\max} = \frac{\sigma_1 - \sigma_3}{2} = \sigma_1 \qquad \text{because } \sigma_2 = 0, \quad \sigma_3 = -\sigma_1$$

$$\sigma_0 = \frac{\sqrt{2}}{2}[(\sigma_1 - \sigma_3)^2 + \sigma_1^2 + \sigma_3^2]^{1/2}$$

$$\sigma_0 = \sqrt{3}\,\sigma_1$$

Therefore,

$$\tau_{max} = \tau_0 = \tau_{rz} = \frac{\sigma_0}{\sqrt{3}}$$

As a result of this,

$$\mu = \frac{\tau_{rz}}{p} = \frac{\tau_0}{\sigma_0} = \frac{1}{\sqrt{3}} = 0.577$$

As expected, the coefficient of friction for sticking conditions is higher than that for sliding conditions.

The flow of material is inhomogeneous in forging; this is particularly true in closed-die forging. Figure 2.31 shows a region in which plastic flow is highly concentrated because of the imposed external forces. Two effects may arise in these regions of localized flow (called shear bands): (1) the material work hardens, with an attendant increase of flow stress and a less pronounced tendency for deformation; and (2) the heat generated by the localized deformation is large enough to upset the flow stress increase and a flow stress decrease results. This flow stress decrease will result in a concentration of deformation in that region, with more and more heat generation (90 to 95% of deformation work is converted into heat). This self-feeding process will create a region of intense flow concentration leading to eventual fracture and/or melting. This effect is more marked at higher strain rates. Under high-energy-rate deformation, the conditions are quasiadiabatic (there is no time for heat flow outside the deformation area)

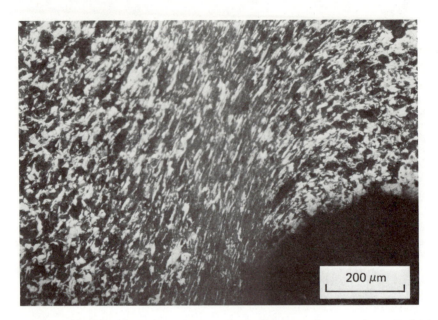

Figure 2.31 Concentration of plastic flow in forging as evidenced by change in shape of individual grains.

Figure 2.32 Shear band of the transformation type produced in AISI (1040) steel quenched and tempered at 400°C by the impact of a projectile. (Courtesy of H. C. Rogers, Drexel University.)

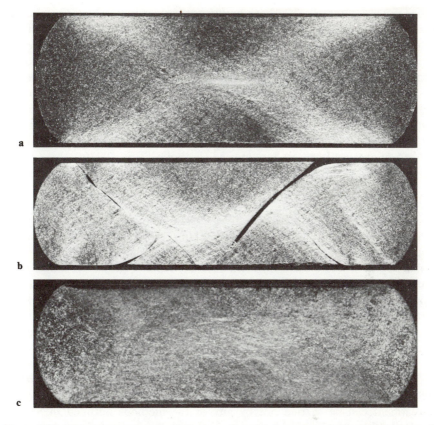

Figure 2.33 Hot sidepressed rods of titanium alloy: (a) formation of shear bands due to heat transfer to dies with associated cooling, creating temperature (and flow stress) differential; (b) same as (a), but shear band formation more drastic because alloy is upset at a slower rate and temperature (and flow stress) differentials become greater; (c) isothermal forging, resulting in homogeneous microstructure. (Courtesy of S. L. Semiatin, Battelle Memorial Institute.)

and these bands are called adiabatic shear bands. They have important implications in ballistic armor penetration, explosive fragmentation, high-velocity shaping and forming, machining, grinding, and erosion. Rogers[43] presents a comprehensive description of adiabatic shear bands and classifies them into *deformed* and *transformed* bands. The first are narrow bands of intense shear without a structural transformation. This is the case for the majority of nonferrous metals. Transformed bands, on the other hand, are clearly distinguishable from the surrounding material since they etch white. Fig. 2.32 shows the tip of a transformed band in an impacted plate of AISI 1040 steel quenched and tempered at 400°C. The white band is the shear band; at its tip, one can see the sheared (but untransformed) material. The mechanisms responsible for the formation of transformation bands are not yet very well understood.[43,44]

Temperature gradients arising in hot forging due to cooling of the workpiece in contact with the colder dies are also an important source of inhomogeneous flow. Figure 2.33(a) shows the cross section of a titanium alloy after being homogeneously upset.[40] The shear bands are clearly noticeable. The alloy in contact with the dies is colder and has, as a consequence, a higher flow stress. This effect is rendered more pronounced when the forging velocity is decreased (by switching from a mechanical to a hydraulic press) and more time is allowed for cooling of the workpiece [Fig. 2.33(b)]. The shear bands become more severe and shear cracks form. However, if the dies are preheated to a temperature at which the temperature sensitivity of the workpiece is small, it is possible to obtain a homogeneous flow of the material. This is a technique that has met good success for difficult-to-forge alloys; it is called *isothermal forging* and is described in greater detail by Semiatin and Lahoti[45] [Fig. 2.33(c)].

2.5.8 Sheet-Metal Forming

Stamping, the most important sheet-metal forming operation, involves essentially two processes: drawing and stretching. Different parts of the blank are subjected to different constraints, but deep drawing and stretching are a rational way to separate them into two different types. Figure 2.12 shows schematically a typical sheet forming operation. The region of the blank where the punch acts will be "stretched" down, because the blank holder will apply a certain pressure on the edges of the blank. On the other hand, the material under the blank holder will be "drawn" inward in the stamping process, because the holder does not hold the blank rigidly in position but allows it to slide inward in a controlled way during the forming operation.

Thus, we can study both theoretically and experimentally these stresses and loads in two schematic situations: stretching and deep drawing. The theoretical treatment can be carried out by any of the methods outlined in Section 2.5.4. Both Blazynski[25] and Hecker et al.[46] provide excellent theoretical treatments. However, they will not be presented and only a brief description of the subject will be given here. It is based on the presentation given by Hecker and Ghosh.[47] Deep drawing and stretching are shown in Fig. 2.34(a) and (b), respectively.

In deep drawing, the blank diameter is reduced from D_0 to D_f. During the action of the punch, the blank is forced to contract circumferentially as it is drawn radially inward; it tends to buckle and wrinkle, and it is the pressure of the holder that counteracts the tendency. Since the volume remains constant, the thickness of the edge of the blank has to increase (in the event of no wrinkles) as it approaches the 90° region. Most of the deformation in deep drawing occurs in the flange region [indicated in Fig. 2.34(a)]. If the hold-down pressure is too high, the load required in the punch will cause the side walls of the cup to deform plastically and eventually neck. Consequently, the necking in the side walls of the cup establishes the maximum value for the punch load. In a biaxial stress state necking at the cup walls is produced by localized thinning. A material that has thickness anisotropy (the strain in the thickness direction being lower than in the two sheet directions) is ideal for that purpose. Section 18.2 treats this aspect in greater detail. Another effect of anisotropy can be seen in Fig. 18.1; here planar anisotropy (different properties on the *plane* of

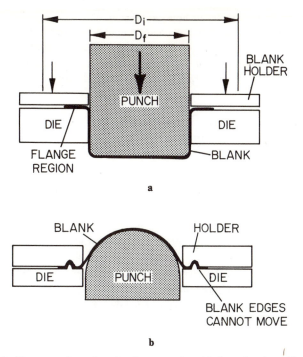

Figure 2.34 Two operations that simulate stamping: (a) deep drawing; (b) stretching.

the sheet) causes the drawing to be more inhibited along certain orientations. The cup height varies; this effect is called "earing."

In stretching, on the other hand, [Fig. 2.34(b)], there is no inward drawing of the blank edges, and tensile stresses act on the entire surface of the sheet.

EXERCISES

2.1. A polycrystalline metal has a plastic stress–strain curve that obeys Hollomon's equation,

$$\sigma = K\epsilon^n$$

Determine n knowing that the flow stresses of this material at 0.1 and 0.2% plastic deformation (offset) are equal to 175 and 185 MPa, respectively.

2.2. You are traveling in an airplane. The engineer who designed it is, casually, on your side. He tells you that the wings were designed using von Mises' criterion. Would you feel safer if he had told you that Tresca's criterion had been used? Why?

2.3. A material is under a state of stress such that $\sigma_1 = 3\sigma_2 = 2\sigma_3$. It starts to flow when $\sigma_2 = 140$ MPa.

(a) What is the flow stress in uniaxial tension?

(b) If the material is used under conditions in which $\sigma_1 = -\sigma_3$ and $\sigma_2 = 0$, at which value of σ_3 will it flow, according to Tresca's and von Mises' criteria?

2.4. A steel with a yield stress of 300 MPa is tested under a state of stress where $\sigma_2 = \sigma_1/2$, $\sigma_3 = 0$. What is the stress at which yielding occurs if it is assumed that:

(a) The maximum normal stress criterion holds?

(b) The maximum shear stress criterion holds?

(c) The distortion-energy criterion holds?

2.5. Determine the maximum pressure that a cylindrical gas reservoir can withstand, using the three flow criteria.

Material: AISI 304 stainless steel—hot finished and annealed, $\sigma_y = 205$ MPa
Thickness: 25 mm
Diameter: 500 mm
Length: 1 m

[*Hint:* Determine the longitudinal and circumferential (hoop) stresses by the method of sections.]

2.6. Determine the value of Poisson's ratio for an isotropic cube being plastically compressed between two parallel plates.

2.7. A low-carbon-steel cylinder, having a height of 50 mm and a diameter of 100 mm, is being forged (upset) at 1200°C and velocity of 1 m/s, until its height is equal to 15 mm. Assuming an efficiency of 60% and that the flow stress at the specified strain rate is 80 MPa, determine the power required to forge the specimen.

2.8. Briefly describe the steps that would be required to go from a steel ingot (AISI 1008) to a 2-mm sheet. [*Hint:* Consult H. E. McGannon (ed.), *The Making, Shaping, and Treating of Steel*, 9th ed., U.S. Steel, Pittsburgh, Pa., 1971.]

2.9. Determine the engineering strain, the true strain, and the percentual change in area q for specimens being plastically deformed under uniaxial stress:

(a) Extension from L to $1.1L$.

(b) Compression from h to $0.9h$.

2.10. **(a)** At a plant, 10-mm wire is being produced, starting with 25 mm-thick stock rod. Calculate the number of stands (dies) required to do the job, assuming that all dies have the same angle $\alpha = 15°$ and that the Coulomb coefficient of friction is equal to 0.25.

(b) An engineer proposed the use of a new lubricant that would reduce the coefficient of friction to 0.17. Would this result in a decrease in the number of stands?

2.11. What are the maximum, mean, and minimum strain rates for a wire being drawn at a speed of 25 m/s with a reduction of 20% assuming that $\alpha = 15°$ and that the initial diameter is 5 mm?

2.12. **(a)** A strip of copper with width 50 mm and initial thickness 10 mm undergoes a 25% reduction by strip drawing. (See figure.) Knowing that the yield stress of copper (under uniaxial tension) is 50 MPa, determine the *force* required to pull the strip through the dies. Use the Tresca criterion.

(b) Assuming that the energy expended to overcome friction is equal to half of the energy required to deform the strip and that the deformation is homogeneous, determine the efficiency of the process.

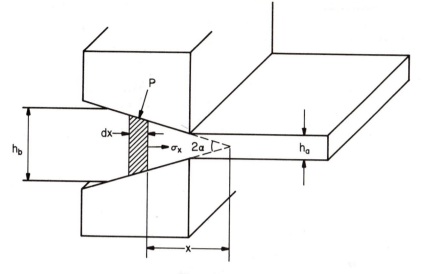

Figure E2.12

2.13. A 10-mm-thick plate initially in the annealed (copper) condition and with $\sigma_0 = 120$ MPa is being hot-rolled. A reduction of 15% is desired; the roll diameters are 100 cm.

  (a) Can this reduction be effected in one pass?

  (b) Plot a pressure versus distance diagram, after making all the calculations, assuming a roll diameter of 100 cm. Repeat the procedure for a roll diameter of 300 cm. Assume that the Coulomb coefficient of friction is 0.225.

  (c) Which situation will require less power?

2.14. Show the difference between hot and cold working by means of hardness versus strain plots.

REFERENCES

[1] H. Tresca, *Compt. Rend. Acad. Sci. Paris, 59* (1864) 754; *64* (1867) 809.

[2] R. von Mises, *Göttinger Nachr. Math. Phys. Klasse,* 1913, p. 582.

[3] E.P. Popov, *Mechanics of Materials*, 2nd ed., Prentice-Hall, Englewood Cliffs, N.J., 1976.

[4] G. Murphy, *Advanced Mechanics of Materials*, McGraw-Hill, New York, 1964, p. 83.

[5] H.A. Kuhn and C.L. Downey, *Int. J. Powder Met.,* 7 (1971) 15.

[6] L.M. Clarebrough, M.E. Hargreaves, and M.H. Loretto, in *Recovery and Recrystallization of Metals*, L. Himmel (ed.), Interscience, New York, 1963, p. 63.

[7] B. de Saint-Venant, *Compt. Rend. Acad. Sci. Paris, 70* (1870) 473; *74* (1872) 1009.

[8] M. Levy, *Compt. Rend. Acad. Sci. Paris, 70* (1870) 1323.

[9] L. Prandtl, *Proc. First Int. Congr. Appl. Mech.*, Delft, 1924, p. 43.

[10] A. Reuss, *Z. Angew. Math. Mech.*, *10* (1930) 266.

[11] H. Hencky, *Proc. First Int. Congr. Appl. Mech.*, Delft, 1924; *Z. Angew. Math. Mech.*, *4* (1924) 323.

[12] A. Nadai, *Plasticity*, McGraw-Hill, New York, 1931, p. 77.

[13] T.E. Davidson, in *Advances in Deformation Processing*, Sagamore Army Materials Research Conf. Proc., Vol. 21, J.J. Burke and V. Weiss (eds.), Plenum Press, New York, 1978, p. 535.

[14] W.A. Backofen, *Deformation Processing*, Addison-Wesley, Reading, Mass., 1972.

[15] H.E. McGannon (ed.), *The Making, Shaping, and Treating of Steel*, 9th ed., U.S. Steel, Pittsburgh, Pa., 1971.

[16] B. Avitzur, *Metal Forming: Processes and Analysis*, Kreiger Publishing Co., Melbourne, Florida, 1968, p. 167.

[17] B.J. Nieber and A.B. Draper, *Product Design and Process Engineering*, McGraw-Hill, New York, 1974, p. 577.

[18] Society for Manufacturing Engineers, *Fundamentals of Tool Design*, Prentice-Hall, Englewood Cliffs, N.J., 1962.

[19] R.N. Orava and R.H. Wittman, in *Advances in Deformation Processing*, Sagamore Army Materials Research Conf. Proc., Vol. 21, J.J. Burke and V. Weiss (eds.), Plenum Press, New York, 1978, p. 485.

[20] A.A. Ezra, *Principles and Practice of Explosive Metal Working*, Garden City Press, London, 1973.

[21] R.N. Orava, in *Metallurgical Effects at High Strain Rates*, R.W. Rohde, B.M. Butcher, J.R. Holland, and C.H. Karnes (eds.), Plenum Press, New York, 1973, p. 126.

[22] J.S. Hirschhorn, *Introduction to Powder Metallurgy*, American Powder Metallurgy Institute, Princeton, N.J., 1969, p. 309.

[23] D.S. Fields, Jr., and J.F. Hubert, cited in [13], p. 441.

[24] T.E. Davidson, cited in [13], p. 535.

[25] T.Z. Blazynski, *Metal Forming: Tool Profiles and Flow*, Wiley, New York, 1976, p. 70.

[26] J.J. Jonas and M.J. Luton, cited in [13], p. 222.

[27] W. Johnson, cited in [13], p. 21.

[28] W. Prager and P.G. Hodge, Jr., *Theory of Perfectly Plastic Solids*, Chapman & Hall, London, 1951.

[29] T.Z. Blazynski, *Int. J. Mech. Sci.*, *13* (1971) 113.

[30] W.A. Backofen, J.J. Burke, L.F. Coffin, Jr., N.L. Reed, and V. Weiss (eds.), *Fundamentals of Deformation Processing*, Sagamore Army Materials Research Conf. Proc., Vol. 9, Syracuse University Press, New York, 1964, p. 11.

[31] E.H. Lipmann (ed.), *Metal Forming Plasticity*, Springer-Verlag, Berlin, 1979, pp. 169, 395.

[32] H. Kudo and S. Matsubara, cited in [31], p. 378.

[33] Y. Yamada, A.S. Wifi, and T. Hirakawa, cited in [31], p. 159.

[34] O. Hoffman and G. Sachs, *Introduction to the Theory of Plasticity for Engineers*, McGraw-Hill, New York, 1953, p. 206ff.

[35] H.E. McGannon (ed), cited in [15], p. 636.

[36] R. Hill and S.J. Tupper, *J. Iron Steel Inst.*, *159* (1948) 333.

[37] P.W. Whitton, *Wire Ind.*, Aug. 1958, p. 735.

[38] P.R. Cetlin, H. Helman, J.M.R. Caccipoli, R.A.N.M. Barbosa, and M.R. Marini, *Proc. 35th Annu. Meet. Braz. Soc. Metals*, Recife, Brasil, July 1981.

[39] G. Langford and M. Cohen, *Trans. ASM*, *62* (1969) 623.

[40] H.J. Rack and M. Cohen, *Met. Trans.*, *1* (1969) 1050.

[41] G. Langford and M. Cohen, *Met. Trans.*, *6A* (1975) 901.

[42] G.E. Dieter, *Mechanical Metallurgy*, 2nd ed., McGraw-Hill, New York, 1976, pp. 84, 564.

[43] H.C. Rogers, *Annu. Rev. Mater. Sci.*, *9* (1979) 283.

[44] G.B. Olson, J.F. Mescall, and M. Azrin, in *Shock Waves and High-Strain-Rate Phenomena in Metals: Concepts and Applications*, M.A. Meyers and L.E. Murr (eds.), Plenum Press, New York, 1981, p. 221.

[45] S.L. Semiatin and C.D. Lahoti, *Sci. Am.*, August 1981, p. 98.

[46] S.S. Hecker, A.K. Ghosh, and H.L. Gegel (eds.), *Formability: Analysis, Modeling, and Experimentation*, TMS-AIME, New York, 1978.

[47] S.S. Hecker and A.K. Ghosh, *Sci. Am.*, Nov., 1976, p. 100.

SUGGESTED READING

Slip-Line Field Analysis

DIETER, G.E., *Mechanical Metallurgy*, 2nd ed., McGraw-Hill, New York, 1976.

JOHNSON, W., and P.B. MELLOR, *Plasticity for Mechanical Engineers*, Van Nostrand, London, 1962.

ROWE, G.W., *An Introduction to the Principles of Metalworking*, Arnold, London, 1965.

THOMPSON, E.G., D.T. YANG, and S. KOBAYASHI, *Mechanics of Plastic Deformation*, Macmillan, New York, 1965.

Finite Element Method

DESAI, C.S., and J.F. ABEL, *Introduction to the Finite Element Method*, Van Nostrand Reinhold, New York, 1972.

LIPPMANN, E.H. (ed.), *Metal Forming Plasticity*, Springer-Verlag, Berlin, 1979.

NORRIE, D.H., and G. DEVRIES, *An Introduction to Finite Element Analysis*, Academic Press, New York, 1978.

ZIENKIEWICZ, O.C., *The Finite Element Method in Engineering Science*, McGraw-Hill, London, 1971.

Deformation Processing

AVITZUR, B., *Metal Forming: Processes and Analysis*, McGraw-Hill, New York, 1968.

BACKOFEN, W., *Deformation Processing*, Addison-Wesley, Reading, Mass., 1972.

BACKOFEN, W.A., J.J. Burke, L.E. Coffin Jr., N.L. Reed, and V. Weiss (eds.), *Fundamen-*

tals of *Deformation Processing*, Sagamore Army Materials Research Conf. Proc., Vol. 9, Syracuse University Press, New York, 1964.

BLAZYNSKI, T.Z., *Metal Forming: Tool Profiles and Flow*, Wiley, New York, 1976.

BURKE, J.J., and V. WEISS (eds.), *Advances in Deformation Processing*, Sagamore Army Materials Research Conf. Proc., Vol. 21, Plenum Press, New York, 1978.

DIETER, G.E., *Mechanical Metallurgy*, 2nd ed., McGraw-Hill, New York, 1976.

HECKER, S.S., A.K. GHOSH, and H.L. GEGEL (eds.), *Formability: Analysis, Modeling and Experimentation*, TMS-AIME, Warrendale, Pa., 1978.

HIRSCHHORN, J.S., *Introduction to Powder Metallurgy*, American Powder Metallurgy Institute, Princeton, N.J., 1969.

HOFFMAN, O., and G. SACHS, *Introduction to the Theory of Plasticity for Engineers*, McGraw-Hill, New York, 1953.

MCGANNON, H.E. (ed.), *The Making, Shaping, and Treating of Steel*, 9th ed., U.S. Steel, Pittsburgh, Pa., 1971.

SUH, N.P., and A.P.L. TURNER, *Elements of the Mechanical Behavior of Solids*, McGraw-Hill, New York, 1975.

UNDERWOOD, L.R., and A. MCCANCE, *The Rolling of Metals*, Wiley, New York, 1950.

UNSKOV, E.P., *An Engineering Theory of Plasticity*, Butterworths, London, 1961.

WISTREICH, J.G., "The Fundamentals of Wire Drawing," *Met. Rev.*, 2 (1958) 10.

Plasticity Theory

DIETER, G.E., *Mechanical Metallurgy*, 2nd ed., McGraw-Hill, New York, 1976.

HILL, R., *The Mathematical Theory of Plasticity*, Clarendon Press, Oxford, 1950.

HOFFMAN, O., and G. SACHS, *Introduction to the Theory of Plasticity for Engineers*, McGraw-Hill, New York, 1953.

KACHANOV, L.M., *Foundations of the Theory of Plasticity*, North-Holland, Amsterdam, 1971.

PRAGER, W., *An Introduction to Plasticity*, Addison-Wesley, Reading, Mass., 1959.

SPENCER, G.C., *Introduction to Plasticity*, Chapman & Hall, London, 1968.

Chapter 3

FRACTURE AND
FRACTURE TOUGHNESS

‡3.1 INTRODUCTION

The separation or fragmentation of a solid body into two or more parts, under the action of stresses, is called fracture. The subject of fracture is vast and involves disciplines as diverse as solid-state physics, materials science, and continuum mechanics. Fracture is a phenomenon, not a discipline. Moreover, fracture of a material by cracking can occur in many ways, the principal ones being:

1. Slow application of external loads
2. Rapid application of external loads (impact)
3. Cyclic or repeated loading (fatigue)
4. Time-dependent deformation under a constant load (creep)

Clearly, a detailed treatment of all these aspects of fracture is outside the scope of this book. This chapter delves into some fundamental aspects of fracture, such as nucleation and propagation of cracks under different conditions, and introduces briefly the discipline known as linear elastic fracture mechanics (LEFM).

The process of fracture can, in most cases, be subdivided into the following categories:

1. Damage accumulation
2. Nucleation of one or more cracks
3. Propagation of a crack to complete separation of the material

The damage accumulation is associated with the material properties, such as its atomic structure, crystal lattice, grain boundaries, and the prior loading history. When the local strength or ductility is exceeded, a crack is formed (two free surfaces). On continued loading, the crack propagates through the section until complete rupture.

The discipline of linear elastic fracture mechanics applies the theory of linear elasticity to crack propagation and does not concern itself with the crack nucleation problem. Defining the fracture toughness of a material as its resistance to crack propagation, LEFM provides a quantitative measure of fracture toughness or tenacity of a material. There exist specifications of ASTM (American Society for Testing and Materials) and BSI (British Standards Institution) for fracture toughness tests (see Chapter 19).

3.2 STRESS CONCENTRATION AND GRIFFITH CRITERION OF FRACTURE

The most fundamental requisite for the propagation of a crack is that the stress at the crack tip must exceed the theoretical cohesive strength of the material. It is indeed the fundamental criterion, but it is not very useful because it is almost impossible to measure the stress at the crack tip. An equivalent criterion, called the Griffith criterion, is more useful and predicts the force that must be applied to a body containing a crack for the propagation of the crack. The Griffith criterion is based on an energy balance and is described in Section 3.2.3. Let us first grasp the basic idea of stress concentration in a solid.

‡3.2.1 Stress Concentrations

The failure of a material is associated with the presence of high local stresses and strains in the vicinity of defects. Thus, it is important to know the magnitude and distribution of these stresses and strains around cracklike defects.

Consider a plate having a through-the-thickness notch, and subjected to a uniform tensile stress away from the hole (Fig. 3.1). We can imagine the applied external force being transmitted from one end of the

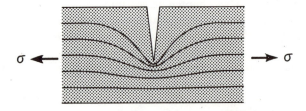

Figure 3.1 "Lines of force" in a bar with a side notch. The direction and density of these lines of force indicate the direction and magnitude of stress in the bar under a uniform stress σ away from the notch. There is a concentration of the lines of force at the notch tip.

plate to the other by means of lines of force (similar to the well-known magnetic lines of force). At the ends of the plate, which is being uniformly stretched, the spacing between the lines is uniform. The lines of force in the central region of the plate are severely distorted by the presence of the hole (i.e., the stress field is perturbed). The lines of force, acting as elastic strings, tend to minimize their lengths and thus group together near the ends of the elliptic hole. This grouping together of lines causes a decrease in the line spacing locally and consequently an increase in the local stress (a stress concentration), there being more lines of force in the same area.

‡3.2.2 Stress Concentration Factor

The theoretical fracture stress of a solid is of the order $E/10$ (see Chapter 4), but the strength of solids (crystalline or otherwise) in practice is orders of magnitude less than this value. The first attempt at giving a rational explanation of this discrepancy was due to A. A. Griffith.[1] His analytical model was based on the elastic solution of a cavity elongated in the form of an ellipse.

Figure 3.2 shows an elliptical cavity in a plate under a uniform stress σ away from the cavity. The maximum stress occurs at the ends of the major axis of this elliptical cavity and is given by the formula due to Inglis:[2]†

$$\sigma_{max} = \sigma\left(1 + 2\frac{a}{b}\right) \tag{3.1}$$

where $2a$ and $2b$ are the major and the minor axes of the ellipse, respectively. The value of stress at the leading edge of the cavity becomes extremely large as the ellipse is flattened. In the case of an extremely flat ellipse or a very narrow crack of length $2a$, having a radius of curvature $\rho = b^2/a$, Eq. (3.1) can be written as

$$\sigma_{max} = \sigma\left(1 + 2\sqrt{\frac{a}{\rho}}\right) \simeq 2\sigma\sqrt{\frac{a}{\rho}} \qquad \text{for } \rho \ll a \tag{3.2}$$

We note that as ρ becomes very small, σ_{max} becomes very large, and in the limit of $\rho \to 0$, $\sigma_{max} \to \infty$. We define the term $2\sqrt{a/\rho}$ as the stress concentration factor, K_t (i.e., $K_t = \sigma_{max}/\sigma$). K_t simply describes the geometric effect of the crack on the local stress (i.e., at the crack tip). Note that the stress concentration factor K_t depends more on the form of the cavity than on its size. There exist a number of texts and handbooks giving a compilation of stress concentration factors K_t for components containing cracks or notches of various configurations.[3,4]

In addition to producing a stress concentration, a notch also produces a local situation of biaxial or triaxial stress. For example, in the case of a plate

---

† The derivation of this equation, which can be found in more advanced texts (e.g., J. F. Knott, Fundamentals of Fracture Mechanics, Butterworths, London, 1973, p. 51) involves the solution of the biharmonic equation (Eq. 1.146), the choice of an appropriate Airy stress function (Eq. 1.145) and complex variables.

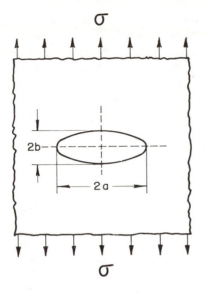

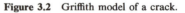

Figure 3.2 Griffith model of a crack.

containing a circular hole and subjected to an axial force, there exist radial as well as tangential stresses. Timoshenko and Goodier[5] have shown that the stresses in a large plate, containing a circular hole (diameter $2a$) and axially loaded (Fig. 3.3), can be expressed as (for $r \gg a$)

$$\sigma_r = \frac{\sigma}{2}\left(1 - \frac{a^2}{r^2}\right) + \frac{\sigma}{2}\left(1 + 3\frac{a^4}{r^4} - 4\frac{a^2}{r^2}\right)\cos 2\theta$$

$$\sigma_\theta = \frac{\sigma}{2}\left(1 + \frac{a^2}{r^2}\right) - \frac{\sigma}{2}\left(1 + 3\frac{a^4}{r^4}\right)\cos 2\theta \tag{3.3}$$

$$\sigma_{r\theta} = -\frac{\sigma}{2}\left(1 - 3\frac{a^4}{r^4} + 2\frac{a^2}{r^2}\right)\sin 2\theta$$

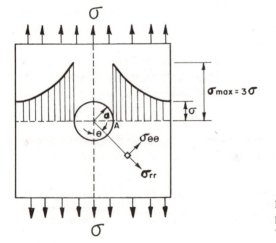

Figure 3.3 Stress distribution in a large plate containing a circular hole. (Adapted with permission from [5], p. 78.)

Mechanics of Deformation   Part I

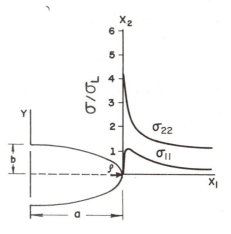

Figure 3.4 Stress concentration at an elliptical hole for $a = 3b$. (Adapted with permission from [6], p. 4.)

The maximum stress occurs at point A, where $\theta = \pi/2$ and $r = a$. In this case

$$\sigma_\theta = 3\sigma = \sigma_{max}$$

where σ is the uniform stress applied at the ends of the plate. Thus, K_t in this case is 3.

In the case of an elliptical hole,[6] for $a = 3b$, Fig. 3.4 shows that σ_{22} falls from its maximum value and attains σ asymptotically. σ_{11} increases to a peak and then falls to zero with the same tendency as σ_{22}. The general result is that the major perturbation in the applied stress state occurs over a distance approximately equal to a from the boundaries of the cavity, with the major stress gradients being confined to a region of dimensions roughly equal to ρ surrounding the maximum concentration position.

Although the exact formulas vary according to the form of the crack, in all cases K_t increases with an increase in the crack length a and a decrease in the root radius at the crack tip, ρ.

Despite the fact that the analysis of Inglis represented a great advance, the fundamental nature of the fracture mechanism remained obscure. If the Inglis analysis was applicable to a body containing a crack, how does one explain that, in practice, larger cracks propagate more easily than smaller cracks? What is the physical significance of the root radius at the crack tip?

‡3.2.3 The Griffith Criterion

Griffith estimated the change in energy that results when an elliptical crack is introduced in an infinite plate. Griffith calculated the decrease in energy due to the presence of a crack in a plate according to the formulation of Inglis and found it to be equal to $\pi\sigma^2 a^2/E$ per unit thickness of the plate for the plane-stress condition. The increase in energy due to the presence of two new surfaces equals $2a \times 2\gamma_s$ per unit thickness of the plate. Thus, the change in potential energy of the plate when a crack is introduced in it can be written as

$$U - U_0 = 4a\gamma_s - \frac{\pi\sigma^2 a^2}{E}$$

where U is the potential energy per unit thickness of the plate in the presence of crack, U_0 the potential energy per unit thickness of the plate in the absence of crack, σ the applied stress, a the half crack length, E the modulus of elasticity, and γ_s the specific surface energy (i.e., surface energy per unit area).

As the crack grows, strain energy is released but additional surfaces are created. The crack is stable when these energy components balance each other. If they are not in balance, we have an unstable crack (i.e., the crack will grow). We can obtain the equilibrium condition by equating to zero the first derivative of the potential energy U with respect to the crack length. Thus, remembering that $\partial U_0/\partial a = 0$, as U_0 is the potential energy in the absence of the crack and does not change with the crack length a, we have

$$\frac{\partial U}{\partial a} = 4\gamma_s - \frac{2\pi\sigma^2 a}{E} = 0 \tag{3.4}$$

or

$$2\gamma_s = \frac{\pi\sigma^2 a}{E} \tag{3.5}$$

The reader can further check the nature of this equilibrium by taking the second derivative of U with respect to a. A negative second derivative would imply that Eq. 3.4 (or 3.5) represents an unstable equilibrium condition and that the crack will advance.

Rearranging Eq. (3.5), we may write for the crack propagation stress

$$\sigma_c = \sqrt{\frac{2E\gamma_s}{\pi a}} \qquad \text{plane stress} \tag{3.6a}$$

and

$$\sigma_c = \sqrt{\frac{2E\gamma_s}{\pi a(1 - \nu^2)}} \qquad \text{plane strain} \tag{3.6b}$$

The length of the crack, preexisting or resulting from a nucleation event, is critical in determining whether or not it will propagate in a given stress field. According to Griffith's thermodynamic analysis, a necessary condition for crack propagation is

$$-\frac{\partial U_e}{\partial a} \geq \frac{\partial U_\gamma}{\partial a}$$

where U_e is the elastic energy of the system (i.e., machine plus test piece) and U_γ is equal to $2\gamma_s$ in the simplest case. This is necessary condition for fracture by rapid crack propagation. But it may not always be sufficient. If the local stress at the crack tip is not sufficiently large to break the atomic bonds, the energy criterion of Griffith will be inadequate.

It is worth pointing out here that the tensile fracture stress σ_c is inversely proportional to $\sqrt{a}$. Moreover, the quantity $\sigma_c\sqrt{a}$ depends only on the material constants. From Eq. 3.2,

$$\sigma_c\sqrt{a} = \tfrac{1}{2}(\sigma_{max})_c\sqrt{\rho} = \text{constant}$$

where the subscript c denotes the fracture condition.

The great value of the Griffith theory is that, instead of the local and difficult-to-measure parameters ($\sigma_{max})_c$ and $\sqrt{\rho}$, it permits the use of the applied stress and the crack length, which are easy to measure. The quantity $\sigma_c\sqrt{a}$ is denoted by K_c and is called the fracture toughness of the material. We treat this in detail later.

3.2.4 Crack Propagation in a Material That Deforms Plastically

If the material in which the crack is propagating can deform plastically, the crack-tip form changes because of plastic strain. A sharp crack tip will be blunted. Another important factor is time. Plastic deformation requires time to occur. Thus, the amount of plastic deformation that can occur at the crack tip will depend on how fast the crack is moving.

In a great majority of materials, localized plastic deformation at and around the crack tip is produced because of the stress concentrations there. In such a case a certain amount of plastic work is done during crack propagation, in addition to the elastic work done in creation of two fracture surfaces. The mechanics of fracture will, then, depend on the magnitude of γ_p, the plastic work done, which in its turn depends on the crack speed, temperature, and the nature of the material. For an inherently brittle material, at low temperatures and at high crack velocities, γ_p is relatively small ($\gamma_p < 10\gamma_s$). In such a case, the crack propagation would be continuous and elastic. Nowadays, such cases are usefully treated by means of linear elastic fracture mechanics, which is dealt with later. In any event, in the case of plastic deformation the work done in the propagation of a crack per unit area of the fracture surface is increased from γ_s to ($\gamma_s + \gamma_p$). Consequently, the Griffith criterion (Eqs. 3.6a and 3.6b) is modified to

$$\sigma_c = \sqrt{\frac{2E}{\pi a}(\gamma_s + \gamma_p)} \qquad \text{plane stress} \qquad (3.7a)$$

and

$$\sigma_c = \sqrt{\frac{2E}{\pi a(1 - v^2)}(\gamma_s + \gamma_p)} \qquad \text{plane strain} \qquad (3.7b)$$

Rearranging Eq. 3.7a, we get

$$\sigma_c = \sqrt{\frac{2E\gamma_s}{\pi a}\left(1 + \frac{\gamma_p}{\gamma_s}\right)}$$

For $\gamma_p/\gamma_s \gg 1$,

$$\sigma_c \simeq \sqrt{\frac{2E\gamma_p}{\pi a}}$$

Thus, the plastic deformation around the crack tip makes it blunt and serves to relax the stress concentration by increasing the radius of curvature of the crack at the tip. Localized plastic deformation at the crack tip therefore improves the fracture toughness of the material.

This is the conventional treatment of the plastic work contribution to the fracture process wherein γ_p is considered to be a constant. However, the reader should be warned that this is not strictly true. As a matter of fact, the value of γ_p increases with the stress intensity factor, K $(= f\sigma\sqrt{a})$. Consider Eq. 3.7a. As pointed out above, in the conventional approach γ_p will be very much larger than γ_s for a ductile material such as polycrystalline copper. Thus, according to this conventional treatment, the fracture stress σ_c should be relatively insensitive to changes in γ_s. However, in the embrittlement of copper with beryllium, all we change is the γ_s part of Eq. 3.7a (along the grain boundaries where the fracture proceeds). The γ_p part in Eq. 3.7a (i.e., the plastic behavior of copper) does not change appreciably by the beryllium addition to copper. Weertman[7] has given a theoretical treatment that explains how fracture stress σ_c does change appreciably if γ_s is changed, without altering appreciably the γ_p contribution.

As pointed out earlier, equations of the type 3.6 or 3.7 are difficult to use in practice. It is not a trivial matter to measure quantities such as surface energy and energy of plastic deformation. Irwin,[8] in a manner similar to that of Griffith, made a fundamental contribution to the mechanics of fracture. He proposed that fracture occurs at a stress that corresponds to a critical value of the crack extension force G (sometimes called the strain energy release rate), where

$$G = \frac{1}{2}\frac{\partial U}{\partial a} = \text{rate of energy change with crack growth}$$

Now $U = \pi a^2 \sigma^2 / E$, energy released by the advancing crack per unit of plate thickness. This is for plane stress. For plane strain a factor of $(1 - \nu^2)$ is introduced in the denominator. Thus,

$$G = \frac{\pi a \sigma^2}{E}$$

At fracture, $G = G_c$, and

$$\sigma_c = \sqrt{\frac{EG_c}{\pi a}} \qquad \text{plane stress} \qquad (3.8a)$$

and

$$\sigma_c = \sqrt{\frac{EG_c}{\pi a(1 - \nu^2)}} \qquad \text{plane strain} \qquad (3.8b)$$

From Eqs. 3.7 and 3.8 we see that

$$G_c = 2(\gamma_s + \gamma_p)$$

We shall come back to this idea of crack extension force later in the chapter.

‡3.3 MICROSTRUCTURAL ASPECTS OF CRACK NUCLEATION AND PROPAGATION

The microstructure of a metal has a great influence on its fracture behavior. We present below a brief description of the various models of nucleation and propagation of cracks. The reader should consult Chapter 6 to understand this section more clearly.

‡3.3.1 Crack Nucleation

Nucleation of a crack in a perfect crystal essentially involves the rupture of interatomic bonds. The stress necessary to do this is the theoretical cohesive stress, which is dealt with in Chapter 4, starting from an expression for interatomic forces. From this expression we see that ordinary materials break at much lower stresses—of the order of $E/10^4$, where E is Young's modulus of the material. The explanation lies in the existence of surface and internal defects that act as preexisting cracks and in the plastic deformation that precedes the fracture. When both plastic deformation and fracture are eliminated, for example in "whiskers," stresses of the order of the theoretical cohesive stresses are obtained.

Crack nucleation mechanisms vary according to the type of the material: brittle, semibrittle, or ductile. The brittleness of a material has to do with the behavior of dislocations in the region of crack nucleation. In highly brittle materials, the dislocations are practically immobile, in semibrittle materials dislocations are mobile but only on a restricted number of slip planes, and in ductile materials there are no restrictions on dislocation movement other than those inherent in the crystalline structure of the material. In Table 3.1 are presented various materials classified according to the above-mentioned criterion regarding dislocation mobility.

The exposed surface of a brittle material can suffer damage by mechanical contact with even microscopic dust particles. If a glass fiber without surface treatment were rolled over a tabletop, it would be seriously damaged mechanically.

Any material heterogeneity that produces a stress concentration can nucleate cracks. For example, steps, striations, depressions, holes, and so on, act as stress raisers on apparently perfect surfaces. In the interior of the material, there can exist voids, air bubbles, second-phase particles, and so on. Crack nucleation will occur at the weakest of these defects, where the conditions would be most favorable. We generally assume that the size as well as the location of defects are distributed in the material according to some function of standard

TABLE 3.1 Materials of Various Degrees of Brittleness[a]

Type	Principal Factors	Materials
Brittle	Bond rupture	Structures of type diamond, ZnS, silicates, alumina, mica, boron, tungsten, carbides, and nitrides
Semibrittle	Bond rupture, dislocation mobility	Structures of type NaCl, ionic crystals, hexagonal compact metals, majority of body-centered cubic metals, glassy polymers
Ductile	Dislocation mobility	Face-centered cubic metals, nonvitreous polymers, some body-centered cubic metals

[a] Adapted with permission from [6], p. 17.

distribution whose parameters are adjusted to conform to experimental data. It should be recognized that in this there is no explicit consideration of nature or origin of the defects.

In the semibrittle materials, there is a tendency for slip initially, followed by fracture on well-defined crystallographic planes. That is, there exists a certain inflexibility in the deformation process and the material, not being able to accommodate localized plastic strains, initiates a crack to relax stresses.

There exist various models based on the idea of crack nucleation at an obstruction site. For example, the intersection of a slip band with a grain boundary, another slip band, and so on, would be obstruction sites.

‡3.3.1.1 Ductile fracture.   In ductile materials the role of plastic deformation is very important. The important point here is the flexibility of slip. Dislocations can move on a large number of slip systems and even cross from one plane to another (cross-slip). Consider the deformation of copper single crystal, a ductile metal, under uniaxial tension. The single crystal undergoes slip throughout its section. There is no nucleation of cracks and the crystal deforms plastically until the start of plastic instability called necking. The deformation from this point onward is concentrated in this region until the crystal separates along a line giving a knife-edge-type fracture. Figure 3.5(a) shows this schematically and Fig. 3.5(b) shows a micrograph of the fracture surface of a copper single crystal. However, if in a ductile material there are microstructural elements such as particles of a second phase, internal interfaces, and so on, then microcavities may be nucleated in regions of high stress concentrations in a manner similar to that of semibrittle materials, except that due to large

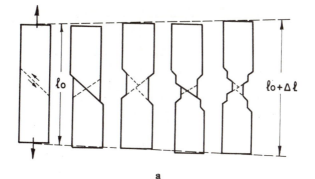

a

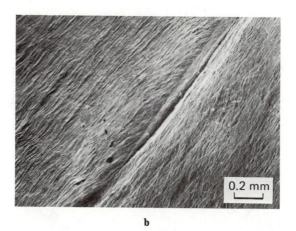

b

Figure 3.5 Failure by shear (glide) in a pure metal. (Reprinted with permission from [9], p. 362.) (b) Failure in a pure copper monocrystal according to the model of Fig. 3.5(a), resulting in a knife-edge type of fracture.

plasticity, cracks generally do not propagate from these cavities. The regions between the cavities, however, behave as small test samples that elongate and break by plastic instability as described for the single crystal.

Zener[10] was the first one to propose that in crystalline solids cracks can be nucleated by grouping up of dislocations piled up against a barrier. High stresses at the head of a pile up are relaxed by crack nucleation, as shown in Fig. 3.6. But this would occur only in case there is no relaxation of stresses by dislocation movement on the other side of the barrier. Depending on the slip geometry in the two parts and the kinetics of dislocation motion and multiplication, this could possibly occur (see Table 3.1). Figure 3.7(a) shows a bicrystal that has a slip band in grain I. The stress concentration at the barrier due to the slip band is completely relaxed by slip on two systems in grain II. Figure 3.7(b) shows the case of only a partial relaxation and the resulting appearance of a crack at the barrier.

Lattice rotation associated with the bend planes and deformation twins can also nucleate cracks (Fig. 3.8). Figure 3.9 shows crack nucleation in zinc

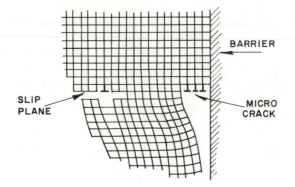

Figure 3.6 Grouping of dislocations piled up at a barrier leading to a microcrack formation.

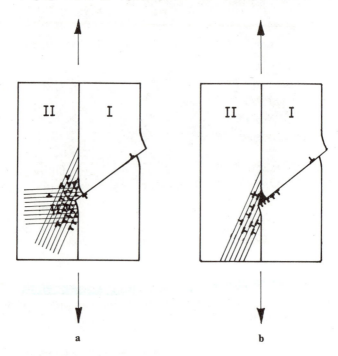

Figure 3.7 Bicrystal with a slip band in grain I: (a) the stress concentration at the barrier boundary due to slip band is fully relaxed by multiple slip; (b) the stress concentration is only partially relaxed, resulting in a crack at the boundary.

Figure 3.8 Crack nucleation by (a) lattice rotation due to the bend planes, and (b) deformation twins.

144

Figure 3.9 Crack nucleation in zinc due to lattice rotation associated with bend planes. (Reprinted with permission from [11], p. 83.)

as in the model of Fig. 3.8(a), and Fig. 3.10 shows crack nucleation in MgO at slip band intersection.

At high temperatures, two other types of cracking are observed:

1. The first one occurs at triple points (where three grain boundaries meet); Fig. 3.11 shows this. Due to grain boundary sliding that occurs at high temperatures, stress concentrations are developed at triple points. Figure 3.12 shows a micrograph of copper showing crack nucleation according to this mechanism. This type of crack is called w-type.

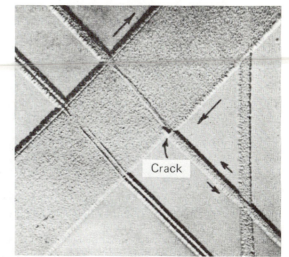

Figure 3.10 Crack nucleation in MgO at a slip-band intersection. (Reprinted with permission from [11], p. 85.)

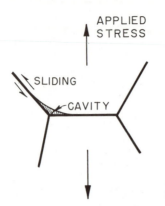

Figure 3.11 *w*-type cavitation at a grain-boundary triple point.

2. Another type of cracking occurs under conditions of low stresses and high temperature. Small cavities form on the grain boundaries (Fig. 3.13) that are, predominantly, at approximately 90° to the stress axis. These are called r-type cavities. Figure 3.14 shows such r-type cavities in copper.

The most familiar example of ductile fracture is the one in uniaxial tension, giving the classic "cup and cone" fracture. When the maximum load is reached, the plastic deformation in a cylindrical tensile test piece becomes macroscopically heterogeneous and is concentrated in a small region. This phenomenon is called necking. The final fracture occurs in this necked region and has the characteristic appearance of a conical region on the periphery resulting from shear, and a central flat region resulting from the voids created there. In extremely pure metal single crystals (e.g., free of inclusions, etc.), the plastic deformation continues until the sample section is reduced to a knife edge, a geometric consequence of slip, as shown in Fig. 3.5.

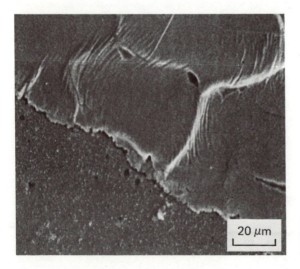

20 μm

Figure 3.12 *w*-type cavities nucleated at grain boundaries in copper. (Scanning electron microscope.)

Mechanics of Deformation   Part I

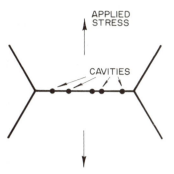

Figure 3.13 *r*-type cavitation at a grain boundary normal to the stress axis.

In practice, materials generally contain a large quantity of dispersed phases. These can be very small particles (1 to 20 nm), as, for example, carbides of alloy elements; particles of intermediate size (50 to 500 nm), as, for example, alloy element compounds (carbides, nitrides, carbonitrides) in steels; or dispersions such as Al_2O_3 in aluminum and ThO_2 in nickel. Precipitate particles obtained by appropriate heat treatment also form part of this class (e.g., Al-Cu-Mg system); and finally, one may include inclusions of large size (on the order of millimeters), for example, oxides and sulfides.

If the second-phase particles are brittle and the matrix is ductile, the former will not be able to accommodate the large plastic strains of the matrix and, consequently, these brittle particles break in the very beginning of plastic deformation. In case the particle–matrix interface is very weak, interfacial separation will occur. In both cases, microcavities are nucleated at these sites (Fig. 3.15). Generally, the voids nucleate after a few percent of plastic deformation, while the final separation may occur around 25%. The microcavities grow with slip and the material between the cavities can be visualized as a small tensile test piece. This material between voids undergoes necking on a microscopic scale and the voids join together. However, these microscopic necks do not contribute significantly to total elongation. This mechanism of initiation, growth, and coalescence of microcavities gives the fracture surface a characteristic appearance. When viewed in the scanning electron microscope (SEM), one observes

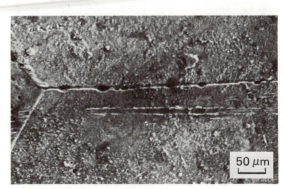

Figure 3.14 *r*-type cavities nucleated at grain boundaries in copper. (Scanning electron microscope.)

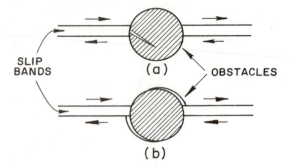

Figure 3.15 Nucleation of a cavity at a second-phase particle in a ductile material. (Adapted with permission from [6], p. 40.)

that such a fracture surface consists of small dimples which represent the micro-cavities after coalescence (Fig. 3.16). In many of these dimples one can see the inclusions that were responsible for the void nucleation (Fig. 3.16). At times, due to unequal triaxial stresses, these voids are elongated in one or the other direction.

It is worth pointing out here that by "ductility" of a given material, we understand its capacity to undergo extensive plastic deformation. As it happens,

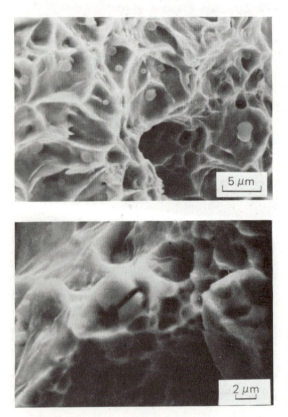

Figure 3.16 Dimple fracture resulting from the nucleation, growth, and coalescence of microcavities. Note the inclusion particles which served as the microcavity nucleation sites. (Scanning electron microscope.)

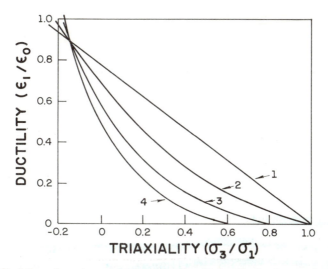

Figure 3.17 Variation of maximum plastic strain (ductility) with the degree of triaxiality: (1) theory of maximum tensile stress failure; (2) plane-strain conditions; (3) von Mises criterion; and (4) power law of plastic strain. (Adapted with permission from [12], p. 265.)

this is not a fundamental property of the material, because the plastic strain before fracture is a function of the state of stress, strain rate, temperature, environment, and prior history. The state of stress is defined by the three-dimensional distribution of normal and shear stresses at a point or by the three principal stresses at a point (see Chapter 1). The multiaxial stresses may be obtained by external multiaxial loading, by geometry of the structure or microstructure under load, by thermal stresses, or by volumetric microstructural changes. One can define[12] a simple "triaxiality" factor by the ratio σ_3/σ_1, where $\sigma_1 > \sigma_2 > \sigma_3$ are the principal stresses. If ϵ_0 is the plastic strain at fracture in uniaxial tension and ϵ_1 is the maximum principal plastic strain, one can define a ductility ratio as ϵ_1/ϵ_0.

This ductility ratio shows, theoretically, a decrease with increasing triaxiality; that is, ϵ_1/ϵ_0 goes to zero as σ_3/σ_1 goes to 1 (Fig. 3.17). Thus, an increase in degree of stress triaxiality results in a decrease in the ductility of the material.

The temperature and the strain rate have contrary effects. A high temperature (or a low strain rate) leads to high ductility, whereas a low temperature (or a high strain rate) leads to low ductility.

‡3.3.1.2 Brittle fracture or cleavage fracture.

The most brittle form of fracture is the cleavage fracture. The tendency for a cleavage fracture increases with an increase in the strain rate or a decrease in the test temperature (see Sec. 19.3). This is shown, typically, by a ductile–brittle transition in steel (Fig. 3.18). The ductile–brittle transition temperature (DBTT) increases with an increase in the strain rate. Above DBTT the steel shows a ductile fracture, while below DBTT it shows a brittle fracture. Ductile fracture needs a lot more energy

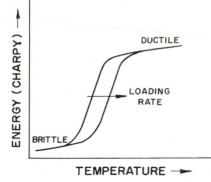

Figure 3.18 Ductile–brittle transition in a steel and the effect of loading rate (schematic).

than the brittle fracture. We deal with these aspects of DBTT in more detail in Chapter 19.

Cleavage occurs by direct separation along specific crystallographic planes by means of a simple rupturing of atomic bonds. Iron, for example, undergoes cleavage along its cubic planes (100). This gives the characteristic flat surface appearance within a grain on the fracture surface (Fig. 3.19).

There is evidence that some kind of plastic yielding and dislocation interaction is responsible for cleavage fracture. Low[13] studied the fracture behavior of a low-carbon steel at 77 K. He compared the yield stress in compression

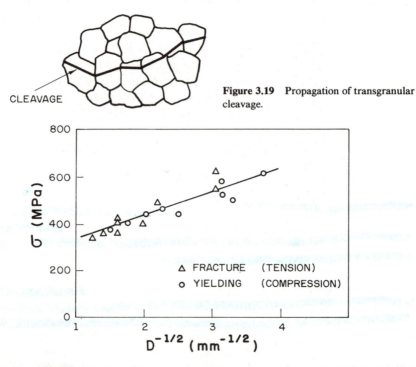

Figure 3.19 Propagation of transgranular cleavage.

Figure 3.20 Effect of grain size on fracture and yield stress of a carbon steel at 77 K. (Adapted from [13], p. 60.)

Mechanics of Deformation   Part I

Figure 3.21 (a) Cleavage facets in 300-M steel. (Scanning electron microscope.) (b) River markings on a cleavage facet in 300-M steel. (Scanning electron microscope.)

(the fracture does not occur in this case) and the stress for cleavage in tension. He did this for a number of samples with different grain sizes and showed (Fig. 3.20) that the variation in grain size in both the cases followed a Hall–Petch type of relationship, which showed that the controlling mechanism in yielding was also the controlling mechanisms for fracture initiation (see also Chapter 14).

We mentioned that cleavage occurs along specific crystallographic planes. As in a polycrystalline material the adjacent grains have different orientations; the cleavage crack changes direction at the grain boundary in order to continue along the specific crystallographic planes. The cleavage facets through the grains have a high reflectivity, which gives the fracture surface a shiny appearance [Fig. 3.21(a)]. Sometimes the cleavage fracture surface shows some small irregu-

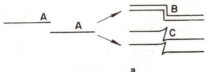

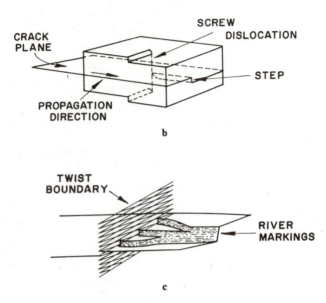

Figure 3.22 Formation of cleavage steps: (a) parallel cracks (A, A) join together by cleavage (B) or shear (C); (b) cleavage step initiation by the passage of a screw dislocation; (c) formation of river markings after the passage of a grain boundary. (Adapted from [9], p. 33.)

larities; for example, the river markings in Fig. 3.21(b). What happens is that within a grain a crack may grow simultaneously on two parallel crystallographic planes [Fig. 3.22(a)]. The two parallel cracks can join together, by secondary cleavage or by shear, to form a step. Cleavage steps can be initiated by the passage of a screw dislocation as shown in Fig. 3.22(b). In general, the cleavage step will be parallel to the crack propagation direction and perpendicular to the plane containing the crack, as this would minimize the energy for the step formation by creating a minimum of additional surface. A large number of cleavage steps can join and form a multiple step. On the other hand, steps of opposite signs can join and disappear. The junction of cleavage steps results in a figure of a river and its tributaries. These river markings can appear by the passage of a grain boundary as shown in Fig. 3.22(c). We know that a cleavage crack tends to propagate along a specific crystallographic plane. This being so, when a crack passes through a grain boundary, it has to propagate now in a grain with a different orientation. Figure 3.22(c) shows the encounter of a cleavage crack and a grain boundary. The crack should propagate now on a cleavage plane that is oriented in a different manner. The crack can do this at various points and spread into the new grain. This process gives rise to the formation of a number of steps that can group together to give a river marking [Fig. 3.21(b)]. The convergence of tributaries is always in the direction of the flow of the river (i.e., downstream). This fact furnishes the possibility of determining local crack propagation direction in a micrograph.

Another characteristic of cleavage fracture is the cleavage tongue, so called because of its form. It is believed that the cleavage tongues are formed by local fracture along a twin-matrix interface (twins form due to high strain rates in front of an advancing crack).

Under normal circumstances, face-centered cubic (FCC) metals do not show cleavage. In these metals, there will occur a large amount of plastic deformation before the stress necessary for cleavage is reached. Cleavage is common in body-centered cubic (BCC) and hexagonal close-packed (HCP) structures, particularly in iron and low-carbon steels (BCC). Tungsten, molybdenum, and chromium (all BCC) and zinc, beryllium, and magnesium (all HCP) are other examples of metals that commonly show cleavage.

Quasicleavage is a type of fracture when cleavage occurs on a very fine scale and on cleavage planes that are not very well defined. Typically, one sees this type of fracture in quenched and tempered steels. These steels contain tempered martensite and a network of carbide particles whose size and distribution can lead to a poor definition of cleavage planes in the austenite grain. Thus, the real cleavage planes are exchanged for small and ill-defined cleavage facets that initiate at the carbide particles. Such small facets can give the appearance of a much more ductile fracture than that of normal cleavage and, generally, the river markings are not observed.

Intergranular fracture is a low-energy fracture mode. The crack follows the grain boundaries as shown schematically in Fig. 3.23, which gives the fracture a bright and reflective appearance on a macroscopic scale. On a microscopic scale, the crack may deviate around a particle and make some microcavities locally. Figure 3.24 shows an example of this in a micrograph of an intergranular fracture in steel. Intergranular fractures tend to occur when the grain boundaries are more brittle than the crystal lattice. This occurs, for example, in stainless steel when it is accidentally sensitized. This accident in the heat treatment produces a film of brittle carbides along the grain boundaries. This film is then the preferred trajectory of the crack tip. Segregation of phosphorus or sulfur to grain boundaries can also lead to intergranular fracture. In many cases, fracture at high temperatures and in creep (see Chapter 20) tends to be intergranular.

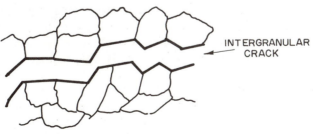

Figure 3.23 Intergranular fracture (schematic).

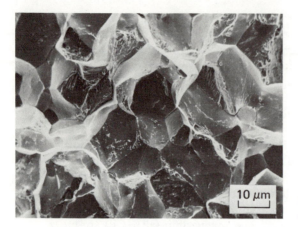

Figure 3.24 Intergranular fracture in a steel. (Scanning electron microscope.)

‡3.3.1.3 Fracture in aggressive environments. The fracture behavior of a material is very often influenced by the environment.

‡*Liquid Metal Embrittlement.* Some metals that fail in a ductile manner under normal conditions fail in a very brittle manner when deformed in liquid mercury. This phenomenon is called liquid metal embrittlement and, generally, in polycrystalline metals this type of fracture occurs in an intergranular manner. However, sometimes there can be beneficial effects, too. For example, ionic crystals can undergo larger plastic deformations in water than in air. The explanation for this lies in the fact that in water the surface defects are blunted by dissolution. When a certain environment has some effect on a certain material, the interaction generally is a particular one for that combination of material and environment. For example, liquid mercury embrittles zinc but has no effect on magnesium.

This phenomenon of liquid metal embrittlement occurs by a direct interaction between the highly strained bonds at the crack tip and the absorbed liquid metal atoms. Brittle fracture of a normally ductile metal occurs when the bond rupture stress is reduced to a level below that of local plastic deformation. Once the condition for bond rupture is satisfied, crack propagation occurs rapidly under tension and is limited only by the rate of liquid metal arrival at the crack tip.

In terms of the Griffith treatment, this embrittlement due to adsorption is explained by a great reduction in γ_s, the specific surface energy. In certain cases this can even be of some advantage. For example, in the drilling of quartz rock, addition of $AlCl_3$ to the water lubricant allows us to double the drilling speed without increasing the wear of the drilling bit.

‡*Hydrogen Embrittlement.* Hydrogen may be dissolved in molten steel from the water vapor or in solid steel when it is in contact with nascent hydrogen. The terrible embrittlement effect of hydrogen may be due to the rapid diffusion of the small hydrogen atoms to regions of high triaxiality (e.g., at the crack

front). This effect can be substantially reduced or practically eliminated by means of an appropriate annealing treatment. Hydrogen can diffuse out of steel at ambient temperature if the section is small enough and if a sufficiently long time is given. Large sections require very long times for hydrogen to diffuse out. Hydrogen levels greater than ~0.0005% produce such effects. Vacuum degasification can reduce the hydrogen content substantially.

In other materials, hydrogen embrittlement has been attributed to the formation of brittle hydrides. In this case the material itself has a low fracture toughness, while in steels hydrogen provides a mechanism of producing such a crack length that at a given fracture toughness of the material, the applied stress causes the material to fail.

‡*Stress Corrosion Cracking.* In many materials the fracture under stress corrosion conditions is intergranular, due, perhaps, to the potential difference between the grain boundary and the grain interior caused by solute segregation. The intergranular fracture may be due to the presence of second-phase particles at the grain boundaries.

This type of cracking is a phenomenon that depends on time, on the presence of a specific environment, and on a tensile stress (may be residual stress). The time to failure decreases with an increase in tension. The principal difference from the liquid metal or hydrogen embrittlement is that in this case electrochemical dissolution is involved.

‡3.3.2 Fracture Mechanism Maps

Data presented in the form of mechanism maps is increasingly gaining popularity.[14-17] This trend is understandable in view of the varied information available about the material behavior. According to Fields and Smith,[16] the idea of mechanism maps is just an extension of the phase diagrams concept in alloy chemistry where different phases coexisting in multicomponent systems are represented as a function of composition and temperature.

Deformation and fracture mechanism maps provide information about mechanical properties in a compact form. Deformation mechanism maps are described in Section 20.4. In fracture mechanism maps, one can plot normalized tensile strength, σ_{UTS}/E, against the homologous temperature, T/T_m. Regions of different fracture types are classified based on fractography or fracture-time or fracture-strain studies. One can also plot stress intensity factor (see Section 3.4) against temperature and obtain information about crack growth during the fracture process. Figure 3.25 shows these two types of fracture mechanism maps.

3.4 LINEAR ELASTIC FRACTURE MECHANICS

When in a material the ductility (i.e., the capacity to deform plastically) is very small, it is not capable of relaxing the peak stresses. The crack then propagates very rapidly with little plastic deformation of the material around the

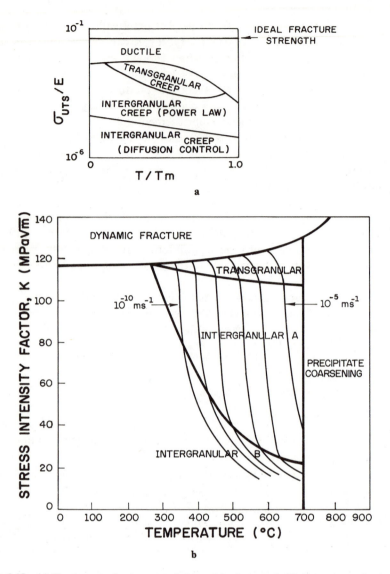

Figure 3.25 (a) Fracture mechanism map (schematic) of a metal. (b) Fracture mechanism map of Inconel 718 alloy. (Adapted with permission from [16], p. 44.)

crack surface, resulting in a brittle fracture. Another characteristic of brittle fracture is that the crack propagation is sudden, rapid, and unstable. In practical terms this definition of brittleness, which refers to the onset of instability under an applied stress less than the general yielding stress, is much more useful.

Numerous brittle fractures have occurred in service and there are abundant examples of them in a great variety of structural and mechanical engineering fields. Details of catastrophic failures in service of ships, bridges, pressure vessels, oil ducts, turbines, and so on, can be found in [18–20]. In view of the great importance of brittle fracture in real life, a discipline called linear elastic fracture mechanics (LEFM) has emerged which enables us to obtain a quantitative measure of the resistance of a material against crack propagation.

3.4.1 Fracture Toughness

LEFM is the most recent treatment in our ever-constant fight against catastrophic fracture. This new approach is based on the concept that the relevant material property, the fracture toughness, is the force necessary to extend a crack through a structural member. Under special circumstances, this crack extension force becomes independent of the specimen dimensions, and this parameter can then be used for a classification of materials in order of their fracture toughness.

LEFM adopts an entirely new approach in designing against fracture. It does not concern itself with the prevention of crack nucleation. On the contrary, it is admitted that defects always will be present in a structural element and an answer is sought to the following question: Given a certain stress, what is the largest size of a defect (crack) that can be tolerated without the failure of the member?

Knowing the answer to this question, it remains only to use appropriate inspection techniques to select a material that does not possess defects larger than the critical size for the given design stress.

3.4.2 Hypotheses of LEFM

We assume that:

1. The cracks are inherently present in a material, because there is a limit to the sensibility of crack-detecting equipment.
2. A crack is a free, internal, and plane surface in a linear elastic stress field. With this hypothesis, the linear elasticity furnishes us stresses near the crack tip as

$$\sigma_{r\theta} = \frac{K}{\sqrt{2\pi r}} f(\theta) \tag{3.9}$$

where r and θ are polar coordinates and K is a constant called the stress intensity factor (SIF).
3. The crack growth leading to the failure of the structural member is then predicted in terms of the tensile stress acting at the crack tip. In other words, the stress situation at the crack tip is characterized by the value of K. It can be shown by elasticity theory[7] that $K = \alpha\sigma\sqrt{\pi a}$, where σ is the applied stress, a the half crack length, and α is a constant that depends on the crack opening mode and the specimen geometry.

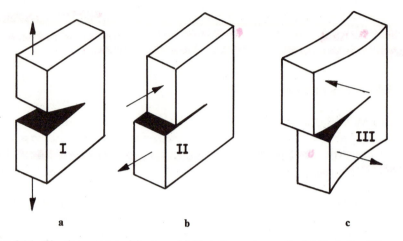

Figure 3.26 The three modes of fracture: (a) Mode I: opening mode; (b) Mode II: sliding mode; (c) Mode III: tearing mode.

3.4.3 Crack-Tip Separation Modes

The three modes of fracture are shown in Fig. 3.26. Mode I [Fig. 3.26(a)], called the opening mode, has tensile stress normal to the crack faces. Mode II [Fig. 3.26(b)] is called the sliding mode or the forward shear mode. The shear stress is normal to the advancing crack front. Mode III [Fig. 3.26(c)] is called the tearing mode or transverse shear mode, with the shear stress in this case being parallel to the advancing crack front.

3.4.4 Stress Field in an Isotropic Material in the Vicinity of the Crack Tip

The stress components for the three fracture modes in an isotropic material are given below. In the case of anisotropic materials, these relations must be modified to permit the asymmetry of stress at the crack tip. K_I, K_{II}, and K_{III} indicate the modes I, II, and III, respectively.†
Mode I:

$$\begin{bmatrix} \sigma_{11} \\\\ \sigma_{22} \\\\ \sigma_{12} \end{bmatrix} = \frac{K_I}{\sqrt{2\pi r}} \cos\frac{\theta}{2} \begin{bmatrix} 1 - \sin\frac{\theta}{2} \sin\frac{3\theta}{2} \\\\ 1 + \sin\frac{\theta}{2} \sin\frac{3\theta}{2} \\\\ \sin\frac{\theta}{2} \cos\frac{3\theta}{2} \end{bmatrix} \tag{3.10}$$

† The derivation of these equations involves the choice of an appropriate Airy stress function which, upon differentiation, gives the stresses (Eq. 1.145); it is provided in more advanced texts, such as J. F. Knott, Fundamentals of Fracture Mechanics, Butterworths, London, 1973, p. 53.

$$\sigma_{13} = \sigma_{23} = 0$$

$$\sigma_{33} = 0 \qquad \text{plane stress}$$

$$\sigma_{33} = \nu(\sigma_{11} + \sigma_{22}) \qquad \text{plane strain}$$

Mode II:

$$
\begin{bmatrix} \sigma_{11} \\ \\ \sigma_{22} \\ \\ \sigma_{12} \end{bmatrix}
= \frac{K_{\text{II}}}{\sqrt{2\pi r}}
\begin{bmatrix}
-\sin\dfrac{\theta}{2}\left(2\cos\dfrac{\theta}{2}\ \cos\dfrac{3\theta}{2}\right) \\ \\
\sin\dfrac{\theta}{2}\ \cos\dfrac{\theta}{2}\ \cos\dfrac{3\theta}{2} \\ \\
\cos\dfrac{\theta}{2}\left(1 - \sin\dfrac{\theta}{2}\ \sin\dfrac{3\theta}{2}\right)
\end{bmatrix}
\tag{3.11}
$$

$$\sigma_{13} = \sigma_{23} = 0$$

$$\sigma_{33} = 0 \qquad \text{plane stress}$$

$$\sigma_{33} = \nu(\sigma_{11} + \sigma_{22}) \qquad \text{plane strain}$$

Mode III:

$$
\begin{bmatrix} \sigma_{13} \\ \\ \sigma_{23} \end{bmatrix}
= \frac{K_{\text{III}}}{\sqrt{2\pi r}}
\begin{bmatrix} -\sin\dfrac{\theta}{2} \\ \\ \cos\dfrac{\theta}{2} \end{bmatrix}
\tag{3.12}
$$

$$\sigma_{11} = \sigma_{22} = \sigma_{33} = \sigma_{12} = 0$$

3.4.5 Details of the Crack-Tip Stress Field in Mode I

Consider an infinite, homogeneous, and elastic plate containing a crack of length $2a$ (Fig. 3.27). The plate is subjected to a tensile stress σ far away from and normal to the crack. The stresses at a point (r, θ) near the crack tip are given by Eq. 3.10. Ignoring the subscript of K, we may write the stress components in expanded form as:

$$\sigma_{11} = \frac{K}{\sqrt{2\pi r}}\cos\frac{\theta}{2}\left(1 - \sin\frac{\theta}{2}\sin\frac{3\theta}{2}\right)$$

$$\sigma_{22} = \frac{K}{\sqrt{2\pi r}}\cos\frac{\theta}{2}\left(1 + \sin\frac{\theta}{2}\sin\frac{3\theta}{2}\right)$$

$$\sigma_{12} = \frac{K}{\sqrt{2\pi r}}\cos\frac{\theta}{2}\sin\frac{\theta}{2}\cos\frac{3\theta}{2} \tag{3.13}$$

$$\sigma_{13} = \sigma_{23} = 0$$

$$\sigma_{33} = 0 \quad \text{(plane stress)},$$

$$\sigma_{33} = \nu(\sigma_{11} + \sigma_{22}) \quad \text{(plane strain)}$$

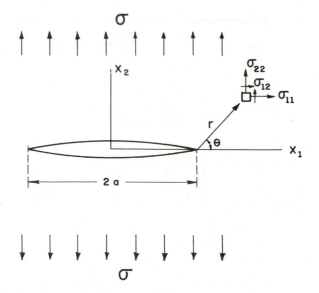

Figure 3.27 Infinite, homogeneous, and elastic plate containing a through-the-thickness central crack of length $2a$, subjected to a tensile stress σ.

where K, the stress intensity factor, for the infinite plate shown above, is given by

$$K = \sigma\sqrt{\pi a} \tag{3.14}$$

The stress intensity factor K has the units $(\text{N/m}^2)\sqrt{\text{m}}$ or $\text{Pa}\sqrt{\text{m}}$ or $\text{Nm}^{-3/2}$.

It should be mentioned that Eq. 3.13 is applicable in the region $r \ll a$ (i.e., in the vicinity of the crack tip). For $r \sim a$, higher-order terms must be included.

For a thin plate, one has the plane-stress conditions and $\sigma_{33} = \sigma_{13} = \sigma_{23} = 0$. For a thick plate (infinite in the thickness direction) there exist plane-strain conditions [i.e., $\sigma_{33} = \nu(\sigma_{11} + \sigma_{22})$ and $\sigma_{13} = \sigma_{23} = 0$].

Consider Eq. 3.13. The right-hand side has three quantities, K, r, and $f(\theta)$. The terms r and $f(\theta)$ describe the stress distribution around the crack tip. These two characteristics of the stress field at the crack tip [i.e., $\sqrt{r}$ dependence and $f(\theta)$] are identical for all cracks in two- or three-dimensional elastic solids. The stress intensity factor K includes the influence of the applied stress σ and the appropriate crack dimensions, in this case the half crack length a. Thus, K will characterize the external conditions (i.e., nominal applied stress σ and the half crack length a) that correspond to fracture when stresses and strains at the crack tip reach a critical value. This value of K is characterized as critical and is designated as K_c. It turns out, as we shall see later, that K_c depends on specimen dimensions. In the case of thin sample (plane stress conditions), K_c depends on the sample thickness, whereas in the case of a sufficiently thick sample (plane strain conditions), it is independent of specimen thickness and is designated as K_{Ic}.

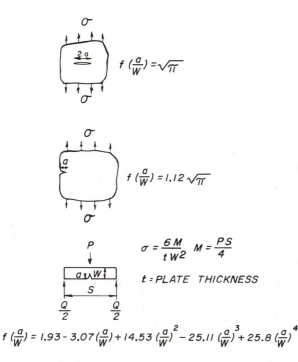

Figure 3.28 Some common load and crack configurations and the corresponding expressions for $f(a/w)$ in $K = \sigma\sqrt{a}\, f(a/w)$.

The stress intensity factor K measures the amplitude of the stress field around the crack tip and should not be confused with the stress concentration factor K_t discussed in Section 3.2.2. It is important to distinguish between K and K_c or K_{Ic}. The stress intensity factor K is a quantity, determined analytically or not, that varies as a function of configuration (i.e., the crack geometry and the manner of application of external load). Thus, the analytical expression for K varies from one system to another. However, once K attains its critical value, K_{Ic}, in plane strain for a given system and material, it is essentially a constant for all the systems made of this material. The difference between K_c and K_{Ic} is the following: K_c depends on the thickness, whereas K_{Ic} is independent of the thickness. The forms of K for various load and crack configurations have been calculated and the reader can find them in [21–25]. Some of the more common configurations and the corresponding expressions for K are presented in Fig. 3.28.

For samples of finite dimensions, the general practice is to consider the solution for an infinite plate and modify it by an algebraic or trigonometric function that would make the surface tractions vanish. Thus, for a central through-the-thickness crack of length $2a$, in a plate of width W, we have

$$K = \sigma\left(w \tan \frac{\pi a}{W}\right)^{1/2} \tag{3.15}$$

For the same crack in an infinite plate, we have

$$K = \sigma\sqrt{\pi a}$$

If we expand Eq. 3.15, we get

$$K = \sigma\, W^{1/2}\left(\frac{\pi a}{W} + \frac{\pi^3 a^3}{3\,W^3} + \cdots\right)^{1/2}$$

$$= \sigma\sqrt{\pi a}\left(1 + \frac{\pi^2 a^2}{3\,W^2} + \cdots\right)^{1/2}$$

Thus, for an infinite solid, $a/W = 0$, and we have $K = \sigma\sqrt{\pi a}$, as expected. For an edge crack in a semi-infinite plate, we have $K = 1.12\,\sigma\sqrt{\pi a}$. The factor 1.12 here takes care of the fact that stresses normal to the free surface must be zero.

At this point it would be appropriate to make some comments on the limitations of LEFM. It was pointed out earlier that the expressions for stress components (Eqs. 3.10–3.12) are valid only in the neighborhood of the crack tip. The reader will have noticed that these stress components tend to infinity as we approach the crack tip (i.e., as r goes to zero). Now, there does not exist a material in real life that can resist an infinite stress. The material in the neighborhood of the crack tip, in fact, would inevitably deform plastically. These expressions for stress components based on linear elasticity theory are not valid in the plastic zone at the crack tip. The deformation process in a plastic zone, as is well known, will be a sensitive function of the microstructure, among other things. However, in spite of ignorance of the exact nature of the plastic zone, the LEFM treatment is valid for low-enough stresses such that the plastic zone size at the crack tip is small with respect to the crack length and dimensions of the sample. We shall see in the next section how to incorporate a correction term for the presence of a plastic zone at the crack tip.

3.4.6 Plastic Zone Size Correction

Equations 3.10–3.12 give a $\sqrt{r}$ singularity; that is, σ_{11}, σ_{22}, and σ_{12} go to infinity when $\sqrt{r}$ goes to zero. For a great majority of materials local yielding will occur at the crack tip which would relax the peak stresses. As we shall see below, the utility of the elastic stress field equations is not affected by the presence of this plastic zone as long as the nominal stress in the material is below the general yielding stress of the material. When yielding occurs at the crack tip, it becomes blunted; that is, the crack surfaces separate without any crack extension (see Fig. 3.29). The plastic zone (radius r_y) will be embedded in an elastic stress field. Outside and far away from the plastic zone, the elastic stress field "sees" the crack and the perturbation due to plastic zone, as if there were present a crack in an elastic material with the leading edge of the crack situated inside the plastic zone. A crack of length $2(a + r_y)$ in an ideal elastic material produces stresses almost identical to elastic stresses in a locally yielded member outside the plastic zone.

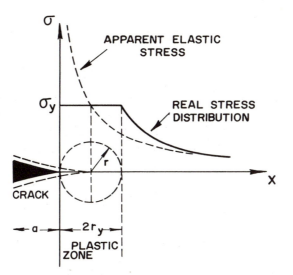

Figure 3.29 Plastic zone correction (after Irwin). The effective crack length is $(a + r_y)$.

If the stress applied is too large, the plastic zone increases in size in relation to the crack length and the elastic stress field equations lose precision. When the whole of the reduced section yields, the plastic zone spreads to the edges of the sample and K does not have any validity as parameter defining the stress field.

When the plastic zone is small in relation to the crack length, it can be visualized as a cylinder (Fig. 3.29) of radius r_y at the crack tip. From Eq. 3.13, for $\theta = 0$, $r = r_y$, and $\sigma_{22} = \sigma_y$, the yield stress, we can write

$$\sigma_y = \frac{K}{\sqrt{2\pi r_y}}$$

and the plastic zone radius to the first approximation will be

$$r_y = \frac{1}{2\pi}\left(\frac{K}{\sigma_y}\right)^2 \tag{3.16}$$

In fact, the plastic zone radius is a little bigger than $(1/2\pi)(K/\sigma_y)^2$ due to redistribution of load in the vicinity of the crack tip. Irwin,[26] taking into account the plastic constraint factor in the case of plane strain, gave the following expressions for the plastic zone size:

$$r_y \approx \frac{1}{2\pi}\left(\frac{K}{\sigma_y}\right)^2 \qquad \text{plane stress}$$

$$r_y \approx \frac{1}{6\pi}\left(\frac{K}{\sigma_y}\right)^2 \qquad \text{plane strain}$$

Thus, the center of perturbation, the apparent crack tip, is located at a distance r_y from the real crack tip. The effective crack length is

$$(2a)_{\text{eff}} = 2(a + r_y)$$

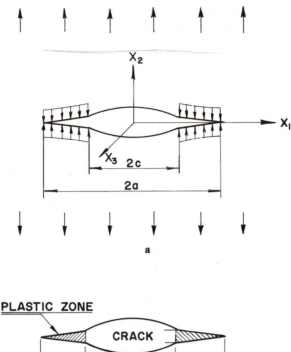

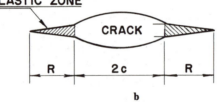

Figure 3.30 Dugdale–Bilby–Cottrell–Swinden model of a crack.

Substitution of a by $(a + r_y)$ in the elastic stress field equations gives an adequate adjustment for the crack-tip plasticity under conditions of small-scale yielding. With this adjustment, the stress intensity factor, K, is useful for characterization of the fracture conditions.

There is another model for the plastic zone at the crack tip for the plane-stress case, called the Dugdale—BCS crack model.[27,28] In this model the plasticity spreads out at the two ends of a crack in the form of narrow strips of length R (Fig. 3.30). These narrow plastic strips in front of the actual crack tips are under the yield stress σ_y that tends to close the crack. Mathematically, the internal crack of length $2c$ is allowed to extend elastically a distance $2a$ and then internal stress is applied to reclose the crack in this region. Dugdale showed, combining the internal stress field surrounding the plastic enclaves with the external stress field associated with the applied stress σ acting on the crack, that

$$\frac{c}{a} = \cos \frac{\pi \sigma}{2\sigma_y}$$

From this relation one notes that as $\sigma \rightarrow \sigma_y$, $c/a \rightarrow 0$ or $a \rightarrow \infty$ (i.e., general yielding occurs). On the other hand, as σ/σ_y decreases, we can write (using the series expansion for cosine),

$$\frac{c}{a} = 1 - \frac{\pi^2 \sigma^2}{8\sigma_y^2} + \cdots$$

Noting that $a = c + R$ and using the binomial expansion, we have

$$\frac{c}{a} = \frac{c}{c+R} = \left(1 + \frac{R}{c}\right)^{-1} = 1 - \frac{R}{c} + \cdots$$

Thus, for $\sigma \ll \sigma_y$, we have

$$\frac{R}{c} \simeq \frac{\pi^2}{8}\left(\frac{\sigma}{\sigma_y}\right)^2$$

or

$$R \approx \frac{\pi}{8}\left(\frac{K}{\sigma_y}\right)^2 \tag{3.17}$$

Comparing Dugdale's equation (3.17) with that of Irwin, Eq. 3.16, we note that there is good agreement between the two ($\pi/8 \approx 1/\pi$). In fact, the plastic zone size varies with θ also. A formal representation of the plastic zone at the crack front through the plate thickness is shown in Fig. 3.31. The reader is also advised to see Exercises 3.10 and 3.11 at the end of the chapter.

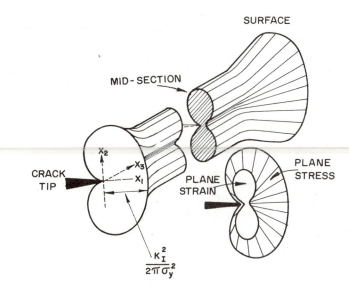

Figure 3.31 Formal representation of the plastic zone at the crack tip in a through-the-thickness crack in a plate.

3.4.7 Variation of Fracture Toughness with Thickness

The elastic stress state is markedly influenced by the plate thickness, as indicated by Eq. 3.13. The material in the plastic zone deforms in such a way that its volume is kept constant. Thus, the large deformations in the x_1 and x_2 directions tend to induce a contraction in the x_3 direction (parallel to the crack front or the plate thickness direction) which is resisted by the surrounding elastic material (Fig. 3.31). We make a dimensional analysis. As the elastic material surrounding the plastic zone is the primary source of constraint, the size of the plastic zone, $2r_y$, will be the significant dimension to be compared with the plate thickness B. The ratio of plate thickness B to plastic zone size $2r_y$ is given by

$$\frac{B}{2r_y} = \pi \frac{B}{(K_c/\sigma_y)^2}$$

and this would be a convenient parameter to characterize the variation of fracture toughness, K_c, with thickness. Data for Al 7075-T6 and H-11 steel are plotted in Fig. 3.32 in the form of K_c/σ_y versus $B/(K_c/\sigma_y)^2$.[29,30] Observe that when $B/(K_c/\sigma_y)^2$ is greater than $1/\pi$ (i.e., $B \gg 2r_y$), the fracture toughness value K_c does not change with B. Apparently, beyond a thickness $B \gg 2r_y$, the constraint in the thickness direction (x_3) is completely effective and additional plate thickness does not change K_c. This particular value of K_c that is independent of specimen thickness is labeled the fracture toughness of the material and the symbol K_{Ic} is used to denote it.

On the other extreme, when the ratio $B/(K_c/\sigma_y)^2$ is much smaller than $1/\pi$ (i.e., $B \ll 2r_y$), we expect the fracture toughness to increase linearly with the plate thickness. In the region of $B/(K_c/\sigma_y)^2 = 1/\pi$ corresponding to $B = 2r_y$, the data for both the materials show a rapid fall to a constant level of K_{Ic}. This decrease in the peak value of K_c (Fig. 3.32) to the K_{Ic} level represents a change in the fracture mode from that of a plane stress type to

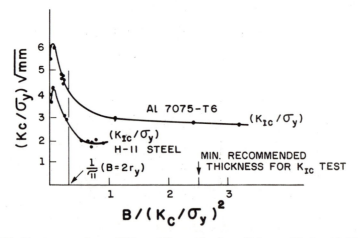

Figure 3.32 Fracture toughness (K_c) variation with plate thickness (B) for Al 7075-T6 and H-11 steel. (Reprinted with permission from [29], p. 144, and [26], p. 418.)

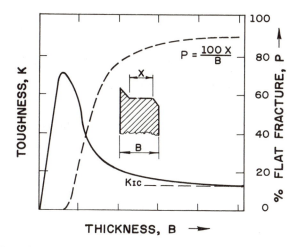

Figure 3.33 Schematic variation of fracture toughness K_c and percentage of flat fracture P with the plate thickness B.

that of a plane strain condition. The fracture in a relatively thin plate (plane stress) usually consists of a certain fraction of slant fracture (high energy) and another fraction of flat fracture (low energy). In general, with increasing specimen thickness, the percentage of slant fracture decreases and the energy necessary for crack propagation also decreases—hence the fall in the K_c value. At a certain critical specimen thickness, the crack propagates under plane strain conditions and the stress intensity factor reaches the minimum value designated as K_{Ic}. Figure 3.33 shows schematically the variation of K_c and the percentage of flat fracture P with the plate thickness B. K_{Ic} is especially relevant in the material evaluation, as it is a constant that is essentially independent of the specimen dimensions. One may say that the relation between K_{Ic} and K_c is the same as that existing between the yield stress σ_y and the strength σ. Table 3.2 gives representative values of the fracture toughness K_{Ic} for some metals and alloys.

TABLE 3.2 K_{Ic} and σ_y of Some of Metals and Alloys at Ambient Temperature

Material	σ_y (MPa)	K_{Ic} (MPa $\sqrt{m}$)
Al–4% Cu	405	26
Al–3% Mg–7% Zn	500	25
Ti–6% Al–1% V	850	60
Ti–6% Al–4% V	1020	52
Stainless steel (18–8)	340	200
A533 steel	450	120
(0.25% C, 1.3% Mn)		
12% Cr steel	1550	50
Maraging steel	1930	74
(18% Ni, 8% Co, 4% Mo)		
Cast iron	250	20

3.4.8 Relation Between *G* and *K*

The elastic strain energy release rate with respect to the crack area or the crack extension force *G* (sec Section 3.2.4) is a parameter that can be used to characterize the fracture condition in a manner similar to the stress intensity factor *K*. From Eqs. 3.8 and 3.14 one can see that these two parameters are related in the following way for elastic, homogeneous, and isotropic materials:

$$GE = K^2 \qquad \text{plane stress}$$
$$GE = K^2(1 - \nu^2) \qquad \text{plane strain} \tag{3.18}$$

The measurement of *G* is described in the next section.

3.5 CRACK EXTENSION FORCE AND ITS MEASUREMENT

The concept of crack extension force *G* due to Irwin is considered in some detail in this section. *G* can be interpreted as a generalized force. One can say that fracture mechanics is the study of the response of a crack (measured in terms of crack velocity) to the application of various magnitudes of this crack extension force. Let us consider an elastic body of uniform thickness *B* containing a through-the-thickness crack of length 2*a*. Let the body be loaded as shown in Fig. 3.34(a). With increasing load *P*, the displacement *e* of the loading point increases. The load-displacement diagram is shown in Fig. 3.34(b). At point 1 we have the load as P_0 and displacement as e_0. Now let us consider a "gedanken" experiment in which the crack extends by a small increment, δa. Due to this small increment, δa, in crack extension, the loading point is

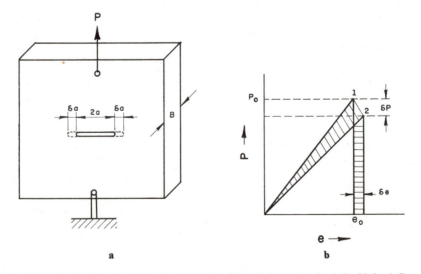

Figure 3.34 (a) Elastic body containing a crack of length 2*a* under load *P*; (b) load *P* versus displacement *e* diagram.

displaced by δe while the load falls by δP. Now, before the crack extension, the potential energy stored in body was

$$U_1 = \tfrac{1}{2}Pe$$

and is represented by the area of the triangle through point 1 in Fig. 3.34. After the crack extension, the potential energy stored in the body is

$$U_2 = \tfrac{1}{2}(P - \delta P)(e + \delta e)$$

and is represented by the area of the triangle passing through the point 2 in Fig. 3.34. In this process of crack extension, the change in potential energy, $U_2 - U_1$, is given by the difference in the areas of the two crosshatched regions in Fig. 3.34. Considering the small increment δa in crack length, we can write G, the crack extension force per unit length, as

$$GB\,\delta a = U_2 - U_1 = \delta U$$

The change in elastic strain energy with respect to the crack area, in the limit of the area going to zero, equals the crack extension force, that is,

$$G = \lim_{\delta A \to 0} \frac{\delta U}{\delta A}$$

where $\delta A = B\,\delta a$.

It is convenient to evaluate G in terms of compliance c of the sample defined as

$$e = cP \tag{3.19}$$

Now

$$\delta U = U_2 - U_1 = \tfrac{1}{2}(P - \delta P)(e + \delta e) - \tfrac{1}{2}Pe$$

or

$$\delta U = \tfrac{1}{2}P\,\delta e - \tfrac{1}{2}e\,\delta P - \tfrac{1}{2}\,\delta P\,\delta e \tag{3.20}$$

Differentiating Eq. 3.19, we have

$$\delta e = c\,\delta P + P\,\delta c \tag{3.21}$$

Substituting Eq. 3.21 in Eq. 3.20, we have

$$\delta U = \tfrac{1}{2}Pc\,\delta P + \tfrac{1}{2}P^2\,\delta c - \tfrac{1}{2}e\,\delta P - \tfrac{1}{2}e(\delta P)^2 - \tfrac{1}{2}P\,\delta P\,\delta c \tag{3.22}$$

Remembering that $e = cP$ and ignoring the higher-order product terms, we can write

$$\delta U = \tfrac{1}{2}Pc\,\delta P + \tfrac{1}{2}P^2\,\delta c - \tfrac{1}{2}Pc\,\delta P$$

or

$$\delta U = \tfrac{1}{2}P^2\,\delta c \tag{3.23}$$

Then

$$G = \lim_{\delta A \to 0} \frac{\delta U}{\delta A} = \lim_{\delta A \to 0} \frac{\frac{1}{2} P^2 \, \delta c}{\delta A}$$

or

$$G = \tfrac{1}{2} \frac{P^2}{B} \frac{\delta c}{\delta a} \tag{3.24}$$

From Eq. 3.24 we see that G is independent of the rigidity of the surrounding structure or the test machine. It depends only on the change in compliance of the cracked member due to crack extension. Thus, to obtain G for a specimen sample, all we need do is to determine the compliance of the specimen as a function of crack length and measure the gradient of the resultant curve, dc/da, at the appropriate initial crack length (Fig. 3.35).

This method is more useful for relatively small test samples where exact measurements can be made in the laboratory. One of the important uses of Eq. 3.24 is that it provides a value of G (or K) for complex structures that have not been (or cannot be) treated analytically. An experimental determination of G_c, the critical crack extension force, using this equation requires the value of fracture load (measured during experiment) and the value of dc/da. The com-

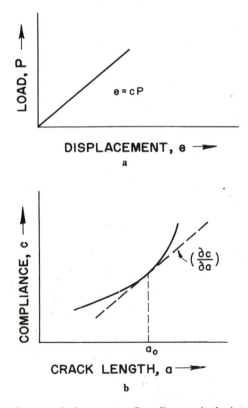

Figure 3.35 (a) Load P versus displacement e. Compliance c is the inverse of the slope of this curve. (b) Compliance c versus crack length a. a_0 is the initial crack length.

pliance can be measured by calibrating a series of samples with different crack lengths. We obtain a c versus a diagram and dc/da is evaluated as the slope at the appropriate initial crack length.

3.6 IMPORTANCE OF K_{Ic} IN PRACTICE

K_{Ic} is the critical stress intensity factor under conditions of plane strain ($\epsilon_{33} = 0$), which is characterized by small-scale plasticity at the crack tip. The material is fully constrained in the thickness direction. When determined under these rigorous conditions, K_{Ic} will be a material constant. Thus, when one needs to characterize materials by their tenacity (in the same way that one characterizes materials by their ultimate tensile strength or tensile yield strength), only valid K_{Ic} data should be considered.

K_c is the critical stress intensity factor under conditions of plane stress ($\sigma_{33} = 0$), which is characterized by large plasticity at the crack tip. In this case, the through-thickness constraint is negligible. K_c values can be up to two times greater than the K_{Ic} values of the same material. K_{Ic} depends on temperature T, strain rate $\dot{\epsilon}$, and on metallurgical variables, in general. For example, dependence of K_{Ic} on tensile strength and on sulfur level is shown in Fig. 3.36. Notice that as expected, K_{Ic} decreases monotonically with increases in tensile strength or sulfur content. Figure 3.37 shows that the same holds for K_{Ic} as a function of the yield strength.[32] K_c depends on these variables,

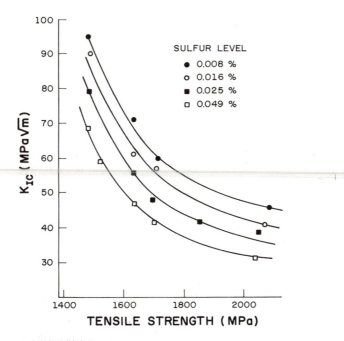

Figure 3.36 Variation of fracture toughness K_{Ic} with tensile strength and sulfur content in a steel. (Adapted with permission from [31], p. 982.)

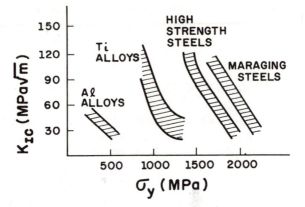

Figure 3.37 Variation of fracture toughness K_{Ic} with yield strength σ_y for a series of alloys. (Adapted with permission from [32], p. 270).

but it also depends on the thickness B of the sample. Besides, for a given set of conditions (T, $\dot{\epsilon}$, B), the K_c value also depends on the initial crack length a_0.

For a number of structural materials used in real structures, service conditions in terms of temperature, strain rate, and thickness are such that they are in a plane-stress condition and not in a plane-strain condition. Thus, there is a great interest in determining the fracture toughness of materials under plane-stress conditions; some of these attempts are crack opening displacement, J integral, and R curve. These are nothing but attempts at extending the LEFM to elastoplastic regimes. We shall describe briefly these frontier areas of fracture mechanics.

3.7 CRACK OPENING DISPLACEMENT[33,34]

The development of a plastic zone at the crack tip results in a displacement of the faces without crack extension. This relative displacement of opposite crack edges is called the crack opening displacement (COD) (Fig. 3.38). Wells[33] suggested that when this displacement at the crack tip reaches a critical value δ_c, fracture ensues.

LEFM is applicable only when the plastic zone is small in relation to the crack length (i.e., well below the yield stress and in plane strain). Consider small cracks in brittle materials. We have

$$\sigma_c = K_{Ic}(\sqrt{\pi a})^{-1} \qquad \text{as } a \longrightarrow 0, \quad \sigma_c \longrightarrow \infty$$

But this, as we very well know, does not occur. Instead, a plastic zone develops and may extend through the section such that

$$\sigma_{\text{net}} = \sigma \frac{W}{W - a} \geq \sigma_y$$

where W is the width of sample and σ_y is the yield stress. In practice, $\sigma_c \leq 0.66\sigma_y$ for the K_{Ic} validity.

In more ductile materials, the critical stress predicted by LEFM will be higher than σ_y. One can use the concept of COD in such cases. In the elastic case (Fig. 3.39)

$$\text{COD} = \Delta = \frac{4\sigma}{E} \sqrt{(a^2 - x^2)} \tag{3.25}$$

At the center of the crack ($x = 0$), the maximum opening is

$$\Delta_{\max} = \frac{4\sigma a}{E}$$

Applying the plastic zone correction, we have, from Eq. 3.25,

$$\Delta = \frac{4\sigma}{E} \sqrt{(a + r_y)^2 - x^2}$$

where $(a + r_y)$ is the effective crack length.

The crack-tip opening displacement (CTOD), δ, is given for $x = a$ and $r_y \ll a$ as

$$\delta = \frac{4\sigma}{E} \sqrt{2ar_y} \tag{3.26}$$

A displacement of the origin to the crack tip gives a general expression for the crack opening:

$$\Delta = \frac{4\sigma}{E} \sqrt{2a_{\text{eff}}r_y}$$

Substituting $r_y = \sigma^2 a / 2\sigma_y^2$ (Eq. 3.16) gives

$$\delta = \frac{4}{\pi} \frac{K_I^2}{E\sigma_y} \tag{3.27}$$

Equation 3.27 is valid in the LEFM regime and fracture occurs when $K_I = K_{Ic}$, which should occur, according to Eq. 3.27, at a constant value of δ_c, a material constant.

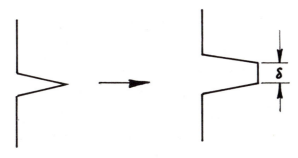

Figure 3.38 Crack opening displacement.

The use of COD criterion demands measurement of δ_c. Direct measurement of δ_c is not easy. An indirect way is the following. We have

$$\Delta = \frac{4\sigma}{E} \sqrt{(a + r_y)^2 - x^2}$$

$$= \frac{4\sigma}{E} \sqrt{a^2 + 2ar_y + r_y^2 - x^2}$$

Ignoring the r_y^2 term and using the relationship of Eq. 3.26, we can write

$$\Delta = \frac{4\sigma}{E} \left(a^2 - x^2 + \frac{E^2}{16\sigma^2} \delta^2 \right)^{1/2} \tag{3.28}$$

According to Eq. 3.28, δ can be measured indirectly from a COD measurement (e.g., at $x = 0$, at the center of the crack) without making any simplifications about the plastic zone size correction. Δ can be measured by means of a clip gage.

Another way of measuring δ is to use the equations for δ, according to Dugdale's model of the crack (see Section 3.4.6). Thus, according to Dugdale's model,

$$\delta = \frac{8\sigma_y a}{\pi E} \log \sec \frac{\pi \sigma}{2\sigma_y}$$

Expanding the log sec function in series, we get

$$\delta = \frac{8\sigma_y a}{\pi E} \left[\frac{1}{2} \left(\frac{\pi \sigma}{2\sigma_y} \right)^2 + \frac{1}{12} \left(\frac{\pi \sigma}{2\sigma_y} \right)^4 + \cdots \right]$$

For $\sigma \ll \sigma_y$, we can write

$$\delta = \frac{\pi \sigma_a^2}{E \sigma_y} = \frac{G_I}{\sigma_y} \tag{3.29}$$

Comparing Eq. 3.29 with Eq. 3.27, we note that the difference is in the factor $4/\pi$, which comes from the plastic zone size correction. In general,

$$\delta = \frac{G_I}{\lambda \sigma_y} = \frac{K_I^2 (1 - \nu^2)}{E \lambda \sigma_y} \qquad \text{for plane strain} \tag{3.30}$$

The factor $(1 - \nu^2)$ should be ignored in the case of plane stress.

In the literature, we encounter various λ values. These depend on the exact location where CTOD is determined (i.e., the exact location of the crack tip). Wells[34] suggested that experimentally $\lambda \approx 2.1$ for compatibility with LEFM (i.e., limited plasticity). For cases involving extensive plasticity, the engineering design application approach is to take $\lambda \approx 1$.

Thus, at unstable fracture, $G_{Ic} = \lambda \sigma_y \delta c$. The important point about COD is that theoretically, δ_c can be computed for both elastic and plastic materials, whereas G_{Ic} is restricted only to the elastic regime. The COD thus allows one to treat fracture under plastic conditions. A word of caution is in

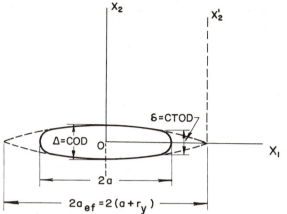

Figure 3.39 Relationship between crack opening displacement (COD, Δ); crack-tip opening displacement (CTOD, δ); crack length (2a); and plastic zone size (r_y).

order here. Figure 3.39 presents a comparison between COD and CTOD. We should realize that the strain fields and crack opening displacements associated with a crack tip will be different for different specimen configurations. Thus, we cannot define a single critical COD value for a given material in a manner equivalent to that of K_{Ic}, as the COD value will be affected by the geometry of the test specimen.[35]

3.8 J INTEGRAL[36-42]

The J integral, first proposed and used for dislocations by Eshelby[36] and applied, independently, to cracks by Cherepanov[37] and Rice,[38] is defined as a line integral, independent of path, along a curve Γ surrounding the crack tip. Mathematically, Rice[38] has shown that

$$J = \int_{\Gamma} \left(W \, dy - T \frac{\partial u}{\partial x} \, ds \right) \tag{3.31}$$

where W is the strain energy density function, T is the traction vector perpendicular to the surface Γ, u is the displacement in the x_1 direction and ds is an arc along Γ (Fig. 3.40).

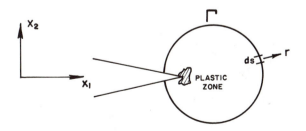

Figure 3.40 Definition of the J integral.

From a physical point of view, the J integral represents the difference in the potential energy of identical bodies containing cracks of length a and $a + da$, that is,

$$J = -\frac{1}{B}\frac{\partial U}{\partial a} \tag{3.32}$$

where U is the potential energy, a the crack length, and B the plate thickness.

U is equal to area under the load versus displacement curve. Figure 3.41 shows this interpretation, where the shaded area is $\partial U = JB\, \partial a$. Like G_{Ic}, J_{Ic} measures the critical energy associated with the initiation of crack growth, but in this case accompanied by substantial plastic deformation. In fact, Begley and Landes[39] showed the formal equivalence of J_{Ic} and G_{Ic} by measuring the former from small fully plastic specimens and the latter from large elastic specimens satisfying the plane strain conditions for the LEFM test.

The path independence of the J integral, together with this interpretation in terms of energy, makes it a powerful analytical tool. The J integral is path independent in the case of materials behaving elastically, linear or nonlinear. When there occurs extensive plastic deformation, the current practice is to assume that the plastic yielding can be described by deformation theory of plasticity. According to this theory, stresses and strains are functions only of the point of measurement and not of the path taken to get to that point. As in the case of slow, stable crack growth, there will be relaxation of stresses at the crack tip, so there will be a violation of this postulate. Thus, the use of J

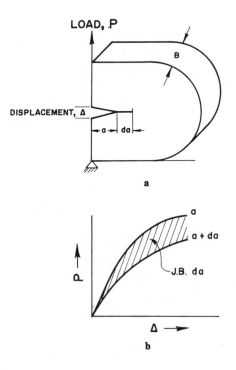

Figure 3.41 Physical interpretation of the J integral. The J integral represents the difference in potential energy (shaded area) of identical bodies containing cracks of length a and $a + da$.

Mechanics of Deformation　Part I

should be limited to the initiation of crack propagation, by stable or unstable processes. Studies using incremental plasticity or flow theories with finite elements indicate the path independence of the J integral, but there is no theoretical proof as yet and the problem remains unresolved.

3.9 *R* CURVE[43]

The R curve characterizes the resistance of a material to fracture during slow and stable propagation of a crack. An R curve graphically represents this resistance to crack propagation of a material as a function of crack growth. With increasing load in a cracked structure, the crack extension force G at the crack tip also increases (see Eq. 3.24). However, the material at the crack tip presents a resistance R to crack growth. According to Irwin, the failure will occur when the rate of change of the crack extension force $(\partial G/\partial a)$ equals the rate of change of this resistance to crack growth in the material $(\partial R/\partial a)$. The resistance of the material to crack growth, R, increases with an increase in plastic zone size. The plastic zone size increases nonlinearly with a; thus, R will also be expected to increase nonlinearly with a. G increases linearly with a. Figure 3.42 shows the instability criterion: the point of tangency between the curves G versus a and R versus a. Figure 3.42(a) shows the R curve for a brittle material and Fig. 3.42(b) shows the R curve for a ductile material. Crack extension occurs for $G > R$. Consider the G line for a stress σ' [Fig. 3.42(b)]. At this stress σ', the crack in this material will grow only from a_0 to a' since $G > R$ for $a < a'$. $G < R$ for $a > a'$ and the crack does not extend beyond a'. As the load is increased, the G-line position changes, as indicated in the figure. When G becomes tangent to R, unstable fracture ensues. The R curve for a brittle material [Fig. 3.42(a)] is a "square" curve and the crack does not extend at all until the contact is reached, at which point $G = G_c$ and the unstable fracture follows.

The R-curve method is another version of the Griffith energy balance. One can conveniently make this kind of analysis if an analytical expression

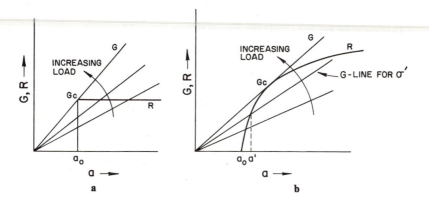

Figure 3.42 R curves for (a) brittle material and (b) ductile material.

for the R curve is available. Experimental determination of R curves, however, is complicated and time consuming.

EXERCISES

3.1. A thin plate is rigidly fixed at its edges (see Fig. E3.1). The plate has a height L and thickness (normal to the plane of figure) t. A crack moves from left to right through the plate. Every time the crack moves a distance Δx, two things happen:
1. Two new surfaces (γ, specific surface energy) are created.
2. The stress falls to zero behind the advancing crack front in a certain volume of the material.
Obtain an expression for the critical stress necessary for crack propagation in this case. Explain the significance of this expression.

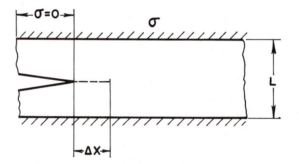

Figure E3.1

3.2. Compute the theoretical strength of iron ($\gamma = 1500$ mJ/m²). Compare this value with the maximum strength obtainable in commercially produced ultra-high-strength steels.

3.3. Explain why FCC metals generally do not fracture in a brittle manner at low temperatures, whereas BCC metals generally do.

3.4. Gray cast iron is a rather cheap material containing between 2.4 and 4% C and 1 and 3% Si, most of the carbon being present in the form of graphite flakes. This material has been used for centuries for engine cylinder blocks, boilers, lamp-posts, and structure and machinery foundation parts. But, invariably, only its good *compressive strength* is taken into account. Where even a reasonable tensile strength and toughness are required, the cast iron is rejected by engineers. Why?

3.5. Consider a brittle material with $\gamma = 1$ J/m² and $E = 100$ GPa. What is the breaking strength of this material if it contains cracklike defects as long as 1 mm? Should it be possible to increase γ to 3000 J/m², what would be the breaking strength for a 1-mm-long crack?

3.6. Explain the cracking of a metal due to thermal shock.

3.7. A microalloyed steel, quenched and tempered at 250°C, has a yield strength (σ_y) of 1750 MPa and a plane-strain fracture toughness K_{Ic} of 43.50 MPa $\sqrt{m}$. What is the largest disk type inclusion, oriented most unfavorably, that can be tolerated inside this steel at an applied stress of $0.5\sigma_y$?

3.8. Consider a maraging steel plate of thickness (B) 3 mm. Two specimens of width (W) equal to 50 mm and 5 mm were taken out of this plate. What is the largest through-the-thickness crack that can be tolerated in the two cases at an applied stress of $\sigma = 0.6\sigma_y$, where σ_y (yield stress) $= 2.5$ GPa? The plane-strain fracture toughness K_{Ic} of the steel is 70 MPa $\sqrt{m}$. What are the critical dimensions in the case of a single-edge notch specimen?

3.9. A steel plate containing a through-the-thickness central crack of length 15 mm is subjected to a stress of 350 MPa normal to the crack plane. The yield stress of the steel is 1500 MPa. Compute the size of plastic zone and the effective stress intensity factor.

3.10. An AISI 4340 steel plate has a width W of 30 cm and has a central crack $2a$ of 3 mm. The plate is under a uniform stress σ. The expression for the stress intensity factor is

$$K = f\left(\frac{a}{w}\right)\sigma\sqrt{\pi a}$$

where the function $f(a/w)$ is as follows:

a/w	$\leqslant 0.074$	0.207	0.275	0.337	0.410	0.466	0.535	0.592
$f(a/w)$	1.00	1.03	1.05	1.09	1.13	1.13	1.25	1.33

This steel has a K_{Ic} value of 50 MPa $\sqrt{m}$ and a service stress of 1500 MPa. Compute the maximum crack size that the steel may have without failure.

3.11. An infinitely large plate containing a central crack of length $2a = 50/\pi$ mm is subjected to a nominal stress of 300 MPa. The material yields at 500 MPa. Compute:
(a) The stress intensity factor at the crack tip.
(b) The size of plastic zone at the crack tip.
Comment on the validity of Irwin's correction for the plastic zone size in this case.

3.12. Compute the approximate size of plastic zone r_y for an alloy that has a Young's modulus $E = 70$ GPa, yield strength $\sigma_y = 500$ MPa, and toughness $G_c = 20$ kJ/m².

3.13. The plastic zone size at the crack tip in the general plane-stress case is given by

$$r_y = \frac{K_I^2}{2\pi\sigma_y^2}\cos^2\frac{\theta}{2}\left(4 - 3\cos^2\frac{\theta}{2}\right)$$

(a) Determine the radius of the plastic zone in the direction of the crack.
(b) Determine the angle θ at which the plastic zone is the largest.

3.14. For the plane-strain case, the expression for the plastic zone size is

$$r_y = \frac{K_I^2}{2\pi\sigma_y^2}\cos^2\frac{\theta}{2}\left\{4[1 - \nu(1 - \nu)] - 3\cos^2\frac{\theta}{2}\right\}$$

(a) Show that the expression above reduces to the one for plane stress.
(b) Make plots of the plastic zone sizes as a function of θ for $\nu = 0$, $\nu = \frac{1}{3}$, and $\nu = \frac{1}{2}$. Comment on the size and form of the zone in the three cases.

3.15. "Leak before break" is an important concept in regard to pressure vessels. It is important that a pressure vessel leaks before it breaks should a through-the-thickness crack develop (i.e., the material should not suffer a catastrophic failure before the crack is long enough to cause a leak). Assume that a crack will cause a leak when its length is twice the plate thickness. Determine:

(a) The maximum tolerable stress in a maraging steel pressure vessel with a 5-mm wall thickness.

(b) The value of K_{Ic} required to permit a stress level of $0.6\sigma_y$.

3.16. A 25-mm² bar of cast iron contains a crack 5 mm long and normal to one face. What is the load required to break this bar if it is subjected to three-point bending with the crack toward the tensile side and the supports 250 mm apart?

3.17. 300-M steel, commonly used for airplane landing gear, has a G_c value of 10 kN/m. A nondestructive examination technique capable of detecting cracks that are 1 mm long is available. Compute the stress level that the landing gear can support without failure.

3.18. Sih[22] has presented a more general treatment of fracture mechanics which allows one to predict the direction in which the crack will propagate at the onset of instability. Sih defines a strain energy density function S as follows:

$$S = a_{11}k_1^2 + 2a_{12}k_1k_2 + a_{22}k_2^2 + a_{33}k_3^2$$

where

$$a_{11} = \frac{1}{16G}\,[(3 - 4\nu - \cos\theta)(1 + \cos\theta)]$$

$$a_{12} = \frac{1}{16G}\,\{(2\sin\theta)[\cos\theta - (1 - 2\nu)]\}$$

$$a_{22} = \frac{1}{16G}\,[4(1 - \nu)(1 - \cos\theta) + (1 + \cos\theta)(3\cos\theta - 1)]$$

$$a_{33} = \frac{1}{4G}$$

Consider a plate containing a through-the-thickness crack and under a shear stress τ (see Fig. E3.18). For this loading condition only k_2 is nonzero and is equal to $\tau\sqrt{a}$. Derive an expression for the critical strain energy density S_c for crack propagation and compute the angle of crack propagation direction.

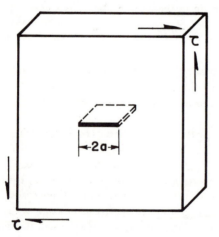

Figure E3.18

REFERENCES

[1] A.A. Griffith, *Phil. Trans. Roy. Soc. London, A221* (1921) 163.

[2] C.E. Inglis, *Proc. Inst. Naval Arch., 55* (1913) 163.

[3] R.E. Peterson, *Stress Concentration Design Factors*, Wiley, New York, 1953.

[4] *Metals Engineering Design*, ASME Proc., McGraw-Hill, New York, 1953.

[5] S. Timoshenko and J.N. Goodier, *Theory of Elasticity*, 2nd ed., McGraw-Hill, New York, 1951, p. 78.

[6] B.R. Lawn and T.R. Wilshaw, *Fracture of Brittle Solids*, Cambridge University Press, Cambridge, 1975, pp. 4, 16.

[7] J.R. Weertman, *Acta Met., 26* (1978) 1731.

[8] G.R. Irwin, *J. Appl. Mech., 24* (1957) 361.

[9] D. Broek, *Elementary Engineering Fracture Mechanics*, 3rd ed. Martinus Nijhoff, The Hague, Netherlands, 1982, p. 33.

[10] C. Zener, in *Fracturing of Metals*, ASM, Metals Park, Ohio, 1948, p. 3.

[11] J.J. Gilman, "Physical Nature of Plastic Flow and Fracture," General Electric Report No. 60-RL-2410M, Apr. 1960.

[12] M.J. Manjoine, in *Fracture: An Advanced Treatise*, Vol. 3, H. Liebowitz (ed.), Academic Press, New York, 1971, p. 265.

[13] L.R. Low, in *Madrid Colloquium on Deformation and Flow of Solids*, Springer-Verlag, Berlin, 1956, p. 60.

[14] M.F. Ashby, *Acta Met., 20* (1972) 887.

[15] M.F. Ashby in *Advances in Research on Strength and Fracture of Materials, 1,* D.M.R. Taplin (ed), Pergamon Press, Elmsford, N.Y., 1978, p. 1.

[16] R.J. Fields and J.H. Smith, *Metal Prog., 118* (Aug. 1981) 39.

[17] M.F. Ashby, C. Gandhi, and D.M.R. Taplin, *Acta Met., 27* (1979) 699; *27* (1979) 1565.

[18] G.M. Boyd, *Brittle Fracture in Steel Structures*, Butterworths, London, 1970.

[19] E.R. Parker, *Brittle Fracture of Engineering Structures*, Wiley, New York, 1957.

[20] C.F. Tipper, *The Brittle Fracture Story*, Cambridge University Press, London, 1962.

[21] G.C. Sih, *Handbook of Stress Intensity Factors for Researchers and Engineers*, Institute of Fracture and Solid Mechanics, Lehigh University, Bethlehem, Pa., 1973.

[22] G.C. Sih (ed.), *Methods of Analysis and Solutions of Crack Problems*, Noordhoff, Leyden, The Netherlands, 1973.

[23] H. Tada, P.C. Paris, and G.R. Irwin, *The Stress Analysis of Cracks Handbook*, Del Research Corp., Hellertown, Pa., 1973.

[24] D.P. Rooke and D.J. Cartwright, *Compendium of Stress Intensity Factors*, HMSO, London, 1976.

[25] A. Kamei and T. Yokobori, "Some Results on Stress Intensity Factors of the Cracks and/or Slip Band Systems," *Ref. Res. Inst. Strength and Fracture of Materials, Tohoku Univ.*, Sendai, Japan, Vol. 10, 1974, p. 29.

[26] G.R. Irwin, in *Encyclopaedia of Physics*, Vol. VI, Springer-Verlag, Heidelberg, 1958; see also *J. Basic Eng., Trans. ASME, 82* (1960) 417.

[27] B.A. Bilby, A.H. Cottrell, and K.H. Swinden, *Proc. Roy. Soc., A272* (1963) 304.

[28] D.S. Dugdale, *J. Mech. Phys. Solids, 8* (1960) 100.

[29] J.E. Srawley and W.F. Brown, ASTM STP 381, ASTM, Philadelphia, p. 133.

[30] W.F. Brown and J.E. Srawley, ASTM STP 410, ASTM, Philadelphia, 1966, p. 1.

[31] A.J. Birkle, R.P. Wei, and G.E. Pellissier, *Trans. ASM, 59* (1966) 981.

[32] D. Broek, *Elementary Engineering Fracture Mechanics*, 3rd ed. Martinus Nijhoff The Hague, Netherlands, 1982, p. 310.

[33] A.A. Wells, *Brit. Weld. J., 13* (1965) 2.

[34] A.A. Wells, *Eng. Fract. Mech., 1* (1970) 399.

[35] D.C. Drucker and J.R. Rice, *Eng. Fract. Mech., 1* (1970) 577.

[36] J.D. Eshelby, *Phil. Trans. Roy. Soc. London, A244* (1951) 87.

[37] G.P. Cherepanov, *Appl. Math. Mech. (Prinkl. Mat. Mekh.), 31*, No. 3 (1967) 503.

[38] J.R. Rice, *J. Appl. Mech., 35* (1968) 379.

[39] J.A. Begley and J.D. Landes, ASTM STP 514, ASTM, Philadelphia, 1972, p. 1.

[40] J.R. Rice, P.C., Paris, and J.G. Merkle, ASTM STP 536, ASTM, Philadelphia, 1973, p. 231.

[41] J.D. Landes and J.A. Begley, ASTM STP 560, ASTM, Philadelphia, 1974, p. 170.

[42] F.M. Burdekin and D.E.W. Stone, *J. Strain Anal., 1* (1966) 145.

[43] *Fracture Toughness Evaluation by R-Curve Methods*, ASTM STP 527, ASTM, Philadelphia, 1973.

SUGGESTED READING

ASTM STP Nos. 381 (1965), 415 (1967), 463 (1970), 486 (1971), 514 (1972), ASTM, Philadelphia.

BROEK, D., *Elementary Engineering Fracture Mechanics*, 3rd ed. Martinus, Nijhoff. The Hague, Netherlands, 1982.

CHELL, G.G. (ed.), *Developments in Fracture Mechanics*, Vol. 1, Applied Science Publishers, Essex, England, 1979.

Fracture Mechanics in Design and Service—Living with Defects, special issue of Phil. Trans. Roy. Soc. London, Vol. A299, 1981.

FRANCOIS, D. (ed.), *Advances in Fracture Research* (Fracture 81), Vols. 1–5, Pergamon Press, Elmsford, N.Y., 1982.

HERTZBERG, R.W. *Deformation and Fracture Mechanics of Engineering Materials*, Wiley, New York, 1976.

KNOTT, J.F., *Fundamentals of Fracture Mechanics*, Butterworths, London, 1973.

LATZKO, D.G.H. (ed.)., *Post-yield Fracture Mechanics*, Applied Science Publishers, Essex, England, 1979.

ROLFE, S.T., and J.M. BARSOM, *Fracture and Fatigue Control in Structures*, Prentice-Hall, Englewood Cliffs, N.J., 1977.

TAPLIN, D.M.R. (ed.), *Advances in Research on Strength and Fracture of Materials*, Vols. 1–4, Pergamon Press, Elmsford, N.Y., 1978.

Part II

METALLURGY
OF DEFORMATION

In Chapter 4, the theoretical resistance of a crystal to shear and cleavage is derived from first principles. Metallic glasses are presented and the relationship between bonding and elastic constants is discussed. The defects responsible for the actual strength of the metals are presented in the next three chapters: point defects (5), line defects (6), and interfacial defects (7). The equilibrium concentration of point defects as a function of temperature is derived in Chapter 5 from thermodynamic considerations; the effects of these defects on the mechanical properties are also described. Chapter 6 provides a comprehensive treatment of dislocations: their nature, stress fields, interactions, reactions, and sources. The various interfacial defects that have important effects on the mechanical properties of metals are presented in Chapter 7: grain boundaries, twin boundaries, antiphase boundaries, and stacking faults. In Chapter 8 additional treatment on dislocations is provided. Included are the Orowan equation, which provides a link between micromechanics and macromechanics by correlating dislocation motion to strain. Treated in detail are the effects of stresses on dislocation velocities and the possible limiting velocity of dislocations. The concept of dislocation motion as a thermally activated phenomenon is presented; it leads to the definition of internal and effective stresses and the activation volume. The geometric relationships between the crystallographic slip systems and the imposed stresses are also presented; in particular, Schmid's law is derived. The treatment of deformation by slip in monocrystals is extended to multi- and polycrystals.

Chapter 4

PERFECT CRYSTALS

4.1 INTRODUCTION

The crystallinity, or periodicity of the structure, does not exist in gases or liquids. Among the solids the metals, ceramics, and polymers may or may not exhibit it, depending on a series of processing and composition parameters. For example, a metal cooled at a superfast rate from the liquid—splat-cooled— is amorphous, in spite of its normal crystalline structure. This subject is treated in greater detail in Section 4.5. SiO_2 can exist as amorphous (fused silica) or as crystal (crystoballite or tridimite). Polymers consisting of molecular chains can exist in various degrees of ordering.

Readers not familiar with structures, lattices, crystal systems, and Miller indices should study this subject before proceeding. Most books on materials science, physical metallurgy, or X-rays treat this subject completely.

4.2 THE STRUCTURE OF METALS

The metallic bond can be visualized, in a very simplified way, as an array of positive ions held together by a glue consisting of electrons. These positive ions, which repel each other, are attracted to the "glue," which is known as electron gas.[1,2] Ionic and covalent bonding, on the other hand, can be visualized as direct attractions between atoms. Hence, their bonding is strongly directional and determines the number of nearest neighbors that one atom will have, as well as their position. This, in turn, determines the type of structure; often,

TABLE 4.1   Crystalline Structure of Most Important Metals

FCC	BCC	HCP
Fe (910–1390°C)	Fe ($T < 910$°C)	Ti
Ni	Be ($T > 1250$°C)	Be ($T < 1250$°C)
Cu	Co ($T > 427$°C)	Co ($T < 427$°C)
Al	W	Ce (-150°C $< T < -10$°C)
Au	Mo	Zn
Pb	Cr	Mg
Pt	V	Zr
Ag	Nb	Hf ($T < 1950$°C)

they are very complicated for ionic and covalent bonding. On the other hand, the directionality is not very important for metals, and the atoms pack into the simplest and most compact forms; they can be visualized as spheres. The structures favored by metals are the face-centered cubic (FCC), body-centered cubic (BCC), and hexagonal close-packed (HCP). In the periodic table, from the 81 elements to the right of the Zindl line, 53 have either the FCC or HCP structure, 21 the BCC structure; the remaining 8 have other structures. The Zindl line defines the boundary of the elements with metallic character, in the periodic table. It is only natural that the analytical efforts of metallurgists concentrate on these three structures, since they encompass the vast majority of metals. Table 4.1 shows the structure of some common metallic elements. Some of them have several structures, depending on the temperature. Perhaps the most complex of the metals is plutonium, which undergoes six polymorphic transformations. The range of problems and challenges posed to the metallurgist is very wide. In addition to the metallic elements, intermediate phases and intermetallic compounds exist in great numbers, with a variety of structures. For instance, the beta phase in the Cu-Mn-Sn system exhibits a special ordering for the composition Cu_2MnSn.[3] The unit cell (BCC) is shown in Fig. 4.1. However, the ordering of the Cu, Mn, and Sn atoms creates a superlattice composed of four BCC cells. This superlattice is FCC. Hence, the unit cell for the ordered

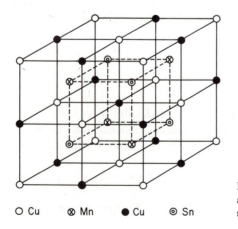

○ Cu   ⊗ Mn   ● Cu   ◉ Sn

Figure 4.1   Beta-ordered phase in Heusler alloys (Cu_2MnSn). (Reprinted with permission from [3], p. 40.)

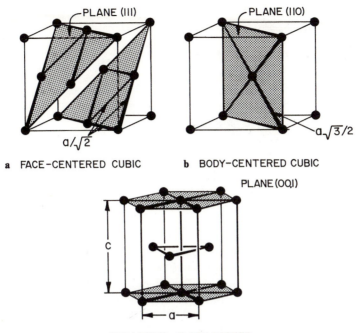

PLANE (III)   PLANE (IIO)

$a/\sqrt{2}$ $a\sqrt{3}/2$

a FACE-CENTERED CUBIC   **b** BODY-CENTERED CUBIC

PLANE (OOI)

c

a

c HEXAGONAL CLOSE PACKED

Figure 4.2 Closest-packed planes in (a) FCC; (b) BCC; (c) HCP.

phase is FCC, whereas for the disordered phase one has a BCC unit cell. This ordering has important effects on the mechanical properties and is discussed in Section 15.5.

Figure 4.2 shows the three main metallic structures. The atom positions are marked by small spheres and the atomic planes by dark sections. The small spheres do not correspond to the scaled-up size of the atoms, which would almost completely fill the available space, touching each other. For the FCC and HCP structures, the coordination number (number of nearest neighbors of an atom) is 12. For the BCC structure, it is eight.

The planes with densest packing are indicated in Fig. 4.2. They are (111), (110), and (00.1) for the FCC, BCC, and HCP structures, respectively. These planes have an important effect on the directionality of deformation, as will be seen in Chapters 6 and 8. The distances between nearest neighbors are also indicated in Fig. 4.2. The reader should try to calculate them, as an exercise. These distances are $a/\sqrt{2}$, $a\sqrt{3}/2$, and a for the FCC, BCC, and HCP structures, respectively.

The similarity between the FCC and HCP structures is much greater than would be expected from looking at the unit cells. Planes (111) and (00.1) have the same packing, as can be seen in Fig. 4.2. This is the densest possible packing of coplanar spheres. It is shown in Fig. 4.3(a). The packing of a second plane similar to the first one on top of the first plane (called A) can be made in two different ways, shown in Fig. 4.3(b); these two planes are indicated by

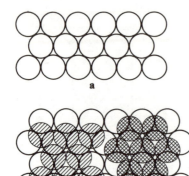

a

LAYER A   LAYER C   LAYER B

b

Figure 4.3 (a) Layer of closest-packed atoms corresponding to (111) in FCC and (00.1) in HCP; (b) packing sequence of densest-packed planes in *AB* and *AC* sequence.

B and *C*. Hence, either alternative *B* or *C* can be used. A third plane, when placed on top of plane *B*, would have two options: *A* or *C*. If the second plane is *C*, the third plane can be either *A* or *B*. If only the first and second layers are considered, the FCC and HCP structures are identical. If the position of the third layer coincides with that of the first (*ABA* or *ACA* sequence), we have the HCP structure. Since this packing has to be systematically maintained in the lattice, one would have *ABABAB* . . . or *ACACAC*. . . . In case the third plane does not coincide with the first, we have one of these two alternatives: *ABC* or *ACB*. Since this sequence has to be systematically maintained, one has: *ABCABCABC* . . . or *ACBACBACB*. . . . This stacking sequence corresponds to the FCC structure. We can conclude that the only difference between the FCC and HCP structures (the latter with a theoretical *c/a* ratio of 1.833) is the stacking sequence of the densest packed planes. The difference resides in the next neighbors and in the greater symmetry of the FCC structure.

4.3 THEORETICAL STRENGTH OF METALS

In the twenties, there were heated arguments due to the enormous differences—two or more orders of magnitude—between the actual and theoretical strength of metals. This divergence was explained with the postulation of dislocations in 1934; later their existence was fully confirmed experimentally. The theoretical strength of a perfect lattice is of great importance. It will be shown that it is very high and that experimental efforts have been made to reach it. This theoretical strength is determined by the nature of the interatomic forces, by the temperature (which causes atoms to vibrate), and by the stress state. The calculations will be conducted in the next sections for two states of stress: uniaxial normal stress and shear stress. These two stress states determine two different types of failure: cleavage and shear, respectively. The stresses required for failure under the two situations will be calculated, and the theoretical strength should be the lower of the two values.

‡4.3.1 Theoretical Tensile Strength by Orowan's[4] Method

A material is said to cleave when it breaks under normal stress and the fracture path is perpendicular to the applied stress. The process involves the separation of the atoms along the direction of the applied stress. Orowan[4] developed a simple method to obtain the theoretical tensile strength of a crystal. It should be noted that no stress concentrations at the crack tip are assumed; it is assumed that all atoms separate simultaneously once their distance has reached a critical value. Figure 4.4 shows how the stress required to separate two planes will vary as a function of the distance between these planes. The distance is initially equal to a_0. Naturally, $\sigma = 0$ for $a = a_0$; σ will also be zero when the separation is infinite. The exact form of the σ versus a curve depends on the nature of the interatomic forces (a more specific case is presented in Section 4.3.2). In Orowan's model the curve is simply assumed to be a sine function—hence the generality of Orowan's model. The area under the curve is the work required to cleave the crystal. This work of deformation—and here there is a certain similarity with Griffith's crack propagation theory presented in Chapter 3—cannot be lower than the energy of the two new surfaces created by the cleavage. If the surface energy per unit area is γ and the cross-sectional area of the specimen is A, the total energy is $2\gamma A$ (two surfaces formed). The stress dependence on plane separation is given by the following equation, admitting a sine function:

$$\sigma = K \sin \frac{\pi}{d}(a - a_0) \tag{4.1}$$

K can be determined by the following artifice: when a is close to a_0, the material responds linearly to the applied loads (Hookean behavior). Assuming that the elastic deformation is restricted to the two planes shown in Fig. 4.4 and that the material is isotropic, we have

$$\frac{da}{a_0} = d\epsilon$$

$$\frac{d\sigma}{d\epsilon} = \frac{d\sigma}{da/a_0} = E \tag{4.2}$$

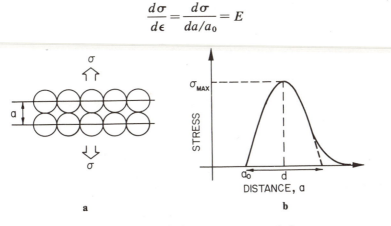

a

b

Figure 4.4 Stress required to separate to atomic layers.

or

$$a_0 \frac{d\sigma}{da} = E$$

Taking the derivative of Eq. 4.1 and substituting it into Eq. 4.2 for $a = a_0$, we have

$$a_0 \frac{d\sigma}{da} = K \frac{\pi}{d} a_0 \cos \frac{\pi}{d} (a - a_0) = E$$

$$K = \frac{E}{\pi} \frac{d}{a_0}$$

(4.3)

However, a is not known; to determine a, the area under the curve has to be equated to the energy of the two surfaces created.

$$\int_{a_0}^{a_0+d} \sigma \, da = 2\gamma$$

(4.4)

Substituting Eq. 4.1 into Eq. 4.4, we get

$$\int_{a_0}^{a_0+d} K \sin \frac{2\pi}{2d} (a - a_0) \, da = 2\gamma$$

(4.5)

From the *Handbook of Physics and Chemistry*, the integral above can be evaluated:

$$\int \sin ax \, dx = -\frac{1}{a} \cos ax$$

(4.6)

A variable substitution is required to solve Eq. 4.5; applying the standard Eq. 4.6, we have $a - a_0 = y$; therefore, $dx = dy$ and

$$K \int_0^d \sin \frac{\pi}{d} y \, dy = 2\gamma$$

$$K \frac{d}{\pi} = \gamma$$

and

$$d = \frac{\pi\gamma}{K}$$

(4.7)

The maximum value of σ is equal to the theoretical cleavage stress. From Eq. 4.1 and making the sine equal to 1, we have

$$\sigma_{max} = K = \frac{E}{\pi} \frac{d}{a_0}$$

(4.8)

Substituting Eq. 4.7 into 4.8 yields

$$K = \sigma_{max} = \frac{E\gamma}{a_0 K}$$

TABLE 4.2   Theoretical Cleavage Stresses According to Orowan's Theory[a]

Element	Direction	Young's Modulus (GPa)	Surface Energy (mJ/m^2)	σ_{max} (GPa)	σ_{max}/E
α-Iron	$\langle 100 \rangle$	132	2000	30	0.23
	$\langle 111 \rangle$	260	2000	46	0.18
Silver	$\langle 111 \rangle$	121	1130	24	0.20
Gold	$\langle 111 \rangle$	110	1350	27	0.25
Copper	$\langle 111 \rangle$	192	1650	39	0.20
	$\langle 100 \rangle$	67	1650	25	0.38
Tungsten	$\langle 100 \rangle$	39	3000	86	0.22
Diamond	$\langle 111 \rangle$	121	5400	205	0.17

[a] Adapted with permission from [6], p. 10.

and

$$K^2 = (\sigma_{max})^2 = \frac{E\gamma}{a_0}$$

or

$$\sigma_{max} = \sqrt{\frac{E\gamma}{a_0}} \tag{4.9}$$

According to Orowan's model, the surface energy is given by

$$\gamma = \frac{Kd}{\pi} = \frac{E}{a_0}\left(\frac{d}{\pi}\right)^2$$

It has been experimentally determined that d is approximately equal to a_0. Hence,

$$\gamma \simeq \frac{Ea_0}{10} \tag{4.10}$$

We can conclude from Eq. 4.9 that a material, in order to have a high theoretical cleavage strength, has to have a high Young's modulus and surface energy and a small distance a_0 between atomic planes. Table 4.2 presents the theoretical cleavage strengths for a number of metals. The greatest source of error is γ; it is not easy to determine γ with great precision in solids and the values used in Table 4.2 come from different sources and were not necessarily determined at the same temperature.

4.3.2 More Exact Calculations of the Theoretical Tensile (Cleavage) Stress

The force between two atoms that are bonded covalently can be described as a Morse function.[5,6] More exact calculations of the theoretical cleavage stress, using Morse's function, have been made for carbon and silicon; the values found are about half of the Orowan cleavage stress, as will be seen below. It can be concluded that Orowan's equation superestimates the cleavage stress.

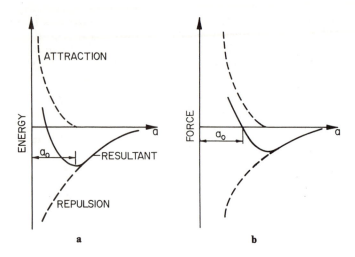

Figure 4.5 Morse curves showing qualitatively the variation with distance a of (a) interaction energy; (b) interaction forces.

The Morse function represents the interaction energy (the potential energy due to the position) of two atoms. The potential energy of repulsion tends to zero at a faster rate than the potential energy of attraction, as shown in Fig. 4.5(a). Therefore, there is a well, called a potential well. The equilibrium separation is a_0. The derivative with respect to a provides the interaction forces, plotted in Fig. 4.5(b). It can be seen that, at the equilibrium separation a_0, the force between the atoms is zero. The binding energy between two covalent atoms can be approximated as

$$U = U_0[e^{-2k(a-a_0)} - 2e^{-k(a-a_0)}] \tag{4.11}$$

The parameter k is related to the curvature of the Morse function. The treatment given by de Boer[7] for bonding in carbon is presented below. This work is described by Kelly.[6] The maximum value of the force corresponds to the separation a at which the energy curve has an inflection point. This is obtained by setting the second derivative of Eq. 4.11 equal to zero. After finding the distance,

$$a = \frac{ka_0 + \ln 2}{k}$$

we substitute it into the first derivative of Eq. 4.11 and find that

$$F_{\text{max}} = \frac{U_0 k}{2}$$

The parameter k can be obtained from the slope of the F versus a curve at the equilibrium separation, since

$$\left(\frac{dF}{da}\right)_{a=a_0} = \left(\frac{\partial^2 U}{\partial a^2}\right)_{a=a_0} = 2k^2 U_0$$

The binding energy between carbon atoms is 5.8×10^{-19} J per atom and a_0 is 1.54×10^{-10} m. The parameter k can be obtained by means of experimental tests. Silver[8] found it to be equal to 2.12×10^{10} m^{-1}. Hence, the maximum force (force required to break the C—C bond) is equal to 6.1×10^{-9} N. In order to determine the theoretical cleavage stress of diamond, we have to know the number of bonds per unit area, 1.82×10^{19} bonds/m^2.[6] Consequently, the theoretical cleavage stress along $\langle 111 \rangle$ is equal to 110 GPa. Orowan's calculation (Table 4.2) predicts approximately 205 GPa, or twice that amount.

Similar calculations[9,10] using the Morse curve have been conducted for NaCl and the theoretical cleavage stress was found to be between 2 and 2.6 GPa.

4.3.3 Theoretical Shear Stress

Frenkel[11] performed a simple calculation of the theoretical shear strength of crystals by considering two adjacent and parallel lines of atoms subjected to a shear stress; this configuration is shown in Fig. 4.6; (a) is the separation between the adjacent planes and (b) is the interatomic distance. Under the action of τ the top line will move in relation to the bottom line; the atoms will pass through successive equilibrium positions A, B, C, for which the stress τ is zero. When the applied shear stress is enough to overcome these barriers, plastic deformation will occur and the atoms will move until a shear fracture is produced. The stress is also zero when the atoms are exactly superimposed; in the latter case, the equilibrium is metastable. Between these values the stress varies cyclically with a period b. Frenkel[11] assumed a sine function, as we would expect:

$$\tau = k \sin \frac{2\pi x}{b} \tag{4.12}$$

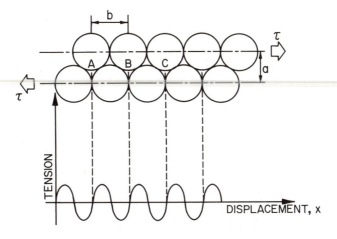

Figure 4.6 Stress required to shear crystal.

For small displacements,

$$\tau \simeq k\frac{2\pi x}{b} \tag{4.13}$$

Since for small displacements one can consider the material as elastically deformed, we have

$$\tau = G\frac{x}{a} \tag{4.14}$$

where x/a is the shear strain and G is the shear modulus. Substituting Eq. 4.14 into Eq. 4.13, we have

$b\left(\frac{x}{a}\right) = K\frac{2\pi x}{b} \qquad K \approx \frac{Gb}{2\pi a}$

$\frac{bb}{a} \approx K 2\pi$

$$k = \frac{Gb}{2\pi a} \tag{4.15}$$

Substituting Eq. 4.15 into 4.12 yields $\quad \tau = K \sin \frac{2\pi x}{b}$

$$\tau = \frac{Gb}{2\pi a}\sin\frac{2\pi x}{a}$$

The maximum of τ occurs for $x = b/4$:

$$\tau_{max} = \frac{Gb}{2\pi a} \tag{4.16}$$

For FCC metals we have $b = a_0/\sqrt{2}$ and $a = a_0\sqrt{3/8}$. Hence,

$$\tau_{max} \simeq \frac{G}{5.4} \tag{4.17}$$

More complex models have been advanced in which the sine function has been replaced by more precise curves expressing the interaction energy. The method developed by Mackenzie[12] is an example. He took into account the distortion of the planes. Table 4.3 shows the stresses calculated by Mackenzie's method. It should be noticed that the τ_{max}/G ratio varies between 0.039 and 0.24. Consequently, it is fairly close to Frenkel's ratio (0.18) obtained by the simpler method.

TABLE 4.3 Theoretical Shear Strength Calculated According to Mackenzie's[12] Method[a]

Element	G (GPa)	τ_{max} (GPa)	τ_{max}/G
Iron	60.0	6.6	0.11
Silver	19.7	0.77	0.039
Gold	19.0	0.74	0.039
Copper	30.8	1.2	0.039
Tungsten	150.0	16.5	0.11
Diamond	505.0	121.0	0.24
NaCl	23.7	2.8	0.12

[a] Adapted with permission from [6], p. 28

Comparing Tables 4.2 and 4.3, we note that the theoretical shear strengths are all much lower than the theoretical cleavage strengths. In a tensile test we have, for the plane of maximum shear, $\tau = \sigma/2$. Even if this adjustment is made, the theoretical shear strength is found to be substantially lower than the theoretical cleavage strength, showing that ideal metals would fail under shear. It should be emphasized, however, that the calculation of the shear strength was made assuming zero normal stress perpendicular to the shear plane. Normal stresses can significantly affect τ_{max}.

4.3.4. Importance of the σ_{max}/τ_{max} Ratio

Kelly et al.[13] proposed that the σ_{max}/τ_{max} ratio had an important bearing on the brittleness of a material; if this ratio is high, a ductile behavior is expected. Conversely, a low ratio is an indication of brittleness. The FCC metals possess a high σ_{max}/τ_{max} ratio; it is 34 for gold, 30 for silver, 22 for nickel, and 28 for copper. Iron and tungsten, on the other hand, have ratios of 6.85 and 5.[6] Qualitatively, the reason for the enhanced ductility when τ_{max} is relatively low is the ease of forming dislocations at the crack tip, blunting it. Based on these concepts, Weertman[14] developed a theoretical model that predicts crack-tip blunting for $\sigma_{max}/\tau_{max} > 7$. In his model, pairs of dislocations of opposite signs are created close to the crack tip by the high stresses. One of the dislocations moves toward the crack and the other away from it. After several such events, a pile-up is created at the crack tip; when the stress is high enough, the piled-up dislocations enter the crack tip and, in so doing, blunt it.

‡4.4 ACTUAL STRENGTH OF METALS

The theoretical strengths derived in previous sections are of the order of gigapascals; unfortunately, the actual strength of metals is orders of magnitude below that. This fact was very worrisome to investigators at the turn of the century. One of the views was that the real metals had small internal cracks, at whose extremities high stress concentrations were set up.[15] Hence, the theoretical strength would be achieved at the crack tip at applied loads that were only a fraction of this stress. But this theory could not account for work hardening: after plastic deformation, the cracks would become longer, and the stress concentration at them would be even higher. Griffith's theory, seen in Chapter 3, explains this situation very clearly. It is now known that this is correct but that these stress concentrations are much lower in ductile materials, since plastic flow can take place at the crack tip, blunting it (Section 3.3). Another accepted theory was that there are slip planes in metals which can slide with perfect ease; however, when they do, portions of the material come loose and get turned around, forming keys that block the slip. In 1934 the dislocations were postulated. This startling breakthrough is one of the foundations of physical/mechanical metallurgy.

In essence, not only dislocations, but the various defects existing in metals and most important, their interactions, are responsible for the difference between real and theoretical strength. These defects are studied in detail in Chapters 5 through 15. According to their dimensionality, they can be classified into:

1. Point defects
2. Line or unidimensional defects
3. Interfacial or bidimensional defects
4. Volume or tridimensional defects

Their dimensions are very different, as can be seen in Table 4.4.

To approach the theoretical strength of metal, there are two possible methods: (1) eliminating all defects, and (2) creating so many defects that their interactions render them inoperative. The first approach has yielded some materials with extremely high strength. Unfortunately, this has only been possible in special configurations that are called "whiskers." The second approach is the one more commonly pursued, because of the obvious dimensional limitations of the first; the strength levels achieved in bulk metals have steadily increased by an ingenious combination of strengthening mechanisms but are still much lower than the theoretical strength. Maraging steels with useful strength up to 2 GPa have been produced, and patented steel wires with strengths of up to 4.2 GPa; these are the highest-strength steels.[16]

Figure 4.7 compares the ambient-temperature strength of tridimensional, filamentary, and "whisker" materials. The whiskers have a cross-sectional diameter of only a few micrometers and are usually monocrystalline (although polycrystalline whiskers have also been developed).[17] Brenner[18-20] has pioneered the study of whiskers. They are the strongest materials developed by man, and SiC whiskers with a strength of 40 GPa have been in fabricated in the Philips Laboratories, Eindhoven, the Netherlands. The dramatic effect of the elimination of two dimensions is shown clearly in Fig. 4.7. The strongest whiskers are ceramics. Table 4.5 provides some illustrative examples. Iron whiskers with 12.6 GPa strength have been produced, compared with the strongest bulk steels (2 GPa). This value is essentially identical to the theoretical shear stress (Table 4.3) because the normal stress is twice the shear stress. In general, FCC whiskers tend to be much weaker than BCC whiskers and ceramics; for instance, Cu

TABLE 4.4   Dimensional Ranges of Different Classes of Defects

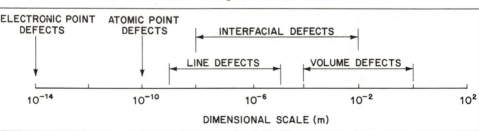

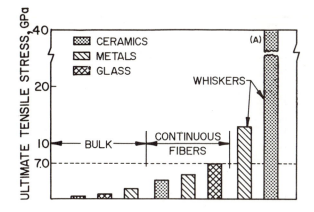

Figure 4.7 Theoretical strength of tridimensional materials, continuous fibers, and whiskers. The strength of the SiC whisker produced by the Philips Eindhoven Laboratory is indicated by (A).

has a strength of 2 GPa. This is consistent with the much lower theoretical shear strength exhibited by them. In Table 4.3, silver, gold, and copper have τ_{max}/G ratios of 0.039. Hence, they are not good whisker materials. Figure 4.8 shows a stress–strain curve for a copper whisker. The specimen had a length between 2 and 3 mm and a cross-sectional diameter of 6.7 μm. The stress drops vertically after the yield point, with a subsequent plateau corresponding to the propagation of a Lüders band.

In the elastic range, the curve deviates slightly from Hooke's law and exhibits some temporary inflections and drops (not shown in Fig. 4.8). In many cases, both for metals and nonmetals, failure occurs at the elastic line, without appreciable plastic strain. When plastic deformation occurs, such as in the case of copper and zinc, a very large yield drop is observed. Although the strength of whiskers is not completely understood, it is connected to the absence of dislocations. It is impossible to produce a material virtually free of dislocations or perfect. However, for whiskers these dislocations can easily escape out of the material, during elastic loading. Their density and mean free path are such that they will not interact and produce other dislocation sources. Hence, the

TABLE 4.5 Tensile Strengths of Whiskers at Room Temperature[a]

Material	Maximum Tensile Strength (GPa)	Young's Modulus (GPa)
Graphite	19.6	686
Al$_2$O$_3$	15.4	532
Iron	12.6	196
SiC	20–40	700
Si	7	182
AlN	7	350
Cu	2	192

[a] Adapted with permission from [6], p. 263.

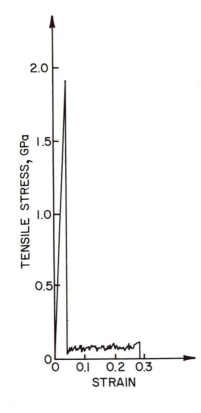

Figure 4.8 Stress–strain curve for a copper whisker with a fiber direction $\langle 100 \rangle$. The whisker diameter is 6.8 μm. (Adapted with permission from [21], p. 826.)

yield point is the stress required to generate dislocations from surface sources. The irregularities observed by Yoshida et al.[21] in the elastic range indicate that existing dislocations move and escape out of the whisker. The whisker becomes, at a certain stress, essentially dislocation-free. When the stress required to activate surface sources is reached, the material yields plastically, or fails. This is shown in the results obtained by Mehan et al.[22] and Grenier and Kelly[23] and reproduced in Fig. 4.9. The experimental results by Mehan et al.[22] show very clearly an increase in yield stress as the diameter of the whisker is reduced. Below 5 μm this increase becomes extremely rapid; this is the range below which the interaction between existing dislocations is minimized. However, the variability of the experimental points is extremely large. Grenier and Kelly,[23] on the other hand, carefully selected and chemically polished the whiskers in order to obtain smooth and regular surfaces and obtained results that were much more self-consistent. This shows that the surface plays a very important role in determining the beginning of plastic deformation and failure. The subject will be treated in greater detail in Section 6.8 (dislocation sources). Once the surface sources are activated, a large enough number of dislocations is generated so that they can interact and multiply by other mechanisms; hence, the crystal looses its "whisker" characteristic (Fig. 4.8).

loses There are a variety of whisker fabrication techniques: stress-induced growth, growth by vapor deposition, reaction with a gaseous phase (such as a

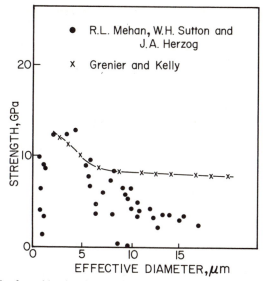

Figure 4.9 Strength of sapphire (α-Al_2O_3) whiskers as a function of the square root of the cross-sectional area. (Adapted with permission from [6], p. 254.)

carbonyl), electrolytic growth, growth from the solution, and unidirectional growth of eutectics. Whiskers have potential applications in composite materials (Chapter 12).

‡4.5 METALLIC GLASSES (OR AMORPHOUS METALS)

Under extreme processing conditions it is possible to obtain solid metals in a noncrystalline structure. This "glassy" structure is characterized by an absence of a regular periodicity in the atomic arrangement. Figure 4.10 shows a crystalline and a glassy alloy of the same composition. The liquid state is frozen in and the structure resembles that of glasses. It is possible to arrive at these special structures by cooling the alloy at such a rate that virtually no reorganization of the atoms into periodic arrays can take place. This requires cooling rates of 10^6 to $10^{8\circ}$ C/s. The original technique to obtain this high cooling rate was called splat cooling and was pioneered by Duwez in 1960.[24] This technique consisted of propelling a drop of liquid metal with a high velocity against a heat-conducting surface such as copper. The interest in these alloys was mainly academic at that time. However, the unusual magnetic properties and high strength exhibited by these alloys triggered worldwide interest and the subsequent research has resulted in thousands of papers.[24] The splat-cooling technique has been refined[25] to the point where 0.07- to 0.12-mm-thick wires can be ejected from an orifice. Production rates as high as 1,800 m/min can be obtained. Sheets and ribbons can be manufactured by the same technique. An alternative technique consists of vapor deposition on a substrate (sputtering). This seems to be the most promising approach and samples with a thickness of several millimeters have been successfully produced. The potential applications of metal-

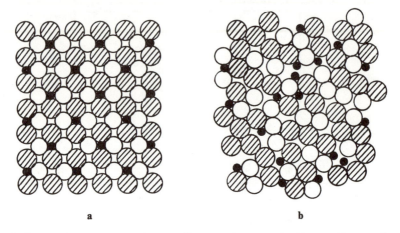

a   b

Figure 4.10 Atomic arrangements in crystalline and glassy metals: (a) crystalline metal section; (b) glassy metal section. (Courtesy of L. E. Murr, Oregon Graduate Center.)

lic glasses are connected with their unique magnetic properties.[25] They have been shown to be suitable for application in magnetic bubble memories. Another area of major interest involves rare earth (RE) transition metal amorphous alloys, with compositions such as $(RE)Fe_2$. These materials can be made over a wide range of composition by the sputtering technique. They can be crystallized to an intermediate structure that is very fine grained and exhibits highly desirable magnetic properties: this makes them excellent candidate materials for permanent magnets. At 4 K, $TbFe_2$ has a coercivity of over 30,000 Oe and a saturation magnetization of over 100 emu/g, leading to one of the highest values of the energy product, BH_{max}, ever measured. Yet another group of glassy metals has a great potential as "soft" magnets, with high magnetic permeability. Examples are 80% (Fe, Co, Ni) and 20% (B, C, Si, P) alloys.

The unique mechanical properties exhibited by metallic glasses are connected to their structure. Table 4.6 lists the hardnesses, yield stresses, and Young moduli for several metallic glasses. The unique compositions correspond to regions in the phase diagram that have a very low melting point. The low melting points aid in the retention of the "liquid" structure. The Metglas group is commercially produced in wire and ribbon form. The Young's modulus varies between 60 and 70% of the Young's modulus of the equilibrium crystalline structure. J. C. M. Li[26] has proposed a relationship between the shear modulus of the glassy and crystalline states:

$$G_g = \frac{0.947}{1.947 - \nu} G_c \qquad (4.18)$$

where G_g and G_c are the shear moduli of the glassy and crystalline states and ν is Poisson's ratio. Megusar et al.[27] show that the crystalline Young's modulus of $Cu_{60}Zr_{40}$ is regained when the material is annealed and crystallinity sets in. The yield stresses of metallic glasses are high, as can be seen in Table 4.6. For Fe-B metallic glasses, strength levels over 3.5 GPa were achieved.[28]

TABLE 4.6   Mechanical Properties of Some Metallic Glasses[a]

Alloy	HV (GPa)	σ_y, (GPa)	H/σ_y	E_g, (GPa)	E_g/σ_y
$Ni_{36}Fe_{32}Cr_{14}P_{12}B_6$ (Metglas 286AA)	6.1	1.9 (tension)	3.16	99.36	52
$Ni_{49}Fe_{29}P_{14}B_6S_2$ (Metglas 286B)	5.5	1.7 (tension)	3.26	91.1	54
$Fe_{80}P_{16}C_3B_1$ (Metglas 2615)	5.8	1.7 (tension)	3.35		
$Pd_{77.5}Cu_6Si_{16.5}$	3.4	1.08 (compr.)	3.17	61.9	57
$Pd_{64}Ni_{16}P_{20}$	3.1	1. (compr.)	3.17	61.9	57
$Fe_{80}B_{20}$ (Metglas 2605)	7.6	2.55 (tension)	2.97	116.6	45

[a] Adapted with permission from [28], p. 369, Table 1.

This is close to the highest yield strengths achieved in polycrystalline metals (see Section 4.4). The yield stresses of the metallic glasses are usually 10 to 30 times higher than the yield stress of the same alloy in the crystalline state. Polk et al.[25] suggest that it should, in the future, be possible to produce amorphous fibers with strengths of at least 4 GPa; these fibers would be excellent candidate materials for composite materials.

The micromechanical deformation mechanisms responsible for the unique mechanical properties are still not very well understood. The absence of crystallinity has a profound effect on the mechanical properties. Grain boundaries, dislocations, mechanical twinning, and other very important components of the deformation of crystalline metals are not directly applicable to metallic glasses. Although the dislocations are not introduced until Chapter 6, the concept is used in this section in an attempt to rationalize the mechanical response of metallic glasses. The lowered Young's modulus is probably due to the less efficient packing of atoms with a consequent larger average interatomic distance. Section 4.6 shows that elastic constants are very sensitive to interatomic distance. The plastic part of the stress–strain curve also differs from the crystalline one. Here we have to distinguish between the behavior of the metallic glass above and below T_g, the glass transition temperature. As in silicate glasses, a temperature is defined above which the glass becomes viscous and deformation occurs by a viscous flow that is homogeneous. Only the deformation at temperatures below T_g will be discussed here. Pampillo and Chen[29] were able to test small cylindrical specimens under compression and found the curves shown in Fig. 4.11. There is little evidence of work hardening and the plastic range is close to horizontal. The surface of the specimens usually exhibits steps produced by shear bands. These shear bands have been found to be in 20 nm thick and the shear offset (step) has been found to be around 200 nm. This shows that deformation is highly inhomogeneous in metallic glasses and that, once shear starts on a certain

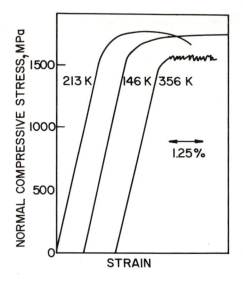

Figure 4.11 Compression stress–strain curves for $Pd_{77.5}Cu_6Si_{16.5}$. (Adapted with permission from [29], p. 182.)

plane, it tends to continue there. This indicates that the plane of shear actually becomes softer than the surrounding regions. We can compute the amount of shear strain in a band by dividing the band height by the offset. For the case above, it is equal to 10. This behavior is termed work softening. The curves of Fig. 4.11 provide macroscopic support for the absense of work hardening. Gilman[30,31] has suggested that the equivalent of a dislocation exists in a metallic glass. The slip vector of this dislocation would fluctuate in direction and magnitude along the dislocation line, but its mean value would be dictated by some structural parameter. Another model, based on the concept of dislocations proposed by Li,[26] is shown schematically in Fig. 4.12. To create an amorphous

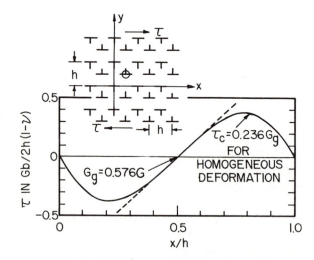

Figure 4.12 Dislocation lattice model for an amorphous (glassy) metal. (Reprinted with permission from [26], p. 535.)

Metallurgy of Deformation   Part II

structure from a crystalline one, Li inserted a series of dislocations in the structure, disrupting the periodicity. Positive dislocations were inserted in the places where a lattice atom was pushed upward to accommodate a solute atom and negative dislocations where the lattice atom was pushed downward. The yield stress of this structure is, according to Li, the stress required to force the dislocation to move through the stress fields. Under an applied shear stress τ the positive edge dislocations would move in a direction opposite to the negative ones. The stress versus distance (x/h) plot has a sine shape and the slope of the curve at $\tau_0 = 0$ (linear range) provides the shear modulus. Li's calculations yield (as can be seen from Eq. 4.18) $G_g = 0.575 G_c$ for $v = 0.3$. If we assume that both the glassy and crystalline metal have the same Poisson's ratio, we would have $E_g = 0.575 E_c$. This is fairly close to what is found experimentally. As an example, we have $E_g = 0.61 E_c$ for $Pd_{80}Si_{20}$. Li's model also allows determination of the stress needed for plastic deformation. This corresponds to the maximum in the plot of Fig. 4.12. Li found that

$$\tau_g \equiv \frac{0.19 G_c b}{h(1 - v)} \tag{4.19}$$

For $h = 2b$ (a realistic estimate of the dislocation density, if b is the distance between nearest neighbors), we find that

$$\tau_g = 0.236 G_g \tag{4.20}$$

This leads to the following ratio between yield stress and the Young's modulus, for $v = 0.3$:

$$\frac{E_g}{\sigma_y} = 0.83 \tag{4.21}$$

Inspection of Table 4.6 shows that this ratio is about 50 times higher than the experimentally observed results. This led Li to postulate the "dislocation in a dislocation lattice" concept. It is shown in Fig. 4.13. It consists of an array of dislocations forming a lattice; this lattice is, in its turn, dislocated. It can be readily seen that the stress required to move the dislocation-in-a-dislocation lattice is much lower than the one calculated previously. All that is needed is a minor readjustment of the dislocation lattice to propagate the deformation. The reader should compare this concept with the conventional dislocations discussed in Chapter 6. Only one dislocation at a time is moving. The plot in Fig. 4.13 shows that we have

$$\frac{\tau_g}{G_g} = \frac{1}{80}$$

For $v = 0.3$, this corresponds to

$$\frac{E_g}{\sigma_y} = 104 \tag{4.22}$$

This value is much closer to those actually observed.

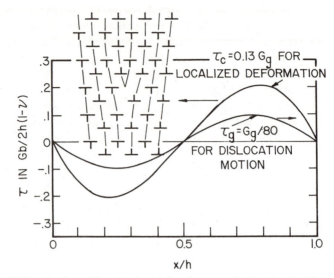

Figure 4.13 Dislocation in a dislocation lattice for glassy metal. (Reprinted with permission from [26], p. 538.)

An alternative theory has been proposed by Spaepen and Turnbull[32] to explain the deformation of glassy metals. They suggested that plastic flow occurs when the local viscosity is decreased due to the applied stress. The lowered viscosity in a region (the shear band) produces the localization of viscous flow. At the Second International Conference on Rapidly Quenched Metals held at MIT in 1975, the relative merits of the two approaches—dislocation and viscous flow—were discussed at length, but no consensus could be reached.

Computer models incorporating interatomic potentials have been used to simulate the deformation of metallic glasses. Initially, two-dimensional models were used (Maeda and Takeuchi[33] used 254 atoms); three-dimensional models using 2067 atoms were used later by Maeda and Takeuchi.[34] Nevertheless, these models did not predict the shear bands. Another approach was used by Argon and Kuo;[35] they made bubble raft experiments. By mixing bubbles of two sizes, they simulated the arrangement of Fig. 4.10(b). The application of a shear stress led to the identification of small regions undergoing shear transformations. These internal regions of about five bubble diameters underwent complex internal rearrangements, producing a net shear. They also observed slipping between two adjacent rows of atoms. Based on these observations, Argon[36] proposed a mechanism for deformation.

Figure 4.14 shows slip lines and steps produced after (a) bending and (b) unbending. We can see the slip lines terminating inside the metallic glass. The slips decrease in height on unbending. These observations and others, as well as the inability of computer simulations and bubble-raft analogs to predict shear bands, tend to confirm the relevance of some kind of dislocation mechanism in the plastic deformation of metallic glasses; J. C. M. Li's[37–39] later work extends his earlier model, described in the preceding paragraphs.

a

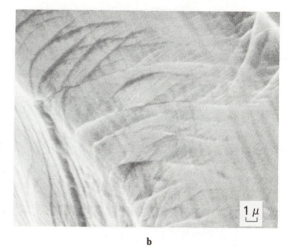

b

Figure 4.14 Shear steps terminating inside material after annealing at 250°C/h produced by (a) bending and decreased by (b) unbending Metglas $Ni_{82.4}Cr_7Fe_3Si_{4.5}B_{3.1}$ strip. (Courtesy of X. Cao and J. C. M. Li, Rochester University.)

4.6 RELATIONSHIP BETWEEN BONDING AND ELASTIC CONSTANTS

The linear theory of elasticity, treated in Part B of Chapter 1 makes the assumption that the material is a *continuum*. This assumption is correct when we are dealing with large bodies; the micromechanics of deformation, on the other hand, take place on a scale where the continuum breaks down into a periodic array of atoms: the crystalline structure. It is theoretically possible to calculate the microscopic elastic constants from the consideration of the interatomic forces. These calculations can be conducted for ionic structures, such as NaCl, considering only electrostatic forces. In metals the situation is more complex. Even approximate quantitative determinations require the use of wave mechanics. The effect of temperature on atomic vibration and/or the lattice parameter is discussed, as well as the attendant changes in elastic properties.

The electrostatic nature of the forces between ionic crystals renders the determination of the elastic constants less arduous. The NaCl structure is a simple cubic structure, with Na and Cl occupying alternate positions in the lattice. Each Na$^+$ ion is surrounded by six Cl$^-$ ions. If we consider one isolated ion (either Na$^+$ or Cl$^-$) and compute all attractive and repulsive forces by neighbors, next-neighbors, next-next-neighbors, and so on, it is possible to determine the resultant electrostatic force.

This force between individual ions is coulombic (i.e., it varies with the square of the distance). Computing all the forces and transforming the resultant force into a stress and the displacement into strain, it can be shown that one obtains an equation of the form

$$E = \frac{Ke^2}{r_0^4} \tag{4.23}$$

where r_0 is the interatomic distance, K a constant, and e the charge of an electron. This very simplified calculation shows that Young's modulus should vary with r_0^{-4}. The same dependence should exist for the bulk modulus B. Figure 4.15 shows that this type of dependence is actually observed. In the log-log plot, the slope of -4 corresponds to the dependence shown in Eq. 4.23. Elements from groups I, II, III, and IV obey that relationship. Elements of the same group were taken together because they have the same valence (electronegativity). Group I elements are monovalent and have the weakest bonding. Hence, their line is the lowest in Fig. 4.15.

In spite of the fact that bonding is more complex in metals than in ionic crystals, Gilman[40] has shown that an r_0^{-4} type of relationship can be found for some metals. This is shown in Fig. 4.16. The alkali metals seem to obey the r_0^{-4} (where r_0 is the interatomic distance) quite well; the transition metals are situated above them. The elastic properties are strongly dependent, obviously, on bonding. Figure 4.17 shows a plot of the bulk modulus versus melting point for a number of transition metals. The melting point is the temperature at

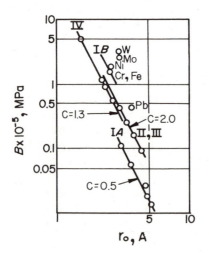

Figure 4.15 Effect of radius of closest approach on bulk modulus for groups I, II, III, IV, and transition elements.

Metallurgy of Deformation Part II

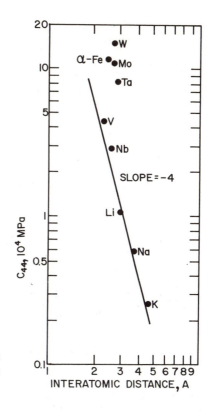

Figure 4.16 Variation of elastic constant C_{44} with r_0 for BCC metals. (Adapted with permission from [40], p. 96.)

which thermal energy is sufficient to disrupt the metallic bonding. Hence, the stronger the bonding, the higher the melting point. This correlation is clearly evident in Fig. 4.17. The lines join elements from the same column in the periodic table. Some of the series of three elements fall remarkably well in a straight line: Cr-Mo-W, V-Nb-Ta, Ag-Cu-Au.

A plot that emphasizes the importance of the electronic structure on elastic constants is the one of Fig. 4.18. The periodicity in the variation of the Young's moduli (lines represent rows in the periodic table) is indicative of the importance of the electronic structure. Transition metals, which are characterized by strong bonding by electrons from the d shell, have particularly high Young's moduli. Os, Ru, and Fe have six d electrons each and are the elements that have the highest melting point for each of the three rows of transition elements in the periodic table. In Fig. 4.16, it can be seen that the transition elements have C_{44} higher than what would have been predicted from the r_0^{-4} relationship. This confirms the indication that the strong d bonding is responsible for additional stiffness. One element stands out in Fig. 4.18: *beryllium*. Having a relatively small atomic number, it has an extremely high stiffness, comparable to that of tungsten and molybdenum. The ratio between Young's modulus and density is extremely high (six to seven times as high as for titanium and aluminum). It has unique applications in the aerospace industry. Its first use[41] was in spacers for the Minuteman missile, and it is used nowadays in space vehicles. A high

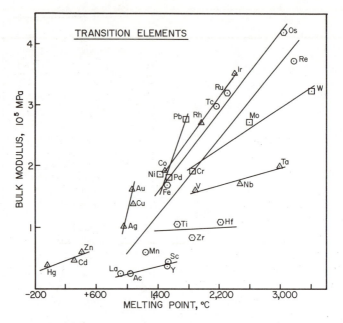

Figure 4.17 Relationship between melting point and bulk modulus for transition elements.

stiffness is required in large satellites because the lowest natural frequency of vibration must exceed a specified value to avoid resonant coupling with the booster control system during powered flight. The higher the stiffness, the higher the natural frequency of vibration. However, the metallurgical problems posed in beryllium production are many, because of its structure (HCP) and high toxicity.

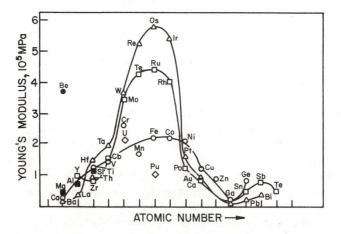

Figure 4.18 Periodic variation of Young's modulus of elements. The three lines represent the rows in the periodic table. (Adapted with permission from [43], p. 373.)

The dependence of elastic properties of a metal on interatomic separation (which can, as a first approximation, be expressed as r_0^{-4}) can be applied to rationalize two different phenomena: the temperature dependence of elastic properties and the effect of magnetic fields on elastic properties. They will be discussed below. As the temperature increases, the metals expand; this lattice expansion is treated in detail by Mott and Jones[42] but is beyond the scope of this book. It suffices to say that as the temperature increases, the amplitude of vibration of the atoms increases (although the frequency remains constant). This amplitude increase will accommodate the thermal energy term (kT) and the expanded lattice will have a larger r_0. This, in turn, will produce a decrease in the elastic constants. The change in elastic constants with temperature is much less pronounced than the change of yield stress, UTS, elongation. The Young's modulus at the melting point is usually between one-half to two-thirds of the low-temperature value. The temperature dependence of the yield point is much more pronounced. It will be seen in Chapter 8 that this is because plastic deformation is a thermally activated process. Figure 4.19 illustrates the changes in E with temperature for some metals. The effect of magnetic fields can be explained by the same rationale. A magnetic field, due to the magnetostrictive effect, changes the lattice parameter slightly; this, in turn, affects the elastic properties. When Young's modulus of nickel in the presence and absence of a magnetic field is measured, appreciable differences are found. Actually, between 200 and 360°C (Curie temperature, where ferromagnetic–paramagnetic transformation takes place) the Young's modulus of nickel increases with temperature.[43] By appropriate alloying it is possible to obtain alloys that have essentially a constant Young's modulus over a certain temperature range. Such an alloy is Elinvar (36% Ni, 12% Cr, 1 to 2% Si, 0.8% C, balance Fe), and it has an essentially constant E between 15 and 40°C; it is ideal for springs in watches and other precision instruments.

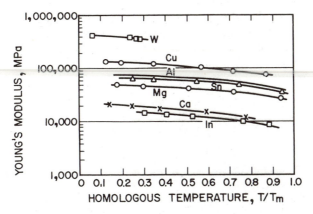

Figure 4.19 Effect of temperature on dynamic Young's modulus. (Adapted with permission from [43], p. 376.)

EXERCISES

4.1. Draw a cubic unit cell, designating the planes of the six faces by their Miller indices.

4.2. Draw a hexagonal unit cell designating the directions connecting the origin to the vertices by their Miller indices.

4.3. Draw a hexagonal unit cell, designating by their Miller indices:
   (a) The direction corresponding to the six parallel sides of the hexagon.
   (b) The directions corresponding to the diagonals of the six rectangles that compose the lateral faces.
   (c) The base planes.
   (d) The planes that compose the lateral faces.

4.4. Prove that the c/a ratio is equal to 1.633 for an HCP unit cell composed of rigid spheres.

4.5. (a) In a cubic cell, draw the planes with indices

$$
\begin{array}{cc}
(112) & (321) \\
(4\bar{1}2) & (01\bar{1})
\end{array}
$$

   (b) In a hexagonal cell, draw the directions with indices

$$
\begin{array}{cc}
[1 \quad 2 \quad \bar{3} \quad 0] & [3 \quad 1 \quad \bar{4} \quad 1] \\
[1 \quad \bar{1} \quad 0 \quad 2] & [0 \quad 1 \quad \bar{1} \quad 2]
\end{array}
$$

4.6. Determine the theoretical cleavage stress for iron (BCC) along $\langle 100 \rangle$ and $\langle 111 \rangle$.

4.7. Along which directions are the atoms in direct contact for the BCC, FCC, and HCP structures?

4.8. Give the Miller indices for the closest-packed planes in the FCC, BCC, and HCP structures.

4.9. Assume that the iron atoms have the same radius (1.26 Å) in both the FCC and BCC structures. What would be the density for both cases?

4.10. The low-temperature dependence of the yield stress exhibited by $Pd_{77.5}Cu_6Si_{16.5}$ indicates that thermal activation does not play a very important role in the onset of deformation. How does this relate to Li's mechanism?

4.11. If the atoms at the surface of an iron crystal are relaxed from their normal positions, with an increase of 10% in the average interatomic spacing, how would the Young's modulus be affected? Calculate the change, using the value of Poisson's ratio from Table 1.5.

REFERENCES

[1] R.C. Evans, *An Introduction to Crystal Chemistry*, 2nd ed., Cambridge University Press, Cambridge, 1964.

[2] S. Raimes, *The Wave Mechanics of Electrons in Metals*, Elsevier/North-Holland, New York, 1970.

[3] M.A. Meyers, C.O. Ruud, and C.S. Barrett, *J. Appl. Cryst.*, 6 (1973) 39.

[4] E. Orowan, *Rep. Prog. Phys.*, 12 (1949) 185.

[5] R. Gomer and C.S. Smith (eds.), *The Structure and Properties of Solid Surfaces*, University of Chicago Press, Chicago, 1952.

[6] A. Kelly, *Strong Solids*, 2nd ed., Clarendon Press, Oxford, 1973.

[7] J. H. de Boer, *Trans. Faraday Soc.*, *32* (1936) 10.

[8] S. Silver, *J. Chem. Phys.*, *8* (1940) 919.

[9] F. Zwicky, *Phys. Z.*, *24* (1923) 131.

[10] W.R. Tyson, *Phil. Mag.*, *14* (1966) 925.

[11] J. Frenkel, *Z. Phys.*, *37* (1926) 572.

[12] J.K. Mackenzie, Ph.D. dissertation, University of Bristol, England, 1949; cited in [6].

[13] A. Kelly, W.R. Tyson, and A.H. Cottrell, *Phil. Mag.*, *15* (1967) 567.

[14] J. Weertman, *Phil. Mag. A*, *43* (1981) 1103.

[15] G.I. Taylor, in *The Sorby Centennial Symposium on the History of Metallurgy*, C.S. Smith (ed.), Gordon and Breach, New York, 1965, p. 355.

[16] W.C. Leslie, *J. Metals*, *23*, No. 12 (1977) 31.

[17] D.S. Lashmore, W.A. Jesser, D.M. Schladitz, H.J. Schladitz, and H.G.F. Wilsdorf, *J. Appl. Phys.*, *48* (1977) 478.

[18] S.S. Brenner, *J. Appl. Phys.*, *27* (1956) 1484.

[19] S. S. Brenner, *Acta Met.*, *4* (1956) 62.

[20] S.S. Brenner, *Growth and Perfection of Crystals*, Wiley, New York, 1958, p. 157.

[21] K. Yoshida, Y. Goto, and M. Yamamoto, *J. Phys. Soc. Jap.*, *21* (1966) 825.

[22] R. L. Mehan, W. H. Sutton, and J.A. Herzog, General Electric Rep. No. R65SD28, 1965.

[23] P. Grenier and A. Kelly, *Compt. Rend. Acad. Sci. Paris, Ser. B*, *266* (1968) 859.

[24] *Mater. Sci. Eng.*, *23* (1976) 81.

[25] D.E. Polk, B.C. Giessen, and F.S. Gardner, *Mater. Sci. Eng.*, *23* (1976) 309.

[26] J.C.M. Li, in *Frontiers in Materials Science—Distinguished Lectures*, L.E. Murr and C. Stein (eds.), Marcel Dekker, New York, 1976, p. 527.

[27] J. Megusar, J.B. Vander Sande, and N.J. Grant, in *Rapidly Quenched Metals,* N.J. Grant and B.C. Giessen (eds.), MIT Press, Cambridge, Mass. 1976, p. 401.

[28] L.A. Davis, cited in [27], p. 369.

[29] C.A. Pampillo and H.S. Chen, *Mater. Sci. Eng.*, *13* (1974) 181.

[30] J.J. Gilman, *J. Appl. Phys.*, *44* (1973) 675.

[31] J.J. Gilman, *J. Appl. Phys.*, *46* (1975) 1625.

[32] F. Spaepen and D. Turnbull, *Scripta Met.*, *8* (1974) 563.

[33] K. Maeda and S. Takeuchi, *Phys. Status Solidi (a).*, *49* (1978) 685.

[34] K. Maeda and S. Takeuchi, *J. Phys. F*, *12* (1978) L283.

[35] A.S. Argon and H.Y. Kuo, *Mater. Sci. Eng.*, *39* (1979) 101.

[36] A.S. Argon, *Acta Met.*, *27* (1979) 47.

[37] J.C.M. Li, in *Metallic Glasses*, ASM, Metals Park, Ohio, 1978.

[38] J.C.M. Li, *Proc. 4th Int. Conf. Rapidly Quenched Metals*, Sendai, Japan, 1981.

[39] J.C.M. Li, *Proc. Mater. Res. Soc. Symp. Rapidly Solidified Amorphous and Crystalline Alloys*, Boston, Nov. 17, 1981.

[40] J.J. Gilman, *Mechanical Behavior of Crystalline Solids*, NBS Monograph 59, 1963, p. 79.

[41] R.J. Switz, *Metals Eng. Quart.*, Nov. 1973, p. 35.

[42] N.F. Mott and H. Jones, *The Theory of the Properties of Materials and Alloys*, Dover, New York, 1958.

[43] O.D. Sherby, in *Nature and Properties of Materials: An Atomistic Interpretation*, J. Pask (ed.), Wiley, New York, 1967, pp. 373, 375, 376.

SUGGESTED READING

CAGLIOTI, G. (ed.), *Atomic Structure and Mechanical Properties of Metals*, Proc. International School of Physics Enrico Fermi, Course LXI, North-Holland, Amsterdam, 1976.

GILMAN, J.J., and H.H. LEAMY (eds.), *Metallic Glasses*, ASM, Metals Park, Ohio, 1978.

GRANT, N.J., and B.C. GIESSEN (eds.), *Rapidly Quenched Metals*, MIT Press, Cambridge, Mass., 1976.

KELLY, A., *Strong Solids*, 2nd ed., Clarendon Press, Oxford, 1973.

LEVITT, A.P. (ed.), *Whisker Technology*, Wiley, New York, 1970.

Chapter 5

POINT DEFECTS

5.1 INTRODUCTION

Point defects are so called in spite of the fact that they have a volume; they also generate a stress field in the crystal lattice. This name is due to the small dimensions of these defects. As Table 4.4 shows, they can have a diameter of approximately 10^{-15} or 10^{-11} nm, depending on whether they are electronic or atomic point defects, respectively. Thus, it can be seen that these defects are confined to the volume of an atom.

The electronic point defects are characterized by the lack (electron hole) or excess of electrons. They have an important effect in modifying the properties of ionically or covalently bonded materials.[1] In this case, a lack or excess of electrons is introduced in the lattice by introducing ions with valence different from the lattice ions; this process is called "doping" and leads to extrinsic semiconductors. In metallic structures, on the other hand, the presence of ions with a different valence does not have such a profound effect on the electronic properties. The most obvious effect would be the alteration of the electrical potential of the metal.

If electronic point defects do not have a significant effect on the properties of metals, such is not the case with atomic point defects. They can be of three types:

1. *Vacancies*. When an atomic position in the Bravais lattice is vacant.
2. *Interstitial point defects*. When an atom occupies a position that would normally not be occupied (an interstitial position). This interstitial position

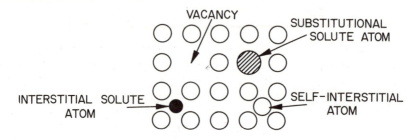

Figure 5.1 Atomic point defects in metals.

can be occupied by an atom of the material or by a foreign atom; it is called a self-interstitial or interstitial impurity for the two cases above, respectively.

3. *Substitutional point defects.* When a regular atomic position is occupied by a foreign atom.

Figure 5.1 shows the three types of atomic point defects. The representation of self-interstitials is oversimplified. Self-interstitials seem to come in all varieties; crowdions (Section 15.3) are an example. The vacancy concentration in pure elements is very low at low temperatures. The probability that an atomic site is a vacancy is of approximately 10^{-6} at low temperatures, rising to 10^{-3} at the melting point. In spite of their low concentration, vacancies have a very important effect on the properties because they control the self-diffusion and substitutional diffusion rates. The movement of atoms in the structure is coupled to the movement of vacancies. Section 5.3 calculates the equilibrium concentrations of vacancies and self-interstitials.

5.2 INTERSTICES IN THE FCC, BCC, AND HCP STRUCTURES

The self-interstitials and interstitial impurities lodge themselves in the "holes" that the structure has. There is more than one type of "hole" in the FCC, BCC, and HCP structures, and their diameters and positions will be determined below.

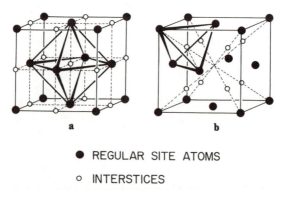

● REGULAR SITE ATOMS

○ INTERSTICES

Figure 5.2 Interstices in FCC structure: (a) octahedral void; (b) tetrahedral void.

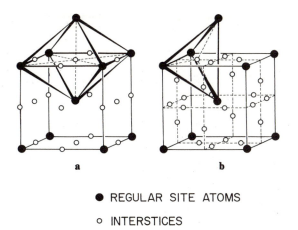

● REGULAR SITE ATOMS

○ INTERSTICES

Figure 5.3 Interstices in the BCC structure: (a) octahedral void; (b) tetrahedral void.

The FCC structure, shown in Fig. 5.2, has two types of voids: the larger, called octahedral, and the smaller, called tetrahedral. The names are derived from the nearest-neighbor atoms; they form the vertices of the polyhedra above. If we consider the atoms as rigid spheres, we can calculate the maximum radius of a sphere that would fit into the void without straining the lattice. The reader is encouraged to engage in this exercise; with some luck, he will find radii of 52 and 28 pm for octahedral and tetrahedral voids, respectively, in γ-iron.[2] Hence, carbon ($r = 80$ pm) and nitrogen ($r = 70$ pm) produce distortions in the lattice, when they occupy the voids.

In BCC metals there are also octahedral and tetrahedral voids, as shown in Fig. 5.3. In this case, however, the greater void is the tetrahedral. Considering rigid spheres for α-iron, the void radii are 36 and 19 pm for tetrahedral and octahedral interstices, respectively. Hence, a solute atom is accommodated in an easier way in FCC than BCC iron, in spite of the fact that the FCC structure is more closely packed.

Analogously, the HCP structure presents tetrahedral and octahedral voids, shown in Fig. 5.4; the reader is reminded of the similarity between the FCC

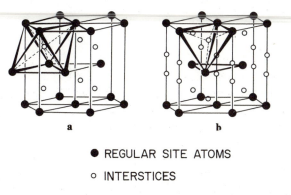

● REGULAR SITE ATOMS

○ INTERSTICES

Figure 5.4 Interstices in the HCP structure: (a) octahedral void; (b) tetrahedral void.

and HCP structures (see Section 4.3), which explains the presence of the same voids.

5.3 EQUILIBRIUM OF POINT DEFECTS

A very important characteristic of vacancies and self-interstitial atoms, in contrast with line and surface defects, is that they can exist in thermodynamic equilibrium at temperatures above 0 K. The thermodynamic equilibrium in a system of constant mass, at a constant pressure and temperature, and that does not execute any work in addition to the work against pressure, is reached when the Gibbs free energy is minimum.[3] The formation of point defects in a metal requires a certain quantity of heat (as there is no work being executed, except the one against pressure). Hence, if $dH = \delta q$,[4] the enthalpy of the system increases. The configurational entropy also increases, because there are a certain number of different ways of putting the defects into the system. Figure 5.5 shows schematically the variations in H, TS, and F with the number n of defects. The enthalpy H and the entropy versus temperature product (TS) are plotted. The Gibbs free energy (F) is, by definition,

$$F = H - TS \qquad (5.1)$$

One can thus see that the free energy will reach a minimum for a certain value of n different from zero; at 0 K the entropic term is zero and the equilibrium concentration is zero.

The equilibrium concentration of point defects can be calculated from statistical considerations[5] and is equal to

$$V = \frac{n}{N} = e^{-H_f/kT} \qquad (5.2)$$

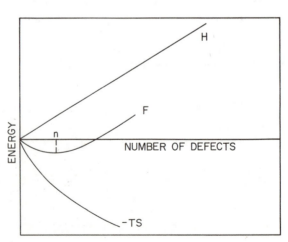

Figure 5.5 Variation of Gibbs free energy with concentration of point defects.

where n and N are the number of point defects and sites, respectively; H_f is the enthalpy of formation of the defects; and k is Boltzmann's constant. It is often energetically favorable for defects to arrange themselves in groups. Divacancies, trivacancies, and vacancy–interstitial complexes are some examples. For instance, the equilibrium concentration of divacancies in an FCC unit cell is

$$(V_2)_{\text{FCC}} = 6(V_1)^2 e^{B/kT} \qquad (5.3)$$

V_1 and V_2 are the concentrations of vacancies and divacancies, and B is the enthalpy of bonding of divacancies. The enthalpies of formation of point defects are presented by Huntington[6,7] and described by Swalin.[8] Indirect measurements allow their determination. For copper, the free energies of formation of vacancies and interstitials were found to be

$$E_v = 125 \text{ kJ/mol} \qquad E_i = 878 \text{ kJ/mol}$$

We have, approximately, the following ratio:

$$\frac{E_i}{E_v} = \frac{H_i}{H_v} \simeq 7$$

Therefore, for copper, the enthalpy of formation of a vacancy is approximately seven times lower than that of a self-interstitial defect. Using Eq. 5.2, we can obtain the ratio between the vacancy (X_v) and interstitial (X_i) concentrations. It is

$$\frac{X_v}{X_i} \simeq \exp\left(\frac{H_i - H_v}{kT}\right)$$

For copper at 1000 K,

$$\frac{X_v}{X_i} \simeq 10^{39}$$

It can be concluded that, at least in close-packed structures, the concentration of interstitials is negligible with respect to that of the vacancies. If the entropic term is neglected, and using Eq. 5.2 for copper at 1000 K, we obtain

$$X_v \simeq 3 \times 10^{-7}$$

Hence, there is only one vacancy for each 3×10^6 copper atoms at 1000 K. This number is very small; in spite of this, it corresponds to approximately 10^{16} vacancies/cm³. The low concentration of self-interstitials in close-packed structures is a consequence of the small diameter of the interstitial voids (Section 5.2). In more open structures these concentrations can be higher. Even so, there are no reports of high interstitial concentrations in equilibrium structures.

5.4 MORE COMPLEX POINT DEFECTS

It was seen in Section 5.3 that point defects can group themselves in more complex arrangements. The enthalpy of formation of divacancies has been determined for several metals. For example, one has, for copper ($H_f = 5.63 \times 10^{-19}$ J), $B = 0.96 \times 10^{-19}$ J. According to calculations conducted by Seeger and Bross,[9] the enthalpy of formation of divacancies in noble metals is of the order of 0.48×10^{-19} J. It is nowadays accepted that divacancies are stable, in spite of the fact that their enthalpies of bonding are not very well known. Larger groups of vacancies are also formed, but their enthalpies of formation are not easily calculated. Some calculations to predict the relative stability of the various configurations have been conducted. The first calculation for a trivacancy has been done using a Morse function,[10] for copper. Tetra- and pentavacancies are also studied by computation.

Di-interstitials also exist and their energies can be calculated by the same processes. Similarly, the vacancies can bind themselves to impurity atoms when the binding energy is positive. The binding energy between a vacancy or self-interstitial and an impurity is determined, to a great extent, by the interaction between the stress fields and electrostatic interactions. If the substitutional solute atom is greater than the matrix atom, the lattice around it is under compression; consequently the presence of a vacancy and its stress field would result in a decrease on free energy. On the other hand, a substitutional atom smaller than the lattice atoms would produce tensile stresses in the lattice and attract interstitial defects. The binding energies can be calculated; Hasiguti[11] showed that, for copper, a small impurity atom (beryllium) has a binding energy with a self-interstitial of 0.5 eV (0.8×10^{-19} J).

5.5 PRODUCTION OF POINT DEFECTS

Intrinsic point defects in a metal—vacancies and self-interstitials—exist in well-established equilibrium concentrations (Section 5.3). By appropriate processing the concentration of these defects can be increased.

Quenching or ultra-high-speed cooling is one of these methods. The concentration of vacancies in BCC, FCC, and HCP metals is greatly superior to that of interstitials and of the order of 10^{-3} close to the melting point; it is only 10^{-6} at temperatures of about half the melting point. Hence, if a specimen is cooled at a high enough rate, the high-temperature concentration can be retained at low temperature. For this to occur, the rate of cooling has to be such that the vacancies cannot diffuse to sinks: grain boundaries, dislocations, surface, and so on. Theoretically, gold would have to be cooled from 1330 K to ambient temperature at a rate of 10^{11} K/s to retain integrally its high-temperature vacancy concentration. The fastest quenching technique to cool thin wires produces cooling rates lower than 10^5 K/s; nevertheless, a significant portion of the high-temperature point defects is retained. Kimura and Maddin[12] treat the subject in detail.

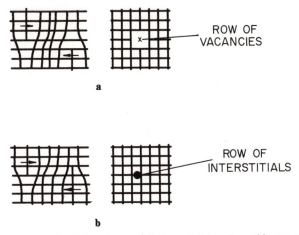

Figure 5.6 Formation of point defects by annihilation of dislocations: (a) row of vacancies; (b) row of interstitials.

Another method of increasing the concentration of point defects is by plastic deformation. The movement of dislocations, as will be seen in Chapter 6, generates dislocations by two mechanisms: nonconservative motion of jogs and annihilation of parallel dislocations of opposite sign, producing a line of vacancies or interstitials. The jogs are created by dislocation intersections; since they cannot glide with dislocations, they have to climb as the dislocation moves. In a screw dislocation, they are small segments having edge character; the slip plane of this segment is not compatible with that of the dislocation. The climb is possible only by continuous emission of vacancies or interstitials. The second mechanism is depicted schematically in Fig. 5.6. When the two dislocations cancel each other, they create rows of interstitials or vacancies if their slip planes do not coincide.

Deformation by shock waves (see Chapter 15) is another means by which high concentrations of point defects are created.[13] Figure 5.7 shows the vacancy

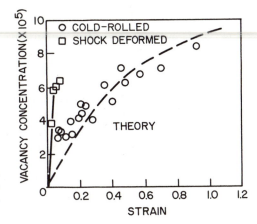

Figure 5.7 Effect of deformation by shock waves and cold rolling on vacancy concentration in nickel. (Reprinted with permission from [13], p. 138.)

concentrations obtained by Kressel and Brown [13] for shock-loaded and conventionally deformed (by cold-rolling) nickel; at the same strain, shock deformation induces more defects than does conventional deformation. Meyers and Murr[14] calculated the concentration of vacancies for a shock loading pressure of 20 GPa and found 5×10^{-5}; this is in fair agreement with the results reported by Kressel and Brown.[13] The calculations are based on the generation of vacancies by nonconservative jog motion (climb).

Quenching produces mostly vacancies and vacancy groups. The concentrations obtained are lower than 10^{-4}. Deformation, on the other hand, can introduce higher concentrations of vacancies and equivalent ones of interstitials; the problem is that it introduces a number of other substructural changes that complicate the situation. Dislocations are introduced, and they interact strongly with point defects. There is a method of producing point defects that does not present the problems outlined above. Radiation of the metal by high-energy particles allows the introduction of a high concentration of point defects. The radiation produces the following effects: (1) displacement of the electrons or ionization, (2) production of displaced atoms by elastic collisions, and (3) production of fission and thermal spikes. This subject is treated in greater detail in Chapter 15. The displacement of atoms is produced by the elastic collision of the bombarding particles with the lattice atoms, transferring their kinetic energy to them. This may cause the atoms to travel through the lattice. In the majority of the cases the atom travels a few atomic distances and places itself in an interstitial site. Consequently, a vacancy is produced, together with a self-interstitial. The energy transferred in the collision has to be well above the energy required to form an interstitial–vacancy pair in a reversible thermodynamic process (3 to 6 eV, or 4.8×10^{-19} to 9.6×10^{-19} J). It is believed that the energy transferred to the atom has to be approximately 25 eV (40×10^{-19} J). Different particles can be used in the bombardment process: neutrons, electrons, γ rays, α particles.

‡5.6 EFFECT OF POINT DEFECTS ON MECHANICAL PROPERTIES

Point defects have a marked effect on the mechanical properties. For this reason, the effect of radiation is of great importance (Chapter 15). The effect of excess vacancy concentration on the mechanical properties had been observed by Li et al.[15] in 1953, among others. Nevertheless, the first systematic study is due to Maddin and Cottrell.[16] They used aluminum monocrystals with various purity levels, observing that the yield stress increased with quenching. Quenching was accomplished by taking the specimens from 600°C and throwing them into a water–ice mixture, while annealed material was slowly cooled in the furnace. The yield stress increased from 550 to 5900 kPa, on average. The effect of impurity atoms could be discarded because the increase in yield stress was consistent throughout the specimens. The effect of possible residual stresses due to quenching was also discounted. With the purpose to obtain evidence that was still more convincing, a crystal was tested immediately after quenching,

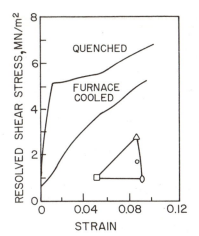

Figure 5.8 Stress versus strain curves for aluminum monocrystals. The crystallographic orientation is shown in the stereographic triangle. (Adapted with permission from [16], p. 737.)

while another was tested after staying a few days at ambient temperature. The yield stress increased from 5.9 GPa to 8.4 GPa in the aged condition. Maddin and Cottrell's[16] work allows the following conclusions to be made. The strengthening in quenching is due to the interaction of dislocations and vacancies or groups thereof. The effect of jogs, formed by the condensation of vacancies on the dislocations, can also be considerable. During aging, the excess concentration of vacancies forms groups and/or annihilates preexisting dislocations.

There are also alterations in the plastic portion of the stress versus strain curve, seen in Fig. 5.8. The initial work-hardening rate of the quenched aluminum is lower. At greater strains, however, the two work-hardening rates become fairly similar. Hence, the effect of quenching disappears at higher strains. This is thought to be due to the fact that the excess concentrations are eliminated during plastic deformation; at the same time, excess vacancies are generated by dislocation motion, so that the concentrations in the quenched and annealed materials become the same.

The increase in hardness in many quenched metals is negligible, in spite of the obvious changes in the stress versus strain curve. This is explained by the fact that the effect of quenching disappears after a certain amount of plastic deformation. Since the indenter deforms plastically (in an extensive way) the metal, the effect of quenching is minimum. Aust et al.[17] were able to obtain hardness differences, after especially designed experiments.

The excess of vacancies also has an effect on the internal friction of metals. The concept of internal friction is explained by Reed-Hill.[18] In these experiments, a metallic specimen (wire) is subjected to a cyclic torsion by a rotational pendulum attached to its bottom. When the frequency of the pendulum is equal to the frequency of an internal phenomenon, considerable damping occurs in the oscillations. This damping can be determined experimentally. Internal friction is affected in two ways by vacancies. First, there is Zener's relaxation, caused by the atomic rearrangement due to the changes in the external state of stress; the concentration and mobility of vacancies determines the relaxation rate. Second, the vacancies interact with the dislocations or are annihilated in the formation of jogs. Consequently, the mobility of the dislocations is decreased and the internal friction due to the movement of the dislocations is decreased in quenched metals. This was observed experimentally by Levy and Metzger.[19]

‡5.7 OBSERVATION OF POINT DEFECTS

The field-ion microscope (FIM) allows the direct imaging of atoms. The tip of a fine tungsten wire is shown in Fig. 5.9(a). The atoms are bright spots and are arranged in circles around a point (the center of the tip). Although

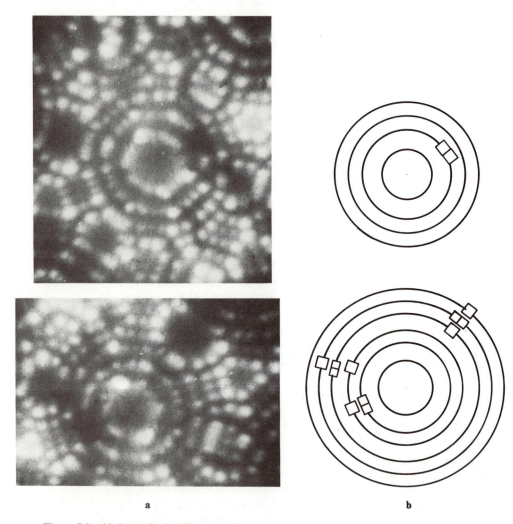

a　　　　　　　　　　　　　　　b

Figure 5.9 (a) Laser-shock-induced vacancy clusters in tungsten. The vacancy clusters can be seen as dark regions. (Courtesy of O. T. Inal, New Mexico Institute of Mining and Technology.) (b) Schematic representation of (a) showing vacancy positions as squares; atoms positioned in concentric rings because emission tip is viewed "head on."

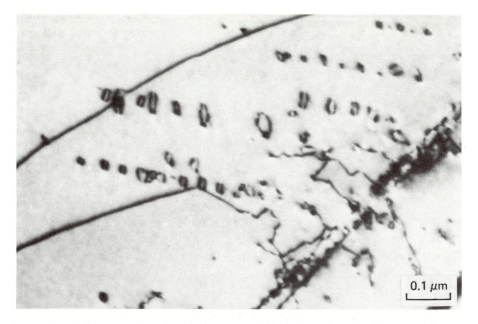

Figure 5.10 Nickel heated at 600°C for 10 min and quenched in liquid nitrogen. Strings of vacancy loops can be clearly seen. (Courtesy of L. E. Murr, New Mexico Institute of Mining and Technology.)

the atomic positions are not exactly reproduced in the FIM, one can clearly see the presence of gaps, sketched as squares in Fig. 5.9(b). These are vacancies. One should be careful, however, in the interpretation of the significance of these vacancies, since one is looking at the surface and not directly into the bulk. These vacancy clusters were produced by laser shock.

Transmission electron microscopy (TEM) does not allow the high magnifications of FIM and one cannot under usual conditions see the individual vacancies. However, groups of vacancies become visible, and Fig. 5.10 shows strings of vacancy loops that were "punched" out from a source. The nickel sheet was heated to 600°C for 10 minutes and quenched in liquid nitrogen.

EXERCISES

5.1. Calculate the radii of the tetrahedral and octahedral holes in BCC and FCC iron; assume lattice parameters of 0.286 and 0.357 nm, respectively.

5.2. Calculate the concentration of monovacancies in gold at 1000 K knowing that $H_f = 1.4 \times 10^{-19}$ J. If the gold is suddenly quenched to ambient temperature, what will be the excess vacancy concentration?

5.3. How many vacancies per cubic centimeter are there in gold, at ambient temperature, assuming a lattice parameter of 0.408 nm?

5.4. What is the effect of vacancies on electrical conductivity?

5.5. Using Le Châtelier's principle, comment on the effect of hydrostatic pressure on vacancy concentration in a cubic material.

5.6. What is the effect of vacancies on the amplitude of vibration of the neighboring atoms?

REFERENCES

[1] C.H. Johansson, *Arch. Eisenhüttenv.*, *11* (1937) 241.

[2] L.W. Strock, *Z. Krist.*, *93* (1936) 285.

[3] L.S. Darken and R.W. Gurry, *Physical Chemistry of Metals*, McGraw-Hill, New York, 1953, p. 189.

[4] L.S. Darken and R.W. Gurry, *Physical Chemistry of Metals*, McGraw-Hill, New York, 1953, p. 160.

[5] N. Davidson, *Statistical Mechanics*, McGraw-Hill, New York, 1962, p. 87.

[6] H.B. Huntington and F. Seitz, *Phys. Rev.*, *61* (1942) 325.

[7] H.B. Huntington, *Phys. Rev.*, *91* (1953) 1092.

[8] R.A. Swalin, *Thermodynamics of Solids*, Wiley, New York, 1962, p. 222.

[9] A. Seeger and H. Bross, *Z. Phys.*, *145* (1956) 161.

[10] A.C. Damask, G.J. Dienes, and V.C. Weizer, *The Physical Review, 113* (1959) 781.

[11] R.R. Hasiguti, *J. Phys. Soc. Jap.*, *15* (1960) 1807.

[12] H. Kimura and R. Maddin, *Quench Hardening in Metals*, in series *Defects in Crystalline Solids*, S. Amelinckx, R. Gevers and J. Nihoul (eds.), North-Holland, Amsterdam, 1971.

[13] H. Kressel and N. Brown, *J. Applied Physics*, *38* (1967) 138.

[14] M.A. Meyers and L.E. Murr, in *Shock Waves and High-Strain-Rate Phenomena in Metals: Concepts and Applications*, M.A. Meyers and L.E. Murr (eds.), Plenum Press, New York, 1981, p. 482.

[15] C.H. Li, J. Washburn, and E.R. Parker, *Trans. AIME, 197* (1953) 1223.

[16] R. Maddin and A.H. Cottrell, *Phil. Mag.*, *46* (1955) 735.

[17] K.T. Aust, A.J. Peat, and J.H. Westbrook, *Acta Met.*, *14* (1966) 1469.

[18] R.E. Reed-Hill, *Physical Metallurgy Principles*, 2nd ed., D. Van Nostrand, New York, 1973, p. 433.

[19] M. Levy and M. Metzger, *Phil. Mag.*, *46* (1955) 10.

SUGGESTED READING

BARRETT, C.S., and T.B. MASSALSKI, *Structure of Metals*, 3rd ed., McGraw-Hill, New York, 1966.

CRAWFORD, J.H., JR., and L.M. SLIFKIN (eds.), *Point Defects in Solids*, Plenum Press, New York, 1972.

DAMASK, A.C., and G.J. DIENES, *Point Defects in Metals*, Gordon and Breach, New York, 1963.

FLYNN, C.P., *Point Defects and Diffusion*, Clarendon Press, Oxford, 1972.

KIMURA, H., and R. MADDIN, *Quench Hardening in Metals*, in series *Defects in Crystalline Solids*, S. Amelinckx, R. Gevers, and J. Nihoul (eds.), North-Holland, Amsterdam, 1971.

NOWICK, A.S., and B.S. BERRY, *Anelastic Relaxation in Crystalline Solids*, Academic Press, New York, 1972.

SWALIN, R.A., *Thermodynamics of Solids*, 2nd ed., Wiley, New York, 1972.

VAN BUEREN, H.G., *Imperfections in Crystals*, North-Holland, Amsterdam, 1961.

Chapter 6

LINE DEFECTS

‡6.1 INTRODUCTION

Bands in the surface of plastically deformed metallic specimens had been reported as early as the nineteenth century.[1,2] With the discovery of the crystalline nature of metals (see Section 4.2), these bands were interpreted as being the result of the shear of one part of the specimen with respect to the other. Similar slip bands (or markings) were observed by geologists in rocks. However, the calculations of the theoretical strength of crystals based on the simultaneous motion of all atoms along the slip band showed systematic deviations of several orders of magnitude with respect to the experimental values (Section 4.4). In the early part of the twentieth century several proposals arose; one of the most popular ones consisted of small cracks that propagated throughout the material, producing shear and reclosing themselves. In 1934 the correct solution to the apparent contradiction was proposed independently by Orowan,[3] Polanyi[4] and Taylor,[5] and the concept of a dislocation was introduced. Orowan[6] described later the mental process that led him to propose dislocations. He apparently had been developing the concept since 1929, when he was working on his doctoral thesis under Becker. He communicated the concept to Polanyi, who worked on it and submitted, almost simultaneously with Orowan, two papers to the German journal *Zeitschrift für Physik*. G. I. Taylor had been involved in studies on the nature of the strength of metals and considered the existing concepts of cracks in the material "very vague and certainly untrue in the case of face-centered cubic metals."[7] He decreased the size of these cracks until they were only one atomic layer wide. In order to study the interactions between these unit cracks, he needed to know the stress fields around them:[7] "As soon as I realized this, I looked up Love's textbook on elasticity to find out whether the stress distributions corresponding with such irregularities, or dislocations, had been calculated. I found that Timpe had given the particular solution I required in 1905 and that the matter had been followed up and generalized to other types of dislocations by Volterra in 1907."

Having postulated dislocations, Taylor proceeded to study the relationships between these stress fields and arrived at a parabolic relationship between plastic stress and plastic strain. His theory on the parabolic work hardening of crystals is still accepted today (see Chapter 9) and predicts the well-known proportionality between the stress and the square root of the dislocation density. Unfortunately, Sir Geoffrey Taylor, a meteorologist and a physicist, did not pursue his ideas: "In 1934 I was developing the statistical theory of turbulence and experimenting with anchors. These activities took up most of my time so I did not attempt to develop the work on dislocation beyond the state described in the two papers published that year in the *Proceedings* of the Royal Society."

The concept of dislocation permitted that one part of a metal could be sheared without the need for *simultaneous* movement of the atoms above the interface; instead, the dislocation would move throughout the crystal. Hence, although the energy expended in deformation is the same for simultaneous shear and dislocation motion, the applied stress is much lower in the latter case. The type of dislocation proposed by Orowan, Polanyi, and Taylor is known as *edge dislocation*. Later, in 1939, the Dutch metallurgist Burgers[8] proposed another type: *screw dislocation*. Consider the perfect cube of Fig. 6.1. Depending on the cut and subsequent shear applied, two types of dislocations are obtained. Line *AA* (the end of the cut) is called dislocation line. If the atoms are displaced perpendicularly to the dislocation line, we have an edge dislocation; if they are displaced parallel to the line, we have a screw dislocation.

There is also a mixed dislocation that possesses both screw and edge character. Figure 6.1(b) shows such a dislocation together with a "cut." It can be seen that the shear direction is neither parallel (screw) nor perpendicular (edge) to the cut direction. The line *AA* defines a mixed dislocation.

There is also another type of dislocation, called "helical dislocation"; it is described by Hull.[9] It forms a large spiral and is sometimes observed in crystals that were heat-treated to produce climb. A mechanism for their generation was proposed by Amelinckx et al.[10] These dislocations are of mixed character; the reader should not confuse them with screw dislocations.

Dislocations are, after X-ray diffraction, the most important conceptual development in physical metallurgy. They will be studied in detail in this chapter since they are the building blocks for the understanding of the mechanical response of metals. Important as they are, the knowledge of their behavior has not given us, alas, the power to control them. Thus, metallurgy is and will remain as much an art as it is a science.

The dislocation treatment given in this chapter is far from comprehensive. It is assumed that the reader is familiar with the concept of dislocations. Any good textbook in physical metallurgy provides background material; dislocations

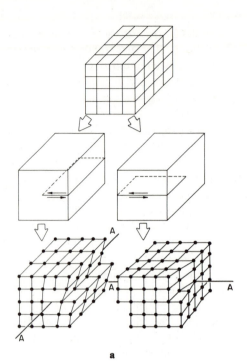

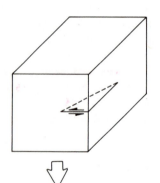

a

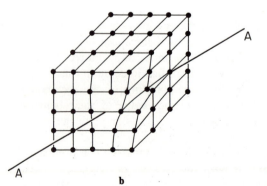

b

Figure 6.1 (a) Edge and screw dislocations obtained from a perfect crystal; (b) mixed dislocation.

and other specific topics should be studied by students who feel unprepared. Especially recommended is the excellent monograph by Weertman and Weertman (see "Suggested Reading").

‡6.2 EXPERIMENTAL OBSERVATION OF DISLOCATIONS

It took 20 years to verify the existence of dislocations experimentally, and this period (1935–1955) was surrounded by skepticism and harsh polemics. It is *arguments about doctrines* said that in the best known mining school this skepticism prolonged itself to the late 1960s. Nevertheless, their existence is nowadays universally recognized, and the "lunatic" theories and models have been proven to be remarkably correct. A number of techniques have allowed the observation of dislocations. Some of them are described and illustrated below.

"Etch pitting" uses the fact that chemical attack is more intense at the region in which a dislocation intersects the free surface; this greater rate of dissolution is due to the stresses generated by the dislocation. Figure 6.2 shows an array of etch pits;[11] the surface plane of the specimen (copper) is (111), which explains the pyramidal shape of the pits. The decoration technique has also been successfully used. Impurity atoms are attracted to the dislocation line due to the stress fields (See Chapter 10); they "decorate" the dislocations and make them visible by light microscopy. Figure 6.3 shows a hexagonal array of dislocations in NaCl doped with silver. The migration of the interstitial silver atoms toward the dislocations renders the latter visible by optical microscopy in transmission of a foil (with thickness of 0.2 mm).

Special X-ray techniques—specifically, Berg–Barrett's technique—have also been used to reveal the presence of dislocations. Figure 6.4 shows X-ray topographs obtained by Newkirk.[13] In Fig. 6.4(a) the same region is viewed by optical microscopy; dislocations were made visible by an appropriate etch. Depending on the diffraction condition, either the edge [Fig. 6.4(b)], the screw [Fig. 6.4(c)], or all dislocations [Fig. 6.4(d)] are seen. The special configurations of dislocations will be explained in the coming sections.

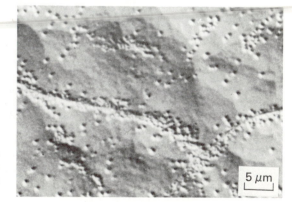

Figure 6.2 Etch pits at dislocations observed on copper by scanning electron microscopy. (Reprinted with permission from [11], p. 55.)

5 μm

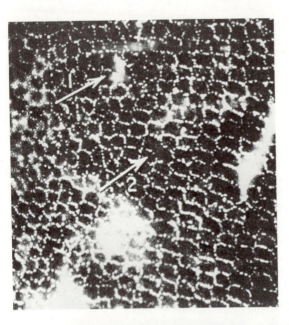

Figure 6.3 Dislocations in NaCl revealed by decoration with silver. (Reprinted with permission from [12], p. 26.)

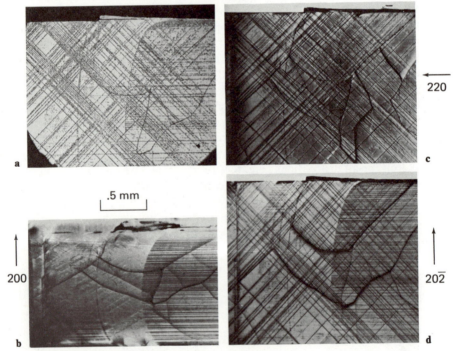

Figure 6.4 Dislocations viewed by X-rays (Berg–Barrett technique). (a) Optical micrograph of lithium fluoride crystal chemically etched. The diagonal lines show emergent edge dislocations. The horizontal lines show emergent screw dislocations. (b) In a (200) reflection, only edge dislocations appear. (c) In a (220) reflection, only screw dislocations appear. (d) In a (20$\bar{2}$) reflection, the two types of dislocations appear. (Reprinted with permission from [13], p. 490.)

Transmission electron microscopy (TEM) has established itself as the main observation technique for dislocations; it was employed for the first time by Heidenreich[14] and Hirsch et al.[15] In TEM the foil has to be thinned to a thickness between 1000 and 3000 Å, becoming transparent to electrons. The dislocations produce distortions of the atomic planes. Hence, for certain orientations of the foil with respect to the beam, the region around a dislocation diffracts the beam. The dislocations can then be seen as dark, thin lines. Nowadays TEMs with higher operating voltages (megavolts) are available and allow observation of thicker specimens. Figure 6.5 shows dislocations in nickel rendered visible by this technique.

It is also possible—but much more difficult—to observe dislocations by field-ion microscopy. This technique allows the observation of individual atoms; the magnification ranges from 10^6 to 10^7. Figure 6.6 shows a field-ion micrograph; the atoms are seen as bright spots. The region observed is the emission tip of a specimen and the configuration of the atoms reproduces the "tip." A screw

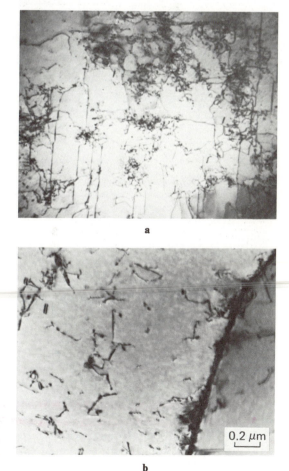

a

0.2 μm

b

Figure 6.5 Dislocations seen by transmission electron microscopy of thin foil (a) nickel shock-loaded at low pressure; (b) annealed nickel.

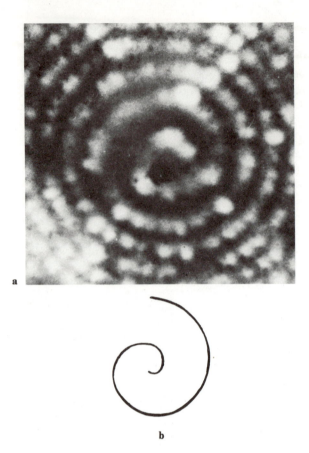

Figure 6.6 (a) Screw dislocation in tungsten seen by field-ion microscopy; (b) sketch of spiral array at field tip around screw dislocation in (a). (Courtesy of O. T. Inal, New Mexico Institute of Mining and Technology.)

dislocation—schematically shown in Fig. 6.6(b)—can be seen if one closely analyzes the micrograph. The line of the dislocation is perpendicular to the plane of the micrograph and passes through the tip of the beam; hence, the atoms exhibit the spiral configuration.

6.3 BEHAVIOR OF DISLOCATIONS

6.3.1 Dislocation Loops

A dislocation line can form a closed loop instead of extending until it reaches an interface or the surface of the crystal. This is the case of Fig. 6.7(a), where a square loop is sketched. Two cuts, along perpendicular sections, were made: *AAAA* and *BBBB*. Figure 6.7(b) and (c) show these sections. It can clearly be seen that the dislocation segments *CF* and *DE* [Fig. 6.7(b)] are of

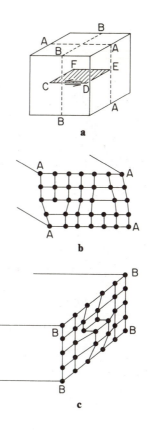

Figure 6.7 Square dislocation loop.

edge character, while segments *CD* and *FE* [Fig. 6.7(c)] are of screw character. This is due to the shear direction. The loop can be imagined as a cut made in the interior of the crystal (impossible feat, of course); the edges of the cut form the dislocation line, after shear is applied to the crystal. Dislocations *CF* and *DE* are of the same type, with opposite signs; the same applies to *CD* and *FE*. The sign convention used for screw dislocations is the following: if the extra semiplane (wedge) is on the top portion, it is positive; if on the bottom, it is negative. Hence, *CF* is positive and *DE* negative. For screw dislocations a similar convention is used. If the helix turns in accord with a normal screw, it is positive. If not, it is negative. According to this convention, *CD* is positive and *FE* is negative.

   The actual dislocation loops are not necessarily square. An elliptical shape would be more favorable energetically than a square. For an elliptical or circular shape, the character of the dislocation changes continuously along the line. Figure 6.8 shows this situation; the regions that are edge and screw are shown by appropriate symbols. The symbols most commonly used are an inverted T (⊥) for a positive edge, and an S (Ꙅ) for a positive screw dislocation. The negative signs can be described by a correct T (T) and by an inverted S (Ꙅ). In Fig. 6.8, all the portions of the loop between the short segments of pure screw and edge character are mixed.

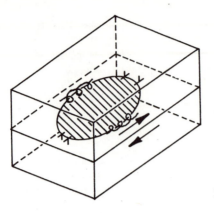

Figure 6.8 Elliptic dislocation loop.

There is another type of loop, called a "prismatic loop"; it should not be confused with a common loop. A prismatic loop is created when a disk of vacancies is either inserted or removed from the crystal. Figure 6.9 shows this situation; cuts *AAAA* and *BBBB* are indicated. A disk having the thickness of one atomic layer was introduced and it can be seen that both sections *AAAA* [Fig. 6.9(b)] and *BBBB* [Fig. 6.9(c)] are identical. They are edge dislocations with opposite signs. This configuration is very different from that encountered in normal loops. One can also remove a disk of atoms, instead of adding it. It will be seen in Section 6.3.2 that these loops do not have the same ability to move as normal loops.

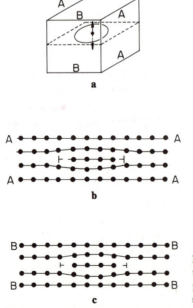

Figure 6.9 Prismatic loop produced by the introduction of a disk into metal: (a) perspective view; (b) section AAAA; (c) section BBBB.

6.3.2 Movement of Dislocations

The plastic deformation of metals is normally accomplished by the movement of dislocations. The elements of dislocation motion are reviewed in this section together with the resulting deformations. In actual deformation and for elevated strains there are complex interactions between dislocations. These complex interactions can be broken down into simple basic mechanisms that will be described here. Two edge dislocations are shown in Fig. 6.10(a). After the passage of one of them, one part of the lattice is displaced in relation to the other part by a distance equal to the Burgers vector. Both a positive and a negative dislocation can generate the same shear. They have to move in opposite directions in order to accomplish this. It should be noted that the shear and dislocation motion directions are the same.

Screw dislocations can produce the same lattice shear [Fig. 6.10(b)]. However, in this case the shear takes place perpendicularly to the direction of motion of the dislocations. Positive and negative screw dislocations have to move in opposite directions in order to produce the same shear strain.

The plane in which a dislocation moves is called slip plane. The slip plane and the loop plane coincide in Fig. 6.11. A loop will eventually be ejected from a crystal upon expanding if there is no barrier to its motion. The expansion of a loop will produce an amount of shear in the crystal equal to the Burgers vector of the dislocation. It is worth noting that the shears of the different dislocations are all compatible; there is no opposition of movement.

The prismatic loops, consisting totally of edge dislocations, cannot expand as the normal loops. Thus, because the plane of the dislocation does not coincide with the loop plane, the coupled movement of the edge dislocations will force the loop to move perpendicularly to its plane, maintaining the same diameter. Upon being ejected from the crystal, a slip will be formed at the surface. Figure 5.10 shows a succession of vacancy loops.

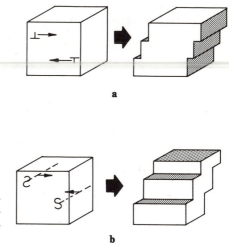

Figure 6.10 Slip produced by the movement of dislocations: (a) positive and negative edge dislocations; (b) positive and negative screw dislocations.

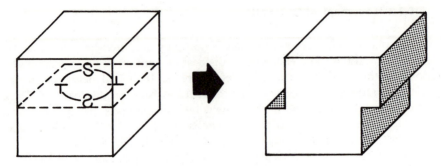

Figure 6.11 Expansion of a dislocation loop.

6.4 STRESS FIELD AROUND DISLOCATIONS

Dislocations are defects; hence, they introduce stresses and strains in the surrounding lattice. The mathematical treatment of these stresses and strains can
be substantially simplified if the medium is considered as isotropic and continuous. The anisotropic treatment is discussed in detail by Bacon et al.[16]; it will
not be introduced here. A dislocation is, under the conditions of isotropy, completely described by the line and Burgers vectors. With this in mind, and considering the simplest possible situation, dislocations are assumed to be straight and
infinitely long lines. Figures 6.12 and 6.13 show hollow cylinders sectioned
along the longitudinal direction. Different deformations are applied in the two
cases. The one in Fig. 6.12 portrays the deformation around a screw dislocation,
while Fig. 6.13 is an idealization of the strains around an edge dislocation.
The cylinders, with external radii R, were longitudinally and transversally displaced by a quantity **b**, called the Burgers vector. **b** is parallel and perpendicular
to the cylinder axis in the representation of screw and edge dislocations, respectively. In both cases an internal hole with radius r_0 is made through the center.
This is done to simplify the mathematical treatment. In a continuous medium
the stresses on the center would build up and become infinite, in the absence
of a hole; in real dislocations the crystalline lattice is periodic and this does

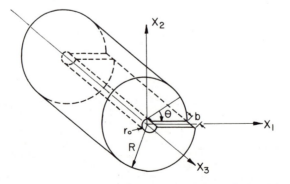

Figure 6.12 Positive screw dislocation with Burgers vector b.

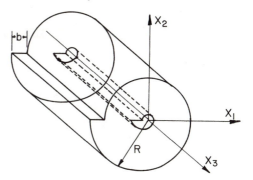

Figure 6.13 Edge dislocation with Burgers vector b.

not occur. Therefore, this is a way of reconciling the continuous medium hypothesis with the periodic nature of the structure. To analyze the stresses around a dislocation, we have to use the formal theory of elasticity (see Section 1.12.5). For that, one has to use the relationships between stresses and strains (constitutive relationships), the equilibrium equations, the compatibility equations, and the boundary conditions. Hence, the problem is somewhat elaborate. The mathematical formalism given by Kuhlmann-Wilsdorf[17] will be followed here.

6.4.1 Screw Dislocations

As can be seen in Fig. 6.12, the displacements are

$$u_1 = 0, \qquad u_2 = 0, \qquad u_3 \neq 0$$

The displacement on the direction Ox_3 can be assumed to be approximately equal to

$$u_3 = f(\theta) = \frac{b}{2\pi} \theta$$

This is so because the displacement is b after a rotation of 2π. An alternative form of u_3 is

$$u_3 = \frac{b}{2\pi} \arctan \frac{x_2}{x_1} \tag{6.1}$$

Using Eqs. 1.121 through 1.126,

$$\epsilon_{11} = 0 \qquad \qquad \epsilon_{22} = 0$$

$$\epsilon_{12} = 0 \qquad \qquad \epsilon_{23} = \frac{1}{2}\frac{\partial u_3}{\partial x_2}$$

$$\epsilon_{13} = \frac{1}{2}\frac{\partial u_3}{\partial x_1} \qquad \epsilon_{33} = \frac{\partial u_3}{\partial x_3} = 0$$

Substituting Eq. 6.1 into the equations above, we obtain

$$\epsilon_{13} = \frac{-bx_2}{4\pi(x_1^2 + x_2^2)} \tag{6.2}$$

$$\epsilon_{23} = \frac{bx_1}{4\pi(x_1^2 + x_2^2)} \tag{6.3}$$

$$\sigma_{33} = 0$$

Now, using Eqs. 1.11, we have

$$\sigma_{13} = 2G\epsilon_{13}$$

$$\sigma_{23} = 2G\epsilon_{23}$$

$$\sigma_{13} = \sigma_{31} = -\frac{Gbx_2}{2\pi(x_1^2 + x_2^2)} \tag{6.4}$$

$$\sigma_{23} = \sigma_{32} = \frac{Gbx_1}{2\pi(x_1^2 + x_2^2)} \tag{6.5}$$

We have to verify whether Eqs. 6.3, 6.4, and 6.5 obey the equilibrium conditions. The equilibrium equations lead to

$$\frac{\partial \sigma_{13}}{\partial x_1} + \frac{\partial \sigma_{23}}{\partial x_2} = 0 \tag{6.6}$$

Taking the partial derivatives of Eqs. 6.4 and 6.5 yields

$$\frac{\partial \sigma_{13}}{\partial x_1} = \frac{Gbx_1x_2}{\pi(x_1^2 + x_2^2)^2} \tag{6.7}$$

$$\frac{\partial \sigma_{23}}{\partial x_2} = -\frac{Gbx_1x_2}{\pi(x_1^2 + x_2^2)^2} \tag{6.8}$$

Summing 6.7 to 6.8, we obtain Eq. 6.6. Hence, equilibrium conditions are satisfied. However, we have to verify the compatibility equations. The compatibility equations (Eqs. 1.131 through 1.136) become:

$$\frac{\partial^2 \epsilon_{12}}{\partial x_1 \partial x_2} = 0 \tag{6.9}$$

$$\frac{\partial^2 \epsilon_{23}}{\partial x_2 \partial x_3} = 0 \quad \text{or} \quad \frac{\partial}{\partial x_2}\left(\frac{\partial \epsilon_{23}}{\partial x_3}\right) = 0 \tag{6.10}$$

$$\frac{\partial^2 \epsilon_{13}}{\partial x_1 \partial x_2} = 0 \quad \text{or} \quad \frac{\partial}{\partial x_1}\left(\frac{\partial \epsilon_{13}}{\partial x_3}\right) = 0 \tag{6.11}$$

$$\frac{-\partial^2 \epsilon_{23}}{\partial x_1^2} + \frac{\partial_2 \epsilon_{13}}{\partial x_1 \partial x_2} = 0 \tag{6.12}$$

$$\frac{\partial^2 \epsilon_{13}}{\partial x_2^2} + \frac{\partial^2 \epsilon_{23}}{\partial x_2 \partial x_1} = 0$$

$$\frac{\partial^2 \epsilon_{13}}{\partial x_3 \partial x_2} + \frac{\partial^2 \epsilon_{23}}{\partial x_3 \partial x_1} = 0 \tag{6.13}$$

Since $\partial/\partial x_3$ is equal to zero, we are left with only Eqs. 6.12 and 6.13. By differentiating Eqs. 6.2 and 6.3 twice, we can show that Eqs. 6.12 and 6.13 are obeyed and all compatibility conditions are satisfied.

Finally, it is necessary to verify whether the boundary conditions are obeyed. For the model to truly represent a screw dislocation, the following requirements have to be met: (1) no normal stresses should be acting on the lateral surfaces of the cylinder; (2) normal stresses should be acting on the surfaces normal to the cylinder axis; and (3) there should be no torsion of the cylinder around its axis. Condition (1) is satisfied because $\sigma_{11} = \sigma_{22} = 0$. Condition (2) is not satisfied because $\sigma_{33} = 0$. Condition (3) is not satisfied because σ_{23} and σ_{13} result in a torsion. There is a deviation from the response of a real dislocation. This difference can be easily seen by the following analogy. If a rubber hose (e.g., a garden hose) is cut along its axis and submitted to a moment, the line of the cut will describe a helical trajectory around the axis. A real dislocation, on the other hand, could not exhibit this behavior. Its slip plane is a "plane" and cannot describe a helix around the dislocation line. As a consequence, the model has to be corrected in such a way that the "cut" stays on one plane and that the equations obey this boundary condition. Eshelby and Stroh[18] performed this task; because of its complexity, it transcends the objectives of this book.

6.4.2 Edge Dislocations

The method used for screw dislocations is not applicable to edge dislocations because u_1 and u_2 are variables, and only $u_3 = 0$. This can be readily seen in Fig. 6.13. The requirement of $u_3 = 0$ establishes a state of plane strain. The general solution for elasticity problems in plane strain was developed in Section 1.12.5; it involves the determination of the Airy stress function. Recalling Section 1.12.5, we have to determine the Airy stress function; once it is known, the stresses and strains are readily found by differentiation (Eqs. 1.145). It is interesting to note that Taylor found the mathematical solution readily available when he proposed dislocations in 1934. The term "distorsioni" had been given to them by Volterra[19] in 1907; Love,[20] in 1927, talked about "dislocations." The mathematical treatment had been developed as early as 1905 by Timpe.[21] Hence, when dislocations were postulated in 1934 as a defect in a crystalline structure, the stresses could readily be obtained. The most difficult part—obtaining a satisfactory Airy stress function—had been worked out. The following Airy stress function had been found:

$$\Phi = -\frac{Gb}{2\pi(1-\nu)}\, x_2\, \ln\, (x_1^2 + x_2^2)^{1/2} \tag{6.14}$$

It should be noticed that the isotropy hypothesis is implicit. The stresses are obtained by applying Eq. 1.145 to Eq. 6.14:

$$\sigma_{11} = \frac{\partial^2 \Phi}{\partial x_2^2} = -\frac{Gbx_2(3x_1^2 + x_2^2)}{2\pi(1-\nu)(x_1^2 + x_2^2)^2} \tag{6.15}$$

$$\sigma_{12} = \frac{\partial^2 \Phi}{\partial x_1 \, \partial x_2} = \frac{Gbx_1(x_1^2 - x_2^2)}{2\pi(1 - \nu)(x_1^2 + x_2^2)^2} \qquad (6.16)$$

$$\sigma_{22} = \frac{\partial^2 \Phi}{\partial x_1^2} = \frac{Gbx_2(x_1^2 - x_2^2)}{2\pi(1 - \nu)(x_1^2 + x_2^2)^2} \qquad (6.17)$$

σ_{33} is then obtained by

$$\sigma_{33} = \nu(\sigma_{11} + \sigma_{22}) = -\frac{Gb\nu x_2}{\pi(1 - \nu)(x_1^2 + x_2^2)} \qquad (6.18)$$

These equations satisfy automatically the equations of equilibrium and compatibility. The reader should verify this, as an exercise. The boundary conditions have to be checked. In the case above, there was resulting bending moment applied on the Ox_3 axis. The real dislocations cannot, however, have a bending moment applied on their line. The Airy stress function above is not entirely satisfactory and corrections have to be introduced. An improved solution was presented by Leibfried and Lücke,[22] but the above one is generally accepted as satisfactory.

Figure 6.14 shows the stress field around an edge dislocation,[23] isostress contours clearly indicate the regions of maximum tension, compression, and shear.

6.4.3 Strain (or Deformation) Energy of Dislocations

The elastic deformation energy of a dislocation can be found by integrating the elastic deformation energy over the whole volume of the deformed crystal. The deformation energy, calculated in Section 1.11.5 is given by

$$U = \tfrac{1}{2}\sigma_{ij}\epsilon_{ij}$$

We have, for an isotropic material, using Eq. 1.11:

$$U = \frac{1}{2G}\left[\frac{1}{2(1 + \nu)}(\sigma_{11}^2 + \sigma_{22}^2 + \sigma_{33}^2) + (\sigma_{12}^2 + \sigma_{13}^2 + \sigma_{23}^2)\right.$$
$$\left. - \frac{\nu}{(1 + \nu)}(\sigma_{11}\sigma_{33} + \sigma_{11}\sigma_{22} + \sigma_{22}\sigma_{33})\right] \qquad (6.19)$$

For a screw dislocation, we have, using Eqs. 6.4 and 6.5,

$$U_s = \frac{1}{2G}\left[\frac{G^2b^2x_2^2}{4\pi^2(x_1^2 + x_2^2)^2} + \frac{G^2b^2x_1^2}{4\pi^2(x_1^2 + x_2^2)^2}\right]$$

$$U_s = \frac{Gb^2}{8\pi^2(x_1^2 + x_2^2)} \qquad (6.20)$$

Substituting $(x_1^2 + x_2^2)$ by r^2 (shown in Fig. 6.12), we find that

$$U_s = \frac{Gb^2}{8\pi^2 r^2} \qquad (6.21)$$

Integrating Eq. 6.21 between r_0 and R, we get

$$U_s = \int_{r_0}^{R} \frac{Gb^2}{8\pi^2 r^2} 2\pi r\, dr = \frac{Gb^2}{4\pi} \ln \frac{R}{r_0} \tag{6.22}$$

In a similar way, the energy of a straight edge dislocation per unit length is equal to

$$U_\perp = \frac{Gb^2}{4\pi(1-\nu)} \ln \frac{R}{r_0} \tag{6.23}$$

It should be observed that the factor $(1 - \nu)$ is approximately equal to $\frac{2}{3}$. Hence, the energy of an edge dislocation is about $\frac{3}{2}$ of that of a screw dislocation.

For a mixed dislocation, one may consider the Burgers vector as the resultant of two components: one parallel and one perpendicular to the dislocation line. If b makes an angle α with the dislocation line, we have

$$b_\perp = b \sin \alpha$$

$$b_s = b \cos \alpha$$

The energy of a mixed dislocation can be expressed as

$$U_M = \frac{Gb^2}{4\pi(1-\nu)} (1 - \nu \cos^2 \alpha) \ln \frac{R}{r_0} \tag{6.24}$$

It can be seen that Eq. 6.24 reduces itself to Eqs. 6.22 and 6.23 for $\alpha = 0$ and $\alpha = 90°$, respectively.

The models used in Section 6.4 present the nuclei as holes to avoid the infinite stresses in the dislocation core. Several methods have been used to estimate r_0. In this book, r_0 will be assumed to be equal to $5b$. It should be noted that the stresses given by the equations above become infinite for infinite R; hence, one has to establish an approximate value for R. Dislocations in a metal never occur in a completely isolated manner; they form irregular arrays with a mean density ρ. This density is derived as the total length of dislocation line per unit volume. The spaghetti analogy can be used here. Imagine a pot with water and spaghetti. The spaghetti density would be obtained by measuring the total spaghetti length and dividing it by the pot volume. The stress fields of the various dislocations interact, as will be seen in the subsequent sections; we generally assume a value of R equal to the average distance between the dislocations. The reader can readily show, by means of a simplified array, that it is equal to $\rho^{-1/2}$. In order to consider the dislocation nucleus, a certain strain energy is added to Eqs. 6.22, 6.23, and 6.24. This energy is taken as $Gb^2/10$, for metals. Hence, the total energy of a dislocation is

$$U_T = U_{\text{nucleus}} + U_{\text{periphery}}$$

Equation 6.24 takes the form

$$U_T = \frac{Gb^2}{10} + \frac{Gb^2}{4\pi(1-\nu)} (1 - \nu \cos^2 \alpha) \ln \frac{\rho^{-1/2}}{5b} \tag{6.25}$$

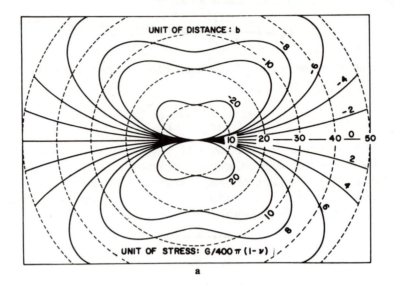

a

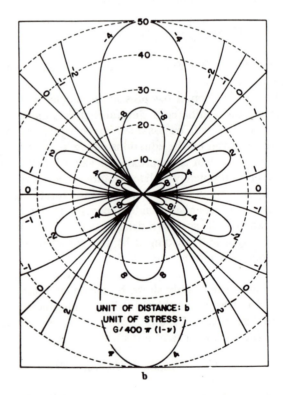

b

Figure 6.14 Stress fields around edge dislocation (dislocation line is Ox_3); (a) σ_{11}; (b) σ_{22}; (c) σ_{33}; (d) σ_{12}. (Adapted with permission from [23], pp. 22–25.)

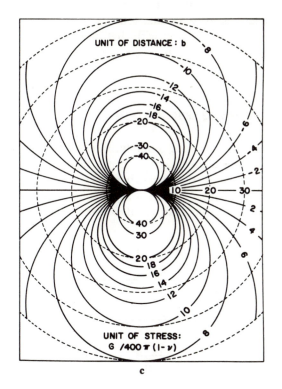

c

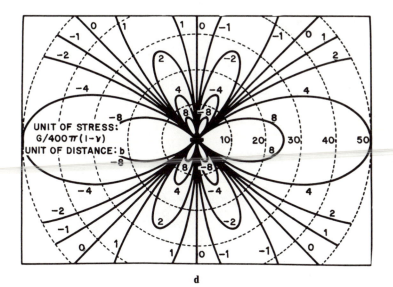

d

Figure 6.14 (Continued)

For typical metals U_T is equal to a few electron volts per atomic plane. The energy of the nucleus is 10% of this total. The energy of a dislocation per atomic plane is high in comparison with that of a vacancy; it is approximately 3 eV (4.8×10^{-19} J) versus about 1 eV (1.6×10^{-19} J) for a vacancy (see Chapter 5).

6.5 ENERGY OF A CURVED DISLOCATION

In a curved dislocation there are, forcibly, interactions between the opposing arms. To determine the change in self-energy of a dislocation as it becomes curved, we have to determine the change in energy as one dislocation approaches one parallel to it and having opposite sign. This configuration is shown in Fig. 6.15(a). The stress field around an isolated dislocation decreases as $Gb/2\pi r$,

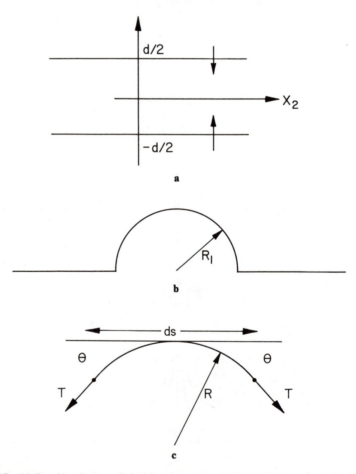

Figure 6.15 (a) Two identical parallel dislocations separated by a distance d and having opposite signs; (b) semicircular configuration for a dislocation segment; (c) forces acting on a dislocation segment.

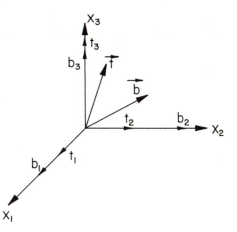

Figure 6.16 Burgers vector b and dislocation line t for most general orientation.

as seen in the preceding section. When the distances are large with respect to the dislocation separation d shown in Fig. 6.16, the stress fields vary according to the equation

$$\sigma \alpha \pm \frac{Gb}{2\pi}\left[\left(x_1 - \frac{d}{2}\right)^{-1} - \left(x_1 + \frac{d}{2}\right)^{-1}\right]$$

For $x_1 \gg d$ we have approximately

$$\sigma \alpha \frac{Gbd}{2\pi x_1^2}$$

Hence, the stress field in the region outside two dislocations drops with the square of x, while the one with an isolated dislocation changes only with x.

Such is always the case with a dislocation loop that can be decomposed into a series of segments placed symmetrically around a center. Hence, the stresses at points removed from the loop, in relation to the loop radius, are negligible.

We proceed in the same fashion if we want to know the line energy of an edge dislocation for the configuration of Fig. 6.15(b). For the straight portion,

$$U = \frac{Gb^2}{4\pi(1-v)}\,\ell n\,\frac{R}{5b}$$

For the semicircular portion,

$$U = \frac{Gb^2}{4\pi(1-v)}\,\ell n\,\frac{R_1}{5b} \tag{6.26}$$

The line tension is defined as the energy per unit length. This definition is given in analogy to soap bubbles that have a surface tension equal to the surface energy per unit area. The tendency of a dislocation is to remain straight in order to minimize its length, and this is due to its self-energy minimization.

The forces acting on a dislocation segment are shown in Fig. 6.15(c). We have a segment of length ds and curvature radius R. The line tension is T. This line tension always acts tangentially to the dislocation line. The vertical force that the dislocation segment undergoes is $2T \sin \theta$.

$$F = 2T \sin \theta$$

and

$$F \simeq 2T\theta \tag{6.27}$$

But

$$\theta \simeq \frac{ds}{2R}$$

Hence, the force is

$$F = \frac{T\, ds}{R} \tag{6.28}$$

This force, per unit length, is T/R. In order for the dislocation to maintain this configuration, an upward vertical force with the same magnitude (T/R) has to act on it. Otherwise, the dislocation will tend to regain its straight configuration. The same applies to a loop. To avoid its collapse, a force has to act on it continuously.

6.6 FORCES ACTING ON DISLOCATIONS

Either external loads, internal stresses, or chemical factors can exert forces on dislocations. Examples of internal stresses are other dislocations, point defects, precipitates, inclusions, thermal gradients, and image forces due to other phases.

The completely general situation is that of a dislocation with a Burgers vector and dislocation line along general directions. Figure 6.16 shows this configuration. The following equation is obtained for the force exerted per unit length of dislocation:[24]

$$\mathbf{F} = (t_2 G_3 - t_3 G_2)\mathbf{i} + (t_3 G_1 - t_1 G_3)\mathbf{j} + (t_1 G_2 - t_2 G_1)\mathbf{k} \tag{6.29}$$

in which G_1, G_2, and G_3 are

$$G_1 = \sigma_{11} b_1 + \sigma_{12} b_2 + \sigma_{13} b_3$$

$$G_2 = \sigma_{21} b_1 + \sigma_{22} b_2 + \sigma_{23} b_3$$

$$G_3 = \sigma_{31} b_1 + \sigma_{32} b_2 + \sigma_{33} b_3$$

Readers who recall vector calculus will see that 6.29 can be expressed as

$$\mathbf{F} = \begin{vmatrix} \mathbf{i} & \mathbf{j} & \mathbf{k} \\ t_1 & t_2 & t_3 \\ G_1 & G_2 & G_3 \end{vmatrix} = \mathbf{t} \wedge \mathbf{G} \tag{6.30}$$

Equation 6.30 is called the general *Peach–Koehler* equation. Its detailed derivation is presented by Peach and Koehler.[24] **F** is perpendicular to the plane
defined by **b** and **G**. Therefore, the force acting on a dislocation is *always* perpendicular to its line.

6.7 RELATIONS BETWEEN DISLOCATIONS

6.7.1 Dislocations in Face-Centered Cubic Crystals

It was seen in Chapter 4 that, among the 80 or so metals, 55 are FCC.
The FCC structure is the closest-packed one, together with the HCP one. Thus,
it is natural that dislocations are more carefully studied for the FCC structure.

When we visualize a dislocation, we generally think of a defect that, upon
passing, recomposes the original structure of the crystal. Hence, in a simple
cubic structure the Burgers vector would have the direction [100] and magnitude
a (lattice parameter). However, there are cases in which the original structure
is not recomposed. This type of dislocation is known as *imperfect* or *partial*.

In FCC crystals, the closest-packed planes are (111). These planes are
usually called A, B, and C, depending on their order in the stacking sequence.
Figure 6.17 shows an atomic plane A. The glide movement of the atoms of
plane A that would recompose the same lattice would be indicated by Burgers
vector b_1. This vector has the direction $[10\bar{1}]$. The magnitude can be determined
from the Eq. 6.31[25]

$$d_{hkl} = \frac{a}{\sqrt{h^2 + k^2 + l^2}} = \frac{a}{\sqrt{2}} \qquad (6.31)$$

$$1^2 + 0 + (-1)^2$$

The magnitude of b_1 is equal to the distance between $(10\bar{1})$; in cubic structures,
planes and directions with the same indices are perpendicular. Vector b_1 is
expressed with respect to unit vectors **i**, **j**, and **k**, of the coordinate system
$Ox_1x_2x_3$ as

$$\mathbf{b}_1 = \frac{a}{2}\mathbf{i} + 0\mathbf{j} - \frac{a}{2}\mathbf{k} = \frac{a}{2}(\mathbf{i} - \mathbf{k})$$

It can be seen that the magnitude is

$$|\mathbf{b}_1| = \frac{a}{\sqrt{2}}$$

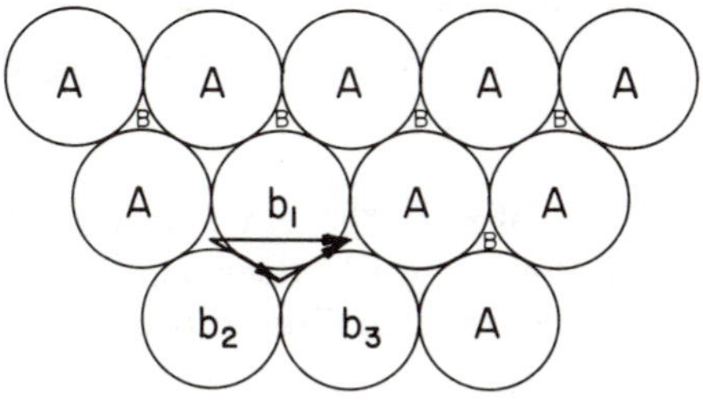

Figure 6.17 Decomposition of dislocation.

This vector is, logically, the same as that of Eq. 6.31. The simplified notation used for Burgers vectors is

$$b_1 = \frac{a}{2}[10\bar{1}] \quad \text{or} \quad b_1 = \frac{1}{2}[10\bar{1}]$$

Hence, the term in brackets gives the direction of the vector, while the term that precedes it is the same fraction as that used in the definition of unit vectors **i**, **j**, and **k**. There is also a graphic method to determine this fraction. First, one draws the vector **b** connecting points $(0, 0, 0)$ to point $(1, 0, -1)$. Then one draws b_1. b_1 will be a fraction of b (in this case, half). The fraction is the term that precedes the bracketed term.

One possibility of decomposition for the dislocation is shown in Fig. 6.17. b_2 and b_3 connect B to C. They both define partial dislocations, because they change the stacking sequence ABC. But, acting together (or sequentially), they would have the same effect as b_1 and maintain the correct stacking sequence. The magnitude of b_2 and b_3 is

$$d_{hkl} = \frac{a}{\sqrt{h^2 + k^2 + l^2}} = \frac{a}{\sqrt{6}}$$

Hence, we have the following possible reaction:

$$\frac{a}{2}[10\bar{1}] \quad \longrightarrow \quad \frac{a}{6}[2\bar{1}\bar{1}] + \frac{a}{6}[11\bar{2}]$$

It can be seen that the total energy decreases with decomposition, taking the square of the magnitude of the Burgers vectors:

$$\frac{a^2}{2} > \frac{a^2}{6} + \frac{a^2}{6}$$

The easiest way of studying the possible orientations is by means of the Thompson tetrahedron.[26] It is shown in Fig. 6.18(a). In this regular tetrahedron the vertices are points (011), (101), (110), (000). They are identified by A, B, C, O. The centers of the faces opposite to these vertices are identified by the corresponding Greek letters α, β, γ, δ. The four sides of the tetrahedron are planes $\{111\}$ (slip planes) and the edges are the directions. $<110>$ are the Burgers vectors of perfect dislocations $(a/2)<110>$; they define these directions. The perfect dislocations decompose themselves according to

$$\frac{a}{2}[1\bar{1}0] \quad \longrightarrow \quad \frac{a}{6}[2\bar{1}1] + \frac{a}{6}[1\bar{2}\bar{1}]$$

There are two alternatives

$$AB \quad \longrightarrow \quad A\gamma + \gamma B$$

$$AB \quad \longrightarrow \quad A\delta + \delta B$$

The same applies to the other dislocations, AC, BC, AO, BO, and CO. We can by this manner easily visualize all slip planes and directions and their associated perfect and partial dislocations, for the FCC structure. Figure 6.18(b) shows this construction, in a plane, with all dislocations indicated.

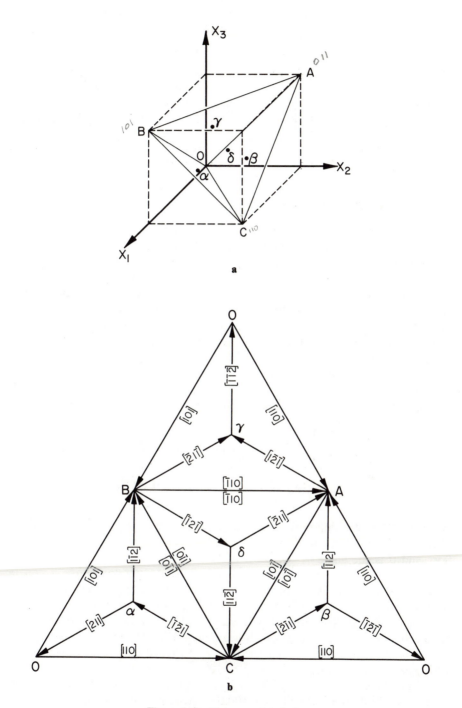

Figure 6.18 Thompson tetrahedron.

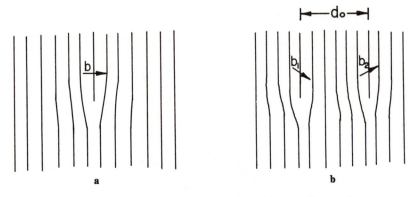

Figure 6.19 Two partial dislocations b_1 and b_2 at a distance d_0.

When a perfect dislocation decomposes itself into partials, a region of faulty stacking is created between the partials. This decomposition is shown in Fig. 6.19. The dislocations generate a region in which the stacking is ABC AC ABC. Hence, we have four planes in which the stacking is $CACA$. This is exactly the stacking sequence of the HCP structure. This structure has a higher Gibbs free energy than the equilibrium FCC structure because it is not thermodynamically stable under the imposed conditions. This specific array of planes is called the *stacking fault* and the energy associated with it determines the separation between the two partial dislocations: the repulsive force between the two partials is balanced by the attraction trying to minimize the region with the stacking fault. The two equations below (from Murr[27] and Kelly and Groves,[28] respectively) allow the calculation of the equilibrium separation between the partial dislocations, d:

$$\gamma_{SF} = \frac{G|b_p|^2}{8\pi d} \left[\frac{2-\nu}{1-\nu} \left(1 - \frac{2\nu \cos 2\theta}{2-\nu} \right) \right]$$

$$\gamma_{SF} = \frac{Gb_1 b_2}{2\pi d} \left(\cos\theta_1 \cos\theta_2 + \frac{\sin\theta_1 \sin\theta_2}{1-\nu} \right)$$

where γ is the stacking-fault free energy (SFE) per unit area (free energy of HCP minus free energy of FCC), b_p is the Burgers vector of the partial dislocation, and θ is the angle of the Burgers vector with the dislocation line. Table 6.1 presents the SFEs for some metals. From the equations above it can be seen that d is inversely proportional to γ. The effect of alloying elements is generally to decrease the SFE. The addition of aluminum to copper has a drastic effect on its SFE. It drops from 78 to 6 mJ/m². Murr[27] gives a detailed treatment of stacking faults. Aluminum, which has a high SFE (166 mJ/m²), does exhibit a very small separation between partials: 10 Å. On the other hand, in certain alloy systems the distance can go up to 100 Å or more.

The stacking-fault energy is very sensitive to composition. For example, Kestenbach[29] showed that different stainless steels exhibited significantly different SFEs: AISI 303, 8.4 mJ/m²; AISI 304, 23.3 mJ/m²; AISI 316, 20.5 mJ/m²; AISI 310, 47 mJ/m². Kestenbach[29] also showed that there is a relation-

TABLE 6.1   Stacking-Fault Free Energies and Separation Between
Shockley Partials for Metals ($\theta = 30°$)[a]

Metal	γ (mJ/m²)	a_0 (nm)	b (nm)	G (GPa)	d (nm)
Aluminum	166	0.41	0.286	26.1	1
Copper	78	0.367	0.255	48.3	3.2
Gold	45	0.408	0.288	27.0	4.0
Nickel	128	0.352	0.249	76.0	2.9
Silver	22	0.409	0.289	30.3	9

[a] Adapted from [27], pp. 145–147.

ship between the orientation of a slip system with respect to the tensile (or compressive) axis and the separation between the partial dislocations and, consequently (as will be seen next), an effect on the dislocation substructure. The SFE can be determined by the node method. Figure 6.20 shows how three sets of partial dislocations acquire special configurations when they encounter at a node. The hatched portion corresponds to the stacking fault. By measuring the radius r and appropriate calculations, one can obtain γ. Figure 6.20(a) shows a material with high γ; Fig. 6.20(b) shows a material with low γ.

Figure 6.21 shows some stacking faults in AISI 304 stainless steels viewed by transmission electron microscopy. The region corresponding to the stacking

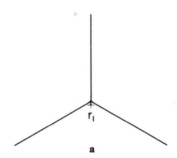

a

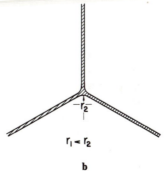

$r_1 < r_2$

b

Figure 6.20 Schematic representation of stacking faults at dislocation nodes. By measuring the node radius one can determine the SFE.

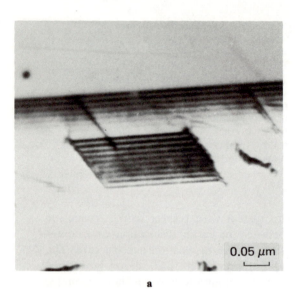

a

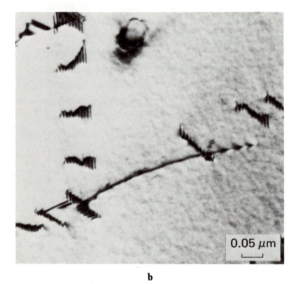

b

Figure 6.21 (a) Short segment of stacking fault in AISI 304 stainless steel overlapping with coherent twin boundary. Differences in the nature of these defects are illustrated by fringe contrast differences. (b) Dislocations in AISI 304 stainless steel splitting into partials bounded by short stacking-fault region. (Courtesy of L. E. Murr, New Mexico Institute of Mining and Technology.)

fault can be clearly seen by the characteristic fringe (////) pattern. The extremities of the fringes are bound by the partial dislocations. In Fig. 6.21(a) the stacking fault lies parallel to a coherent twin boundary, which is much longer. It can be distinguished from the coherent twin boundary by the differences in fringe contrast. While all the fringes of the stacking fault are dark, the ones in the twin are dark at the top and become successively lighter. Figure 6.21(b) shows a number of dislocations (probably emitted from the same source) whose segments are trapped on the foil. These segments have decomposed into partials and one can clearly distinguish the stacking-fault regions by the characteristic fringe contrast.

The effect of the stacking-fault energy on the deformation substructure can be seen in Fig. 6.22. This figure shows (a) an Fe–34.5% Ni and (b) an FE–15% Cr–15% Ni alloy after deformation by shock loading under identical conditions (7.5 GPa peak pressure; 1.2 μs pulse duration).[30] The Fe–15% Cr–15% Ni alloy has a significantly larger stacking-fault energy than does the Fe–34.5% Ni alloy, and the resultant deformation substructures seem to be strongly affected by this difference. This subject is discussed in Chapters 8 and 9 and it suffices to say here that low-SFE metals tend to exhibit a deformation substructure characterized by banded, linear arrays of dislocations, whereas high-SFE metals tend to exhibit dislocations arranged in tangles or cells.

There are other types of dislocations on FCC structures. They are called sessile or Frank[31] dislocations and they are immobile. They appear under two specific conditions that are shown in Fig. 6.23. In Fig. 6.23(a) a disk was removed in plane (111); in Fig. 6.23(b) a disk was added. It can be seen that in both cases the stacking sequence was changed: *ABCBCA* and *ABCBABC* for Fig. 6.23(a) and (b), respectively. The magnitude of the Burgers vectors is determined applying Eq. 6.31. We have

$$d_{hkl} = \frac{a}{\sqrt{h^2 + k^2 + l^2}} \qquad ; \qquad \mathbf{b} = \frac{a}{3}[111]$$

We have a sample of an *intrinsic* stacking fault in Fig. 6.23(a) and an *extrinsic* or double stacking fault in Fig. 6.23(b). In Thompson's tetrahedron, the Frank dislocations are represented by $A\alpha$, $B\beta$, $C\gamma$, and $D\delta$. Since the Burgers vector is not in the slip plane, they are immobile. Another type of immobile dislocation that can occur in FCC metals is the Lomer[32]–Cottrell[33] lock. Let us consider two (111) and (11$\bar{1}$) planes that are represented by δ and γ, respectively, in Thompson's tetrahedron. The three perfect dislocations in (111) are

$$\mathbf{b}_1 = \frac{a}{2}[1\bar{1}0] \qquad \scriptstyle 1-1+0=0$$

$$\mathbf{b}_2 = \frac{a}{2}[\bar{1}01] \qquad \scriptstyle -1+0+1=0$$

$$\mathbf{b}_3 = \frac{a}{2}[01\bar{1}] \qquad \scriptstyle 0+1-1=0$$

For plane (11$\bar{1}$),

$$\scriptstyle -1+1+0=0 \qquad \mathbf{b}_4 = \frac{a}{2}[\bar{1}10]$$

$$\scriptstyle 1+0-1=0 \qquad \mathbf{b}_5 = \frac{a}{2}[101]$$

$$\scriptstyle 0+1-1=0 \qquad \mathbf{b}_6 = \frac{a}{2}[011]$$

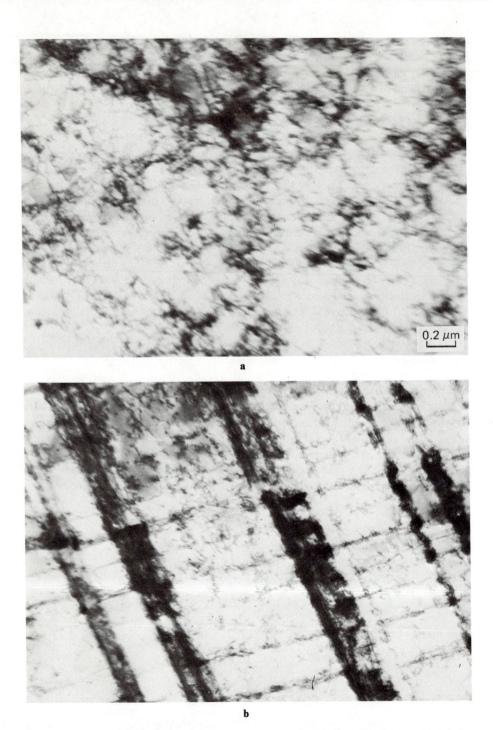

a

b

Figure 6.22 Effect of stacking fault energy on dislocation substructure: (a) higher-stacking-fault-energy alloy (Fe–34% Ni); (b) lower-stacking-fault-energy alloy (Fe–15% Cr–15% Ni). (Both alloys shock-loaded at 7.5 GPa pressure and 2 μs pulse duration.)

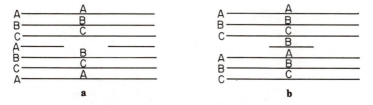

Figure 6.23 Frank sessile dislocations: (a) intrinsic; (b) extrinsic.

One good rule to determine whether a direction belongs to a plane is: the scalar product between the direction **b** and the normal to the plane has to be zero (in a cubic structure). This rule comes from vector calculus. $\mathbf{b}_1$ and $\mathbf{b}_4$ have the same direction and opposite senses; the common direction is also that of the intersection of the two planes. Hence, both dislocations will cancel when they encounter each other. The combination of $\mathbf{b}_2$ and $\mathbf{b}_5$ would result in

$$\mathbf{b}_2 + \mathbf{b}_5 = \frac{a}{2}[\bar{1}01] + \frac{a}{2}[101] = \frac{a}{2}[002] = a[001]$$

The energy of these dislocations is

$$\frac{a^2}{2} + \frac{a^2}{2} = a^2$$

Therefore, this reaction will not occur, because it would not result in a reduction of the energy. The sole combinations that would result in a decrease in the overall energy would be of the type

$$\mathbf{b}_3 + \mathbf{b}_5 = \frac{a}{2}[01\bar{1}] + \frac{a}{2}[101]$$

$$= \frac{a}{2}[110]$$

$$\frac{a^2}{2} + \frac{a^2}{2} > \frac{a^2}{2}$$

This reaction is energetically favorable; it is shown in Fig. 6.24. This dislocation is not mobile in either the (111) or (11$\bar{1}$) plane; hence, it acts as a barrier for

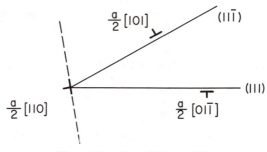

Figure 6.24 Cottrell–Lomer lock.

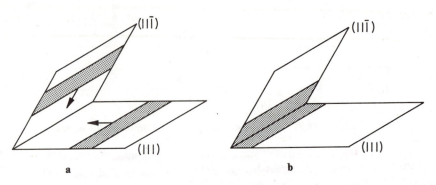

Figure 6.25 Stairway dislocation.

any additional dislocation moving in these planes. It impedes slip and is therefore called a "lock." It was initially proposed by Lomer.[32] Cottrell[33] later showed that the same reasoning could be applied to partial dislocations (also known as Shockley partials[34]). The resultant configuration is shown in Fig. 6.25; it resembles a stair and is therefore called "stair-rod" dislocation. The leading partials react and immobilize the partials coupled to them (trailing partials). The bands of stacking faults form a configuration resembling steps on a stairway. These steps are barriers to further slip on the atomic planes involved, as well as in the adjacent planes.

6.7.2 Dislocations in Hexagonal Close-Packed Crystals

In HCP crystals the stacking sequence of the most densely packed planes is $ABAB$ (see Section 4.2). These planes are known as basal planes [Fig. 4.2(c)]. Figure 6.26 shows the main planes in the HCP structure. Perfect dislocations moving in the basal plane can decompose into Shockley partials, just as in the FCC structure. Stacking faults are also formed (only intrinsic stacking faults). This analogy can be easily understood if one realizes the similarity between the two structures. The (111) planes in the FCC structure are the equivalent

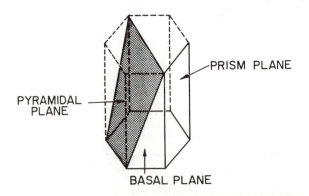

Figure 6.26 Basal, pyramidal, and prism plane in HCP structure.

to the basal planes in the HCP structure. A perfect dislocation in the basal plane has the following Burgers vector:

$$b = \frac{a}{3}[2 \ \bar{1} \ \bar{1} \ 0]$$

In an ideal crystal the c/a ratio is 1.633. However, in real crystals this never happens. It has been experimentally observed that, for crystals with $c/a >$ 1.633, slip occurs mainly in the basal plane, while the pyramidal and prism planes are preferred in crystals with $c/a <$ 1.633. This is due to the dependence on the distance between the atoms upon c/a; it is well known that the dislocations prefer to move in the highest packed planes. A detailed treatment of dislocations in HCP metals is given by Teutonico.[35]

6.7.3 Dislocations in Body-Centered Cubic Crystals

In BCC crystals the atoms are closest to each other along the <111> directions [see Fig. 4.2(b)]. Any plane in the BCC crystal that contains this direction is a suitable slip plane. Slip has been experimentally observed in (110), (112), and (123) planes. The slip markings in BCC metals are usually wavy and ill defined. The following reaction has been suggested for a perfect dislocation having its Burgers vector along <111>:

$$\frac{a}{2}[\bar{1}\bar{1}1] \quad \longrightarrow \quad \frac{a}{8}[\bar{1}\bar{1}0] + \frac{a}{4}[\bar{1}\bar{1}2] + \frac{a}{8}[\bar{1}\bar{1}0]$$

This corresponds to the equivalent of Shockley partials. Apparently, the stacking-fault energy is very high because they cannot be observed by transmission electron microscopy. The wavyness of the slip markings is also indicative of the high stacking-fault energy. If the partials were well separated, slip would be limited to one plane. Cross-slip, which will be treated in Chapter 8, is much easier when the stacking-fault energy is high.

6.8 DISLOCATION SOURCES

It is experimentally observed that the dislocation density increases with plastic deformation; specifically, the relationship $\tau \alpha \rho^{1/2}$ (see Chapter 9) has been found to be closely obeyed. While the dislocation density of an annealed polycrystalline specimen is typically 10^7 cm^{-2}, a plastic strain of 10% raises this density to 10^{10} cm^{-2} or more. This is an increase of three orders of magnitude. This is an apparent paradox, because one would think that the existing dislocations would be ejected out of the crystalline structure by the applied stress. If one calculates the strain that the existing dislocations in an annealed metal would be able to produce by their motion until they would leave the crystal, one would arrive at very small numbers. Consequently, the density of dislocations has to increase with plastic deformation, and internal sources have to be activated. In the next sections, some possible dislocation-generation mechanisms will be discussed.

6.8.1 Homogeneous Nucleation

The homogeneous nucleation of a dislocation consists in the rupture of the atomic bonds along a certain line. Figure 6.27 shows schematically the sequence of steps leading to the formation of a pair of edge dislocations (one negative, one positive). In Fig. 6.27(a) the lattice is elastically stressed, until (b) an atomic plane is sheared; this generates two dislocations that move in opposite senses. This mechanism allows the formation of dislocations from an initially perfect lattice. It can be intuitively seen that the stress required would be extremely high. The calculations were done by Hirth and Lothe.[36] For copper, this stress is of the order of

$$\frac{\tau_{\text{hom}}}{G} = 7.4 \times 10^{-2}$$

Comparing these values with the theoretical strength of crystals described in Chapter 4, one can see that the difference is not very large. Hence, these values would be obtained only if either the applied stresses are very high or there

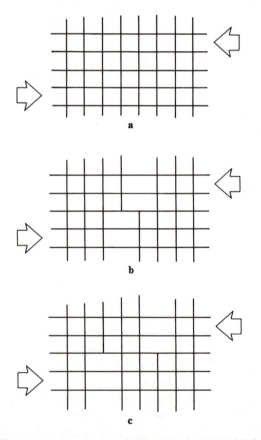

a

b

c

Figure 6.27 Homogeneous nucleation of dislocations in conventional deformation.

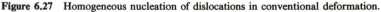

are internal regions of high stress concentration. In conventional deformation other dislocation-generation mechanisms should become operational at much lower stresses, rendering homogeneous nucleation highly unlikely.

There is one special case, however, where it is believed that dislocations are homogeneously nucleated; shock-wave deformation (or shock loading), which is described in greater detail in Chapter 15. In shock-wave deformation, a plastic wave having a steep front and an amplitude many times higher than the yield stress of the material propagates through it at a velocity slightly below or above the velocity of longitudinal elastic waves. The deformation substructure is characterized by a high density of dislocations (see Chapter 15). There are serious limitations to dislocations moving at sonic speeds; this subject is treated in detail in Section 8.4. The stress generated by the shock has shear components that can easily exceed the stress required for homogeneous dislocation generation. Figure 6.28 shows the sequence of what is thought to happen as the shock front travels through the material.[37–39] The shock wave travels downward; its front is represented by arrows in Fig. 6.28. Figure 6.28(a) shows the distortion of a simple cubic lattice if the wave were elastic. The region that is compressed is distorted from its cubicity. If the amplitude of the wave is such that the stress required for homogeneous dislocation generation is exceeded, dislocations are "pinched out" at the interface, accommodating the shear stresses [Fig. 6.28(b)]. Since these dislocations cannot move with the front, the latter advances, leaving behind successive layers of dislocations, that reorganize themselves into more stable arrays.

Figure 6.29 is a transmission electron micrograph of nickel subjected to a shock wave of amplitude 15 GPa, from an initial temperature of 77 K. We can see that the dislocation distribution is very homogeneous; this agrees with the fact that the dislocations moved very little.

6.8.2 Interfacial and Surface Nucleation of Dislocations

J. C. M. Li[40,41] introduced the concept of grain boundaries as dislocation sources. According to him, irregularities at the grain boundaries (steps, or ledges) could be responsible for emission of dislocations into the grains. Murr,[42] among others, showed that grain boundaries could actually emit dislocations. Figure 6.30 shows the emission of dislocations from a grain-boundary source; dislocations are seen as they are being generated at the ledge. The stress due to heating produced by the electron beam generates the force on the dislocations. It is presently thought that dislocation emission from grain boundaries is an important source of dislocations in the first stages of plastic deformation of a polycrystal. Additional information is provided in Section 14.3.1. Price and Hirth[43] proposed a specific mechanism for screw dislocation generation from grain boundaries.

In monocrystals the surfaces can act as dislocation sources. Small steps at the surfaces act as stress concentration sites; hence, the stress can be several times higher than the average stress. At these regions dislocations can be generated and "pumped" into the monocrystals. Nixon et al.[44] showed that the majority of dislocations in monocrystals deformed in tension were generated at the surface. Pangborn et al.[45] determined the bulk and surface dislocation

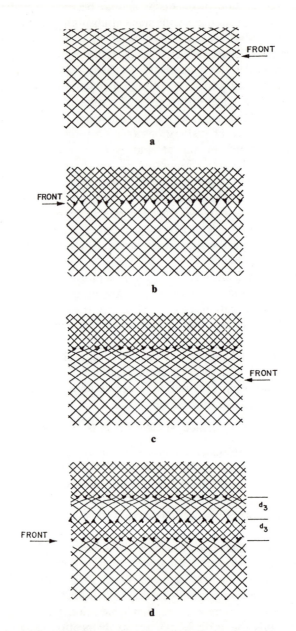

Figure 6.28 Successive stages in the homogeneous nucleation of dislocations during shock loading: (a) elastic distortion of the lattice; (b) dislocation interface coinciding with shock front; (c) shock front advances, leaving behind dislocation interface; (d) as shear stresses build up, new dislocation interfaces are homogeneously nucleated. (Reprinted with permission from [37], p. 550.)

Figure 6.29 Uniform distribution of dislocations in nickel shock-loaded at 15 GPa, 2 μs, and 77 K. Dislocations have moved very little from their original positions.

density in silicon, aluminum, and gold monocrystals after tensile deformation. The dislocation density close to the surface was up to six times higher than that in the bulk. The dislocation surface layer (with higher dislocation density) extended for approximately 200 μm into the material at the surface. The surface sources cannot have a significant effect on polycrystal deformation because the majority of the grains would not be in contact with the free surface. Since dislocation activity is restricted to the grains, the surface sources would not

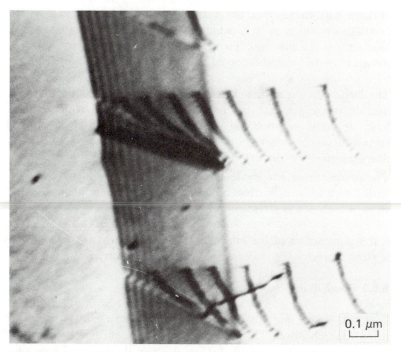

Figure 6.30 Emission of dislocations from ledges in grain boundary as observed in transmission electron microscopy during heating by electron beam. (Courtesy of L. E. Murr, New Mexico Institute of Mining and Technology.)

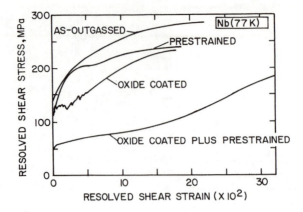

Figure 6.31 Effect of oxide layer on the tensile properties of niobium. (Reprinted with permission from [47], p. 531.)

be able to affect the internal grains. Incoherent interfaces between the matrix and precipitates, dispersed phases, or reinforcing fibers (composites) are also sources of dislocations.

The importance of interfaces in the production of dislocations is seen in the results obtained by Gibala and coworkers.[46-48] The low-temperature tensile response of BCC metals was dramatically affected by the presence of an oxide layer. Figure 6.31 exemplifies this response for niobium. The flow stress of monocrystalline niobium at 77 K is highly dependent on the state of the surface. The oxide softens the material. Two effects are responsible for the lowering of the flow stress by the introduction of an oxide layer:

1. The oxide puts the surface layers under tensile stresses, because the introduction of oxygen into the lattice expands it. The oxide, on the other hand, is under compression. The resultant resolved shear stress at the surface is much higher (in the presence of the oxide layer) than the one due exclusively to the externally applied load.
2. The predeformed and oxide-coated specimen (lowest curve in Fig. 6.31) has an even lower flow stress because the predeformation introduces surface steps, which act as stress-concentration sites.

Hence, the joint action of the internal stresses generated by the oxide and the surface steps activates the dislocation sources at the surface.

6.8.3 Frank–Read Sources

In 1950, Frank and Read[49] proposed their now classic mechanism for dislocation multiplication. This mechanism, later confirmed experimentally, is shown schematically in Fig. 6.32. In Fig. 6.32(a), there is a dislocation *ABCD* with Burgers vector *b*. Only the segment *BC* is mobile in the slip plane α. Segments *AB* and *CD* do not move under the imposed stress. The applied

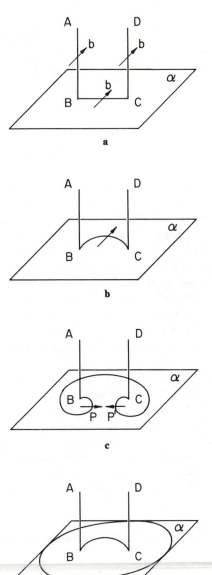

Figure 6.32 Sequence of the formation of dislocation loop by Frank–Read mechanism.

stress will generate a force on segment BC; this force is equal to (see Equation 6.28)

$$F = \frac{T\,ds}{R}$$

The radius of curvature of the dislocation segment decreases until it reaches its minimum, equal to $BC/2$. At this point, the force is maximum (and so is the stress). Hence, the dislocation reaches a condition of instability beyond

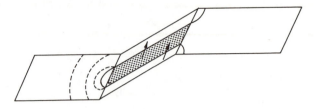

Figure 6.33 Frank–Read source formed by cross-slip.

this point. The critical position is shown in Fig. 6.32(c). When P approaches P', the dislocation segments have opposite signs; they attract each other, forming a complete loop when they touch each other and are pinched off. The stress required to activate a Frank–Read source is equal to the one needed to curve the segment BC into a semicircle with radius $BC/2$; beyond this point, the stress is decreased. However, as loops are formed, they establish a back stress, so that the stress to generate successive loops increases steadily. If the loops are expelled from the material, they cease to exert a back stress.

Only few Frank–Read sources have been observed in metals. However, in a tridimensional array of dislocations, nodes define segments. These segments can bow and effectively act as Frank–Read sources. Another possibility is the source formed when a screw dislocation cross-slips and returns to a plane parallel to the original slip plane (Fig. 6.33). Incidentally, edge dislocations cannot cross-slip because their Burgers vector could not be contained in the cross-slip plane. The Burgers vector of a screw dislocation, on the other hand, is parallel to its line and will be in the cross-slip plane if the intersection of the two dislocations is parallel to it. After the segment in the cross-slip plane advances a certain extent, the stress system applied might force it into a plane parallel to the original slip plane. At this point, a Frank–Read source is formed. Although it is thought that the original formulation of the Frank–Read source is not common, its modifications cited above—the node and the cross-slip case—might be the principal mechanism of dislocation generation, after the first few percent of plastic strain.

6.8.4 Other Sources of Dislocations

The growth of crystals from the gaseous phase or dilute liquid solutions was, for a long time, considered as a mystery. Two unexplained effects occurred: first, the observed growth rate was much higher than the calculated one, and second, the crystals often exhibited a pyramidal shape.

Both effects were explained by Frank[50] when he suggested that the growth of crystals was associated with screw dislocations. Figure 6.34 shows how this process takes place. The sequence (a), (b), (c) shows how the crystalline phase grows around the irregularity, which is a screw dislocation. A gaseous or liquid atom has a much greater tendency to attach itself to the solid in the dislocation region than in a flat surface. This is due to the fact that it will be bonded on two or three surfaces. There is no longer a need to nucleate a new plane once the deposition on one plane has been concluded, since the helicoid renders

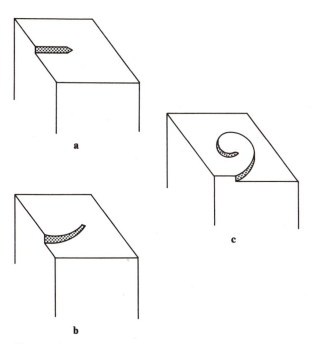

Figure 6.34 Growth of a crystal around a screw dislocation.

the process continuous. As the growth proceeds in the helicoidal fashion, a screw dislocation will form along the central line. Since the growth also proceeds in the direction perpendicular to the dislocation line, a pyramidal morphology results.

Crystals formed by the growth over a substrate (a technique commonly employed in the production of thin films) show dislocations whose formation can be easily explained. The substrate never has exactly the same lattice parameter as the crystal overgrowth. Figure 6.35 shows the sequence of formation of dislocations as the crystal grows over the substrate. If a_s and a_o are the lattice parameters of the substrate and overgrowth, respectively, the separation between the dislocations is

$$d = \frac{a_s^2}{|a_s - a_o|}$$

Often the impurity content of a crystal varies cyclically due to solidification; this is called segregation. The periodic change in composition is associated with changes in lattice parameter; these variations in lattice parameter can be accommodated by dislocation arrays.

Vacancies can condense and form disks and prismatic loops if they are present in a "supersaturated" concentration. In FCC crystals, these disks and loops occur in {111} planes. As seen in Section 6.3.1, the dislocations that form the edges of these features are called Frank dislocations. Kuhlmann-Wilsdorf[17] proposed, and this was later experimentally confirmed, that they can act as Frank–Read sources.

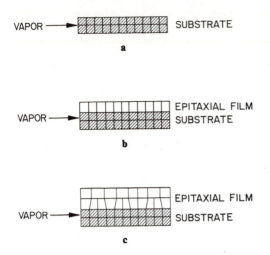

Figure 6.35 Epitaxial growth of thin film: (a) substrate; (b) start of epitaxial growth; (c) formation of dislocations.

EXERCISES

6.1. Derive Eqs. 6.2 and 6.3 from Eq. 6.1.

6.2. Derive Eqs. 6.15 to 6.17 from Eq. 6.14.

6.3. Derive Eq. 6.19 from the knowledge you gained in Chapter 1.

6.4. Derive the equation for the energy of an edge dislocation. If you have difficulties, consult a source from "Suggested Reading."

6.5. **(a)** What stress is required to render operational a Frank–Read source in iron, knowing that the distance between points B and C is 200 Å and that the Goldschmidt radius of the iron atoms is 0.14 nm?
(b) What stress is required to form the second loop?
(c) And the third loop?

6.6. Make all possible reactions between dislocations (perfect) in $(11\bar{1})$ and $(1\bar{1}\bar{1})$ in an FCC crystal. Among them, which ones are Lomer locks?

6.7. Consider all possible reactions between partial Shockley dislocations (only the front dislocation, from the pair) in (111) and $(11\bar{1})$ in an FCC crystal. Among them, which ones will form a stair-rod dislocation?

6.8. **(a)** Show that the reaction

$$\frac{a}{2}[10\bar{1}] \longrightarrow \frac{a}{6}[21\bar{1}] + \frac{a}{6}[11\bar{2}]$$

is either vectorially correct or incorrect.
(b) Is it energetically favorable?

6.9. 10^7 and 10^{11} cm^{-2} are typical values for the dislocation density of annealed and deformed nickel, respectively. Calculate the average space among dislocation lines (assuming a random dislocation distribution), as well as the line energy for edge and screw dislocations, in both cases. Material: nickel ($E = 210$ GPa; $\nu = 0.3$, lowest distance between atom centers: 0.25 nm).

6.10. Calculate the dislocation density for Fig. 6.5(a); assume a foil thickness of 0.3 μm.

6.11. The concentration of vacancies in aluminum at 600°C is 9.4×10^{-4}; by quenching this concentration is maintained at ambient temperature. These vacancies tend to form disks, with Frank partials at the edges. Determine the loop concentration and dislocation density, assuming that:

(a) Disks with 5 nm radius are formed.

(b) Disks with 50 nm radius are formed.

For aluminum, assume that the radius of the atoms is 0.143 nm. (*Hint:* The length of the Frank dislocation corresponding to a disk is equal to the length of the circle.)

6.12. The flow stress of monocrystals is of the order of 10^{-4} G. Using the concept of Frank–Read sources, determine the length of segments required for this stress level. If the length of the segments is determined by dislocations on a second slip plane (tree dislocations), obtain an estimate for the dislocation density in annealed monocrystals. Assume that the dislocations are equally distributed on the slip planes of an FCC crystal.

6.13. In what planes of a BCC structure can the $a/2$ [111] move?

6.14. An edge dislocation, upon encountering an obstacle, stops. A second edge dislocation with identical Burgers vector and moving in the same plane approaches the first dislocation, driven by a stress equal to 140 MPa.

(a) What will be the equilibrium separation between the two dislocations? Assume that the metal is nickel ($E = 210$ GPa, $\nu = 0.3$, $r = 2.49$A).

(b) What would be the equilibrium separation if they were both screw dislocations?

6.15. LiF is an ionic crystal with a NaCl-type structure (cubic). The Li atoms occupy the vertices and the centers of the faces of the unit cell, while the F atoms occupy the edges and one F atom is in the body-centered position. There are eight atoms per unit cell. Knowing that the slip plane for LiF is {110}, determine the Burgers vector of a perfect dislocation. Remember that one has an ionic crystal and that there is a strong repulsion between ions of the same sign. Explain your results.

6.16. Draw a unit cell for an HCP crystal. Show the perfect dislocations in the base plane. Can they decompose into partials? If so, represent them on the special notation for dislocations.

6.17. Nickel sheet is being rolled at ambient temperature in a rolling mill (roll diameter 50 cm; velocity 200 rpm). The initial thickness is 20 mm and the final thickness is 10 mm (one pass).

(a) Calculate the average strain rate.

(b) Calculate the energy that will be stored in the material, assuming that the final density is 10^{11} cm^{-2}.

(c) Assuming that the process is adiabatic, calculate the temperature increase ($C_p = 0.493$ J/g °C). Calculate this taking into account dislocation motion and using the Peach–Koehler equation. [*Hint:* Use Orowan's equation; consult R. W. Rohde and C. H. Pitt, *J. Appl. Phys.*, *38* (1967) 876.]

(d) Determine the total energy expenditure per unit volume from part (c).

(e) Why does the energy stored represent only a fraction of the energy expended?

6.18. The response of copper to plastic deformation can be described by Hollomon's equation $\sigma = \kappa \epsilon^{0.7}$.

It is known that for $\epsilon = 0.25$, $\sigma = 120$ MPa. The dislocation density varies with flow stress according to the well-known relationship

$$\sigma = K' \rho^{1/2}$$

(a) If the dislocation density at a plastic strain of 0.4 is equal to 10^{11} cm^{-2}, plot the dislocation density versus strain.
(b) Calculate the work performed to deform the specimen.
(c) Calculate the total energy stored in the metal as dislocations after a plastic deformation of 0.4 and compare this value with the one obtained in part (b). Explain the difference.

6.19. There are two alternative ways of drawing a Thompson tetrahedron. One is indicated in Fig. 6.18. The other one consists of taking the points (1, 0, 0), (0, 1, 0), (0, 0, 1), and (1, 1, 1) as the vertices. Draw this tetrahedron and project the four faces on a plane, indicating all perfect and partial dislocations.

REFERENCES

[1] O. Mügge, *Neues Jahrb. Min.*, *13* (1883).

[2] A. Ewing and W. Rosenhain, *Phil. Trans. Roy. Soc. London*, *A193* (1889) 353.

[3] E. Orowan, *Z. Phys.*, *89* (1934) 604, 634.

[4] M. Polanyi, *Z. Phys.*, *89* (1934) 660.

[5] G.I. Taylor, *Proc. Roy Soc. (London)*, *A145* (1934) 362.

[6] E. Orowan, in *The Sorby Centennial Symposium on the History of Metallurgy*, C.S. Smith (ed.), Gordon and Breach, New York, 1965, p. 359.

[7] G.I. Taylor, in *The Sorby Centennial Symposium on the History of Metallurgy*, C.S. Smith (ed.), Gordon and Breach, New York, 1965, p. 355.

[8] J.M. Burgers, *Proc. Kon. Ned. Akad. Wetenschap.*, *42* (1939) 293, 378.

[9] D. Hull, *Introduction to Dislocations*, Pergamon Press, Elmsford, N.Y., 1969, p. 66.

[10] S. Amelinckx, W. Bontinck, W. Dekeyser, and F. Seitz, *Phil. Mag.*, *2* (1957) 355.

[11] K.K. Chawla, *J. Fiber Sci. Technol.*, *8* (1975) 49.

[12] S. Amelinckx, in *Dislocations and Mechanical Properties of Crystals*, J.C. Fisher, W.G. Johnston, R. Thomson, and T. Vreeland, Jr. (eds.), Wiley, New York, 1947, p. 3, 26.

[13] J.B. Newkirk, *Trans. TMS-AIME*, *215* (1959) 483.

[14] R.D. Heidenreich, *J. Appl. Phys.*, *20* (1949) 993.

[15] P.B. Hirsch, R.W. Horne, and M.J. Wheelan, *Phil. Mag.*, *1* (1956) 677.

[16] D.J. Bacon, D.M. Barnett, and R.O. Scattergood, *Prog. Mater. Sci.*, *23* (1979) 51.

[17] D. Kuhlmann-Wilsdorf, in *Physical Metallurgy*, R.W. Cahn (ed.), Elsevier/North-Holland, New York, 1970, p. 787.

[18] J.D. Eshelby and A.N. Stroh, *Phil. Mag.*, *42* (1951) 1401.

[19] V. Volterra, *Ann. Ecole Norm. Sup.*, *24* (1907) 400.

[20] A.E.H. Love, *The Mathematical Theory of Elasticity*, Cambridge University Press, Cambridge, 1927, p. 221.

[21] A. Timpe, *Z. Math. Phys.*, *52* (1905) 348.

[22] G. Leibfried and K. Lücke, *Z. Physik*, *126* (1949) 450.

[23] J.C.M. Li, in *Electron Microscopy and Strength of Crystals*, G. Thomas and J. Washburn (eds.), Interscience Publishers, New York, 1963, p. 713.

[24] M. Peach and J.S. Koehler, *Phys. Rev.*, *80* (1950) 436.

[25] B.D. Cullity, *Elements of X-Ray Diffraction*, Addison-Wesley, Reading, Massachusetts, 1956, p. 459.

[26] N. Thompson, *Proc. Phys. Soc.*, *B66* (1953) 481.

[27] L.E. Murr, *Interfacial Phenomena in Metals and Alloys*, Addison-Wesley, Reading, Massachusetts, 1975, p. 142.

[28] A. Kelly and G.W. Groves, *Crystallography and Crystal Defects*, Addison-Wesley, Reading, Massachusetts, 1970, p. 233.

[29] H.J. Kestenbach, *Phil. Mag.*, *36* (1977) 1509.

[30] M.A. Meyers, C.Y. Hsu, L.E. Murr, and G.A. Stone, *Mater. Sci. & Eng.*, 57 (1983) 113.

[31] F.C. Frank, *Proc. Phys. Soc.*, *A62* (1949) 202.

[32] W.M. Lomer, *Phil. Mag.*, *42* (1951) 1327.

[33] A.H. Cottrell, *Phil. Mag.*, *43* (1952) 645.

[34] R.D. Heidenreich and W. Shockley, in *Report of a Conference on Strength of Solids*, Physical Soc., London, 1948, p. 57.

[35] L.J. Teutonico, *Mater. Sci. & Eng.*, 6 (1970) 27.

[36] J.P. Hirth and J. Lothe, *Theory of Dislocations*, McGraw-Hill, New York, 1968, p. 688.

[37] M.A. Meyers, *Scripta Met.*, *12* (1978) 21.

[38] M.A. Meyers, in *The Strength of Metals and Alloys*, P. Haasen, V. Gerold, and G. Kostorz, (eds.) Pergamon Press, Elmsford, N.Y., 1980, p. 547.

[39] M.A. Meyers and L.E. Murr, in *Shock Waves and High-Strain-Rate Phenomenon Metals: Concepts and Applications*, M.A. Meyers and L.E. Murr (eds.), Plenum Press, N.Y., 1981, p. 487.

[40] J.C.M. Li, *Trans. TMS-AIME*, *227* (1963) 239.

[41] J.C.M. Li, *J. Appl. Phys.*, *32* (1961) 525.

[42] L.E. Murr, *Met. Trans.*, *6A* (1975) 505.

[43] C.W. Price and J.P. Hirth, *Mater. Sci. & Eng.*, 9 (1972) 15.

[44] W.E. Nixon, M.H. Massey, and J.W. Mitchell, *Acta Met.*, *27* (1979) 943.

[45] P.N. Pangborn, S. Weissmann, and I.R. Kramer, *Met. Trans.*, *12A* (1981) 109.

[46] V.K. Sethi and R. Gibala, *Thin Solid Films*, *39* (1976) 79.

[47] V.K. Sethi and R. Gibala, *Scripta Met.*, 9 (1975) 527.

[48] V.K. Sethi and R. Gibala, *Acta Met.*, *25* (1977) 321.

[49] F.C. Frank and W.T. Read, in *Symposium on Plastic Deformation of Crystalline Solids*, Carnegie Institute of Technology, Pittsburgh, Pa., 1950, p. 44.

[50] F.C. Frank, *Proc. Roy. Soc. (London)*, *A62* (1949) 202.

SUGGESTED READING

COTTRELL, A.H., *Dislocations and Plastic Flow in Crystals*, McGraw-Hill, New York, 1953.

COTTRELL, A.H., *Theory of Crystal Dislocations*, Gordon and Breach, New York, 1964.

FISHER, J.C., W.G. JOHNSTON, R. THOMSON, and T. VREELAND, JR. (eds.), *Dislocations, and Mechanical Properties of Crystals*, Wiley, New York, 1957.

FRIEDEL, J., *Dislocations*, Pergamon Press, Elmsford, N.Y., 1967.

HIRTH, J.P., and J. LOTHE, *Theory of Dislocations*, 2nd ed., J. Wiley, New York, 1981.

HULL, D., *Introduction to Dislocations*, Pergamon Press, Elmsford, N.Y., 1965.

KELLY, A., and G.W. GROVES, *Crystallography and Crystal Defects*, Addison-Wesley, Reading, Mass., 1974.

KOVACS, I., and L. ZSOLDOS, *Dislocations and Plastic Deformation*, Pergamon Press, Elmsford, N.Y., 1973.

KUHLMANN-WILSDORF, D., in *Physical Metallurgy*, 2nd ed., R.W. Cahn (ed.), North-Holland, Amsterdam, 1970, p. 787.

NABARRO, F.R.N., *Dislocations*, Oxford University Press, Oxford, 1967.

NABARRO, F.R.N. (ed.), *Dislocations in Solids*, Elsevier/North-Holland, New York, 1979.

READ, W.T., JR., *Dislocations in Crystals*, McGraw-Hill, New York, 1953.

WEERTMAN, J., and J.R. WEERTMAN, *Elementary Dislocation Theory*, Macmillan, New York, 1964.

Chapter 7

PLANAR DEFECTS

7.1 INTRODUCTION

In Chapters 5 and 6 we dealt with point and line defects, respectively. There is another class of defects called interfacial or planar defects. These imperfections, as the name signifies, occupy an area or surface and so are bidimensional, and are of great importance in mechanical metallurgy. As examples, we may cite free surfaces of material, grain boundaries, twin boundaries, domain, and anti-phase boundaries. Of all these, the grain boundaries are the most important from the mechanical properties point of view. In what follows, we consider in detail the structure of grain and twin boundaries and their importance in various deformation processes, and, very briefly, the structure of other interfacial defects. The details regarding the strengthening by grain boundaries are given in Chapter 8.

7.2 GRAIN BOUNDARIES

Crystalline solids generally consist of a large number of grains separated by boundaries. Each grain (or subgrain) is a single crystal and the grain boundaries are, thus, transition regions between the neighboring crystals. These transition regions may consist of various kinds of dislocation arrangements. When the misorientation between two grains is small, the grain boundary can be described by a relatively simple configuration of dislocations (e.g., an edge dislocation wall) and is, fittingly, called a low-angle boundary. When the misorientation is large (high-angle grain boundary), more complicated structures are involved (as in a configuration of soap bubbles simulating the atomic planes in crystal lattices). A general grain boundary has five degrees of freedom. We need three degrees to specify the orientation of one grain with respect to the other and two degrees to specify the orientation of the boundary with respect to one of the grains. Grain structure is usually specified by giving the average diameter or using a procedure due to ASTM according to which grain size is specified by a number n in the expression $N = 2^{n-1}$, where N is the number of grains per square inch when the sample is examined at 100×.

7.2.1 Tilt and Twist Boundaries[1]

The simplest grain boundary consists of a configuration of edge dislocations between two grains. The misfit in the orientation of the two grains (one on each side of the boundary) is accommodated by a perturbation of the regular arrangement of crystals in the boundary region.

Figure 7.1 shows some vertical atomic planes terminating in the boundary and each termination is represented by an edge dislocation. The misorientation at the boundary is related to spacing between dislocations, D, by the following relation:

$$D = \frac{b}{2 \sin (\theta/2)} \simeq \frac{b}{\theta} \qquad \text{(for } \theta \text{ very small)} \qquad (7.1)$$

where b is the Burgers vector.

Thus, as the misorientation θ increases, the spacing between dislocations is reduced, until, at large angles, the description of the boundary in terms of simple dislocation arrangements does not make sense. Theta becomes so large that the dislocations are separated by one or two atomic spacings; for such

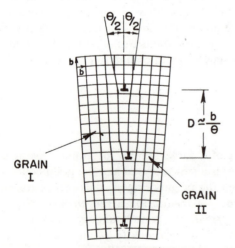

Figure 7.1 Low-angle tilt boundary.

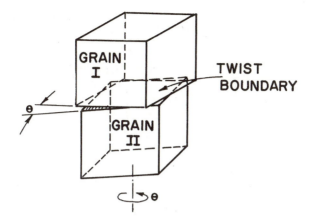

Figure 7.2 Low-angle twist boundary.

small separations the dislocation core energy becomes important and the linear elasticity does not hold. In such cases, the grain boundary is a region of severe localized disorder.

Boundaries consisting entirely of edge dislocations are called tilt boundaries, because the misorientations, as can be seen in Fig. 7.1, can be described in terms of a rotation about an axis normal to the plane of the paper and contained in the plane of dislocations. The example shown in Fig. 7.1 is called the symmetrical tilt wall as the two grains are symmetrically located with respect to the boundary. A boundary consisting entirely of screw dislocations is called twist boundary, because the misorientation can be described by a relative rotation of two grains about an axis. Figure 7.2 shows a twist boundary consisting of two groups of screw dislocations.

It is possible to produce misorientations between grains by combined tilt and twist boundaries. In such a case, the grain boundary structure will consist of a network of edge and screw dislocations.

7.2.2 Calculation of the Energy of a Grain Boundary

The dislocation model of grain boundary can be used to compute the energy of low-angle boundaries ($\theta \leq 10°$). For such boundaries the distance between dislocations in the boundary is more than a few interatomic spaces, as

$$\frac{b}{D} \simeq \theta \leq 10° \simeq \tfrac{1}{6} \text{ rad} \qquad \text{or} \qquad D \simeq 6b$$

Consider a tilt boundary consisting of edge dislocations with spacing D. Let us isolate a small portion of dimension D, as in Fig. 7.3, with a dislocation at its center. The energy associated with such a portion, E, includes contributions from the regions marked I, II, and III in Fig. 7.3. E_I is the energy due to the material inside the dislocation core of radius r_I. E_{II} is the energy contribution

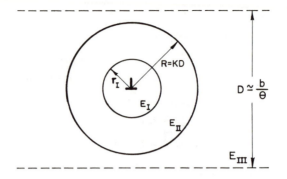

Figure 7.3 Model for the computation of grain boundary energy.

of the region outside the radius r_I and inside the radius $R = KD > b$, where K is a constant less than unity. In this region II, the elastic strain energy contributed by other dislocations in the boundary is very small. E_{II} is mainly due to the elastic strain energy associated with the dislocation in the center of this portion. E_{III}, the rest of the energy in this portion, depends on the combined effects of all dislocations.

The total strain energy per dislocation in the boundary is, then,

$$E_\perp = E_I + E_{II} + E_{III}$$

Consider now a small decrease, $d\theta$, in the boundary misorientation. The corresponding variation in the strain energy is

$$dE_\perp = dE_I + dE_{II} + dE_{III}$$

and

$$-\frac{d\theta}{\theta} = \frac{dD}{D} = \frac{dR}{R} \qquad \text{(as } R = KD\text{)}$$

(i.e., a decrease in θ increases R and D).

The new dimensions of this crystal portion are shown in Fig. 7.4. The region immediately around the dislocation, contributing an energy E_I, does not change. This region does not change because E_I, the localized energy of atomic misfit in the dislocation core, is independent of the disposition of other dislocations. Thus, $dE_I = 0$. E_{II} increases by a quantity dE_{II}, corresponding to an increase in R by dR. E_{III}, however, does not change with an increase in D, because although the volume of region III increases, the number of dislocations contributing to the strain energy of this region decreases. Therefore,

$$dE_{III} \simeq 0$$

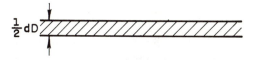

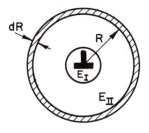

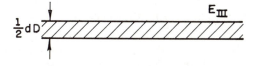

Figure 7.4 New dimensions of a portion of crystal after a decrease $d\theta$ in the boundary misorientation.

Thus, $dE_\perp$ corresponds only to an increase in the energy of region II, dE_{II}, caused by an increase of dR in R. The increase in strain energy per volume element between the cylinders of radii R and $(R + dR)$ is

$$dE_\perp = dE_{II} = \frac{Gb^2}{4\pi(1 - \nu)}\frac{dR}{R} = -\frac{Gb^2}{4\pi(1 - \nu)}\frac{d\theta}{\theta}$$

The energy per dislocation is

$$E_\perp = \int dE_\perp = \frac{Gb^2}{4\pi(1 - \nu)}(A - \ln \theta) \tag{7.2}$$

where A is a constant of integration.

The energy per unit area of the boundary is obtained by multiplying the dislocation energy per unit length by the number of dislocations per unit length of the boundary measured in a direction normal to the dislocation lines:

$$E = E_\perp\left(\frac{1}{D}\right) = E_\perp\left(\frac{\theta}{b}\right) = \frac{Gb}{4\pi(1 - \nu)}\theta(A - \ln \theta)$$

or

$$E = E_0\theta(A - \ln \theta) \tag{7.3}$$

where

$$E_0 = \frac{Gb}{4\pi(1 - \nu)} \tag{7.4}$$

7.2.3 Variation of Grain-Boundary Energy with Misorientation

Consider Eq. 7.3. Because of the $(-\ln\theta)$ term, a merger of two low-angle boundaries, forming a high-angle boundary, always results in a net decrease in the total energy of the interface. Thus, low-angle boundaries have a tendency to combine and form boundaries of large misorientation.

A plot of E versus θ gives a curve with a maximum at $\theta_{max} \simeq 0.5$ rad ($\approx 30°$). However, the dislocation model of grain boundaries loses validity at much smaller orientations ($\theta \simeq 10°$). Some recent studies, using field-ion microscopy, have shown that the high-angle grain boundaries consist of rather large regions of atomic fit separated by regions of misfit, to which are associated the grain boundary ledges. The boundary thickness is not more than two to three atomic diameters. Low-angle grain boundaries have a dislocation density that increases proportionally to the misorientation angle (see Eqs. 7.1 and 7.3), and, consequently, the energy of a low-angle boundary increases linearly with θ near $\theta = 0°$. After this, the energy increases slowly as the stress fields of adjacent dislocations interact more strongly. This behavior is shown in Fig. 7.5. A surface tension, γ_{gb}, can be associated with an ordinary (high-angle) grain boundary, which consists of a mixture of various types of dislocations. As the value is relatively high, it is instructive to determine the stable forms assumed by the grains of a given material. As it happens, there are certain special boundaries for which a particular high angle between two adjacent crystals produces a low value of γ. These special boundaries can be divided into two categories: coincidence boundaries and coherent twin boundaries. A coincidence boundary (Fig. 7.6) is incoherent as an ordinary grain boundary; that is, a majority of the atoms of one crystal, in the boundary, do not correspond to the lattice sites of the other crystal. On an average, however, this noncorrespondence in a coincidence boundary is less as the density of coincidence sites increases. For example, in Fig. 7.6, one atom in seven, in a boundary, is in a lattice position for both the crystals. We call this boundary a one-seventh coincidence boundary and the atomic sites (black atoms in Fig. 7.6) in question form

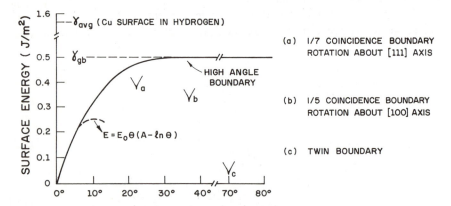

Figure 7.5 Variation of grain boundary energy γ_{gb} with misorientation θ. (Adapted with permission from [2], p. 212.)

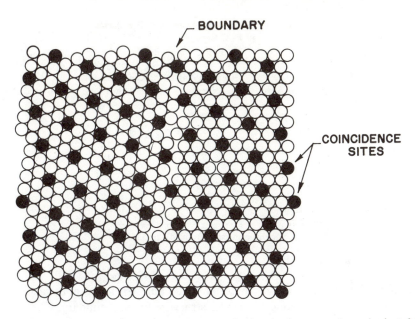

Figure 7.6 Coincidence lattice made by every seventh atom in the two grains, misoriented 22° by a rotation around the ⟨111⟩ axis. (Adapted from [3], p. 506.)

a coincidence lattice for the two grains.[3] Coincidence lattices occur in all common crystalline structures and have a density of sites varying from $\frac{1}{3}$ to $\frac{1}{9}$ and less.

A twin boundary is frequently a kind of coincidence boundary, but it is convenient to treat it separately. The energy of a twin boundary γ_{twin} is generally about $0.1\gamma_{gb}$ (see Fig. 7.5), whereas the energy of a coincidence boundary is only slightly less than γ_{gb}. The two most common twin orientations are (1) rotation twins (coincidence), produced by a rotation about a direction $[hkl]$ called the twinning axis; and (2) reflection twins, in which the two lattices maintain a mirror symmetry with respect to a plane $[hkl]$ called the twinning plane.

Some of the orientations that give the highest density of coincidence lattice sites in crystals are given in Table 7.1.[4] These boundaries have lower energies than those of random high-angle boundaries. Contrary to the great majority of low-energy boundaries, coincidence site boundaries have greater mobility than that of the random boundaries. Twin boundaries, even with low energies, have lower mobility because they are coherent.

7.2.4 Grain-Boundary Dislocations

Various experimental observations of the structure of grain boundaries[5-7] have demonstrated the existence of grain-boundary dislocations (GBDs) when the orientation relations deviate from the ideal coincidence lattice site orientations. A grain-boundary dislocation belongs to the grain boundary and is not

**TABLE 7.1 Some Coincidence Site Boundaries
in FCC Crystals**

Rotation Axis	Rotation Angle (deg)	Density of Coincidence Sites
$\langle 111 \rangle$	38	1 in 7
	22	1 in 7
	32	1 in 13
	47	1 in 19
$\langle 110 \rangle$	39	1 in 9
	$50\frac{1}{2}$	1 in 11
	$26\frac{1}{2}$	1 in 19
$\langle 100 \rangle$	37	1 in 15

Reprinted with permission from [4], p. 326.

a common lattice dislocation. In order to be able to define a Burgers vector of a GBD, one must assume a periodic structure in the boundary. The possible Burgers vectors for perfect GBDs are equal to the difference of the two lattice vectors. Thus,

$$B = \mathbf{b}_A \pm \mathbf{b}_B$$

where $\mathbf{b}_A$ and $\mathbf{b}_B$ are the Burgers vectors of grains A and B, respectively, that join at the boundary. A GBD model is shown in Fig. 7.7. Figure 7.7(a) shows the boundary without GBD and Fig. 7.7(b) shows the same boundary with a GBD of Burgers vector b. A GBD can be imagined as consisting of two planes inserted into the lattice (indicated by arrows). The dashed line indicates the grain boundary.

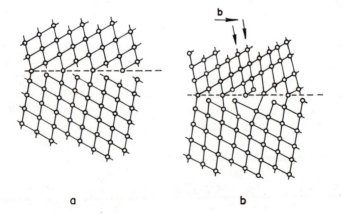

a   b

Figure 7.7 Atomistic model of a grain boundary dislocation (GBD): (a) boundary without GBD; (b) boundary with GBD. (Reprinted with permission from [7], p. 1056.)

7.2.5 Grain-Boundary Ledges

Grain-boundary dislocations (GBDs) can acquire, by grouping together, the geometry of a grain-boundary ledge. This agglomeration that leads to the formation of a step is shown in Fig. 7.8. Figure 7.8(a) shows the movement of GBDs along the grain-boundary plane in the direction indicated by the arrow. Figure 7.8(b) shows the coalescence of GBDs to make a grain-boundary ledge. Another way of ledge formation is shown in Fig. 7.8(c) and (d). Lattice dislocations, under the applied tension, can move from grain A, through the boundary plane, to grain B [Fig. 7.8(c)]. The passage through the boundary results in heterogeneous shear of the boundary, forming a ledge.

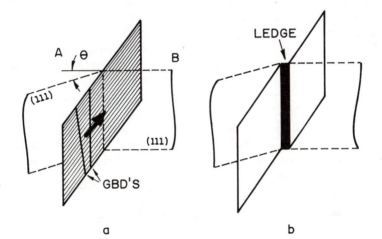

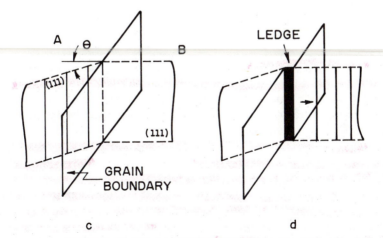

Figure 7.8 Models of the ledge formation in a grain boundary. (Reprinted with permission from [8], p. 255.)

In the simplified situation shown in Fig. 7.8 the (111) of the neighboring grains intersect along the boundaries. Ledges in the grain boundaries constitute an important structural characteristic of the high-angle boundaries. It has been observed that the density of ledges increases with an increase in the boundary misorientation. One of the important aspects of this structure of boundaries is that the ledges function as effective sources of dislocations, a fact that has important implications on the mechanical properties of polycrystals. These are described in Sections 6.8 and 14.2.3.

7.2.6 Grain Boundaries as a Packing of Polyhedral Units

Ashby et al.[9] have described the grain-boundary structure in terms of a packing of polyhedral units. If equal spheres are packed to form a shell, such that all spheres touch their neighbors, then the sphere centers are at the vertices of a "deltahedron," a polyhedron with equilateral triangles as faces. Ashby et al. regard a crystal as a regular packing of polyhedral holes. The FCC structure, for example, consists of a regular packing of tetrahedra and octahedra. The main advantage of such a description of the structure is that it remains valid even when the structure becomes completely disordered (i.e., an amorphous structure). Any grain boundary between metallic crystals can be described as per this scheme as a packing of eight basic deltahedra. This model, according to the authors, is able to describe a number of properties associated with grain boundaries; for example, segregation of certain elements to the boundaries, the characteristically high diffusion rates in the boundaries, and grain-boundary faceting in the presence of impurities.

7.2.7 Role of Grain Boundaries in Plastic Deformation

Grain boundaries have a very important role in plastic deformation of polycrystalline materials, and this is treated in detail in Chapter 14. We outline below the important aspects of the role of grain boundaries:

1. At low temperatures ($T < 0.5 T_m$, where T_m is the melting point in K), the grain boundaries act as strong obstacles to dislocation motion. Mobile dislocations can pile up against the grain boundaries (see Section 14.2) and thus give rise to stress concentrations that can be relaxed by initiating locally multiple slip.

2. There exists a condition of compatibility among the neighboring grains during the deformation of polycrystals (see Section 8.5); that is, if the development of voids or cracks is not permitted, the deformation in each grain must be accommodated by its neighbors. This accommodation is realized by multiple slip in the vicinity of the boundaries which leads to a high strain hardening rate. It can be shown, following von Mises, that for each grain to stay in contiguity with others during deformation, there must be operating at least five independent slip systems (see Section 8.5.1). This condition of strain compatibility leads a

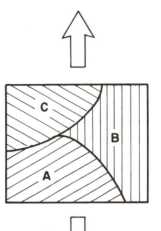

Figure 7.9 Strain compatibility problem among three grains of a hexagonal close-packed material. The lines represent the basal plane traces.

polycrystalline sample to have multiple slip in the vicinity of grain boundaries. The smaller the grain size, the larger will be the total boundary surface area per unit volume. In other words, for a given deformation in the beginning of the stress–strain curve, the total volume occupied by the work-hardened material increases with the decreasing grain size. This implies a greater hardening due to dislocation interactions induced by multiple slip.

Let us consider the case of a solid material that does not possess five independent slip systems in operation. Figure 7.9 shows three grains of a polycrystalline HCP material. The lines indicate the basal plane traces. The resolved shear stress in grain *B* is zero; that is, the grain *B* constitutes a barrier to slip in grains *A* and *C*. Symmetrically oriented grains such as *A* and *C* can deform easily. In a polycrystal with grains oriented randomly, there exist barriers of strength varying from type *A-C* to that of type *B-C*.

3. At high temperatures the grain boundaries function as sites of weakness. Grain boundary sliding may occur, leading to plastic flow and/or opening up of voids along the boundaries (see Chapters 3 and 20).

4. Grain boundaries can act as sources and sinks for vacancies at high temperatures, leading to diffusion currents as, for example, in the Nabarro–Herring creep mechanism (see Chapter 20).

5. In polycrystalline materials, the individual grains usually have a random orientation with respect to one another. We call this grain structure as randomly oriented. Frequently, however, the grains of a metal may be preferentially oriented. For example, an Fe–3% Si solid solution alloy, used for electrical transformer sheets because of its excellent magnetic properties, has grains with their [110] planes nearly parallel and their $\langle 100 \rangle$ direction along the rolling direction of the sheet (Fig. 7.10). This material is said to have a texture or preferred orientation. Preferred orientation of grains is also frequently observed in drawn wires.

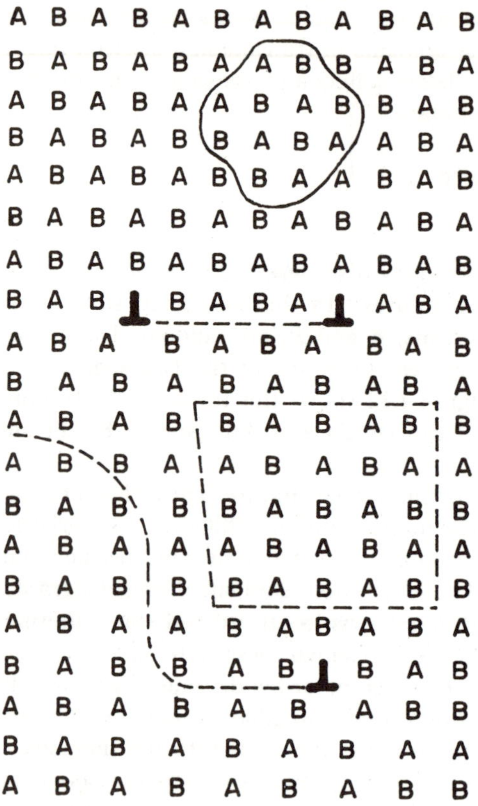

Figure 7.13 Ordered bidimensional structure with two superlattice domains completely surrounded by a straight and a curved antiphase boundary. Also shown are two antiphase boundaries limited by dislocations. (Reprinted with permission from [10], p. 364.)

small region of HCP stacking in an FCC crystal. Although the material presents the same orientation on the two sides of the defect, one lattice is translated with respect to the other by a fraction of interplanar distance.

7.4 TWINNING AND TWIN BOUNDARIES

There are two types of twin boundaries: deformation twins and annealing twins. A brief description of these two types of boundaries follows.

7.4.1 Deformation Twins

Deformation or mechanical twinning is the second most important mechanism of plastic deformation in metals after slip, although it is not as common as slip. The crystallographic nature of deformation twins is shown in Fig. 7.14. When a crystal deforms plastically by twinning, there occur atomic displacements, as shown in Fig. 7.14, which give rise to crystal bands within the grain that are twin-oriented. Hexagonal metals, such as Zn and Mg, behave in this way when deformed at ambient temperatures, while BCC metals, such as iron,

Figure 7.15 Deformation twins in shock-loaded AISI 1010 steel. (Reprinted with permission from [11], p. 1737.)

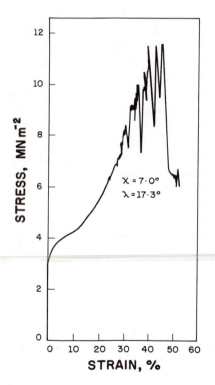

$$\chi = 7 \cdot 0°$$
$$\lambda = 17 \cdot 3°$$

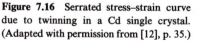

Figure 7.16 Serrated stress–strain curve due to twinning in a Cd single crystal. (Adapted with permission from [12], p. 35.)

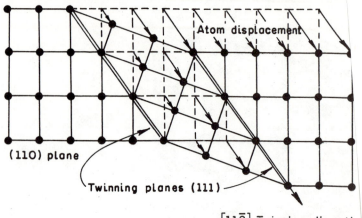

Figure 7.14 Schematic of twinning in FCC metals.

show this behavior when deformed at subambient temperatures. This mec[l]
is not of great importance in the deformation of FCC metals. The twin [ ]
twin vectors, and the shear produced by them are given in Table 7.2 for[ ]
BCC, and HCP crystals. Figure 7.15 shows deformation twins in AIS[I]
steel deformed by shock loading at a pressure of approximately 12 GPa[ ]
can see that the twins (dark lines) do not traverse the grain boundaries. Co[ ]
tional deformation induces twins in iron only at very low temperatures.

The mechanism of plastic deformation by twinning is very different [ ]
that of slip. First, the twinned region of a grain is a mirror image of the or[i]
lattice, while the slipped region has the same orientation as that of the or[i]
unslipped grain. Second, slip consists of a shear displacement of an entire [b]
of crystal, while twinning consists of uniform shear strain. Third, the slip d[ ]
tion can be positive or negative (i.e., in tension or compression), while [ ]
twinning direction is polar. Twinning results in a change of shape of def[ ]
type and magnitude as determined by the crystallographic nature of the twin[i]
elements.

The stress necessary to form twins is, generally, greater than but [ ]

**TABLE 7.2 Twinning Planes, Directions,
and Shears**

Structure	Twin Plane and Direction	Shear
FCC	$(111)\,[112]$	0.707
BCC	$(112)\,[111]$	0.707
HCP	$(10\bar{1}2)\,[10\bar{1}1]$	Cd: 0.171
		Zn: 0.139
		Mg: 0.129
		Ti: 0.189
		Be: 0.199

sensitive to temperature than that necessary for slip. This stress for initiation of twinning is much larger than the stress necessary for its propagation. Deformation twinning occurs when the applied stress is high due to work hardening or low temperatures or, in the case of HCP metals, when the resolved shear stress on the basal plane is low. Copper and other FCC metals can be made to deform by twinning at very low temperatures. Deformation twins, however, play an important role in the straining of HCP metals. The "cry of tin" heard when a polycrystalline sample of tin is bent plastically is caused by the sudden formation of deformation twins. The bursting of twins during straining can lead to a serrated form of stress strain curve (Fig. 7.16). In many HCP metals, the slip is restricted to basal planes. Thus, twinning can contribute to plastic deformation by the shear that it produces, but this is generally small (see Table 7.2). More important, however, the twinning process serves to reorient the crystal lattice to favor further basal slip. In HCP metals, the common twinning elements are $(10\bar{1}2)$ plane and $[10\bar{1}1]$ direction. Twinning results in a compression or elongation along the c axis depending on the c/a ratio. For $c/a > \sqrt{3}$ (the case of Zn and Cd), twinning occurs on $(10\bar{1}2)$ $[\bar{1}011]$ when compressed along the c axis. When $c/a = \sqrt{3}$ the twinning shear is zero. For $c/a < \sqrt{3}$ (the case of Mg and Be), twinning occurs under tension along the c axis. Figure 7.17 shows this dependence on the c/a ratio.

In terms of the dislocation theory, the following mechanism has been proposed for deformation twin formation.[13] It is called the pole mechanism (Fig. 7.18). A partial dislocation approaches a tree dislocation that has, at least, partially, a screw character. The atomic planes form a ramp around the screw or partially screw dislocation. The free dislocation serves as a pole for the spiral ramp. When the partial dislocation meets the pole dislocation, an arm of the

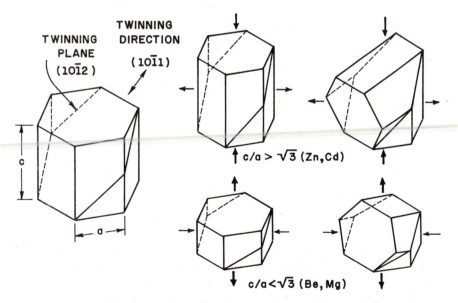

Figure 7.17 Twinning in HCP metals with c/a ratio more than or less than $\sqrt{3}$.

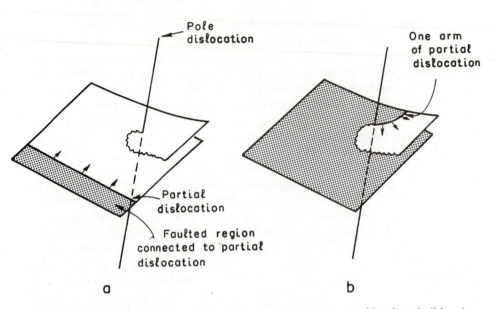

Figure 7.18 Pole mechanism for twinning: (a) partial dislocation approaching the pole dislocation; (b) two arms of partial dislocation start the spiral around the pole dislocation.

partial dislocation will move a plane up and the other arm will move a plane down. The two arms, thus, will move in spiral around the pole dislocation. Eventually, one will have moved a partial dislocation through all the planes in succession and, thus, one will have a deformation twin. This is the pole mechanism proposed by Cottrell and Bilby[14] for BCC metals and extended by Venables[15] to FCC metals. The principal deterrent to the pole mechanism is the maximum rate of growth predicted by this model; it is of the order of 1 m/s. This is about three orders of magnitude below values reported by Bunshah,[16] Reid et al.,[17] and Takeuchi.[18] Takeuchi[18] found that a twin propagated at 2500 m/s in iron, and that this velocity was virtually independent of temperature in the interval −196 to +126°C. The latter observation is very important and indicative of the fact that growth is not a thermally activated process. Hornbogen[19] suggested that the propagation of twins in Fe-Si alloys occurs at such a rate as to generate shock waves. It is the velocity limitation that led Cohen and Weertman[20] to propose a much simpler model for FCC metals, involving the production of Shockley partials at Cottrell–Lomer locks and their motion through the material. The velocity of propagation of a twin is in this case simply established by the velocity of motion of the Shockley partials. Hirth and Lothe[21] proposed a yet simpler model in which the dislocations are simply homogeneously nucleated; while the stress required to homogeneously nucleate the first dislocation is of the order of 10% of the shear modulus, the subsequent loops would require stresses that are much lower (1% of G). This "homogeneous nucleation" concept was forwarded first by Orowan.[22]

By comparison, the mechanism of twin formation proposed by Sleeswyk[23] for BCC materials is phenomenologically identical to that of Cohen and

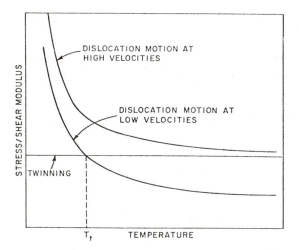

Figure 7.19 Effect of temperature on the stress required for twinning and slip (at low and high strain rates).

Weertman[20] for FCC materials, except that it involves systematic glide of dislocations on [112] planes and takes into account the slip-plane multiplicity in BCC.†

One may regard slip and twinning as competing mechanisms; it is experimentally found that either an increase in strain rate or a decrease in temperature tend to favor twinning over slip. In this context, the graphical scheme proposed by Thomas for martensite can be generalized.[24] This generalization is extended here to high strain rates, showing that they favor twinning. It is shown in Fig. 7.19. The low temperature dependence of the stress required for twin initiation is a strong indication that it is not a thermally activated mechanism. Hence, τ/G for twinning is not temperature dependent. On the other hand, the thermally activated dislocation motion becomes very difficult at low temperatures; T is the temperature below which the material will yield by twinning in conventional deformation. However, at high strain rates and in shock loading, dislocation generation and dynamics are such that the whole curve is translated upward, while the twinning curve is stationary, for reasons that will be given later. As a consequence, the intersection of the two curves takes place at a higher temperature.

7.4.2 Annealing Twins

The formation of annealing twins that are commonly seen in FCC metals and alloys has been explained in various ways. Figure 7.20 shows the three morphologies that such twins frequently assume. A is a grain edge twin, B is

† S. Mahajan [*Acta Met. 23* (1975) 671] carefully studied the crystallography and dislocation configurations in twinned Mo–3.5% Re alloy by transmission electron microscopy. His observations seem to support Sleeswyk's model over that of Cottrell and Bilby. The initiation of twinning seems to occur by the dissociation of a $\frac{1}{2}\langle 111 \rangle$ screw dislocations into three $\frac{1}{6}\langle 111 \rangle$ dislocations on adjacent layers. The parallel movement of these dislocations produces a three-layer twin. The broadening (increase in thickness) is accomplished by additional dissociations on parallel planes.

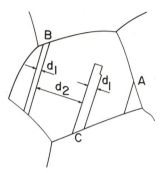

Figure 7.20 Three types of annealing twins observed in FCC metals and alloys. A is a grain-edge twin; B is a complete twin of parallel faces; C is an incomplete twin of parallel faces.

a complete twin of parallel sides, and *C* is an incomplete twin of parallel sides. These parallel sides are coherent twin boundaries and are (111) planes. Twins of type *C* also have an incoherent (or semicoherent) twin boundary joining the parallel faces.

Annealing twins, as the name suggests, are formed during heat treatments (recovery, recrystallization, and grain growth). They have certain importance in determining the mechanical behavior of metals. Babiak and Rhines[26] showed that they should be included in grain size determination. They made experiments taking them into account or not, and came to the conclusion that the Hall–Petch relationship (see Section 14.1) was obeyed only when twin boundaries were included. There are also cases where fracture under creep conditions, at high temperatures, propagates preferentially along coherent twin boundaries.

The formation of annealing twins in FCC metals and alloys has been the object of great speculation and there have been various attempts at explaining them. These attempts can be classified into four groups, described below.

1. *Growth accident.*[27,28] A coherent twin boundary forms accidentally at a grain boundary when it migrates during grain growth.

2. *Grain encounter (Burgers–Nielsen).*[29,30] Different grains, initially separated, are by sheer force of probability in twin orientation. Assuming that these two grains grow, there will come a moment when they will touch. The boundary between these will be a coherent twin boundary. This boundary thus formed will have a low-energy orientation such that it coincides with the "mirror plane" and is coherent.

3. *Dash and Brown model.*[31] This involves nucleation by means of stacking faults in the migrating grain boundaries, growth by means of grain boundary motion leaving behind a twin of parallel faces, or growth by combined movement of the incoherent (or semicoherent) twin boundary and the grain boundary in opposite directions.

4. *"Pop-out" model.*[25] According to this model, the twin formation occurs in two stages: initiation and propagation. The initiation takes place at a grain-boundary ledge. After "popping out" of the grain boundary, the twin grows in the grain interior by means of migration of the incoherent (or semicoherent) twin boundary. Figure 7.21 shows how this occurs. This process does not need an increase in the interfacial energy, in view of the fact that the energy of a

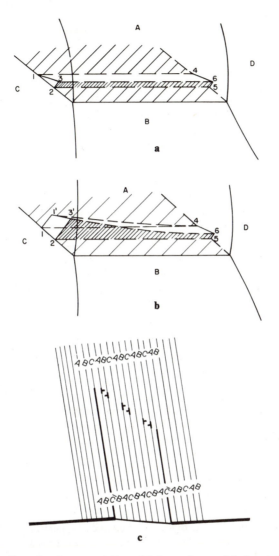

Figure 7.21 Three-dimensional representation of twin nucleus and incipient formation of grain which appears as a step in the grain boundary: (a) small nucleus of a triangular section; (b) after "jumping" outside the boundary, the incoherent twin boundary advances toward the grain interior; (c) dislocation array forming semicoherent boundary in "pop-out" model. (Reprinted with permission from [25], p. 955.)

coherent twin boundary is only a small fraction of random high-angle boundaries. The energy of incoherent (or semicoherent) twins is also less than the random grain boundaries. The model proposed by Meyers and Murr[25] explains the formation of the three types of twins: *A*, *B* and *C*. Figure 7.21(c) shows the special array of Shockley partials (with total Burgers vector zero, since annealing twins do not produce strain) that is supposed to generate, upon advancing, an annealing twin.

EXERCISES

7.1. Calculate the dislocation spacing in a symmetrical tilt boundary ($\theta = 0.5°$) in a copper crystal.

7.2. Starting from the equation $E = E_0\theta(A - \ln\theta)$ for a low-angle boundary, show how one can obtain graphically the values of E_0 and A.

7.3. Taking $A = 0.3$, compute the value of θ_{max}.

7.4. Show that for a low-angle boundary, we have

$$\frac{E}{E_{max}} = \frac{\theta}{\theta_{max}}\left(1 - \ln\frac{\theta}{\theta_{max}}\right)$$

where E_{max} and θ_{max} correspond to the maximum in the E versus θ curve.

7.5. Consider two parallel tilt boundaries with misorientations θ_1 and θ_2. Show that, thermodynamically, we would expect the two boundaries to join and form one boundary of misorientation $(\theta_1 + \theta_2)$.

7.6. Can you suggest a quick technique to check whether lines observed in an optical microscope on the surface of a polished sample after deformation are slip lines or twin markings?

7.7. A twin boundary separates two crystals of different orientations; however, we do not necessarily need dislocations to form a twin. Why?

7.8. Let m be the total length of dislocations per unit area of the grain boundary. Assume that at yield, all the dislocations in the grain interiors (ρ) are the ones emitted by the boundaries. Assume also that the grains are spherical (diameter d). Derive the Hall–Petch relation ($\sigma = \sigma_0 + kd^{-1/2}$) for this case and give the expression for k.

7.9. Consider a piano wire which has a 100% pearlitic structure. When this wire undergoes a reduction in diameter from D_0 to D_ϵ, the pearlite interlamellar spacing normal to the wire axis is reduced from d_0 to d_ϵ, that is,

$$\frac{d_0}{d_\epsilon} = \frac{D_0}{D_\epsilon}$$

where the subscript 0 refers to the original dimensions while the subscript ϵ refers to the dimensions after a true plastic strain of ϵ. If the wire obeys a Hall–Petch type of relationship between the flow stress and the pearlite interlamelar spacing, show that the flow stress of the piano wire can be expressed as

$$\sigma = \sigma_i + \frac{k'}{\sqrt{d_0}}\exp\left(\frac{\epsilon}{4}\right)$$

REFERENCES

[1] W.T. Read, *Dislocations in Crystals*, McGraw-Hill, New York, 1953.

[2] A.G. Guy, *Introduction to Materials Science*, McGraw-Hill, New York, 1972, p. 212.

[3] M.L. Kronberg and H.F. Wilson, *Trans. AIME*, 85 (1949) 501.

[4] J.W. Christian, *The Theory of Transformations in Metals and Alloys*, Pergamon Press, Elmsford, N.Y., 1965, p. 326.

[5] G. Bäro and H. Gleiter, *Mater. Sci. Eng.*, *3* (1968) 92.

[6] H. Gleiter and G. Bäro, *Mater. Sci. Eng.*, *2* (1967) 224.

[7] H. Gleiter, E. Hornbogen, and G. Bäro, *Acta Met.*, *16* (1968) 1053.

[8] L.E. Murr, *Interfacial Phenomena in Metals and Alloys*, Addison-Wesley, Reading, Mass., 1975, p. 255.

[9] M.F. Ashby, F. Spaepen, and S. Williams, *Acta Met.*, *26* (1978) 1647.

[10] J.P. Hirth and J. Lothe, *Theory of Dislocations,* McGraw-Hill, New York, 1968, p. 364.

[11] M.A. Meyers, C. Sarzeto, and C.Y. Hsu, *Met. Trans.*, *11A* (1980) 1737.

[12] W. Boas and E. Schmid, *Z. Phys.*, *54* (1929) 16.

[13] R.E. Reed-Hill, J.P. Hirth, and H.C. Rogers (eds.), *Deformation Twinning* TMS-AIME Conf. Proc., Gordon and Breach, New York, 1965.

[14] A.H. Cottrell and B.A. Bilby, *Phil. Mag.*, *42* (1951) 573.

[15] J.A. Venables, *Phil. Mag.*, *6* (1961) 379.

[16] R.F. Bunshah, cited in [13], p. 390.

[17] C.N. Reid, G.I. Hahn, and A. Gilbert, cited in [13], p. 386.

[18] T. Takeuchi, *J. Phys. Soc. Jap.*, *21* (1966) 2616.

[19] E. Hornbogen, *Trans. AIME*, *221* (1961) 712.

[20] J.B. Cohen and J. Weertman, *Acta Met.*, *11* (1963) 997.

[21] J.P. Hirth and J. Lothe, *Theory of Dislocations*, McGraw-Hill, New York, 1968, p. 364.

[22] E. Orowan, *Dislocations in Metals*, AIME, New York, 1954, p. 116.

[23] A.W. Sleeswyk, *Acta Met.*, *10* (1962) 803.

[24] G. Thomas, University of California at Berkeley, *Personal Communication*, 1978.

[25] M.A. Meyers and L.E. Murr, *Acta Met.*, *26* (1978) 951.

[26] W.J. Babiak and F.N. Rhines, *Trans. AIME*, *218* (1960) 21.

[27] H. Carpenter and S. Tamura, *Proc. Roy. Soc.* (*London*), *A213* (1926) 161.

[28] R.F. Fullman and J.C. Fisher, *J. Appl. Phys.*, *22* (1951) 1350.

[29] W.G. Burgers, *Nature*, *157* (1946) 76.

[30] J.P. Nielsen, *Acta Met.*, *15* (1967) 1083.

[31] S. Dash and N. Brown, *Acta Met.*, *11* (1963) 1067.

SUGGESTED READING

CHAUDHARI, P., and J.W. MATTHEWS (eds.), *Grain Boundaries and Interfaces*, North-Holland, Amsterdam, 1972.

GLEITER, H., "On the Structure of Grain Boundaries in Metals," *Mater. Sci. Eng.*, *52* (1982) 91.

GLEITER, H., and B. CHALMERS, *High-Angle Grain Boundaries*, Progress in Materials Science, Vol. 16, B. Chalmers, J.W. Christian, and T.B. Massalski (eds.), Pergamon Press, Elmsford, N.Y., 1972.

HU, H. (ed.), *The Nature and Behavior of Grain Boundaries*, Plenum Press, New York, 1972.

MAHAJAN, S., and D.F. WILLIAMS, *Int. Met. Rev.*, *18* (1973) 43.

MURR, L.E., *Interfacial Phenomena in Metals and Alloys*, Addison-Wesley, Reading, Mass., 1975.

REED-HILL, R.E., J.P. HIRTH, and H.C. ROGERS (eds.), *Deformation Twinning*, TMS-AIME Conf. Proc., Gordon and Breach, New York, 1965.

WALTER, J.L., J.H. WESTBROOK, and D.A. WOODFORD (eds.), *Grain Boundaries in Engineering Materials*, Proc. 4th Bolton Landing Conf., Claitor's, Baton Rouge, La., 1975.

Chapter 8

ADDITIONAL DISLOCATION EFFECTS
AND GEOMETRY OF DEFORMATION

8.1 DISLOCATION PILE-UP

All dislocations generated by a Frank–Read source are in the same slip plane if they do not cross-slip. In low stacking-fault energy metals the large separation between the partials renders cross-slip more difficult. In case one of the dislocations encounters an obstacle (a grain boundary, a precipitate, etc.), its motion will be hampered. The subsequent dislocations will "pile up" behind the leading dislocation, after being produced by the Frank–Read source. Figure 8.1 shows schematically a pile-up. The distance between the dislocations increases as their distance from the obstacle increases. On the other hand, if the metal has a very high stacking-fault energy, cross-slip will easily occur and the planar array will be destroyed; edge dislocations cannot, obviously, cross-slip because of their Burgers vector.

Figure 8.2 shows an example, obtained by etch pitting in copper. It should be observed that the dislocation configurations for a pile-up and a grain-boundary source are similar and that many grain-boundary sources have in the past been mistaken for pile-ups. Figure 6.30 shows a grain-boundary source.

Each dislocation in a pile-up is in equilibrium under the effect of the applied stress and of the stresses due to the other dislocations (in the pile-up). Assuming that the dislocations are of edge character and parallel, the resulting force acting on the nth dislocation is obtained by applying the equation that gives the forces between dislocations:

$$\tau b - \sum_{\substack{j=0 \\ i \neq j}}^{n} \frac{Gb^2}{2\pi(1-\nu)(x_i - x_j)} = 0 \tag{8.1}$$

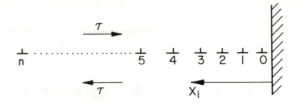

Figure 8.1 Pile-up of dislocations against a barrier.

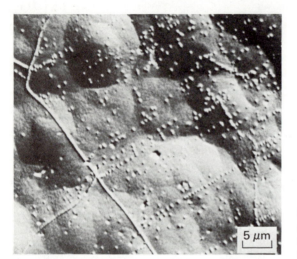

Figure 8.2 Pile-up of dislocations against grain boundaries (or dislocations being emitted from grain-boundary sources?) in copper observed by etch pitting.

Solving the n equations with n unknowns $(x_i - x_j)$ for the dislocations behind the lead dislocation, we obtain the positions. This derivation was introduced by Eshelby et al.[1] and is not reproduced here, because of its complexity.

The stress acting on the lead dislocation due to the presence of the other dislocations and due to the applied stress is found to be

$$\tau^* = n\tau \qquad (8.2)$$

So, the effect of the n dislocations in the pile-up is to create a stress at the lead dislocation n times greater than the applied stress. For this reason the dislocation pile-up is sometimes treated as a superdislocation with a Burgers vector nb. The calculations above can also be applied to screw dislocations, removing the term $(1 - \nu)$.

8.2 DISLOCATION INTERSECTION

A dislocation, when moving in its slip plane, encounters other dislocations, moving along other slip planes. If we imagine the first dislocation moving in a horizontal plane, it will "see" the other dislocations as trees in a forest. The latter name designates dislocations in other slip planes. When the dislocation intersects another dislocation and, since it shears the material equally (by a quantity b) on the two sides of the slip plane, it will form one or more steps. These steps are of two types: "jogs" if the tree dislocation was transferred to another slip plane, "kink" if the tree dislocation remains in the same slip system. There are various possible outcomes from dislocation intersections; they are shown in Fig. 8.3. Figure 8.3(a) shows an edge dislocation traversing a forest composed of two edge and two screw dislocations. A good rule to determine the direction of "jogs" and "kinks" is the following: The direction of the segment is the same as the Burgers vector of the dislocation that is traversing the forest; on the other hand, the Burgers vector of the jog or kink is the Burgers vector of the dislocation in which it is located, because the Burgers vector is always the same along the length of a dislocation. Figure 8.3(b) shows a screw dislocation after traversing a forest. The reader is asked to verify the directions of dislocation segments and Burgers vectors; he should also verify whether they are jogs or kinks.

The ability of these segments to slip with a dislocation is of great importance in determining the work hardening of metal. It should be noted that some authors use the name "jog" for both types of segments. Jogs and kinks can have either screw or edge character. From Fig. 8.4 it can be seen that segments on an edge dislocation cannot impede their motion; they can slip with the dislocation. On the other hand, in screw dislocations there are segments that can slip with the dislocations and segments that cannot. When the segment can move with the dislocation, the motion is called "conservative." When it cannot move by slip, the motion is called "nonconservative." Figure 8.5 shows some interactions. At the left there is a conservative motion by slip, and at the right a nonconservative motion. The nonconservative motion of a jog is, in essence, a climb process and requires thermal activation. Vacancies or interstitials are produced as the segment moves. If the temperature is not high enough to provide sufficient thermal activation, the jog does not move, and loops are formed as the dislocation advances; this is shown in Fig. 8.6. The dislocation forms a dipole, upon advancing, because the jog stays back. At a certain point, the dipole will be pinched out, producing a loop.

8.3 DEFORMATION PRODUCED BY DISLOCATION MOTION (OROWAN'S EQUATION)

A dislocation produces, upon moving, a certain deformation in the material. This deformation is inhomogeneous. Figure 6.7 shows the step generated by the passage of a dislocation. If we consider a large number of dislocations acting on different systems, we can consider the association of a large number of small steps as creating a homogeneous state of deformation. The deformation is related to both the number of dislocations that move and the distance traveled by them. This equation is known as Orowan's or Taylor–Orowan's equation and is derived below. Figure 8.7 shows a cube with dimensions dx_1, dx_2, and dx_3

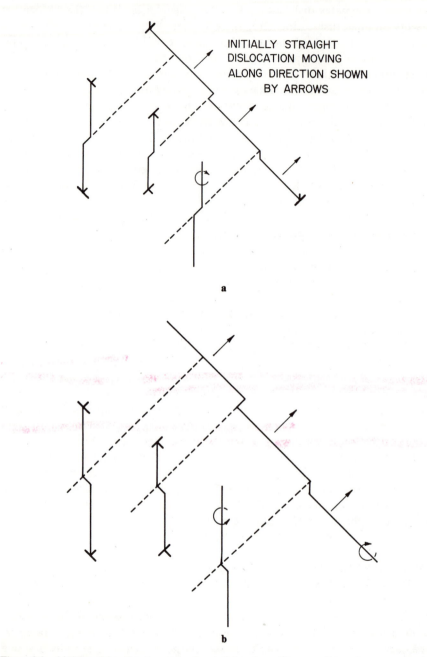

INITIALLY STRAIGHT
DISLOCATION MOVING
ALONG DIRECTION SHOWN
BY ARROWS

a

b

Figure 8.3 (a) Edge dislocation traversing forest dislocations; (b) screw dislocation traversing forest dislocations.

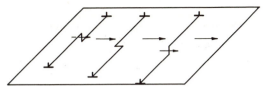

Figure 8.4 Kink and jog in edge dislocation.

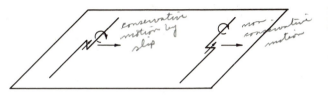

Figure 8.5 Kink and jog in screw dislocation.

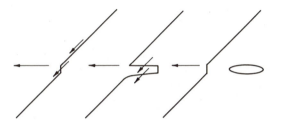

Figure 8.6 Loop being pinched out when jog is left behind by dislocation motion.

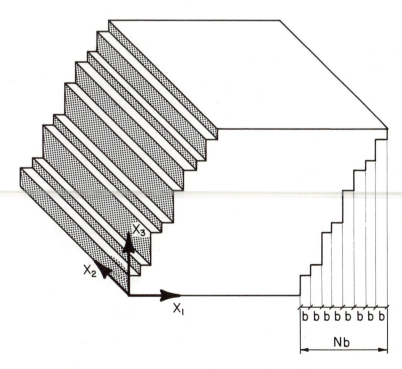

Figure 8.7 Shear produced by the passage of parallel dislocations.

that was sheared by the passage of N dislocations moving along plane Ox_1x_2. The plastic shear strain can be expressed as

$$\gamma_{13} = \frac{Nb}{dx_3} \tag{8.3}$$

This is so because all dislocations are of edge character and have the same sign, with identical Burgers vector b. The density of dislocations, ρ, is the total length $N\, dx_2$ in the volume $dx_1\, dx_2\, dx_3$. Therefore,

$$\rho = \frac{N\, dx_2}{dx_1\, dx_2\, dx_3} \qquad \text{and} \qquad N = \rho\, dx_1\, dx_3 \tag{8.4}$$

Substituting Eq. 8.4 into Eq. 8.3 yields

$$\gamma_{13} = \rho b\, dx_1$$

A cube isolated in space was considered in the situation above; dislocations were generated in one face and "popped out" at the opposite face. In real situations dislocations remain within the material and the deformation generated by each dislocation is related to the distance traveled by it. Assuming that dislocations travel an average distance $\bar{l}$, we have

$$\gamma_{13} = \rho b \bar{l}$$

But in a general case of deformation, five independent slip systems are activated. The deformation is not perfectly aligned with the movement of dislocations and it is necessary to introduce a correction parameter ϕ that takes this into account:

$$\gamma_p = \phi \rho b \bar{l} \tag{8.5}$$

If one assumes that the density of mobile dislocations is not affected by the rate of deformation (strain rate), one would have, taking the time derivative of both sides of Eq. 8.5,

$$\frac{d\gamma_p}{dt} = \phi \rho b \frac{d\bar{l}}{dt} \qquad \text{or} \qquad \dot{\gamma}_p = \phi \rho b \bar{v} \tag{8.6}$$

where $\bar{v}$ is the mean velocity of the dislocations. We can also use the longitudinal strain ϵ_{11} if applying the situation to a tensile test. It can be shown that $\gamma = 2\epsilon$ (Section 8.6.3) for an ideal orientation for slip.

As an illustration, if iron ($b \approx 2.5$ Å) is being deformed at 10^{-3} s^{-1}, and the density of mobile dislocations is around 10^{10} cm^{-2}, their approximate velocity would be 4×10^{-6} cm/s.

Attention should be called to the fact that the density of mobile dislocations is lower than the total density of dislocations in the material. As the dislocation density increases in a deformed material, a greater and greater number of them are locked at various types of barriers, such as grain boundaries, cell walls, or by the action of a great number of jogs. The actual density of mobile dislocations is only a fraction of the total dislocation density. Gilman[2] treats the subject in greater detail.

8.4 DISLOCATION DYNAMICS

8.4.1 Relationship between Dislocation Velocity and Applied Stress

"Most of our knowledge on the dynamic behavior of materials is largely empirical and somewhat questionable. This arises because it is analytically impossible to extract the dynamic behavior of materials from evidence based on conventional plastic wave-propagation experiments." The opening sentence of a review article by Dorn et al.[3] on dislocation dynamics is indicative of the uncertainties about the dynamics of dislocations and, especially, of high-velocity dislocations.

Gilman and Johnston,[4,5] in their now classic experiments, measured the velocities of dislocations in LiF as a function of stress. They used an ingenious etch-pit technique, and their results are depicted schematically in Fig. 8.8. This is, to the authors' knowledge, the first study on dislocation dynamics. This pioneering work was followed by a succession of other papers. Some of these are: Stein and Low[6] for Fe-Si, Ney et al.[7] for Cu and Cu-Al, Schadler[8] for tungsten, Chaudhuri et al.[9] for germanium and silicon, Erickson[10] for Fe-Si, Gutmanas et al.[11] for NaCl, Rohde and Pitt[12] for nickel, Pope et al.[13] for

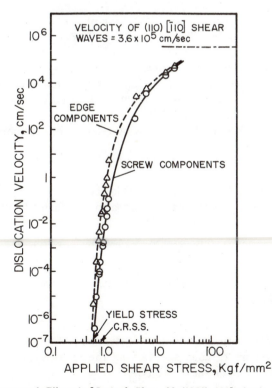

Figure 8.8 Johnston and Gilman's [*J. Appl. Phys., 33* (1959) 129] classic experiments showing dislocation velocity versus resolved shear stress for as-grown LiF single crystal. Units left unchanged. (1 Pa = 9.8 × 10⁶ kgf/mm²). (Adapted with permission from [5], p. 132.)

zinc, Greenman et al.[14] for copper, Suzuki and Ishii[15] for copper, Blish and Vreeland[16] for zinc, Gorman et al.[17] for aluminum, Parameswaran and Weertman[18] for lead, and Parameswaran et al.[19] for aluminum.

There is no experimental report, to the authors' knowledge, of supersonic dislocation. The stress dependence of dislocation velocity has been fitted into different types of equations, some empirical, some with a theoretical backing. Johnston and Gilman[5] expressed the dislocation velocity as

$$v \alpha \sigma^m e^{-E/kT}$$

where m varied between 15 and 25 when v was expressed in cm/s and σ in kgf/mm². At a constant temperature, we should have

$$v = K\sigma^m \tag{8.7}$$

Stein and Low,[6] on the other hand, found their data to fit more closely the expression

$$v = A \exp\left(-\frac{A}{\tau}\right) \tag{8.8}$$

where τ is the resolved shear stress. Rohde and Pitt[12] found a good correlation with the expression

$$v = \frac{kT}{h} K \exp\left(-\frac{\Delta H}{kT}\right) \exp\left[\frac{B(\tau_a - \tau_l)}{kT}\right] \tag{8.9}$$

where h is Plank's constant, K a numerical constant, B an activation volume, ΔH the enthalpy of activation, τ_a the applied resolved shear stress, and τ_l the long-range internal stress. This equation is based on the theory of reaction rates as presented by Glasstone et al.[20]

Greenman et al.[14] found the following relationship for copper:

$$v = v_0 \left(\frac{\tau}{\tau_0}\right)^m \tag{8.10}$$

where m was equal to 0.7, v_0 was unit velocity, and τ_0 a material constant ($= 0.25 \times 10^3$ N/m²). But the data could also satisfactorily fit the equation with $m = 1$ and $\tau_0 = 2.7 \times 10^3$ N/m²:

$$v = \frac{v_0}{2.7 \times 10^3} \tau \qquad \text{or} \qquad v = K\tau \tag{8.11}$$

For the basal plane of zinc (and edge dislocations), Pope et al.[13] obtained a good fit by assuming that

$$v = 3.4 \times 10^{-5}\tau \tag{8.12}$$

where v was expressed in cm/s and τ in dyn/cm². For the $\{\bar{1} \ 2 \ \bar{1} \ 2\} > < \bar{1}\bar{2} \ 13 >$ slip systems of zinc, and varying the temperature, they were able to fit the data into an equation of the type

$$v = K\tau \exp\left(-\frac{E}{kT}\right) \tag{8.13}$$

They found different K values for edge and screw dislocations. Notice that, at constant temperature, Eqs. 8.11 and 8.12 are special cases of Eq. 8.13. For aluminum, Gorman et al.[17] were also able to obtain a linear relationship between the dislocation velocity and stress. Similarly, Weertman and co-workers[18,19] were able to fit the data into Eq. 8.11, for aluminum and lead.

As a conclusion it can be said that, although the correct form over a wide range of stresses is not a power function or a proportionality, these expressions can be used as an approximation. Gilman,[21] in his book *Micromechanics of Flow in Solids*, presents the following equation, which should, according to him, describe the behavior of almost any material:

$$v = v_s^*(1 - e^{-\tau/S}) + v_d^* e^{-D/\tau} \tag{8.14}$$

where D is the characteristic drag stress and v_s and v_d are limiting velocities. For $\tau \to \infty$, one would have

$$v = v_s^* + v_d^*$$

So there is an upper limit for the velocity. On the other hand, the other models (Eqs. 8.7 to 8.13) do not necessarily predict an upper limit for v.

The approximation of $v\alpha\tau$ is valid at moderate stresses and is taken advantage of in the determination of the dislocation-damping constant B. If we apply Newton's second law, we have

$$F = ma \tag{8.15}$$

The corresponding applied stress is given by the Peach–Koehler equation (Eq. 6.30):

$$\mathbf{F} = \mathbf{t} \wedge \mathbf{G}$$

If the stress is ideally oriented for dislocation motion, we have the simplified form

$$F = \tau b \tag{8.16}$$

Substituting Eqs. 8.15 and 8.16, we would have

$$\tau b = ma \tag{8.17}$$

If we assume a certain dislocation mass, the dislocation would undergo acceleration under a constant applied stress. However, the equations previously described predict a constant velocity at a certain stress. Therefore, we would have to have something "holding back" the dislocations once a certain velocity is reached. There are two possibilities:

1. The mass m (or equivalent mass of a dislocation that will be defined later) increases with increasing velocity, causing the acceleration to decrease, as the velocity increases (for the same applied stress).
2. In analogy with the movement of fluids, there is a drag stress, or damping of the dislocation motion.

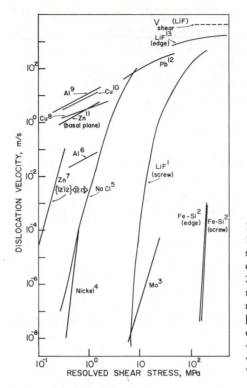

Figure 8.9 Compilation of results in the literature relating dislocation velocities and stress. Numbers refer to references indicated below. 1 (Johnston and Gilman [5]); 2 (Stein and Low [6]); 3 (Preckel and Conrad [22]); 4 (Rohde and Pitt [12]); 5 (Gutmanas et al. [11]); 6 (Parameswaran et al. [19]); 7 (Pope et al. [13]); 8 (Greenman et al. [14]); 9 (Gorman et al. [17]); 10 (Ney et al. [7]); 11 (Pope et al. [13]); 12 (Parameswaran and Weertman [18]); 13 (Cotner and Weertman [23]).

These two effects are analyzed in Sections 8.4.2 and 8.4.3.

The various mathematical equations (Eqs. 8.7 to 8.14) used to describe the dynamic behavior of dislocations can be confusing; for this reason, the log-log plot of Fig. 8.9 was made. It is a compilation of data from several sources, and some trends are readily visible. The general conclusion at which one can arrive is: *The slope decreases as the velocity increases.* One can also see that the experimental results of Rohde and Pitt,[12] in spite of the fact that they were adjusted to Eq. 8.9, could very well be represented by Eq. 8.7 (the log-log relationship of Gilman and Johnston).[4] If one wants to predict the response of nickel (or of any metal, for that matter) over the full velocity range, one should observe that in the medium-velocity range the velocity and stress have been found by several investigators to be linearly dependent. In the log-log plot of Fig. 8.9, this corresponds to a unit slope found by Vreeland and co-workers,[13,14,16,17] Weertman and co-workers,[18,19,23] and Haasen and co-workers.[7] This can be clearly seen from Eq. 8.7, making $m = 1$. Accordingly, Fig. 8.10 establishes a hypothetical stress dependence of dislocations velocities in nickel, in accord with the aforementioned observations. Three regions of response are defined: I, II, and III. One has, for the coefficients of Eq. 8.7, $m_I > m_{II} > m_{III}$ and $m_{II} = 1$. The limiting velocity of the dislocation is set as the velocity of propagation of elastic shear waves. This was established as the limiting velocity because both edge and screw dislocations generate shear stresses (and strains). These disturbances, on their turn, cannot propagate at

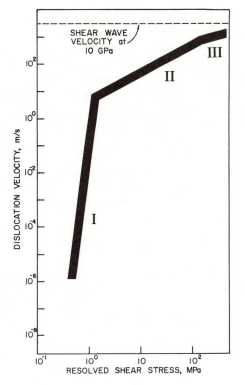

Figure 8.10 Schematical representation of the stress versus velocity behavior of nickel. Three regions of response were established. Region I ($m_I > 1$), region II ($m_{II} = 1$), region III ($m_{III} < 1$). Region I was obtained from data replotted from Rohde and Pitt [12].

velocities higher than the velocity of elastic shear waves. Therefore, the dislocations, which are assemblies of stresses and strains, cannot travel faster than this.

8.4.2 Velocity Dependence of Dislocation Self-Energy

It is appropriate to open this section with a quote from Friedel:[24] "A dislocation moving under a constant applied external stress cannot exceed the speed of sound C, for it is a signal, which can be Fourier analyzed into plane waves of strain (sound waves or phonons). And one knows that a signal cannot travel faster than the waves that carry it." It is well known that an isolated dislocation is surrounded by a stress field and associated strain field. The displacements generated by a dislocation have to precede it if this dislocation is in motion. But disturbances have well-determined velocities in metals; this can be seen from Section 1.13. If the displacement is parallel to the direction of motion of the dislocation, a situation where u, the particle displacement, is parallel to the wave direction is attained, and one has a longitudinal wave. On the other hand, if the disturbance is caused by a displacement parallel to the dislocation line and perpendicular to the direction of dislocation motion, a shear wave is produced. Friedel's rationale is the following: How can a dislocation, being composed basically of a set of displacements, move faster than the displacements?

In this section it will be shown analytically that this is not possible. The derivations will be conducted for a screw dislocation and was originated by Eshelby.[25] Similar treatments can be found in other sources.[26-29] The treatment given here is based on Weertman.[26] Weertman and Weertman[29] have recently presented a more comprehensive treatment. The equations for the stresses and strains around a screw dislocation (see Section 6.4) are used in the development. The equation of equilibrium (see Section 1.10) expressed as $\Sigma F = 0$ is now replaced by Newton's second law, since one does have a dynamic situation. From Eq. 1.40, we have

$$\frac{\partial \sigma_{13}}{\partial x_1} + \frac{\partial \sigma_{23}}{\partial x_2} + \frac{\partial \sigma_{33}}{\partial x_3} = \rho \frac{\partial^2 u_3}{\partial t^2} \tag{8.18}$$

From Eqs. 1.11 and 1.75, one obtains the stresses in terms of the displacements:

$$\sigma_{13} = G \frac{\partial u_3}{\partial x_1} \tag{8.19}$$

$$\sigma_{23} = G \frac{\partial u_3}{\partial x_2} \tag{8.20}$$

Substituting the displacements for the stresses (Eqs. 8.19 and 8.20 into 8.18) yields

$$\frac{G}{\rho} \left(\frac{\partial^2 u_3}{\partial x_1^2} + \frac{\partial^2 u_3}{\partial x_2^2} \right) = \frac{\partial^2 u_3}{\partial t^2} \tag{8.21}$$

The equation for a wave disturbance ω traveling at a velocity v along x is given in most elementary textbooks and is expressed as

$$v^2 \frac{\partial^2 \omega}{\partial x^2} = \frac{\partial^2 \omega}{\partial t^2} \tag{8.22}$$

From the consideration of Eq. 8.22 it can be seen, by analogy, that Eq. 8.21 represents a disturbance u_3 traveling at a velocity

$$v = \left(\frac{G}{\rho} \right)^{1/2}$$

It has been shown in Section 1.13 (Eq. 1.148) that this is the velocity of an elastic shear wave. If the dislocation is at rest, Eq. 8.21 reduces itself to

$$\frac{\partial^2 u_3}{\partial x_1^2} + \frac{\partial^2 u_3}{\partial x_2^2} = 0 \tag{8.23}$$

This is the Laplace equation, and the solution for a screw dislocation is known to be (see Section 6.4.1)

$$u_3 = \frac{b}{2\pi} \tan^{-1} \frac{x_2}{x_1} \tag{8.24}$$

Actually, the procedure used in Section 6.4.1 was to assume the displacement function given by Eq. 6.1 and to apply the equilibrium condition (Eq. 6.6) to it, showing that it satisfied it. We will now determine the solution for Eq. 8.21 when the dislocation is moving. The solution of this equation was found to be very similar to the treatment employed in the special relativity theory. By an appropriate transformation—similar to the one made in special relativity theory—Eq. 8.21 is transformed into the Laplace equation. This transformation is made

$$x_1^* = \frac{x_1 - vt}{(1 - v^2/v_s^2)^{1/2}}; \qquad x_2^* = x_2; \quad x_3^* = x_3 \tag{8.25}$$

This transformation is such that the system of reference moves with the dislocation; v_s is the velocity of the shear wave. This transformation is called a Lorentz transformation in special relativity theory. Calling $(1 - v^2/v_s^2)^{1/2} = \beta$, we have

$$x_1^* = \frac{x_1 - vt}{\beta} \qquad \text{and} \qquad \partial x_1^* = \frac{\partial x_1}{\beta}$$

so

$$\frac{\partial}{\partial x_1^*} = \beta \frac{\partial}{\partial x_1} \qquad \text{and} \qquad \frac{\partial}{\partial x_1^*}\left(\frac{\partial}{\partial x_1^*}\right) = \beta \frac{\partial}{\partial x_1^*}\left(\frac{\partial}{\partial x_1}\right)$$

so

$$\frac{\partial^2}{\partial x_1^{*2}} = \beta^2 \frac{\partial^2}{\partial x_1^2} \tag{8.26}$$

For the derivation with respect to $\partial/\partial t^2$, we have

$$v = \frac{\partial x_1}{\partial t} \qquad \text{and} \qquad v\frac{\partial}{\partial x_1} = \frac{\partial}{\partial t} \tag{8.27}$$

at constant v; applying the operator $\partial/\partial t$, we have

$$\frac{\partial}{\partial t}\left(v\frac{\partial}{\partial x_1}\right) = \frac{\partial^2}{\partial t^2}; \qquad v\frac{\partial}{\partial t}\left(\frac{\partial}{\partial x_1}\right) = \frac{\partial^2}{\partial t^2}$$

from

$$v^2 \frac{\partial^2}{\partial x_1^2} = \frac{\partial^2}{\partial t^2} \tag{8.28}$$

Substituting 8.21 for 8.26 and 8.28, we have

$$\frac{G}{\rho}\left(\frac{1}{\beta^2}\frac{\partial^2 u_3}{\partial x_1^{*2}} + \frac{\partial^2 u_3}{\partial x_2^2}\right) = v^2 \frac{\partial^2 u_3}{\partial x_1^2} = \frac{v^2}{\beta^2}\frac{\partial^2 u_3}{\partial x_1^{*2}}$$

Recalling that

$$\beta = \left(1 - \frac{v^2}{v_s^2}\right)^{1/2}, \qquad \text{therefore,} \qquad -(\beta^2 - 1) = \frac{v^2}{v_s^2}$$

so

$$\frac{v^2}{-(\beta^2-1)}\left(\frac{1}{\beta^2}\frac{\partial^2 u_3}{\partial x_1^{*2}}+\frac{\partial^2 u_3}{\partial x_2^{*2}}\right)=\frac{v^2}{\beta^2}\frac{\partial^2 u_3}{\partial x_1^{*2}}$$

Simplifications will lead to

$$\frac{\partial^2 u_3}{\partial x_1^{*2}}+\frac{\partial^2 u_3}{\partial x_2^{*2}}=0 \tag{8.29}$$

This is the Laplace equation, whose solution is given by Eq. 8.24 with the appropriate changes:

$$u_3=\frac{b}{2\pi}\tan^{-1}\frac{x_2^*}{x_1^*}=\frac{b}{2\pi}\tan^{-1}\frac{x_2}{(x_1-vt)/\beta} \tag{8.30}$$

Differentiation of Eq. 8.30 will yield

$$\epsilon_{13}=-\frac{b}{4\pi}\frac{x_2}{x_2^2+\left(\dfrac{x_1-vt}{\beta}\right)^2}$$

$$\epsilon_{23}=\frac{b}{4\pi}\frac{(x_1-vt)/\beta}{x_2^2+\left(\dfrac{x_1-vt}{\beta}\right)^2} \tag{8.31}$$

The corresponding stresses are

$$\sigma_{13}=-\frac{Gb}{2\pi}\frac{x_2}{x_2^2+\left(\dfrac{x_1-vt}{\beta}\right)^2}$$

$$\sigma_{23}=\frac{Gb}{2\pi}\frac{(x_1-vt)/\beta}{x_2^2+\left(\dfrac{x_1-vt}{\beta}\right)^2} \tag{8.32}$$

Figures 8.11 and 8.12 show the stress fields around a screw dislocation as a function of its velocity. These plots were computer-generated for three velocities: 0, 0.5v_s, and 0.995v_s, where v_s is the velocity of elastic shear waves. The dislocation axis is assumed to be along Ox_3. The situation is similar to the one shown in Section 6.4.2. (see Fig. 6.14). Figure 8.11 shows the shear stress σ_{13} as a function of dislocation velocity, while the same is shown for σ_{23} in Fig. 8.12; the dislocation is moving along the horizontal axis Ox_1. The isostress lines clearly show that the stress fields (initially circular) are compressed along the Ox_1 axis. At 0.5v_s the compression is barely visible. However, as v_s is approached, the compression becomes very pronounced and the stresses will eventually be reduced to the Ox_2x_3 plane at $v=v_s$. Concurrently with the deformation of the stress fields, there is an increase in the dislocation self-energy, as will be demonstrated below. So there seems to be a shear velocity wall, just as there is a sound wall for airplanes at Mach 1. For edge dislocations the situation is

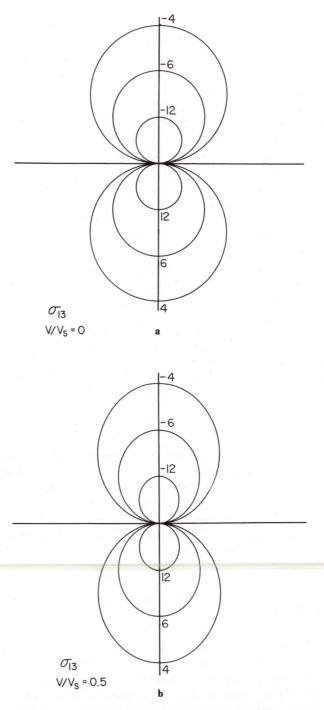

Figure 8.11 Variation of σ_{13} of a screw dislocation with velocity; (a) $v = 0$; (b) $v = 0.5v_s$; (c) $v = 0.995v_s$ (v_s is the velocity of elastic of shear waves). (Courtesy of A. R. Pelton and L. K. Rabenberg, University of California at Berkeley.) (Continued on next page)

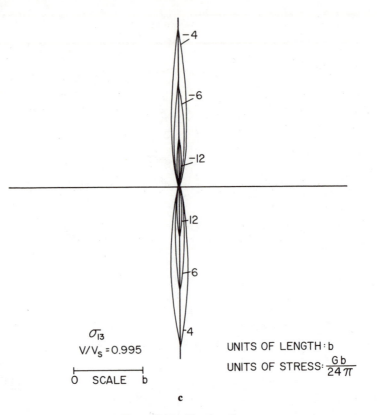

$\sigma_{\overline{13}}$

$V/V_s = 0.995$

├─────────┤
0   SCALE   b

UNITS OF LENGTH: b

UNITS OF STRESS: $\dfrac{Gb}{24\pi}$

c

Figure 8.11 (Continued)

similar. The shear stresses and strains cannot propagate faster than v_s. However, the edge dislocations also generate normal stresses σ_{33} (and the associated longitudinal strains ϵ_{33}; see Eq. 6.18). These longitudinal strains can travel at velocities less or equal to the velocity of longitudinal waves, v_l. To a first approximation, v_l is twice as large as v_s. So, edge dislocations are subjected to two sonic walls: at v_s and v_l. It can also be proven that their self-energy becomes infinite at $v = v_s$ and $v = v_l$. Weertman defines three ranges of velocity for dislocations:

$$v < v_s \quad \longrightarrow \quad \text{subsonic dislocation}$$

$$v_s < v < v_l \quad \longrightarrow \quad \text{transonic dislocation}$$

$$v > v_l \quad \longrightarrow \quad \text{supersonic dislocation}$$

It should be noticed that a supersonic screw dislocation was proposed by Eshelby;[30] it is different from a normal dislocation and is known as an "Eshelby" dislocation. According to Eshelby,[30] it should propagate along planes that give up finite amounts of energy. Weertman[31] extended Eshelby's mathematical treatment to gliding and climbing edge dislocations (both in the trans- and supersonic range) and mixed dislocations. According to the mathematical treatment presented, these dislocations are possible. Weiner and Pear[32] and

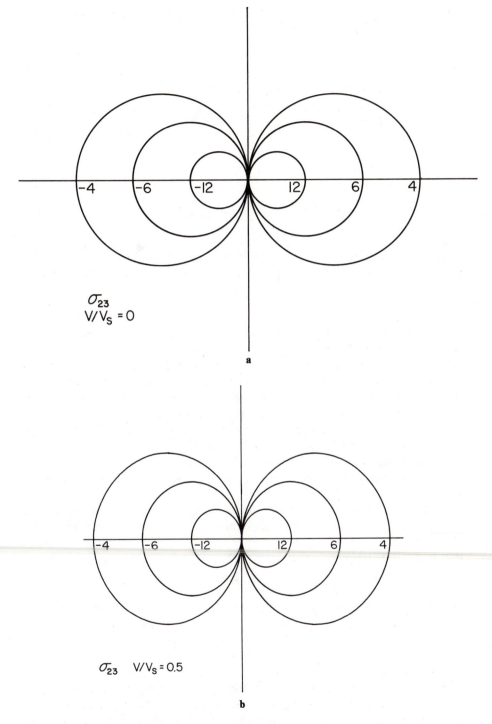

Figure 8.12 Variation of σ_{23} of a screw dislocation with velocity (a) $v = 0$; (b) $v = 0.5v_s$; (c) $v = 0.995v_s$ (v_s is the velocity of elastic of shear waves). (Courtesy of A. R. Pelton and L. K. Rabenberg, University of California at Berkeley.) (Continued on next page)

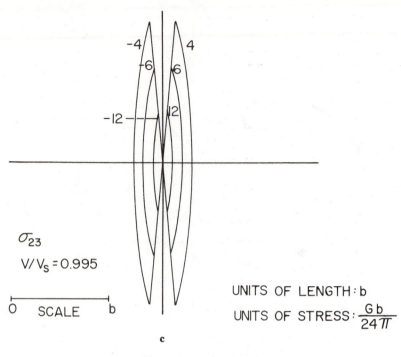

σ_{23}

$V/V_s = 0.995$

|—————————————|
O   SCALE   b

UNITS OF LENGTH: b

UNITS OF STRESS: $\dfrac{Gb}{24\pi}$

c

Figure 8.12 (Continued)

Earmme and Weiner,[33] instead of treating the metal as a continuum, have used the concept of atoms vibrating. The simplest visualization is to consider the atoms connected by springs. Vibrations are transferred throughout the lattice as a propagating wave. Using atomistic models and lattice vibrations, they found, mathematically, that a dislocation breakdown was observed when $v = v_s$. These studies are, however, inconclusive because, at the present moment, experimental support is lacking.

The total energy of a moving dislocation is calculated in a manner similar to the total energy of a stationary dislocation. Now, however, both kinetic and potential terms have to be added. The energy per unit volume of a screw dislocation is (see Eq. 1.86)

$$U = \tfrac{1}{2}\sigma_{ij}\epsilon_{ij}$$

Substituting Eqs. 8.31 and 8.32 and integrating with respect to r and θ, we obtain

$$U_p = \frac{Gb^2}{4\pi}\ln\left(\frac{R}{r_0}\right)\frac{1+\beta^2}{2\beta} \tag{8.33}$$

The kinetic energy, given by $\tfrac{1}{2}mv^2$, is expressed as (per unit volume)

$$U_k = \frac{1}{2}\rho\left(\frac{\partial u_3^*}{\partial t}\right)^2$$

Integration yields

$$U_k = \iint \frac{1}{2} \rho \left(\frac{\partial u_3^*}{\partial t} \right)^2 2\pi r \, dr \, d\theta$$

$$U_k = \left(\frac{v}{v_s} \right)^2 \frac{1}{2\beta} \frac{Gb^2}{4\pi} \ell n \frac{R}{r_0}$$

(8.34)

The total energy is

$$U_T = U_p + U_k = \frac{U_0}{\beta}$$

where U_0 is the energy of the stationary dislocation. It can be seen that, as v approaches v_s, U_T approaches infinity.
So

$$\lim_{v \to v_s} U_T = \infty$$

The analogy to Einstein's relativity theory can be extended to the mass. Einstein's famous equation $E = mc^2$ stated that the energy depended on the mass. The mass at a velocity v is related to the mass at rest by

$$m = \frac{m_0}{(1 - v^2/c^2)^{1/2}}$$

(8.35)

where c is the velocity of light. In our case, we can define an "equivalent mass" for a dislocation.

Defining, for a stationary screw dislocation, $U_0 = m_0 v_s^2$, we have

$$\frac{Gb^2}{4\pi} \ell n \frac{R}{r_0} = m_0 \left(\frac{G}{\rho} \right)$$

and

$$m_0 = \frac{\rho b^2}{4\pi} \ell n \frac{R}{r_0}$$

Usually, $\ell n \, (R/r_0)$ can be assumed to be $\sim 4\pi$, so that

$$m_0 \simeq \rho b^2$$

(8.36)

Since $U_T \simeq U_0/\beta$, we would have

$$U_T = \frac{1}{\beta} m_0 v_s^2$$

If $U_T = mv_s^2$,

or

$$m = \frac{m_0}{(1 - v^2/v_s^2)^{1/2}}$$

(8.37)

This equation is in all respects analogous to Eq. 8.35 from Einstein's relativity theory. Recalling Eq. 8.17, it can be clearly seen that the acceleration caused by a certain stress would decrease at higher velocities.

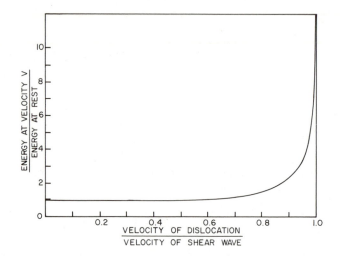

Figure 8.13 Effect of velocity on self-energy of a dislocation. This is a relativistic effect.

Figure 8.13 presents, for nickel, a plot of the self-energy of a screw dislocation versus its velocity. In the calculations, the energy for the dislocation core
was included, in addition to the energy for the surrounding strain field. The
energy for the core was taken as $Gb^2/10$ and $ln\ (R/r_0)$ was assumed to be
equal to 4π. It can be seen that, beyond $0.5v_s$, the energy changes very rapidly.
For an edge dislocation, the energy becomes infinite for $v = v_s$, and again for
$v = v_l$. It is not known whether the edge dislocation traverses the first barrier
and becomes transonic. This is treated in detail by Weertman.[31]

8.4.3 Damping of Dislocation Motion

It is well known that dislocation motion causes a temperature increase.
The cold rolling of a piece of nickel increases its temperature substantially. It
is also known that the energy stored in the material after deformation (as defects)
is only a small portion of the energy spent to deform it. The energy of the
substructure (residual) is usually only 5 to 10% of the total energy. So 95%
of the energy is dissipated as the dislocations move in the material. Hence,
there are dissipative forces opposing the applied stresses. They can be expressed
as a viscous behavior of the solid. To a first approximation, the solid can be
assumed to act as Newtonian viscous material with respect to the dislocation.
Hence,

$$f_v = Bv$$

B, the viscous damping coefficient, is independent of v for a Newtonian fluid.
Under the application of a certain external stress, the dislocation will accelerate
until it reaches a steady-state velocity. Recalling Eq. 8.16, equilibrium will be
reached when

$$F = f_v$$

$$\tau b = Bv$$

(8.38)

We should remember that this is exactly the form under which dislocation motion has been studied by Greenman et al.[14], Pope et al.,[13] Gorman et al.,[17] and Parameswaran et al.[18,19] Their studies actually allow obtaining B. For instance, it is possible to estimate B from Eq. 8.11 comparing it with Eq. 8.38:

$$K = \frac{b}{B} \quad \text{and} \quad B = \frac{b}{K}$$

$$B = \frac{\tau_0 b}{v_0} = \frac{2.7 \times 10^3 \times 2.55 \times 10^{-10}}{10^{-2}}$$

$$= 7 \times 10^{-5} \text{ Ns/m}^2$$

Some investigators (e.g., Gillis et al.)[34] consider B to be dependent on the velocity of the dislocation. This allows the use of B over a very wide range of velocities. The expression proposed by Gillis et al.[34] is

$$B = \frac{B_0}{1 - v^2/v_s^2} \tag{8.39}$$

It can be seen that viscous drag decreases as the velocity increases.

The mechanisms responsible for viscous drag can be classified into two groups. These are mechanisms involving thermally activated processes and mechanisms not involving them. In the first group, one can include the overcoming of Peierls–Nabarro stresses, the interaction of dislocations with point defects, and with forest dislocations. The mechanisms that are not thermally activated are the interaction of the dislocation with thermal vibrations (phonon† drag) and with electrons (electron viscosity); additionally, they also include relaxation effects in the dislocation core. They will be described briefly below. They are discussed in detail by Gorman et al.,[17] Dorn et al.,[3] and especially Granato.[35]

1. *Phonon viscosity.* A change in the compressive stress will trigger an increase in the radiation pressure of the phonon gas, causing an increase in the modulus of rigidity. This increase in modulus relaxes with time, as the phonons approach equilibrium.

2. *Phonon scattering.* There are several ways by which phonon scattering mechanisms can operate. One of these is the scattering of phonons by dislocation strain field in a manner similar to the refraction of light. Another is the absorption of energy from phonons by dislocations, with the subsequent vibrations by dislocations.

3. *Thermoelastic effect.* A moving dislocation alternatively strains adjacent regions in tension and compression; these regions show temperature decreases and increases, respectively. An irreversible process of heat flow ensues, increasing the entropy and depleting the dislocation of some of its energy.

---

† A phonon is an elastic vibration propagating through the crystal. It is quantized because the lattice is discrete and not continuous.

4. *Electron viscosity*. The free electrons in a metal will affect the dislocation motion in the same way as the phonon viscosity.

5. *Anharmonic radiation*. A dislocation, when undergoing positive or negative acceleration, emits elastic waves. This corresponds to an outflow of energy. Thus, the dislocation, while accelerating between two equilibrium positions, gives off energy.

6. *Glide-plane viscosity*. There seems to be a certain consensus with respect to the relative importance of the various drag mechanisms under various conditions. Gilman's[2] ideas are, in this respect, corroborated by Granato's.[35] They are given below (according to Gilman[2]):

   a. *Covalent bonding*. At low stress levels ($\leq G/100$) the motion is thermally activated (e.g., overcoming of Peierls stresses) and is stopped at low temperatures. However, stresses alone, when high enough ($>G/100$), cause motion at low temperatures.

   b. *Ionic bonding (salts)*. At low stress levels and high temperature, phonons are responsible for the drag.

   c. *Metals*. At high temperatures, phonons cause drag; at low temperatures, electrons cause drag. At high stresses, for both metals and salts, the viscosity increases because of relativistic effects.

Granato adds that, for materials not containing an electronic cloud (covalent and ionic bonding), drag by radiation is the mechanism operating at low temperatures. They also comment on the interaction between dislocations and point defects, if these are moving slow enough. This belongs to the group of thermally activated mechanisms.

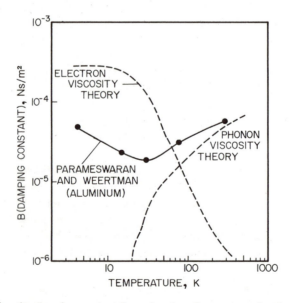

Figure 8.14 Dislocation damping constant B as a function of temperature for aluminum, compared with theoretical calculations from Mason's theories. (Adapted with permission from [19], p. 2985.)

In this context it is instructive to look at the results obtained by Parameswaran and Weertman.[18] For lead,[18] lead–indium,[18] and aluminum,[19] they found the temperature dependence of B shown in Fig. 8.14. It can be seen that phonon viscosity is the drag mechanism above 20 K; below 20 K, B does not tend to zero, however. It increases, as would be predicted by the electron-viscosity theory.

8.4.4 Dislocation Acceleration

Gillis and Kratochvil[36] treated the problem of dislocation acceleration and concluded that it is, under most conditions, extremely high. So the time interval and distance required to bring the dislocation to its steady-state velocity is very small and can, in many applications, be neglected. The acceleration of a dislocation can be calculated without too much difficulty. Figure 8.15 shows a dislocation segment; the applied force (F_{AP}) generated a certain displacement, which is opposed by the frictional force (F_{FR}) and by the back forces due to the bowing [$F(r)$]. Applying Newton's second law, we have

$$F_{AP} - F_{FR} - F(r) = \frac{d}{dt}(mv)$$

Assuming a straight dislocation ($r = 0$), the back stresses vanish. Applying Peach–Koehler's equation (Eq. 8.16) yields

$$b\tau_{AP} - b\tau_{FR} = m\frac{dv}{dt} + v\frac{dm}{dt} \tag{8.40}$$

But the frictional stress is given by (Eq. 8.38):

$$b\tau_{FR} = Bv$$

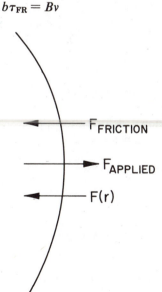

Figure 8.15 Forces on moving dislocation: applied force, frictional force, and force due to back-stress produced by bowing.

Equations 8.36 and 8.37 provide

$$m = \frac{m_0}{(1 - v^2/v_s^2)^{1/2}} = \rho b^2 \left(\frac{v_s^2}{v_s^2 - v^2}\right)^{1/2} \tag{8.41}$$

$$\frac{dm}{dt} = \rho b^2 v_s \frac{d}{dt} (v_s^2 - v^2)^{-1/2}$$

$$= \rho b^2 v_s v (v_s^2 - v^2)^{-3/2} \frac{dv}{dt} \tag{8.42}$$

Substituting Eqs. 8.41 and 8.42 into Eq. 8.40, we have

$$\rho b^2 \left(\frac{v_s^2}{v_s^2 - v^2}\right)^{1/2} a + v^2 \rho b^2 v_s (v_s^2 - v^2)^{-3/2} a = b \tau_{AP} - Bv \tag{8.43}$$

Substituting the value for B given in Eq. 8.39 into Eq. 8.42, and remembering that $\beta^2 = 1 - v^2/v_s^2$, we obtain

$$a = \frac{v_s^2 \tau_{AP} \beta^2 - B_0 v_s^2 v b^{-1}}{\rho b v_s^2 \beta + v^2 \rho b/\beta}$$

$$a = \frac{v_s^2 \tau_{AP} \beta^2 - B_0 v_s^2 v b^{-1}}{\rho b (v_s^2 \beta + v^2/\beta)} \tag{8.44}$$

$$a = \frac{v_s^2 \tau_{AP} \beta^3}{\rho b} - \frac{B_0 v \beta}{\rho b^2} \tag{8.45}$$

Dimensional analysis provides a quick check of the correctness of this equation. It might be somewhat tricky, but the astute student will know how to untie the Gordian knot.

If only relativistic effects are considered, Eq. 8.45 reduces itself to

$$a = \frac{v_s^2 \tau_{AP}}{\rho b} \left(1 - \frac{v^2}{v_s^2}\right)^{3/2} \tag{8.46}$$

As an illustration, the acceleration as a function of velocity was plotted for copper, at an applied stress level of 10 MPa in Fig. 8.16. This is a reasonable level in conventional deformation processes; the yield stress of copper single crystal is a few megapascals. The following values were used:

$$\rho = 8.92 \times 10^3 \text{ kg/m}^3$$

$$v_s = 2.92 \times 10^3 \text{ m/s (from Eq. 1.148)}$$

$$b = 2.55 \times 10^{-10} \text{ m}$$

$$B_0 = 7 \times 10^{-5} \text{ N s/m}^2$$

Since the value of B obtained by Greenman et al.[14] (7×10^{-5} N s/m²) was obtained at low dislocation velocities, it was assumed to be equal to B_0. It

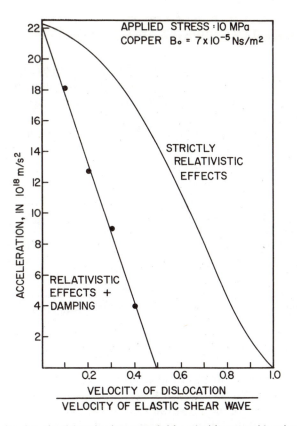

Figure 8.16 Acceleration of a dislocation in copper (with and without consideration of drag effects), as a function of velocity, for an applied stress of 10 MPa.

does not have to be corrected because B, defined as the drag force per unit length of dislocation per unit velocity, was obtained per centimeter, and the unit velocity was taken as 1 cm/s; the two "cm" cancel each other. Figure 8.16 shows that the initial acceleration of a dislocation is extremely high: 2.2×10^{19} m/s². If only relativistic effects were considered, the acceleration would steadily decrease until it becomes zero, when the dislocation velocity becomes equal to the shear-wave velocity. On the other hand, the drag stress has an important effect: it establishes a steady-state velocity of about $0.5 v_s$ at the applied stress level. The steady-state velocity is reached when the drag stress becomes equal to the applied stress. In any event, the acceleration to the steady-state velocity is extremely high. If one takes a mean acceleration of 10×10^{19} m/s², a dislocation will reach the steady-state velocity of $0.5 v_s$ in approximately 1.5×10^{-16} s. This corresponds to a distance traveled of less than 1 Å. So for all intents it can be assumed that dislocations reach their steady-state velocity instantaneously.

8.5 THE MOVEMENT OF DISLOCATIONS
AS A THERMALLY ACTIVATED PHENOMENON

The resistance of crystals to plastic deformation is determined by the resolved shear stress that is required to make the dislocations glide in their slip planes. If no obstacles were present, the dislocations would move under infinitesimal small stresses. However, in real metals the nature and distribution of obstacles determines their mechanical response. Becker[37] was the first to point out the importance of thermal energy in helping the applied stress in overcoming existing obstacles. Seeger[38-40] divided the stress required for deformation, τ, into two parts: τ^*, temperature dependent, and τ_G, essentially temperature independent. The relative importance of τ^* and τ_G can be studied by determining the temperature (or strain-rate) dependence of the flow stress. In stage I of the stress–strain curve (easy glide region; see Chapter 9) one frequently observes that the τ^*/τ_G ratio is close to unity. As the plastic strain increases, one goes to stage II (linear hardening); the ratio τ^*/τ_G decreases. This is the Cottrell–Stokes law.

Let U_0 be the total energy required to free a dislocation from an obstacle. It is represented by the area of the force versus distance (moved by the dislocation) plot. Since this is the extra force needed to move the dislocation through the barrier, one has:

$$\text{extra force} = f = (\tau - \tau_G)bl$$

where τ is the applied stress, τ_G the stress required to move the dislocation in the absence of a barrier, τ_G is due to the long-range stress fields caused by the other dislocations in the lattice, and l is the length of the dislocation segment. The plot shown in Fig. 8.17 is a general one; we have

$$U_0 = \int_{x=0}^{x=d} f \, dx$$

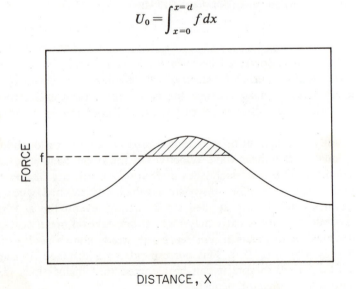

Figure 8.17 Variation of the force required for dislocation motion with distance.

The exact form of the curve depends on the nature of the barrier as well as on the dislocation. If the applied stress is not sufficiently high—for example, force f in Fig. 8.17—slip can still occur if there is a thermal fluctuation. If this thermal fluctuation provides a thermal energy U_T sufficient to overcome the force peak, the dislocation will be able to move. Its magnitude has to be equal to the crosshatched area in Fig. 8.17.

$$U_T = U_0 - \text{work performed by force } f$$

$$U_T = A$$

8.5.1 Simple Case of a Constant-Height Barrier

Figure 18.18 shows a simple case of a square barrier of width d. The total energy required to free a dislocation is

$$U_0 = (\tau_0 - \tau_G)bld$$

If the applied stress is τ and $\tau < \tau_0$, the thermal energy required to overcome the obstacle is

$$U_T = U_0 - (\tau - \tau_G)bld$$

The distance d in which the extra force is needed is called "activation distance"; the product bld is called "activation volume":

$$v^* = bld \tag{8.47}$$

Hence,

$$U_T = U_0 - (\tau - \tau_G)v^* \tag{8.48}$$

The plastic shear strain is given by

$$\gamma = bAN$$

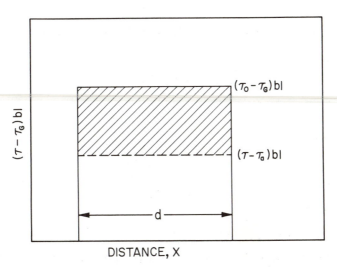

Figure 8.18 Square barrier with width d.

where N is the number of dislocations per unit volume that sweep an area A and b is their Burgers vector. Let ν be the frequency of vibration of a dislocation (i.e., the number of times it tries to move per unit time). The plastic strain rate will be

$\dot{\gamma} = A \cdot b \cdot$ (number of dislocations moving per unit time)

$\dot{\gamma} = A \cdot b \cdot N \cdot$ (number of attempts/unit time) $\cdot$ (fraction of the attempts that is successful)

The theory of thermal activation and thermally activated reactions can be applied to the equations above; it is the basis of chemical kinetics. One knows that the probability of overcoming the obstacle increases exponentially with temperature. Hence, we have

$$\dot{\gamma} = AbN\nu \, \exp\left(-\frac{U_T}{kT}\right) \tag{8.49}$$

where k is Boltzmann's constant and T is the temperature in K. For the simple square barrier considered above, we just substitute Eq. 8.48 into Eq. 8.49:

$$\dot{\gamma} = bAN\nu \, \exp\left[-\frac{U_0 - (\tau - \tau_G)\nu^*}{kT}\right] \tag{8.50}$$

Equation 8.50 expresses the stress dependence of strain rate.

8.5.2 Variation of Flow Stress with Temperature

Solving Eq. 8.50 for τ, we have

$$\tau = \tau_G + \frac{U_0 - kT \ln(bAN\nu/\dot{\gamma})}{\nu^*}$$

$$\tau = \tau_G + \frac{U_0}{\nu^*}\left(1 - \frac{mkT}{U_0}\right) \tag{8.51}$$

where

$$m = \ln\frac{bAN\nu}{\dot{\gamma}}$$

Designating the term

$$\frac{U_0}{\nu^*}\left(1 - \frac{mkT}{U_0}\right)$$

by τ^*, we have

$$\tau = \tau_G + \tau^* \tag{8.52}$$

τ_G and τ^* are the athermal and thermal components of stress, respectively. τ^* can obviously not be negative, but it tends to zero when

$$T = T_0 = \frac{U_0}{mk}$$

Hence, when

$$T \gg T_0, \qquad \tau = \tau_G$$

then

$$T < T_0, \qquad \tau = \tau_G + \tau^*$$

T_0 is about 173 K for aluminum and 373 K for magnesium (when it deforms by basal slip). The variation in flow stress with the temperature is schematically shown in Fig. 8.19. It can be concluded that, for $T \geqslant T_0$, the flow stress exhibits the same temperature response as τ_G. The temperature response of τ_G is that of G, the shear modulus. It is very low, as can be seen from the analysis conducted in Chapter 4.

It should be emphasized that the analysis above, due to Seeger,[38] assumes the existence of a single thermally activated process controlling the deformation over the whole temperature range. A more detailed analysis is described in the next section.

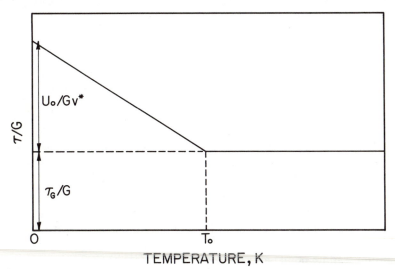

Figure 8.19 Variation of flow stress with temperature.

8.5.3 Different Barriers

Conrad and Wiedersich[41] extended Seeger's analysis, considering different types of barriers. It is described in detail by Conrad[42,43] and will be succinctly presented here. Figure 8.20 shows the variation of the internal stress fields encountered by a dislocation, as it moves through crystal lattice. The long-range stresses have a wavelength λ; the dislocation motion is actually aided when they are negative. Thermal energy cannot aid the dislocation to overcome

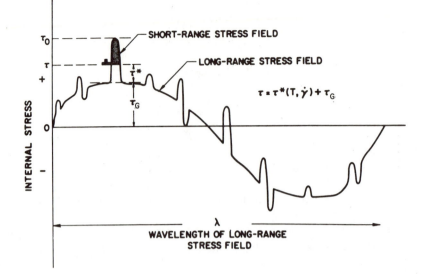

Figure 8.20 Internal stress fields encountered by a dislocation moving through the crystal lattice. (Reprinted with permission from [42], p. 584.)

the long-range stress fields. These long-range fields are due to precipitates, grain boundaries, other dislocations, image forces due to other phases, and so on. Riding on the long-range stress fluctuations in Fig. 8.20, one can see the short-range spikes, due to short-range obstacles. Thermal energy can and does help the dislocations to overcome them. As one can see from Fig. 8.20, different short-range barriers have different heights. Usually, at a specific temperature one barrier is the rate-limiting step (the highest). There are several types of short-range obstacles. The Peierls–Nabarro barriers are the most fundamental ones, and they exist in any lattice. They represent the resistance that the crystalline lattice offers to the movement of a dislocation. Figure 8.21 shows the stress that one has to apply to a dislocation to make it move by a distance b. When the extra plane is moved away from its equilibrium position (either to the right or left), one has to overcome a barrier. The difference in energy between the equilibrium (saddle point) and the most unstable position is called the Peierls–Nabarro energy, and the stress required to overcome this energy barrier is the Peierls–Nabarro (P–N) stress. Theoretical considerations predict a fairly low P–N stress for metals at ambient temperature, while ceramics have a very big P–N stress. FCC and HCP metals have a relatively low P–N stress at low temperatures. Iron, tungsten, niobium, vanadium, molybdenum, and tantalum exhibit an exponential yield stress increase as 0 K is approached. On the other hand, FCC metals do not exhibit this strong dependence. The low-temperature plastic deformation of BCC metals is controlled by the Peierls–Nabarro stresses (see Table 8.1); the activation volume is small ($<100b^3$). Krausz and Eyring[44] calculate the temperature dependence of yield stress as a function of activation volume and find it to be much higher for small values of v. The Peierls–Nabarro

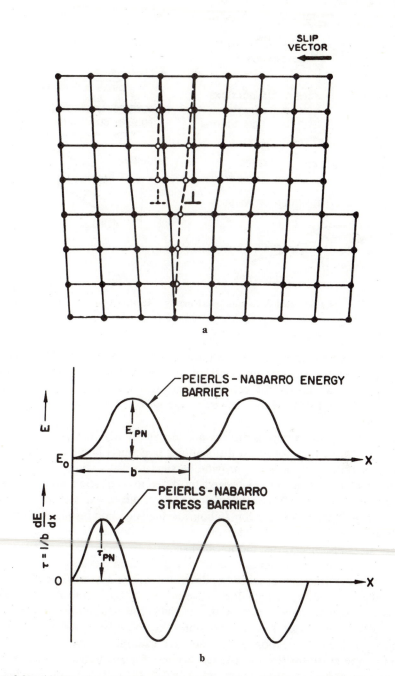

Figure 8.21 (a) Movement of dislocation away from its equilibrium position; (b) variation of Peierls–Nabarro stress with distance. (Reprinted with permission from [42], p. 583.)

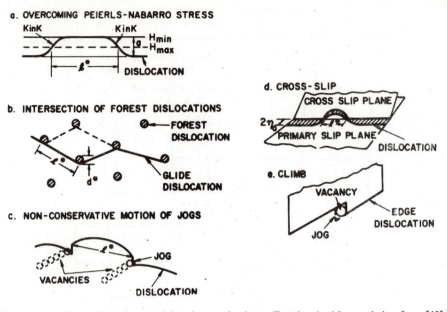

a. OVERCOMING PEIERLS-NABARRO STRESS

KinK   KinK
H_{min}
H_{max}
ℓ^*
DISLOCATION

b. INTERSECTION OF FOREST DISLOCATIONS

FOREST DISLOCATION
ℓ^*
d^*
GLIDE DISLOCATION

c. NON-CONSERVATIVE MOTION OF JOGS

ℓ^*
JOG
VACANCIES
DISLOCATION

d. CROSS-SLIP

CROSS SLIP PLANE
$2\eta_0$
PRIMARY SLIP PLANE
DISLOCATION

e. CLIMB

VACANCY
EDGE DISLOCATION
JOG

Figure 8.22 Thermally activated dislocation mechanisms. (Reprinted with permission from [42], p. 583.)

stresses for edge dislocations are known to be high in BCC metals. The difficulty of movement of the edge components of dislocations at low temperatures is thought to play an important role in the strong temperature dependence of the flow stress of BCC metals; however, the effect of interstitials cannot be ignored.

Additional short-range obstacles are shown in Fig. 8.22, which is self-explanatory. Intersection of forest dislocations, nonconservative motion of jogs, cross-slip, and dislocation climb provide barriers to dislocation motion. These different mechanisms are characterized by different activation volumes (bld). The calculation of these activation volumes by indirect experiments provides a way of establishing the rate-controlling mechanism. For instance, climb has an activation volume of the order of b^3, while it varies between $10^2 b^3$ and $10^4 b^3$ for the nonconservative motion of jogs.

A number of experimental techniques have been devised to obtain the fundamental deformation parameters described above (internal and effective stress, activation volume). Mechanical tests in which the strain rate is suddenly changed at a constant temperature, or in which the temperature is suddenly changed at constant strain rate, are examples thereof. The stress-relaxation test (described in detail in Section 6.5) is another example. Table 8.1 shows the effect of temperature on the short-range barriers for a number of metals. The temperature was divided into three ranges. For $T > 0.5 T_m$, climb and nonconservative motion of jogs tend to be the most important mechanisms. At lower temperatures, different metals tend to react differently.

TABLE 8.1 Rate-Controlling Mechanisms in Metals[a]

Crystal Structure	Materials	Most Likely Mechanism	Alternative Mechanism
		I. Low Temperatures, $T < 0.25T_m$	
HCP	Zn, Cd, Mg	Intersection of dislocations	Nonconservative motion of jogs
FCC	Al, Cu, Ag, Au, Ni	Intersection of dislocations	Conservative motion of glissile jogs
BCC	V, Nb, Ta, Cr, Mo, W, Fe	Overcoming the Peierls–Nabarro stress	Nonconservative motion of jogs Overcoming interstitial atoms or precipitates Cross-slip
		II. Intermediate Temperatures, $0.25T_m < T < 0.5T_m$	
HCP	Zn, Cd, Mg	Intersection of dislocations	Nonconservative motion of jogs
FCC	Al	Cross-slip	Breakdown of Cottrell-Lomer barriers
BCC	V, Nb, Ta, Cr, Mo, W, Fe	Interaction of dislocations and interstitials	
		III. High Temperatures, $T > 0.5T_m$	
HCP	Be, Ti, Zn, Mg	Climb ($0.5T_m < T < 0.8T_m$) Overcoming Peierls–Nabarro stress for prismatic slip ($T < 0.8T_m$)	Nonconservative motion of jogs Cross-slip
FCC	Al, Cu, Au, Pb	Climb	Nonconservative motion of jogs
BCC	Nb, Mo, Ta, Fe	Climb	Nonconservative motion of jogs

[a] Reprinted with permission from [42], p. 587.

8.6 GEOMETRY OF DEFORMATION

8.6.1 Stereographic Projections

The mechanical properties of crystals are anisotropic and slip only occurs in certain planes, along certain directions (see Section 6.7). For this reason it is important to define the *orientation* of a crystal before studying it. The most common technique is the stereographic projection. It will be presented here in an abbreviated way; greater details are provided by Barrett and Massalski.[45] The stereographic projection is a geometric representation of the directions and planes of a crystal. From the stereographic projections one can determine the angles between planes, planes and directions, and directions. The stereographic projection is the projection of a sphere on a plane. We imagine a unit cell of a certain crystalline structure at the center of the sphere. The directions and plane poles (normals to the planes passing through the origins) intercept a sphere at points; these points are projected on a plane. Figure 8.23 shows a standard cubic projection. This projection is known as a [100] standard projection because the [001] direction corresponds to the center. There are a series of other standard projections: [110], [111], [112], and so on. The angles between directions and/or plane poles are theoretically measured on the sphere; in practice, however, these angles are measured on the standard projection, making use of a special chart, the Wulff net. It is the projection of a plane of a sphere in which all the meridians and parallels are marked at regular degree intervals. The sphere has the same diameter as the standard projection. By inserting a tack at the center and rotating the standard projection around it, we can easily find all desired angles. Barrett and Massalski[45] describe the exact procedure.

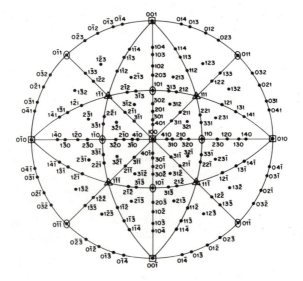

Figure 8.23 Standard [100] stereographic projection. (Reprinted with permission from [45], p. 39.)

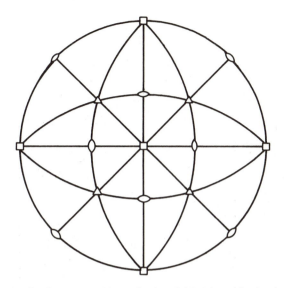

Figure 8.24 Standard [001] stereographic projection divided into 24 triangles. (Reprinted with permission from [50], p. 84.)

In the stereographic projection the crystalline symmetry can be clearly seen. For instance, the ⟨100⟩ directions form a cross in Fig. 8.23. The crystalline symmetry was indicated in Fig. 8.23; two-, three-, four-, and six-field symmetry axes are indicated. The symbols have been introduced in Section 1.12.3 and the reader is referred to Table 1.1. For ⟨111⟩, ⟨110⟩, and ⟨100⟩ the symmetry is four-, two-, and threefold, respectively, in the cubic system. As a consequence of this, the standard projection can be divided, by means of great circles, into 24 spherical triangles that are crystallographically equivalent. Their vertices are ⟨100⟩, ⟨110⟩, and ⟨111⟩. This can be seen in Fig. 8.24. Comparing Figs. 8.23 and 8.24, it can be seen that the directions on the sides and within the spherical triangles are also equivalent. Consequently, one single triangle is sufficient to specify any crystallographic orientation in cubic system; the [100], [110], [111] triangle is used most commonly. The reader should be warned, however, that this simplification is not applicable to the other crystal systems.

8.6.2 Stress Required for Slip

The flow stresses of crystals are highly anisotropic. For instance, the yield stress of zinc under uniaxial tension varies by a factor of at least 6 (see Fig. 8.26). Consequently, it is very important to specify the orientation of the load. In shear or torsion tests, the shear plane and directions are precisely known. As dislocations can only glide under the effect of shear stresses (see the Peach–Koehler relationship in Section 6.6), these shear stresses have to be determined. In uniaxial tensile and compressive tests (the most common tests) one has to determine mathematically the shear component of the applied stress acting on the plane in which slip is taking place. Figure 8.25 shows a crystal with a

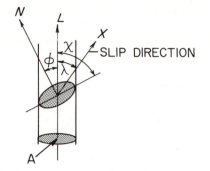

Figure 8.25 Relationship between stress axis and slip plane and direction.

normal cross-sectional area A; a tensile load acting on it generates a uniaxial stress P/A. The slip plane and direction are also indicated by the angles ϕ and λ that they make with the tensile axis, respectively. Notice that, for the plane, the normal to it was taken in the determination of the angle ϕ. The shear component of stress, in the slip plane and along the slip direction, is given by

$$\tau = \sigma \sin \chi \cos \lambda = \sigma \cos \phi \cos \lambda \qquad (8.53)$$

This equation shows that τ can be zero when either λ or ϕ are equal to 90°. On the other hand, the shear component is maximum when both ϕ and λ are equal to 45°. We have, in this case,

$$\tau_{max} = \sigma \cos 45° \cos 45° = \frac{\sigma}{2} \qquad (8.54)$$

Schmid and co-workers[46,47] used the variation in the resolved shear stress to explain the great differences in the yield stresses of monocrystals of certain metals. Known as the *Schmid law*, the following rationalization was proposed by them: Metal flows plastically when the resolved shear stress acting in the plane and along the direction of slip reaches a critical value.

$$\tau_c = \sigma_0 \sin \chi \cos \lambda = M\sigma_0 \qquad (8.55)$$

The factor M is usually known as *Schmid factor*:

$$M = \sin \chi \cos \lambda = \cos \phi \cos \lambda \qquad (8.56)$$

Schmid's law has found experimental confirmation principally in hexagonal crystals. Figure 8.26 shows the experimental results compared with Schmid's prediction for high-purity zinc; these results were obtained by Jillson[48] and show an excellent correlation. The full line shows the hyperbola obtained by use of Eq. 8.55, assuming a critical resolved shear stress of 18.4 kPa. It is worth noting that the yield stress is minimum for $M = 0.5$.

For cubic crystals the correspondence between Schmid's law and experimental results is not as good. This is mainly due to the great number of slip systems in these structures. Haasen[49] observed that for nickel the critical resolved shear stress is practically orientation independent. On the other hand, for copper the critical resolved shear stress is dependent on orientation, being

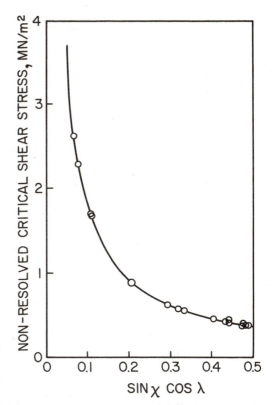

Figure 8.26 Comparison of Schmid law prediction with experimental results. (Adapted with permission from [48], p. 1130.)

constant in the center of the stereographic triangle and assuming higher values close to the sides. Figure 8.27 shows the inverse of Schmid's factor in stereographic triangle based on {111} ⟨110⟩ slip.[50] This is the situation for FCC crystals. The orientation for which FCC crystals are softest is $M = 0.5$, or $M^{-1} = 2$. This occurs approximately at the center of the triangle. The dependence of τ_c on the orientation for cubic systems is thought to be due to the fact that the components of compressive stresses acting normal to the slip planes are different for different orientations at a same applied stress level. These compressive stresses should have an effect on τ_c. Additional information on the Schmid factor for BCC and FCC crystals can be found in [50–54].†

---

† Easy glide in FCC crystals is greatest in the center of the stereographic projection in the region closer to (but not coinciding with) the ⟨110⟩ corner. It is affected by a number of parameters, the most notable being:

1. *Specimen size.* Specimens with a smaller cross-sectional area tend to have a more extended easy glide region.

2. *Temperature.* Easy glide is more pronounced at lower temperature, and may vanish completely at high temperature.

3. *Stacking-fault energy.* FCC metals with high stacking-fault energy tend to have less pronounced easy glide region.

4. *Solute atoms.* If they pin the dislocations, they will shorten their mean free path and the

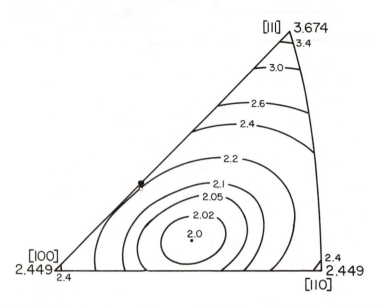

Figure 8.27 Effect of orientation on the inverse of Schmid factor for FCC metal. (Adapted with permission from [50], p. 85.)

8.6.3 Deformation Shear

In the same way as a tensile test does not provide directly the shear stress in the slip plane and along the slip direction, it does not directly provide the corresponding deformation. The latter one has to be determined taking into account the relative orientations of tensile axis and slip system. If a tensile specimen is attached to the grips of a tensile testing machine by means of universal joints, it can be seen that the slip plane will rotate with respect to the tensile axis as deformation proceeds. Therefore, it is important to know the deformation and, consequently, the change in orientation with the attendant alteration in Schmid's factor. This will be done next.

Consider a crystal, initially cylindric, that is deformed in a limited portion by means of two glide planes, as shown in Fig. 8.28. Assume that this crystal underwent homogeneous slip along its whole extension, in such a way that the cylinder still has a central axis after deformation. The glide of the planes between P_1 and P_2, occurring along the direction CD results in the displacement of the crystal toward the left. Since the ends of the crystal are gripped, the

---

extent of stage I. If they contribute to the lowering of the stacking-fault energy or to ordering primarily, they increase the easy glide range.

Some of the controversy surrounding the critical resolved shear stress in BCC metals was resolved by R. Ayres and D. F. Stein [*Mater. Sci. Eng. 13* (1974) 223]. They found that variations of 300% in the CRSS of molybdenum found by previous investigators for different orientation of the tensile axis were due to impurities.

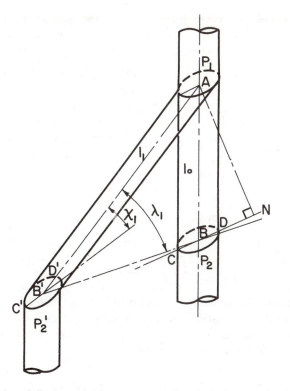

Figure 8.28 Rotation of slip system with respect to tensile axis during extension in uniaxial tensile test.

slip plane and direction change; it will be shown that they rotate toward the tensile axis as deformation takes place.

The angles χ_0 and λ_0, defined in Section 8.6.2, change to χ_1 and λ_1. The relationship between the amount of extension and the rotation of the crystal slip planes can be obtained from geometric considerations on triangles ABB', ABN, and $AB'N$. In triangle ABB' we have

$$\frac{l_1}{l_0} = \frac{\sin \lambda_0}{\sin \lambda_1}$$

λ_0 and λ_1 are the angles between the tensile axis and the slip direction in the beginning and end of the deformation, respectively.

From equilateral triangles ABN and $AB'N$, we have

$$AN = l_0 \sin \chi_0 = l_1 \sin \chi_1$$

$$\frac{l_1}{l_0} = \frac{\sin \chi_0}{\sin \chi_1} \tag{8.57}$$

χ_0 and χ_1 are the angles between the tensile axis and the slip plane at the start and finish of deformation, respectively.

So both χ and λ decrease during the test, and this is tantamount to saying that the plane and direction of slip rotate toward the tensile axis. For the determination of the shear strain in the slip plane, the following equation is used:

$$\gamma = \frac{BB'}{AN}$$

From triangle ABB' we have

$$BB' = \frac{l_1 \sin(\lambda_0 - \lambda_1)}{\sin \lambda_0}$$

From Eq. 8.57, we obtain AN,

$$\gamma = \frac{1}{l_0 \sin \chi_0} \frac{l_1 \sin(\lambda_0 - \lambda_1)}{\sin \lambda_0}$$

or

$$\gamma = \frac{1}{\sin \chi_0} \left\{ \left[\left(\frac{l_1}{l_0} \right)^2 - \sin^2 \lambda_0 \right]^{1/2} - \cos \lambda_0 \right\} \tag{8.58}$$

Equation 8.58 establishes the shear strain in the slip plane from the extension l_1/l_0 and the initial orientations of the slip plane and direction. If we consider a small strain increment, we can neglect the rotation of the slip system and have, at the instantaneous value of χ and λ,

$$d\gamma = \frac{d\epsilon}{\sin \chi \cos \lambda} = \frac{d\epsilon}{M} \tag{8.59}$$

The reader should prove this equation. Therefore, when $M = 0.5$, we have $\tau = 0.5\sigma$ and $\gamma = 2\epsilon$.

8.6.4 Slip Systems

Equations 8.55 and 8.59 establish the stress and strain in the plane and direction of shear and are therefore important from a point of view of dislocation motion. In HCP structures the slip is more easily maintained in one plane. However, in BCC and FCC structures other slip systems are easily activated. The rotation of the slip plane and direction will easily put other systems in a favorable position. This situation is shown in the stereographic projection of Fig. 8.29. A certain crystal has the tensile axis within the crosshatched stereographic triangle. The first slip system to be activated will be the one with the highest Schmid factor (see Eq. 8.55). There are eight slip systems around axis P in Fig. 8.29. The reader can check by using great circles, whether the directions given below really belong to the planes. These are the systems

$(11\bar{1})$	$[101]$	$(11\bar{1})$	$[1\bar{1}0]$
(111)	$[1\bar{1}0]$	(111)	$[10\bar{1}]$
$(1\bar{1}1)$	$[10\bar{1}]$	$(1\bar{1}1)$	$[110]$
$(1\bar{1}\bar{1})$	$[110]$	$(1\bar{1}\bar{1})$	$[101]$

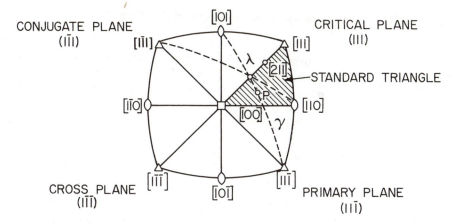

Figure 8.29 Stereographic projection showing the rotation of slip plane during deformation.

The maximum value of Schmid's factor, $M = 0.5$, is obtained for $\chi = \lambda = 45°$. The angles between P and the $\langle 100 \rangle$ directions are determined by means of a Wulff net, passing a great circle through the two poles. The slip system having highest Schmid factor is, among the eight systems above, $(11\bar{1})$ [101]; slip will initially take place in this system. Plane $(11\bar{1})$ is therefore called *primary slip plane*. As deformation proceeds, χ and λ will rotate. In the stereographic projection, this is indicated by the rotation of the axis P (actually, the specimen rotates with respect to the axis). The axis P will tend to align itself with direction [101], decreasing λ in the process; this is shown in Fig. 8.29. However, when the great circle passing through [100] and (111) is reached, the primary system and the *conjugate slip system* $(1\bar{1}1)$ [110] will have the same Schmid factor. The typical behavior in this case is double slip in both systems; the axis P will tend toward the direction [211], as shown in Fig. 8.29. In reality there are deviations from this behavior, and there is a tendency for "overshoot" and subsequent correction. The two other slip systems are called the *cross* system and the *critical* system. The reader should be warned that this nomenclature, due to Clarebrough and Hargreaves,[51] is not universal. Often the term "cross-slip" is used to describe a different situation: small slip segments in a secondary slip system joining slip lines in a primary slip system receive this name. McGregor Tegart[55] favors the term "connective slip" instead of cross-slip for the latter situation.

As a conclusion it can be said that a cubic crystal will initially undergo slip in one system if P is within the stereographic triangle. If P is on the sides of the triangle, two systems have the same Schmid factor. On the other hand, if P coincides with one of the edges, the situation is more complicated: eight systems will have the same Schmid factor if P coincides with [100], four if it coincides with [110] and six if it coincides with [111]. The term "polyslip" refers to a crystal oriented in such a way that more than one system is activated.

8.7 DEFORMATION OF MULTICRYSTALS AND POLYCRYSTALS

The mechanical properties of polycrystals are very different from single crystals. Some of the factors responsible for these differences will be studied here. First, the definition of a polycrystal will be established. McGregor Tegart[55] defines multicrystals (bi-, tri-, tetra-, penta-, etc.) as specimens having between 2 and 20 grains per section normal to the tensile (or compressive) axis. Polycrystals have more than 20 grains per cross section. However, more recent studies have shown that the free surface can be a source of dislocations. Nixon et al.[56] have shown, in the deformation of a Cu–10.5 at% Al alloy, that all dislocations initially observed could be accounted for by surface sources and that surface-generated dislocations could travel distances at least 3.2 mm into the crystal. McGregor Tegart's definition does not take this into account, and if we had the 20 grains arranged in a linear array, we would have a polycrystal in spite of the fact that all grains would have free surfaces. A careful investigation of the effect of specimen thickness on the tensile properties of metals, conducted by Miyazaki et al.,[57] disclosed that below some t/d ratios (d is the grain diameter and t is the specimen thickness) the yield stress and flow stress at various plastic strains were strongly affected. Figure 8.30 shows this effect for Al, Cu, Fe, and alloy Cu–13 at% Al. Changes in these properties start occurring below a critical t/d ratio, which is dependent on the material (and, incidentally, on the grain size also). For instance, for Fe we should stay above $t/d = 10$ to assume polycrystallinity. In view of the above, it is felt that the definition presented by Kocks[58] is more acceptable. He calculates that a wire that has 100 grains per cross section has about one-third of its grains at the free surface. This is indeed a significant value, which could have an important effect on mechanical properties. A good criterion, according to him, would be to have a *fraction of surface grains lower than 10%*. For a cylindrical specimen and isoaxed grains, the specimen diameter has to be at least 30 to 40 times the grain diameter.

The deformation of polycrystals is much more complex than that of monocrystals, and great analytical effort has been devoted over the past years in order to bridge the gap between mechanical behavior of mono- and polycrystals. There are some basic differences in their response, due, in part, to the following factors:

1. Monocrystals are elastically and plastically anisotropic; polycrystals are, in the absence of texture, macroscopically treated as isotropic materials.
2. Monocrystals can undergo deformation in one single slip system if the load axis is appropriately oriented. Slip on a single system cannot take place in a polycrystal because deformation in the various grains has to be compatible.
3. Deformation in polycrystals is inherently inhomogeneous; that is, it varies from grain to grain and even in a single grain (see [59]).
4. The movement of dislocations in a polycrystal is much more hampered because it is restricted to a single grain.

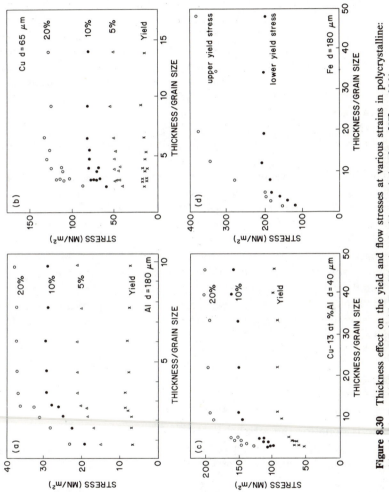

Figure 8.30 Thickness effect on the yield and flow stresses at various strains in polycrystalline: (a) Al; (b) Cu; (c) Cu–13% Al; (d) Fe. (Reprinted with permission from [57], p. 856.)

5. Grain boundaries play an important role in polycrystalline deformation, while in single crystals the free surfaces act as dislocation sources.

8.7.1 Independent Slip Systems in Polycrystals

For any FCC crystal with the tensile axis inside the stereographic triangle, the deformation should start at the primary system. However, if this crystal is surrounded by other crystals with different crystallographic orientations— as is the case in the polycrystalline aggregate—it could not start deforming in the same manner. The strain taking place in the first crystal would have to be compatible with the strain in the other crystals (grains). In other words, it is not possible to form discontinuities along the grain boundaries. Deformation has to propagate from one grain to another if continuity at the boundary is to be maintained. Von Mises[60] noted, for the first time, that five independent slip systems are required to produce a general homogeneous strain in a crystal by slip. The argument developed by von Mises, as presented by Groves and Kelly,[61] is given below. The same argument is presented, in a slightly modified way, by Chin.[50]

The slip along several parallel systems produces, macroscopically, a translation of one part of the crystal with respect to the other and, consequently, a certain shear. Since the plastic flow generally occurs without appreciable volume change, we have $\epsilon_{11} + \epsilon_{22} + \epsilon_{33} = 0$. This relationship reduces the components of strain from six (ϵ_{11}, ϵ_{22}, ϵ_{33}, ϵ_{12}, ϵ_{13}, ϵ_{23}) to five; von Mises observed that the operation of one slip system produces only one independent component of the strain tensor. Therefore, one can conclude that five independent slip systems are required for the deformation of one grain in a polycrystalline aggregate.

To check whether five slip systems are really independent, we determine all the components of strain introduced by one system. So we have, for slip system 1:

$$(\epsilon_{11})_1, \ (\epsilon_{22})_1, \ (\epsilon_{33})_1, \ (\epsilon_{12})_1, \ (\epsilon_{23})_1, \ (\epsilon_{13})_1$$

For system 2,

$$(\epsilon_{11})_2, \ (\epsilon_{22})_2, \ (\epsilon_{33})_2, \ (\epsilon_{12})_2, \ (\epsilon_{23})_2, \ (\epsilon_{13})_2$$

The same is done for the other three systems, always referring the strains to the same system of axis. Since the three longitudinal strains ϵ_{11}, ϵ_{22}, ϵ_{33} are not independent, the following matrix can be built:

$$
\begin{bmatrix}
(\epsilon_{11} - \epsilon_{22})_1 & (\epsilon_{11} - \epsilon_{33})_1 & (\epsilon_{12})_1 & (\epsilon_{23})_1 & (\epsilon_{13})_1 \\
(\epsilon_{11} - \epsilon_{22})_2 & (\epsilon_{11} - \epsilon_{33})_2 & (\epsilon_{12})_2 & (\epsilon_{23})_2 & (\epsilon_{13})_2 \\
(\epsilon_{11} - \epsilon_{22})_3 & (\epsilon_{11} - \epsilon_{33})_3 & (\epsilon_{12})_3 & (\epsilon_{23})_3 & (\epsilon_{13})_3 \\
(\epsilon_{11} - \epsilon_{22})_4 & (\epsilon_{11} - \epsilon_{33})_4 & (\epsilon_{12})_4 & (\epsilon_{23})_4 & (\epsilon_{13})_4 \\
(\epsilon_{11} - \epsilon_{22})_5 & (\epsilon_{11} - \epsilon_{33})_5 & (\epsilon_{12})_5 & (\epsilon_{23})_5 & (\epsilon_{13})_5
\end{bmatrix}
$$

If the determinant of this matrix is different from zero, the five slip systems are independent. This is a consequence of the following property of matrices: If one row (or column) can be expressed as a linear combination of other rows (or columns), the determinant is zero. If one slip system produces a strain that is a linear combination of the strain produced by other systems, it is not an independent slip system. For instance, if x_1 (row 1) $= x_2$ (row 2) $+ x_3$ (row 4), the strain produced by the operation of systems 2 and 4 and system 1 can be the same, and system 1 would not be an independent slip system:

$$\text{row } 1 = \frac{x_2}{x_1} \text{ (row 2)} + \frac{x_3}{x_1} \text{ (row 4)}$$

Groves and Kelly[61] present a method for the determination of independent slip systems using Thompson's tetrahedron (see Section 6.7). Physically speaking, the fact that the determinant has to be different from zero means that each slip system has to be able to contribute with a change in shape of the crystal that cannot be produced by any combination of the other four systems. As an illustration it can be said that a piece of monocrystalline metal can be forged into any possible shape by the operation of only five independent slip systems.

8.7.2 Multicrystals

Figure 9.4 shows a stress–strain curve typical of an FCC monocrystal in which only one slip system operated initially. The curves for polycrystals do not show these various stages (they are discussed in greater detail in Section 9.7.1); for polycrystals the curve can be approximated as a parabola or an exponential. The study of the effect of grain boundaries on the stress–strain curves can be made by studying bi-, tri-, tetra-, and other multicrystals. Of these, the bicrystals are the simplest. Among the bicrystals, the isoaxial ones have the same crystallographic orientation with respect to the tensile axis, the boundary being parallel to the tensile axis [Fig. 8.31(a)]. One crystal could be

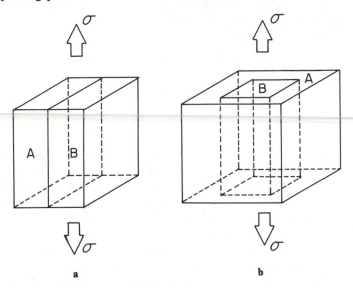

Figure 8.31 Isoaxial (a) and totally surrounded (b) bicrystals being pulled in tension.

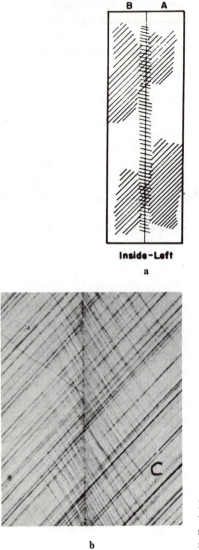

B A

Inside-Left

a

C

b

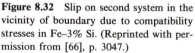

Figure 8.32 Slip on second system in the vicinity of boundary due to compatibility stresses in Fe–3% Si. (Reprinted with permission from [66], p. 3047.)

made to coincide with the other by means of a rotation around the tensile axis. There are certain relative orientations between the two crystals for which the independent operation of the slip systems in the two crystals is compatible; in other words, the two monocrystals could be deformed separately and later glued together and reconstruct the same situation as when the deformation is applied simultaneously to the two sides. In this case the bicrystal is known as compatible and the boundary does not have any effect. But for most of the relative orientations, an isoaxial bicrystal is *incompatible*. Therefore, compatible bicrystal deformation can be accomplished by the operation of one system per grain; for incompatible bicrystals this is not possible. These bicrystals have been studied by, among others, Livingston and Chalmers,[62] Elbaum,[63] Hauser and Chalmers,[64] and Hirth.[65,66] Another type of bicrystal that requires the operation of a larger number of slip systems is the totally surrounded one. As shown in Fig. 8.31(b), one crystal completely surrounds the other. This bicrystal simulates a polycrystal much better than an isoaxial bicrystal. Analytic calculations show that for a totally surrounded bicrystal a total of six independent slip systems have to be activated in the two crystals. Therefore, the situation is fairly close to a polycrystal. Investigations in incompatible and compatible tri- and tetracrystals have also been conducted.

Figure 8.32 illustrates the noncompatibility of strains in a bicrystal of Fe–3% Si after axial loading. Along the interface between the two crystals a second slip system was activated in order to accommodate the additional constraints due to compatibility. The two crystals are oriented in such a way that $M = 0.5$. However, the grain boundary region is subjected to a much higher stress due to elastic incompatibility. The orientation of this stress is such that another slip system is activated.

8.7.3 Polycrystals

Figure 8.33 shows stress–strain curves for aluminum mono- and polycrystals (0.2 mm grain size, giving 15% surface grains). Three single crystals were tested. The lowest curve corresponds to a tensile axis in the central region of the stereographic triangle. This curve is characteristic of all orientations inside the triangle. For the specific orientation, $M = 0.5$ and one can clearly see an easy glide plateau, corresponding to slip in the primary system. The two other single crystals are oriented in such a way that the tensile axis coincides with $\langle 100 \rangle$ and $\langle 111 \rangle$. Recalling Section 8.6.4, it can be seen that at these orientations

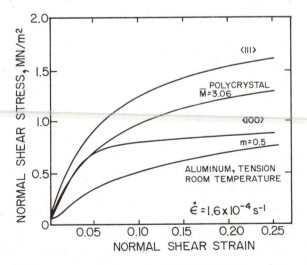

Figure 8.33 Engineering stress–strain curves for 99.99% Al for three single crystals of different orientations and polycrystal with grain size of 0.2 mm. (Reprinted with permission from [58], p. 1121.)

one has several slip systems with identical Schmid factors and polyslip from the onset of plastic deformation. Both flow stresses and work-hardening rates are naturally much higher for these cases. The polycrystalline curve is shown in the same plot. This section is devoted to the attempts at calculating polycrystalline responses from monocrystalline ones.

Sachs[67] was the first to extrapolate the monocrystalline mechanical response to polycrystals. He assumed that all grains underwent the same strain and that they all were activated initially on one slip system. For a uniaxial stress in the x_3 direction (from Eqs. 8.55 and 8.59),

$$\tau_s = \overline{M}\sigma_{33} \qquad \text{and} \qquad \epsilon_{33} = \overline{M}\gamma_s$$

Assuming a random orientation distribution for all grains, he obtained an average inverse Schmid factor of 2.24 for FCC crystals. From a shear stress–shear strain curve for a single crystal oriented for slip in the primary slip plane (τ_s versus γ_s) he obtained the stress–strain curve for the polycrystal (σ_{33} versus ϵ_{33}). However, his model failed to recognize (1) the fact that there are compatibility constraints to be satisfied and (2) that polycrystals do not slip by the activation of one single slip system. Consequently, Sachs' analysis underestimates the polycrystalline stress–strain curve.

The theory accepted nowadays is the result of Taylor's[68] and Bishop and Hill's[69] work. In particular, Taylor's work is admirable because it was proposed in 1938, only four years after he had postulated the existence of dislocations. Both Taylor[68] and Bishop and Hill[69] recognized the importance of considering the compatibility of deformations in the different grains in order to assure intergranular cohesion. They both assumed that each grain underwent the same strain than the macroscopic average strain undergone by the polycrystalline specimen:

$$d\epsilon = d\bar{\epsilon}$$

They also assumed that this strain was imparted to each grain by the operation of five independent slip systems. The FCC structure has 12 slip systems. Using probability theory, one finds the number of combinations of five distinguishable numbers, out of a total of 12. Of these, it is possible to obtain the following combinations of five slip systems:

$$C = \frac{12 \times 11 \times 10 \times 9 \times 8}{5!} = \frac{12!}{5! \, 7!} = 798$$

Of these, it is found that 384 are independent, by applying the criterion shown in Section 8.7.1. For the BCC structure, there are 384 groups of five slip systems of the $\{1\bar{1}0\} \langle 111 \rangle$ and more of the different types of slip systems. In the HCP crystals, there is a much lower number of combinations of five independent slip systems and various types of systems can operate, depending on the material. Greater details are provided by Groves and Kelly.[61] Taylor[68] used the following criterion: The combination of five independent slip systems that would be able to perform a certain deformation with the least amount of shear was the one that would be operative. He searched among the 384 combinations for the

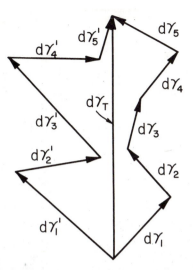

Figure 8.34 Generation of strain $d\gamma_t$ by the activation of two alternative sets of slip system. Both $d\gamma_1 + \cdots + d\gamma_5$ and $d\gamma_1^1 + \cdots + d\gamma_5^1$ produce $d\gamma_T$.

FCC crystal and obtained a specific Schmid factor. Figure 8.34 shows schematically Taylor's hypothesis. We could put it:

$$d\epsilon = \sum_{i=1}^{5} M_i \, d\gamma_i \qquad d\epsilon = \sum_{i=1}^{5} M_i^1 \, d\gamma_i^1 \qquad (8.60)$$

M_i and M_i^1 are Schmid factors for the various slip systems. The strains on the right of Fig. 8.34 can perform the deformation with a lower amount of total shear than the ones on the left. The normal stress that would activate all five slip systems would be

$$\sigma \geq \frac{\tau_c}{M_i}$$

Multiplying this equation by Eq. 8.60 gives us

$$\sigma \, d\epsilon = \sigma \sum_{i=1}^{5} M_i \, d\gamma_i \geq \sum_{i=1}^{5} \tau_c \, d\gamma_i$$

and

$$\sigma \, d\epsilon = \sigma \sum_{i=1}^{5} M_i^1 \, d\gamma_i^1 \geq \sum_{i=1}^{5} \tau_c \, d\gamma_i^1$$

Since the sum of $d\gamma_i$ is lower than the sum of $d\gamma_i^1$, we have

$$\Sigma\tau_c \, d\gamma_i \leq \Sigma\tau_c \, d\gamma_i^1$$

These products provide the deformation work. Therefore, Taylor's criterion is equivalent to saying that the combination of slip systems that will be operative will be the one minimizing the deformation energy (or work). Taylor[68] averaged the value of M over all orientations of the tensile axis, after obtaining for each orientation the set of five M_i giving the least shear. For a FCC crystal, he found that $1/\overline{M} = 3.06$ and

$$\sigma_0 = \frac{\tau_c}{M} = 3.06\tau_c$$

In BCC metals, the slip direction is well established: $\langle 111 \rangle$. However, the slip plane can be $\{1\bar{1}0\}$, $\{11\bar{2}\}$, or $\{12\bar{3}\}$. If deformation is such that the slip direction is crystallographically prescribed, but the slip plane is that of maximum resolved shear stress, the term *pencil glide* is used to describe the situation. This is a common situation in BCC metals, and accounts for the wavyness of the slip markings. For this situation it is found that

$$\frac{1}{M} \simeq 2.75$$

Bishop and Hill[69] used a slightly different approach. They chose the combination of five independent slip systems based on the principle of maximum external work. This name is slightly misleading; the principle comes from a theory of plasticity that derives its name from the fact that the work executed corresponds to the maximum of a certain expression, meaning physically that the flow stress is reached (but not surpassed) in all slip systems involved. It was later shown[50,70] that both Taylor's[68] and Bishop and Hill's[69] methods are equivalent.

Taylor[68] compared the results of a stress–strain curve experimentally obtained for aluminum with a curve calculated from a monocrystal with tensile axis in the center of the stereographic triangle. The correlation was rather poor, and he was able to improve it remarkably by taking a single crystal with a tensile axis oriented along [100] or [110]. This is due to the fact that these orientations produce polyslip. Chin and co-workers[71,72] used computers for the Taylor analysis in the determination of the active slip system for axysymmetric flow in the cases where slip occurred on $\{110\}$ $\langle 111 \rangle$, $\{112\}$ $\langle 111 \rangle$, and $\{123\}$ $\langle 111 \rangle$ systems as well as combinations of these.

The Taylor–Bishop–Hill theory neglects several effects. The most important is the effect of the grain size on the yield stress and work hardening. Grain size is known to have a consistent effect on these parameters. This subject is treated separately in Chapter 14. Another effect is elastic anisotropy. If we calculate the Young moduli of an FCC metal along different orientations, we find remarkable differences. For instance, we have, for nickel (see Section 1.12.4):

$$E[100] = 137 \text{ GPa}$$
$$E[110] = 233 \text{ GPa}$$
$$E[111] = 303 \text{ GPa}$$

Young's modulus in the [111] direction is twice the one in [100] direction, and this will set up stress gradients and concentrations in the material during extension. This will, in turn, generate inhomogeneities in the plastic deformation.

As one can see in Fig. 8.35, for pure nickel a Taylor-corrected curve for $\langle 111 \rangle$ closely matches the polycrystalline curve for large grain sizes; on the other hand, the specimen with a grain size of 2 μm is not well represented by the Taylor model. Thompson,[73,74] applying and modifying Ashby's[75,76] con-

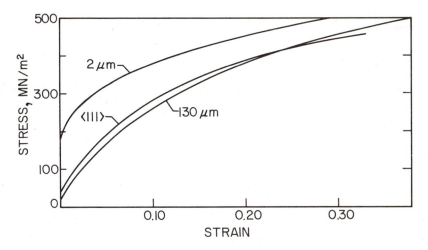

Figure 8.35 Stress–strain curves for pure nickel, as 2-μm and 130-μm grain-size polycrystalline and as ⟨111⟩-oriented single crystal. (Reprinted with permission from [74], p. 111.)

cepts of inhomogeneous plastic flow, was able to develop a model that predicted a grain size dependence of work hardening.

EXERCISES

8.1. An FCC monocrystal (nickel) is being sheared by $\gamma_{12} = 0.1$. Assuming that the dislocation density is equal to 10^8 cm^{-2} and that it remains constant, what is the average distance each dislocation will have to move? If the shear strain rate is 10^{-4} s^{-1}, what is the mean velocity of the dislocations?

8.2. Using the plot of velocity versus stress for nickel, determine the stress required to move the dislocations in Exercise 8.1. If the strain rate were increased to 10^{-3} s^{-1}, what would be the percentual increase in stress required?

8.3. The stress axis in an FCC crystal makes angles of 31° and 62° with the normal to the slip plane and with the slip direction, respectively. The applied stress is 10 MN/m².
  (a) Determine the resolved stress in the shear plane.
  (b) Is it larger when the angles are 45° and 32°, respectively?
  (c) Determine, using a stereographic projection, the resolved stresses on the other slip systems.

8.4. Magnesium oxide is cubic (NaCl structure). The slip planes and directions are {110} and ⟨110⟩, respectively. Along which directions, if any, can a tensile (or compressive) stress be applied without producing slip?

8.5. A Cu monocrystal (FCC) of 10 cm length is pulled in tension. The stress axis is [$\bar{1}23$].
  (a) Which is the stress system with highest resolved shear stress?
  (b) If the extension of the crystal continues until a second slip system becomes operational, what will this system be?
  (c) What rotation will be required to activate the second system?
  (d) How much longitudinal strain is required to activate this second system?

8.6. The flow stress varies with strain rate; one equation that has been used to express this dependence is

$$\sigma = c \dot{\epsilon}^{m'}{}_{\epsilon,T}$$

where m' is the strain-rate sensitivity. It is generally lower than 0.1. Some metals, known as superplastic, can undergo elongations of up to 1000% in uniaxial tension. Assuming that these tests are performed at a uniform velocity of the cross-head, will the metals have a very high or a very low value of m'? Explain, in terms of the formation and inhibition of the neck.

8.7. W. G. Johnston and J. J. Gilman [*J. Appl. Phys.*, *30* (1959) 129] experimentally determined the relationship between dislocation velocity and applied stress:

$$\nu = K\sigma^m$$

Assuming that the mobile dislocation density does not depend on their velocity, obtain a relationship between m and m' (from Exercise 8.6).

8.8. The following results were obtained in an ambient temperature tensile test, for an aluminum monocrystal having a cross-sectional area of 9 mm² and a stress axis making angles of 27° with [100], 24.5° with [110], and 29.5° with [111].

Load (N)	Length (cm)
0	10.000
12.40	10.005
14.30	10.040
16.34	10.100
18.15	10.150
21.10	10.180
23.60	10.200
26.65	10.220

(a) Plot the results in terms of true stress versus true strain.
(b) Determine the resolved shear stress on the system that will slip first.
(c) Determine the longitudinal strain at the end of the easy glide stage (when a second slip system becomes operative).

8.9. Take a stereographic triangle for a cubic metal. Indicate, if the FCC slip systems are operative, the number of slip systems having the same Schmid factor if the stress axis is
(a) [111]
(b) [110]
(c) [100]
(d) [123]
Use the stereographic projection to show this.

8.10. A copper bicrystal is composed of two monocrystals separated by a coherent twin boundary (111). This bicrystal is being compressed in a homogeneous upset test in such a way that the twin boundary is perpendicular to machine plates. The compression direction is the same for both crystals: [134].
(a) Is this crystal isoaxial?
(b) Is deformation compatible or incompatible?

8.11. A monocrystal (diameter 4 mm, length 100 mm) is being pulled in tension.

    (a) What is the elongation undergone by the specimen if 1000 dislocations on slip planes making 45° with the tension axis cross the specimen completely? $b = 0.25$ nm.

    (b) What would the elongation be if all dislocations that exist in the crystal (10^6 cm^{-2}) were ejected by the applied stress? Assume a homogeneous dislocation distribution. Assume that the crystal is FCC and all the dislocations are in the same slip system.

8.12. A long crystal with a square cross section (1×1 cm) is bent to form a semicircle with radius $R = 25$ cm.

    (a) Determine the total number of dislocations generated if it is assumed that all bending is accommodated by edge dislocations.

    (b) Determine the dislocation density ($b = 0.3$ nm).

REFERENCES

[1] J.D. Eshelby, F.C. Frank, and F.R.N. Nabarro, *Phil. Mag.*, 42 (1951) 351.

[2] J.J. Gilman, *Micromechanics of Flow in Solids*, McGraw-Hill, New York, 1969, p. 195.

[3] J.E. Dorn, J. Mitchell, and F. Hansen, *Exp. Mech.*, No. 5 (1965), 353.

[4] J.J. Gilman and W.G. Johnston, *Dislocations and Mechanical Properties of Crystals*, Wiley, New York, 1957, p. 116.

[5] W.G. Johnston and J.J. Gilman, *J. Appl. Phys.*, 33 (1959) 129.

[6] D.F. Stein and J.R. Low, Jr., *J. Appl. Phys.*, 32 (1960) 362.

[7] H. Ney, R. Labusch, and P. Haasen, *Acta Met.*, 25 (1977) 1257.

[8] H.W. Schadler, *Acta Met.*, 12 (1964) 861.

[9] A.R. Chaudhuri, J.R. Patel, and L.G. Rubin, *J. Appl. Phys.*, 33 (1962) 2736.

[10] J.S. Erickson, *J. Appl. Phys.*, 33 (1962) 2499.

[11] E.Y. Gutmanas, E.M. Nadgornyi, and A.V. Stepanov, *Sov. Phys.-Solid State*, 5 (1963) 743.

[12] R.W. Rohde and C.H. Pitt, *J. Appl. Phys.*, 38 (1967) 876.

[13] D.P. Pope, T. Vreeland, Jr., and D.S. Wood, *J. Appl. Phys.*, 38 (1967) 4011.

[14] W.F. Greenman, T. Vreeland, Jr., and D.S. Wood, *J. Appl. Phys.*, 38 (1967) 3595.

[15] T. Suzuki and T. Ishii, *Suppl. Trans. Jap. Inst. Met.*, 9 (1968) 687.

[16] R.C. Blish II and T. Vreeland, Jr., *J. Appl. Phys.*, 40 (1969) 884.

[17] J.A. Gorman, D.S. Wood, and T. Vreeland, Jr., *J. Appl. Phys.*, 40 (1969) 833.

[18] V.R. Parameswaran and J. Weertman, *Met. Trans.*, 2 (1971) 1233.

[19] V.R. Parameswaran, N. Urabe, and J. Weertman, *J. Appl. Phys.*, 43 (1972) 2982.

[20] S. Glasstone, K.J. Laidler, and H. Eyring, *The Theory of Rate Processes*, McGraw-Hill, New York, 1941.

[21] J.J. Gilman, cited in [2], p. 179.

[22] H.L. Preckel and H. Conrad, *Acta Met.*, 15 (1967) 955.

[23] J. Cotner and J. Weertman, *Disc. Faraday Soc.*, No. 38 (1964) 225.

[24] J. Friedel, *Dislocations*, Pergamon/Addison-Wesley, Elmsford, N.Y./Reading, Mass., 1964, p. 63.

[25] J.D. Eshelby, *Proc. Phys. Soc. (London)*, *A12* (1949) 307.

[26] J. Weertman, in *Response of Metals to High Velocity Deformation*, AIME Proc., P.G. Shewmon and V.F. Zackay (eds.), Interscience, New York, 1961, p. 205.

[27] J.J. Gilman, cited in [2], p. 169.

[28] D. Kuhlmann–Wilsdorf, in *Physical Metallurgy*, 2nd ed., R.W. Cahn (ed.), Elsevier/North-Holland, New York, 1970, p. 787.

[29] J. Weertman and J.R. Weertman, in *Dislocations in Solids*, Vol. 3, F.R.N. Nabarro (ed.), Elsevier/North-Holland, New York, 1980, p. 1.

[30] J.D. Eshelby, *Proc. Phys. Soc. (London)*, *B69* (1956) 1013.

[31] J. Weertman, *J. Appl. Phys.*, *38* (1967) 5293.

[32] J.H. Weiner and M. Pear, *Phil. Mag.*, *31* (1975) 679.

[33] Y.Y. Earmme and J.H. Weiner, *J. Appl. Phys.*, *48* (1977) 3317.

[34] P.P. Gillis, J.J. Gilman, and J.W. Taylor, *Phil. Mag.*, *20* (1969) 279.

[35] A.V. Granato, in *Metallurgical Effects at High Strain Rates*, AIME Proc., R.W. Rohde, B.M. Butcher, J.R. Holland, and C.H. Karnes (eds.), Plenum Press, New York, 1973, p. 255.

[36] P.P. Gillis and J. Kratochvil, *Phil. Mag.*, *21* (1970) 425.

[37] R. Becker, *Z. Phys.*, *26* (1925) 919.

[38] A. Seeger, *Z. Naturforsch.*, *9A* (1954) 758, 819, 856.

[39] A. Seeger, *Phil. Mag.*, *1* (1956) 651.

[40] A. Seeger and P. Schiller, *Acta Met.*, *101* (1962) 348.

[41] H. Conrad and H. Wiedersich, *Acta Met.*, *8* (1960) 128.

[42] H. Conrad, *J. Metals*, *16* (1964) 582.

[43] H. Conrad, *Mater. Sci. Eng.*, *6* (1970) 265.

[44] A.S. Krausz and H. Eyring, *Deformation Kinetics*, Wiley, New York, 1975, p. 156.

[45] C.S. Barrett and T.B. Massalski, *The Structure of Metals*, 3rd ed., McGraw-Hill, New York, 1966, p. 39.

[46] E. Schmid and G. Siebel, *Z. Elektrochem.*, *37* (1931) 447.

[47] E. Schmid and W. Boas, *Kristalplasticzitat*, Springer-Verlag, Berlin; *Plasticity of Crystals*, F.A. Hughes and Co., 1950.

[48] D.C. Jillson, *Trans. AIME*, *188* (1950) 1129.

[49] P. Haasen, *Phil. Mag.*, *3* (1958) 384.

[50] G.Y. Chin, in "The Role of Preferred Orientation in Plastic Deformation," *Inhomogeneities of Plastic Deformation*, ASM, Metals Park, Ohio, 1973, p. 83–111.

[51] L.M. Clarebrough and M.E. Hargreaves, *Prog. Mater. Sci.*, *8* (1959) 1.

[52] J. Diehl, *Z. Metallk.*, *47* (1956) 331.

[53] F.R.N. Nabarro, Z.S. Basinsky, and D.B. Holt, *Adv. Phys.*, *13* (1964) 193.

[54] J.W. Christian, *Proc. Second International Conf. on Strength of Metals and Alloys*, ASM, Metals Park, Ohio, 1970, p. 31.

[55] W.J. McGregor Tegart, *Elements of Mechanical Metallurgy*, Macmillan, New York, 1966, p. 120.

[56] W.E. Nixon, M.H. Massey, and J.W. Mitchell, *Acta Met.*, *27* (1979) 943.

[57] S. Miyazaki, K. Shibata, and H. Fujita, *Acta Met.*, *27* (1979) 855.

[58] U.F. Kocks, *Met. Trans.*, *1* (1976) 1121.

[59] W. Boas and M.E. Hargreaves, *Proc. Roy. Soc. (London)*, *A196* (1948) 89.

[60] R. von Mises, *Z. Angew. Math. Mech.*, *8* (1928) 161.

[61] G.W. Groves and A. Kelly, *Phil. Mag.*, *8* (1963) 876.

[62] J.D. Livingston and B. Chalmers, *Acta Met.*, *5* (1957) 322.

[63] C. Elbaum, *Trans. TMS-AIME*, *218* (1960) 444.

[64] J.J. Hauser and B. Chalmers, *Acta Met.*, *9* (1961) 802.

[65] R.E. Hook and J.P. Hirth, *Acta Met.*, *15* (1967) 535.

[66] J.P. Hirth, *Met. Trans.*, *3* (1972) 3047.

[67] G. Sachs, *Z. Ver. Dtsch. Ing.*, *72* (1928) 734.

[68] G.I. Taylor, *J. Inst. Metals*, *62* (1938) 307.

[69] J.F. Bishop and R. Hill, *Phil. Mag.*, *42* (1951) 414, 1298.

[70] G.Y. Chin, E.A. Nesbitt, and A.J. Williams, *Acta Met.*, *141* (1966) 467.

[71] G.Y. Chin and W.L. Mammel, *Trans. TMS-AIME*, *239* (1967) 1400.

[72] G.Y. Chin, W.F. Hosford, and D.R. Menford, *Proc. Roy. Soc. (London)*, *A309* (1969) 433.

[73] A.W. Thompson, *Acta Met.*, *25* (1977) 83.

[74] A.W. Thompson, in *Work Hardening in Tension and Fatigue*, AIME, New York, 1977, p. 111.

[75] M.F. Ashby, *Phil. Mag.*, *21* (1970) 399.

[76] M.F. Ashby, in *Strengthening Methods in Crystals*, A. Kelly and R.B. Nicholson (eds.), Wiley, New York, 1971, p. 137.

SUGGESTED READING

Dislocation Dynamics

GILMAN, J.J., *Micromechanics of Flow in Solids*, McGraw-Hill, New York, 1969.

ROHDE, R.W., B.M. BUTCHER, J.R. HOLLAND, and C.H. KARNES (eds.), *Metallurgical Effects at High Strain Rates*, Plenum Press, New York, 1973.

VREELAND, T., JR., "Dislocation Velocity Measurements," in *Techniques of Metals Research*, R.F. Bunshah (ed.), Wiley-Interscience, New York, 1968, Chap. 12.

WEERTMAN, J., and J.R. WEERTMAN, in *Dislocations in Solids*, Vol. 3, F.R.N. Nabarro (ed.), Elsevier/North-Holland, New York, 1979, p. 1.

Geometry of Deformation

GIL SEVILLANO, J., P. VAN HOUTTE, and E. AERNOUDT, *Large Strain Work Hardening and Textures*, Progress in Materials Science, Vol. 25, J.W. Christian, P. Haasen, and T.B. Massalski (eds.), Pergamon Press, Elmsford, N.Y., 1981, p. 69.

HIRTH, J.P., "The Influence of Grain Boundaries on Mechanical Properties," *Met. Trans.*, *3* (1972) 3047.

HONEYCOMBE, R.W.K., *The Plastic Deformation of Metals*, St. Martin's Press, New York, 1968.

Inhomogeneities of Plastic Deformation, ASM, Metals Park, Ohio, 1970.

KOCKS, U.F., "The Relation Between Polycrystal Deformation and Single-Crystal Deformation," *Met. Trans.*, *1* (1970) 1121.

MCLEAN, D., "Mechanical Behavior of Metals," *Proc. 1971 Int. Conf. Mech. Behavior Mater.*, Society of Materials Science, Japan, 1972, p. 93.

Part III

STRENGTHENING
MECHANISMS

Metallurgists have devised—or fortuitously discovered—many ways of strengthening metals. The ultimate objective of bringing the real strength to coincide with the theoretical strength of metals can be achieved by (1) eliminating all dislocations, or (2) creating as many and as powerful barriers for dislocation motion as possible. The latter route is the one chosen in most strengthening mechanisms. The approach used here is to present, in a quantitative way, the theories that explain the response of metals being strengthened by the different mechanisms. The multiplicity of behavior and compositions of commercial alloys is so great that they are used only to illustrate the effects discussed. Chapter 9 discusses the various work-hardening theories, ranging chronologically from Taylor's to Kuhlmann–Wilsdorf's. The shapes of the plastic range of polycrystalline stress–strain curves are mathematically expressed. Solid-solution hardening is due to the interactions between dislocations and solute atoms; these interactions are analyzed in Chapter 10, together with the various mechanical effects they are responsible for. Chapter 11 presents the major theoretical concepts proposed to explain the mechanical properties of metals containing second phases. These are Mott's theory of hardening by precipitates, Orowan's theory of precipitate bypass by dislocations, and the concepts of inhomogeneous plastic flow leading to statistically stored and geometrically necessary dislocations advanced by Ashby. Chapter 12 presents composite materials and the methods used to calculate their mechanical properties from the properties of the matrix and fibers. The martensitic transformation and its strengthening effect are presented in Chapter 13, which also discusses the shape-memory effect. The yield stress of metals is significantly affected by grain size; the various proposals accounting for this effect are presented in Chapter 15. Thermomechanical processing, dual-phase steels, radiation hardening, shock hardening, order hardening, and texture are other microstructural ways of increasing the strength of metals; they are treated sequentially in Chapter 15. Thermomechanical processing, dual-phase steels, and shock hardening are not strengthening mechanisms in a rigorous sense; several mechanisms contribute to the strengthening, and "strengthening processes" would be a better term.

Chapter 9

WORK HARDENING

COLD WORKING - applying a stress that exceeds the y.s. while deforming the material to a more useful shape.

‡9.1 INTRODUCTION

The relaxation times for the molecular processes in gases and in a majority of liquids are so short that they are almost always in a well-defined state of complete equilibrium. Consequently, the structure of a gas or liquid does not depend on its past history. In contrast, the relaxation times for some of the significant atomic processes in crystals are so long that a state of equilibrium is rarely, if ever, achieved. It is for this reason that the metals show the generally desirable characteristic of work hardening with strain, also called strain hardening.

In fact, hardening by plastic deformation (rolling, drawing, etc.) is one of the most important methods of strengthening metals, in general. Certain metals, in particular (e.g., copper), do not have many precipitation hardening systems but are ductile and can be appreciably hardened by cold working. If the relaxation times were short, the structure would return almost immediately to its state of equilibrium and a constant stress for plastic deformation would result independent of the extent of deformation. This is shown in Fig. 9.1 as the elastic ideally plastic solid. However, when a real crystalline solid is deformed plastically, it turns more resistant to deformation and a greater stress is required for additional deformation (Fig. 9.1). This is called work hardening. Basically, the hardening in a crystalline material occurs because these materials deform plastically by movement of dislocations and these interact directly among themselves and with other imperfections, or indirectly with the internal stress field (short range or long range) of various imperfections and obstacles. These interactions lead to a reduction in the mean mobility of dislocations which is, thus,

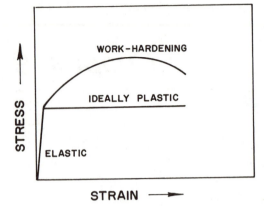

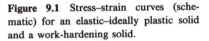

Figure 9.1 Stress–strain curves (schematic) for an elastic–ideally plastic solid and a work-hardening solid.

accompanied by the necessity of a greater stress for accomplishing further movement of dislocations (i.e., with continuing plastic deformation we need to apply an ever greater stress for further plastic deformation): hence the phenomenon of work hardening.

Many theories have been advanced to explain the phenomenon of work hardening. The most important and difficult part in the attempt to predict the work-hardening behavior is to determine how the density and distribution of dislocations vary with the plastic strain. The problem is that stress is a state function in the thermodynamic sense (i.e., it depends only on its position, not how it got there). Plastic strain, on the other hand, is a path function of its position (i.e., it depends on the actual path traversed in reaching a certain strain value). In other words, it is history dependent. Thus, the presence or absence of dislocations and their distributions can tell us nothing about how a certain amount of strain was accumulated in the crystal because we do not know the path that dislocations traversed to obtain that strain. Thus, one constructs models that recreate the processes by means of which the various dislocation configurations emerge and try to correlate them with the configurations observed experimentally. Both the density and the distribution of dislocations are very sensitive functions of the crystal structure, stacking fault energy, temperature, and rate of deformation. In view of all this it is not surprising that a unique theory of work hardening that would explain all the aspects does not exist. In fact, Cottrell has observed that work hardening was the first problem that dislocation theory tried to solve and will be the last one to be solved.

In what follows, we first briefly review the various phenomenological attempts at describing the stress–strain curves, and then try to get a general view of the situation as it exists today in terms of the various work-hardening theories.

9.2 TAYLOR'S THEORY[1]

This is one of the oldest theories. At the time when this theory was postulated (1934), the stress–strain curve for metallic crystals such as aluminum was considered to be parabolic [the stress–strain curve consisting of three stages was un-

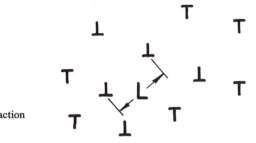

Figure 9.2 Taylor model of interaction among dislocations in a crystal.

known (Section 9.4)]. This being so, Taylor proposed a model that would predict the parabolic curve. The principal idea, which, incidentally, still is used in one form or another by modern theories, was that the dislocations, on moving, elastically interact with other dislocations in the crystal and become trapped. These trapped dislocations give rise to internal stresses that, generally, increase the stress necessary for deformation (i.e., the flow stress).

Let l be the average distance that a dislocation moves before it is stopped. Let ρ be the dislocation density after a certain strain; then the shear strain γ is given by (see Section 8.3)

$$\gamma = K\rho bl \tag{9.1}$$

where K is an orientation-dependent factor and b is the Burgers vector.

Taylor considered only edge dislocations and assumed that the dislocation distribution was uniform, and thus the separation between dislocations, L, will be equal to $\rho^{-1/2}$ (Fig. 9.2). The effective internal stress, τ, as a result of these interactions among dislocations, is the stress necessary to force two dislocations past each other and can be written as

$$\tau = \frac{kGb}{L} \tag{9.2}$$

where k is a constant, L the distance between dislocations and G the shear modulus.

Or, considering that $L = \rho^{-1/2}$,

$$\tau = kGb\sqrt{\rho} \tag{9.3}$$

From 9.1 and 9.3, we get

$$\tau = kGb\sqrt{\frac{\gamma}{Kbl}} = k'G\sqrt{\frac{\gamma}{l}}$$

$$\tau \; \alpha \; \sqrt{\gamma} \tag{9.4}$$

Equation 9.4 is a parabolic relation between stress τ and strain γ. It describes, approximately, the behavior of many materials at large deformations.

Among the criticisms of the Taylor theory, one may include:

1. Such regular configurations of dislocations are rarely observed in cold-worked crystals.

2. Screw dislocations are not involved, and thus the cross-slip is excluded.

3. Two dislocations on neighboring planes may be trapped due to the stress fields of each other, and thus become incapable of moving independent of each other. But the pair of dislocations may be pushed by a third dislocation.

4. We know now that stress–strain curves for hexagonal crystals as well as the stage II of cubic crystals are linear (Section 9.4). Taylor's theory does not explain this linear hardening.

5. The parabolic relation derives from the supposition of uniform distribution of deformed regions inside the crystal. In fact, the distribution is not uniform and, experimentally, we observe slip bands, cells, and other non-uniform arrangements.

9.3 MOTT'S THEORY[2,3]

Mott's theory considers groups of piled-up dislocations, in lieu of individual dislocations, as the sources of internal stress. As was seen earlier (Section 8.1), a dislocation pile-up containing n dislocations (each of Burgers vector, b) can be regarded as a superdislocation with a Burgers vector of nb, for the purposes of computing its stress field at large distances. Mott's theory is nothing but a modification of Taylor's theory. It permits a nonuniform configuration of dislocations. The essential features of Mott's theory may be enumerated as follows:

1. The dislocations are piled up against Lomer–Cottrell barriers and thus are trapped.

2. The piled-up groups of dislocations act as superdislocations of Burgers vector nb.

3. The dislocation distribution is nonuniform, which is in accord with the experimental observation of slip bands in real crystals.

Consider two active slip planes, each of length $2L$ and separated by a distance d. The dislocations pile up against obstacles (mainly Lomer–Cottrell barriers) at each extremity of the slip plane (Fig. 9.3). Each group of piled-up dislocations can be considered to be a superdislocation of Burgers vector nb. The density of superdislocations is

$$\rho = \frac{1}{Ld}$$

and the average distance between the superdislocations is $\sqrt{Ld}$. Thus, each superdislocation is subjected to an average internal stress due to its neighbors,

$$\tau_i = \frac{G(nb)}{2\pi\sqrt{Ld}} \tag{9.5}$$

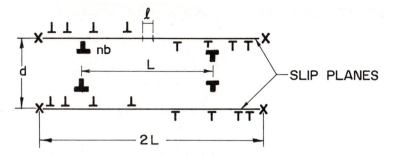

Figure 9.3 Mott model of hardening.

Now, the plastic deformation is given by (ignoring the orientation constant, K)

$$\gamma = \rho(nb)L$$

$$= \frac{1}{Ld}(nb)L = \frac{nb}{d} \qquad (9.6)$$

From 9.5 and 9.6, we have

$$\tau_i = \frac{Gnb}{2\pi\sqrt{L\dfrac{nb}{\gamma}}}$$

or

$$\tau_i = \frac{G}{2\pi}\sqrt{\frac{nb\gamma}{L}} \qquad (9.7)$$

Equation 9.7 is a parabolic relation between stress and strain, similar to that of Taylor. L is still an unknown parameter. To eliminate L, Mott assumed that dislocation loops were generated dynamically by a Frank–Read source (Chap. 6); that is, the kinetic energy of the loop is sufficient for the source to continue operating until the back stress of the dislocations generated becomes equal to the applied stress on the Frank–Read source. We need, for the operation of a Frank–Read source of length l, a stress

$$\tau = \frac{Gb}{l} \,\dagger$$

† This equation is obtained from eq. 6.28, by inserting Peach-Koehler's equation (eq. 6.30) in its simplest form ($F = \tau b$) and making the following simplification in the equation for the dislocation self-energy (line tension; eq. 6.22):

$$T = U \approx \frac{Gb^2}{2}$$

Hence, $\tau b = F = \dfrac{Gb^2}{2R}$ and $\tau = \dfrac{Gb}{l}$

The opposing stress at the source, in a slip plane of half length L, of a superdislocation nb, is (Eq. 8.1)

$$\tau_B = \frac{Gnb}{2\pi L}$$

or

$$n = \frac{2\pi \tau_B L}{Gb}$$

Now, $\tau (= Gb/l)$ is the applied stress on the source and, for stopping of dynamic operation of source, we have, $\tau_B = \tau$, that is,

$$\frac{Gb}{l} = \frac{Gnb}{2\pi L}$$

or

$$\frac{n}{L} = \frac{2\pi}{l} \tag{9.8}$$

Thus, from Eqs. 9.7 and 9.8, we obtain

$$\tau_i = G\sqrt{\frac{b\gamma}{2\pi l}} \tag{9.9}$$

The following comments can be made about Mott's model of work hardening:

1. When the length l of the Frank–Read source is constant, we obtain a parabolic relation between the stress and strain.
2. For $l \simeq 10^{-6}$ m, which is reasonable for Al, the calculated and the observed yield stress are of the same order of magnitude (2 MPa).
3. The relation between stress and strain is independent of L, the obstacle spacing. This is consistent with the parabolic hardening of the crystal.
4. We need about a thousand dislocations per pile-up, an extremely large number.

9.4 MORE RECENT THEORIES

Before presenting the current theoretical situation, it is in order to present the present experimental situation in the light of which the modern theories have been formulated. Figure 9.4 shows the generic shear stress–shear strain curve for FCC metal single crystals. The curve can be divided, conveniently, into three regions I, II, and III: θ_I, θ_{II}, and θ_{III} being the respective inclinations $(d\tau/d\gamma)$ of the regions. In what follows, we shall describe the salient points of the various stages and the current theories that try to explain them.

The stage I starts after elastic deformation at the critical stress τ_0. This

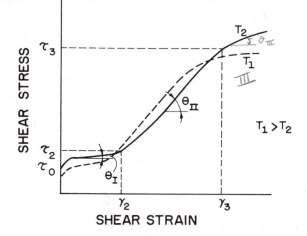

Figure 9.4 Generic shear stress–shear strain curves for FCC single crystals for two different temperatures.

stage, called "easy glide," is a linear region of low strain hardening rate. θ_I is approximately one-tenth of θ_{II}. This stage is characterized by long slip lines (100 to 1000 μm),† straight and uniformly spaced (10 to 100 nm). Stage I does not exist in polycrystals. The extent of this stage depends strongly on the crystal orientation. The strain at the end of stage I (γ_2) has a maximum value when the crystal orientation is located in the center of the standard stereographic triangle (see Section 8.6.2 footnote). The end of stage I is considered to be the start of secondary slip. A more detailed description is provided in Section 8.6.4.

The stage II or linear hardening stage has the following important characteristics:

1. A linear hardening regime with high θ_{II}.
2. $\theta_{II}/G \approx 1/300$. This parameter is relatively constant for a great majority of metals (maximum variation being by a factor of about 2).

θ_{II} is approximately equal to $10\theta_I$ and is, relatively, independent of temperature, although it has a significant effect on the extent of stage II.

3. Rapid and linear hardening is associated with the secondary slip system activity, although a great part of deformation occurs by slip on the primary system.

4. Experimental observations on dislocation distributions, by means of transmission electron microscopy, X-ray topography, or chemical etch pitting, indicate that the arrangement of dislocations is very heterogeneous, a majority of dislocations being concentrated in small regions and separated by regions

---

† We adopt the nomenclature used by A. Seeger (in J. C. Fischer, W. G. Johnston, and T. Vreeland, *Dislocations and Mechanical Properties of Crystals*, John Wiley, New York, 1957, p. 243). *Slip lines* are the "elementary structure" of slip and can only be observed in the electron microscope. In the optical microscope, one observes *slip bands*; they occur at the higher strains and are made up of clusters of slip lines. On the other hand, *slip markings* are observed as steps at the specimen surface.

of low dislocation density. It has also been observed that the secondary dislocation density in the single crystal is comparable to the primary dislocation density throughout the stage II, in spite of the fact that macroscopic plastic deformation on secondary systems is very small.

5. Although the primary system is more active, the secondary systems are also operating. Dislocations on active systems interact with those on other systems and form barriers to the continuing movement of dislocations. The barriers increase with increasing deformation and the slip lines on the surface become ever shorter, closely and not so regularly spaced. It has been shown by means of transmission electron microscopy that the slip-line length, L_s, is inversely related to the shear strain in stage II, that is,

$$L_s = \frac{\Lambda}{\gamma - \gamma_2}$$

where Λ is a constant ($\approx 4 \times 10^{-4}$ cm), γ is the total shear strain and γ_2 is the shear strain at the start of stage II.

6. The flow stress τ is proportional to the square root of the dislocation density ρ. If τ_0 is the stress necessary for moving a dislocation in the absence of other dislocations, then

$$\tau = \tau_0 + \alpha G b \sqrt{\rho}$$

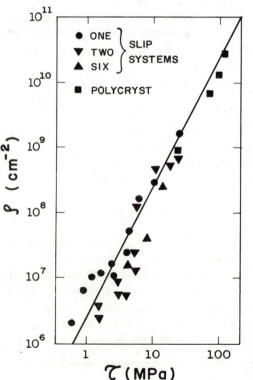

Figure 9.5 Average dislocation density ρ as a function of resolved shear stress τ for copper. (Adapted with permission from [19], p. 427.)

where α is a constant with a value between 0.3 and 0.6. This relationship has been observed to be valid for a majority of the cases (Fig. 9.5).

The stage III starts at the point (γ_3, τ_3) and the hardening rate decreases continuously; the change in stress can be described by the equation

$$\tau = \theta_{III}(\gamma - \gamma')^{1/2}$$

where γ' is a constant. This third stage of hardening is parabolic. The stress τ_3 decreases exponentially with increase in temperature and the θ_{III} shows a similar tendency. This stage is characterized by the presence of wavy slip lines as the slip is not restricted to one unique plane.

The low stacking fault energy (SFE) metals show all three work-hardening stages at room temperature. The high-SFE metals show only stages II and III. For example, aluminum (a high-SFE metal) shows, at room temperature and above, a poorly developed stage II and stage III. We must test aluminum at 77 K to see all three stages. Copper, on the other hand, exhibits all three stages at room temperature. This difference between copper and aluminum is due to their different stacking fault energies which affect the width of extended dislocations in them.

The flow stress of a metal is affected by a change of temperature or strain rate. It is convenient to think of the flow (shear) stress (τ) as consisting of two parts (see equation 8.52):

$$\tau = \tau^* + \tau_G$$

where τ^* is the temperature-dependent part of flow stress while τ_G is the temperature-independent part of flow stress. τ_G depends on temperature only through variation of shear modulus with temperature (see Section 8.5). One can conveniently study the relative importance of τ^* and τ_G by measuring the flow stress dependence on temperature or strain rate (i.e., one measures the change in flow stress on changing the temperature or strain rate).

The ratio τ^*/τ_G is observed to be approximately unity in stage I. In stage II, this ratio decreases and reaches a constant value of about 0.1. This constancy of the τ^*/τ_G ratio is often referred to as the Cottrell–Stokes law.

9.4.1 Long-Range Theories

Stage I.[4,5] According to the school of thought belonging to Seeger and his co-workers, the stage I hardening is due to long-range interactions between well-spaced dislocations. These dislocation loops are blocked by unspecified obstacles, all on the primary system. One assumes that there exist N dislocation sources per unit volume and that this number N remains constant throughout stage I. When a stress τ is reached, each of these sources has emitted n dislocation loops. It was observed experimentally that each loop moves a distance L in the crystal, much greater than the distance between the dislocations on adjacent planes which are a distance d apart (i.e., $L \gg d$). The stress level being low in stage I, the dislocations generated by a given source are well spaced such that one can consider their stress field to be that of a superdislocation.

An increment of stress, $\delta\tau$, will result in an increase, δn, in the number of loops. This will give an increment of strain, $\delta\gamma$, which may be written as

$$\delta\gamma = bNL^2\,\delta n \qquad (9.10)$$

where $L^2\delta n$ is the incremental area δA, that is, an increment in shear strain $\delta\gamma = bN\,\delta A$, a variation of Eq. 9.1. Square loops were assumed.

Now, a volume of L^2d contains one source. Thus, in a unit volume one will have $1/L^2d$ number of sources. This, by definition, is N:

$$N = \frac{1}{L^2d}$$

Therefore, Eq. 9.10 becomes

$$\delta\gamma = \frac{b\,\delta n}{d} \qquad (9.11)$$

The generation of δn new dislocations will also increase the back stress, τ_B, at the dislocation sources by

$$\delta\tau_B = \frac{Gb}{2\pi L}\,\delta n \qquad (9.12)$$

When the increment $\delta\tau_B$ becomes equal to the increment $\delta\tau$, no more dislocations are generated.

Then, from Eqs. 9.11 and 9.12, we have

$$\delta\tau = \frac{Gb}{2\pi L}\frac{\delta\gamma d}{b}$$

or

$$\theta_I = \frac{\delta\tau}{\delta\gamma} = \frac{G}{2\pi}\frac{d}{L} \qquad (9.13)$$

A more detailed treatment gives the expression

$$\theta_I = \frac{8G}{9\pi}\left(\frac{d}{L}\right)^{3/4}$$

Experimentally it was observed that this expression described the hardening behavior in stage I of Cu, Ni, Ni-Co, and Zn.

In terms of criticism of this theory one can point out the following points:

1. It ignores completely the existence of dislocation dipoles and their effect on hardening.
2. Nothing is said about the obstacles responsible for blocking the dislocations.

Stage II.[5,6,7] The increase in hardening in the beginning of stage II is due to the decrease in the mean slip distance as a result of slip activity on the secondary systems. It is thought that this activation of the secondary slip systems gives rise to the formation of obstacles whose density increases throughout the stage II. The only important function of the secondary activity is to furnish such barriers. It is assumed that the number of dislocation sources operating increases continuously.

It is thought that the principal obstacles that block the slip are Lomer–Cottrell barriers. The dislocations pile up against these barriers and give rise to long-range internal stresses which control the flow stress.

Consider, for the sake of convenience, square loops of dislocations generated by a Frank–Read source. Let the length of a slip line (equal to a side of the square loop) be L. Then, an increment of deformation, $d\gamma$, will be given by

$$d\gamma = L^2 bn \, dN \tag{9.14}$$

where dN is the increase in the number of sources, per unit volume, in operation during the incremental deformation, $d\gamma$; n is the number of dislocations per slip line (i.e., per source); and b is the Burgers vector. The reader should note that N was considered to be constant in stage I, while in stage II it is considered to increase continuously.

The relation between the slip-line length, L, and deformation, γ, is introduced as an experimental fact, that is,

$$L = \frac{\Lambda}{\gamma} \tag{9.15}$$

where Λ is a constant to be determined experimentally and γ is measured from the starting point of stage II, that is, γ equal to $(\gamma - \gamma_2)$ in Eq. 19.15.

Substituting 9.15 into 9.14 and integrating (assuming n constant), we obtain

$$\gamma^3 = 3\Lambda^2 bnN \tag{9.16}$$

We assume again that the dislocation groups piled up against obstacles can be considered to be superdislocations with Burgers vector equal to nb and with an average distance l of separation, where

$$l = \frac{1}{\sqrt{NL}} \tag{9.17}$$

Now, the stress required to move a dislocation past another one, on a parallel slip plane, is proportional to the Burgers vector of a dislocation and inversely proportional to the separation distance. The superdislocation will furnish an effective barrier to the dislocation motion and this strength of barrier will increase with diminishing spacing of superdislocations.

The stress required to move a dislocation through the stress fields is given by an expression of the form

$$\tau = \frac{\alpha Gbn}{2\pi l} \tag{9.18}$$

where α is a constant of the order of unity. From Eqs. 9.17 and 9.18, we get

$$\tau = \frac{\alpha Gbn}{2\pi} (NL)^{1/2} \tag{9.19}$$

Manipulating Eqs. 9.15, 9.16, and 9.17, one gets a linear hardening expression and the work-hardening rate is given by

$$\frac{d\tau}{d\gamma} = \theta_{II} = \frac{\alpha G}{2\pi}\left(\frac{nb}{3\Lambda}\right)^{1/2} \tag{9.20}$$

For $n \sim 25$ and $\Lambda \sim 5 \times 10^{-4}$ cm we find $\theta_{II}/G \simeq \frac{1}{300}$ for all FCC metals.

The following comments may be made on this model of stage II:

1. Although dislocation pile-ups have been observed in thin-film samples in transmission electron microscopes, they are by no means universal characteristic of deformed FCC metals. The samples are generally examined in an unloaded state. Thus, one would expect, on unloading, some reverse plastic deformation as the dislocations should run back to the source. Also, dislocation pile-ups have been observed primarily at grain boundaries.

2. One would expect that the high stresses at the heads of dislocation pile-ups will be relaxed by slip on other systems.

3. This model does not explain the influence of temperature on the flow stress.

9.4.2 Other Theories

Other researchers have proposed other mechanisms as responsible for this phenomenon of resistance to deformation. Hirsch[8] and Mott,[9] for example, at one time advanced the theory based on the presence of sessile jogs in the dislocations as being responsible for hardening. Others[10,11] have pointed out that elastic stresses that must be overcome in forest dislocation intersections are the rate-controlling factor in the work-hardening rate.

Kuhlmann–Wilsdorf[12] proposed the so-called "mesh-length" theory, which is based on the stress necessary for dislocation bowing. In stage I the dislocations multiply into certain restricted regions and penetrate into regions as yet substantially free of mobile dislocations, until a quasiuniform distribution of dislocations is obtained. The only resistance to deformation is the dislocation line tension. Thus, hardening occurs due to the fact that free segments of dislocations become ever smaller. The stage II starts when there are no more "virgin" areas left for penetration by new dislocations. The stress required to bow segments of dislocation is responsible for a great part of the stage II hardening. We discuss later (Section 9.5) Kuhlmann–Wilsdorf's recent work on hardening due to cell structure.

Schoeck[13] has criticized the mesh-length theory because of the absence of a real mechanism to block the dislocations, as the line tension is not a blocking mechanism. If one segment of a dislocation bends, other neighboring segments must be locked in some way. Such a blocking mechanism may be provided by (1) presence of internal stresses due to the presence of dislocations on parallel slip planes (theories of Taylor, Mott, and Seeger), or (2) formation of large jogs in screw dislocations. It is worth emphasizing that dislocation movement becomes ever more difficult due to the generation of an ever larger number of obstacles, and the only real obstacle suggested until now is the Lomer–Cottrell lock. The various theories based on the intersection with the forest dislocations and jogs are basically inadmissible because of the absence of an effective blocking mechanism.

Another important criticism of the various models or theories of work hardening is that all of them require specific distributions of dislocations; these specific distributions exist rarely. The distributions, in practice, are mixed and complex, varying from metal to metal and even with orientation in a given metal. For example, dislocations in aluminum tend to align in the form of cells, whereas in copper, they tend to form tangles. However, there does not exist such a great difference in the hardening behavior of these two metals. Thus, it would appear that a specific distribution of dislocations is not crucial, and that strain hardening in practice is a statistical result of some factor which remains more or less the same for various distributions.

9.4.3 Theories of Stage III

This stage is characterized by a decreasing work-hardening rate. Wavy slip lines appear because the slip is not restricted to one particular slip plane.

With increasing deformation in stage II, the average dislocation density increases and the dislocation distribution is ever more characterized by regions (cells) free of dislocations, surrounded by dislocation cell walls. The dislocation configuration acquired in stage II is practically independent of the temperature and the strain rate. At higher temperatures and stresses, dislocations are capable of undergoing processes that are not possible at lower temperatures and stresses. For example, dislocations can have the capacity to circumvent barriers developed in stage II; dislocations of opposite sign may mutually annihilate (e.g., the partials of a Lomer–Cottrell can recombine) and thus reduce the internal stress field.

Seeger[7] considers the phenomenon of cross-slip to overcome the barriers, which is, perhaps, most in conformity with the experimental observations. That is, the dynamic recovery occurs by cross-slip of thermally activated dislocations and the phenomenon is facilitated by the high applied stress. The stage III is also characterized by thick slip bands, which are nothing but groups of finely spaced slip lines. It is thought that cross-slip is responsible for the formation of these bands.

A perfect dislocation in an FCC metal dissociates into two partial dislocations with a ribbon of stacking fault between them. In this way, a screw dislocation is confined to a plane containing the stacking-fault ribbon. It could move to another slip plane only if it were possible to transfer the stacking fault ribbon to the new plane. Figure 9.6 shows a method for accomplishing this. A constriction is formed in the dislocation along the line of intersection of the primary and the cross-slip planes. At the constriction, the dislocation dissociates itself again into partials, but this time on the cross-slip plane.

We expect the work-hardening rate to decrease with the onset of cross-slip of screw dislocations, as this would relieve the stress concentrations due to pile-ups. The phenomenon of cross-slip requires an expenditure of energy

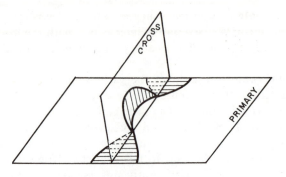

Figure 9.6 Model of cross-slip.

(i.e., it is an activated process). It is difficult to occur at a low level of stresses, and its operation is aided by high temperatures. Thus, one expects that the stress necessary at the start of stage III, τ_3, would depend on temperature. Such, indeed, is the case in practice. τ_3 increases with a decrease in temperature.[14]

The start of stage III is also markedly dependent on the stacking-fault energy (SFE) of the metal. Metals with relatively low SFE, for example, brasses, bronzes, and austenitic steels, have a rather wide stacking-fault ribbon and consequently, need a higher activation energy for the cross-slip. This is so because for cross-slip to occur in these metals it is necessary to form a constriction over a wide ribbon of the stacking fault, in order to have a certain length of perfect dislocation. Thus, in low-SFE metals and alloys cross-slip will be difficult at normal stress levels. This makes it difficult for the screw dislocations to change their slip plane. The dislocation density is high and the transition from stage II to stage III is retarded. Aluminum, on the other hand, has a higher SFE. Thus, the stress necessary for cross-slip in aluminum, at a given temperature, is much less than that in, say, Cu or brass. In fact, aluminum shows the stage III at extremely low applied stress levels and as the dislocations in Al are not dissociated, the cross-slip can occur easily. A method of calculating the separation between the partial dislocations is presented in Section 6.7.1.

9.5 KUHLMANN–WILSDORF THEORY OF WORK HARDENING[15-17]

Kuhlmann–Wilsdorf and co-workers have advanced another theory of work hardening based on dislocation cell structure observed in deformed metals. We shall present the fundamentals of this theory. Kuhlmann–Wilsdorf (hereafter simply K–W) does not agree with the conventional three-stage work-hardening curve as shown in Fig. 9.4. She points out that the stage II is not linear in a simple manner, particularly for not very symmetrical orientations, and that stage III follows stage II without discontinuity in either shear stress, τ, or slope, $d\tau/d\gamma$. Figure 9.7 shows this for Cu single crystals. This condition of

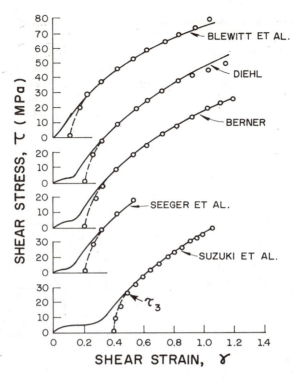

Figure 9.7 Experimental shear stress–shear strain curves for copper (from different authors as indicated) with Bell's calculated points. (Reprinted with permission from [18], p. 112.)

continuity in τ, as well as in $d\tau/d\gamma$ at τ_3, implies that neither ρ, the dislocation density, nor L, the dislocation mean free path, change discontinuously at τ_3. This argument, according to her, rules out the theory due to Seeger and co-workers because according to the Seeger school of thought (see Section 9.4.3) stage III signifies the onset of cross-slip, which allows mutual annihilation of previously trapped dislocations and circumventing obstacles by piled-up screw dislocations, and so on. In Fig. 9.7 the experimental curves are interpolated through points calculated by Bell. According to Bell, the experimental stress–strain curves follow the following relations:

$$\tau = \theta_{\mathrm{III}}\sqrt{\gamma - \gamma_b} = \sqrt{2\theta_{\mathrm{II}}\tau_3(\gamma - \gamma_b)} \tag{9.21}$$

with

$$\theta_{\mathrm{III}} = \sqrt{2\theta_{\mathrm{II}}\tau_3} \tag{9.22}$$

The symbols in these relations have familiar meanings. The only new symbol is γ_b, which is the strain at the midpoint of stage II counting it from its linear extrapolation back to the γ axis. Thus, the stage III is the parabola centering on the strain axis and joining stage II at τ_3 without discontinuity in τ or $d\tau/d\gamma$.

K–W focuses attention on cell structure, commonly found in deformed metals in which dislocations are quite mobile and where one has available "several mutually independent noncoplanar slip systems." The detailed dislocation configurations are indeed quite complex. So K–W and co-workers have resorted to some simplifications based on experimental work, the most important one being that cell walls are low-angle boundaries. The average cell wall misorientation is ≲10°, which is within the range of misorientations associated with low-angle boundaries. K–W and co-workers propose development of dislocation cells from dislocation arrays in the following manner (Fig. 9.8). The basic building block is a so-called terminated loop hexapole composed of edge and screw dislocations [Fig. 9.8(a)] which has stable threefold nodes. To minimize stored

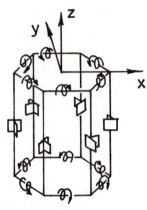

(a)

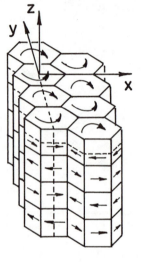

(b)

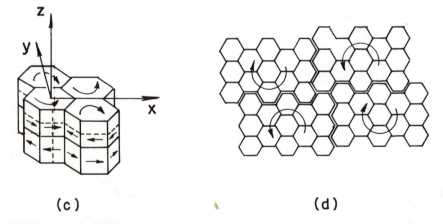

(c)

(d)

Figure 9.8 Development of dislocation cells: (a) terminated loop hexapole consisting of edge and screw dislocations; (b) cell combinations to minimize net rotational stress; (c) two rows of hexapolar units of alternating rotations; (d) 13-unit groups with alternating rotation forming an extended array. (Adapted with permission from [17], pp. 13, 19, 20.)

energy, adjacent cells of opposite rotations (i.e., tilt and twist cell walls of equal and opposite dislocation density) are combined [Fig. 9.8(b)]. Figure 9.8(c) shows this canceling of rotational stress fields extended to two adjoining rows of stacked cells. This pattern, however, cannot be extended beyond two rows. Accepting a larger net rotation per group leads to the extendable arrays shown in Fig. 9.8(d). These arrays have parallel misorientation axes and equal and opposite senses of rotation. K–W recognizes that these are idealized structures, but they are in accord with a number TEM observations on cell structures, having low long-range and short-range stresses.

Consider a cell structure of a deformed metal where the cell walls occupy a volume fraction f of the material, with an average link length $\bar{l}$. Then the overall dislocation density in the material is

$$\rho = \frac{fm}{\bar{l}^2} \qquad (9.23)$$

where m is a numerical factor that increases with the irregularity of the network; K–W estimates it to be about 5.

The instantaneous flow stress, τ, has an important and crucial component, according to this theory, called "source stress," τ_s, which is the stress required to make dislocation links grow beyond their critical bowing-out radius. The bowing-out dislocation links, called "source links," will be the longest links of the network. K–W[15] shows that the ratio of source link length, l_s, to average link length, $\bar{l}$, called n is $\simeq 3$.

The total flow stress is the sum of this source stress, and the sum of all other contributions, τ_0. That is,

$$\tau = \tau_0 + \tau_s \simeq \tau_0 + \frac{Gb}{2\pi l_s} \ln \frac{l_s}{b}$$

or

$$\tau = \tau_0 + \frac{Gb}{2\pi n\bar{l}} \ln \frac{n\bar{l}}{b} \qquad (9.24)$$

The logarithmic term in this equation results from more refined estimates of dislocation line tension, specifically, taking into account its orientation dependence (see Section 6.4.3). Substituting for $\bar{l}$ from Eq. 9.23, we may rewrite Eq. 9.24 as

$$\tau = \tau_0 + \frac{Gb\sqrt{\rho}}{2\pi n\sqrt{mf}} \ln \frac{n\sqrt{mf}}{b\sqrt{\rho}} \qquad (9.25)$$

This is nothing but our familiar relation

$$\tau = \tau_0 + \alpha Gb\sqrt{\rho}$$

where

$$\alpha = \frac{\ln(n\sqrt{mf}/b\sqrt{\rho})}{2\pi n\sqrt{mf}} \qquad (9.26)$$

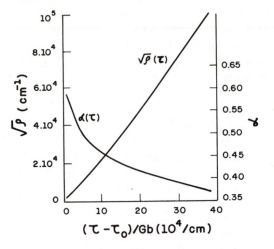

Figure 9.9 Variation of parameter α and dislocation density ρ with the temperature-independent part of flow stress $(\tau - \tau_0)$, for $n \simeq 3$ and $mf \simeq 1$ (see Eq. 9.26). (Reprinted with permission from [17], p. 6.)

Figure 9.9 shows the variation of α and ρ with $(\tau - \tau_0)$, the temperature-independent part of the flow stress, for $n \simeq 3$, $f = \frac{1}{5}$, $m \simeq 5$, and thus $mf \simeq 1$. One notes the square-root dependence of the flow stress on the dislocation density over a range of ρ (from $\sim 10^6$ to 10^{10} cm^{-2}) and $\alpha = 0.47 \pm 0.1$.

Now, to obtain the shear stress–shear strain relationship in stage II, we proceed as follows. Assume that the cells have roughly circular section with a diameter $L = g\bar{l}$. Then the generation of one primary dislocation loop per unit volume by one link bowing out on the primary slip plane will result in a shear strain increment

$$dy = \frac{\pi}{4} L^2 b \tag{9.27}$$

and an increment in dislocation density

$$d\rho = \beta \pi L \tag{9.28}$$

where β represents the fraction of the initial primary dislocation length that has been added to the crystal during this bowing-out process. β will, generally, be smaller than unity due to mutual dislocation annihilation.

From Eq. 9.23 we can write

$$L = g\bar{l} = g \frac{\sqrt{fm}}{\sqrt{\rho}} \tag{9.29}$$

Then, we can rewrite Eq. 9.27 using Eqs. 9.28 and 9.29 as follows:

$$dy = \frac{\pi}{4} bLL = \frac{\pi}{4} b \frac{g\sqrt{fm}}{\sqrt{\rho}} \frac{d\rho}{\beta \pi}$$

or

$$dy = \frac{bg\sqrt{fm}}{4\beta} \frac{d\rho}{\sqrt{\rho}} \tag{9.30}$$

Integrating Eq. 9.30, we get

$$\gamma - \gamma_0 = \frac{bg\sqrt{fm}}{2\beta} \sqrt{\rho} \tag{9.31}$$

Substituting for ρ from Eq. 9.31 into Eq. 9.25, we get the shear stress–shear strain relation for stage II:

$$\tau - \tau_0 = \left[\frac{G\beta}{\pi nmfg} \ln \frac{nmfg}{2\beta(\gamma - \gamma_0)} \right] (\gamma - \gamma_0) \tag{9.32}$$

$$= \theta_{II}(\gamma - \gamma_0)$$

Equation 9.32 does not strictly describe a linear stage II. The dependence of θ_{II} on γ is very small. It would be independent of γ if the logarithmic factor were neglected in Eq. 9.32. The value of θ_{II} cannot be determined without knowing the value of g. In the case of $g = 200$ and $\beta = 1$, $\theta_{II}/G \simeq \frac{1}{300}$.

Stage III. According to K–W, stage III arises because the cells cannot go on shrinking indefinitely. Dislocation cells shrink not by network motion but by subdivision of the largest cells. Here K–W makes the following assumption (lacking experimental confirmation but logically compelling, according to her): The process of subdivision of cells ceases when the cells become so small that no more new cell walls are nucleated because of the negligible chance encounters of slip dislocations. Let the critical cell diameter at which the subdivision ceases be L_c. Then, from Eqs. 9.29 and 9.30, we get

$$dy = \frac{bL_c}{4\beta} d\rho \tag{9.33}$$

Integrating, we get

$$\gamma - \gamma_b = \frac{bL_c}{4\beta} \rho$$

or

$$\sqrt{\rho} = \left(\frac{4\beta(\gamma - \gamma_b)}{bL_c} \right)^{1/2} \tag{9.34}$$

where γ_b is the strain at the midpoint of stage II, counting it from its linear extrapolation back to strain axis.

Substituting for ρ from Eq. 9.34 in Eq. 9.25, we obtain for stage III,

$$\tau - \tau_0 = \left[\frac{G\sqrt{\beta b}}{\pi n \sqrt{mfL_c}} \ln \frac{n\sqrt{mfL_c}}{\sqrt{4\beta b(\gamma - \gamma_b)}} \right] \sqrt{\gamma - \gamma_b}$$

or

$$\tau - \tau_0 = \theta_{III}\sqrt{\gamma - \gamma_b} \tag{9.35}$$

One notes that this equation also does not represent a parabolic stage III. However, Kuhlmann–Wilsdorf points out that the deviation from Bell's parabola is small, within the accuracy of interpolation. In fact, K–W points out that Eq. 9.35 satisfies Eqs. 9.21 and 9.22. As to the deviation from linearity in stage II, K–W points out that computed stress–strain curves based on Eqs. 9.32 and 9.35 are close to experimental curves in the literature.

Despite the rather satisfactory accounting of many aspects of work hardening by this theory it is questionable to attribute work hardening to stress necessary for the initiation of new dislocation sources of appropriate length. The presence of long-range stresses is also completely ignored.

9.6 THE INTRACTABLE PROBLEM[19]

Experimental results show that strain hardening is related in a singular way to the dislocation density after deformation. This is indicated by the general validity of a linear relation between τ and $\sqrt{\rho}$ (Fig. 9.6). In a very general way, one may write

$$\tau = \tau_0 + \alpha Gb\sqrt{\rho} \tag{9.36}$$

where τ_0 is the shear stress required to move a dislocation in the absence of others. In order to obtain a relationship between shear strain, γ, and the dislocation density, ρ, one needs an expression of the form

$$d\gamma = b\pi[r(\gamma)]^2 \, dN \tag{9.37}$$

where r is the radius of the dislocation loop generated by a source at the point where it was stopped by some obstacle. Thus, r should be a function of the accumulated strain. dN is the number of loops emitted per unit volume during the strain increment, $d\gamma$. Then the total perimeter of the new dislocation loops will be equal to the total increment in the length of dislocation,

$$d\rho = 2\pi r(\gamma) \, dN \tag{9.38}$$

From 9.37 and 9.38, we get

$$d\rho = \frac{2}{br(\gamma)} \, d\gamma \tag{9.39}$$

What makes this problem of hardening intractable is the nonexistence of a theoretical treatment that allows one to derive the function $r(\gamma)$. This is understandable in view of the remarks made in the introductory section of this chapter about the nature of plastic strain in metals: the plastic deformation is history dependent. However, should r be independent of γ, then the equations (9.36) and (9.39) will give a parabolic stress–strain curve (characterized by FCC polycrystals). Should $r(\gamma) \propto 1/\gamma$, we get a linear relationship between stress and strain (stage II of FCC single crystals).

‡9.7 WORK HARDENING IN POLYCRYSTALS

In preceding sections work hardening in single crystals was attributed to the interaction of dislocations with other dislocations and barriers which impede the dislocation motion through the crystal lattice. In polycrystals, too, this basic idea remains valid. However, due to the mutual interference of neighboring grains and the problem of compatible deformations among the adjacent grains (see Section 8.7), multiple slip occurs rather easily and, consequently, there occurs an appreciable work hardening right in the beginning of straining.

In a manner similar to that in the single crystals, primary dislocations interact with secondary dislocations, giving rise to dislocation dipoles and loops which result in local dislocation tangles and eventually a three-dimensional network of subboundaries. Generally, the size of these cells decreases with increasing strain. The structural differences between one metal and another are mainly in the sharpness of these cell boundaries. In BCC metals and high-SFE FCC metals like Al, the dislocation tangles rearrange into a well-defined cell structure while in the low-SFE metals or alloys (e.g., brasses, bronzes, austenitic steels, etc.) where the cross-slip is rather difficult and the dislocations are extended, the sharp subboundaries do not form even at very large strains.

The plastic deformation and the consequent work hardening results in an increase in the dislocation density. An annealed metal, for example, will have about 10^6 to 10^8 dislocations per cm^2, while a plastically cold-worked metal may contain up to 10^{12} dislocations per cm^2. The relationship between the flow stress and the dislocation density is the same as that observed for single crystals, that is,

$$\tau = \tau_0 + \alpha G b \sqrt{\rho}$$

A similar relationship exists between the flow stress, τ, and the mean grain diameter or the cell diameter. This linear relationship between τ and the square root of the grain size (D) is known as the Hall–Petch relationship and is dealt with in Chapter 14. A comparison of the stress–strain curves of a few polycrystalline metals is shown in Fig. 9.10. One notes that stages I and II are absent in the polycrystalline metals. Metals with an HCP structure (e.g., zinc) possess a much smaller number of slip systems than cubic metals (e.g., aluminum). Therefore, HCP metals are not capable of deforming easily under the restrictions imposed by a polycrystalline structure. Other deformation processes, such as mechanical twinning (see Section 7.4.1), which require higher stresses, must operate. Consequently, HCP polycrystals are much more difficult to deform and show a much smaller ductility than the appropriately oriented monocrystals of the same metals (Fig. 9.10).

‡9.7.1 Stress–Strain Curves of Polycrystals

The most commonly used equation during many decades to describe the stress–strain curve of a polycrystalline metal is due to Ludwik[20]

$$\sigma - \sigma_0 = K\epsilon^n \tag{9.40}$$

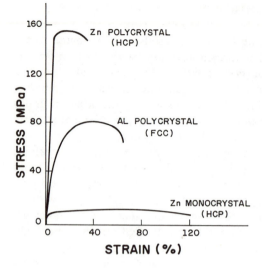

Figure 9.10 Comparison of stress–strain curves (engineering) for single and poly-crystals.

and to Hollomon[21]

$$\sigma = K\epsilon^n \tag{9.41}$$

where K is a constant and the exponent n depends on material, temperature, and strain. The exponent n varies generally between 0.2 and 0.5, while the value of K varies between $G/100$ and $G/1000$, G being the shear modulus. In Eq. 9.40, ϵ is the true plastic strain while in Eq. 9.41, ϵ is true total strain. Equations 9.40 and 9.41 describe parabolic behavior. However, such a description is valid only in a narrow stretch of the stress–strain curve, for the following reasons. First, these equations predict a slope of infinity for $\epsilon = 0$ which does not conform with the experimental facts. Second, these equations imply that $\sigma \rightarrow \infty$ when $\epsilon \rightarrow \infty$. We know that this is not correct and that, experimentally, at higher strains there occurs a saturation of stress.

Voce[22] introduced a much different equation,

$$\frac{\sigma_\infty - \sigma}{\sigma_\infty - \sigma_0} = \exp\left(-\frac{\epsilon}{\epsilon_c}\right) \tag{9.42}$$

where σ_∞, σ_0, and ϵ_c are empirical parameters which depend on the material, temperature, and strain rate. This equation says that the stress reaches exponentially an asymptotic value of σ_∞ at higher strain values. Furthermore, it gives a finite slope to the stress–strain curve at $\epsilon = 0$ or $\sigma = \sigma_0$. As it happens, Voce's equation does not describe the entire stress–strain curve, only the region of high strains.[23] Reed–Hill et al.[24] have critically analyzed Hollomon's, Ludwik's, and Swift's equations and found serious drawbacks in all of them. The parameters in these equations depend on the choice of the initial stress and/or strain. For instance, if one prestrains a material, one would affect K in Ludwik's equation. Guimarães and Valeriano Alves[25] suggested that one should eliminate the effect of choosing the origin (prestrain) by defining an equation of the type

$$\sigma - \sigma_0 = \phi(\epsilon - \epsilon_0)$$

where σ_0 and ϵ_0 are adjustable parameters and ϕ is a functional relationship. Applying this concept to Voce's equation we arrive, after rearranging terms, at

$$\sigma - \sigma_0 = \exp\left[-k(\epsilon - \epsilon_0)\right]$$

This equation is more general; k, a work-hardening parameter, would not be dependent on prestraining the specimen.

In case the material is subjected to a triaxial state of stress, we can substitute the uniaxial stresses and strains by effective stresses and strains (see Sections 1.10.7 and 1.11.4). Thus, Eq. 9.41 becomes

$$\sigma_{\text{eff}} = K\epsilon_{\text{eff}}^{n} \tag{9.43}$$

It is clear that Eq. 9.43 reduces to Eq. 9.41 for the uniaxial stress if one takes the Poisson ratio to be equal to 0.5 (which is reasonable for plastic deformation, as the volume remains constant).

The fact that some equations reasonably approximate the stress–strain curves does not imply that they are capable of describing the curves in a physically satisfactory way. There are two reasons for this: (1) in the different positions of stress–strain curves, different microscopic processes predominate; and (2) plastic deformation is a complex physical entity, and it is worth reminding the reader again that plastic deformation depends on the path taken and is not a thermodynamic state function. This is to say that the accumulated plastic deformation is not uniquely related to the dislocation structure of the material. This being so, it is not very likely that simple expressions could be derived for the stress–strain curves in which the parameters would have definite physical significance.

9.8 WORK SOFTENING

In spite of the fact that metals generally strain harden with plastic deformation, there are some notable exceptions that deserve a brief description here. It is the phenomenon of work softening, discovered by Polakowsky[26] and studied by many others.[27-30] This phenomenon has been observed in a variety of metals; however, for the sake of simplicity, we shall describe here the specific case of nickel. This phenomenon should not be confused with the softening due to adiabatic heating of the samples at high strain rates or with the softening that can occur in single crystals when the rotation suffered by the slip planes during deformation is such that there occurs a reduction in the Schmid factor.

Work softening is a manifestation of dynamic recovery. It can be produced in the following ways:

1. Prestrain a sample of nickel at a very low temperature (e.g., 77 K).[29] This creates a substructure that is characterized by dislocation tangles and cells with thick walls. One can obtain a similar structure by deforming the sample

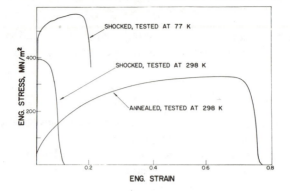

Figure 9.11 Phenomenon of work softening in preshocked Ni. (Reprinted with permission from [30], p. 1583.)

at ambient temperature but at an exceptionally high strain rate: 10^9 s^{-1}. This can be attained by application of shock waves (see Section 15.4).

2. Stress the sample at or about ambient temperature. Once the yield stress is attained, the nominal stress decreases instead of increasing. After a certain deformation, the equilibrium can be established with the formation of a plateau. Finally, the nominal stress returns to its descending trajectory. As the work-hardening coefficient is zero or negative, the necking starts soon after the yield stress is attained. This can be explained by Considère's criterion (see Section 16.1.4). Figure 9.11 shows the work softening in nickel that has been preshocked to a 20-GPa pressure.[30] Also shown are the nominal stress strain curves for annealed Ni, tested at ambient temperature, and preshocked Ni tested at 77 K. The softening with continuing deformation is due to the dynamic recovery of the sample during the test. The substructure generated by shock (dislocation tangles and thick cell walls) is not characteristic of the conventional deformation at ambient temperature (thin-walled cells). Thus, when the test sample reaches the yield stress, at ambient temperature, the dislocations start moving, reorganizing themselves into configurations that are more stable under conditions imposed by the tensile test. This movement is a case of dynamic recovery. Therefore, instead of generating new dislocations with the consequent hardening, there simply occurs a reorganization of the already existing dislocations. On the other hand, the preshocked material does not show softening when tested at 77 K. This is because the shock-generated substructure and the stable substructure at this temperature are similar; in this case, there exist conditions for work hardening to take place.

9.9 DEFORMATION AT LARGE STRAINS[31]

In most metalworking processes the total effective strain imparted to the workpiece is greater than 1. The models described in Sections 9.1 to 9.6 apply to idealized situations in which the total effective strain is not very large. Hence, they apply mostly to the range of strains encountered in the tensile test prior

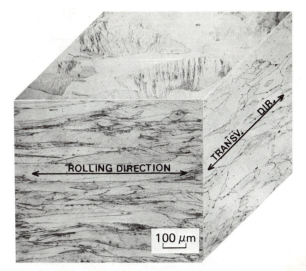

Figure 9.12 Perspective view of microstructure of Nickel-200 cold-rolled to a reduction of thickness of 60%. (Courtesy of D. Jaramillo, New Mexico Institute of Mining and Technology.)

to necking. These strains usually are in the range 0.2 to 0.5. Figure 9.12 shows, as an illustration, three views of a nickel specimen cold-rolled to a reduction in thickness of 60%. This corresponds to an effective strain (see Eq. 1.81) of 0.6. The distortion of the grains in the rolling direction can be clearly seen. They increase in length while decreasing in thickness. Their dimensions in the transverse direction remain unchanged, as one would expect from the condition of plane strain imposed by rolling: the width of the plate is unchanged, and the reduction of thickness is accommodated by an attendant increase in length. The substructures developed during metal deformation processes resemble the idealized models only in the first stages. As the imposed deformation increases, the cell diameters (in medium- and high-stacking-fault-energy alloys) decrease; additionally, the cells become elongated in the general direction of the deformation. The cell walls tend to become progressively sharper as the misorientation between two adjacent cells increases. A cell wall is essentially a low-angle grain boundary, and when the misorientation between adjacent cell walls reaches a certain critical value, we can no longer refer to the boundary as a low-angle grain boundary. The boundary between two cells becomes freer of dislocations, and "subgrains" are formed. This process is called polygonization. This transition from cells to subgrains occurs at different effective strains for different materials: 0.80 to 99.97 for aluminum, 1 to 1.20 for copper. A detailed treatment of the work hardening and formation of texture at large imposed plastic strains is given by Gil Sevillano et al.[31] For metals with low-stacking-fault energies the development of fine lamellar substructure consisting of microtwins, twin bundles, shear bands, and stacking faults is the characteristic feature of high-strain deformation.

The collective behavior of dislocations which undoubtedly is responsible for these deformations is not very well understood. As an illustration, Fig.

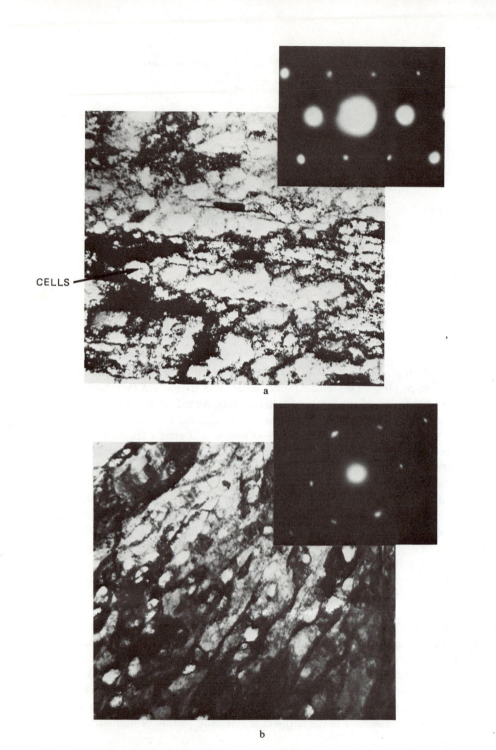

CELLS

a

b

Figure 9.13 Development of substructure of Nickel-200 as a function of plastic deformation by cold rolling: (a) 20% reduction; (b) 40% reduction; (c) 80% reduction. (Continued on next page)

378

SUBGRAIN
BOUNDARIES

Figure 9.13   (Continued)

9.13 shows the changes in substructure observed in nickel rolled at room temperature. At reductions up to 40%, we clearly have a cellular structure. We can
see that at 40% [Fig. 9.13(b)] we already have a large dislocation density. At
80% reduction, we can clearly see that many of the cell walls have disappeared
and are replaced by well-defined boundaries. The observation is made more
difficult because of the large density of dislocations. The electron diffraction
patterns show the effect very well (right-hand corner of photomicrographs).
Up to 40% reductions, the diffraction spots are fairly clear, with little asterism
(elliptical distortion). At 80%, the asterism is very pronounced, and elongated
spots break down into smaller spots, indicating that a distorted grain has broken
down into subgrains, which have relatively little distortion. At 90% reduction,
this effect is still more pronounced. Truckner and Mikkola[32] studied the evolution of substructure and tensile properties in copper cold-rolled and found that
at around 95% reduction in thickness the yield stress started to drop [annealed
value: $\sigma_y = 60$ MPa; σ_y (90%) = 225 MPa]. They attributed this drop to
dynamic recovery (recovery occurring simultaneously with the deformation process). Additional phenomena that take place at large strains are shear instabilities.
These shear bands propagate in a non-crystallographic way (they are described
in Section 2.5.7) and could account for a substantial portion of the total plastic
strain at large deformations.

EXERCISES

9.1. What objections can you point out against the use of thin-film transmission electron microscopy techniques for studying the dislocation behavior in bulk materials? How do the other methods of observing dislocations compare? (See A. Seeger, in *Work Hardening*, TMS-AIME Conf., Vol. 46, 1966, p. 27.)

9.2. (a) The interaction stress, τ, between an obstacle and a dislocation can be related to the distance between the two as follows:

$$\tau \propto \frac{1}{(d)^n}$$

where n is an integer. Plot the graph of τ versus d for (1) $n = 1$ and (2) $n = 2$. In which of the two cases will the interaction be athermal and in which case thermal?

(b) Give three examples each for thermal and athermal obstacles.

9.3. What is the effect of presence of solute atoms on the critical resolved shear stress and the extent of easy glide in an FCC single crystal?

9.4. If we strain an FCC and an HCP single crystal, which of the two will have a larger amount of easy glide, and why?

9.5. Consider dislocations blocked with an average spacing of l in a copper crystal. If the flow stress is controlled by the stress necessary to operate a Frank–Read source, compute the dislocation density ρ in this crystal when it is deformed to a point where the resolved shear stress in the slip plane is 42 MPa. Take $G = 50$ GPa.

9.6. For a copper single crystal, we can write the flow stress as

$$\tau = \frac{1}{2} Gb\sqrt{\rho}$$

where G is the appropriate shear modulus for the orientation, b Burgers vector, and ρ the dislocation density. It was experimentally observed that the dislocation mean free path, L, in this crystal was equal to $100l$, where l is the average dislocation spacing. Show that

(a) The rate of dislocation storage $d\rho$ in a strain interval of $d\gamma$ is given by $d\rho/d\gamma = 10^{-2}/bl$.

(b) The strain hardening rate, $\theta(= d\tau/d\gamma)$, is $G/400$.

9.7. Explain why a metal like lead does not work harden when deformed at room temperature, whereas a metal such as iron would.

9.8. What is the effect of cold work and annealing on the Young's modulus of a metal? Why?

9.9. The stress–strain curve of polycrystalline metals is at times represented by

$$\sigma = K\epsilon^n$$

where σ is the true stress, ϵ the true strain, and K and n are constants. Show that if this equation is obeyed, the strain at necking is equal to n. What is the usual upper value of n? What is the significance of this upper value of n?

9.10. For the single crystal of an FCC metal, the work-hardening rate $d\tau/d\gamma = 0.3$ GPa. Compute the work-hardening rate $d\sigma/d\epsilon$ for a polycrystal of this metal. Take the Taylor factor M_p to be 3.1.

REFERENCES

[1] G.I. Taylor, *Proc. Roy. Soc. (London)*, *A145* (1934) 362.

[2] N.F. Mott, *Proc. Phys. Soc. (London)*, *B64* (1951) 729.

[3] N.F. Mott, *Phil. Mag.*, *43* (1952) 1151.

[4] A. Seeger, H. Kronmüller, S. Mander, and H. Trauble, *Phil. Mag.*, *6* (1961) 939.

[5] A. Seeger, S. Mader, and H. Kronmüller, in *Electron Microscopy and Strength of Crystals*, G. Thomas and J. Washburn (eds.), Wiley-Interscience, New York, 1963, p. 665.

[6] S. Mader, A. Seeger, and H.M. Thieringer, *J. Appl. Phys.*, *34* (1963) 3376.

[7] A. Seeger, in *Dislocations and Mechanical Properties of Crystals*, Wiley, New York, 1957, p. 243.

[8] P.B. Hirsch, *Phil. Mag.*, *7* (1962) 67.

[9] N.F. Mott, *Trans. AIME*, *218* (1960) 962.

[10] W.T. Brydges, *Phil. Mag.*, *15* (1967) 1079.

[11] Z. Basinski and S. Basinski, *Phil. Mag.*, *9* (1964) 51.

[12] D. Kuhlmann–Wilsdorf, *Trans. AIME*, *224* (1962) 1047.

[13] G. Schoeck, *Phil. Mag.*, *9* (1964) 335.

[14] T.E. Mitchell, *Prog. Appl. Mater. Res.*, *6* (1964) 117.

[15] D. Kuhlmann–Wilsdorf, in *Work Hardening*, J.P. Hirth and J. Weertman (eds.), Gordon and Breach, New York, 1968, p. 97.

[16] D. Kuhlmann–Wilsdorf, *Met. Trans.*, *1* (1970) 3173.

[17] D. Kuhlmann–Wilsdorf, in *Work Hardening in Tension and Fatigue*, A.W. Thompson (ed.), TMS-AIME, New York, 1977, p. 1.

[18] J.F. Bell, *Phil. Mag.*, *10* (1964) 107.

[19] H. Wiedersich, *J. Metals*, *16* (1964) 425.

[20] P. Ludwik, *Elemente der Technologischen Mechanik*, Springer, Berlin, 1909, p. 32.

[21] J.H. Hollomon, *Trans. AIME*, *162* (1945) 268.

[22] E. Voce, *J. Inst. Met.*, *74* (1948) 537.

[23] R.W. Swindeman, *J. Eng. Mater. Technol.*, *Trans. ASME*, *97* (1975) 98.

[24] R.E. Reed-Hill, W.R. Cribb, and S.N. Monteiro, *Met. Trans.*, *4* (1973) 2665.

[25] J.R.C. Guimarães and D.L. Valeriano Alves, *Scripta Met.*, *9* (1975) 1147.

[26] N.H. Polakowsky, *J. Iron Steel Inst.*, *169* (1952) 337.

[27] A.H. Cottrell and R.J. Stokes, *Proc. Roy. Soc. (London)*, *A233* (1955) 17.

[28] W.P. Longo and R.E. Reed-Hill, *Scripta Met.*, *4* (1970) 765.

[29] W.P. Longo and R.E. Reed-Hill, *Metallography*, *4* (1974) 181.

[30] M.A. Meyers, *Met. Trans.*, *8A* (1977) 1581.

[31] J. Gil Sevillano, P. van Houtte, and E. Aernoudt, *Prog. Mater. Sci.*, *25* (1981) 69.

[32] W.G. Truckner and D.E. Mikkola, *Met. Trans.*, *8A* (1977) 45.

SUGGESTED READING

BASINSKI, S.J., and Z.S. BASINSKI, "Plastic Deformation and Work Hardening," in *Dislocations in Solids*, F.R.N. Nabarro (ed.), Elsevier/North-Holland, New York, 1979, p. 261.

CLAREBROUGH, L.M., and M.E. HARGREAVES, "Work Hardening of Metals," in *Progress in Metal Physics*, Vol. 8, B. Chalmers and W. Hume–Rothery (eds.), Pergamon Press, N.Y., 1959, p. 1.

COTTRELL, A.H., *Dislocations and Plastic Flow in Crystals*, Oxford University Press, London, 1953.

HIRSCH, P.B. (ed.), *The Physics of Metals*, Vol. 2: *Defects*, Cambridge University Press, Cambridge, 1975.

HIRTH, J.P., and J. LOTHE, *Theory of Dislocations*, McGraw-Hill, New York, 1968.

LAVRENTEV, F.F., *Mater. Sci. Eng.*, *46* (1980) 191.

MITCHELL, T.E., *Prog. Appl. Mater. Res.*, 6 (1964) 117.

SEEGER, A., in *Work Hardening*, TMS-AIME Conf., Vol. 46, 1966, p. 27.

THOMAS, G., and J. WASHBURN (eds.), *Electron Microscopy and Strength of Crystals*, Wiley-Interscience, New York, 1959.

THOMPSON, A.W. (ed.), *Work Hardening in Tension and Fatigue*, TMS-AIME, New York, 1977.

Chapter 10

SOLID SOLUTION HARDENING

‡10.1 INTRODUCTION

Dislocations are moderately mobile in pure metals and plastic deformation occurs by means of dislocation motion (i.e., by shear). A very versatile method of obtaining high strength levels in metals would be to restrict this rather easy motion of dislocations. We saw earlier that grain boundaries (Chapter 8) and stress fields of other dislocation (Chapter 9) can play this role at low temperatures and increase the strength of the material. Another method of restricting the dislocation mobility is to introduce heterogeneity such as solute atoms or precipitates or to add hard particles in a ductile matrix (i.e., dispersion hardening). When the mechanical properties of a solid are modified by the introduction of solute atoms, the resulting strengthening is called solid solution hardening and the alloy is called a solid solution. Figure 10.1 shows the increase in yield stress of steels as a function of the solute content. One notes that carbon and nitrogen have much larger effects than other elements. This will be explained later. In order to analyze the phenomenon of hardening due to the presence of solute atoms, we must consider the increase in the stress necessary to move a dislocation in its slip plane in the presence of these discrete barriers. Conceptually, it is useful and easier to think in terms of an energy of interaction between the dislocation and the barrier (e.g., a solute atom or a precipitate). For a stationary dislocation, the interaction energy U is the change in energy of the system crystal-dislocation when a solvent atom is removed and substituted by a solute atom, in the case of substitutional solutions. Knowing the interaction energy U, the force necessary to move a dislocation a distance dx normal to

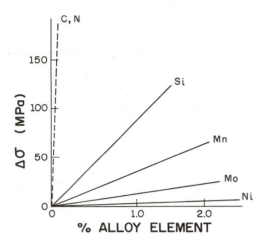

Figure 10.1 Increase in strength $\Delta\sigma$ of steel as a function of solute content. (Adapted with permission from [1], p. 11.)

its length is given by dU/dx. Specifically, in the case of solid solution hardening, this interaction leads to a migration of solute atoms to dislocation where they form an atmosphere around it. This solute atmosphere, called the Cottrell atmosphere, has the effect of locking the dislocation, making it necessary to apply more force to free the dislocation from the atmosphere. This results in the well-known phenomenon of pronounced yield drop in annealed low-carbon steels. A word of caution is in order here. Temperature is an important variable in this solute migration to dislocation. If the temperature is too low, there will be very little solute diffusion to allow a redistribution of solute atoms to dislocations. This solute redistribution, thus, may be thermodynamically expected, but if the temperature is too low, will not occur in a reasonable time. At high enough temperatures ($>0.5T_m$) the mobility of foreign atoms will be much higher than that of dislocations, with the result that they will not restrict dislocation motion. In the range of temperature where solute atoms and dislocations are about equally mobile, there occur strong interactions with dislocations. The serrated stress–strain curve (or the Portevin–Le Châtelier effect) is a manifestation of this.

To evaluate the total interaction energy, we must consider the various contributions. We assume that various contributions to the interaction energy (misfit interaction, electrostatic interaction, chemical interaction, etc.) are mutually independent. This supposition is a reasonable one for dilute solid solutions. For not-so-dilute solid solutions, a statistical treatment becomes necessary. We analyze below the various contributions individually.

10.2 DILATATION MISFIT OR ELASTIC MISFIT INTERACTION ENERGY

We treat the host lattice (matrix) as an isotropic and elastic continuum. The substitution of a regular atom in the matrix lattice by a larger impurity atom is then treated as a classical elasticity problem in which a spherical cavity of radius r_0 is made in the matrix and a sphere of radius R $(R > r_0)$ is lodged in it. The principal effect of such an insertion of substitutional impurity, with a radial misfit of $\epsilon = \Delta r_0/r_0$, is one of a radial displacement, u_r (positive in this case). Thus,

$$R = r_0 + \Delta r_0 = r_0(1 + \epsilon)$$

Starting from the equilibrium equations and the equations linking strains to displacements in spherical coordinates, we can show[2] that the differential equation in this case of spherical symmetry is

$$\frac{d^2 u_r}{dr^2} + \frac{2}{r} \frac{du_r}{dr} - \frac{2u_r}{r^2} = 0 \tag{10.1}$$

where u_r is the radial displacement and r is the distance from the center of the sphere.

Integrating, we find that the displacements should be of the form

$$u_r = \frac{c}{r^2} + c'r \tag{10.2}$$

where c and c' are integration constants. Applying the boundary conditions, we get the displacement of the matrix around a misfitting, spherically symmetrical solute atom. The boundary conditions are

$$u_r = 0 \qquad \text{at } r = \infty \tag{10.3}$$

and

$$u_r = \Delta r_0 \qquad \text{at } r = r_0 \tag{10.4}$$

From Eq. 10.3, it follows that $c' = 0$; otherwise, u_r will be become infinite. Thus, Eq. 10.2 becomes

$$u_r = \frac{c}{r^2} \tag{10.5}$$

Applying now the condition given by Eq. 10.4, we get for $r = r_0$,

$$u_r = \Delta r_0 = \frac{c}{r_0^2}$$

or

$$c = \Delta r_0 r_0^2 \tag{10.6}$$

Thus, from Eqs. 10.5 and 10.6, we obtain

$$u_r = \Delta r_0 \left(\frac{r_0}{r}\right)^2 \tag{10.7}$$

But the radial misfit ϵ is by definition equal to $\Delta r_0/r_0$. Thus,

$$u_r = \epsilon r_0 \left(\frac{r_0}{r}\right)^2 \tag{10.8}$$

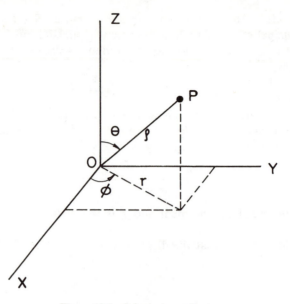

Figure 10.2 Spherical coordinates.

Note that the radial displacement decreases rapidly with increasing r.

We can make use of the symmetry of the problem and use spherical coordinates to represent the displacements in the r, θ, and ϕ directions by u_r, u_θ, and u_ϕ, respectively (Fig. 10.2). For our special case where displacements occur only in the radial directions (i.e., $u_\theta = u_\phi = 0$), it can be shown that the strain components are given by

$$\epsilon_{rr} = \frac{\partial u_r}{\partial r}, \qquad \epsilon_{\phi\phi} = \epsilon_{\theta\theta} = \frac{-\partial u_r}{2\partial r}$$

$$\gamma_{r\theta} = \gamma_{r\phi} = \gamma_{\theta\phi} = 0$$

(10.9)

For our problem of an oversized or undersized sphere in a cavity, we have, from Eqs. 10.8 and 10.9,

$$-\frac{\epsilon_{rr}}{2} = \epsilon_{\theta\theta} = \epsilon_{\phi\phi} = \epsilon \left(\frac{r_0}{r}\right)^3$$

(10.10)

This is a general result and gives the strain components when a sphere is inserted in a larger or smaller cavity. Knowing these displacements, we can compute the total force acting on a dislocation.[3]

10.3 INTERACTIONS OF SOLUTE ATOMS WITH DISLOCATIONS

A dislocation has associated with it a stress field (Chapter 6). Solute atoms, especially when their sizes are too large or too small in relation to the host atom size, are also centers of elastic strain (see Section 10.2). Consequently, the stress fields from these sources can interact and can mutually exert force. This is the elastic interaction due to misfit. Other types of interactions, such as electrical and chemical, are also possible.

10.3.1 Elastic Interactions

In the case of a positive edge dislocation, there is a compressive stress above the slip plane and a tensile stress under it. As a solute atom placed randomly in a crystal produces a stress field around it, this stress field will be minimized if the solute atom were to move to the dislocation. For the case of interstitial atom carbon in iron, the minimum energy position at an edge dislocation is the dilated region near the core. A substitutional atom tends to move to the compressive side if it is smaller than the solvent atom, and, if it is bigger than the solvent atom, it is expected to move to the tensile side.

Substitutional atoms such as Zn in Cu give rise to a completely symmetrical distortion in the lattice which corresponds exactly to the elasticity problem of a ball in a bigger or smaller hole (Section 10.2); that is, the substitutional solute atom acts as a point source of dilatation of spherical symmetry. It is important to note that in view of the spherically symmetric stress fields due to substitutional impurity atoms, they can interact only with defects that have a hydrostatic component in their stress fields, as happens to be the case of an edge dislocation. Screw dislocations, on the other hand, have a stress field of a pure shear character, and, therefore, there is no interaction between screw dislocations and substitutional atoms, such as Zn in Cu.

Interstitial atoms such as carbon or nitrogen in α-iron, however, produce not only a dilatational misfit (in volume), but also induce a tetragonal distortion. Both, carbon and nitrogen, occupy interstitial positions at the face centers and/or the midpoints of the edges of the body-centered cubic structure (Fig. 10.3).

There is evidence indicating that the carbon atoms are located in α-Fe at the midpoints of $\langle 001 \rangle$ edges and that the cubic cell changes into a tetragonal one along the $\langle 001 \rangle$ axis (Section 5.2). The tetragonal distortion thus produced will interact with hydrostatic as well as shear stress fields. The important result of this tetragonal distortion is that the impurity atoms will interact and form atmospheres at both types of dislocations, edge and screw, and lead to a more effective impediment to dislocation movement than that in the case of substitutional atoms (see Fig. 10.1).

Let σ_p be the hydrostatic component of the stress field of a dislocation and let ΔV be the change in volume induced by the introduction of a solute atom of radius $r_0 (1 + \epsilon)$ in a cavity of radius r_0, ϵ being positive. Then, for ϵ very small,

$$\Delta V = 4\pi \epsilon r_0^3$$

The stress field of an edge dislocation in cylindrical coordinates is given by

$$\sigma_{rr} = \sigma_{\theta\theta} = -\frac{Gb}{2\pi(1-\nu)}\frac{\sin\theta}{r}$$

$$\sigma_{zz} = -\frac{\nu Gb}{\pi(1-\nu)}\frac{\sin\theta}{r}$$

$$\sigma_{r\theta} = \frac{Gb}{2\pi(1-\nu)}\frac{\cos\theta}{r}$$

$$\sigma_{\theta z} = \sigma_{zr} = 0$$

and the hydrostatic pressure is, by definition, equal to $-\frac{1}{3}(\sigma_{rr} + \sigma_{\theta\theta} + \sigma_{zz})$. Thus,

$$\sigma_p = \frac{1+\nu}{1-\nu}\frac{Gb}{3\pi}\frac{\sin\theta}{r} \tag{10.11}$$

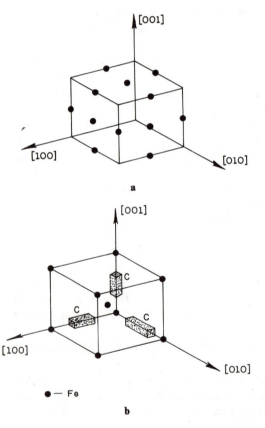

a

b

● — Fe

Figure 10.3 (a) Positions of interstitial atoms in the cube; (b) carbon atom shown as a producer of a tetragonal effect.

To convert this into rectangular coordinates, we use the relations $r = (x^2 + y^2)^{1/2}$ and $\sin\theta = y/(x^2 + y^2)^{1/2}$.

Figure 10.4 shows the coordinates of a solute atom in the strain field of a dislocation. The elastic interaction energy due to misfit, U_{misfit}, for a solute atom at (r, θ) and the origin $(0, 0)$ being at the dislocation, is given by

$$U_{\text{misfit}} = \sigma_p \, \Delta V = \frac{1+v}{1-v} \frac{Gb}{3\pi} \frac{\sin\theta}{r} \, 4\pi \epsilon r_0^3$$

$$= A \frac{\sin\theta}{r} \tag{10.12}$$

where

$$A = \frac{4}{3} \frac{1+v}{1-v} Gb\epsilon r_0^3$$

Equation 10.12 is derived on the basis of linear elasticity theory; thus, it will not be valid at the dislocation core region where linear elasticity does not apply. It is a great omission, as the binding energy U will be a maximum precisely at the dislocation core. Therefore, the reader is forewarned that the interaction energy determined above is only an estimate.

Consider again Eq. 10.12 and Fig. 10.4. U is positive in the region above the slip plane $(0 < \theta < \pi)$ and is negative below the slip plane $(\pi < \theta < 2\pi)$ for large solute atoms $(\Delta V$ positive). This means that a solute atom larger in size than the matrix atom (i.e., ΔV positive) will be repelled by the compressive side of the edge dislocation and will be attracted by the tensile side, as the interaction energy will be negative there. For a solute atom of smaller size than the matrix atom (i.e., ΔV negative), the interaction energy will be negative in the upper part $(0 < \theta < \pi)$ and will be attracted there. In both cases, this migration of solute atoms results in a reduction of the free energy of the system.

The tetragonal strain field around an interstitial atom like C or N in iron can be written as

$$\epsilon_{ij} = \begin{pmatrix} \epsilon_{11} & 0 & 0 \\ 0 & \epsilon_{22} & 0 \\ 0 & 0 & \epsilon_{33} \end{pmatrix}$$

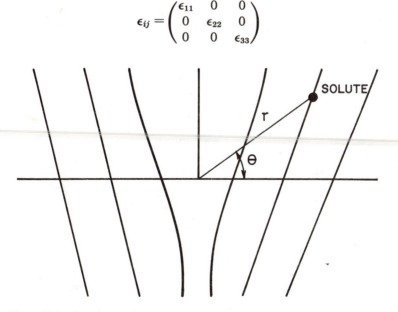

Figure 10.4 Coordinates of a solute atom in the strain field of an edge dislocation.

This tetragonal strain field interacts with both edge and screw dislocations. The interaction energy in this case between a screw dislocation and the strain field due to the interstitial can be written as[4]

$$U_{\text{misfit}} = \epsilon_{ij} \sigma_{ij}^{\text{screw}} \Omega$$

where $\sigma_{ij}^{\text{screw}}$ represents the stress field associated with a screw dislocation and Ω is the specific volume given by $\Delta V = 3\Omega\epsilon$ (ϵ being the misfit parameter). The resultant force exerted by solute atom on dislocation is similar in form to that in the case of a substitutional solute atom (see Section 10.5) but with the misfit parameter replaced by $(\epsilon_{11} - \epsilon_{22})/3$.[4]

A substitutional solute atom producing an isotropic strain field, $\epsilon_{11} = \epsilon_{22}$, does not interact with a screw dislocation. However, for C in α-Fe one has the extensional strain in the [100] direction, $\epsilon_{11} = 0.38$, while the contractional strains along the two orthogonal directions, [010] and [001], are -0.026. Therefore, C in α-Fe hinders all dislocations. This is understandable in terms of the Hume–Rothery solubility rules that require $\epsilon \leq 0.14$ for a finite solubility of the substitutional atoms. However, interstitial atoms with $(\epsilon_{11} - \epsilon_{22})$ as much as 1 show solubility in BCC metals. The reason for this is that metals can accommodate a greater uniaxial distortion than isotropic distortion by solute atoms since the electron energy depends mainly on the specific volume.

The solute atoms thus segregate to the dislocations because this results in a decrease in the free energy. A complete segregation of solute atoms to dislocations will occur under easy conditions for atomic diffusion. In iron, atoms of C and N diffuse easily at ambient temperatures, but in many substitutional alloys, one has to resort to treatments at higher temperatures. The solute concentration, C, at a point where the binding energy is U, can be written[5] as

$$C = C_0 \exp\left(\frac{U}{kT}\right) \tag{10.13}$$

where C_0 is the average concentration, U the elastic interaction energy, k Boltzmann's constant, and T the temperature in K. The concentration C is given without units as it is the ratio of number of sites occupied by solute atoms to the total number of sites. This type of equation would give a Maxwell type of distribution around the dislocation, but in the case of carbon and nitrogen in iron, at ambient temperature, the elastic interaction is very strong (i.e., $U \gg kT$); therefore, the Cottrell atmosphere condenses completely in a line at the atoms along the dislocation core. We can use Eq. 10.13 to estimate the temperature below which this type of condensation occurs. In a completely condensed atmosphere $C = 1$, $T = T_{\text{crit}}$, and $U = U_{\text{max}}$. Then,

$$T_{\text{crit}} = \frac{U_{\text{max}}}{k \ln (1/C_0)}$$

Typically, U_{max} for carbon in iron is 0.08 aJ (0.5 eV) and $C_0 = 10^{-4}$, which gives a T_{crit} of 700 K. The existence of a well-defined yield point in low-carbon steels is explained by this interaction. We treat this phenomenon of well-defined yield point in Section 10.5.

There is another effect, called the Snoek effect, which results because of the anisotropic distortion caused by the carbon atom in the iron lattice.[6] When a uniaxial stress is applied to the iron lattice along a cube direction, this anisotropic distortion causes the carbon atom to change its position. This change of site leads to a reduction in strain energy. On the other hand, a stress applied in the [111] direction does not result in a site change because all three cube directions are equally stressed and, on an average, equally occupied by the carbon atoms. This phenomenon of changing sites by carbon atoms gives rise to an elastic effect called the Snoek effect. The Snoek effect can and is used to measure the carbon content of α-Fe. Would you expect to observe the Snoek effect in γ-Fe?

10.3.2 Interaction Due to the Modulus Difference

Even when the size difference is zero (i.e., $\epsilon = 0$), a contribution to the binding energy between solute and dislocation can result due to difference in the moduli of the two. The solute atom behaves as an elastic heterogeneity in the dislocation strain field.

Since the elastic strain energy of a dislocation is proportional to the shear modulus, it will depend on the distance between dislocation and solute atom. To the extent that linear elasticity is applicable, the two contributions due to difference in size and modulus can be treated separately.

Figure 10.5 shows a solute atom as a sphere in the matrix continuum. If the solute is softer (smaller shear modulus) than the matrix, the energy of the strain field of dislocation can be reduced by distortion of solute. This is to say that U will be negative and thus there will be an attraction between solute

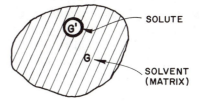

BEFORE DEFORMATION

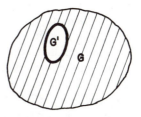

Figure 10.5 Solute of shear modulus G' embedded in a matrix of shear modulus G.   **AFTER DEFORMATION**

and matrix. For a solute that is harder than the matrix, the contrary will occur; U being positive, there will be a repulsion between the two.

The change in energy due to the difference in moduli has been treated by a number of authors.[7-9] In order to avoid treating solute atoms as if they possessed the elastic properties of the bulk material, Kröner's[7] idea of a dielastic polarization of a solvent medium by solute atoms is attractive, as it relates the change in energy, U_{mod}, to the macroscopic modulus change $\eta = d \ln G/dc$, where c is the solute concentration. For a screw dislocation[7]

$$U_{mod} = \frac{Gb^2\Omega\eta}{8\pi^2 r^2} \tag{10.14}$$

Comparing U_{mod} with U_{misfit}, we see that the modulus interaction is of a second order and decreases as $1/r^2$ while the elastic misfit interaction goes as $1/r$. Hence, the interaction force in the modulus interaction is smaller than that in the elastic misfit case. The parameter η, on the other hand, is about 20ϵ. Thus, we would expect both interactions to contribute significantly to solid solution hardening. The modulus interaction, of course, applies to edge dislocations also [with a $1/(1 - \nu)^2$ factor]. However, the dielastic interaction or the interaction due to the modulus difference is symmetrical in x, while the misfit interaction has a different sign above and below the slip plane of an edge dislocation. Thus, one cannot simply add the effects of solute atoms on different sides of the slip plane.

10.3.3 Electrical Interaction

The electrostatic contribution to the interaction energy is significant in ionic solid solutions. The electrical interaction will tend to make the solute atom move to a position of minimum electrostatic energy.[10] In ionic solids, an edge dislocation will have along its entire length an excess of alternately positive and negative ions, and thus the dislocation will attract positively and negatively charged atoms. In order to make the dislocation move in such solids, extra work must be done against this electrostatic binding energy.

10.3.4 Chemical Interaction (or Suzuki Interaction)

In the discussions of elastic and electrical contributions above, we tacitly assumed that the dislocations were perfect; that is, they were not dissociated into partials (see Section 6.7.1). However, we know that in close-packed crystal structures it is energetically favorable for dislocations to dissociate into partials. In an FCC structure, the region within the stacking fault has an HCP stacking sequence. Suzuki[11] pointed out that the equilibrium solute concentration in the stacking fault region must be different from the rest of the material that is free of the fault. This would be so because the cohesive energy among the atoms in the faulted regions (i.e., HCP) is different from that in the fault-free FCC region. Thus, when a dissociated dislocation moves in an alloy, there must occur a transfer of solute from the faulted region to fault-free region.

The work done in this transfer of atoms corresponds to the energy of chemical interaction between the dislocation and the solute atoms.

10.3.5 Local Order Interaction

The solute distribution in the solvent is, in general, not random. The ideal solutions of thermodynamics with random distribution of solute represent only a theoretical model. In reality, many times the total energy is minimized by some preferential arrangement of solute atoms in solvent. For example, there may exist a long-range order in which dislocations occur in pairs (superdislocations) and are joined by antiphase domains (see Chapter 7). There also exists a short-range order wherein, locally, a given solute atom will be surrounded by a nonrandom number of solvent atoms. In short, a locally ordered material would consist of groups of preferentially arranged atoms. For dislocation movement to occur, these groups of preferentially arranged atoms must be partially disordered, which would cause an increase in the energy of the material. In order to sustain this energetically unfavorable dislocation motion, additional work, representing this interaction energy, must be done.

10.4 THE STATISTICAL PROBLEM OF SOLID SOLUTION HARDENING

In Section 10.3 we considered various interactions between a dislocation and a single solute atom (a point obstacle). It is easy to see that this is far from the situation existing in a real solid solution that is not dilute where a dislocation will, in all probability, interact with many solute atoms at any time. This is a formidable problem, as it involves a calculation of the overall effect of forces exerted by all the solute atoms on a dislocation line length.

Consider a dislocation of finite line tension, T, having many solute atoms on both sides of the slip plane. With no applied stress, the dislocation will occupy a position of zero net interaction force. As the stress is increased from zero, the dislocation will experience the force peaks due to the solute atoms. For small concentrations of solute, Fleischer's[8] treatment wherein a dislocation either touches the point obstacles at full force, F_{max}, or not at all, gives the following expression of the critical shear stress:

$$\tau_y b = \frac{F_{max}^{3/2} c^{1/2}}{(6T)^{1/2}} \tag{10.15}$$

where F_{max} is the maximum obstacle strength and c is the solute concentration.

To arrive at this expression, Fleischer defines a mean separation, l, between solute obstacles. It is not the spacing between the nearest solute neighbors in the slip plane. l is the Friedel distance for the average spacing between obstacles touched by a dislocation of line tension T and under a stress τ, and is given by[12] $l = (6T/\tau bc)^{1/3}$. Computer simulations by Foreman and Makin[13] and by Kocks[14] show Eq. 10.15 to be correct for moderately strong point obstacles.

However, Fleischer's theory does not explicitly treat the statistical problem of solution hardening. Rather, it deals with average parameters (τ_y, l). Hanson and Morris[15] and Labusch[16-19] have dealt with the statistical aspects of a dislocation moving through a random array of point obstacles. Labusch takes into account the fact that, due to the restraining effects of the dislocation line tension, all point obstacles do not act with their maximum force, F_{max}, on the dislocation line at the same time. In other words, when the solid solution is not dilute, the dislocation will contact point obstacles of all interaction strengths F, not only those with $F = F_{max}$ or zero as considered by Fleischer. Thus, Labusch takes into account the statistics of distribution of point obstacles such as solute atoms. Labusch defines a distribution function $\tilde{\rho}(F)\,dF$ as the number of solute atoms touched by a unit length of dislocation with interaction forces between F and $F + dF$. Knowing the force profile, one can redefine $\tilde{\rho}(F)$ as another distribution function $\tilde{\rho}(x)$ of the distances x from the point obstacles. $\tilde{\rho}$ will depend on τ and at the critical shear stress, τ_y, it becomes stationary [i.e., $(\partial\tilde{\rho}/\partial\tau)_y = 0$]. Labusch's derivation of the distribution function $\rho_c(x)$, the density of dislocation elements in front of the solute atom, results in the following relation:

$$\tau_y b = \frac{F_{max}^{4/3} c^{2/3} z^{1/3}\alpha}{T^{1/3}} \tag{10.16}$$

where α is a numerical factor of the order of unity. Let us compare this result of Labusch (Eq. 10.16) with that of Fleischer (Eq. 10.15), valid for very dilute solid solutions. Equation 10.16 has different exponents and contains the obstacle range, measured by its distance z from the slip plane. Thus, Labusch's theory takes into account the various obstacles present at different distances z from the slip plane. For greater details, the reader is referred to [16–20].

‡10.5 MECHANICAL EFFECTS ASSOCIATED WITH SOLID SOLUTIONS

There are certain technologically important and scientifically interesting aspects associated with the presence of solute atoms. Among these, the most obvious, of course, is the increase in the yield stress. Some of these effects are described below.

‡10.5.1 Well-Defined Yield Point in the Stress–Strain Curves

Figure 16.5(b) shows a typical stress–strain curve of this type of behavior. Characteristically, annealed low-carbon steels show this. There are two main theories that explain this type of behavior. The first one, applicable to steels, is due to Cottrell and Bilby.[21] According to it, the dislocations in annealed steels ($\sim 10^7$ cm^{-2}) are locked by solute atoms. Of the various solute atoms, the interstitials carbon and nitrogen are the ones that have great mobility. When the stress is increased in a tensile test, one reaches a point where dislocations

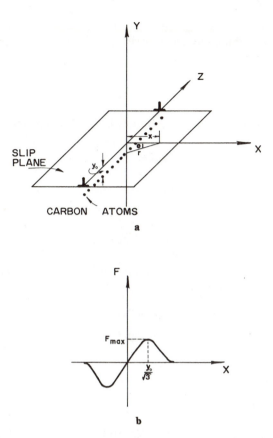

Figure 10.6 (a) Interaction between an edge dislocation and a row of carbon atoms; (b) the force of interaction for the case shown in (a).

become unlocked. The stress necessary to move these unlocked dislocations is less than the stress necessary to free them: hence the phenomenon of yield drop and the appearance of a lower yield point in the tensile stress–strain curve. Let us estimate the force necessary to free the dislocation from the associated row of carbon atoms. Consider a row of carbon atoms condensed at an edge dislocation in the position of maximum interaction energy (Fig. 10.6). This is the Cottrell atmosphere. The interaction energy at a distance x, in the slip plane and normal to the row of carbon atoms is (Eq. 10.12)

$$U_i = \frac{A \sin \theta}{r} = -\frac{A y_0}{r^2} = -\frac{A y_0}{x^2 + y_0^2}$$

The force required to separate the dislocation a distance x from carbon atoms in given by

$$F = \frac{\partial U_i}{\partial x} = \frac{2A y_0 x}{(x^2 + y_0^2)^2} \tag{10.17}$$

Figure 10.6 shows this force graphically. The separation corresponding to the maximum force is obtained by differentiating and equating to zero the expression for F (Eq. 10.17) and is

$$x = \frac{y_0}{\sqrt{3}}$$

The value of the corresponding maximum force F_{max} per atom plane is

$$F_{max} = \frac{3\sqrt{3}\,A}{8y_0^2}$$

However, a well-defined yield point has been observed for a series of metals, for example, molybdenum, titanium, and even in materials of very high purity, such as germanium. Thus, there are cases where the explanation above does not work. For such cases, the ideas of dislocation dynamics (Section 8.4) are used. Let ρ_1 and ρ_2 be the mobile dislocation densities immediately before and after the yielding, respectively. Then, applying the Orowan equation (Section 8.3) and differentiating with respect to time, we get

$$\dot{\epsilon}_1 = \phi\rho_1 b_1 v_1$$

$$\dot{\epsilon}_2 = \phi\rho_2 b_2 v_2$$

But the strain rate is constant; therefore,

$$\rho_1 v_1 = \rho_2 v_2$$

Applying Eq. 8.7 to both sides, we have

$$\rho_1 \tau_1^{\,m} = \rho_2 \tau_2^{\,m}$$

or

$$\frac{\rho_1}{\rho_2} = \left(\frac{\tau_2}{\tau_1}\right)^m$$

where τ_1 and τ_2 are the stresses immediately before and after the yielding. As we must have $\rho_2 > \rho_1$, we get $\tau_2 < \tau_1$ for m positive. Thus, the stress immediately before yielding (upper yield point) will be higher than the stress immediately after yielding (lower yield point). It should be pointed out that, in general, the yield drop due to the Cottrell atmosphere in low-carbon steels is much sharper than that caused by dislocation multiplication.

‡10.5.2 Plateau in the Stress–Strain Curve After the Well-Defined Yield Point or Lüders Bands

Generally, as indicated in Fig. 16.5(b), there follows a plateau region, after the load drop, in which the stress fluctuates around a certain value. The elongation that occurs in this plateau [region of extent H in Fig. 16.5(b)] is called the yield-point elongation. It corresponds to a region of nonhomogeneous deformation. In a portion of the tensile sample where there is a stress concentra-

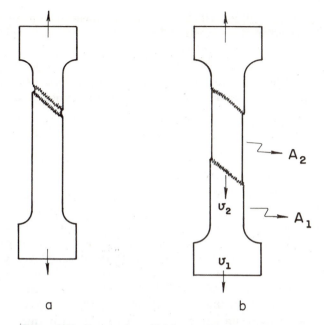

a   b

Figure 10.7 Propagation of Lüders band in a tensile sample.

tion, a deformation band appears such as that indicated in Fig. 10.7(a). As the material is deformed, this deformation band propagates through the test sample. An intermediate position is indicated in Fig. 10.7(b). The deformation is restricted to the interface. This deformation band is known as the Lüders band. There could be, in the plateau region of stress–strain curve, two or more such bands. Sometimes these Lüders bands are visible to the naked eye. After the formation of the last band, the stress–strain curve resumes its normal trajectory of strain hardening. Knowing the cross-sectional areas A_1 and A_2 in Fig. 10.7, one can determine from the yield point elongation the number of Lüders bands. One can also determine, from the strain rate of the sample, the speed of propagation of these bands.

An aspect of great technological importance is the formation of these bands during the stamping of low-carbon steels, with the consequent irregularities in the final sheet thickness. This problem is tackled, in practice, in two ways:

1. By changing the alloy composition to eliminate the yield point. The addition of aluminum, vanadium, titanium, niobium, and boron to steel, leads to the formation of carbides and nitrides, which may remove the interstitial atoms from the solid solution.
2. By prestraining the sheet to a strain greater than the yield point strain such that the strains during the stamping operations occur in the strain-hardening region.

The explanation for the Lüders bands formation is intimately related to the cause of the appearance of the well-defined yield point (Section 10.5.1).

The unlocking of dislocations which occurs at the upper yield point is, initially, a localized phenomenon. These unlocked dislocations move at a very high velocity, as the stress to unlock them is much higher than the stress required to move them, until they are stopped at grain boundaries. The joint effect of the applied stress and the stress concentration due to the dislocations that accumulated at grain boundaries unlocks the dislocations in the neighboring grains.

‡10.5.3 Strain Aging

Figure 10.8 shows the result of experiments done with an annealed austenitic alloy of composition Fe–31% Ni–0.1% C.[22] The tensile test was stopped three times, for 3 h each time, after three different strains: ϵ = 0.08, 0.18, and 0.27. The test was stopped by simply turning off the machine. Initially, the sample did not show a well-defined yield point. However, on reloading after the 3-h rest, the stress–strain curve showed clearly the appearance of a yield point followed by a plateau, a horizontal load drop region, and finally a return to the original trajectory. This can be seen clearly in Fig. 10.8. The dashed lines indicate the stress at which the test was stopped. Note that on reloading, the yield stress of the alloy increased for the three strains. This phenomenon is known as strain aging. The term "aging" is normally used when there is precipitate formation (see Section 11.1). However, this is not necessarily the case in the example given above. As the test and its interruption were carried out at ambient temperature, there occurs a migration of interstitial atoms to dislocations during the test stoppage, with its consequent dislocation locking. Thus, on reloading, these dislocations have to be unlocked with the appearance of the well-defined yield point. Meyers and Guimarães[22] repeated these experiments under identical conditions but maintaining the test sample unloaded during 3 h. The well-defined yield point reappeared, but it was less marked. This indicates that the applied stress has an accelerating effect on the aging process.

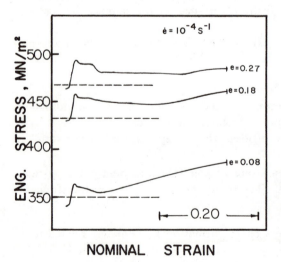

Figure 10.8 Reloading curves after stopping the test for 3 h at nominal strains of 0.08, 0.18, and 0.27. The dashed lines indicate the stresses at which the test was stopped. Note the formation of a well-defined yield point in the three cases. (Reprinted with permission from [22], p. 708.)

Low-carbon steels show strain aging. Guimarães and Meyers[23] also observed strain aging in commercial-purity titanium.

‡10.5.4 Serrated Stress–Strain Curve

The serrations in stress–strain curves shown in Fig. 16.6 are manifestations of the Portevin–Le Châtelier effect. Irregularities in a stress–strain curve can be caused by the interaction of solute atoms with dislocations, by mechanical twinning, or by stress-assisted ("burst"-type) martensitic transformations. The first type (i.e., due to solute–dislocation interaction) receives the name of Portevin–Le Châtelier effect. It generally occurs within a specific temperature and strain rate range. The solute atoms, being able to diffuse in the test sample at a speed greater than the displacement speed of the dislocations (imposed by the applied strain rate), "persecute" the dislocations, locking them eventually. With increasing load, the unlocking of dislocations causes the load drop with the formation of small irregularities in the stress–strain curve.

‡10.5.5 Blue Brittleness

Carbon steels heated in range 230 and 370°C show a notable reduction in elongation. This phenomenon is due to the interaction of dislocations in motion with the solute atoms (carbon or nitrogen). This phenomenon is intimately connected with the Portevin–Le Châtelier effect.[24] We classify it separately merely due to its distinct importance. When the temperature and the strain rate are such that the speed of interstitial atoms is more than that of the dislocations, the latter are continually captured by the former. This results in very high strain hardening rate and strength with a reduction in the elongation (i.e., it results in an embrittlement of the steel). In fact, the use of the word "embrittlement" is not correct as the fracture mode is still ductile. With increasing strain rates, this effect occurs at ever-higher temperatures as the diffusivity increases with the temperature. The word "blue" in the name of this effect refers to the coloration that the steel acquires due to the oxide layer formed when submitted to this temperature range.

In the range of temperature and strain rate in which the material is subjected to dynamic aging (Portevin–Le Châtelier effect and brittleness), there occurs another effect worth mentioning. The strain rate sensitivity m (see Section 16.1.5) is also affected; m tends to increase linearly with temperature. However, in the presence of dynamic aging, m becomes very small and the yield stress becomes practically independent of the strain rate.

EXERCISES

10.1. Why are substitutional solid solutions more common than interstitial solid solutions?

10.2. The interaction energy U between an edge dislocation (at the origin) and a solute atom (at r, θ) is given by

$$U = \frac{A}{r} \sin \theta$$

where A is a constant. Transforming into Cartesian coordinates, plot lines of constant energy of interaction for different values (positive and negative) of $A/2U$. Plot on the same graph the curves for the interaction force. Indicate by arrows the direction in which the solute atoms, with ΔV positive, will migrate.

10.3. Consider a metal with shear modulus $G = 40$ GPa and atomic radius $r_0 = 0.15$ nm. It has a solute that results in a misfit of $\epsilon = (R - r_0)/r_0 = 0.14$. Compute the elastic misfit energy per mole of solute.

10.4. Show that in Fe (BCC), the interaction energy between a carbon atom and a dislocation (edge or screw) is of the order of 0.08 aJ (0.5 eV).

10.5. Estimate the amount of solute (atomic percent) necessary to put one solute atom at each site along all the dislocations. Assume that 1 mm³ of metal contains about 10^6 mm of dislocation lines.

10.6. The substitutional zinc atom is about 12% larger than the solvent copper atom. Take the radius of the Cu atom as $b/2$, where b is the Burgers vector. Compute the interaction energy with the dilatation component of an edge dislocation due to this size misfit. Take appropriate values of G, b, and ν for copper.

10.7. Compute the condensation temperature, T_c, for the following cases:
   (a) Carbon in iron with C_0 (average concentration) = 0.01% and U_i (interaction energy) = 0.08 aJ (0.5 eV).
   (b) Zinc in copper with $C_0 = 0.01\%$ and $U_i = 0.019$ aJ (0.12 eV).

10.8. One of the Hume–Rothery rules for solid solutions (W. Hume–Rothery and G. V. Raynor, *The Structure of Metals and Alloys*, Institute of Metals, London, 1956, p. 97) is that the solubility of solute B in solvent A becomes negligible when the atomic radii of A and B differ by more than 15%. Plot the maximum solubility (atomic percent) of Ni, Pd, Pt, Au, Al, Ag, and Pb as a function of the ratio of solute and solvent (Cu) radii and verify that the solid solubility in Cu drops precipitously at a size ratio of about 1.15.

REFERENCES

[1] F.B. Pickering and T. Gladman, *ISI Special Report 81, Iron and Steel Inst.*, London, 1963, p. 10.

[2] A.E.H. Love, *A Treatise on the Mathematical Theory of Elasticity*, 4th ed., Dover, New York, 1944, p. 142.

[3] J. Weertman and J.R. Weertman, *Elementary Dislocation Theory*, Macmillan, New York, 1964, p. 173.

[4] A.W. Cochardt, G. Schöeck, and H. Wiedersich, *Acta Met.*, 3 (1955) 533.

[5] A.H. Cottrell, *Dislocations and Plastic Flow in Crystals*, Oxford University Press, London, 1953, p. 134.

[6] G. Schöeck and A. Seeger, *Acta Met.*, 7 (1959) 469.

[7] E. Kröner, *Phys. Kond. Mater.*, 2 (1964) 262.

[8] R.L. Fleischer, *Acta Met.*, *9* (1961) 996.

[9] J.D. Eshelby, *Acta Met.*, *3* (1955) 487.

[10] A.H. Cottrell, S.C. Hunter, and F.R.N. Nabarro, *Phil. Mag.*, *44* (1953) 1064.

[11] H. Suzuki, *Sci. Rep.: Tohoku Univ. Res. Inst.*, *44* (1952) 455.

[12] J. Friedel, *Dislocations*, Pergamon Press, Oxford, 1964.

[13] A.J.E. Foreman and M.J. Makin, *Phil. Mag.*, *13* (1966) 13.

[14] U.F. Kocks, *Phil. Mag.*, *13* (1966) 541.

[15] K. Hanson and J.W. Morris, *J. Appl. Phys.*, *46* (1975) 983.

[16] R. Labusch, *Cryst. Lattice Defects*, *1* (1969) 1.

[17] R. Labusch, *Phys. Status Solidi*, *41* (1970) 659.

[18] R. Labusch, *Acta Met.*, *20* (1972) 917.

[19] R. Labusch, *J. Appl. Phys.*, *48* (1977) 4550.

[20] U.F. Kocks, *Mater. Sci. Eng.*, *27* (1977) 291.

[21] A.H. Cottrell and B.A. Bilby, *Proc. Phys. Soc.* (*London*), *A62* (1949) 49.

[22] M.A. Meyers and J.R.C. Guimarães, *Metalurgia-ABM*, *34* (1978) 707.

[23] J.R.C. Guimarães and M.A. Meyers, *Scripta Met.*, *11* (1977) 193.

[24] A. Portevin and F. Le Châtelier, *Compt. Rend. Acad. Sci. Paris*, *176* (1923) 507.

SUGGESTED READING

COTTRELL, A.H., *Dislocations and Plastic Flow in Crystals*, Oxford University Press, London, 1953.

FIORE, N.F., and C.L. BAUER, *Binding of Solute Atoms to Dislocations*, Progress in Materials Science, Vol. 13, B. Chalmers and W. Hume–Rothery (eds.), Pergamon Press, Oxford, 1967, p. 85.

HAASEN, P., in *Dislocations in Solids*, Vol. 4, F.R.N. Nabarro (ed.), Elsevier/North-Holland, New York, 1979, p. 155.

PECKNER, D. (ed.), *Strengthening of Metals*, Reinhold, New York, 1965.

Strengthening Mechanisms in Solids, ASM, Metals Park, Ohio, 1960.

SUZUKI, H., in *Dislocations in Solids*, Vol. 4, F.R.N. Nabarro (ed.), Elsevier/North-Holland, New York, 1979, p. 191.

Chapter 11

PRECIPITATION AND
DISPERSION HARDENING

‡11.1 INTRODUCTION

Precipitation of a second phase from a supersaturated solid solution is, in practice, a very versatile and common strengthening technique. Figure 11.1(a) shows a typical example of an Al alloy, specifically Al 6061, with Mg_2Si precipitates at the boundaries and in the grain interiors, and Fig. 11.1(b) shows γ' precipitates and aged carbides in a superalloy. The supersaturated solid solution is obtained by sudden cooling from a sufficiently high temperature. The heat treatment that causes precipitation of solute is called aging. The process can be applied to a number of alloy systems, although the specific behavior varies. In general, the system must obey the following requisites:

1. Form a supersaturated solid solution at high temperatures.
2. Reject a finely dispersed precipitate during aging (i.e., the phase diagram must show a declining solvus line).

Figure 11.2 shows a part of the phase diagram of the Al-Cu system where precipitation hardening can occur. The precipitation treatment consists of the following steps:

1. *Solubilization*. This involves heating of the alloy to the monophase region and maintaining it there for a sufficiently long time to dissolve any soluble precipitates.

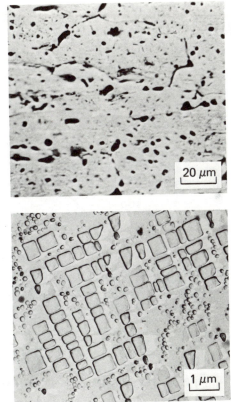

20 μm

b

Figure 11.1 (a) Aluminum alloy, Al6061, with Mg₂Si precipitates distributed in the grain boundaries and grain interiors; (b) γ′ precipitates and aged carbides in superalloy. (Courtesy of R. N. Orava, South Dakota School of Mines and Technology.)

1 μm

2. *Quenching.* This involves cooling very rapidly to room temperature or lower so that the formation of stable precipitates is avoided. Thus, one obtains a supersaturated solid solution.

3. *Aging.* This treatment consists in leaving the material at room temperature or above, and it results in rather fine scale transition structures (≈10 nm).

For example, in the Al-Cu system, transitory structures (metastable) GP1, GP2, and θ' form before the formation of equilibrium phase $CuAl_2$. The name "GP zones" is given to honor the researchers Guinier and Preston, who were the first ones to detect them. These precipitates in the Al matrix give rise to local distortions and strain fields which restrict the dislocation mobility. Table 11.1 presents some of the precipitation-hardening systems, with the precipitation sequence and the equilibrium precipitate.

Although the behavior of different systems varies in detail, one may write the general aging sequence as follows:

$$\begin{array}{ccc}
\text{supersaturated} & \longrightarrow & \text{transition} & \longrightarrow & \text{aged} \\
\text{solid solution} & & \text{structures} & & \text{phase} \\
& & \text{(metastable)} &
\end{array}$$

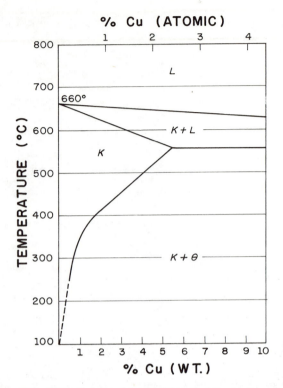

Figure 11.2 Phase diagram of the Al-rich end of the Al-Cu system.

The precipitate produced during the aging treatment can be coherent, semicoherent, or incoherent with the matrix (Fig. 11.3). Coherency signifies that there exists a one-to-one correspondence between the precipitate lattice and that of the matrix [Fig. 11.3(a)]. Semicoherent signifies that there exists only a partial correspondence between the two sets of lattice planes. Dislocations form at the noncorrespondence sites [Fig. 11.3(b) and (c)]. Incoherent signifies

TABLE 11.1 Some Precipitation-Hardening Systems

Base Metal	Alloy	Sequence of Precipitates			
Al	Al-Ag	Zones (spheres)	→ γ' (plates)	→	$\gamma(Ag_2Al)$
	Al-Cu	Zones (disks)	→ θ'' (disks) → θ' →		$\theta(CuAl_2)$
	Al-Zn-Mg	Zones (spheres)	→ M′ (plates)	→	$(MgZn_2)$
	Al-Mg-Si	Zones (rods)	→ β'	→	(Mg_2Si)
	Al-Mg-Cu	Zones (rods or spheres)	→ S′	→	$S(Al_2CuMg)$
Cu	Cu-Be	Zones (disks)	→ γ'	→	$\gamma(CuBe)$
	Cu-Co	Zones (spheres)		→	β
Fe	Fe-C	ϵ-Carbide (disks)		→	Fe_3C("laths")
	Fe-N	α'' (disks)		→	Fe_4N
Ni	Ni-Cr-Ti-Al	γ' (cubes)		→	$\gamma(Ni_3Ti, Al)$

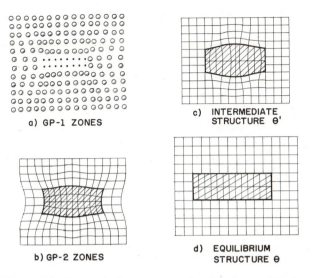

Figure 11.3 Sequence of structures produced during precipitation.

the inexistence of any correspondence between the two lattices [Fig. 11.3(d)]. This situation of incoherency is also present in dispersion-hardened systems, in which one introduces hard particles such as alumina or silica in a ductile matrix, by compaction and sintering of powders, or by internal oxidation, or by addition of solid particles to the alloy in the molten state.

Dispersion-hardened systems have one advantage over those hardened by precipitation: the former have high strength at high temperatures where precipitates tend to dissolve in the matrix. Can the reader cite some advantage of strengthening by precipitation over that by dispersion of strong, inert particles?

The strengthening in these systems has its origin in the dislocation interaction with the precipitates or the dispersed phase. This interaction, in general, depends on the dimensions of the precipitate, its strength, spacing, and fraction present. The detailed behavior, of course, differs from system to system.

11.2 DISLOCATION–PRECIPITATE INTERACTION

Finely distributed precipitates represent an effective barrier to dislocation motion. The following important models have been put forth to explain this interaction:

1. Long-range interactions between the particles and the dislocations. Particle shearing or not by the dislocations is not considered.
2. Dislocations cut through the particles in the slip plane.
3. Dislocations circumnavigate around the particles in the slip plane.

Mott and Nabarro[1] were the first ones to give an explanation, based on long-range interactions between dislocations and precipitates, of precipitation

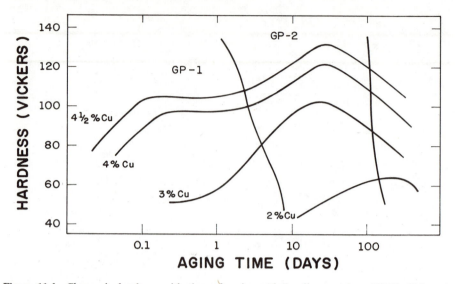

Figure 11.4 Change in hardness with time of various Al-Cu alloys aged at 130°C. (Adapted with permission from [2], p. 195.)

hardening (which includes, in a latent way, the solution hardening discussed earlier). We consider now the interaction between dislocation and precipitates (rather than solute atoms).

Substitutional solute atoms can produce three basic hardening effects:

1. Solid solution, which was discussed in Chapter 10
2. Zones or coherent precipitates
3. Incoherent precipitates

In each case, the mobile dislocations have to overcome the resistance of a barrier. The precipitation from solid solution occurs in the sequence 1–2–3 cited above. The hardening obtained in the Al-Cu system as a function of aging time is shown in Fig. 11.4. The peak in the hardening curve is due to an optimum distribution and size of the precipitates, and the coherency strains in matrix. The fall in strength with overaging is due to the formation of incoherent precipitates of rather large size.

11.2.1 Long-Range Particle–Dislocation Interactions

The most important of the long-range interactions is treated in the theory due to Mott and Nabarro.[1] Their theory is based on the analysis of internal stresses resulting from the difference in average atomic volume of the matrix and of that of the precipitate.

Consider a spherical particle of precipitate embedded coherently in an infinite matrix. Let $(1 + \delta)^3$ be the atomic volume of the precipitate, while the matrix volume is 1. In case δ is positive, the precipitate material will be under hydrostatic pressure and its lattice parameter will be reduced to ϵ, where $\epsilon < \delta$. Thus, ϵ is the dilatation misfit between the precipitate and the matrix, $\Delta r_0 / r_0$ where r_0 is the radius of the cavity. Then, under pressure,

$$\frac{\text{atomic volume of precipitate}}{\text{atomic volume of matrix}} = \left(\frac{1 + \epsilon}{1}\right)^3$$

Let u_r be the displacement in the matrix at a distance r from the center of the precipitate particle. We derived, in Chapter 10 (Eq. 10.2), the equation for such a displacement in the case of spherical symmetry:

$$u_r = cr + \frac{c'}{r^2} \tag{11.1}$$

and the solution was (see Section 10.2):

$$u_r = \epsilon r \qquad \text{when } r < r_0$$

$$u_r = \epsilon r_0 \left(\frac{r_0}{r}\right)^2 \qquad \text{when } r > r_0$$

Let B be the bulk modulus, E be the Young's modulus, and ν be the Poisson's ratio of the matrix. The strain in precipitate is one of uniform hydrostatic compression:

$$\frac{\Delta V}{V} = \frac{(1 + \delta)^3 - (1 + \epsilon)^3}{1} \approx 3(\delta - \epsilon) \tag{11.2}$$

where we have neglected the higher-order terms because δ and ϵ are very small.
The pressure in the precipitate is given by

$$\sigma_p = B \frac{\Delta V}{V} \approx 3B(\delta - \epsilon) \tag{11.3}$$

We also derived, in Chapter 10 (Eqs. 10.9 and 10.10), the following equations for the radial and circumferential strains in the matrix:

$$\epsilon_{rr} = \frac{\partial u_r}{\partial r} = -2\epsilon \left(\frac{r_0}{r}\right)^3 \tag{11.4}$$

$$\epsilon_{\theta\theta} = \epsilon_{\phi\phi} = -\frac{1}{2} \frac{\partial u_r}{\partial r} = \epsilon \left(\frac{r_0}{r}\right)^3$$

Let σ_r, σ_θ, and σ_ϕ be the corresponding stress components. Then, by Hooke's law for isotropic materials (Eq. 1.119), one can write

$$\epsilon_{rr} = \frac{1}{E}[\sigma_r - \nu(\sigma_\theta + \sigma_\phi)] \qquad \text{etc.} \tag{11.5}$$

$$-2\epsilon \left(\frac{r_0}{r}\right)^3 = \frac{1}{E}(\sigma_r - 2\nu\sigma_\theta)$$

and

$$\epsilon \left(\frac{r_0}{r}\right)^3 = \frac{1}{E}\left[\sigma_\theta - \nu(\sigma_r + \sigma_\theta)\right] \tag{11.6}$$

But at the surface of the precipitate (i.e., at $r = r_0$) the radial stress must be equal to pressure in precipitate (i.e., $\sigma_p = \sigma_r$ at $r = r_0$).

Then, from Eqs. 11.3, 11.5, and 11.6, we have

$$\epsilon = \frac{3B\delta}{3B + [2E/(1+\nu)]} \tag{11.7}$$

Note that strain in the matrix is one of shear without dilatation (i.e., $\Delta V = \epsilon_{rr} + \epsilon_{\theta\theta} + \epsilon_{\phi\phi} = 0$). The shear in the matrix at a certain distance R from the precipitate is $\epsilon(r_0/R)^3$, where ϵ depends on the materials and is called the misfit strain.

Let us consider now a monocrystal containing N particles per unit volume, each of radius r_0. We can calculate an average shear strain, γ_m, assuming R to be half the distance between particles [i.e., $R = \frac{1}{2}(N^{-1/3})$]. Thus,

$$\gamma_m = \frac{\epsilon r_0^3}{[\frac{1}{2}(N^{-1/3})]^3} = 8\epsilon r_0^3 N \tag{11.8}$$

Let f be the precipitate volume fraction,

$$f = \frac{\frac{4}{3}\pi r_0^3 N}{1} \tag{11.9}$$

Then, from Eqs. 11.8 and 11.9, we have

$$\gamma_m = 8\epsilon r_0^3 \frac{f}{\frac{4}{3}\pi r_0^3} \approx 2\epsilon f \tag{11.10}$$

The critical shear stress for yielding in the alloy is then given by

$$\tau_y = \delta_m G \approx 2G\epsilon f \tag{11.11}$$

According to Eq. 11.11, the critical shear stress is independent of the size and of the spacing between the particles (Λ), and depends only on the amount of the precipitate (f). However, a dislocation is a flexible line with tension T (see Section 6.5). The radius of curvature of such a flexible dislocation line can be reduced by an applied stress, and is approximately given by (using Eq. 11.11)

$$r \approx \frac{Gb}{2\tau_y} = \frac{b}{4\epsilon f} \tag{11.12}$$

Equation 11.12 is a modified form of Eq. 6.28 (see footnote in Section 9.3).

Thus, the dislocation line, being flexible, can avoid the obstacles by bending around the precipitate stress fields and the radius of curvature must be of the same order of magnitude as the interparticle spacing for the Mott–Nabarro model to be valid (Fig. 11.5). Consider the case in which the precipitate is dispersed on a scale finer than the radius of curvature of dislocation (i.e., r

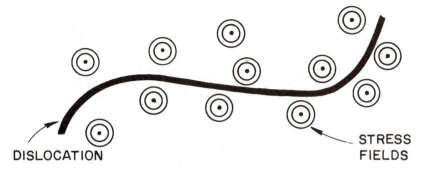

DISLOCATION   STRESS
FIELDS

Figure 11.5   Dislocation bowing around stress fields.

$\gg \Lambda$). In such a case, the dislocation will not be able to bend sufficiently to be entirely in the low-internal-energy regions and the algebraic sum of all the interaction energies will be zero. The alloy in this case will be soft and ductile, as there will be little resistance to dislocation movement. The theory due to Mott and Nabarro thus predicts a critical interparticle spacing for maximum hardening. When the separation between the particles, Λ, is greater than this critical value, the yield stress will only depend on f and ϵ.

In reality, when the precipitate distribution has roughly the critical spacing, the dislocations will probably shear the precipitate particles and the short-range interactions involved in this shearing will affect the yield stress in shear, τ_y. Thus, in practice, the probability is very little that τ_y will be controlled only by the magnitude of ϵ. The theory of Mott and Nabarro also does not take into account the long-range interaction due to difference in the moduli of elasticity of the dispersed phase and the matrix and the exact nature of the coherency strains. Notice that Mott and Nabarro[1] did not consider the intersection of a dislocation with the precipitates; this is an important factor, as will be seen below.

11.2.2 Dislocation–Particle Interactions with and without Particle Shear

Depending on the nature of the precipitate and on the crystallographic relationship between the precipitate and the matrix, we can have two limiting cases.

1. The precipitate particles are impenetrable to the dislocations. Orowan[3] pointed out that if a ductile matrix has second-phase particles interpenetrating the slip plane of dislocations, an additional stress will be necessary to make a dislocation expand between the particles. Should the stress be sufficiently high to bend the dislocations in roughly semicircular form between the particles, the dislocations go around the particles, leaving dislocation loops around them, as by the mechanism shown schematically in Fig 11.6(a) and evidenced by a transmission electron micrograph in Fig. 11.6(b). Figure 11.6(b) illustrates the phenomenon of obstruction of dislocation motion by uniformly distributed non-shearing precipitates. The material is an Al–0.2% Au alloy, solution annealed,

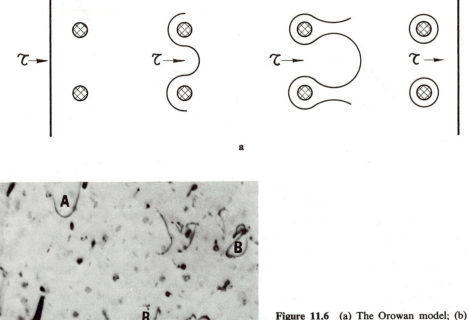

a

Figure 11.6 (a) The Orowan model; (b) obstruction of dislocation motion by uniformly distributed nonshearing particles in an aluminum alloy. (Transmission electron microscope.) (Courtesy of M. V. Heimendahl, Univ. of Erlangen.)

0.1 μm

b

followed by 60 h at 200°C and 5% plastic deformation. At points marked A, one can see dislocations being pinned by the precipitates and Orowan bowing of dislocation segments. At points B, the dislocations have left the slip plane and have formed prismatic dislocation loops. The dislocations in this micrograph are characteristically very short and have been severely impeded in their movement.

It is convenient at this point to define some additional terms such as τ_m, the critical shear stress for matrix yielding in the absence of the precipitate and τ_{LR}, the shear stress necessary for overcoming the long-range barriers of the kind described in Mott and Nabarro's theory. Now, the stress necessary to bend a dislocation to a radius r is given roughly by Eq. 11.12,

$$\tau \simeq \frac{Gb}{2r} \tag{11.13}$$

Let x be the average separation between two particles in the slip plane. Then a dislocation must be bent to a radius of the order of $x/2$ for it to be extruded between the particles instead of cutting them. The shear stress to do this is given by making $r = \dfrac{x}{2}$ in eq. 11.13, that is,

$$\tau \simeq \frac{Gb}{x} \qquad (11.14)$$

Thus, the stress necessary to make the dislocation move as per Orowan bowing must exceed the sum $(\tau_m + \tau_{LR})$. Should the stress necessary to cause the particle shear be greater than Gb/x (rigorously speaking, $2T/bx$, where T is the dislocation line tension) the dislocation will bow between the particles rather than shearing them. This, in essence, is the Orowan model of strengthening due to dispersion or incoherent precipitates. The increase in the yield stress due to the presence of particles is given by Eq. 11.14, so that the yield stress for an alloy strengthened by a dispersed phase or an incoherent precipitate is given by (if there is no particle shear)

$$\tau_y = \tau_m + \tau_{LR} + \frac{Gb}{x} \qquad (11.15)$$

More precise formulations of the Orowan stress have been made[4,5] involving more accurate expressions for the dislocation line tension T and taking into account the effect of the finite particle size on the average interparticle spacing (see Section 11.10).

2. The precipitate particles are penetrable to dislocations; that is, the particles are sheared by dislocations in their slip planes. If the extra stress (i.e., in addition to $\tau_m + \tau_{LR}$) necessary for particle shear is less than that for bending the dislocation between the particles ($=Gb/x$), the particles will be sheared by dislocations during yielding, and

$$\tau_y < \tau_m + \tau_{LR} + \frac{Gb}{x}$$

Thus, we see that the strength of the particle will determine whether or not dislocation will cut it. In internally oxidized alloys (e.g., $Cu + SiO_2$) where the obstacles to dislocation motion are small, very hard (virtually free of mobile dislocations perhaps) ceramic particles with a very high shear modulus, the initial flow stress is controlled by the stress necessary to extrude the dislocations between the hard and impenetrable particle, as per Orowan mechanism. As the shear strength of obstacle is generally very much higher than that of the matrix, very large stresses will be required for particle shear to occur. The initial yield stress is then controlled by the interparticulate spacing. Such behavior is also shown by precipitation hardened alloys when the equilibrium precipitate is an intermetallic compound (e.g., the system Al-Cu). However, in the initial stages of aging, the small precipitates or zones are coherent with the matrix and thus are sheared by dislocations. A vivid example of particle (Ni_3Al) shear by dislocation is illustrated in a Ni–19% Cr–6% Al alloy (aged at 750°C for 540 h and strained 2%) in Fig. 11.7.

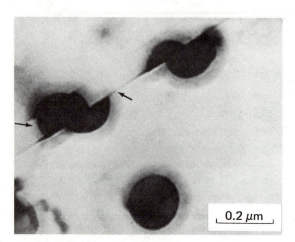

Figure 11.7 γ'-precipitate particles sheared by dislocations in a Ni–19% Cr–6% Al alloy (aged at 750°C for 540 h and strained 2%). The arrows indicate the two slip-plane traces. (Transmission electron microscope.) (Courtesy of H. Gleiter, Univ. des Saarlandes.)

Consider a dislocation such as that at A in Fig. 11.8 which encounters spherical particles (radius r_0 in the slip plane) interpenetrating the slip planes. Let x be the separation between particles in the slip plane. The stress necessary to bend a dislocation between particles is (Eq. 11.14)

$$\tau_{\text{Orowan}} \simeq \frac{Gb}{x}$$

If the particles are not strong enough to support the Orowan stress, the dislocation will cut the particles (radius r_0) by moving from position A to position B, as shown in Fig. 11.8. The passage of a matrix dislocation, which generally will not be a slip dislocation for the precipitate, will result in a faulted plane or an interface, say, of specific energy γ. The increase in the energy of

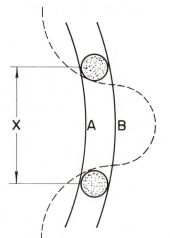

Figure 11.8 Particle shear by dislocations.

Strengthening Mechanisms   Part III

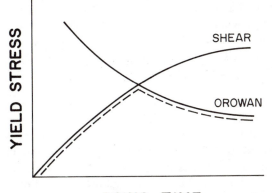

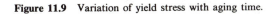

Figure 11.9 Variation of yield stress with aging time.

the particle is $\pi r_0^2 \gamma$. Then, in the absence of any thermally activated process, the stress τ_{shear} necessary to move a dislocation (length x) from A to B by particle shear is given by

$$\tau_{shear} b x 2 r_0 = \pi r_0^2 \gamma$$

or

$$\tau_{shear} = \frac{\pi \gamma r_0}{2xb} \qquad (11.16)$$

If $\tau_{shear} > \tau_{Orowan}$, the dislocation will expand between the precipitate particles and if $\tau_{shear} < \tau_{Orowan}$, the particles will be cut. Whether or not the particles will be sheared depends on r_0, the particle size, and on γ, the specific interface energy. For coherent precipitates, for example, the GP zones in Al alloys, the values of γ are expected to be of the order of magnitude of antiphase domain boundaries, rarely reaching 100 mJ/m². Thus, we estimate that for such values of γ, only very small particles ($2r_0 \leq 50$ nm) will be cut.

With aging, the second-phase particles grow in size so that the average spacing between them also increases (for a given precipitate volume fraction) and τ_{Orowan} for expanding the dislocation between particles decreases monotonically (Fig. 11.9). But the stress necessary to cut the particles increases with the particle size (Fig. 11.9). The yield stress of the alloy, then, will follow the dashed curve as a function of aging time in Fig. 11.9.

11.3 PLASTICALLY NONHOMOGENEOUS CRYSTALS

The stress–strain curve of an alloy containing completely coherent particles which can be cut by dislocations is very much different from that of a matrix containing inherently strong particles, say, oxides that do not deform together with the matrix, at least in the initial stages.[6]

In the case of coherent precipitates or zones, the strain hardening of the alloy is similar to that of the pure metal, as shown by the Cu-Be curve in Fig. 11.10. There may, however, be a small increase in the strain hardening due to change in the stacking fault energy by any solute left in solid solution; there could be small effects due to shear of the particles by dislocations which may alter the form of the precipitates and the nature of the precipitate–matrix interface.

In the case of hard, incoherent particles that do not deform plastically, the initial yield stress will be controlled by interparticle spacing ($\tau \simeq Gb/x$) and the strain-hardening rate will be much greater than that shown by isolated matrix, as shown by the Cu-BeO curve in Fig. 11.10. The dislocation density increases very rapidly in this case.

We define, following Ashby,[7] a crystalline material containing inclusions of a second phase that do not deform together with the matrix as plastically nonhomogeneous. Although the plastic deformation in the matrix may be quite large, the inclusions only deform elastically. We assume throughout that the inclusions adhere strongly to the matrix. These nondeforming inclusions change the slip distribution in the matrix. A homogeneous and monophase single crystal (free of inclusions), appropriately oriented and uniformly stressed, will deform principally on one slip system (i.e., the primary system; see Chapter 8). The GP zones or coherent precipitates which deform together with the matrix in-

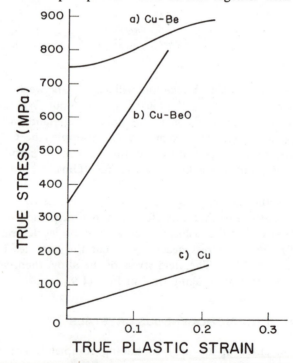

Figure 11.10 Stress–strain curves at 77 K. (a) For a precipitation hardened Cu-Be single crystal (precipitate volume fraction = 20%). These precipitates are cut by dislocations. (Adapted with permission from [6], p. 69.) (b) For a Cu crystal containing 2.8% of BeO which does not deform plastically. (c) For a pure Cu single crystai. (Adapted with permission from [6], p. 69.)

crease the flow stress of the crystal, but the general slip behavior is about the same as that of a crystal free of precipitates (see the Cu-Be curve in Fig. 11.10). As in pure metals, the slip is still sufficiently uniform. Thus, the deformable particles have little influence on strain hardening of the crystal.

Inclusions distributed in the matrix that do not deform or deform only elastically, while the matrix deforms plastically, lead to nonuniform deformation in the matrix. There does not occur any slip (i.e., plastic deformation) in inclusion or in the adjacent matrix region. Far away from the inclusions, however, the slip can be quite large. Thus, the inclusions introduce, microscopically, gradients of deformation in the crystalline matrix, although macroscopically the deformation appears uniform. The smaller the spacing between the inclusions, the larger the strain gradients will be, and it is these strain gradients that lead to extraordinarily fast work hardening.

11.4 THEORIES OF THE EFFECT OF INCLUSIONS ON STRAIN HARDENING OF THE SYSTEM

It is well recognized that rather small volume fractions of strong inclusions increase the yield stress as well as the strain-hardening rate. The form of stress–strain curve changes from that of a single crystal to a roughly parabolic one. In this section we intend to review briefly various theories that have attempted to explain this behavior; Ashby's theory based on the idea of geometrically necessary dislocations will be developed in Sections 11.5 through 11.8.

11.4.1 The Continuum Model

In this theory, the crystal structure of the components is ignored. The matrix is assumed to be a continuum containing a small quantity of inclusions. This leads to a simple result that only the volume fraction f of inclusions is a significant parameter (i.e., a few large inclusions are expected to have the same influence as many small ones). We know that this is not true. This theory is also unable to explain the large increases in the strain-hardening rate that a small inclusion volume fraction can produce.

11.4.2 Fisher–Hart–Pry Theory[8]

In this theory the crystallographic nature of slip is taken into account. The following assumptions are made:

1. The dislocations taking part in deformation are confined to one set of parallel slip planes.
2. The dislocations expand between the strong particles by means of Orowan bowing mechanism.

Thus, slip is confined to a group of parallel slip planes that are penetrated by inclusions. Inclusions act as "elastic pegs" that impede slip locally. When

N dislocations sweep an area of the slip plane, N concentric loops will be left around each particle. These dislocation loops exert long-range stresses which oppose the continued operation of the slip process. The increase in strain hardening rate in plastically nonhomogeneous alloys is attributed, entirely, to this back stress.

The possibility of slip in systems (more than one) is ignored, and thus, any contribution to flow stress due to interference among various slip systems is not considered. For small strains ($\leq 1\%$), this model explains the work-hardening behavior adequately, but it fails at larger strains where stresses around an inclusion become so large that very large local plastic deformation around the inclusion or fracture of inclusion must occur.

11.4.3 Ashby's Theory[9]

The occurrence of secondary slip causes the accumulation of dislocations in the matrix much more rapidly than in a homogeneous, inclusion-free material. These additional dislocations serve to accommodate the strain gradients due to the presence of inclusions. Their presence makes further slip more difficult, thus contributing to extra work hardening of matrix. These ideas are developed in detail in Sections 11.5 to 11.7. Section 11.8 presents the predictions of Ashby's theory.

11.5 GEOMETRICALLY NECESSARY AND STATISTICALLY STORED DISLOCATIONS[7,10]

There are two reasons for accumulation of dislocations during straining of a material:

1. Due to the necessity of strain compatibility among various parts of crystal
2. Due to random mutual capturing of dislocations

The nonuniform slip in a crystal requires that dislocations be stored so that all parts of crystal undergo a compatible strain. Void formation or material transport by means of diffusion is not permitted. It is also assumed that long-range stresses in the crystal do not exist, or that they can be made arbitrarily small. Under such conditions, the number of dislocations required for a compatible deformation can be obtained by means of simple geometrical reasoning. We give two examples.

1. *Plastic bending*. A crystal can be bent plastically without long-range stresses by introducing a dislocation density, ρ. We have (Fig. 11.11(a))

$$R\theta = 1$$

and curvature

$$K = \frac{1}{R}$$

Thus

$$\theta = \frac{1}{R} = K$$

Also

$$(R + 1)\theta = 1 + \rho b$$

or

$$1 + K = 1 + \rho b$$

Consequently,

$$K = \rho b \qquad (11.17)$$

where K is the curvature $(=1/R)$. This is shown in Fig. 11.11(a).

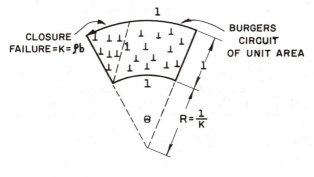

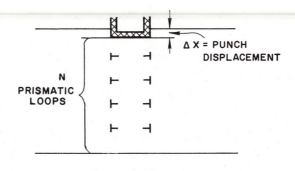

Figure 11.11 Geometrically necessary dislocations: (a) plastic bending; (b) prismatic punching. (Adapted with permission from [7], p. 413.)

2. *Compression by prismatic punching.* In this case, the relation between the punch displacement Δx and the number of punched prismatic loops, N, is given by

$$Nb = \Delta x \tag{11.18}$$

Figure 11.11(b) shows this. The internal stress can be minimized by making N large.

In both cases above, a simple geometric relation between deformation (described by K or Δx) and number of dislocations (ρ or N) is obtained. These dislocations are necessary to accommodate deformation with minimum internal stress and are called geometrically necessary dislocations or compatibility dislocations, ρ^G.

In contrast to what has been said above, dislocations are not geometrically necessary when a homogeneous crystal is deformed under simple tension. But, of course, dislocations accumulate in this case, too; their presence gives the characteristic work hardening of the crystal (Chapter 9). However, as no geometrical argument is involved and these dislocations accumulate due to casual encounters, we call them statistically stored dislocations.

It should be noted that plastic bending introduces both types of dislocations in a crystal: geometrically necessary as well as statistically stored ones. The density of geometrically necessary dislocations, ρ^G, will rarely dominate the total dislocation density, ρ, as the radius of curvature cannot be made arbitrarily small. However, such a limitation would not exist for crystals containing inclusions. This is the main difference. In crystals containing strong and small inclusions, we can introduce very severe strain gradients such that $\rho^G \simeq \rho$.

11.5.1 Accumulation of Dislocations Due to Large Platelike Inclusions

The deformation of a crystal containing inclusions is nonuniform even when the crystal as a whole deforms uniformly, such as under simple tension. The basic difference between this and, say, plastic bending of a homogeneous crystal is that in the former one can obtain extremely large curvatures (i.e., very small radii) and, consequently, one can have the density of geometrically necessary dislocations, ρ^G, so large that it becomes the dominant part of total dislocation density.

Let us consider an ideal case of a crystal containing very strong platelike inclusions which adhere very strongly to the deformable matrix (Fig. 11.12). Dislocations are not allowed to penetrate the particle–matrix interface. Suppose now that this composite is deformed on a single group of slip planes in a direction normal to the plate surface. Let γ be the shear strain in the soft matrix in regions far away from the plates. The strain in an element of matrix near a plate is zero because the ends of each slip plane adhere strongly to the plates, which do not undergo shear (being nondeformable). This implies a rotation of the plates through an angle $\phi = \tan^{-1} \gamma$ with respect to NN', normal to the

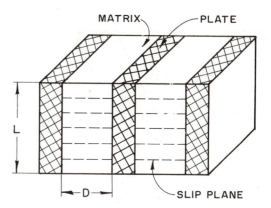

Figure 11.12 Nondeforming plates adhering to the matrix. (Adapted with permission from [7], p. 409.)

original slip plane, and bending of slip planes from ϕ to zero, and again from zero to ϕ, thus reestablishing the original orientation (Fig. 11.13).

The matrix lattice acquires a curvature of magnitude $2\phi\,D$. (*Exercise*: Show that the average curvature is $2\phi\,D$.) This lattice rotation can be easily measured by means of X-rays. A Laue back-reflection spot from the slip planes of composite which was well defined before the deformation will show asterism of angular width equal to ϕ about an axis in the slip plane and normal to the slip direction. The measured lattice rotation is directly related to the density of geometrically necessary dislocations, as shown below.

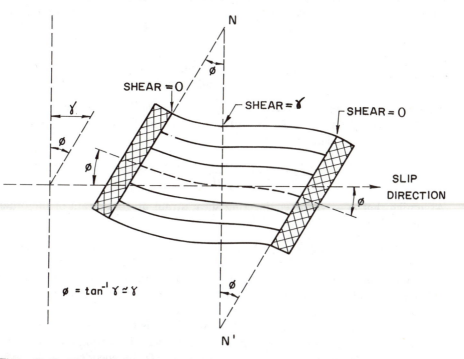

Figure 11.13 The matrix has sheared on a slip system. As it cannot shear near the plate, it suffers a rotation there. (Adapted with permission from [7], p. 409.)

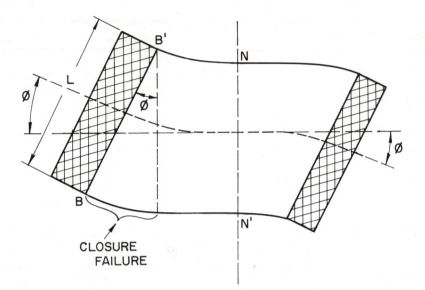

Figure 11.14 Burgers circuit showing closure failure.

Construct a Burgers circuit $BB'NN'$ in the matrix (Fig. 11.14). The absence of any slip at the interface means that a crystallographic direction that was parallel to the interface before straining remains parallel to the interface after straining, too. Thus, the crystallographic direction BB' of matrix that was parallel to NN' (normal to the glide plane) before straining now makes an angle ϕ with it. We can write, then,

$$\text{area of Burgers circuit} = \tfrac{1}{2}DL$$
$$\text{closure failure} = \phi L$$

Thus, the circuit encloses $\phi L/b$ dislocations, where b is the Burgers vector ($\rho b = $ closure failure $= \phi L$) and D is the spacing between the plates. The average dislocation density in the central part of the circuit is

$$\overline{\rho^G} = \frac{\phi L/b}{\tfrac{1}{2}DL} = \frac{2\phi}{Db}$$

Assuming that $\phi \simeq \gamma$, the engineering shear strain up to a strain of the order of 0.4, we have

$$\overline{\rho^G} \simeq \frac{2\gamma}{Db} \tag{11.19}$$

The other half of the crystal (Fig. 11.13) contains an equal number of dislocations, but of opposite Burgers vector. Ignoring the sense of the Burgers vector, Eq. 11.19 describes the dislocation density, taking average over the specimen volume. The reader should note that this density is independent of the detailed manner by which the slip planes assumed the curvature. The total number depends on the shear γ and the spacing D, independent of the local dislocation distribution.

The slip direction, too, in general, will not be normal to the plate face. Besides, two or three slip systems may operate simultaneously. Although a complex pattern may be necessary, the closure failure of Burgers circuit will be the same. The lattice curvature will develop as before and dislocations will accumulate. The dislocation density will be higher than that in the case of single slip, say $4\gamma/Db$, but will vary with D and γ in the same way.

These are the dislocations geometrically necessary to solve the compatibility problem which arises because the inclusions do not deform when the crystal as a whole deforms. By making D sufficiently small (i.e., reducing the plate spacing), we can make the density of geometrically necessary dislocations, ρ^G, to be the dominant factor in the total density, ρ.

11.5.2 Accumulation of Dislocations Due to Equiaxial Particles

Consider an element of matrix in the form of a cube containing a nondeformable particle [Fig. 11.15(a)]. Let the cube be sheared plastically through γ. This plastic shear strain, γ, can be decomposed into the following operations:

1. Remove the particle from the cubic element and deform the cube uniformly with the cavity inside it [Fig. 11.15(b)].

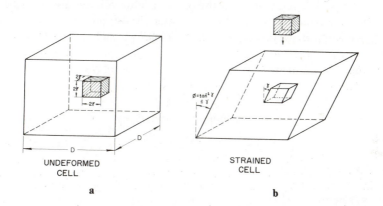

UNDEFORMED CELL

STRAINED CELL

a

b

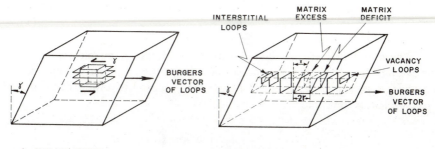

c) SHEAR MODE

d) PRISMATIC MODE

Figure 11.15 Accumulation of dislocations due to equiaxial particles. (Adapted with permission from [7], p. 405.)

2. Replace the nondeformed particle in the cavity. The compatibility requires that the hole be redeformed to bring it back to its original shape.
3. There are various possible groups of displacements that can be used to deform the cavity, returning it to its original shape, such as:
    a. Shear back part of the matrix immediately surrounding the hole by shear of γ [Fig. 11.15(c)]. These displacements are accomplished by inserting n shear loops surrounding the particle. The geometry requires that

$$n = \frac{2r\gamma}{b}$$

where $2r$ is the diameter or size of the particle and b is the Burgers vector.

    b. Prismatic dislocation loops that can restore the original form of the hole by removing material from one side and adding it to the other side [Fig. 11.15(d)]. From the geometry of the situation, we find that the volume of the material on the left-hand side of the particle, which must be removed by prismatic slip, is

$$\Delta V = \tfrac{1}{2} V_p \gamma$$

where V_p is the volume of the particle. This volume ΔV must be added into the matrix in the form of interstitial-type prismatic loops with the same Burgers vector as that of the original shear of primary slip. On the right-hand side of the particle, an equal volume of the matrix must be furnished by the matrix by means of vacancy-type prismatic loops.

Thus, the total number of loops per particle is

$$n = \frac{V_p \gamma}{2r^2 b}$$

where $2r^2$ is the average area of loops which are not of equal size, and $V_p = (2r)^3 = 8r^3$. Thus,

$$n = \frac{4r\gamma}{b} \tag{11.20}$$

These n loops are associated with the volume of an elementary cube. The average density of all types of prismatic loops (i.e., number per unit volume of matrix) depends on the number of particles N_p per unit volume of the matrix. Thus, we can write

$$N_p = \frac{f}{\tfrac{4}{3}\pi r^3} \qquad \text{for the spherical case}$$

and

$$N_p = \frac{f}{8r^3} \qquad \text{for the cubic case}$$

where f is the particle volume fraction.

The number of loops geometrically necessary is

$$\overline{N}^G = nN_p = \frac{4r\gamma}{b}\frac{f}{8r^3}$$

or

$$\overline{N}^G = \frac{1}{8r}\frac{4f\gamma}{br} \tag{11.21}$$

Since the loops are not equal in size, it will be incorrect to take the loop perimeter to equal $16r$. Instead we can approximate the average loop perimeter to equal $8r$; then, the dislocation density in form of prismatic loops is

$$\bar{\rho}^G \simeq \frac{4f\gamma}{br} \tag{11.22}$$

These dislocation loops are not fundamentally different from the ones discussed in the earlier section. They represent only an alternative way of assuring compatibility on a microscopic scale. However, when prismatic loops are introduced, the compatibility is realized without introducing a large-scale curvature. Thus, Laue reflection spots will not show excessive asterism when this mechanism is operating. Hirsch and Humphreys[11] have observed by transmission electron microscopy, in a Cu–30% Zn alloy containing Al_2O_3 particles, shear loops around the alumina particles at low strains ($\gamma \approx 1\%$) and prismatic loops at higher strains.

11.6 STRESS–STRAIN CURVE EQUATION FOR CRYSTALS CONTAINING INCLUSIONS

We derive below the equation for the stress–strain curve of plastically nonhomogeneous crystals.

‡11.6.1 Relation Between Flow Stress and Dislocation Density

Consider the following nondimensional analysis. The shear flow stress, τ, must depend on:

1. Shear modulus, G
2. Magnitude of Burgers vector, b
3. Average dislocation density, $\bar{\rho}$

The only nondimensional combination among these quantities is

$$\frac{\tau}{G} = c(b^2\bar{\rho})^m \tag{11.23}$$

where both the sides are pure numbers and c and m are constants. The applied stress τ exerts a force τb per unit length of the moving dislocation when ideally oriented (Peach–Koehler force; see Section 6.6). The movement of this dislocation is opposed by the nearest obstacle among the dislocation groups which act as obstacles.

The interaction force between any two dislocations, be they parallel or interpenetrating, is proportional to b^2 (see Section 6.6):

$$\tau b \propto b^2$$

or

$$\tau \propto b \tag{11.24}$$

Comparing Eqs. 11.23 and 11.24, we get

$$m = \tfrac{1}{2}$$

or

$$\tau = cGb\sqrt{\bar{\rho}}$$

In general, a friction stress (roughly the matrix flow stress) and internal stresses of constant magnitude are also present and oppose the movement of a dislocation. These may be included by writing

$$\tau = \tau_0 + cGb\sqrt{\bar{\rho}} \tag{11.25}$$

This is a familiar expression of flow stress varying as the square root of the dislocation density (see Chapter 9). Almost any detailed analysis of the interaction between a moving dislocation and another serving as an obstacle gives this square-root dependence of τ on ρ. The constant c is roughly 0.25.

11.6.2 Relation Between Flow Stress and Prismatic Loop Density

A similar analysis predicts the hardening due to a distribution of prismatic loops which also act as obstacle to slip. Thus,

$$\frac{\tau}{G} = c'(\bar{N}rb^2)^m$$

where $\bar{N}r$ has the units of dislocation length and $m = \tfrac{1}{2}$, by the same argument. Including the friction stress term, we can write

$$\tau = \tau_0 + c'Gb\sqrt{\bar{N}r} \tag{11.26}$$

11.6.3 Stress–Strain Relations

It was concluded in Section 11.5.1 that in the general case of dislocation accumulation due to platelike nondeforming particles,

$$\bar{\rho}^G = \frac{4\gamma}{Db}.$$

Taking

$$\tau = \tau_0 + 0.25 Gb\sqrt{\bar{\rho}}$$

and when $\bar{\rho}^G \simeq \bar{\rho}$, we have

$$\tau = \tau_0 + 0.50G \sqrt{\frac{b\gamma}{D}} \tag{11.27}$$

where γ is the engineering shear strain, D is the spacing between the inclusions, and G is the matrix shear modulus.

For equiaxial inclusions, we have (see Section 11.5.2)

$$\overline{N}^G = \frac{f\gamma}{2br^2}, \qquad \bar{\rho}^G = \frac{4\gamma f}{br}$$

and

$$\tau = \tau_0 + c'Gb\sqrt{\overline{N}^G r} = \tau_0 + c'Gb\sqrt{\bar{\rho}}$$

Taking $c = c' = 0.25$ and $\bar{\rho} = \bar{\rho}^G$, we have

$$\tau = \tau_0 + 0.50\, G \sqrt{\frac{f\gamma b}{2r}} \tag{11.28}$$

where $2r$ is the average particle diameter and f is the particle volume fraction.

11.7 THE GEOMETRIC SLIP DISTANCE AND THE RELATIVE MAGNITUDES OF ρ^G AND ρ^S

The parameter geometric slip distance, λ^G, describes the degree of efficiency of obstacles during dislocation storage. The total dislocation density of a crystal can be written as

$$\rho = \rho^G + \rho^S$$

When $\rho^G \gg \rho^S$, properties such as flow stress that depend on the quantity of dislocations will reflect the dominant presence of ρ^G.

11.7.1 Geometric Slip Distance, λ^G

This describes the effect of particles, plates, or even grain boundaries on dislocation accumulation. λ^G is a characteristic of the microstructure. Nondeforming plates, for example, lead to a dislocation density (see Section 11.5.1)

$$\bar{\rho}^G = \frac{1}{D}\frac{4\gamma}{b}$$

Equiaxial particles lead to (see Section 11.5.2)

$$\bar{\rho}^G = \frac{f}{r}\frac{4\gamma}{b}$$

The quantity D for the plates and r/f for the equiaxial particles is called the geometric slip distance. In an analogous manner, we define λ^S as the slip distance for the statistically stored dislocations; thus,

$$\delta\rho^S = \frac{1}{\lambda^S}\frac{4\,\delta\gamma}{b} \tag{11.29}$$

In a very general way, the storage of geometrically necessary dislocations can be described by

$$\bar{\rho}^G = \frac{1}{\lambda^G}\frac{4\gamma}{b} \tag{11.30}$$

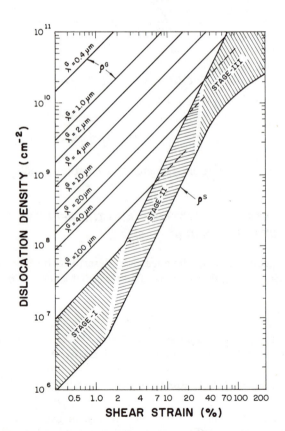

Figure 11.16 Dislocation density (ρ^s and ρ^G) versus shear strain. (Adapted with permission from [7], p. 416.)

It should be noted that λ^G in polycrysta... proportional to the grain size or phase separation... in Chapter 14. Thus, λ^G is a microstructural charac... of deformation. λ^S, on the other hand, varies with... obtained from the slip-line length measurements. Fo... monocrystal, λ^S in stage I is about 100 μm, whereas in 10 μm.

Systems in which the microstructure imposes the con... λ^G or microstructure will have little effect on hardening. If, λ^S, which in practice means that λ^G is less than 10 μm, th... will have a strong effect on dislocation storage and, consequ... hardening.

11.7.2 Relative Magnitudes of ρ^G and ρ^S

The shaded band in Fig. 11.16 shows that ρ^S increases with strai... plotted in the same figure is the variation of ρ^G with strain for a series values. For small strains, ρ^G can predominate the total dislocation den... but at large strains, it can be overtaken by ρ^S. In general, for $\lambda^G < 2$ μm, will predominate, whereas for $\lambda^G > 100$ μm, ρ^G will be negligible.

11.8 WORK HARDENING IN PLASTICALLY NONHOMOGENEOUS MATERIALS

Basically work hardening occurs because stored groups of dislocations, ρ^G and ρ^S, impede the movement of other mobile dislocations. The stress–strain curves for pure copper single crystal and copper containing single crystal of the same orientation but with varying volume fractions of SiO_2 particles that is, nondeforming particles are shown in Fig. 11.17.[12] In pure copper monocrystal, the work hardening is entirely due to ρ^S (Fig. 11.17). Increasing the volume fraction of SiO_2 particles, we manage to reduce λ^G from 20 μm to 2 μm. In the last curve (No. 4) only λ^G controls the work hardening and the stress–strain curve is parabolic.

The work-hardening rates of a series of copper monocrystals strengthened by dispersions[13,15] are plotted as a function of geometric slip distance, λ^G in Fig. 11.18. One notes that the work-hardening rate increases rapidly in relation to the characteristic work-hardening rates of pure FCC crystals as λ^G is reduced to values less than those of λ^S.

‡11.9 MICROALLOYED STEELS

Steels form the most important group of engineering materials. And there has been over the years a continual evolution in the physical and process metallurgy of steels to meet newer demands and challenges. The development of microal-

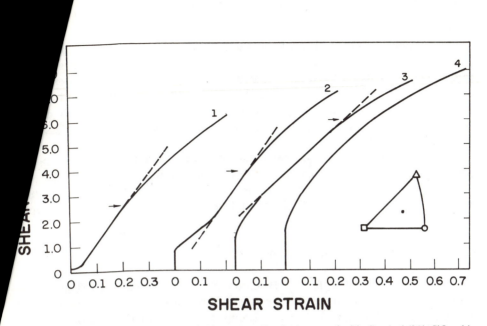

Figure 11.17 Stress–strain curves of (a) a pure Cu single crystal; (b) Cu + 1/3% SiO₂; (c) Cu + 2/3% SiO₂; (d) Cu + 1% SiO₂. (Adapted with permission from [12], p. 815.)

loyed steels in the sixties and seventies can be regarded as one of the greatest metallurgical achievements. One can safely say that it was the result of a clearer understanding of structure–property relations in the low-carbon steels. Of course, the final product resulted from a fruitful combination of physical, mechanical, and process metallurgy. Microalloyed steels are successfully substituting the mild steel as the basic structural materials.

There is some confusion about the terminology in the literature. Earlier, the commonly accepted term was high-strength low-alloy (HSLA) steels; in

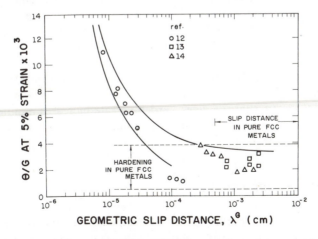

Figure 11.18 Strain hardening rate (θ/G) at 5% of shear strain versus geometric slip distance, λ^G. (Adapted with permission from [7], p. 419.)

TABLE 11.2 Important Precipitates in High-Strength Low-Alloy Steels

Element(s)	Main Precipitates
Niobium	$Nb(C,N)$, Nb_4C_3
Vanadium	$V(C,N)$, V_4C_3
Niobium + molybdenum	$(Nb,Mo)C$
Vanadium + nitrogen	VN
Copper + niobium	Cu, $Nb(C,N)$
Titanium	$Ti(C,N)$, TiC
Aluminum + nitrogen	AlN

the 1970's, though, the term "microalloyed steels" has gained wider acceptance. A microalloyed steel is a low-carbon steel (0.05 to 0.2% C, 0.6 to 1.6% Mn) that contains about 0.1% of elements such as Nb, V or Ti. Some other elements, such as Cu, Ni, Cr, and Mo, may also be present in small proportions (up to about 0.1%). Elements such as Al, B, O, and N also have important effects in microalloying additions. Table 11.2 lists some important second phases generally encountered in HSLA steels.

Microalloyed steels are generally subjected to what is called a "controlled rolling" treatment. Controlled rolling is nothing but a sequence of deformations by hot rolling at certain specific temperatures and controlled cooling. The main objective of this is to obtain a fine ferritic grain size. The ferritic grain size obtained after austenitization and cooling depends on the initial austenitic grain size.[16] This is so because ferrite nucleates preferentially at the austenite grain boundaries. The ferrite grain size also depends on the austenite (γ) $\rightarrow$ ferrite (α) transformation temperature. Lower transformation temperatures favor the nucleation rate which results in a large number of ferritic grains, and consequently in a very small ferritic grain size (5 to 10 μm). Thus, to obtain a maximum of grain refinement, the controlled rolling procedure modifies the hot-rolling process with a view to exploiting the capacity of the microalloying elements to retard the recrystallization of the deformed austenite grains. The microalloyed additions by means of precipitation of second-phase particles during the austenitization treatment impede the austenite grain growth. This precipitation of second phase (e.g., carbides or carbonitrides of Nb, V, or Ti) can inhibit or retard the austenite grain growth, resulting in a posterior ferrite grain refinement. The Hall–Petch relationship between the yield stress and the ferrite grain size (Fig. 11.19) shows the strengthening possible by grain size refinement. Notice the much finer grain size obtainable in the C-Mn-Nb steel compared to the C-Mn steel. The reader will have also noticed that the Nb steel curve in Fig. 11.19 is higher than the C-Mn steel curve. This implies that besides grain size strengthening there is another strengthening mechanism associated with the presence of Nb, strengthening due to carbide precipitation. Thus, in summary, during hot rolling the fine carbide particles form in austenite and control its recrystallization and, consequently, result in a fine ferritic grain size. Second, carbides of Nb precipitate during and soon after the $\gamma \rightarrow \alpha$ transforma-

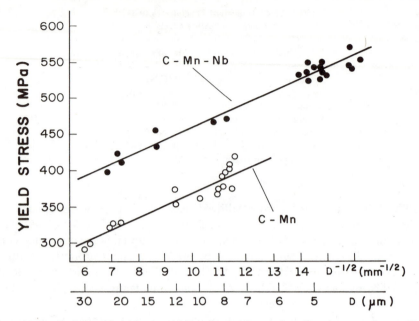

Figure 11.19 Variation of yield stress with grain size for a microalloyed steel and an ordinary steel. (Adapted with permission from [17], p. 91.)

tion lead to a precipitation strengthening of the ferrite. We discuss the precipitation hardening in these steels in the next section. Suffice it here to say that these two strengthening methods together lead to steels with yield strengths in the range 400 to 600 MPa and with good toughness. A word about the toughness of these steels is in order here. One of the measures of toughness is the ductile–brittle transition temperature (see Chapter 3). The variation in this ductile–brittle transition temperature (DBTT) with an increase in the yield stress ($\Delta\sigma_{\text{yield}}$) due to some strengthening mechanisms is shown in Fig. 11.20. We note from this figure that the only microstructural characteristic that improves the strength as well as the toughness of a steel is the grain size refinement. The Hall–Petch relation between the yield stress and inverse square root of the grain size is well known. According to Petch,[18] there exists the following relation between T_c (ductile brittle transition temperature) and the grain diameter D:

$$\beta T_c = \ln\beta - \ln c - \ln D^{-1/2}$$

where β and c are constants. Little et al.[19] showed that the ultrafine grain size was mainly responsible for superior mechanical properties of controlled rolled steels. In particular, they demonstrated (Fig. 11.21) a clear relationship between 50% fracture appearance transition temperature (FATT) and $D^{-1/2}$ (D is the grain size as obtained by mean linear intercept measured parallel to the rolling and transverse directions.)

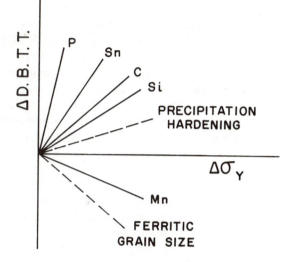

Figure 11.20 Change in ductile brittle transition temperature (ΔDBTT), with some strengthening mechanisms.

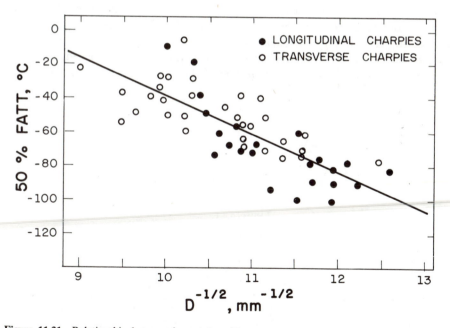

Figure 11.21 Relationship between impact transition temperature as measured by 50% FATT and $D^{-1/2}$ (*D*, grain size). (Adapted with permission from [19], p. 84.)

‡11.10 PRECIPITATES IN STEELS

We saw in Section 11.2.3 that for precipitates with very small spacing (≤10 nm), the dislocations cut them. However, there is little evidence of such shear of precipitates in steels. The carbides, nitrides, or carbonitrides are very hard (Diamond Pyramid Hardness 2500 to 3000). The presence of such hard particles in a matrix means that dislocations will only be able to cut them when they are extremely small. The critical particle size, which corresponds to a transition between the Orowan mechanism and the particle shear mechanism, decreases with an increase in the particle hardness, as shown schematically in Fig. 11.22.

The generally accepted theory for precipitation strengthening is the one proposed by Orowan (Section 11.2.2) and modified by Ashby.[20] According to this theory, the dislocations do bow through the well-spaced particles following the original model of Orowan. Ashby's[20] principal modification stems from the fact that he does not make the approximation of dislocation line tension, T, being equal to the dislocation line energy, E; instead, he uses an expression for T that takes into account the dependence of line tension and energy on the character of the dislocations (i.e., on the angle θ between the Burgers vector and the tangent to the dislocation line). The expression is[20-22]

$$T(\theta) = E(\theta) + \frac{d^2E(\theta)}{d\theta^2}$$

Combining this fact with other modifying factors such as the effect of scale of bending on dislocation energy and the effect of random particle spacing, he arrives at a very simple conclusion: that the yield stress is inversely proportional to the spacing 'l' between the particles when they are well spaced, which is the case in steels, where the precipitation of the second phase occurs heteroge-

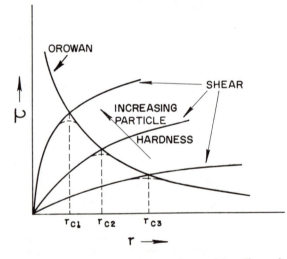

Figure 11.22 Change in shear stress (τ) with increasing particle radius and particle hardness.

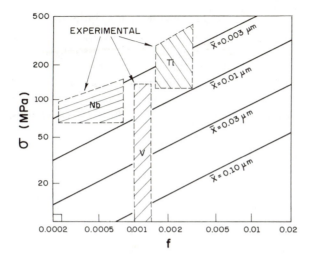

Figure 11.23 Experimental results from commercial steels superimposed on the predictions of the Ashby–Orowan equation. (Adapted with permission from [24], p. 8.)

neously at the dislocations. The final expression for the Ashby–Orowan model can be described by[20]

$$\tau = \frac{Gb}{2\pi \mathit{l}} \mathit{l}n \frac{\overline{x}}{2b}$$

where τ is the resolved shear stress necessary to overcome the effect of precipitates: $\overline{x}$ is the average precipitate diameter measured by planar intercept, l is the spacing (surface to surface) between precipitates and defined as $(1/\sqrt{n_s})$ $-\ \overline{x}$, where n_s is the number of precipitates per unit area of the slip plane, G the shear modulus of ferrite (80.3 GPa), and b the Burgers vector.

Figure 11.23 shows the experimental results from commercial steels superimposed on the predictions of the Ashby–Orowan equation.[23] One notes that the precipitation hardening increases with diminishing precipitate size and increasing volume fraction. However, concomitant with the improvement in the strength of steel with precipitation, there occurs a decrease in toughness. Irvine et al.[25] indicate that for each 4 MPa of increase in the yield stress due to the precipitation of carbides in steels, there occurs an increase of 1°C in the ductile brittle transition temperature.

‡11.11 MECHANICAL ALLOYING

By mechanical alloying it is possible to form true mixtures of metals and metal oxides which, thermodynamically, would not form. Mechanical alloying has been successfully applied to superalloys. It consists of mechanical blending very finely ground powders (e.g., tungsten carbide and cobalt) and then applying pressure and heat, either sequentially or simultaneously. The amount of deformation is considerable. During mechanical alloying, the metal powders are subjected to repetitive fracturing and welding, resulting in an intimate mixing of the constituents. It has been possible, by mechanical alloying, to produce superalloys[26] having the high-temperature strength of dispersion-strengthened alloys and the low-temperature strength of precipitation-strengthened superalloys. Mechanical alloying has also been applied to aluminum alloys, and IN-9052 and IN-9021 are two alloys possessing attractive combinations of tensile strength, fatigue strength, toughness, and stress corrosion cracking (SCC) resistance.[27]

EXERCISES

11.1. The stress required to bend a segment of dislocation between two particles is given by

$$\tau = \frac{2T}{bl}$$

where T is the dislocation line tension, b the Burgers vector, and l is the distance between the two particles. What is the effect on τ of the following changes:
(a) Dissociation of a dislocation into two partials?
(b) Association of two dislocations to form a superdislocation?

11.2. Derive Eq. 11.7.

11.3. Show that, for a given applied shear stress, the radius of curvature of an edge dislocation, r_{edge}, is approximately one-third of the radius of curvature of a screw dislocation, r_{screw}, that is,

$$\frac{r_{edge}}{r_{screw}} \simeq \frac{1}{3}$$

If we take into account the fact that the dislocation self-energy E, or line tension T, varies with the character of dislocation (i.e., θ), what will be the equilibrium form of a dislocation loop under a constant stress τ?

11.4. Consider a unit cube of a matrix containing uniform spherical particles (radius r) of a dispersed second phase.
(a) Show that the average distance between the particles is given by

$$\Lambda = \frac{1}{(N)^{1/3}} - 2r$$

where N is the number of particles per unit volume.
(b) Compute Λ for a volume fraction f of particles equal to 0.001 and $r = 10^{-6}$ cm.

11.5. For a precipitation-hardenable alloy, estimate the maximum precipitate size that can undergo shear by dislocations when plastically strained. Given: matrix shear modulus = 35 GPa, Burgers vector = 0.3 nm, and specific energy of precipitate–matrix interface = 100 mJ/m².

11.6. SAP (sintered aluminum powder) alloy is a trade name alloy of aluminum containing a dispersion of alumina particles. If the volume fraction of Al_2O_3 is 1% and if the mean particle diameter is 20 nm, compute the density of geometrically necessary loops in this alloy when it is deformed to a shear strain of 10%.

11.7. An FCC alloy has a lattice parameter $a = 0.34$ nm and a shear modulus $G = 40$ GPa.

(a) Considering that the grain boundaries act as nondeformable obstacles to slip, compute the flow stress of this alloy when it is deformed 10% in shear. The grain size is 10 μm.

(b) If this alloy contains equiaxial particles of diameter 100 nm and a volume fraction of 1%, what would be the flow stress corresponding to a shear strain of 10%?

Take the friction stress to be zero in both the cases.

11.8. Show that the lattice of a matrix containing platelike inclusions acquires an average curvature of $2\Phi/D$, where D is the spacing between the plates and $\Phi = \tan^{-1} \gamma$, γ being the shear strain. Compute the density of geometrically necessary dislocations, ρ^G, in an alloy containing platelike dispersions with $D = 1$ μm and $\gamma = 10^{-2}$.

11.9. An aluminum alloy contains 2% volume fraction of a precipitate that results in $\epsilon = 5 \times 10^{-3}$. Determine the average spacing (l) between precipitates above which there will be a significant contribution to strength due to the difference in atomic volume of the matrix and the precipitate. Below this critical value of l, what will be the mechanism controlling yielding?

11.10. In metallic alloys where the difference between the coefficient of linear thermal expansion of the matrix and that of the dispersed phase is large, thermal stresses and strains can be generated during fabrication. In order to relax the thermal stresses generated, dislocations may be generated in the matrix volume around each dispersoid. Derive an approximate expression for the number of dislocations generated in an alloy which has particles of diameter d, which was submitted to cooling from a temperature of fabrication T_f to ambient temperature T_a.

11.11. J. Dundurs and T. Mura [*J. Mech. Phys. Solids*, **12** (1964) 177] showed that the interaction between a screw dislocation and an infinite cylindrical inclusion, with the cylinder axis parallel to the dislocation, is given by

$$F = \frac{Gm\, b^2}{2d\pi} \frac{G_i - G_m}{G_i + G_m} \frac{R^2}{d^2 - R^2}$$

where F is the force on dislocation per unit of length; G_m and G_i the shear moduli of the matrix and inclusion, respectively; d the distance between the dislocation and the inclusion axis; and R the radius of the infinite inclusion cylinder.

(a) Develop the expressions for this force when

$$d \gg R$$
$$d = R + x$$

where $x \ll R$ or d.

(b) When will this force be attractive and when will it be repulsive?

(c) For an immensely hard inclusion, derive the expression for this interaction force. What is the significance of this force with respect to the Orowan stress in the alloy containing inclusions; that is, what will be the effect of this force on the effective inclusion size?

(d) Estimate the interaction force between a dislocation and a large void.

11.12. One can write the yield stress of an alloy as

$$\sigma_y = \sigma_f + 0.25\, Gb\sqrt{\rho}$$

where σ_f is the friction stress and the other symbols have the normal significance. Also, in an alloy containing extremely finely distributed nondeforming precipitates, one has $\rho^G \gg \rho^S$. One can consider that in an extremely fine grained (i.e., control rolled) microalloyed steel, the grain boundaries, at low temperatures, act as nondeformable plates. In such a case, derive the Hall–Petch relation

$$\sigma_y = \sigma_f + kD^{-1/2}$$

where D is the average grain diameter and k is a constant. Show that k depends on strain. Explain the significance of this expression.

REFERENCES

[1] N.F. Mott and F.R.N. Nabarro, *Proc. Phys. Soc.* (*London*), *52* (1940) 86.

[2] H.K. Hardy and T.J. Heal, *Prog. Metal Phys.*, 5 (1954) 195.

[3] E. Orowan, in *Internal Stresses in Metals and Alloys*, Institute of Metals, London, 1948, p. 451.

[4] A. Kelly and R.B. Nicholson, *Prog. Mater. Sci.*, *10* (1963) 149.

[5] L.M. Brown and R.K. Ham, in *Strengthening Methods in Crystals*, A. Kelly and R.B. Nicholson (eds.), Elsevier, Amsterdam, 1971, p. 9.

[6] A. Kelly, *Proc. Roy. Soc.* (*London*), *A282* (1964) 63.

[7] M.F. Ashby, *Phil. Mag.*, *21* (1970) 399.

[8] J.C. Fisher, E.W. Hart, and R.H. Pry, *Acta Met.*, *1* (1953) 336.

[9] M.F. Ashby, *Phil. Mag.*, *14* (1966) 1157.

[10] M.F. Ashby, in *Strengthening Methods in Crystals*, A. Kelly and R.B. Nicholson (eds.), Elsevier, Amsterdam, 1971, p. 137.

[11] P.B. Hirsch and F.J. Humphreys, in *Physics of Strength and Plasticity*, A.S. Argon (ed.), MIT Press, Cambridge, Mass., 1969, p. 189.

[12] R. Ebeling and M.F. Ashby, *Phil. Mag.*, *18* (1966) 805.

[13] R.L. Jones and A. Kelly, in *Oxide Dispersion Strengthening*, Proc. 2nd Bolton Landing Conf., G.S. Ansell (ed.), Gordon and Breach, New York, 1968, p. 000.

[14] F.J. Humphreys and J.W. Martin, *Phil. Mag.*, *16* (1967) 927.

[15] R.L. Jones, *Acta Met.*, *17* (1969) 229.

[16] K.J. Irvine and F.B. Pickering, *J. Iron Steel Inst.*, *201* (1963) 944.

[17] A.B. Le Bon and L.N. de Saint-Martin, in *Microalloying '75*, Union Carbide, New York, 1977, p. 90.

[18] N.J. Petch, *Proc. Fracture*, Swampscott Conf., Wiley, New York, 1959, p. 54.

[19] J.H. Little, J.A. Chapman, W.B. Morrison, and B. Mintz, in *The Microstructure and Design of Alloys*, Vol. 1, Metals Society, London, 1976, p. 80.

[20] M.F. Ashby, in *Oxide Dispersion Strengthening*, Proc. 2nd Bolton Landing Conf., G.S. Ansell (ed.), Gordon and Breach, New York, 1968, p. 143.

[21] A.J.E. Foreman, *Acta Met.*, *3* (1955) 322.

[22] G. de Wit and J.S. Koehler, *Phys. Rev.*, *116* (1959) 1113.

[23] L.M. Brown, *Phil. Mag.*, *10* (1964) 441.

[24] T.J. Gladman, D. Dulieu, and I.D. McIvor, in *Microalloying '75*, Union Carbide, New York, 1977, p. 32.

[25] K.J. Irvine, F.B. Pickering, and T.J. Gladman, *J. Iron Steel Inst.*, *205* (1967) 161.

[26] S. Benjamin, *Sci. Am.*, *234*, No. 5 (1976) 40.

[27] D.L. Erich and S.J. Donachie, *Metal Prog.*, Feb. 1982, p. 22.

SUGGESTED READING

ANSELL, G.S. (ed.), *Oxide Dispersion Strengthening*, Proc. 2nd Bolton Landing Conf., Gordon and Breach, New York, 1968.

ARGON, A. (ed.), *Physics of Strength and Plasticity*, MIT Press, Cambridge, Mass., 1969.

FINE, M.E. *Phase Transformations in Condensed Systems*, Macmillan, New York, 1964.

FLETCHER, E.E. *High-Strength, Low-Alloy Steels*, Battelle Press, Columbus, Ohio, 1979.

GEROLD, V., in *Dislocations in Solids*, Vol. 4, F.R.N. Nabarro (ed.), Elsevier/North-Holland, New York, 1979, p. 219.

HONEYCOMBE, R.W.K. *Steels: Microstructure and Properties*, Arnold, London, 1981.

KELLY, A., and R.B. NICHOLSON (eds.), *Strengthening Methods in Crystals*, Elsevier, Amsterdam, 1971.

MARTIN, J.W., *Precipitation Hardening*, Pergamon Press, Oxford, 1968.

Microalloying '75, Union Carbide, New York, 1977.

PICKERING, F.B., *Physical Metallurgy and the Design of Steels*, Applied Science Publishers, Essex, England, 1978.

Chapter 12

FIBER REINFORCEMENT
(COMPOSITE MATERIALS)

‡12.1 INTRODUCTION

Precipitation or dispersion of a hard second phase in a metallic matrix results in a dramatic increase in the yield stress and/or the work-hardening rate. However, the influence of these obstacles on the elastic modulus is negligible. The concept of fiber reinforcement is just the opposite. The idea is not to use the fibers as obstacles in a matrix to modify the properties of latter (such as yield stress) but to use the matrix as a medium to bind the fibers together and to transfer the applied load to the fibers. Thus, one uses the high load-bearing capacity of the fibers. In materials that contain nondeformable particles, the intrinsic properties of the particles (i.e., the high elastic modulus) are not utilized. The principal function of the inclusions is to impede the movement of dislocations and act as a source of secondary slip. Figure 12.1 shows the effect of a series of nondeformable particles on the yield stress of an aluminum matrix as a function of the volume fraction of the particles. Any dependence on the elastic modulus of the particles is conspicuous by its absence. Any kind of particle serves the function: to obstruct dislocation movement in the matrix. The load-bearing capacity of the various strong particles is not exploited. In fiber reinforcement we precisely exploit this greater load-bearing capacity of the second phase, and it occurs because the hard second phase is in the form of fibers. Table 12.1 compares dispersion hardening and fiber reinforcement.

The reinforcement of resins by glass fibers has been in use since the 1920s. Fiber-reinforced resins are light and very strong materials, although the rigidity or the modulus is not very high, mainly because the glass fibers do not possess

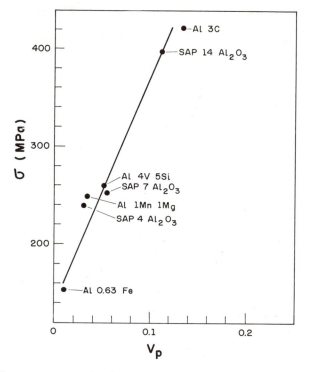

Figure 12.1 Flow stress σ as a function of particle volume fraction V_p at room temperature. (Adapted with permission from [1], p. 321.)

high modulus. Nowadays, there exist various advanced fibers that have extremely high modulus, for example, boron, carbon, and silicon carbide, which are being used for reinforcing resins, metals, and ceramics.[2,3] Table 12.2 lists properties of some of the fibers currently in use. Examples of fiber-reinforced materials are shown in Fig. 12.2. Figure 12.2(a) shows the transverse section of boron-

TABLE 12.1 Comparison of Dispersion Hardening with Fiber Reinforcement

	Dispersion Hardening	Fiber Reinforcement
Role of matrix	Principal load-bearing component	Medium for transferring load to the fibers
Matrix work hardening	Major hardening mechanism; work-hardening rate depends on the particle forms and the spacing between them	Minor hardening factor
Role of the dispersed phase	Impedes the dislocation movement	Principal load-bearing component; obstruction of dislocation movement by indirect means is of secondary importance

TABLE 12.2 Properties of Some Reinforcement Fibers[a]

Fiber	Melting Point (°C)	Density (g/cm³)	Strength (GPa)	Young's Modulus (GPa)	Specific Strength[b]	Specific Modulus[c]
E-glass	700	2.55	1.8-3.5	72	0.7-1.4	28
S-glass	840	2.50	4.6	84	1.8	34
Silica	1660	2.19[d]	6.0	72	2.7	33
Alumina (single crystal)	2072	3.96	2.0	470	0.5	118
Carbon type I	3000	1.90	2.0	390	1.1	204
Carbon type II	3000	1.90	2.6	240	1.4	126
Boron nitride	2980	1.90	1.4	90	0.7	47
Boron[e]	2040	2.63	2.8	380	1.1	145
Silicon carbide[e]	2200	3.35	2.3	470	0.7	140
Boron[e] carbide	2450	2.36	2.3	470	1.0	200
Titanium diboride	2980	4.48	1.0	510	0.2	114
Tungsten	3390	19.3	4.0	410	0.2	210
Beryllium	1284	1.83	1.3	240	0.7	131
Kevlar	—	1.5	2.8	133	1.9	90

[a] Adapted with permission from [4], p. 297.

[b] Strength divided by the numerical value of density.

[c] Modulus divided by the numerical value of density.

[d] Silica bulk density, not in the fiber form.

[e] With tungsten core.

reinforced aluminum, one of the advanced systems. The boron fiber is prepared by vapor deposition of boron on a heated tungsten wire substrate. Thus, the core that one sees in each boron fiber is nothing but a series of tungsten borates. Figure 12.2(b) shows a transverse section of a carbon fiber-reinforced polyester composite. Figure 12.2(c) shows a deeply etched transverse section of a unidirectionally solidified eutectic composite showing a uniform array of NbC rods in a Ni-Cr matrix.

12.2 COMPOSITE PROPERTIES

We describe below some of the important properties of fiber-reinforced composites and their prediction in terms of the component properties and their geometric configuration.

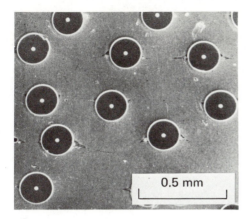

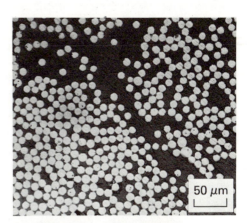

a

b

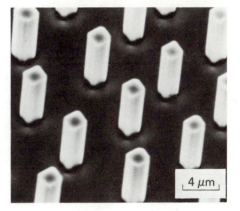

Figure 12.2 (a) Transverse section of a boron fiber-reinforced aluminum composite. $V_f = 10\%$. (Scanning electron microscope.) (b) Transverse section of a carbon fiber-reinforced polyester resin. $V_f = 50\%$. (Optical.) (c) Deeply etched transverse section of a eutectic composite showing NbC fibers in a Ni-Cr matrix. (Courtesy of S. P. Cooper and J. P. Billingham, GEC Turbine Generators Ltd., UK.)

c

12.2.1 Elastic Moduli

The simplest model for predicting the elastic properties of a fiber-reinforced composite is shown in Fig. 12.3. In the longitudinal direction, the composite is represented by a system of "action in parallel" [Fig. 12.3(a)].

For loads applied in the direction of the fibers, assuming equal deformations in the components, the two (or more) phases are viewed as being deformed in parallel. This is the classic case of Voigt's average.[5] In this case, one has

$$P_c = \sum_{i=1}^{n} P_i V_i \tag{12.1}$$

where P is a property, V the volume fraction, and the subscripts c and i indicate, respectively, the composite and the ith component of the total n components.

For the case under study, $n = 2$ and the property P is the Young's modulus. We can write in the extended form

$$E_c = E_f V_f + E_m V_m \tag{12.2}$$

where the subscripts f and m indicate fiber and matrix, respectively.

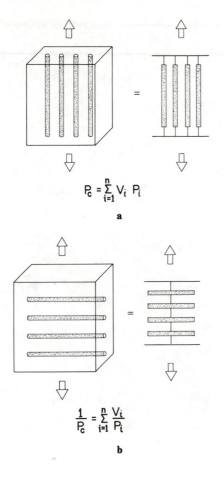

$$P_c = \sum_{i=1}^{n} V_i \, P_i$$

a

$$\frac{1}{P_c} = \sum_{i=1}^{n} \frac{V_i}{P_i}$$

b

Figure 12.3 Simple composite models: (a) longitudinal response (action in parallel); (b) transverse response (action in series).

The elastic properties of such unidirectional composites in the transverse direction can be represented by a system of "action in series" [Fig. 12.3(b)]. On loading in a direction transverse to the fibers, then, we have equal stress in the components. This model is equivalent to the classic treatment of Reuss.[6] We may write

$$\frac{1}{P_c} = \sum_{i=1}^{n} \frac{V_i}{P_i} \tag{12.3}$$

Once again, for the case $n = 2$ and taking the property P to be the Young's modulus, we obtain, for the composite,

$$\frac{1}{E_c} = \frac{V_f}{E_f} + \frac{V_m}{E_m} \tag{12.4}$$

These simple relations (Eqs. 12.2 and 12.4) are commonly referred to as the "rule of mixtures." The reader is warned that this rule is nothing more than a first approximation. More elaborate models have been proposed. A critique of these models and methods has been presented by Chamis and Sendeckyj.[7] A summary of various methods of obtaining composite properties is given below.

1. *The mechanics of materials method*. This deals with specific geometric configuration of fibers in a matrix, for example, hexagonal, square, and rectangular, and we introduce large approximations in the resulting fields.

2. *The self-consistent field method*. This method introduces approximations in the geometry of the phases. We represent the phase geometry by a single fiber embedded in a material whose properties are equivalent of a matrix or an average of composite. The resulting stress field is thus simplified.

3. *The variational calculus method*. This method focuses on the upper and lower limits of the properties and does not predict properties directly. Only when the upper and the lower bounds coincide is the property determined. Frequently, however, the upper and lower bounds are well separated.

4. *The numerical techniques method*. Here we use series expansion, numerical analysis, and computer techniques.

The variational calculus method gives exact results which are also easy to interpret. But, as mentioned above, these results can only be used as indicators of the material behavior when the upper and lower bounds are close enough. Fortunately, this is the case for longitudinal properties. Hill[8] has put rigorous limits on the value of E in terms of bulk modulus in plane strain B_p, Poisson ratio v, and shear modulus G of the two phases. One notes that B_p is the modulus for lateral dilatation with zero longitudinal strain and is given by

$$B_p = \frac{E}{2(1 - 2v)(1 + v)}$$

According to Hill, the bounds on E_c are

$$\frac{4 V_f V_m (v_m - v_f)^2}{V_f/B_{pm} + V_m/B_{pf} + 1/G_m} \leqslant E_c - (E_f V_f + E_m V_m) \leqslant$$

(12.5)

$$\frac{4 V_f V_m (v_m - v_f)^2}{V_f/B_{pm} + V_m/B_{pf} + 1/G_f}$$

It is worth noting that this treatment of Hill does not have restrictions about the form of the fiber, the packing geometry, and so on. We can see, by substituting in Eq. 12.5, that the deviations from the rule of mixtures (Eq. 12.2) are rather small for all practical purposes. For example, take $E_f/E_m = 100$, $v_f = 0.25$, $v_m = 0.6$ (which would be the case of boron or carbon fibers in an epoxy resin) and $V_f = 0.5$. Then the deviation of the Young's modulus of the composite from that predicted by the rule of mixtures is, at most, 2%. For a metallic fiber (e.g., tungsten in a copper matrix), the deviation is less than 1%. Of course, the rule of mixture becomes exact when $v_f = v_m$.

Similarly, Hill showed that for a unidirectionally aligned fiber composite,

$$v_c \gtrsim v_f V_f + v_m V_m \qquad \text{according to} \qquad (v_f - v_m)(B_{pf} - B_{pm}) \gtrsim 0 \qquad (12.6)$$

In real composites, v_f is generally smaller than v_m, and E_f is much greater than E_m. Thus, one has the real value of v_c less than that given by the rule of mixtures ($= v_f V_f + v_m V_m$). The reader should note that the bounds on v_c are not as close as the ones on E_c, because in the former case, the Poisson ratio difference ($v_f - v_m$) appears to the first order (Eq. 12.6), while in the case of the latter, this difference term appears squared (Eq. 12.5).

Once again, if the difference ($v_f - v_m$) is small, which would be the case of metallic matrix reinforced with metallic fibers, the bounds are closer and one can write

$$v_c \simeq v_f V_f + v_m V_m \qquad (12.7)$$

Using these results, Schapery[9] showed that the bounds on the coefficients of thermal expansion are quite narrow and that the expansion coefficient of a composite in the longitudinal direction may be written as

$$\alpha_{cL} \simeq \frac{E_f \alpha_f V_f + E_m \alpha_m V_m}{E_f V_f + E_m V_m} \qquad (12.8)$$

Behrens[10] has shown that the thermal conductivity k of a fiber composite in the direction of the fibers may be written as

$$k_{cL} = k_f V_f + k_m V_m \qquad (12.9)$$

The transverse properties and the shear moduli are not amenable to such simple reductions. They do not obey the rule of mixtures, even to the first approximation. The bounds on these are well spaced. The numerical analysis results show that the composite behavior depends on the fiber form, packing, and spacing between fibers.

One may conclude from the discussion above that the rule of mixtures predicts quite adequately the longitudinal modulus of the composite, and this has been verified experimentally for various systems. Such is not the case for the transverse modulus of the composite. In fact, the simple rule of mixtures gives a conservative estimate for the transverse elastic modulus. In short, composite elastic or thermoelastic properties in the longitudinal direction are adequately given by expressions given in Table 12.3.

**TABLE 12.3 Some
Longitudinal Properties
of Composites**

$$E_c = E_f V_f + E_m V_m$$

$$v_c = v_f V_f + v_m V_m$$

$$\alpha_c = \frac{E_f \alpha_f V_f + E_m \alpha_m V_m}{E_f V_f + E_m V_m}$$

$$k_c = k_f V_f + k_m V_m$$

12.2.2 Strength

In contrast to the elastic or thermoelastic properties, the rule of mixtures does not work as well for the prediction of the strength of a composite given the strengths of the individual components. Specifically, for a composite containing continuous fibers, unidirectionally aligned, loaded in the fiber direction, the stress in the composite is written as

$$\sigma_c = \sigma_f V_f + \sigma_m V_m \qquad (12.10)$$

where σ is the axial stress, V the volume fraction, and the subscripts c, f, and m refer to composite, fiber, and matrix, respectively. The reason that rule of mixtures does not work for properties such as strength, compared to its reasonable application for predicting properties such as Young's modulus in the longitudinal direction, is the following: Elastic modulus is a relatively structure insensitive property and, this being so, the response to an applied stress in the composite state is nothing but the volume-weighted average of the individual responses of the isolated components. Strength, on the contrary, is an extremely structure-sensitive property. Thus, there can occur synergism in the composite state. Let us consider the factors that may influence, in one way or the other, composite properties. First, matrix or fiber structure may be altered during fabrication; and second, composite materials generally consist of two components whose thermomechanical properties are quite different, and thus suffer residual stresses and/or structure alteration due to the internal stresses. The differential contraction during cooling from the fabrication temperature to ambient temperature can lead to rather large thermal stresses, which, in turn, lead the soft matrix to undergo extensive plastic deformation.[11-13] The deformation mode may also be influenced by rheological interaction between the components.[14,15] The plastic constraint on the matrix due to the large difference in the Poisson's ratio of matrix and that of fiber, especially in the stage wherein the fiber deforms elastically while the matrix deforms plastically, can alter the stress state in the composite. Thus, the alteration in the microstructure of one or both the components[11-13] or interaction between the components during straining[14,15] can lead to the phenomenon of synergism in the strength properties. In view of this, the rule of mixtures would be, in the best of the circumstances, a lower bound on the maximum stress of a composite.

Having made these observations about the applicability of the rule of mixtures to the strength properties, it is instructive to consider this lower bound on the composite mechanical behavior. We ignore any negative deviations from the rule of mixtures due to any fiber misalignment or due to the formation of a reaction product between fiber and matrix. Also, we assume that the components do not interact during straining and that these properties in the composite state are the same as those in the isolated state. Then, a series of composites with different fiber volume fractions would show σ_c linearly dependent on V_f. Since $V_f + V_m = 1$, we can rewrite Eq. 12.10 as

$$\sigma_c = \sigma_f V_f + \sigma_m (1 - V_f) \qquad (12.11)$$

We can put certain restrictions on V_f in order to have a real reinforcement. For this, a composite must have a certain minimum fiber (continuous) volume fraction, $V_{\min}$. Assuming that the fibers are identical and uniform (that is, all of them have the same ultimate tensile strength), the composite ultimate strength will be attained, ideally, at a strain equal to the strain corresponding to the fiber ultimate stress. Then, we have

$$\sigma_{cu} = \sigma_{fu} V_f + \sigma'_m (1 - V_f) \qquad V_f \geqslant V_{\min} \qquad (12.12)$$

where σ_{fu} is the fiber ultimate tensile stress in the composite and σ'_m is the matrix stress at the strain corresponding to fiber ultimate tensile stress. The reader should note that σ'_m is to be determined from the stress–strain curve of the matrix alone, that is, it is the matrix flow stress at a strain in the matrix equal to the fiber breaking strain. As indicated above, we are assuming that matrix stress–strain behavior in the composite is the same as in isolation. At low volume fractions, if a work-hardened matrix can counterbalance the loss of load-carrying capacity as a result of fiber breakage, the matrix will control the composite strength. Assuming that all the fibers break at the same time, then, in order to have a real reinforcement effect, one must satisfy the relation

$$\sigma_{cu} = \sigma_{fu} V_f + \sigma'_m (1 - V_f) \geqslant \sigma_{mu} (1 - V_f) \qquad (12.13)$$

where σ_{mu} is the matrix ultimate tensile stress. The equality of this expression serves to define the minimum fiber volume fraction, $V_{\min}$, that must be surpassed in order to have real reinforcement.

$$V_{\min} = \frac{\sigma_{mu} - \sigma'_m}{\sigma_{fu} + \sigma_{mu} - \sigma'_m} \qquad (12.14)$$

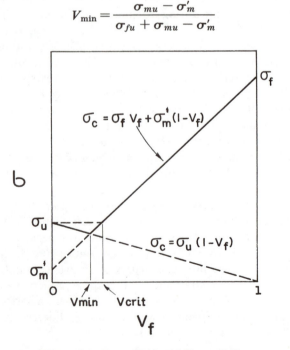

Figure 12.4 Determination of $V_{\min}$ and V_{crit}.

The value of $V_{\min}$ increases with decreasing fiber strength or increasing matrix strength.

In case we require that the composite strength should surpass the matrix ultimate stress, we can define a critical fiber volume fraction, V_{crit}, that must be exceeded. V_{crit} is given by the equation

$$\sigma_{cu} = \sigma_{fu} V_f + \sigma'_m (1 - V_f) > \sigma_{mu} \qquad (12.15)$$

In the case of equality, Eq. 12.15 gives

$$V_{\text{crit}} = \frac{\sigma_{mu} - \sigma'_m}{\sigma_{fu} - \sigma'_m} \qquad (12.16)$$

V_{crit} increases with increasing degree of matrix work hardening $(\sigma_{mu} - \sigma'_m)$. Figure 12.4 shows graphically the determination of $V_{\min}$ and V_{crit}. One notes that V_{crit} will always be greater than $V_{\min}$.

12.3 LOAD TRANSFER FROM MATRIX TO FIBER

The matrix has the important function of transmitting the applied load to the fiber. Remember that we emphasized the idea that in fiber-reinforced composites the fibers are the principal load-carrying members. No direct loading of fibers from the ends is admitted. One imagines each fiber to be embedded inside a matrix continuum; the state of stress (and, consequently, that of strain) of the matrix is perturbed by the presence of the fiber (Fig. 12.5). When the composite is loaded axially, the axial displacements in the fiber and in the matrix are locally different due to the different elastic modulus of the components. Macroscopically, the composite is deformed homogeneously.

The difference in the axial displacements in the fiber and the matrix implies that shear deformations are produced on planes parallel to the fiber axis and

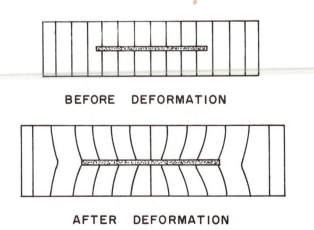

BEFORE DEFORMATION

AFTER DEFORMATION

Figure 12.5 Perturbation of the matrix stress state due to the presence of fiber.

in the direction of this axis. These shear deformations are the means by which the applied load is distributed between the two components.

Let us consider the distribution of the longitudinal stress along the fiber–matrix interface. There are two distinct cases: (1) matrix elastic and fiber elastic, and (2) matrix plastic and fiber elastic.

12.3.1 Fiber Elastic–Matrix Elastic[16]

Consider a fiber of length l embedded in a matrix subjected to a strain. Focus at a point distant x from one end of the fiber. It is assumed that (1) there exists a perfect contact between fiber and matrix (i.e., there is no sliding between them); and (2) Poisson's ratios of fiber and matrix are equal. Then the displacement of this point distant x from one extremity of the fiber can be defined in the following manner: u is the displacement of point x in the presence of fiber, and v is the displacement of the same point in the absence of the fiber.

The transfer of load from the matrix to the fiber may be written as

$$\frac{dP}{dx} = H(u - v) \tag{12.17}$$

where P is the load on the fiber and H is a constant to be defined later (it depends on the geometric arrangement of fibers, the matrix, and their moduli).

Differentiating Eq. 12.17, we obtain

$$\frac{d^2P}{dx^2} = H\left(\frac{du}{dx} - \frac{dv}{dx}\right) \tag{12.18}$$

Now, it follows from the definition that

$$\frac{dv}{dx} = \text{strain in matrix} = e \tag{12.19}$$

$$\frac{du}{dx} = \text{strain in fiber} = \frac{P}{A_f E_f}$$

where A_f is the transverse-sectional area of the fiber. From 12.18 and 12.19, we obtain

$$\frac{d^2P}{dx^2} = H\left(\frac{P}{A_f E_f} - e\right) \tag{12.20}$$

A solution of this differential equation is

$$P = E_f A_f e + S \sinh \beta x + T \cosh \beta x \tag{12.21}$$

where

$$\beta = \left(\frac{H}{A_f E_f}\right)^{1/2} \tag{12.22}$$

The boundary conditions to evaluate the constants S and T are

$$P = 0 \qquad \text{at } x = 0 \quad \text{and} \quad x = l$$

Putting in these values and using the "half-angle" trigonometric formulas, we get the equation

$$P = E_f A_f e \left\{ 1 - \frac{\cosh \beta[(l/2) - x]}{\cosh \beta(l/2)} \right\} \qquad \text{for } 0 < x < \frac{l}{2} \qquad (12.23)$$

or

$$\sigma_f = \frac{P}{A_f} = E_f e \left\{ 1 - \frac{\cosh \beta[(l/2) - x]}{\cosh \beta(l/2)} \right\} \qquad \text{for } 0 < x < \frac{l}{2} \qquad (12.24)$$

The maximum possible value of strain in fiber is the imposed strain e, and thus the maximum stress is eE_f. Thus, as long as we have a sufficiently long fiber, the stress in fiber will increase from the two ends to a maximum value, $\sigma_f^{\max} = E_f e$. It can be shown readily that the average stress in fiber will be

$$\bar{\sigma}_f = E_f e \left[1 - \frac{\tanh(\beta l/2)}{\beta l/2} \right] \qquad (12.25)$$

The variation of shear stress, τ, along the fiber–matrix interface is obtained by considering the equilibrium of forces acting over an element of fiber (radius r_f). Thus,

$$\frac{dP}{dx} dx = 2\pi r_f \, dx \, \tau \qquad (12.26)$$

P is the tensile load on the fiber and is equal to $\pi r_f^2 \sigma_f$. Then

$$\tau = \frac{1}{2\pi r_f} \frac{dP}{dx} = \frac{r_f}{2} \frac{d\sigma_f}{dx} \qquad (12.27)$$

or

$$\tau = \frac{E_f r_f e \beta}{2} \frac{\sinh \beta[(l/2) - x]}{\cosh \beta(l/2)} \qquad (12.28)$$

The variation of τ and σ_f with x is shown in Fig. 12.6.

One must emphasize here that the shear stress τ in Eq. 12.28 will be the smaller of the two shear stresses:

1. Strength of fiber–matrix interface in shear
2. Shear yield stress of matrix

Of these two shear stresses, the one that has a smaller value will control the load transfer phenomenon and should be used in Eq. 12.28.

The constant H still remains to be determined. An approximate value of H is derived below for a particular geometry. Let the fiber length l be much greater than the fiber radius r_f, and let $2R$ be the average fiber spacing (center

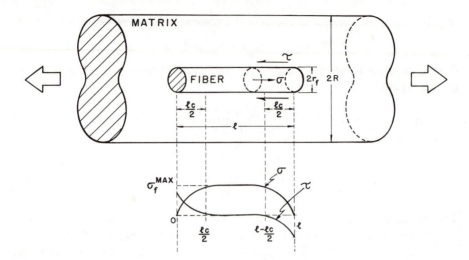

Figure 12.6 Load transfer to fiber. Variation of tensile stress σ in fiber and shear stress τ along the interface with the fiber length l.

to center). Let $\tau(r)$ be the shear stress in the direction of the fiber axis at a distance r from the axis. Then, at the fiber surface ($r = r_f$),

$$\frac{dP}{dx} = -2\pi r_f \tau(r_f) = H(u - v)$$

Thus,

$$H = -\frac{2\pi r_f \tau(r_f)}{u - v} \tag{12.29}$$

Let w be the real displacement in the matrix. At the fiber–matrix interface, not being permitted sliding, $w = u$. At a distance R from the center of a fiber, $w = v$. Considering equilibrium of matrix between r_f and R, we get

$$2\pi r \tau(r) = \text{constant} = 2\pi r_f \tau(r_f)$$

or

$$\tau(r) = \frac{\tau(r_f) r_f}{r} \tag{12.30}$$

The shear strain γ in the matrix is given by $\tau(r) = G_m \gamma$, where G_m is the matrix shear modulus. Then

$$\gamma = \frac{dw}{dr} = \frac{\tau(r)}{G_m} = \frac{\tau(r_f) r_f}{G_m r} \tag{12.31}$$

Integrating from r_f to R, we get

$$\Delta w = \frac{\tau(r_f) r_f}{G_m} \ln\left(\frac{R}{r_f}\right) \tag{12.32}$$

But, by definition,

$$\Delta w = v - u = -(u - v) \tag{12.33}$$

Then

$$\frac{\tau(r_f)r_f}{u - v} = -\frac{G_m}{\ln(R/r_f)} \tag{12.34}$$

From Eqs. 12.29 and 12.34, we get

$$H = \frac{2\pi G_m}{\ln(R/r_f)} \tag{12.35}$$

and from Eq. 12.22, we obtain an expression for the load transfer parameter β:

$$\beta = \left(\frac{H}{E_f A_f}\right)^{1/2} = \left[\frac{2\pi G_m}{E_f A_f \ln(R/r_f)}\right]^{1/2} \tag{12.36}$$

Note that the greater the value of G_m/E_f, the more rapid is the increase in fiber stress from the two ends.

The analysis above is an approximate one and, more specifically, the method of evaluation of load transfer parameter β. More exact analyses[17,18] give results similar to the one above and differ only in the value of β. In all the analyses, however, β is proportional to $\sqrt{G_m/E_f}$ and differences occur only in the term involving fiber volume fraction, $\ln(R/r_f)$, in the equation above.

12.3.2 Fiber Elastic–Matrix Plastic

It should be clear from the discussion above that in order to load high-strength fibers to their maximum strength in the matrix, shear strength must correspondingly be large. A metallic matrix will flow plastically in response to the high shear stress developed. Should the fiber–matrix interface be weaker, it will fail first. Plastic deformation of a matrix implies that the shear stress at the fiber surface, $\tau(r_f)$, will never go above τ_y, the matrix shear yield strength (ignoring any work-hardening effects). In such a case, we get from an equilibrium of forces,

$$\sigma_f \pi \frac{d^2}{4} = \tau_y \pi d \frac{l}{2}$$

or

$$\frac{l}{d} = \frac{\sigma_f}{2\tau_y}$$

We consider $l/2$ and not l because the fiber is being loaded from both the ends. If the fiber is sufficiently long, it should be possible to load it to its breaking stress, σ_{fb}, by means of load transfer through the matrix flowing plastically around it. Let $(l/d)_c$ be the minimum fiber length-to-diameter ratio necessary to accomplish this. We call this ratio l/d as the aspect ratio of a fiber and $(l/d)_c$ as the critical aspect ratio necessary to attain the breaking stress of the fiber, σ_{fb}. Then we can write

$$\left(\frac{l}{d}\right)_c = \frac{\sigma_{fb}}{2\tau_y} \tag{12.37a}$$

Or we can think of a critical fiber length, l_c, for a given fiber diameter d:

$$\frac{l_c}{d} = \frac{\sigma_{fb}}{2\tau_y} \tag{12.37b}$$

Thus, the fiber length l must be equal or greater than l_c for it to be loaded to its maximum stress. If $l < l_c$, the matrix will flow plastically around the fiber and will load it to a stress in its central portion given by

$$\sigma_f = 2\tau_y \frac{l}{d} < \sigma_{fb} \tag{12.38}$$

This is shown in Fig. 12.7. An examination of this figure shows that even for $l/d > (l/d)_c$, the average stress in fiber will be less than the maximum stress to which it is loaded in its central region. In fact, we can write for the average fiber stress,

$$\bar{\sigma}_f = \frac{1}{l} \int_0^l \sigma_f \, dx$$

$$= \frac{1}{l} [\sigma_f(l - l_c) + \phi\sigma_f l_c]$$

$$= \frac{1}{l} [\sigma_f l - l_c(\sigma_f - \phi\sigma_f)]$$

or

$$\bar{\sigma}_f = \sigma_f \left(1 - \frac{1 - \phi}{l/l_c}\right) \tag{12.39}$$

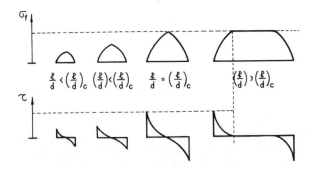

Figure 12.7 Variation of the fiber load transfer length as a function of the aspect ratio l/d.

where $\phi\sigma_f$ is the average stress in the fiber over a portion $l_c/2$ of its length at both the ends. We can thus regard ϕ as a load transfer function. Its value will be precisely 0.5 for an ideally plastic matrix (i.e., the increase in stress in the fiber over the portion $l_c/2$ will be linear). The composite stress can then be written as

$$\sigma_c = \sigma_f V_f \left(1 - \frac{1-\phi}{l/l_c}\right) + \sigma_m(1 - V_f) \tag{12.40}$$

Taking $l/l_c = 10$ and $\phi = 0.5$, we find that the stress in a composite containing discontinuous fibers will be 95% of that of a composite containing continuous fibers. That is, as long as the fibers are reasonably long compared to the load transfer length, there is not much loss of strength due to their discontinuous nature.

12.4 FRACTURE IN COMPOSITES

Fracture is a complex subject even in monolithic materials (see Chapter 3). Undoubtedly, it is even more complex in composite materials. A great variety of deformation modes can lead to composite failure. The operative failure mode will depend, among other things, on loading conditions and the particular composite system. The microstructure has a very important role in the mechanics of composite rupture. For example, fiber diameter, its volume fraction and alignment, damage due to thermal stresses that may develop during fabrication and/or service—all these factors can contribute and influence directly the crack initiation and propagation. A multiplicity of failure modes can exist in a composite under different loading conditions.

12.4.1 Single and Multiple Fracture

In general, the two components of a composite will have different values of strain to fracture. When the component that has the smaller breaking strain fractures, the load carried by this component is thrown onto the other one. If this component, which has a higher strain to fracture, can bear this additional load, the composite will show multiple fracture of the brittle (smaller fracture strain) component; eventually, a particular transverse section of composite becomes so weak that the composite is unable to carry load any further and it fails.

Let us consider the case of a fiber-reinforced composite in which the fiber fracture strain is less than that of the matrix. Then the composite will show a single fracture[19] when

$$\sigma_{fu} V_f > \sigma_{mu} V_m - \sigma'_m V_m$$

where σ'_m is the matrix stress corresponding to the fiber fracture strain and σ_{fu} and σ_{mu} are ultimate tensile stresses of fiber and matrix, respectively. This equation says that when the fibers break, the matrix will not be in a

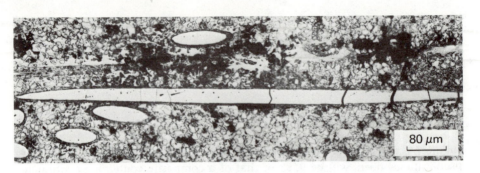

Figure 12.8 Multiple fracture of tungsten fibers in an Fe-Cu matrix. (Optical.)

condition to support the additional load. This is commonly encountered in composites of high V_f, brittle fibers, and ductile matrix. All the fibers break in more or less one plane and the composite fails in that plane.

If, on the other hand, we have a system that satisfies the condition

$$\sigma_{fu} V_f < \sigma_{mu} V_m - \sigma'_m V_m$$

the fibers will be broken into small segments until the matrix fracture strain is reached. An example of this is shown in Fig. 12.8, an optical micrograph of the system Fe-Cu matrix containing small volume fraction of W fibers.

In case the fibers have a fracture strain greater than that of the matrix (an epoxy resin reinforced with metallic wires), we would have a multiplicity of fractures in the matrix and the condition for this may be written as[19]

$$\sigma_{fu} V_f > \sigma_{mu} V_m + \sigma'_f V_f$$

where σ'_f is now the fiber stress corresponding to matrix fracture strain.

12.4.2 Failure Modes in Composites

Two failure modes are commonly encountered in composites:

1. The fibers break in one plane and, the soft matrix being unable to carry the load, the composite failure will occur in the plane of fiber fracture. This mode is more likely to be observed in composites that contain relatively high fiber volume fractions and fibers that are strong and brittle. The latter condition implies that the fibers do not show a distribution of strength with a large variance but show more a strength behavior that can be characterized by Dirac delta function.
2. When the adhesion between fibers and matrix is not sufficiently strong, the fibers may be pulled out of the matrix before the composite failure. This fiber pull-out results in the fiber failure surface being nonplanar.

More commonly, a mixture of these two modes is found: fiber fracture together with fiber pull-out. Fibers invariably have defects distributed along their lengths and thus can break in regions above or below the crack tip. This leads to separation between the fiber and the matrix and, consequently, to fiber pull-out with the crack opening up. Examples of this mixed fracture mode are shown in Fig. 12.9: boron–aluminum [Fig. 12.9(b)] and carbon–polyester [Fig. 12.9(a)].

One of the attractive characteristics of composites is the possibility of obtaining an improved fracture toughness behavior together with high strength. Fracture toughness can be defined loosely as resistance to crack propagation. In a fibrous composite containing a crack transverse to the fibers, the crack propagation resistance can be increased by doing additional work by means of:

1. Plastic deformation of the matrix
2. Fiber pull-out
3. Presence of weak interfaces, fiber–matrix separation, and deflection of the crack

Figure 12.9 Fracture in composites showing the fiber pull out phenomenon: (a) carbon–polyester; (b) boron–aluminum 6061. (Scanning electron microscope.)

b

a

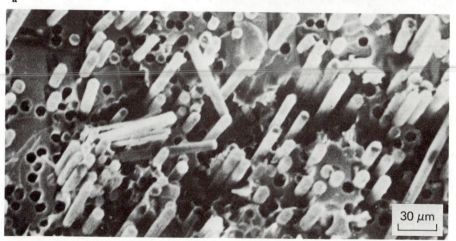

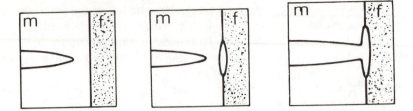

Figure 12.10 Fracture of weak interface in front of crack tip due to transverse tensile stress; m and f indicate matrix and fiber, respectively. (Adapted with permission from [23], p. 514.)

Cooper and Kelly[20] have shown that, for a metallic matrix, the work of fracture is the work done during plastic deformation of matrix to failure, and is proportional to $d\,(V_m/V_f)^2$. Presumably, for fibers of a large diameter for a given V_f, the advancing crack will have to pass through a greater plastic zone of matrix and will result in large work of fracture.

Fiber pull-out contributes to the work of fracture by leading to a large deformation to fracture. In this case, the controlling parameter for work of fracture is the ratio d/τ_i, where d is the fiber diameter and τ_i is the interface shear strength. In the case of discontinuous fibers, the work of fracture due to fiber pull-out also increases with the fiber length, reaching a maximum at $l = l_c$. In the case of continuous fibers, the work to fracture increases with increase in spacing between the defects.[21,22] It would appear that the simplest and most convenient method to increase the work of fracture would be to increase the fiber diameter.

The deflection of the crack along an interface or the separation of the fiber–matrix interface is an interesting mechanism of augmenting the resistance to crack propagation in composites. Cook and Gordon[23] have analyzed the stress distribution in front of a crack tip and concluded that the maximum transverse tensile stress, σ_{11}, is about one-fifth of the maximum longitudinal tensile stress, σ_{22}. They suggested, therefore, that when the ratio σ_{22}/σ_{11} is greater than 5, the fiber–matrix interface in front of the crack tip will fail under the influence of the transverse tensile stress and the crack would be deflected 90° from its original direction. This way the fiber–matrix interface would act as a crack arrester. This is shown schematically in Fig. 12.10. This improvement in fracture toughness due to the presence of weak interfaces has been confirmed qualitatively.[22]

12.5 SOME FUNDAMENTAL CHARACTERISTICS OF COMPOSITES

Composite materials are not like any other common type of material. They are inherently different from the monolithic materials and, consequently, these basic differences must be taken into account when designing and fabricating any article from composite materials. In what follows we give a brief description of some of these fundamental characteristics of composites.

12.5.1 Heterogeneity

Composite materials are inherently heterogeneous, consisting as they do of two components of different elastic moduli, different mechanical behavior, different expansion coefficients, and so on. For this reason, the analysis and the design procedures for composite materials are quite intricate and complex compared to those for ordinary materials. The structural properties of composites are functions of:

1. Component properties
2. Geometric arrangement of components in the composite
3. Interface between the components

Given two components, we can obtain a great variety of properties due to functions 1 and 2.

12.5.2 Anisotropy

The monolithic materials, in general, are reasonably isotropic; that is, their properties do not show any marked preference for any particular direction. The unidirectional composites are anisotropic due to their very nature. Once again, the analysis and design of composites should take into account this strong directionality of properties—properties that cannot be specified without any reference to some direction. Figure 12.11 shows, schematically, elastic moduli of a monolithic material and a composite as a function of fiber orientation, θ. As the monolithic material (e.g., Al) is an isotropic material, its moduli do not vary with the angle of test and the graphs are horizontal.

For the ordinary materials, the designer only needs to open a manual and find one unique value of strength or one unique value of modulus of the material (say, aluminum). But for composite materials, the designer has to consult "performance charts" representing the specific strength and the specific rigidity of the various composite systems.

Ordinary materials, such as aluminum or steel, can be represented by a fixed point (see Fig. 12.12). For the composite material, however, there does not exist a unique combination of these properties. The composite contains a system of properties and must be represented by an area instead of a point. We call these "carpet plots." The highest point on the graph would represent the longitudinal properties, while the lowest point would represent quasi-isotropic properties. The important point to make is that, depending on the construction of a laminate and the appropriate quantity of fiber, the characteristics of the composite can be varied. In other words, the composites can be tailor-made, in accord with the final objective.

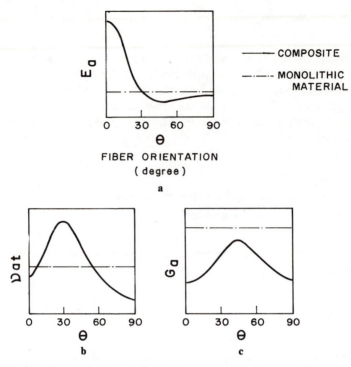

Figure 12.11 Variation of elastic moduli of a fiber composite and a monolithic material with the angle of reinforcement (schematic). E_a is the axial Young's modulus, ν_{at} is the principal Poisson's ratio, and G_a is the axial shear modulus.

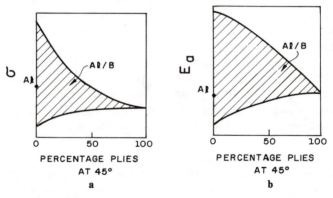

Figure 12.12 Performance chart of a composite (schematic).

12.5.3 Shear Coupling

The composite properties are very sensitive functions of the fiber orientation. They display what is called shear coupling; shear strains are produced by an axial stress, and axial strains are produced by shear stress (see Fig. 12.13).

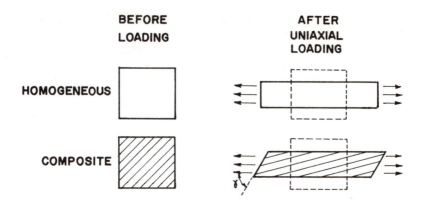

BEFORE LOADING

AFTER UNIAXIAL LOADING

HOMOGENEOUS

COMPOSITE

Figure 12.13 Shear coupling phenomenon in a fiber composite.

In response to a uniaxially applied load, an isotropic material produces only axial and transverse strains. In fiber-reinforced composites, however, there is also produced, in response to an axial load, a shear strain γ due to the fact that the fibers tend to align themselves in the direction of the applied load. This shear distortion can be eliminated if one makes a "cross-ply" composite: A composite containing an equal number of parallel fibers, aligned at a given angle with respect to the loading axis and at a complementary angle to this (Fig. 12.14). That is, we have the various layers in a composite arranged at $\pm\theta$ degrees to the loading axis, and thus the shear distortion due to one layer is compensated by an equal and opposite shear distortion due to the other. However, this balance occurs only in two dimensions. The real-life composites are three-dimensional materials. This leads to an "edge effect." The individual layers deform differently under tension and in the neighborhood of the free edges, giving rise to out-of-plane shear and bending. In this respect, Pagano

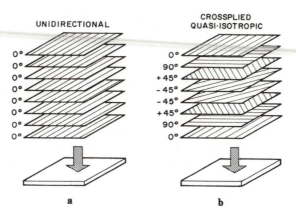

UNIDIRECTIONAL

CROSSPLIED QUASI-ISOTROPIC

0°
0°
0°
0°
0°
0°
0°
0°

0°
90°
+45°
−45°
−45°
+45°
90°
0°

a

b

Figure 12.14 Unidirectional and cross-plied composites.

and Pipes[24] have shown that the stacking sequence of the various layers in the composite is important. For example, in a laminate composite consisting of fibers at +90°, +45°, −45°, −45°, +45°, +90° and subjected to a tensile stress (in plane), there occur compressive stresses in the thickness direction, in the vicinity of the edges. Should the same composite have the sequence +45°, −45°, +90°, +90°, −45°, +45°, however, these stresses in the thickness direction are of a tensile nature and thus tend to delaminate the composite, clearly an undesirable effect.

‡12.6 APPLICATIONS OF COMPOSITES

Although major applications of composite materials have involved resin-based composites, metal matrix composites have also steadily found applications in many areas. Especially in terms of high-temperature applications, metal matrix composites become very important. Metal matrix fiber composites possess high-temperature capability, high thermal conductivity, low thermal expansion in the fiber direction, and high specific stiffness and strength. Also, they do not pick up moisture or outgas as do resin matrices.

Boron fiber–aluminum matrix struts helped support the fuselage frame of the Space Shuttle. Boron–Al has been used experimentally in jet engine fan blades. Boron fiber having a surface barrier coating of silicon carbide (SiC) is used in Al or Ti for high-temperature capability. Boron carbide (B_4C) is another diffusion barrier coating.

Graphite fibers have been used for reinforcing aluminum. Generally, a vapor-deposited titanium boron (Ti-B) coating on graphite fibers is required before being used in an aluminum matrix. This coating promotes wettability of the graphite fibers by Al and prevents adverse reaction between Al and C.

Other important reinforcements include Dupont's FP alumina fiber for Al, Pb, or Mg; SiC-coated tungsten or carbon fibers for Al, Ti, or superalloy matrices; Mo and W fibers for superalloy matrices; and graphite fibers for Pb. The last one, involving the use of graphite lead composites for plates in nuclear submarine lead–acid batteries in a U.S. Navy program, doubled battery lifetime, mainly because of the superior creep properties of the reinforced plates.

One of the ways of obtaining higher efficiencies in the performance of gas turbine engines is by obtaining materials that allow higher turbine inlet temperatures. Conventionally, Ni- and Co-based superalloys, strengthened mainly by coherent, ordered γ' precipitates, are used (see Chapter 20). However, these precipitates tend to dissolve in the matrix at temperatures $>0.7 T_m$, where T_m is the melting point. Directionally solidified eutectic composites show a considerable potential in this regard and increases of 50 to 100°C in turbine inlet temperature have been predicted.[25] Directionally solidified eutectic composites contain a high-melting-point strong phase aligned in the form of fibers or lamellae in a matrix. The fibers are generated in situ and possess low-energy interfaces with the matrix. They show excellent isothermal stability at temperatures as high as $0.95\,T_m$, with little attendant change in microstructure or mechanical properties,[25] although there is some concern regarding thermal stability when cycled in a temperature gradient.[26,27] Nevertheless, such eutectic alloys have been reported to show improved creep properties, low cycle fatigue, low thermal fatigue, and good impact properties compared to conventional superalloys.[25] The close control required in the production of these alloys makes them expensive. Also, they show rather poor oxidation resistance. This problem can be attacked by employing coatings.[28] Multifilamentary composite superconductor is yet another example. An Nb_3Sn (A-15 structure) fiber composite superconductor is prepared by first producing a ductile billet consisting of Nb rods in a Cu-Sn alloy matrix and then extruding this billet down to a wire diameter. This is followed by an annealing treatment that makes Sn diffuse to Nb to form a layer of Nb_3Sn around each filament by solid-state reaction. This is the "bronze method" of fabricating filamentary composite structure. The thickness of the Nb_3Sn layer and the grain size in this layer are important factors that control the current-carrying capacity of the superconductor. Table 12.4 summarizes some of the important metal matrix fiber composite systems and their applications.

TABLE 12.4   Some Metal Matrix Composite Systems and Their Applications

Fiber (Substrate)	Matrix	Applications
Boron (W)	Al, Mg, Ti	Compressor blades, jet engine
Borsic	Al, Ti	fan blades, antennae
Boron (C)	Al, Mg, Ti	structures, high-temperature
Boron carbide-coated boron	Al, Mg, Ti	structures
Rayon T50		Satellite, missile, and helicopter
PAN HTS T300		parts; space and satellite
PAN HM	Al, Mg, Pb, Cu	structures; storage battery
Pitch P		plates; electrical contacts
Pitch LIHM		
FP alumina	Al, Mg, Pb	Storage battery plates, helicopter components
Silicon carbide (W)	Al, Ti	High-temperature structures,
Silicon carbide (C)	Al, Ti	high-temperature engine
Silicon carbide	Al, Ti	components
Tungsten		
WReHfC	Superalloy	High-temperature turbine
W1.5ThO$_2$	Superalloy	components
Molybdenum	Superalloy	
Ni$_3$Al-Ni$_3$Nb	Superalloy	
TaC, Cr$_7$C$_3$,		
Nb$_3$Sn, V$_3$Ga, NbTi	Cu	Superconductors

EXERCISES

12.1. Describe some composite materials that occur in nature. Describe their structure and properties.

12.2. List some nonstructural applications of composite materials. [See M. B. Bever, P. E. Duwez, and W. A. Tiller, *Mater. Sci. Eng.*, 6 (1970) 149–155.]

12.3. Consider a fiber (radius r) embedded up to a length l in a matrix (see Fig. E12.1). When the fiber is pulled, the adhesion between the fiber and the matrix produces a shear stress τ at the interface. In a composite system containing a fiber of fracture stress σ_f equal to eight times the maximum shear stress τ_{max} that the interface can bear, what fiber aspect ratio is required to break it rather than pull it out?

12.4. A steel wire (diameter 1.25 mm) has an aluminum coating such that the composite wire has a diameter of 2.50 mm. Some other pertinent data are:

	Steel	Aluminum
Elastic modulus E	210 GPa	70 GPa
Yield stress σ_y	200 MPa	70 MPa
Poisson ratio ν	0.3	0.3
Coefficient of thermal expansion (linear)	$11 \times 10^{-6} \, K^{-1}$	$23 \times 10^{-6} \, K^{-1}$

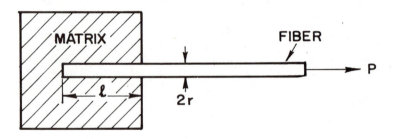

Figure E12.1

(a) If the composite wire is loaded in tension, which of the two components will yield first? Why?

(b) What tensile load can the composite wire support without undergoing plastic strain?

(c) What is the elastic modulus of the composite wire?

(d) What is the coefficient of thermal expansion of the composite wire?

12.5. A boron–aluminun composite has the following characteristics:
Unidirectional reinforcement
Fiber volume fraction $V_f = 50\%$
Fiber length $l = 0.1$ m
Fiber diameter $d = 100 \, \mu m$
Fiber ultimate stress $\sigma_{fu} = 3$ GPa

Fiber strain corresponding to σ_{fu}, $e_{fu} = 0.75\%$ (uniform elongation)

Fiber Young's modulus $E_f = 415$ GPa

Matrix shear yield stress $\tau_{ym} = 75$ MPa

Matrix stress at $e = e_{fu}$, $\sigma'_m|_{e=e_{fu}} = 93$ MPa

Matrix ultimate stress $\sigma_{mu} = 200$ MPa

Compute:

(a) The critical fiber length l_c for the load transfer.

(b) The ultimate tensile stress of the composite.

(c) The $V_{\min}$ and V_{crit} for this composite system.

12.6. In matrices such as Cu and Ni (ignoring work hardening), compute the critical fiber aspect ratio $(l/d)_c$ for a fiber having an ultimate stress of 7 GPa. What will be the value of $(l/d)_c$ for this fiber for an Al matrix? For a polymer matrix?

12.7. For applications of fibrous composites at high temperatures, how does the aspect ratio $(l/d)_c$ vary in comparison with its value for low-temperature applications?

12.8. To promote the wettability and avoid interfacial reactions, protective coatings are sometimes applied to the fibers. Any improvement in the composite behavior will depend on the stability of this layer. The maximum time t for the dissolution of this layer can be estimated by the diffusion distance x:

$$x \approx \sqrt{Dt}$$

where D is the diffusivity of the matrix in the protective layer. Making an approximation that the matrix diffusion in the protective layer can be represented by self-diffusion, compute the time required for a 0.1-μm-thick protective layer on the fiber to be dissolved at T_m and $0.75 T_m$, where T_m is the matrix melting point in Kelvin. Assume a reasonable D value for self-diffusion in metals, taking into account the variation of D with temperature.

12.9. A composite material with aligned fibers is an extremely anisotropic material. An isotropic material needs only two elastic constants to characterize its elastic behavior. In the case of aligned fiber composite plate (i.e., a plane stress and transversely isotropic state; see Fig. E12.2) one can write Hooke's law as

$$\begin{bmatrix} \epsilon_1 \\ \epsilon_2 \\ \epsilon_6 \end{bmatrix} = \begin{bmatrix} S_{11} & S_{12} & 0 \\ S_{12} & S_{22} & 0 \\ 0 & 0 & S_{66} \end{bmatrix} \begin{bmatrix} \sigma_1 \\ \sigma_2 \\ \sigma_6 \end{bmatrix}$$

where

$$S_{11} = \frac{1}{E_{11}}$$

$$S_{22} = \frac{1}{E_{22}}$$

$$S_{12} = \frac{-\nu_{12}}{E_{11}} = \frac{-\nu_{21}}{E_{22}}$$

$$S_{66} = \frac{1}{G_{12}}$$

Transform the elasticity matrix and derive the following equation giving the variation of elastic properties with fiber orientation:

$$\frac{1}{E_\theta} = S_{11} \cos^4 \theta + S_{22} \sin^4 \theta + (S_{66} - 2S_{12}) \cos^2 \theta \sin^2 \theta$$

Plot E_θ versus θ for any fiber composite system of your choice.

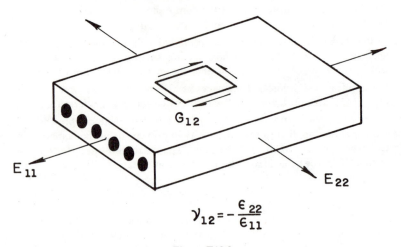

Figure E12.2

12.10. One can obtain a two-dimensional isotropy in a fiber composite by having randomly oriented fibers in the plane of the plate. Show that the average in-plane modulus is

$$\overline{E}_\theta = \frac{\displaystyle\int_0^{\pi/2} E_\theta \, d\theta}{\displaystyle\int_0^{\pi/2} d\theta}$$

Plot $\overline{E}_\theta/E_{11}$ versus V_f for fiber-reinforced composites with $E_f/E_m = 1$, 10, and 100.

12.11. A fiber composite is also highly directional in its strength properties. If σ_l is the composite longitudinal tensile strength, σ_t is the transverse tensile strength, and τ is the in-plane shear strength, show that

$$\frac{1}{\sigma_\theta^2} = \frac{\cos^4 \theta}{\sigma_l^2} + \frac{\sin^4 \theta}{\sigma_t^2} + \cos^2 \theta \sin^2 \theta \left(\frac{1}{\tau^2} - \frac{1}{\sigma_l^2} \right)$$

Plot σ_θ versus θ for any composite system of your choice.

12.12. Discuss the problem of thermal stability and thermal fatigue in metal matrix composites.

12.13. Describe the various fabrication processes for fiber-reinforced metals.

12.14. Fiber composites in general and eutectic composites, in particular, are characterized by a rather large surface-to-volume ratio interfaces. This is likely to be a source of physical and/or chemical instability. Discuss ways and means of reducing this.

12.15. How would you measure the interfacial energy between the Ta matrix and Ta_2C fibers in the eutectic system $Ta-Ta_2C$?

REFERENCES

[1] F. Schuh and P. Polatidis, *Z. Metallk.*, *69* (1978) 320.

[2] K.K. Chawla, *J. Eng. Mater. Technol., Trans. ASME*, *97* (1975) 371.

[3] K.K. Chawla, *Mater. Sci. Eng.*, *48* (1981) 137.

[4] J. Aveston, *Composites*, *1* (1970) 296.

[5] W. Voigt, *Lehrbuch der Kristallphysik*, Teubner, Leipzig, 1910.

[6] A. Reuss, *Z. Angew. Math. Mech.*, *9* (1929) 49.

[7] L.C. Chamis and G.P. Sendeckyj, *J. Composite Mater.*, *2* (1968) 332.

[8] R. Hill, *J. Mech. Phys. Solids*, *12* (1964) 199.

[9] R.A. Schapery, *J. Composite Mater.*, *2* (1968) 380.

[10] E. Behrens, *J. Composite Mater.*, *2* (1968) 2.

[11] K.K. Chawla and M. Metzger, *J. Mater. Sci.*, *7* (1972) 34.

[12] K.K. Chawla and M. Metzger. *Met. Trans.*, *8A* (1977) 1681.

[13] K.K. Chawla, *Metallography*, *6* (1973) 155.

[14] L.J. Ebert and J.D. Gadd, in *Fiber Composite Materials*, ASM, Metals Park, Ohio, 1965, p. 89.

[15] A. Kelly and H. Lilholt, *Phil. Mag.*, *20* (1971) 175.

[16] H.L. Cox, *Brit. J. Appl. Phys.*, *3* (1952) 72.

[17] N.F. Dow, General Electric Rep. No. R63-SD-61, 1963.

[18] B.W. Rosen, in *Fiber Composite Materials*, ASM, Metals Park, Ohio, 1965, p. 37.

[19] D.K. Hale and A. Kelly, in *Annu. Rev. Mater. Sci.*, *2* (1972) 405.

[20] G.A. Cooper and A. Kelly, *J. Mech. Phys. Solids*, *15* (1967) 279.

[21] A. Kelly, in *The Properties of Fibre Composites*, IPC Science and Technology Press, Guilford, Surrey, 1971, p. 5.

[22] G.A. Cooper, *J. Mater. Sci.*, *5* (1970) 645.

[23] J. Cook and J.E. Gordon, *Proc. Roy. Soc. (London)*, *A228* (1964) 508.

[24] N.J. Pagano and R.B. Pipes, *J. Composite Mater.*, *5* (1971) 50.

[25] Proc. Conf. In Situ Composites, Sept. 1972; NMAB-308-II, National Academy of Sciences and National Academy of Engineering, Washington, D.C., Jan. 1973.

[26] D.R.H. Jones, *Metal. Sci.*, *8* (1974) 37.

[27] D.R.H. Jones and G.J. May, *Acta Met.*, *23* (1975) 29.

[28] D.E. Graham and D.A. Woodford, *Met. Trans.*, *12A* (1981) 329.

SUGGESTED READING

BROUTMAN, L.J., and R.H. KROCK, eds., *Composite Materials*, Vols. 1–8, Academic Press, New York, 1971–1975.

BROUTMAN, L.J., and R.H. KROCK, *Modern Composite Materials*, Addison-Wesley, Reading, Mass., 1967.

CALCOTE, L.R., *Analysis of Laminated Composite Structures*, Van Nostrand Reinhold, London, 1967.

CHAWLA, K.K., *Composite Materials–Some Recent Developments,* J. of Metals, 35 (March, 1983).

SCALA, E., *Composite Materials for Combined Functions*, Hayden, Rochelle Park, N.J., 1973.

TSAI, S.W., and H.T. HAHN, *Introduction to Composite Materials*, Technomic, Westport, Conn., 1968.

VINSON, J.R., and T.W. CHOU, *Composite Materials and Their Use in Structures*, Halsted Press (Wiley), New York, 1975.

Chapter 13

MARTENSITIC STRENGTHENING

13.1 CLASSIFICATION OF MARTENSITE

Quenching of steel has been known for over 3000 years and is, up to this day, the single most effective strengthening mechanism known. However, it is only fairly recently that the underlying mechanisms have become understood and been studied in a scientific manner. Initially attributed to a beta phase supposedly existing in the Fe-C system, the strengthening effect is now known to be due to a metastable phase: martensite. The investigations leading to the understanding of the mechanisms governing and factors affecting martensitic transformations have posed a great challenge to metallurgists over the past 50 years. Out of a confusing maze of apparently contradictory phenomena, order is starting to appear; martensitic-like transformations have been identified in a great number of systems, including pure metals, solid solutions, intermetallic compounds, and even ceramics. In order to assess the mechanical behavior of martensite and to take advantage of its unique responses in technological applications, one has to understand the fundamental aspects; this will be done in this chapter. Table 13.1 presents a number of systems in which martensitic-like transformations have been observed.

The original use of martensitic transformation was exclusively to harden steel. More recent developments have led to the use of this transformation in different contexts. In transformation-induced plasticity (TRIP) steels[2-4] the martensitic transformation occurs during deformation and strengthens the regions ahead of a crack or near a neck; the ductility of the material is enhanced, while the strength level remains high. This results in great toughness. Another

TABLE 13.1   Systems in Which Martensitic or
              Quasimartensitic Transformation
              Occurs[a]

Alloy	Structural Change
Co, Fe-Mn, Fe-Cr-Ni	FCC → HCP
Fe-Ni	FCC → BCC
Fe-C, Fe-Ni-C, Fe-Cr-C, Fe-Mn-C	FCC → BCT
In-T*l*, Mn-Cu	FCC → FCT
Li, Zr, Ti, Ti-Mo, Ti-Mn	BCC → HCP
Cu-Zn, Cu-Sn	BCC → FCT
Cu-A*l*	BCC → distorted HCP
Au-Cd	BCC → orth.

[a] Adapted with permission of [1], p. 180.

example is the shape-memory effect. The material, upon being plastically deformed, undergoes internal changes in the configuration of the martensite plates. Heating will recompose the initial shape. This effect is discussed in detail in Section 13.6. Recently, a number of more complex processing procedures for steels have been developed, involving the martensitic transformation; ausforming, maraging, and dual-phase steels are the best known examples (Section 15.1).

Cohen et al.[6] have proposed a classification scheme for martensitic transformations. They start from the broader group of displacive transformations, defined as "a structural change in the solid state which occurs by coordinated shifts of atoms." These coordinated atom shifts can be described as a combination of (1) *homogeneous lattice deformation* (or distortion) and (2) *shuffles*. While the lattice deformation (or Bain strain) converts one lattice into another, the shuffle is a coordinated shift of atoms in unit cell which, by itself, does not produce a strain. Based on these two types of coordinated atom shifts the classification scheme of Fig. 13.1 was proposed by Cohen et al.[6] The first branching occurs depending on whether lattice distortions or shuffles dominate. The distortive-dominated transformation, in its turn, generates stresses in the lattice that can be decomposed into a deviatoric (shear) and a hydrostatic (or dilatational) component. An example of a dilatation-dominant transformation is the one in tin. There is a 12% volume expansion upon cooling below room temperature. When the deviatoric strains are dominant, on the other hand, an undistorted and unrotated plane is made possible. In this class of transformations, Cohen et al.[6] propose that one has a martensitic transformation if the kinetics and morphology are dominated by the strain energy. To this classification scheme, the martensitic transformation subdivision can be added. One has a "reversible" martensitic transformation if the martensite-parent phase interface moves reversibly, without generating plastic deformation; on the other hand, if substructural damage is produced (dislocations, vacancies, etc.), the interface cannot move reversibly and one has an "irreversible" martensitic transformation. Examples

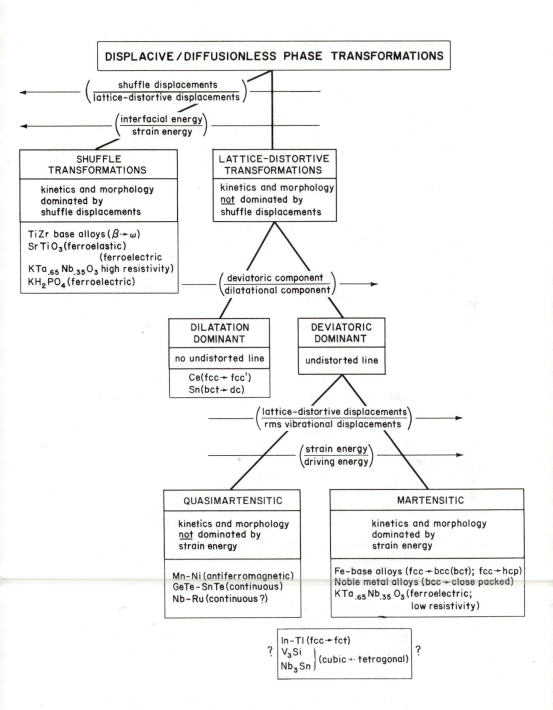

Figure 13.1 Proposed classification scheme for displacive–diffusionless phase transformations. (Reprinted with permission from [6], p. 5.)

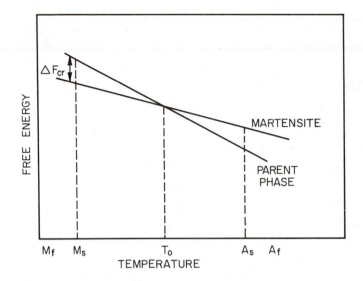

Figure 13.2 Free energy versus temperature for austenitic and martensitic phases. M_s, M_f, A_s, and A_f marked in abscissa.

of the former are shape-memory alloys and of the latter are steels and Fe-Ni alloys.

Consistently with the foregoing scheme, we have:[6] a martensitic transformation is a lattice-distortive, virtually diffusionless structural change having a dominant deviatoric component and associated shape change such that strain energy dominates the kinetics and morphology during the transformation. The requirement of no diffusion stems from thermodynamic requirements. The driving energy required for martensitic transformation is much higher (in the case of irreversible martensites, especially) than the one needed for diffusional decomposition (such as precipitation or spinodal decomposition). Hence, as the alloy is cooled the latter transformations would take place at a higher temperature, when the free-energy difference between the two phases is not very large. Figure 13.2 shows the free energies of the parent and martensitic phase as a function of temperature. At T_0 the two phases have the same free energy and this is the equilibrium temperature. M_s is the highest temperature at which martensite forms spontaneously. The critical free energy required for the martensitic transformation is ΔF_{cr}; it is around 300kcal/mol for Fe-Ni and Fe-C alloys. Hence, if a diffusion-induced transformation competes with the martensitic transformation, the cooling in the $T_0 \rightarrow M_s$ region has to be fast enough to avoid it. On the other hand, if T_0 is low enough, there is essentially no diffusion and slow cooling will produce martensite. Upon heating above A_s the martensite reverts back to austenite. For irreversible martensites the $M_s \rightarrow A_s$ gap is of a few hundred K; for reversible martensite, it is of a few tens of K. Thermodynamics of the martensitic transformation are described in detail by Kaufman and Cohen.[7]

13.2 MORPHOLOGIES AND STRUCTURES OF MARTENSITE

Martensite can exhibit a variety of morphologies, depending on the composition of the alloy, the conditions in which it is formed and its crystalline structure. The three most common morphologies are the lenticular (lens-shaped), the lath (large number of blocks juxtaposed in a shingle-like arrangement), and the acicular (needle-shaped). They are shown in Figs. 13.3, 13.4, and 13.5, respectively. Lenticular martensite occurs in Fe-Ni and Fe-Ni-C alloys with ~30%

Figure 13.3 Lenticular martensite. (Courtesy of J. R. C. Guimarães, Instituto Militar de Engenharia.)

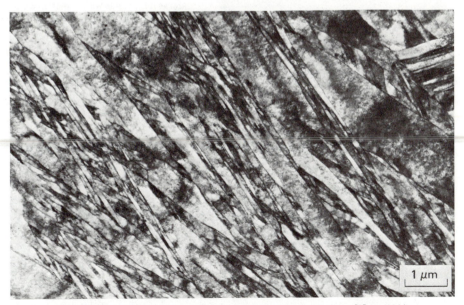

Figure 13.4 Lath martensite. (Reprinted with permission from [8], p. 594.)

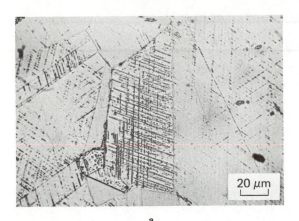

a

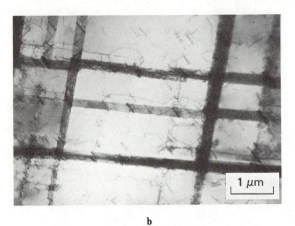

b

Figure 13.5 Needle-shaped α-martensite in shock-loaded AISI 304. The intersections of needles with plane of polish in (a) are seen as dark spots due to heavy etching. Transmission electron micrograph in (b) shows dark bands with α-martensite forming at their intersections. (Reprinted with permission from [9], p. 1948.)

Ni and Fe-C alloys with over 0.6% C. The center region is called the midrib and etches preferentially. The substructure is characterized by twins, dislocations, or both. In the particular case of Fig. 13.3, the region adjacent to the midrib is twinned, and the external parts are dislocated. Lath martensite, on the other hand, is quite different. It consists of small, juxtaposed blocks. Narasimha Rao and Thomas[10] determined the orientation of the different units and found that they are arranged in packets separated by low-angle grain boundaries. Each packet is composed of blocks with a thickness varying between a few tens and a few micrometers; these blocks have specific angles with their neighbors. There is a repetitive pattern in each packet, leading to a 360° rotation and resultant periodicity. Low-carbon steels and Fe-Ni alloys with less than about 30% nickel exhibit this morphology. Acicular martensite is shown in Fig. 13.5. It occurs in austenitic stainless steels after deformation; Fig. 13.5(a) shows the optical micrograph. The dark spots occur in specific arrays and correspond to the intersection of the needles with the plane of the section; they etch more heavily than the austenite. Figure 13.5(b) shows the transmission electron micrograph of a typical region; the needles form at the intersection of the shear bands[9] (either dislocations, stacking faults, twins, ϵ-martensite, or a combination thereof). Since the intersection of these bands is a thin "tube," the martensite forming in it has this specific shape. The acicular martensite has the BCC or BCT structure and has a marked effect on the strength and work-hardening ability of the alloy. Other martensite morphologies have been observed. The ϵ-martensite is HCP and forms in plates. It can be produced in steel by subjecting it to a high pressure[11] ($>$13 GPa), or in austenitic stainless steels by deformation. After substantial plastic deformation, sheaves of fine parallel laths were observed to form along the austenite slip bands in austenitic Fe-Ni-C alloys.[12,13] Yet another morphology is the butterfly martensite, so called because two lenses form in a coupled way; the resultant microstructure resembles a number of butterflies.

In spite of these differences in morphology, there are some unique features common to all martensites. The most important is the existence of an *undistorted and unrotated plane*. The crystallographic orientation relationship between parent and martensite phases is such that there *always* is a plane which has the same indices in the two structures. This undistorted and unrotated plane is called the habit plane; it is usually a plane with irrational indices. Kurdjumov and Sachs[14] found, for a steel with 1.4% carbon, the following relationships:

Habit plane: (225)

$$(111)_A \| (011)_M$$

$$[10\bar{1}]_A \| [1\bar{1}1]_M$$

Steels with less than 1.4% carbon exhibit this relationship. This specific martensite is known as (225) or K-S. Nishiyama[15] investigated the Fe-Ni-C alloys and steels with carbon content greater than 1.4% and obtained:

Habit plane: (259)

$$(111)_A \| (011)_M$$

$$[11\bar{2}]_A \| [01\bar{1}]_M$$

Mehl and Derge[16] were able to obtain the two relationships above, varying the nickel content of alloys with low M_s. Wechsler et al.[17] and Bowles and Mackenzie[18,19] independently developed a theory by which the habit plane and orientation relationships could be predicted from the structure and lattice parameters of the parent and martensitic phase. This theory is considered a very significant step in the theoretical framework of metallurgy. Bilby and Christian[20,21] review it critically, as does Wayman.[22] It will be described only very briefly here, since the total development is rather involved and requires matrix algebra. According to the Wechsler et al. and Bowles–Mackenzie theories, we can arrive at the martensitic structure by the following three operations:

1. *A Bain strain*. Figure 13.6(a) shows an FCC lattice represented by two unit cells. If we take the center points of the top and bottom faces of

the two unit cells and join them, we can form a body-centered unit cell (shown in thicker lines of Fig. 13.6(a)). This body-centered structure has sides a_0 and $a_0/2$; to become body-centered cubic a distortion will have to take place, as shown in Fig. 13.6(b). This contraction along one direction and expansion along the other two receives the name Bain strain; it can be expressed by a matrix (B).

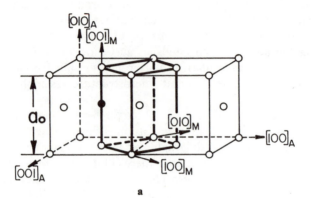

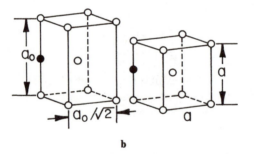

Figure 13.6 Lattice correspondence for the formation of martensite in steels: (a) body-centered tetragonal cell of axial ratio 2 outlined in austenite structure of cell size a_0; (b) deformation carrying this structure into martensite cell with parameter a. Open circles are iron atoms; filled circles are one possible carbon site.

2. *Invariant lattice deformation*, which does not change the lattice structure, but produces shear. In matrix notation: (P).
3. *Rigid lattice rotation* (R), by which the lattice is simply rotated without deformation or distortion.

As a result, the shape deformation P_1 is expressed as

$$P_1 = RBP \qquad (13.1)$$

Martensite plates cannot propagate from one grain to another; the smaller the grain size, the smaller will be the plates formed.

13.3 NUCLEATION AND GROWTH OF MARTENSITE

The generation of a martensite lens is divided into two distinct stages: nucleation and growth. The driving energy for this phase transformation is the decrease in free energy (see Fig. 13.2). However, there is a rather large energy barrier that has to be overcome in order for a nucleus of critical size to be formed. Growth proceeds, in ferrous martensites, at a very high rate once nucleation has taken place, and the activation energy for nucleation is much greater than the one for growth. The large activation energy for nucleation is due to the fact that the formation of a nucleus requires shear, the formation of an interface, and expansion of the lattice. Kaufman and Cohen[7] make a detailed analysis of these factors and arrive at the conclusion that thermal fluctuations in the lattice are not enough for nucleation to take place. Indeed, an undercooling of about 200 K below T_0 is required and, when M_s is low, some other entities have to be invoked. There are essentially two schools of thought regarding nucleation of martensite: the lattice softening and the Olson–Cohen dislocation concepts.

The lattice-softening concept* is based on the propagation of phonons through the lattice. Phonons are quantized lattice vibrations; the movement of the atoms is not random, but coordinated throughout the lattice. The thermal vibrations propagate as waves through the metal; these waves are quantized and called phonons. In the lattice-softening theory it is assumed that the parent phase becomes "soft" in one specific plane and direction, so that the phonon amplitude (vibrational amplitude of atoms) is much greater in that direction. Hence, the vibration amplitude reaches a critical level beyond which the lattice collapses into the new structure. There is an analogy that may be helpful: that of a ripe wheat field over which the wind passes. If the wind exceeds a certain amplitude, the wheat stems will successively lay down, in a domino effect. The result is a whole wheat field flattened.

The Olson–Cohen[23] dislocation model, on the other hand, starts from the precept that there are existing defects in the lattice from which an embryo, and later a nucleus, will form. No collapse of the lattice is required, and the motion of specific dislocations transforms one lattice into another. Olson and Cohen established the following postulates as the basis of their mechanism:

1. Nucleation can be treated as a sequence of steps which take the particle from maximum to minimum coherency.

2. The first nucleation step consists of faulting on the closest-packed planes of the parent phase, the fault displacement being derived from an existing defect.

3. The subsequent steps in the nucleation process take place in such a way as to leave the fault plane unrotated.

* A specific mechanism involving the interactions of soft phonon modes with existing defects was proposed by P. C. Clapp [Phys. Stat. Sol., *57* (1973) 561]. An excellent treatment of nucleation is given by M. Cohen and C. M. Wayman, in "Metallurgical Treatises," J. K. Tien and J. F. Elliott, (eds.), AIME, Warrendale, PA, 1981, p. 445.

For the specific case of the FCC → BCC transformation, the sequence of steps is given in Figure 13.7. A dislocation that already exists is required. The splitting of a total dislocation into partials obeys the reaction

$$\frac{a}{2}[1\bar{1}0] \longrightarrow \frac{a}{6}[1\bar{2}1] + \frac{a}{6}[2\bar{1}\bar{1}]$$

These partial dislocations, on their turn, dissociate themselves, spreading their core over three atomic layers [Fig. 13.7(b)]. Hence, each corresponds to one-third of the original partial dislocation, and one has $a/18$ [$1\bar{2}1$] and $a/18$ [$2\bar{1}\bar{1}$]. An additional set of dislocations moving along different slip planes brings the structure closer to BCC [Fig. 13.7(c)]. The final adjustment is shown in Fig. 13.7(d). One $a_{fcc}/2$ [$\bar{1}10$] on each eight planes completes the process. Meyers[24] has found that the time required for nucleation could be below 55 ns.

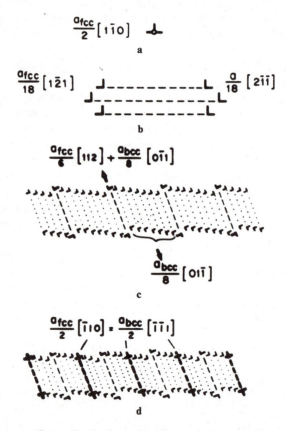

Figure 13.7 Sequence of steps for the formation of a semicoherent BCC embryo from an FCC lattice dislocation: (a) existing $a_{FCC}/2$ [$1\bar{1}0$] screw dislocation; (b) dissociation of dislocation with spreading of cores over consecutive planes to produce a fault with the structure of the first *Bogers–Burgers shear configuration*; (c) relaxation of fault of (b) to BCC structure, producing new partial dislocations in the fault interface; (d) fault structure after addition of lattice screw dislocations which cancel remaining long-range field of partial dislocations of (c). (Reprinted with permission from [23], p. 1907.)

Once an embryo has reached a critical size upon which further increase in size results in a decrease of free energy (top of the energy hill), it is called a nucleus. From this moment on, growth takes place. The different morphologies exhibited by different martensites are the result of different growth patterns. The discussion presented here restricts itself to the classic lenticular martensite characteristic of ferrous alloys. Growth velocity has been found to be extremely high. Bunshah and Mehl[25] obtained values of the order of 1000 m/s; the velocities could be even higher than this. Kulin and Cohen[26] showed that the growth velocity was not significantly affected by temperature. This led them to propose the existence of a strain wave; if the growth mechanism was thermally activated, it would produce low velocities at low temperatures and an exponential temperature dependence. Meyers[27] proposed that growth proceeds in two directions: longitudinal and transverse. His model is schematically shown in Fig. 13.8 and predicts a lenticular shape for martensite.

According to it, the growth of lenticular martensite typically occurring in the Fe-Ni and Fe-C systems takes place by the propagation of waves throughout the material. Two different types of waves are postulated: longitudinal transformation waves and transverse transformation waves, propagating at velocities of the order of elastic waves. The longitudinal transformation wave initiates the transformation process once an embryo has reached a critical size whereupon it becomes a nucleus; it propagates radially along directions contained in the habit plane, forming the midrib. The martensitic disk generated by the longitudinal transformation wave acts as a second-order nucleus for the transverse transformation. The term "second-order nucleus" is used to distinguish it from the "first-order nucleus" that gives rise to the start of the transformation. The transverse transformation waves propagate perpendicularly to the habit plane, starting at the midrib. Accordingly, different defect generation mechanisms operate along the longitudinal and transverse propagation directions, due to the differences in stress state and substructure at the fronts and propagation velocities. The model allows the determination of the shape of a growing martensite plate, which closely resembles lenticular martensite. If xz is the habit plane, the direction of transverse propagation is oy and the major shear direction is ox, the martensite lens can be described, at time t, by the equation

$$\frac{(x^2 + z^2)^{3/2}}{x^2 v_{ed} + z^2 v_{es}} - \frac{1}{k} \ln\left(1 - \frac{k|y|}{v_{es}}\right) = t \tag{13.2}$$

where v_{ed} and v_{es} are the velocities of longitudinal and shear elastic waves and k is a parameter. The arrest of growth takes place by uncoupling between the transformation front and the plastic waves that precede it. Figure 13.9 shows the shape of a martensite lens, as predicted by Eq. 13.2.

‡13.4 STRENGTH OF MARTENSITE[28]

The martensitic transformation has the ability to confer a great degree of strength to steels; other alloys do not seem to have such strong martensites. The strength of martensite in steel is dependent on a number of factors, the most important

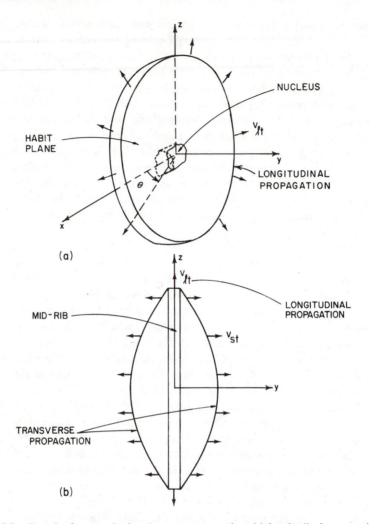

Figure 13.8 Growth of martensite lens by wave propagation: (a) longitudinal propagation along plane *xz* at velocity v_{lt}, starting at nucleus; (b) transverse propagation with velocity v_{st}, perpendicular to habit plane (direction *oy*). (Reprinted with permission from [27], p. 761.)

being the carbon content. While the Rockwell C hardness of iron increases from 5 to 10 when it is transformed to martensite, it increases from 15 to 65 when the carbon content is 0.80 (eutectoid steel). The origin of the high hardness of martensite has been the object of great controversy in the past. It is now fairly well established that there is no single and unique mechanism responsible for it. Rather, a number of strengthening mechanisms operate; most of the mechanisms are described in Part III of this book. Nevertheless, the relative importance of these strengthening mechanisms and their interactions are still the object of controversy. It seems that interstitial solution hardening and sub-structure strengthening (work hardening) are the most important ones.

10 mm

Figure 13.9 Computer-generated perspective view of martensite lens at $t_i = 10^{-5}$ s. (Reprinted with permission from [27], p. 767.)

Most metals exhibit a yield stress dependence on grain size; the martensite lenses divide and subdivide the grain when they form. Hence, a small-grain alloy produces small martensitic plates, whereas a large-grain alloy produces a distribution of sizes whose mean is much larger. Grange[29] confirmed experimentally this effect; it is shown in Fig. 13.10. Three commercial steels (AISI 4310, 4340, and 8650) exhibit a dependence of yield stress on prior austenitic grain size. The slope of the Hall–Petch plot seems to be the same for the three. However, for the range of grain sizes usually encountered the contribution of grain size is not very important: the grain sizes are equal to 0.1 mm or more. Only in steels that have undergone thermomechanical processing to reduce the austenitic grain size is this strengthening mechanism of significant importance.

The contribution of substitutional solid solution elements to the strength of ferrous martensites is relatively unimportant; additionally, it is difficult to separate it from other indirect effects such as the change in M_s and stacking-fault energy due to the addition of these elements.

On the other hand, interstitial solutes (carbon and nitrogen, for instance) can play an important effect. If we regard martensite as a supersaturated solution

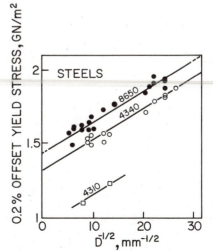

Figure 13.10 Effect of grain size on the yield stress of three commercial martensitic steels. (Adapted with permission from [29], p. 44.)

of carbon in ferrite, a great portion of its strength could be ascribed to solution hardening. Foreman and Makin,[30] based on Fleischer's theory (see Section 10.4) developed an equation of the form below to express the effect of the solute concentration c on the O K yield stress of the alloy, if only the interaction between dislocations and single-atom obstacles is considered:

$$\tau_0 = \left(1 - \frac{\varphi'}{5\pi}\right) G \left(\frac{F_{\max}}{2T}\right)^{3/2} (3c)^{1/2} \tag{13.3}$$

where $F_{\max}$ is the maximum force exerted by the obstacle on the dislocation, T the tension of the dislocation line, G the shear modulus, and φ' the angle turned through by the dislocation immediately before it frees itself from the obstacle. The interesting aspect of Eq. 13.3 is that the yield stress should increase with the square root of the solute concentration. Results obtained by Roberts and Owen[31] confirm Eq. 13.3, as can be seen in Fig. 13.11. They used alloys with very low M_s (below 77 K) to avoid any secondary effect of the carbon atoms, such as precipitation hardening or Cottrell atmosphere formation. The fact that the room-temperature tests exhibit the same slope as the ones conducted at 77 K shows that even at room temperature, solid solution hardening is operating and effectively strengthening martensite.

Snoek ordering[32] consists of the reorientation of a system of point defects of tetragonal or lower symmetry that are randomly distributed in the stress

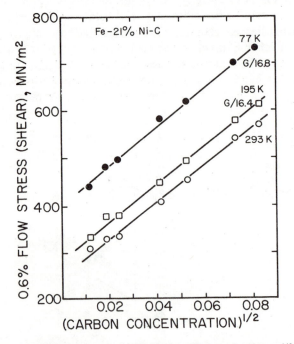

Figure 13.11 Plot of 0.6% proof stress (one-half of tensile stress) versus $c^{1/2}$ for Fe-Ni-C lath martensite at various temperatures. The slopes are shown as fractions of the shear modulus, which is denoted G. (Adapted with permission from [31], p. 377.)

field of a dislocation. Single jumps of carbon atoms can organize them in such a way as to minimize their energy. Owen and Roberts[33,34] found experimental evidence for static aging produced by Snoek ordering. Snoek ordering can take place in a much lower time interval than Cottrell atmosphere formation, because no long-range diffusion is required.

Cottrell atmosphere formation, on the other hand, requires diffusion of the atoms toward regions in the dislocation in which their strain energy will be minimized. Carbon atoms produce tetragonal distortions and shear stresses; hence, they seek regions around both edge and screw dislocations in which the shear strains cancel each other. Cottrell atmospheres produce both static and dynamic aging. A manifestation of the latter is the serrated flow (Portevin–Le Châtelier effect), which was observed by Owen and Roberts.[33,34]

The carbon atoms have also been shown to exhibit a clustering behavior. Carbon-rich regions have been identified by transmission electron microscopy in steels that had been exposed to temperatures no higher than ambient temperature. These clusters do not change the crystalline structure of the martensite, but produce periodic strain fields, resulting in a "modulated" structure. In this sense, the clustering is closer to a spinodal decomposition than to a precipitation reaction. If the martensite is aged at higher temperatures, cementite and other metal carbides are precipitated. The latter process is called tempering.

Frequently, precipitation is observed in martensite. Winchell and Cohen[35] showed how this can take place. Quenched carbon steels with M_s above room temperature may contain precipitates that form during cooling. In certain low-carbon steels these precipitates have been identified as cementite. It seems that carbon is a more efficient strengthener as a precipitate than in solid solution. The calculations of Kelly and Nutting[36] and those of Leslie and Sober[37] agree that the contribution of precipitates in ferrous martensites exceeds that of a solid solution. A very important contribution is that of strain hardening. In twinned martensite a very fine array of twins 50 to 90 Å thick present a very effective barrier for additional deformation. Kelly and Nutting[38] considered these fine twins the most important factor in the strength of martensite. When martensite is dislocated, the density of dislocations is typically 10^{10} to 10^{11} cm^{-2}; the substructure resembles that of heavily deformed BCC steel, by conventional means.

Williams and Thompson[39] estimate the contributions to the strength of the martensite in a 0.4% carbon steel to be distributed as follows:

Boundary strengthening	620 MPa
Dislocation density	270 MPa
Solid solution of carbon	400 MPa
Rearrangement of carbon in quench (Cottrell, Snoek, clustering, precipitation)	750 MPa
Other effects	200 MPa
Total	2240 MPa

They point out that these effects are not necessarily additive; however, this simplified scheme shows the various contributions.

Yet another source of strengthening is the intrinsic resistance of the lattice to dislocation motion (Peierls–Nabarro stress). Christian[28] believes that this intrinsic resistance to dislocation motion accounts for the temperature dependence of yield stress of martensite. Iron exhibits a strong temperature dependence of yield stress at low temperatures, as do other BCC metals. This same behavior is observed in martensite, independent of the existence of precipitates and solutes.

13.5 MECHANICAL EFFECTS

A martensite lens introduces macroscopic strains in the lattice surrounding it. This is best seen by making fiducial marks on the surface and transforming the material. The fiducial marks will be distorted by these strains. The strains introduced by a martensite lens can be decomposed into a dilatational and a shear strain. There is a dilatative strain perpendicular to the midrib plane and a shear strain parallel to the midrib plane. In ferrous alloys, the dilatation is ~0.05 and the shear strain γ is ~0.20. Figure 13.12 shows a fiducial mark made on the surface of the hypothetical alloy. The shear direction is such that the plane is not distorted. Hence, $\tan \theta = \gamma$, and θ is equal to 11°. Machlin and Cohen[40] systematically investigated these strains for the different martensite variants. The strain matrix can be expressed as

$$\begin{pmatrix} \epsilon_{11} & \epsilon_{12} & \epsilon_{13} \\ \epsilon_{12} & \epsilon_{22} & \epsilon_{23} \\ \epsilon_{13} & \epsilon_{23} & \epsilon_{33} \end{pmatrix} = \begin{pmatrix} 0 & 0 & 0 \\ 0 & 0 & 0.10 \\ 0 & 0.10 & 0.05 \end{pmatrix} \qquad (13.4)$$

We must recall that $\epsilon_{23} = \gamma_{23}/2$. These strains are well beyond the elastic limit of the matrix, and there is plastic deformation in the region surrounding the martensitic lens. This is reflected in Fig. 13.12 by the distortion of the fiducial line.

The dilatational and shear stresses and strains imposed by the martensitic transformation interact with externally applied stresses and very special responses

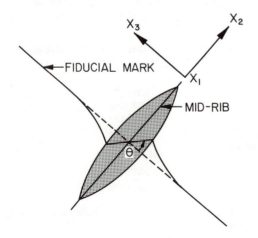

Figure 13.12 Distortion produced by martensite lens on fiducial mark on surface of specimen.

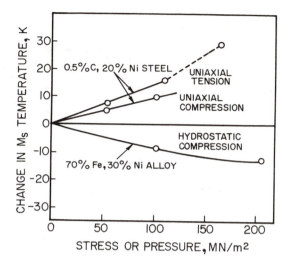

Figure 13.13 Change in M_s temperature as a function of loading condition. (Adapted with permission from [41], p. 532.)

ensue. The effect of externally applied tensile, compressive, and hydrostatic stresses were studied by Patel and Cohen[41] and are shown in Fig. 13.13. The uniaxial tension and compression increase M_s, whereas hydrostatic compression lowers it. The explanation provided by Patel and Cohen[41] is that under the effect of the applied stress, the mechanical work done by the transformation, which can be decomposed into the dilatational and shear components, $\sigma\epsilon$ and $\tau\gamma$, is either increased or decreased. The hydrostatic stress counters the lattice expansion produced by martensite, but does not affect the shear stress. Hence, a greater amount of free energy is required to trigger the transformation. Referring to Fig. 13.2 we can see that a greater ΔF will require a lowering of M_s. For the tensile test, the applied stress can be decomposed into a normal (positive) and shear stress, both of which aid the transformation. The shear stress aids the martensite variants aligned with the direction of maximum shear (45° to tensile axis). These variants will form preferentially. Hence, the free-energy requirement is decreased and M_s is increased. In the compressive test, the normal portion of the stress is negative and counters the dilatational stress of the transformation, whereas the shear stress favors it (there are always favorably oriented variants). Since the shear stress term dominates the expression (because of the greater shear strain γ), the tensile stress should be more effective in increasing M_s than the compressive stress. This is exactly what Patel and Cohen[41] observed (Fig. 13.13).

While Patel and Cohen[41] prestressed their specimens to desired stress levels and lowered the temperature until martensite formed, a slightly different experimental procedure can also be used. It consists of conducting a tensile test in specimens at temperatures above M_s. When the stress level reaches the value at which martensite forms at the test temperature, a significant load drop is observed. Figure 13.14 shows this effect. The load drop is due to the shear

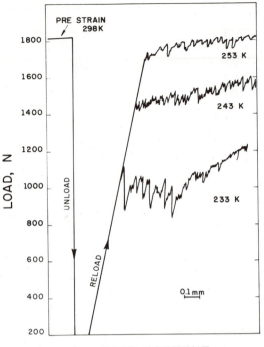

Figure 13.14 Tensile curves for Fe-Ni-C alloy above M_s showing martensite forming in elastic range (stress-assisted). (Courtesy of J. R. C. Guimarães, Instituto Militar de Engenharia.)

strain of the martensite, which produces an instantaneous increase of length of the specimen. This aspect is discussed in more detail in Section 16.2. As the difference between the test temperature and M_s increases, the stress at which martensite starts forming increases; this can be directly inferred from Fig. 13.14. In Fig. 13.15 this yield stress is plotted as a function of temperature; when martensite forms in the elastic line, the stress at which it forms is equal to the yield stress [Fig. 13.14, for instance]. The temperature dependence of the stress for martensite transformation is clearly shown by the three straight lines in Fig. 13.15. At M_s, as expected, martensite forms spontaneously, without any stress. The yield stress versus temperature plots for three different conditions have inverted V shapes. At the point marked M_s^{σ} the slope of the curve changes and above this temperature the yield stress is produced by conventional dislocation motion; hence, it shows the regular increase with decreasing temperature. Between M_s and M_s^{σ}, on the other hand, we have *stress-assisted martensite establishing* yield, and the temperature dependence is inverted, leading to a yield stress of zero at M_s. It is worth noting that the three alloys in Fig. 13.15 have the same composition but different processing histories.[42] M_s temperature is affected by grain size (M_s increases with increasing grain size and predeformation); in the case shown in Fig. 13.15, predeformation was accomplished by shock loading (this is discussed in greater detail in Chapter 15).

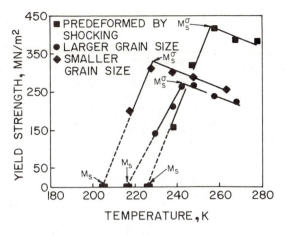

Figure 13.15 Temperature dependence of the yield strength of Fe–31% Ni–0.1% C. (Adapted with permission from [42], p. 413.)

The formation of strain-induced martensite occurs in the temperature range above M_s^σ in Fig. 13.15. Substantial plastic deformation is required before the first martensite forms. The substructure has to be sensitized by plastic deformation. This latter martensite is called *strain-induced,* to differentiate it from the former (*stress-assisted*). Guimarães[12] and later Maxwell et al.[13] investigated both martensites and found that the strain-induced one has a unique morphology resembling thin parallel laths less than 0.5 μm wide.

Strain-induced martensite is responsible for a very beneficial property: the transformation-induced plasticity (TRIP) effect. TRIP steels were developed by Zackay, Parker, and coworkers.[2,43,44] Maxwell et al.[13] ascribe the TRIP effect to strain-induced martensite, not to the stress-assisted one. Remarkable combinations of high strength and toughness have been obtained in TRIP steels. The high strength is due to work hardening, carbide precipitation, and dislocation pinning by solutes during the thermomechanical treatment (described in greater detail in Section 15.1). The high toughness comes from a combination of high strength and ductility. The latter is a direct consequence of the strain-induced martensite transformation. If a certain region in the metal is severely deformed plastically, strain-induced transformation takes place, increasing the local work-hardening rate and inhibiting an incipient neck from further growth. On the other hand, if a crack has already formed, martensitic transformation at the crack tip will render its propagation more and more difficult.

Another mechanical aspect of importance is the fracture of martensite. Fracture is usually initiated in a martensitic alloy along the martensite–austenite or martensite–martensite boundaries. Indeed, Chawla et al.,[45] upon investigating the fracture surfaces of Fe–31% Ni–0.1% C alloy, found that the density of dimples increased as the amount of martensite in the cross section; the same result was obtained by decreasing the grain size. Hence, the dimple size was tied to the density of interfaces. In carbon-free or low-carbon steels, martensite is fairly soft and the fracture is, consequently, ductile. In high-carbon steels, on the other hand, martensite is hard and brittle, and the fracture surface takes a cleavage appearance, with the fracture path traversing the plates (or laths). Of great importance in the initiation of fracture is the existence of microcracks in the structure. Marder et al.[46] found a great number of microcracks in Fe-C martensites; when the grain size was decreased, the incidence of microcracks decreased. These microcracks were formed when one lens impinged on another. Figure 13.16 shows how these cracks occur. These microcracks act as stress-concentration sites when the specimen is loaded. They are initiation sites for macrocracks.

Tempering of martensite in steels is performed to improve the toughness. However, the tempering process might induce embrittlement.[48] Temper martensite embrittlement (TME) results from the segregation of impurities to the previous austenitic grain boundaries, providing a brittle path for fracture propagation. The fracture takes the intergranular morphology. Temper embrittlement (TE) is caused by the impurities antimony, phosphorus, tin, and arsenic (less than 100 ppm required) or larger amounts of silicon and manganese. TME and TE occur in different ranges of temperatures; TME is a much more rapid process.

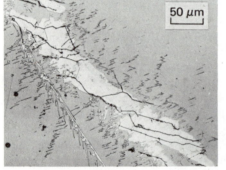
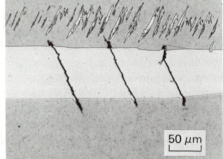

a

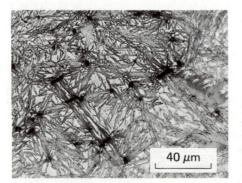

b

Figure 13.16 Microcracks generated by martensite. (a) Fe–8% Cr–1% C (225 martensite sectioned parallel to habit plane). (Courtesy of J. S. Bowles, University of South Wales.) (b) Carburized steel. (Reprinted with permission from [47], p. 1197.)

Strengthening Mechanisms   Part III

13.6 SHAPE-MEMORY EFFECT

The shape-memory effect (SME) is the unique property that some alloys possess according to which, after being deformed at one temperature, they recover the original shape upon being heated to a second temperature. The built-in memory is produced by the martensitic transformation. The effect was first discussed by the Russian metallurgist Kurdjumov. In 1951, Chang and Read[49] reported it for an In-Tl alloy. However, wide exposure of the properties came only after the development of the nickel-titanium alloy by the Naval Ordnance Laboratory (NiTiNOL) in 1968.[50] Since then, the research activity in this field has been intense, and the following β-phase SME alloys have been investigated:[51] AgCd, AgZn, AuCd, CuAl, CuZn, FeBe, FePt, NbTi, NiAl, and ternary alloys. The Nitinol family of alloys has found wide technological applications, and adjustments in the composition can be made to produce M_s temperatures between -273 and $100°C$. This is an extremely helpful feature, and alloys are tailored for specific applications. In the majority of SME alloys the high-temperature phase is a disordered β phase (body-centered cubic), while the martensitic phase is an ordered BCC structure with a superlattice or orthorhombic structure. Two separate mechanical effects characterize the response of SME alloys: pseudoelasticity and strain memory effect. They will be described in connection with tensile tests.

Pseudoelasticity is the result of stress-induced martensitic transformation in a tensile test (see Section 13.5) which reverts back to the parent phase upon unloading. The individual martensite plates do not grow explosively, as in the ferrous martensites, and little irreversible damage is done to the lattice. The shear strain of one plate is accommodated by neighboring plates. The complex motion of the interfaces between the martensite plates along the various variants and within the same martensite plate, takes place by the displacement of the interfaces between the different twins. Figure 13.17(a) shows the pseudoelastic effect for a Cu-Zn-Sn alloy with $M_s = -48°C$. The test is being conducted at $24°C$ ($76°C$ above M_s). At A, stress-induced martensite starts to form. At B, the martensitic transformation has been completed and any straining beyond that point will produce irreversible plastic deformation or fracture. Upon unloading, the martensite reverts to the parent phase between C and D. Further unloading results in the return to the original length of the specimen. The pseudoelastic strain exceeds 6%. The magnitude of the pseudoelastic strain can be calculated from a knowledge of the habit plane of the martensite (and its orientation with respect to the tensile axis) and the magnitude of the shear strain for the transformation. Since the habit plane of martensite is irrational, it has a multiplicity of 24, and there is always a habit plane oriented very closely to the plane of maximum shear.

It is not sufficient for the testing temperature to be above M_s to obtain the pseudoelastic effect, as in Fig. 13.17(b). These tests were conducted on a Cu-Al-Ni alloy. The A_s and A_f temperatures (austenite start and finish, respectively) are also important. These temperatures are defined in Fig. 13.2. If the testing temperature is below A_s, the martensite will not revert back to austenite upon

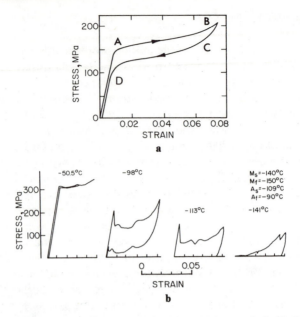

Figure 13.17 (a) Pseudoelastic stress–strain curve for a single-crystal Cu-Zn-Sn alloy at 24°C (76°C above M_s). (Reprinted with permission from [51], p. 31.) (b) Temperature dependence of stress–strain characteristics along the characteristic transformation temperatures. Strain rate: 2.5 × 10⁻³ min⁻¹. (Reprinted with permission from [52], p. 68.)

unloading; the tests conducted at −141°C and −113°C show this irreversibility. For the test conducted at −50.5°C and −98°C, total reversibility is obtained, since this temperature is above A_f (−90°C). Another observation that can be made in Fig. 13.17(b) is that the stress at which martensite forms increases with increasing temperature; this response is similar to the one described in Section 13.5.

When the deformation is irreversible (at −131°C and −141°C in Fig. 13.17), the effect receives the name *strain-memory effect*. Additional heating is required to revert the martensite, since the deformation temperature is below A_s. Upon heating, the original dimensions will be regained, as the martensite interfaces move back to retransform the lattice. The sequences in which the plates form and in which they disappear are inverted. The first plate to form is the last to disappear.

The strain-memory effect is also obtained when deformation is imparted at temperatures below M_s. This is actually the procedure used in most technological applications. In this case, the structure consists of thermally induced martensite; it is present in such a way that all variants occur. When the external stress is applied, the variants that have shear strains aligned with the applied shear strain tend to grow and the unfavorably oriented variants shrink. Figure 13.18 shows schematically how this takes place. Only two variants are shown, for simplicity. The variant that favors the applied tensile strain grows at the expense of the unfavorably oriented one. Hence, all unfavorable variants disappear and the favorable variant takes over the structure. On heating, the structure reverts back to the original one, composed of equal distribution of the two variants, giving the strain recovery.

The shape-memory effect can be described from the point of view of technological applications as a "solution looking for a problem."[53] It has found some unique uses. One is as a tight coupling for pneumatic and hydraulic lines. The F-14 jet fighter tube couplings are made of Nitinol that is fabricated at room temperature with a diameter 4% less than that of the tubes that will be joined. Then, the couplings are cooled below M_s ($-120°C$) and expanded mechanically until their diameter is 4% larger than those of the tubes. They are held at this temperature until they are placed over the tube ends. Allowed to warm, they will shrink to the initial diameter; impeded by the tube, they will provide

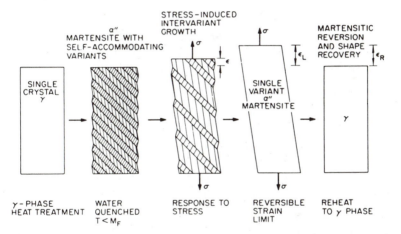

Figure 13.18 Sequence showing how growth of one martensite variant and decrease of others results in strain ϵ_L. (Courtesy of R. Vandermeer, University of Tennessee.)

a tight fit. Another potential application is shown in Fig. 13.19. It is an antenna for a spacecraft. The sequence of steps is self-explanatory. Electrical connectors that are opened and closed by temperature changes are another application. Orthopedic and orthodontic aids have also been tested, and Nitinol seems to react well in the body fluid environment. The pen-drive mechanism in recorders is a very successful application of the SME; over 600,000 of these new drives are now in service. Another application that has received a lot of attention is the utilization of the SME in the direct conversion of heat to mechanical work. The advantage is that such an engine can utilize low-temperature gradient thermal sources ($\sim 20°C$). Salzbrenner[54] measured a maximum efficiency at less than 2% between 25 and 190°C for the polycrystalline alloy Ni-45 at percent

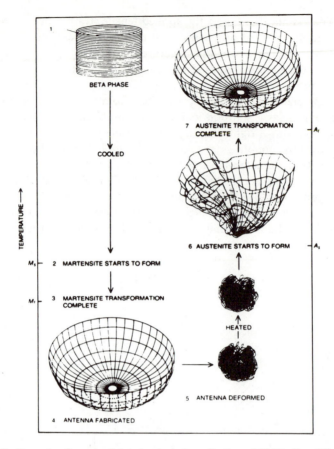

Figure 13.19 Example of a potential technological application of SME alloy. Wire hemisphere for use as antenna in spacecraft; unfolding occurs after spacecraft has been launched and is in orbit. (Reprinted with permission from [53], p. 76.)

Ti operating in a tensile cycle; the calculated Carnot efficiency was 35.6%. The efficiency depends on a number of parameters, including grain size, type of stress cycle, alloy composition, and temperature. Salzbrenner[54] concluded that the practicality of such an engine would be very limited under most circumstances; this is clearly seen by the large difference between the efficiencies. Other potential benefits of the shape-memory effect that are being investigated are the increased damping capacity.[55] Damping capacity seems to be very large because of the work required to form the martensite.

EXERCISES

13.1. Plot the hydrostatic strain versus carbon content for the martensitic transformation in steel from the plot shown below.

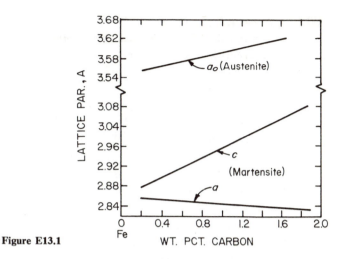

Figure E13.1

13.2. From the data of Fig. 13.14, estimate the M_S temperature of the alloy at zero stress.

13.3. Design a pen-drive system for a recorder using Cu-Zn-Al. Based on the plot presented by L. McDonald Shetky [*Sci. Am., 241* (Nov. 1979)], what composition would you choose for the alloy?

13.4. The AISI 8620 steel shown in Fig. 13.16(b) has a plane strain fracture toughness of 110 MPa$\sqrt{m}$ and a yield stress of 320 MPa. Will the cracks shown in the Figure have a catastrophic effect if a specimen is stressed to 180 MPa?

13.5. Write down all the possible martensite variants for the Kurdjumov–Sachs orientation.

13.6. Verify the correctness of the dislocation reactions in the Olson–Cohen model. Would they be energetically favorable under usual conditions?

13.7. Calculate the total strain energy associated with a martensite lens having a volume of 10 μm^3, assuming that all the energy is elastically stored. Specify the assumptions made; include both shear and longitudinal strain components from Eq. 13.4.

REFERENCES

[1] V.F. Zackay, M.W. Justusson, and D.J. Schmatz, in *Strengthening Mechanisms in Solids*, ASM, Metals Park, Ohio, 1962, p. 179.

[2] V.F. Zackay, E.R. Parker, D. Fahr, and R. Bush, *Trans. ASM, 60* (1967) 252.

[3] L. Delaey, R.V. Krishnan, H. Tas, and H. Warlimont, *J. Mater. Sci., 9* (1974) 1521, 1536, 1545.

[4] J.C. Shyne, V.F. Zackay, and D.J. Schmatz, *Trans. ASM, 52* (1960) 348.

[5] S. Floreen, *Met. Rev., 13* (1968) 115.

[6] M. Cohen, G.B. Olson, and P.C. Clapp, *Proc. Int. Conf. Martensitic Transformations*, M.I.T., Cambridge, Mass., May 1979, p. 1.

[7] L. Kaufman and M. Cohen, "Thermodynamics and Kinetics of Martensitic Transformations," in *Progress in Metal Physics*, Vol. 7, B. Chalmers and R. King (eds.), Pergamon, Elmsford, 1958, p. 165.

[8] C.A. Apple, R.N. Caron, and G. Krauss, *Met. Trans.*, 5 (1974) 593.

[9] H.-J. Kestenbach and M.A. Meyers, *Met. Trans.*, 7A (1976) 1943.

[10] B.V. Narasimha Rao and G. Thomas, *Proc. Int. Conf. Martensitic Transformations*, M.I.T., Cambridge, Mass., 1979, p. 12.

[11] G.E. Duvall and R.A. Graham, *Rev. Mod. Phys.*, 49 (1977) 523.

[12] J.R.C. Guimarães, Ph.D. dissertation, Stanford University, Stanford, Calif., 1972.

[13] P.C. Maxwell, A. Goldberg, and J.C. Shyne, *Met. Trans.*, 5 (1974) 1305.

[14] G. Kurdjumov and G. Sachs, *Z. Phys.*, 64 (1930) 325.

[15] Z. Nishiyama, *Sci. Rep., Tohoku Univ.*, 28 (1934) 627.

[16] R.F. Mehl and G. Derge, *Trans. AIME*, 125 (1937) 482.

[17] M.S. Wechsler, D.S. Lieberman, and T.A. Read, *Trans. AIME*, 197 (1953) 1503.

[18] J.S. Bowles and J.K. Mackenzie, *Acta Met.*, 2 (1954) 138.

[19] J.S. Bowles and J.K. Mackenzie, *Acta Met.*, 2 (1954) 224.

[20] B.A. Bilby and J.W. Christian, *Inst. Met. Monogr. No. 18* (1955) p. 121.

[21] B.A. Bilby and J.W. Christian, *J. Iron Steel Inst.*, 197 (1961) 122.

[22] C.M. Wayman, *Introduction to the Crystallography of Martensitic Transformations*, Macmillan, New York, 1964.

[23] G.B. Olson and M. Cohen, *Met. Trans.*, 7A (1976) 1897, 1905, 1915.

[24] M.A. Meyers, *Met. Trans.*, 10A (1979) 1723.

[25] R.F. Bunshah and R.F. Mehl, *J. Metals*, 5 (1953) 1251.

[26] S.A. Kulin and M. Cohen, *Trans. AIME*, 188 (1950) 1139.

[27] M.A. Meyers, *Acta Met.*, 28 (1980) 757.

[28] J.W. Christian, "The Strength of Martensite," in *Strengthening Methods in Crystals*, A. Kelly and R.B. Nicholson (eds.), Elsevier, Amsterdam, 1971, p. 261.

[29] R.A. Grange, *Trans. ASM*, 59 (1966) 26.

[30] A.J.E. Foreman and M.J. Makin, *Phil. Mag.*, 14 (1966) 191.

[31] M.J. Roberts and W.S. Owen, *J. Iron Steel Inst.*, 206 (1968) 375.

[32] G. Schoeck and A. Seeger, *Acta Met.*, 7 (1959) 469.

[33] W.S. Owen and M.J. Roberts, in *Int. Conf. Strength of Alloys*, Jap. Inst. Metals, Tokyo, 1968, p. 911.

[34] W.S. Owen and M.J. Roberts, in *Dislocation Dynamics*, A.R. Rosenfield, G.T. Hahn, A.L. Bement, Jr. and R.I. Jaffee (eds.), McGraw-Hill, New York, 1968, p. 357.

[35] P.G. Winchell and M. Cohen, *Trans. ASM*, 55 (1962) 347.

[36] P.M. Kelly and J. Nutting, *Iron Steel Inst. Spec. Rep.*, 93 (1965) 166.

[37] W.C. Leslie and R.J. Sober, *Trans. ASM*, 60 (1967) 99.

[38] P.M. Kelly and J. Nutting, *Proc. Roy. Soc. (London)*, A259 (1960) 45.

[39] J.C. Williams and A.W. Thompson, in *Metallurgical Treatises*, J.K. Tien and J.F. Elliot (eds.), TMS-AIME, Warrendale, Pa., 1981 p. 487.

[40] E.S. Machlin and M. Cohen, *Trans. AIME*, 191 (1951) 1019.

[41] J.R. Patel and M. Cohen, *Acta Met.*, 1 (1953) 531.

[42] J.R.C. Guimarães, J.C. Gomes, and M.A. Meyers, *Suppl. Trans. Jap. Inst. Met.*, 17 (1976) 411.

[43] J.A. Hall, V.F. Zackay, and E.R. Parker, *Trans. ASM*, 62 (1969) 965.

[44] W.W. Gerberich, P.L. Hemmings, M.D. Monz, and V.F. Zackay, *Trans. ASM*, *61* (1968) 843.

[45] K.K. Chawla, J.R.C. Guimarães, and M.A. Meyers, *Metallography*, *10* (1977) 201.

[46] A.R. Marder, A.O. Benscoter, and G. Krauss, *Met. Trans.*, *1* (1970) 1545

[47] C.A. Apple and G. Krauss, *Met. Trans*, *4* (1973) 1195.

[48] G. Krauss, *Principles of Heat Treatment of Steel*, ASM, Metals Park, Ohio, 1980, pp. 212, 215.

[49] L.C. Chang and T.A. Read, *Trans. Met. Soc. AIME*, *191* (1951) 47.

[50] W.J. Buehler and F.E. Wang, *Ocean Eng.*, *1* (1968) 105.

[51] C. Rodriguez and L.C. Brown, in *Shape Memory Effects*, J. Perkins (ed.), Plenum Press, New York, 1975, p. 29.

[52] K. Shimizu and K. Otsuka, cited in [51], p. 59.

[53] L. McDonald Schetky, *Sci. Am.*, 232, No. 11 (1974) 24.

[54] R.J. Salzbrenner, unpublished research, 1981.

[55] L. Kaufman, S.A. Kulin, P. Nesche, and R.J. Salzbrenner, cited in [51], p. 547.

SUGGESTED READING

CHRISTIAN, J.W., The Strength of Martensite, in *Strengthening Methods in Crystals*, A. Kelly and R.B. Nicholson (eds.), Elsevier, Amsterdam, 1971, p. 261.

CHRISTIAN, J.W., *The Theory of Transformations in Metals and Alloys*, Pergamon Press, Elmsford, N.Y., 1965.

KRAUSS, G., *Principles of Heat Treatment of Steel*, ASM, Metals Park, Ohio, 1980.

LIEBERMAN, D.S., "Crystal Geometry and Mechanisms of Phase Transformations in Crystalline Solids," in *Phase Transformations*, ASM, Metals Park, Ohio, 1970, p. 1.

MAGEE, C.L., "The Nucleation of Martensite," in *Phase Transformations*, ASM, Metals Park, Ohio, 1970, p. 115.

"New Aspects of Martensitic Transformation," *Proc. First Jap. Inst. Met. Int. Symp.*, *Suppl. Trans. Jap. Inst. Metals*, 1976, Vol. 17.

NISHIYAMA, Z., *Martensitic Transformation*, Academic Press, New York, 1978.

PERKINS, J. (ed.), *Shape Memory Effects in Alloys*, Plenum Press, New York, 1975.

Proc. Int. Conf. Martensitic Transformations, Cambridge, Mass., May 1979.

WAYMAN, C.M., *Introduction to the Crystallography of Martensite Transformation*, Macmillan, New York, 1964.

Chapter 14

GRAIN-SIZE STRENGTHENING

‡14.1 INTRODUCTION

Ever since Hall[1] and Petch[2] introduced their well-known relationship between the lower yield point of low-carbon steels and grain size, a great deal of effort has been devoted to (1) explaining it from a fundamental point of view and (2) applying it to the yield and flow stress of different metals and alloy systems. The Hall–Petch equation has the form

$$\sigma_y = \sigma_0 + kD^{-1/2} \tag{14.1}$$

where σ_y is the yield stress, σ_0 a frictional stress required to move dislocations, and D the grain size. This equation has been applied to many systems, with varying degrees of success. It seems to be a satisfactory description of the grain-size dependence of yield stress when a somewhat limited range of grain sizes is being investigated. Figure 14.1 illustrates the Hall–Petch equation for several metals. BCC and FCC metals exhibiting smooth elastic–plastic transitions and yield points are represented. Table 14.1 presents the parameters for a number of metals.

It should be noted that Hall–Petch data reported in the literature should be analyzed with great care. The scatter in the results reported by different investigators is usually large. The large differences can be due to several reasons:

1. *Specimen dimensions*. True polycrystallinity requires that the specimen size be about 30 to 40 times greater than grain size. This requirement is described in greater detail in Section 8.7. All too often, this requirement is met for the small-grain-sized specimens but not for the large-grain-sized ones. This condition

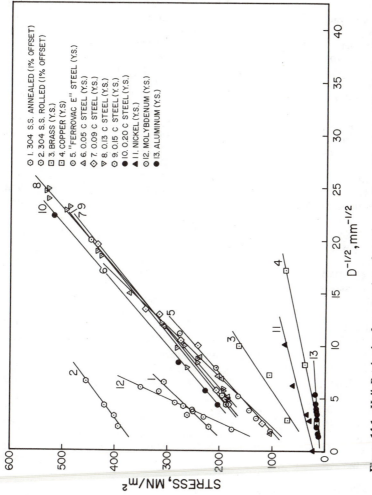

Figure 14.1 Hall–Petch plot for a number of metals and alloys. (Courtesy of K. C. Hsu, New Mexico Institute of Mining and Technology.)

1. 304 S.S. ANNEALED (1% OFFSET)
2. 304 S.S. ROLLED (1% OFFSET)
3. BRASS (Y.S.)
4. COPPER (Y.S)
5. "FERROVAC E" STEEL (Y.S.)
6. 0.05 C STEEL (Y.S.)
7. 0.09 C STEEL (Y.S.)
8. 0.13 C STEEL (Y.S.)
9. 0.15 C STEEL (Y.S.)
10. 0.20 C STEEL (Y.S.)
11. NICKEL (Y.S.)
12. MOLYBDENUM (Y.S.)
13. ALUMINUM (Y.S.)

$D^{-1/2}$, mm$^{-1/2}$

STRESS, MN/m^2

TABLE 14.1 Tabulation of σ_0 and k Values for BCC, FCC, and HCP Structures[a]

Material Specification[b]	σ_0 (MPa)	k MN/m$^{3/2}$
Body-Centered Cubic		
Mild steel, y.p.	70.60	0.74
Mild steel, $\epsilon = 0.10$	294.18	0.39
Swedish iron, y.p.	47.07	0.71
Swedish iron, no y.p.	36.28	0.20
Fe–3% Si, y.p. $-196°C$	505.99	1.54
Fe–3% Si, twinning $-196°C$	284.37	3.32
Fe–18% Ni, $\epsilon = 0.002$	650.14	0.22
Fe–18% Ni, twinning $-196°C$	843.32	1.30
FeCo, ordered, $\epsilon = 0.004$	50.01	0.90
FeCo, disordered, $\epsilon = 0.004$	319.68	0.33
Chromium, y.p.	178.47	0.90
Chromium, twinning $-196°C$	592.52	4.37
Molybdenum, y.p.	107.87	1.77
Molybdenum, $\epsilon = 0.10$	392.24	0.53
Tungsten, y.p.	640.33	0.79
Vanadium, y.p.	318.70	0.30
Niobium, y.p.	68.64	0.04
Tantalum, with O_2, y.p. $0°C$	186.31	0.64
Face-Centered Cubic		
Copper, $\epsilon = 0.005$	25.50	0.11
Cu–3.2% Sn, y.p.	111.79	0.19
Cu–30% Zn, y.p.	45.11	0.31
Aluminum, $\epsilon = 0.005$	15.69	0.07
Aluminum, fracture, 4K	539.33	1.67
Al–3.5% Mg, y.p.	49.03	0.26
Silver, $\epsilon = 0.005$	37.26	0.07
Silver, $\epsilon = 0.002$	23.53	0.17
Silver, $\epsilon = 0.20$	150.03	0.16
Hexagonal Close-Packed		
Cadmium, $\epsilon = 0.001$, $-196°C$	17.65	0.35
Zinc, $\epsilon = 0.005$, $0°C$	32.36	0.22
Zinc, $\epsilon = 0.175$, $0°C$	71.58	0.36
Magnesium, $\epsilon = 0.002$	6.86	0.28
Magnesium, $\epsilon = 0.002$, $-196°C$	14.71	0.47
Titanium, y.p.	78.45	0.40
Zirconium, $\epsilon = 0.002$	29.42	0.25
Beryllium, y.p.	21.57	0.41

[a] Adapted with permission from [3], p. 108.

[b] y.p., yield point.

is not met by the largest grain sizes of Fujita and Tabata[4] and Shiroor et al.[5] and invalidates their results. This is shown in the discussions published by Thompson[6] and Meyers.[7] Recent experiments by Miyazaki et al.[8] show that the yield stress and work hardening of sheet specimens are affected by the specimen thickness, below a certain ratio between grain size and specimen thickness (see Fig. 8.30).

2. *Purity of the material and texture.* Impurities often segregate at grain boundaries, changing their elastic constants and flow stress. This, in turn, should have a strong effect on σ_0 and k.

3. *Method of measuring grain size.* Annealing twins should be included in measurement of grain size, as was shown by Babiak and Rhines.[9] This fact is often overlooked and leads to erroneous interpretations. The measurement of grain size is usually conducted by the linear intercept technique. The mean linear intercept does not really provide the grain size, but is related to a fundamental size parameter, the grain-boundary area per unit volume, S_v.[10]

$$\bar{l} = \frac{1}{N_l} = \frac{2}{S_v} \qquad (14.2)$$

where $\bar{l}$ is the mean linear intercept and N_l the number of intersections per unit length. If we assume that the grains are spherical to a first approximation, we have the following relationship between the grain-boundary area and volume:

$$S_v = \frac{1}{2}\frac{4\pi r^2}{\frac{4}{3}\pi r^3}$$

$$S_v = \frac{3}{2r} = \frac{3}{D} \qquad (14.3)$$

where D is the average grain diameter. The factor 1/2 was introduced in the expression above because each surface is shared between two grains. From Eqs. 14.2 and 14.3,

$$D = \frac{3}{2}\bar{l} \qquad (14.4)$$

Hence, this is the most correct way to express the grain size from linear intercept measurements.

4. *Difference in grain-size range.* If two investigators study two widely different ranges of grain size and the σ_y versus $D^{-1/2}$ approximation is valid only over a narrow range, they are bound to find different values of σ_0 and k. This is studied in greater detail in Section 14.3.

14.2 THEORIES PREDICTING A σ_y VERSUS $D^{-1/2}$ RELATIONSHIP

The principal theories advanced to explain the Hall–Petch relationship are presented here. The first two theories have lost a lot of their credibility because dislocation pile-ups are not thought to be as important as they used to be especially in high-stacking-fault-energy materials.

14.2.1 Hall–Petch Theory

The basic idea behind the separate propositions of Hall[1] and Petch[2] is that a dislocation pile-up can "burst" through a grain boundary due to stress concentrations at the head of the pile-up. If τ_a is the resolved shear stress applied on the slip plane, then the stress acting in the head of the pile-up containing n dislocations is $n\,\tau_a$ (Eq. 8.1). The number of dislocations in a pile-up depends on the length of the pile-up, which, in turn, is proportional to the grain diameter D. According to Eshelby et al.,[11] (Eq. 17).

$$n = \frac{\alpha d \tau_a}{Gb/\pi}$$

(14.5)

where d is the length of the pile-up, that is, for a pile-up with two ends (generated, for instance, by a Frank–Read source in the center of the grain) equal to half the grain diameter D; and α is a geometrical constant equal to 1 for the pile-up of screw dislocations and equal to $(1 - \nu)$ for edge dislocations. If τ_c is the critical stress required to overcome the grain-boundary obstacles, then the dislocations of the pile-up will be able to traverse the grain boundary if

$$n\tau_a \geqslant \tau_c$$

(14.6)

From Eq. 14.5,

$$\frac{\alpha D \tau_a}{2Gb/\pi}\,\tau_a \geqslant \tau_c \qquad \text{or} \qquad \frac{\alpha \pi D \tau_a^2}{2Gb} \geqslant \tau_c$$

In order to take into account the friction stress τ_0 needed to move the dislocations in the absence of any obstacle, we have to add this term:

$$\tau_a \geqslant \tau_0 + kD^{-1/2}$$

(14.7)

Equation 14.7 is essentially identical to Eq. 14.1 once the shear stresses are converted into normal stresses. It should be noted that Eshelby's equation is valid only for a large number of dislocations. Hence, the equation is not applicable to grain sizes below a few micrometers.

14.2.2 Cottrell's[12] Theory

Cottrell[12] used a somewhat similar approach; however, he recognized that it is virtually impossible for dislocations to "burst" through boundaries. Instead, he assumed that the stress concentration produced by a pile-up in one grain activated dislocation sources in the adjacent grain. Figure 14.2 shows how a Frank–Read source at a distance r from the boundary is activated by the pile-up produced by a Frank–Read source in the adjacent grain. The slip band blocked in the boundary was treated by Cottrell as a shear crack. The maximum shear stress at a distance r ahead of a shear crack is given by

$$\tau = (\tau - \tau_0)\left(\frac{D}{4r}\right)^{1/2}$$

τ_0 is the frictional stress required to move dislocations and $r < D/2$. The stress required to activate the Frank–Read source in the neighboring grain, τ_c, would be given by

$$\tau_c = (\tau_a - \tau_0)\left(\frac{D}{4r}\right)^{1/2}$$

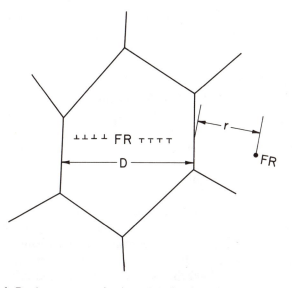

Figure 14.2 Frank–Read source operating in center of grain and producing two pile-ups at grain boundaries.

or

$$\tau_a = \tau_0 + 4\tau_c r^{1/2} D^{-1/2} \tag{14.8}$$

This equation is of a Hall–Petch form.

14.2.3 Li's[13] Theory

J. C. M. Li[13] used a different approach to obtain a relationship. Instead of using pile-ups, he considered the grain boundary to be a source of dislocations. The concept of grain-boundary dislocation sources is discussed in Section 6.8 and it is thought that the onset of yielding in polycrystals is associated with the activation of these sources. Li suggested that the grain-boundary ledges generated dislocations, "pumping" them into the grain. These dislocations act as Taylor (Section 9.2) forests in regions close to the boundary. The yield stress is, according to Li, the stress required to move dislocations through these forests. For many metals the flow stress is related, under most conditions, to the dislocation density by the relationship (Section 9.7)

$$\tau = \tau_0 + \alpha G b \sqrt{\rho} \tag{14.9}$$

where τ_0 is the friction stress, α is a numerical constant and ρ the dislocation density. At this point, use is made of the experimental observation that was invoked by Li: ρ was taken as proportional to the grain diameter D. Li rationalized this as follows. The ledges "pump" dislocations into the grains. The number of dislocations generated per unit deformation is proportional to the number of ledges, or to the grain-boundary surface per unit volume assuming the same ledge density per unit area for different grain sizes.

$$\rho \propto S_v \tag{14.10}$$

Equation 14.3 shows that the grain-boundary surface per unit volume is inversely proportional to D. Hence,

$$\rho \propto \frac{1}{D} \tag{14.11}$$

Substituting Eq. 14.11 into Eq. 14.9, we obtain

$$\tau = \tau_0 + \alpha' GbD^{-1/2}$$

Again, this is a Hall–Petch equation.

14.2.4 Conrad's[14,15] Theory

Also known as the work-hardening theory, Conrad's theory starts from the assumption that the small-grain-sized specimens have a higher dislocation density than the larger-grain-sized specimens at a given value of plastic strain. This is shown schematically in Fig. 14.3. The higher density generates a higher internal stress (internal and effective stresses are discussed in Section 8.5). The overall stress required to move a dislocation is the sum of two components: the thermal stress τ^* (short-range obstacles) and the athermal stress τ_G (long-range obstacles):

$$\tau = \tau^* + \tau_G$$

The athermal stress can be regarded as the sum of a component independent of dislocation density, τ_{G0}, and a component that shows the increase in dislocation density, $\tau_{G\rho}$. Hence,

$$\tau = \tau^* + \tau_{G0} + \tau_{G\rho}$$

Conrad assumes that

$$\tau_{G\rho} = \alpha Gb\rho^{1/2}$$

where ρ is the dislocation density. Thus,

$$\tau = \tau^* + \tau_{G0} + \alpha Gb\rho^{1/2} \tag{14.12}$$

If we consider the effect of grain size on the average free slip distance, it is possible to obtain a relationship between ρ and D. Orowan's equation states (Section 8.3) that

$$\gamma = K_1 \rho b \bar{l}$$

If we assume to a first approximation that the average distance traveled by a dislocation is a certain fraction of the grain diameter,

$$\bar{l} = K_2 D$$

we have

$$\rho = \frac{\gamma}{K_1 K_2 bD} \tag{14.13}$$

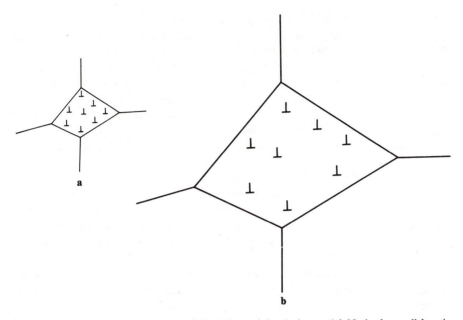

a

b

Figure 14.3 Schematic representation of Conrad's work-hardening model. Notice larger dislocation density for small grain size.

Substituting Eq. 14.13 into Eq. 14.12, we obtain

$$\tau = \tau^* + \tau_{G0} + \alpha Gb\left(\frac{\gamma}{K_1 K_2 bD}\right)^{1/2}$$

or

$$\tau = (\tau^* + \tau_{G0}) + K_3 D^{1/2}$$

where

$$K_3 = \frac{\alpha G\gamma^{1/2}}{K_1 K_2}$$

This is the familiar Hall–Petch equation. Conrad's model emphasizes the motion of dislocations through the grain rather than the behavior in the immediate vicinity of the grain boundary.

14.3 CRITICISM OF HALL–PETCH RELATION

In spite of the fact that the Hall–Petch equation is widely known and accepted, it is by no means free of controversy. For instance, Baldwin[16] showed that in many cases the scatter in the data was such that σ_y versus D^{-1} or σ_y versus $D^{-1/3}$ yielded as good a fit to the experimental results as σ_y versus $D^{-1/2}$. Another strong source of criticism for the Hall–Petch and Cottrell theories is that grain-boundary dislocation sources (and not pile-ups) are being increasingly accepted as the cause of the regular dislocation arrays (see Chapter 6). Conrad's theory fails to recognize that the onset of plastic flow is a highly localized process where only the grain boundaries and the regions closely adjoining them participate. Large deviations from the Hall–Petch equation at small grain sizes were obtained by Thompson[17,18] and Abrahamson.[19] One important flaw of the Hall–Petch equation is that it predicts an infinite yield stress at an infinitely small grain size. Hence, it leads to distorted predictions of the yield stress at very small grain sizes. Grain-size reduction is an important strengthening mechanism; Anderson et al.[20] made a systematic comparison of the slopes k of the Hall–Petch plots of steels published in the literature and were able to rationalize the differences obtained by assuming that the σ_y versus $D^{-1/2}$ plot was not a straight line but a curve; different investigators, working in relatively small grain-size ranges, would fit their curves into a straight line. The slope k tended to decrease for small grain sizes, indicating that the curve was concave toward the $D^{-1/2}$ axis. These results are consistent with the ones obtained by Abrahamson.[19]

It is felt by the authors that the Hall–Petch equation and the theoretical assumptions used in its derivation should be revised in view of the experimental observations. One such attempt is presented in Section 14.3.1.

14.3.1 More Recent Proposals

Experimental evidence by transmission electron microscopy indicating that grain boundaries are sources of dislocations[21,22] has steadily been amassing. Grain-boundary sources are discussed in Chapter 6. Figure 14.4 shows very clearly, for AISI 304 stainless steel, that the onset of plastic deformation is associated with dislocation activity in proximity of the grain boundary; two total strains are shown.

The initiation of plastic flow takes place at stresses much lower than the conventional macroyield stress. Thus, the name "microyield stress" is commonly used to designate the stress at which the first sights of dislocation activity are noticeable, while the macroyield stress is characterized by massive dislocation generation and motion and a substantial deviation from the elastic deformation regime. There is a remarkable group (in their self-consistency) of papers devoted to microyielding in the Fe–3% Si alloy and its study by etch-pitting techniques. Suits and Chalmers,[23] Worthington and Smith,[24] Brentnall and Rostoker,[25] and Douthwaite and Evans[26] incontrovertibly showed that the stresses are higher in the grain-boundary region than in the center of the grains. They made a scratch on the surface of a specimen, subsequently pulling it in tension. Dislocation emission from the grain boundaries started at a stress below that at which the dislocations generated by the scratch could move. This proves that the stress at the grain boundaries was higher than that at the scratch (representative of the bulk). Margolin and Stanescu[27] found similar results for β-titanium. They report the onset of plastic deformation at the grain boundaries, proceeding inward. Based on their observations they proposed an expression for polycrystalline strengthening at constant strain as a function of grain size.

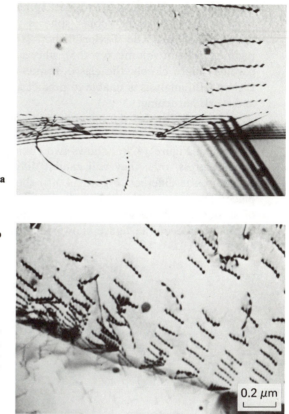

Figure 14.4 Dislocation activity at grain boundaries in AISI 304 stainless steel ($\dot{\epsilon} = 10^{-3}$ s^{-1}): (a) typical dislocation profiles after a strain of 0.15%; (b) same, after a strain of 1.5%. (Courtesy of L. E. Murr, New Mexico Institute of Mining and Technology.)

0.2 μm

The experimental evidence presented in [21–27], among other sources, and the concepts of inhomogeneous plastic flow developed by Ashby,[28] Thompson and coworkers,[17,18] and Margolin and coworkers[27,29] clearly show that the early theories on the effect of grain size on the yield stress are not realistic. Based on these concepts, Meyers and Ashworth[30] proposed a model that will be described briefly here. The fundamental assumption on which the model rests is that the *elastic incompatibility stresses* between adjacent grains cause localized plastic flow at the grain boundaries, at stresses much lower than the macroyield stress. For instance, the elastic moduli for nickel along the [100], [110], and [111] directions are 137, 233, and 303 GPa. These values can be obtained readily from Eq. 1.107. It can readily be seen that Young's modulus in [111] is twice as high as that in [100].

An important assumption used is that the *stress* acting on all the grains is the same. The assumption made in both Taylor's[31] and Bishop and Hill's[32] analyses of the plastic deformation of polycrystalline aggregates is that all individual grains undergo the same *strain* as the tensile specimen. This assumption is not applicable here, as will be shown by the following analogy. The grains can be imagined as rubber and steel spheres filling a box. If a ram compresses the top of the box, the rubber spheres will undergo more deformation than the steel spheres. The Taylor–Bishop–Hill analysis is realistic in the plastic regime (constant volume, $\nu = 0.5$), whereas the assumption of the present model parallels more closely the elastic response of metals ($\nu \simeq 0.3$). The Taylor–Bishop–Hill analysis is unable to predict a grain-size dependence of yield stress and work hardening.

Hirth[33] divided the incompatibility effects into two types. This same approach will be used here and a longitudinal and a shear incompatibility will be treated. Figure 14.5(a) shows two cubic grains: when subjected to the same applied stress, σ_{AP}, they will exhibit different strains if their crystallographic orientation is different. Figure 14.5(b) shows how these distortions might take place. The interfacial bonding, however, establishes the condition of equality of strains across the grain boundary. Figure 14.5(c) shows this situation. The left-hand side of the figure shows the longitudinal incompatibility, in which

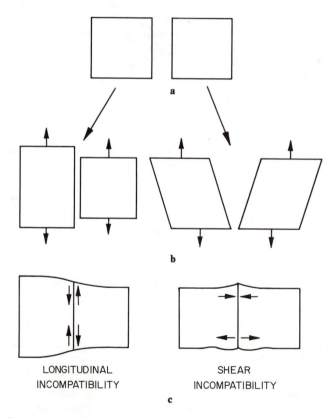

Figure 14.5 Development of incompatibility stresses along interface between two grains: (a) initially cubic grains having identical dimensions; (b) deformation of two grains under the same applied stress; (c) incompatibility stresses: on the left, longitudinal incompatibility; on the right, shear incompatibility.

the crystallographic orientation is such that the faces of the cube remain perpendicular to each other. Crystallographic orientations such as [100] and [111] are characterized by this type of response. The incompatibility stresses were calculated by Meyers and Ashworth[30] and found to be

$$(\tau_I)_{long} = 1.37\sigma_{AP}$$

$$\tau_H = \sigma_{AP}/2$$

Hence, the interfacial shear stress due to the longitudinal incompatibility is almost three times higher than the resolved shear stress homogeneously applied on the grain ($\tau_H = \sigma_{AP}/2$). This means that dislocation activity at the grain boundary starts before dislocation activity at the center of the grains.

When the stress reaches the critical level required for emission, localized plastic deformation will start [Fig. 14.6(a)]. These dislocations do not propagate throughout the grain for two reasons:

1. The stress decreases rapidly with distance from the grain boundary.
2. The center of the grains is under the homogeneous shear stress control, which is maximum at 45° with the tensile axis. On the other hand, the interfacial and homogeneous shear stresses have different orientations. Figure 14.6 shows how the dislocations emitted from the grain boundaries

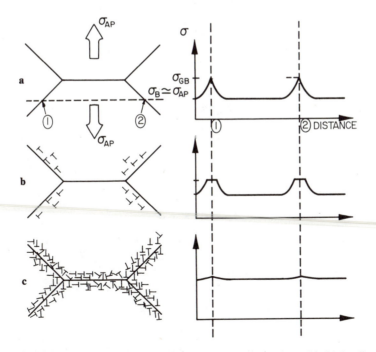

Figure 14.6 Sequence of stages in polycrystalline deformation, starting with (b) localized plastic flow in the grain-boundary regions (microyielding) forming a work-hardened grain-boundary layer (c) that effectively reinforces the microstructure.

will undergo cross-slip. Extensive cross-slip and the generation of dislocation locks will result in a localized layer with high dislocation density.

The plastic flow of the grain-boundary region attenuates the stress concentration; geometrically necessary dislocations accommodate these stresses [Fig.
14.6(b) and (c)]. This marks the onset of microyielding. The dislocations do
not propagate throughout the whole grain because of cross-slip induced by
the difference of orientation between the maximum shear stress (due to the
applied load) and the stress concentration due to elastic incompatibility. The
work-hardened grain-boundary layer has a flow stress σ_{fGB}, while the bulk
has a flow stress σ_{fB} ($\sigma_{fGB} > \sigma_{fB}$). The material behaves, at increasing applied
loads, as a composite made out of a continuous network of grain-boundary
film with flow stress σ_{fGB} and of discontinuous "islands" of bulk material with
flow stress σ_{fB}. The increasing applied stress σ_{AP} does not produce plastic flow
in the bulk in spite of the fact that $\sigma_{AP} > \sigma_{fB}$, because the continuous grain-
boundary network provides the rigidity to the structure. The total strain in
the continuous grain-boundary network does not exceed 0.005, since it is elastic;
hence, plastic deformation in the bulk is inhibited. This situation can be termed
"plastic incompatibility."

When the applied load is such that the stress in the grain-boundary region
becomes equal to σ_{fGB}, plastic deformation reestabishes itself in this region.
The plastic deformation of the continuous matrix results in increases in stress
in the bulk with plastic flow [Fig. 14.6(c)]. This marks the onset of *macroyielding*.
After a certain amount of plastic flow, dislocation densities in the bulk and
grain-boundary regions become the same; since both regions have the same
flow stress, plastic incompatibility disappears, and we have $\sigma_{AP} = \sigma_{GB} = \sigma_B$.

14.4 OTHER INTERNAL OBSTACLES

There are other internal obstacles to dislocation motion that may have an effect
analogous to grain boundaries. Examples are cell walls and deformation twins.
These barriers were studied by several investigators and their effect on flow
stress may be represented by the general equation

$$\sigma_f = \sigma_0 + K\Delta^{-m} \tag{14.14}$$

The coefficient m has been found to vary between $\frac{1}{2}$ and 1. If we want to
include both the effect of grain size and substructure refinement due to the
internal barriers, we can use the following overall equation (e.g., Moin and
Murr[34]) which describes reasonably well the response of the material:

$$\sigma_f = \sigma_0 + K_1 D^{-1/2} + K_2 \Delta^{-m} \tag{14.15}$$

Figure 14.7 shows an example of substructural refinement in nickel. These twins
were induced by shock loading (see Section 15.4) at 45 GPa and 2 μs. It is
easy to understand why these obstacles strengthen the metal. Dislocation movement occurring in subsequent deformation by, say, tensile testing is severely

Figure 14.7 Deformation twins in shock-loaded nickel (45 GPa peak pressure; 2 μs pulse duration). Plane of foil (100); twinning planes (111) making 90°. (Courtesy of L. E. Murr, New Mexico Institute of Mining and Technology.)

hampered by all these planar obstacles. Internal cells are also very effective barriers. Langford and Cohen,[35,36] and later Rack and Cohen,[37,38] systematically investigated the effects of the cell size on the flow stress of highly cold-worked low-carbon steel wire. The straining to high levels was accomplished by wire drawing (see Section 2.5.6) and the material was recovered and showed thin cell walls and virtually dislocation-free cell interiors. Figure 14.8 shows some representative results. The group of data on the left side shows the alloys recovered at 300 and 400°C. The slope in the log-log plot is −1 and we have, consequently,

$$\log(\sigma_f - \sigma_0) - \log(\sigma_1 - \sigma_0) = -1(\log \bar{d} - \log \bar{d}_1) \qquad (14.16)$$

where the equation above expresses the straight line passing through ($\sigma_f - \sigma_0$, $\bar{d}$) and ($\sigma_1 - \sigma_0$, $\bar{d}_1$). Notice that the ordinate in Fig. 14.8 is $\sigma - \sigma_0$. Manipulation of Eq. 14.16 will yield

$$\log \frac{(\sigma_f - \sigma_0)}{(\sigma_1 - \sigma_0)} = \log \left(\frac{\bar{d}}{\bar{d}_1} \right)^{-1}$$

Hence,

$$\sigma_f - \sigma_0 = \frac{\sigma_1 - \sigma_0}{\bar{d}_1^{-1}} \cdot \bar{d}^{-1} = K\bar{d}^{-1}$$

$$\sigma_f = \sigma_0 + K\bar{d}^{-1}$$

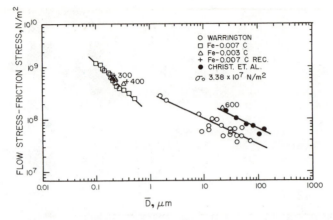

Figure 14.8 Strength of wire-drawn and recovered Fe–0.003% C as a function of transverse linear-intercept cell size. Recovery temperatures as indicated. (Adapted with permission from [38], p. 371.)

On the other hand, if the anneals were done at 600°C and above, recrystallization took place and the group of points on the right side of the plot were found. The slope was decreased to $-\frac{1}{2}$, leading to a regular Hall–Petch relationship.*

‡14.5 EFFECT OF GRAIN SIZE ON OTHER PROPERTIES

A number of other mechanical properties, in addition to the yield stress, are affected by grain size; impact, fatigue, creep, work hardening, stress corrosion, twinning, and work-hardening response are dependent on it. Some of these effects will be discussed here.

* In low-carbon steels the yield stress is strongly dependent on grain size; a steel with a grain size of 0.5 mm and σ_y of 104 MN/m² has its yield stress increased to ~ 402 MN/m² when the grain size is reduced to 0.005 mm. As the carbon content is increased and the steel tends more and more toward eutectoid, other effects, such as the ferrite/pearlite ratio, the spacing of cementile layers in the pearlite, and the size of the pearlite colonies, become important parameters. T. Gladman, I. D. McIvor, and R. F. Pickering [*J.I.S.I.*, *210* (1972) 916] developed an expression for pearlite-ferrite mixtures.

$$\sigma_y(\text{ksi}) = f_\alpha^{1/3} [2.3 + 3.81(\% \text{ Mn}) + 1.13 D^{-1/2}] \\ + (1 - f_\alpha^{1/3})[11.6 + 0.25 \, S_0^{-1/2}] + 4.1(\% \text{ Si}) + 27.6(\sqrt{\% \text{ N}})$$

where f_α is the ferrite fraction, D is the ferrite grain size (in mm), S is the interlamellar spacing in pearlite (in mm), and % Mn, Si, and N are the weight percentages of manganese, silicon, and nitrogen, respectively.

J. M. Hyzak and I. M. Bernstein [*Met. Trans.*, *7A* (1976) 1217] proposed the following equation for fully pearlitic steels.

$$\sigma_y(\text{MPa}) = 2.18 S^{-1/2} - 0.40 P^{-1/2} - 2.88 \, D^{-1/2} + 52.30$$

S is the pearlite interlamellar spacing, P is the pearlite colony size, and D is the austenite grain size (units of S, P, and D are not given in reference, but should be cm).

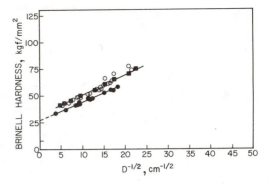

Figure 14.9 Effect of grain size on Brinell hardness number for brass (○ 68–32 brass; ● 70–30 brass; ■ 69–31 brass). (Adapted with permission from [3], p. 102.)

‡14.5.1 Effect on Hardness

The relationship between hardness and grain size has been shown to often be of the Hall–Petch type. This is illustrated for three different brasses in Fig. 14.9. This relationship is easy to rationalize. The hardness represents a very rough measure of the resistance of a body to plastic deformation. Assuming that the work-hardening rates of small- and large-grain-size specimens are the same, their hardnesses will have the same relationship as their yield stresses. So to a first approximation the hardness should obey the same relationship as the yield stresses. Care has to be exercised, however, to ensure that the diameter of the impression is much greater than the grain size. Otherwise, the results loose significance. Hence, microhardness cannot be used for comparison purposes; often, the hardness of the bulk metal is different from that at the vicinity of the grain boundary (see Chapter 17).

‡14.5.2 Effect on the Twinning Stress

Twinning and slip (see Chapter 7) are often competing deformation mechanisms. As the temperature is decreased, so is the thermal activation energy available for dislocation motion. This is especially true for BCC metals, which exhibit a very high rise in the friction stress as the temperature is decreased. Apparently, this is connected to the reduced mobility of the screw (and not the edge) dislocations.[39] The stress required for twinning does not show a great temperature dependence, as shown by the rationale of Thomas in Section 7.4.1. Hence, twinning becomes an increasingly attractive deformation mechanism at low temperatures. When the stress required for twinning becomes less than the stress required for dislocation motion, plastic deformation takes place by the former mechanism. In addition to the temperature dependence, several investigators determined a grain-size dependence for twinning. This is shown in Fig. 14.10 for iron–3% silicon. For grain sizes larger than 30 μm, Fe-Si yields by twinning (at 77 K), while for smaller grain sizes the yielding is initiated by dislocation motion. It can be seen that the stress required for twinning obeys a Hall–Petch relationship, but that the slope is higher than that of the region where yielding by dislocation motion takes place. Twinning and fracture are

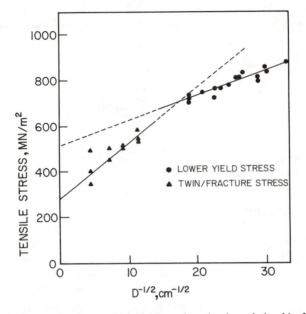

Figure 14.10 Yield stress (or fracture/twinning stress)–grain size relationship for Fe–3% Si at 77 K. (Adapted with permission from [41], p. 194.)

closely associated. Apparently, twinning nucleates cracks, which results in fracture. The samples with grain sizes smaller than 30 μm show a considerable amount of plastic deformation prior to fracture. For tests conducted at higher temperatures (195 and 293 K) Hull[40] found the onset of plastic flow to be connected with dislocation motion for all grain sizes.

Similar results were obtained by Marcinkowski and Lipsitt[41] for chromium (another BCC metal). At one temperature, the specimens with large grain sizes had a greater preference for twinning than the specimens with small grain size. Figure 14.11 presents some of their results. At 77 K all grain sizes deformed by twinning. As the deformation temperature increased, the slip-twinning boundary shifted toward the larger grain sizes; above 148 K only slip was observed. Here, the slope of the twin Hall–Petch curve is about the same as the slip Hall–Petch. Marcinkowski and Lipsitt[42] were able to inequivocally associate crack nucleation with twinning at low temperatures. It should be noticed that the results reported in Fig. 14.11 refer to compression tests.

‡14.5.3 Effect on Ductile-to-Brittle Transition Temperature

While FCC metals retain a reasonable degree of ductility up to very low temperatures, BCC metals suffer a marked decrease in ductility and impact resistance below a specific temperature, the ductile-to-brittle transition temperature (DBTT). The DBTT is determined by competition between shear and cleavage crack initiation processes, the important factors being the stress level and the time delay required. Owen and Hull[43] analyze in detail the DBTT behavior

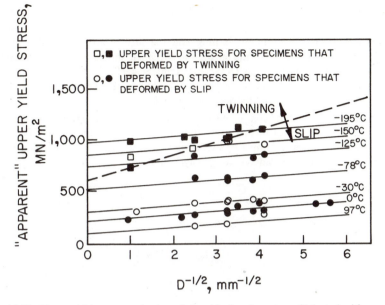

Figure 14.11 Upper yield stress–grain size relationship for chromium. (Adapted with permission from [42], p. 100.)

in group VA (Va, Nb, Ta) and VIa (Cr, Mo, W) BCC metals and find the grain size and level of impurities at the grain boundaries to have an important effect on DBTT. All other factors remaining constant, the smaller the grain size, the lower the DBTT. Armstrong reports that the DBTT obeys a Hall–Petch relationship for steel. The conclusions reached by Owen and Hull[42] are that there are two very important factors that produce brittleness in BCC refractory transition metals. The first is the segregation of impurities at the grain boundaries, reducing their effective surface energy and creating an easy path for the onset of fracture propagation. This embrittlement can be eliminated by either removing the impurities or preventing them from segregating at the boundaries. The second factor is *twinning*. As seen in the preceding section, it is grain-size dependent. It is this grain-size dependence of twinning that has as a corollary the grain-size dependence of DBTT. Twinning does not cause cleavage per se, but induces microcracks (at twin-grain boundary and twin-twin intersections, and twin-matrix interface), which, in their turn, will propitiate the condition for cleavage fracture propagation.

‡14.5.4 Effect on Fatigue

The effect of grain size on fatigue is not the same in all metals. On the one hand, the endurance limit is increased as a consequence of the increased strength due to grain-size reduction. On the other hand, the nature of slip in a specified metal also has a strong bearing on the fatigue life. For greater details, the reader is referred to Pelloux.[44]

EXERCISES

14.1. (a) Determine the mean linear intercept, the surface area per unit volume, and the estimated grain diameter for the specimen shown in the figure below.

(b) Estimate its yield stress (AISI 304 stainless steel).

(c) Estimate the parameters of part (a), excluding the annealing twins. By what percentage is the yield stress going to differ?

14.2. Professor M. I. Dum conducted a study on the effect of grain size on the yield stress of a number of metals using thin foil specimens (thickness of 0.1 mm and width of 6.25 mm). He investigated grain sizes 5 of 5, 25, 45, and 100 μm. Which specimens can be considered truly polycrystalline?

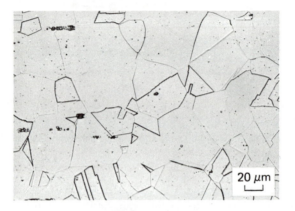

Figure E14.1

14.3. (a) Calculate the Young moduli for aluminum along [100], [110], and [111]. According to the more recent viewpoints, would pure aluminum exhibit a large or small grain-size dependence on yield stress? Why?

(b) Comment on the effect of segregation at the grain boundaries on the Hall–Petch slope of aluminum.

14.4. (a) A. W. Thompson [*Acta Met.*, *25* (1977) 83] obtained the following results for the yield stress of nickel:

Grain Sizes (μm)	Yield Stress (MN/m²)
0.96	251
2	185
10	86
20	95
95	33
130	25

(b) Find the parameters in the Hall–Petch equation. Plot the yield stress versus D^{-1}, $D^{-1/2}$, and $D^{-1/3}$. Which plot shows the best linearity?

(c) Show how you can determine the correct exponent by another plot (not by trial and error).

14.5. If the grain size of a metal is doubled by an appropriate anneal, by what percentage is its surface area per unit volume changed?

14.6. J. L. Nilles and W. S. Owen [*Met. Trans.*, *3* (1972) 1877] found a strong grain-size dependence of the stress required for twinning when deforming an Fe–25% Ni alloy at 4 K. From what you learned in the text, is this behavior expected? Compare the ratio of the Hall–Petch slopes of the twinning and yield stresses for Fe–25% Ni with the ratio found for chromium and Fe-Si. From this, try to develop a model for the initiation of twinning.

REFERENCES

[1] E.O. Hall, *Proc. Roy. Soc.* (*London*), *B64* (1951) 474.

[2] N.J. Petch, *J. Iron Steel Inst.*, *174* (1953) 25.

[3] R.W. Armstrong, in *Advances in Materials Research*, Vol. 5, R.F. Bunshah (ed.), Wiley-Interscience, New York, 1971, p. 101.

[4] H. Fujita and T. Tabata, *Acta Met.*, *21* (1973) 355.

[5] V.S. Shiroor, A.G. Kulkarni, P.P. Rao, and V.R. Parameswaran, *Scripta Met.* *9* (1975) 671.

[6] A.W. Thompson, *Scripta Met.*, *9* (1975) 1069.

[7] M.A. Meyers, *Scripta Met.*, *10* (1976) 159.

[8] S. Miyazaki, K. Shibata, and H. Fujita, *Acta Met.*, *27* (1979) 855.

[9] W.J. Babiak and F.N. Rhines, *Trans. TMS-AIME*, *218* (1960) 21.

[10] R.T. De Hoff and F.N. Rhines (eds.), *Quantitative Microscopy*, McGraw-Hill, New York, 1968.

[11] J.D. Eshelby, F.C. Frank, and F.R.N. Nabarro, *Phil. Mag.*, *42* (1951) 351.

[12] A.H. Cottrell, *Trans. TMS-AIME*, *212* (1958) 192.

[13] J.C.M. Li, *Trans. TMS-AIME*, *227* (1963) 239.

[14] H. Conrad, *Acta Met.*, *11* (1963) 75.

[15] H. Conrad, in *Ultra-fine Grain Metals*, J.J. Burke and V. Weiss (eds.), Syracuse University Press, Syracuse, N.Y. 1970, p. 213.

[16] W.M. Baldwin, Jr., *Acta Met.*, *6* (1958) 14.

[17] A.W. Thompson, *Acta Met.*, *23* (1975) 1337.

[18] A.W. Thompson, *Acta Met.*, *25* (1977) 83.

[19] E.P. Abrahamson II, in *Surfaces and Interfaces II*, Syracuse University Press, Syracuse, N.Y., 1968, p. 262.

[20] E. Anderson, D.W.W. King, and J. Spreadborough, *Trans. TMS-AIME*, *242* (1968) 115.

[21] K. Tangri and T. Malis, *Surface Sci.*, *31* (1972) 101.

[22] L.E. Murr, *Met. Trans.*, *6A* (1975) 427.

[23] J.C. Suits and B. Chalmers, *Acta Met.*, *9* (1961) 854.

[24] P.J. Worthington and E. Smith, *Acta Met.*, *12* (1964) 1277.

[25] W.D. Brentnall and W. Rostoker, *Acta Met.*, *13* (1965) 187.

[26] R.M. Douthwaite and J.T. Evans, *Acta Met.*, *21* (1973) 525.

[27] H. Margolin and M.S. Stanescu, *Acta Met.*, *23* (1975) 1141.

[28] M.F. Ashby, *Phil. Mag.*, *21* (1970) 399.

[29] J. Jinoch, S. Ankem, and H. Margolin, *Mater. Sci. Eng.*, *34* (1978) 203.

[30] M.A. Meyers and E. Ashworth, *Phil. Mag.*, *46* (1982) 737.

[31] G.I. Taylor, *J. Inst. Metals*, *62* (1938) 307.

[32] J.F. Bishop and R. Hill, *Phil. Mag.*, *414* (1951) 1298.

[33] J.P. Hirth, *Met. Trans.*, *3* (1972) 3047.

[34] E. Moin and L.E. Murr, *Mater. Sci. Eng.*, *37* (1979) 249.

[35] G. Langford and M. Cohen, *Trans. ASM*, *62* (1969) 623.

[36] G. Langford and M. Cohen, *Met. Trans.*, *6A* (1975) 901.

[37] H.J. Rack and M. Cohen, *Met. Trans.*, *1* (1970) 1050.

[38] H.J. Rack and M. Cohen, in *Frontiers in Materials Science: Distinguished Lectures*, L.E. Murr (ed.), Marcel Dekker, New York, 1976, p. 365.

[39] J.E. Talia, L. Fernandez, V.K. Sethi, and R. Gibala, in *Strength of Metals and Alloys*, P. Haasen, V. Gerold, and G. Kostorz (eds.), Pergamon Press, Elmsford, N.Y., 1980, p. 127.

[40] D. Hull, *Acta Met.*, *9* (1961) 191.

[41] M.J. Marcinkowski and H.A. Lipsitt, *Acta Met.*, *10* (1962) 95.

[42] W.S. Owen and D. Hull, in *Refractory Metals and Alloys: II*, M. Semchyshen and I. Perlmutter (eds.), Interscience, New York, 1963, p. 1.

[43] R. Pelloux, in *Ultra-fine Grain Metals*, Sagamore Army Materials Research Conf. Proc., Vol. 16, J.J. Burke and V. Weiss (eds.), Syracuse University Press, Syracuse, N.Y., 1970, p. 231.

SUGGESTED READING

ARMSTRONG, R.W., in *Advances in Materials Research*, Vol. 4, R.F. Bunshah (ed.), Wiley-Interscience, 1971, p. 101.

ARMSTRONG, R.W., in *Ultra-fine Grain Metals*, Sagamore Army Materials Research Conf. Proc., Vol. 16, J.J. Burke and V. Weiss (eds.), Syracuse University Press, Syracuse, N.Y., 1970, p. 1.

EMBURY, J.D., in *Strengthening Methods in Crystals*, A. Kelly and R.B. Nicholson (eds.), Elsevier, Amsterdam, 1971, p. 331.

HIRTH, J.P., *Met. Trans.*, *3* (1972) 3047.

THOMPSON, A.W., in *Work Hardening in Tension and Fatigue*, A.W. Thompson (ed.), TMS-AIME, New York, 1978, p. 67.

Chapter 15

OTHER MECHANISMS

‡15.1 THERMOMECHANICAL PROCESSING (OR TREATMENT)

‡15.1.1 Introduction

Thermomechanical processing (TMP) or treatment (TMT) can be defined as the treatment by means of which plastic deformation is introduced in the thermal treatment cycle of the material in a way that alters the processes occurring during thermal treatment and improves the properties.[1] In general, the objective is to obtain a higher mechanical strength; however, frequently, one obtains an improvement in fatigue, creep, corrosion resistance, and fracture toughness. Some authors[2] treat thermomechanical processing or treatment (TMT) separately from mechanical-thermal processing or treatment (MTT). MTT refers to materials that do not undergo phase transformations. Thus, the mechanical thermal treatment (MTT) involves the introduction of a high dislocation density by means of mechanical processing; the distribution of these dislocations is then modified by some kind of recovery treatment which results in a pinning of dislocations by solute atoms or a rearrangement of dislocations in a more favorable configuration from the mechanical point of view. In the thermomechanical treatments (TMT) one uses the synergistic effect of the mechanical treatments (interstitial atoms, line defects, stacking faults, stress- or strain-induced martensite, etc.) and thermal treatments (precipitates, martensite) to obtain better properties. Thus, TMT is not a strengthening mechanism per se, but is a combination of them.

Due to the complexity of alloys and the thermal and mechanical effects to which they can be subjected, and due to the possible multiple combinations of these, TMTs constitute a formidable challenge to the modern metallurgist and require a lot of ingenuity.

The gas turbine is an excellent example of a material application that represents this challenge to the metallurgist. The higher the turbine temperature, the higher will be its efficiency. This explains the constant search for materials having high mechanical strength at elevated temperatures which would permit turbine operation at higher temperatures. While it is difficult to conceive of strengthening techniques or mechanisms that are fundamentally different from the ones employed at present, the simple utilization of combined strengthening techniques together with innovations in processing would permit significant improvements in the mechanical response and thermal stability of these alloys. Thermomechanical treatment is one such option. In recent years, this has been an active research area and is slowly passing from the laboratory stage to real applications.[3] Table 15.1 presents the alloys that were investigated in the decade of the 1960s. In the following sections, we describe briefly the TMTs for ferrous alloys (Section 15.1.2), superalloys (15.1.3), and other alloys (Section 15.1.4).

‡15.1.2 Ferrous Alloys

The great technological interest generated by TMT of ferrous alloys has led to an intense research activity in this area, with the attendant highly confusing terminology and treatment combinations. With a view to clarify some of this confusion, Radcliffe and Kula[5] classified the various steels that undergo phase transformations according to whether deformation is introduced before, during, or after the phase transformation.

Class I: deformation before the austenite transforms; martensite forms in work-hardened austenite.

Class II: deformation during the transformation of austenite; martensite forms during deformation of metastable steels.

Class III: deformation after the transformation of austenite; strain aging of the austenite transformation products.

Figure 15.1 presents a temperature–time–transformation (TTT) curve and the schematic position of mechanical processing with respect to the nose of the curve.

The classification scheme presented in Table 15.2 is not universally accepted. In particular, the Soviets do not seem to follow it. They have applied a great deal of effort to MTT, which has been described by Dunleavy and Spretnak.[6] Basically, one differentiates between the low-temperature treatment (TMT LT) and the high-temperature treatment (TMT HT) in spite of the fact that both belong to class I. There is also the combined treatment (TMCT), which consists of high-temperature treatment followed by low-temperature treatment. Table 15.2 presents the various TMT possibilities for steels according to Araki.[7] It incorporates the classification described above. A variety of possible

TABLE 15.1 Materials Subjected to TMT in the 1960s[a]

Material	Description of TMT	Objective	Observations
1. Carbon steels and low alloy steel for construction purposes	Strain aging by warm finishing	Greater mechanical strength	
2. Low- and medium-alloy high-strength steels	Hot-cold working and/or ausforming	Greater mechanical strength	
3. Maraging steels	Warm working with controlled reductions	Increase in ductility, grain-size reduction	TMT controls grain size as well as precipitation reactions
4. Tool steels	Hot work and cold working and/or ausforming	Increase in toughness	
5. High-speed steels	Hot work and cold work above the TTT curve nose	Increase in abrasion resistance and dimensional stability	
6. Nickel-base superalloys	Controlled reductions at specific temperatures to change metallurgical substructure and precipitation hardening	Increase in mechanical strength and low-cycle fatigue resistance	
7. Titanium alloys	Controlled deformation to control microstructure (and phases)	Optimization of mechanical properties	Properties of all alloys depend on warm working conditions
8. Aluminum alloys	Cold or warm work plus aging to decrease internal stresses	Increase in stress corrosion resistance	

[a] Adapted with permission from [4].

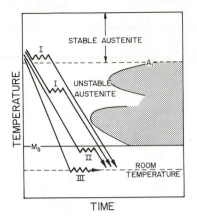

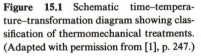

Figure 15.1 Schematic time–temperature–transformation diagram showing classification of thermomechanical treatments. (Adapted with permission from [1], p. 247.)

combinations are shown. The different TMTs are described in detail by Radcliffe and Kula.[5]

By way of illustration, we describe below some of the TMTs. Maraging treatment allows one to attain a yield strength of up to 2100 MPa. A typical composition is 18% Ni, 5% Mo, 0.5% Ti, 8% Co, and the rest Fe. The martensite produced by cooling the steel is relatively ductile, as it does not contain carbon and thus can be mechanically processed rather easily. This process is followed by an aging treatment which hardens the alloy by precipitation of compounds

TABLE 15.2   Thermomechanical Treatments for Steel[a]

Classification		Nomenclature
I. Deformation before transformation, $T > M_d$	A—of stable austenite 1—to martensite (and/or bainite) 2—to ferrite + pearlite (and/or bainite) B—of metastable austenite 1—to martensite (and/or bainite)	HTTMT "Ausforging" Controlled rolling of HSLA steels LTTMT
II. Deformation during transformation (one of the relevant phenomena is "TRIP")	C—of less stable austenite 1—to martensite 2—to ferrite + pearlite	Strain-induced transformation "Zerolling" (cold working of austenite) "Isoforming" (dynamic recovery)
III. Deformation after transformation and aging	D—of martensite without carbon E—of transformed structure 1—of martensite, transformed martensite 2—of ferrite + pearlite, bainite, etc. F—of martensite	Marforming, maraging Warm working E.T.D., etc. Warm working at higher temperatures (dynamic recovery)

[a] Adapted with permission from [7], p. 2.

of Ni, Mo, and Ti. Zackay et al.[8,9] developed in 1967 a new class of high-strength metastable austenitic steels making use of strain-induced martensitic transformation. These steels are known as TRIP steels (from "transformation-induced plasticity"). Their composition is such that they are austenitic at room temperature. The processing is started by deforming the austenite, taking the yield stress to 1400 MPa or more. This requires a 70 to 80% reduction in thickness during rolling. The rolling temperature must, of course, be above M_d so that the steel continues to be austenitic during deformation. Deformation below M_d will induce the martensitic transformation. In uniaxial tensile tests, the martensite formation during deformation provides a large amount of plasticity to the material. This results in rather large elongations (of the order of 50%). This plasticity, as expected, has a beneficial effect on the fracture toughness of the material. One of the reasons why martensite formation provides a large ductility during straining is an increase in volume (about 3%) associated with its formation. This volumetric expansion tends to reduce the three-dimensional tensile stresses that develop at cracks or in the beginning of neck during the plastic deformation. Figure 15.2 compares the performance of an AISI 4340 steel, quenched and tempered, an 18% Ni maraging steel, and a TRIP steel. Note that for a given yield stress, the fracture toughness of TRIP steel is much higher than that of 4340 steel.

Another TMT is the one developed for the high-strength low-alloy (HSLA) or microalloyed steels. The sheets are deformed (rolled) in the austenitic or in

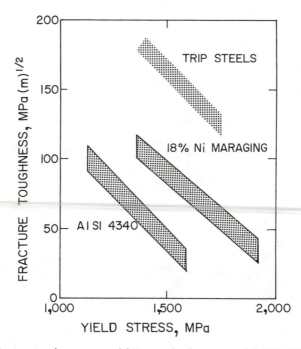

Figure 15.2 Fracture toughness versus yield stress plot for commercial AISI 4340 steel, maraging steel, and TRIP steel. (Adapted with permission from [9], p. 210.)

the $(\alpha + \gamma)$ region, and by means of a proper control of deformation and cooling rates the ferritic grain size can be controlled. The cooling is not fast enough to form martensite; the austenite decomposes into ferrite and pearlite (see also Chapter 11). From Table 15.2 one can see that this TMT falls in class I. Thin sheets of such steels can be coiled, cut, conformed, and welded without any problem. Main applications are for tubes and reservoirs. A related recent development are the dual-phase steels. Because of their great importance and promise, we discuss them separately in Section 15.2.

TMT is currently used in the armament industry, and some of the possible military applications are discussed below.

1. *Armor plates*. Modern armor plates tend to be composed of an outer high-hardness sheet and an inner sheet of low hardness and higher ductility. The objective of this composite armor plate is to better resist ballistic impact, as shown in Fig. 15.3. The outer harder sheet serves to fragment the projectile. The inner sheet avoids spalling due to its larger ductility. Another advantageous use of composite armor plates occurs where projectiles loaded with plastic explosives are encountered. These operate by internal spalling, which is avoided by means of a ductile posterior sheet. The TMT can be incorporated in the production of these composite sheets, as shown in Fig. 15.3. The two sheets are roll-bonded together to obtain a metallurgical bond. The roll bonding is done at about 1420 K, followed by ausforming at 1030 K and finishing at 860 K. The reduction suffered by the material is about 60%. The material is quenched from 590 K, followed by tempering.

2. *Projectiles with a conic tail*. A new type of armament has been developed.[1] This projectile has a conical flange, made from a TRIP steel, at the tail. In passing through the tube, the projectile is conformed. Figure 15.4

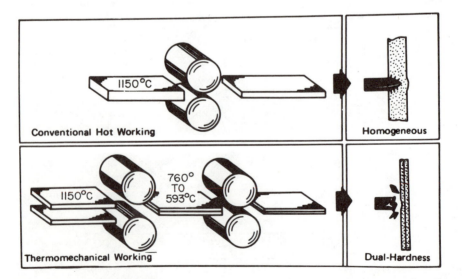

Figure 15.3 Schematic diagram showing fabrication of conventional and composite armor plate by TMT. (Adapted with permission from [1], p. 262.)

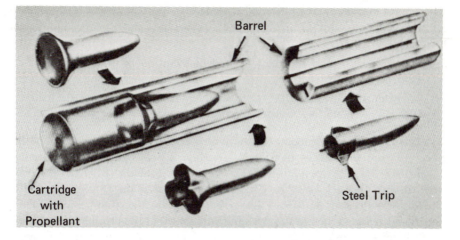

Barrel

Cartridge
with
Propellant

Steel Trip

Figure 15.4 TRIP steel projectile as its fin is being formed during its movement in barrel. (Adapted with permission from [1], p. 273.)

shows this projectile schematically. For a given exit velocity at the mouth of the tube, this type of projectile requires one-fourth of the gaseous pressure required in conventional weapons. This results in large savings in the propellant and a considerable reduction in the thickness of the armament tube.

‡15.1.3 TMT of Superalloys

Alloys meant to resist appreciable stresses at high temperatures (mainly Ni and Co based) can also benefit from TMT. These alloys are mainly used in industries such as aerospace (turbines), chemical (reactors), and other high-temperature applications. Ni-based superalloys are not subject to martensitic transformations and as such, the quench and temper treatment is not possible. Ni-based superalloys have an FCC structure and do not undergo allotropic transformations until the melting point. They possess high toughness and ductility over a large temperature range. For this reason, they can be cold, warm, or hot worked. TMT can be effective up to temperatures of the order of 1000 K; above this temperature the substructure produced by deformation is not stable. There exist a whole class of superalloys (e.g., Inconel 718, Udimet 700) designed for service in the temperature range 700 to 1000 K. For these temperatures, the mechanical properties (strength, toughness, low and high cycle fatigue) are the factors that determine the extent of superalloy life. It is precisely for these alloys that TMT can contribute. Disks and axles in gas turbines are examples of components where use is made of TMT.

The work of Kear et al.[10] and Oblak and Owczarski[11] is very important, as it established the basics of a theory for TMT of superalloys. In contrast with most other works, which treat the subject rather superficially and are based only on results, Kear et al. established a methodology for more conclusive TMT studies. The various phases of such a study can be divided into the following steps:

1. Choice and characterization of a predeformation structure
2. Determination of the deformation temperature and the optimum aging cycles
3. Characterization of morphology and distribution of precipitates and the deformation substructure
4. An exhaustive evaluation of the mechanical properties over a large temperature range, including tensile, creep, and low and high cycle fatigue tests

Examples of experimental programs where use has been made of this philosophy are the TMTs developed by Meyers and Orava[12,13] for Inconel 718 and by Orava[14] for Udimet 700. We describe below, briefly, the results obtained with Inconel 718. The TMTs developed used the results obtained by Conserva et al.[15] with Al 7075. These researchers observed that the introduction of an aging treatment before the straining would benefit the mechanical behavior. The following sequence of treatments was chosen: solubilization + preaging + straining + postaging. The temperatures of pre- and postaging treatments were varied and an optimization of these parameters was sought by means of hardness, tensile, and stress-rupture tests. Among all the combinations of pre- and post-aging treatments, the following produced the best group of mechanical properties:

Solution: 1230 K, 1 h
Pre-aging: 980 K, 4 h
Straining: rolling (19% reduction) or shock (51 GPa)
Post aging: 950 K, 8 h, furnace cooling to 895 K, total of 18 h

Processing by rolling or by shock produced substantial improvements in the mechanical properties. The introduction of preaging has basically two functions. First, it establishes a homogeneous distribution of precipitates which cannot be destroyed by subsequent deformation. Without preaging, the nucleation of precipitates on dislocations during the final aging treatment results in a non-uniform distribution of precipitates which would not favor an optimum response. Second, the preexistent precipitates anchor the dislocations, thus avoiding recovery during the final aging treatment(s). A schematic illustration of what is believed to occur is given in Fig. 15.5. The experimental program described is a good illustration of the factors that must be taken into account in the development of a TMT. Figure 15.6 shows the improvements in stress-rupture life induced by TMP of Inconel 718 at 922 K.

‡15.1.4 TMT of Other Alloys

In aluminum alloys, the objective of TMT is to increase the mechanical strength as well as corrosion resistance. From this point of view, TMT has been a success in Al alloys. One of the more notable developments was the demonstration that an aging treatment before "warm" working is essential to the development of better properties. Ni-based alloys also benefit from this, as indicated in Section 15.1.3. This development is due to Conserva et al.[15] Of

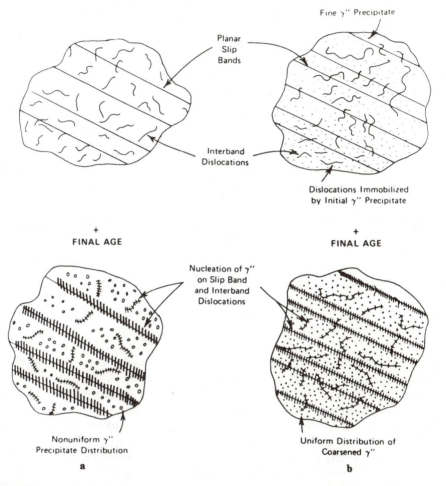

SOLUTION TREATMENT
+
COLD WORK

SOLUTION TREATMENT
+
PARTIAL AGE
+
COLD WORK

Fine γ″ Precipitate

Planar
Slip
Bands

Interband
Dislocations

Dislocations Immobilized
by Initial γ″ Precipitate

+
FINAL AGE

+
FINAL AGE

Nucleation of γ″
on Slip Band
and Interband
Dislocations

Nonuniform γ″
Precipitate Distribution

a

Uniform Distribution of
Coarsened γ″

b

Figure 15.5 Schematic representation of how precipitation interacts with dislocations: (a) solution + deformation + aging; (b) solution + preaging + deformation + aging. TMP schedule (b) provides a more uniform precipitate distribution. (Reprinted with permission from [14], p. 524.)

course, Al alloys do not benefit from martensitic transformations. One resorts to warm working to avoid planar slip (dislocation banding) and a very heterogeneous dislocation distribution. On the other hand, warm working makes it possible to obtain a more homogeneous structure.

In titanium, an increase in strength by developing special textures has been successfully studied. Thus, the mechanical working in the TMT of titanium alloys has been applied in such a way as to produce favorable mechanical textures. One of the significant studies has been that of Kalish and Rack[16] in a beta

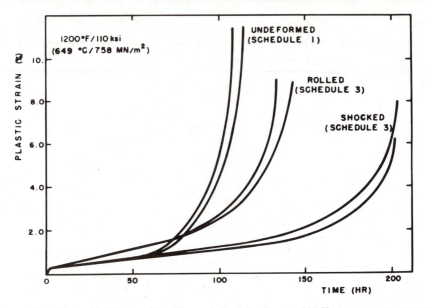

Figure 15.6 Effect of TMP by cold rolling and shock loading on 922-K stress-rupture response of Inconel 718. (Adapted with permission from [12], p. 138.)

III titanium alloy (Ti, 11.5% Mo, 5.5% Zr, 4.5% Sn). They obtained improvements in mechanical properties by means of TMT of this alloy.

‡15.2 DUAL-PHASE STEELS

Dual-phase steels, so called because they consist essentially of a mixture of ferrite and martensite, are now available commercially. These steels offer, for a given strength level, the highest formability (Fig. 15.7). These steels are mostly finding weight-reducing applications in the automobile industry. The excellent combination of strength and ductility displayed by dual-phase steels (Fig. 15.7) results from its composite structure: a strong phase (martensite) dispersed in a strong (extremely fine grained) and ductile (low interstitial content) ferrite matrix. These steels have compositions close to those of HSLA or microalloyed steels, and are quenched from a suitable temperature in the intercritical range A_1 to A_3 such that the typical ferrite–pearlite microstructure of such steels is replaced by a ferrite–martensite structure. In fact, the microstructure is a lot more complex due to the presence of some retained austenite, bainite, and carbides.[18] Figure 15.8 shows schematically how dual-phase steels are produced and the respective microstructures before and after treatment in the intercritical temperature range.* Figure 15.9 shows an optical micrograph of a dual-phase steel showing islands of martensite (m) in a matrix of ferrite (f).

---

* Commercialization of dual-phase steel is most advanced in Japan. One probable reason for this is the availability in Japanese steel plants of large continuous annealing lines, which are so important in dual-phase sheet and strip production in large sizes and quantities.

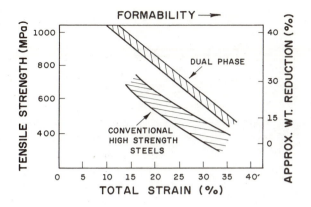

Figure 15.7 Dual-phase steels show a combination of strength and ductility superior to that of the conventional microalloyed steels. (Adapted with permission from [17], p. 26.)

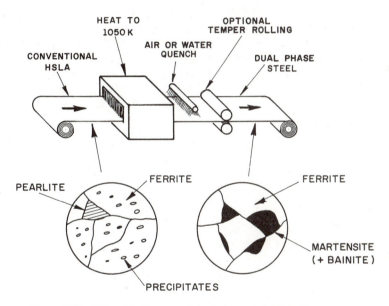

Figure 15.8 Schematic diagram of dual-phase steel fabrication process.

‡15.2.1 Microstructural Aspects

Among the factors that are considered to be critical for obtaining an appropriate microstructure after the intercritical temperature treatment are: ferrite grain size, temperature of intercritical treatment, and cooling rate. The importance of a fine ferritic grain size is obvious. The finer the grain size, the better will be the ductility for a given strength level. The temperature of intercritical treatment is of great importance as it will determine the amount of martensite that will form as well as its composition (i.e., carbon content). According to some initial work,[18,19] the strength of these ferrite–martensite mixtures increased

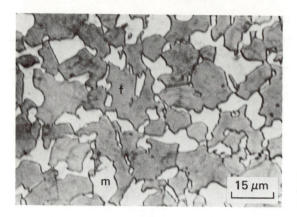

Figure 15.9 Typical microstructure of a dual-phase steel; m and f indicate martensite and ferrite, respectively.

linearly with the martensite volume fraction, independent of the carbon content of martensite. This seems to have been a fortuitous result, as the strength of martensite is a very sensitive function of its carbon content.[18,20,21] The amount of carbon dissolved in martensite also influences its microstructure. Above 0.35%C, twinned martensite is observed, while a dislocated martensite (i.e., containing dislocations) is observed for C < 0.35%.[22,23] The former has a very low toughness and is therefore undesirable. The cooling rate is an important factor and is related to the problem of hardenability. The problem of hardenability becomes critical for sheet thicknesses greater than about 2 to 3 mm, making it necessary to add alloying elements that retard austenite decomposition by diffusional processes. A common addition is that of Mn (~1.5%), due to its low diffusivity at the intercritical annealing temperatures. Additions of Si and Cr in sufficient quantities can permit box annealing treatment.[24]

‡15.2.2 Mechanical Response

Dual-phase steels show the following characteristics in their stress–strain behavior:

1. A low tensile yield stress
2. An absence of discontinuous yielding
3. A very high initial strain-hardening rate

Figure 15.10 shows these aspects of a dual-phase steel in comparison with a conventional microalloyed steel. The absence of a well-defined yield point is generally attributed to the presence of free dislocations in ferrite. These could be introduced by the martensitic transformation into the neighboring ferritic grains.[19]

As to the work-hardening behavior of these steels, there exist various explanations in the literature. According to Rigsbee and Van der Arend,[25] transformation of retained austenite is responsible for the high work-hardening

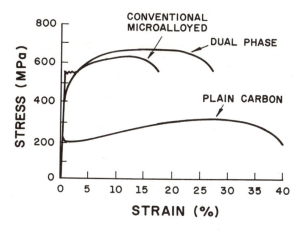

Figure 15.10 Stress–strain curves of dual phase steel and a conventional microalloyed steel. The dual phase steel shows a low yield stress, no yield drop, and has a high initial work hardening rate.

rate at low strains. Davies,[26] on the other hand, attributes this characteristic of dual-phase steels to a clean, interstitial-free ferrite. In any case, this high strain-hardening rate in the initial stages of deformation is of great practical importance, as it permits the development of high strength during forming, even in regions suffering small deformations.

‡15.2.3 Formability

The combination of a low yield stress, high tensile stress, and good total elongation in dual-phase steels (Fig. 15.10) leads to good formability.

A high work-hardening rate is very important for good press formability. At any strain level, a high work-hardening rate results in a more uniform strain distribution throughout the formed piece. A high initial work-hardening rate also implies a low yield stress at low strains. This reduces springback and allows better shape control in large shallow panels (such as automobile skin panels). Figure 15.11 shows that dual-phase steels have a 10% smaller springback than

Figure 15.11 Relationship between tensile strength and springback in pure bend forming for three steels: SH, solution hardened; PH, precipitation hardened; DP, dual phase. All steels had a thickness, $t = 1.2$ mm. (Adapted with permission from [27], p. 37.)

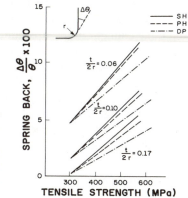

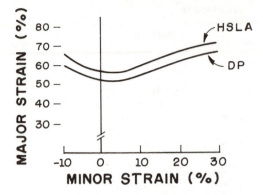

Figure 15.12 Forming limit diagram of an HSLA and a dual-phase (DP) steel. (Adapted with permission from [28], p. 7.)

do conventional high-strength steels for a given strength level.[28] This lower springback implies a considerable "shape-holding" advantage.

The forming limit curves of a conventional HSLA steel and a dual-phase steel are shown in Fig. 15.12. The dual-phase steel curve lies below the HSLA curve.

15.3 RADIATION DAMAGE

15.3.1 Introduction

Irradiation of solids by high-energy particles may produce one or more of the following effects:

1. Production of displaced electrons (i.e., ionization)
2. Production of displaced atoms by elastic collision
3. Production of fission and thermal spikes

Ionization has a much more important role in nonmetals than it has in metals. The high electrical conductivity in metals leads to a very quick neutralization of ionization and there is no observable change in properties due to this phenomenon. Electronic excitations in metals are also eliminated almost instantaneously. Such would not be the case in semiconductors and dielectrics, where electronic excitation configurations are almost permanent ones. Thus, in the case of metals only collisions among incident particles and atomic nuclei are of importance. The basic mechanism in all processes of radiation damage is one of transfer of energy and motion from the incident particle beams to the atoms of the material. The incident particle beam may consist of positive particles (protons, for example), negative particles (which are invariably electrons), or neutral particles (X-rays, γ rays, neutrons, etc.). The spectra of energy transfer by bombardment of neutrons, positive ions, and electrons are shown in Fig. 15.13. Irradiation by neutrons results in a large spectrum of constant energy until T_m, the maximum energy that a particle can transmit to an atom that

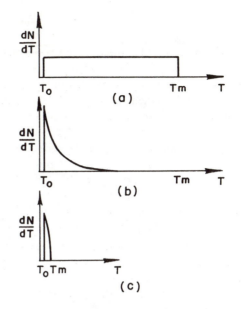

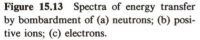

Figure 15.13 Spectra of energy transfer by bombardment of (a) neutrons; (b) positive ions; (c) electrons.

suffered the impact. The term dN/dT represents the number of atoms that receive an energy between T and $T + dT$. A neutron of 1 MeV(0.16pJ) can transfer about 10^5 eV(0.016pJ) to an atom. High-energy transfers can also be obtained by means of positive particles, but such energy transfers are less common. In the case of electrons, only low-energy transfers are possible. We shall consider here mainly the effects of neutron radiation on metals. The primary collision has the function of transferring energy to the atomic system. The subsequent events that occur are as follows:

1. Displacement of an atom from its normal position in the lattice to a position between the normal lattice sites
2. Creation of defects by displacements and their migrations and interactions

When an atom is displaced from its normal lattice site, two defects are created: an interstitial atom referred to as auto-interstitial or self-interstitial, and a vacant lattice site called vacancy. More complex configurations can be regarded as having started from this fundamental step. When an atom receives an energy impulse greater than a certain value T_e, called effective displacement energy, an atom is displaced from its normal position to an interstitial position. In the most simple case, if an atom receives the primary impact of energy T_e, it is displaced. This, however, is not inevitable; sometimes another atom, a neighboring one, is displaced. With an increase in the energy imparted to the impacted atom, various events can occur. At low energies, but higher than T_e, only an interstitial and its connected vacancy are possible. At high energies, the impacted atom becomes an important particle for creating more damage. This leads to cascade elements.

15.3.2 Mechanism of Damage Production

Brinkman[29] formulated the concept of a displacement spike wherein an energetic atom, near the end of its trajectory, displaces all the atoms that it encounters, producing a region similar to a void. Seeger[30] presented a radiation damage model which best represents current understanding of the distribution of defects in a cascade. This model is shown in Fig. 15.14. Seeger pointed out that many atoms that spread about by displacement spikes will locate themselves along the atomic packing lines, and thus this will be a most efficient manner of transporting energy far away from the spike. The impact transferred along a crystallographic direction is called FOCUSON (analogous to photon and phonon). If the energy is not well above the energy required for atomic displacement, it will be transferred into a chain of exchange collisons which makes the atom travel far away from the spike before it comes to a stop as an interstitial. The efficiency of this process is much higher in the close-packed directions ($\langle 110 \rangle$

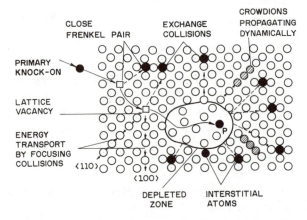

Figure 15.14 Schematic representation of damage produced by irradiation. P indicates the position where the first "knock-on" terminates. (Reprinted with permission from [30], p. 105.)

directions in FCC crystals). The atomic configuration in the $\langle 110 \rangle$ direction in which the interstitial is propagated along a line is called a dynamic crowdion. The efficiency of these focusing processes is directly proportional to the interatomic potential, being higher for heavy metals and low for light metals (such as Al). According to the Seeger model, at zero Kelvin, for each initially displaced atom, one would have one or more regions in which a good fraction of atoms (about 30%) disappear. These regions are surrounded by interstitial clouds that extend for noble metals a few hundreds of atomic distances, while for a metal such as Al, perhaps a few atomic distances. Seeger called the region of lost atoms in the center of a cascade a "depleted zone" and estimated that its typical size would be less than 1 nm.

15.3.3 Radiation Hardening in Metals

In any event, a major portion of radiation damage in common metals caused by neutrons in reactors consists of a large number of interstitials and vacancies produced in a cascade process that follows after a primary knock-on impact. These point defects act as small obstacles to dislocation movement and result in a hardening of the metals. Besides this direct effect on mechanical properties, some indirect effects are possible. These indirect effects arise from the fact that irradiation by neutrons changes the rates and mechanisms of atomic interchange. Damask[31] summarized these effects in the following way:

1. Order destruction
2. Fractionating of precipitates
3. Acceleration of nucleation
4. Acceleration of diffusion

These processes have their origin, directly or indirectly, in the kinetic energy exchanges between energetic neutrons and atoms. According to Seeger's model, atoms can be transported long distances by cooperative focalization along the more densely packed directions and the collision processes create simple defects, such as interstitials and vacancies, and complex defects, such as displacement spikes. If an alloy is ordered, focalization and displacement spikes may destroy the order. If the alloy contains precipitates, a displacement spike may break the precipitates if they are smaller than the spike, and thus revert the precipitate into solution. In an alloy that can have precipitates, the damaged regions caused by spikes can serve as nucleation sites. The excess of vacancies produced by irradiation can accelerate the diffusion rate. All these effects influence significantly the mechanical properties as described in earlier chapters. At ordinary temperatures (i.e., ambient or slightly above) one or both the defects (interstitials and vacancies) are mobile, and thus the ones that survive the annihilation, due to recombination or loss of identity at sinks such as dislocations or interfaces, group together. It is well established that in a majority of metals, irradiation at low temperature ($<0.2\,T_m$, where T_m is the melting point in Kelvin) results in joining of vacancies and interstitials to form groups that are surrounded by dislocations (i.e., loops and tetrahedral packing defects). These groups impede dislocation motion, increase the strength, and reduce the ductility of the material. At high temperatures, the vacancies can group together to form voids. The formation of such groups of defects can cause important and undesirable changes in mechanical properties and result in a dimensional instability of the material. Damage accumulated during irradiation by neutrons (and other particles) can cause significant changes in important properties. For example, the yield stress or the flow stress increases and frequently there is a loss of ductility.[32] In the case of copper, the increase in yield stress in a function of the square root of dose for neutron doses less than 10^{17} n/cm^2.[32] Hull and Mogford[33] found a linear relation between the increase in the yield stress $\Delta\sigma_{ys}$ of a low-carbon

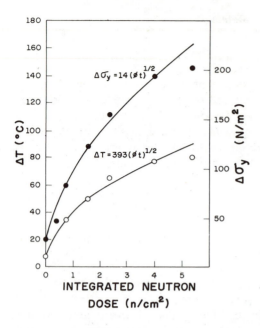

Figure 15.15 Increase in ductile–brittle transition temperature and yield stress as a function of neutron dose for a carbon steel. (Adapted with permission from [34], p. 177.)

steel and the cube root of the neutron yield Φt, where Φ is the neutron flux and t is the time. The units of neutron yield Φt are n/cm². Nichols and Harries,[34] however, observed a linear relation between $\Delta\sigma_{ys}$ and the square root of the integrated neutron dose. This is shown in Fig. 15.15. The big disadvantage of such power relations is that they predict an infinite hardening for an infinite yield of neutrons. In practice, one observes a saturation in the property changes with irradiation. Makin and Minter[35] derived the following expression:

$$\Delta\sigma_{ys} = A[1 - B\exp(-C\Phi t)]^{1/2} \qquad (15.1)$$

where A, B, and C are constants. This expression is found to be observed for various materials.[35-37] The ductile–brittle transition temperature also increases with irradiation, as shown in Fig. 15.15.

The problem of mechanical and dimensional stability is a very serious one for structural components in fast reactors. Cawthorne and Fulton[38] discovered in 1967 that fuel cladding consisting of austenitic stainless steel, when exposed to high doses of fast neutrons, showed internal cavities (~10 nm). These cavities, called voids, result in an increase in the dimensions of the material. It is estimated that the possible maximum dilatation in the structural components can be of the order of 10%. However, as neutron flux and temperature of the sodium coolant are not uniform in the core, the swelling of the component will be nonuniform. This nonuniformity can influence the component in-service behavior.

Irradiation by neutrons causes marked changes in the properties of zirconium alloys Zircaloy-2 and Zircaloy-4 (very much used in light water reactors) and 304 and 316 stainless steels (used in liquid metal fast breeder reactors).[38-43] Figure 15.16 shows the increase in strength (yield strength and ultimate tensile

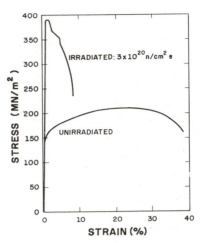

Figure 15.16 Stress–strain curves for irradiated and unirradiated Zircaloy. (Adapted with permission from [39], p. 2223.)

strength) of Zircaloy after neutron radiation. The exact nature of defects introduced by radiation that are responsible for these changes in Zircaloy are not yet very well characterized. There is a considerable variation in the observed microstructures. One of the few observations about which there exists general agreement is the absence of radiation-induced vacancies in Zircaloy, which is a significant difference when compared with, say, the behavior of stainless steels. Stainless steels show swelling due to neutron irradiation. The dilatation induced by neutron irradiation in stainless steel depends on neutron flux and temperature, as shown in Fig. 15.17. It is believed that the vacancies introduced by irradiation combine to form voids, while the interstitials are preferentially attracted to dislocations. According to Shewmon,[40] this dilatation of stainless steel does not affect the viability or security of breeder-type reactors but will have a significant effect on core design and economy of reproduction. Preliminary results indicate that, in spite of not being able to eliminate the effect completely, cold work, heat treatments, or composition changes can reduce the swelling by a factor of 2 or more. Figure 15.18 shows the change in dilatation of stainless steel as a function of Cr and Ni contents.

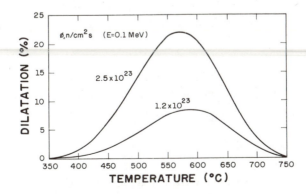

Figure 15.17 Stress-free dilatation in AISI 316 stainless steel (20% cold worked). (Adapted with permission from [39], p. 2223.)

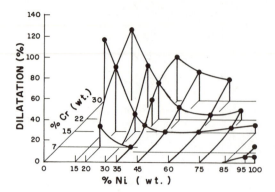

Figure 15.18 Dependence of fast neutron-induced dilatation in stainless steel (Fe-Cr-Ni) as a function of Ni and Cr amounts. (Adapted with permission from [41], p. 735.)

15.3.4 Ion Implantation

An interesting application of this kind of interaction between a solid and charged particles is called *ion implantation*. Charged ions are accelerated in an electric field (e.g., in a linear accelerator) to very high energies (~200 keV) and allowed to strike the target solid in moderate vacuum (~1 mPa). It is worth emphasizing that the selected species of ions gets *implanted* into the target surface and not deposited on the target surface. The technique, originally developed for preparing semiconductor devices in a controlled fashion, has been developed into a sophisticated tool for altering composition and structure of surfaces for any number of purposes,[42] for example, modifying surface chemistry for better corrosion and oxidation resistance, tribological properties, and super-conductivity. The reader can well imagine the powerfulness of the technique by the fact that it allows one to introduce elements into a surface, which may not be possible in conventional heat treatment because of low diffusivity. B^+, N^+, and Mo^+ ions implanted into steel can, depending on the dose, reduce the wear of the tool by an order of magnitude. A good deal of this work has been done by Dearnaley and coworkers at the U.K. Atomic Energy Research Establishment.[43-45] Figure 15.19 shows, as an illustration, a gear being ion-implanted.

The ion implantation technique of modifying composition and structure of surfaces has a number of advantages over conventional techniques.

1. The process is essentially a cold one; therefore, there is no loss of surface finish and dimensions (i.e., the process can be applied to finished particles).
2. One can implant a range of metallic and nonmetallic ions, individually or combined.
3. One can implant selected critical areas.

Ion implantation is particularly suited for this kind of selected modification of small, critical parts. A typical example is given by Dearnaley and Hartley.[46] Oil burners used for injecting a fuel oil and air mixture into boilers of oil-

Figure 15.19 Ion implantation of a gear. (Courtesy of G. Dearnaley, Harwell, England.)

fired power plants face rather severe erosion conditions. Ti and B implantation of oil-burner tips improved erosion properties and increased the service life.

Another very important aspect has to do with the fact that ion implantation is basically a nonequilibrium process. There are thus no thermodynamic constraints, such as solubility limits. In other words, we are able to produce metastable alloys with new and unusual characteristics, amorphous alloys, and so on.[47,48] The technique thus offers a novel way of producing surfaces, in a controlled manner, for scientific studies. This aspect of the technique, despite the obvious technological importance described above, should not be disregarded.

15.4 SHOCK HARDENING

Any material is capable of transmitting, duly transformed, the vibrations that it receives. For example, a person in a completely closed room is capable of hearing external noises. This is because the vibrations of the vocal cords are transmitted to the air, from there to the walls of the room, and from the internal walls again to the air, finally reaching the person's eardrum. Metals are no exception to this rule. Any impact received propagates through them. Initially, the disturbance is elastic, moving with the velocity of sound in the metal. When the amplitude of the disturbance reaches a critical value, large shear stresses lead to the yielding of the metal at this large strain rate. Depending on the geometry of the metal, there are two states of stress for which a great majority of dynamic testing is carried out: uniaxial stress (when the wave propagates along a small-diameter bar) and uniaxial strain (when the wave propagates in a medium in which the lateral dimensions are infinite). In the first case, there is a residual strain, while in the second case, this is a minimum. In the first case, the typical strain rates are of the order of 10^3 to 10^4 s^{-1}, and we call it a plastic wave. It is considerably slower than the elastic precursor wave. Commonly, experiments are carried out with the Hopkinson pressure bar. When the strains are uniaxial, we obtain very high strain rates (10^4 to 10^6 s^{-1}). The plastic wave is slower than the elastic wave for small amplitudes, but at higher pressures it "swallows" the elastic wave. The plastic wave is characterized by an abrupt front, thus receiving the name "shock wave." Figure 15.20(a) shows how the pressure, temperature, and density are changed by the passage of the shock front. The fundamental characteristic of a shock wave, the abrupt shock front, is shown in Fig. 15.20(b). The wave velocity increases with pressure, so that the high-pressure portion overcomes the low-pressure region ($U_{s2} > U_{s1}$), producing the steep front. The passage of shock waves through metals produces very specific structures. The strengthening of metals by shock waves has been known and studied in the last 20 years.[49-60] Taking pressure P as the hydrostatic component of stress and remembering that the state of deformation is uniaxial, that is,

$$\begin{bmatrix} \epsilon_1 & 0 & 0 \\ 0 & 0 & 0 \\ 0 & 0 & 0 \end{bmatrix} \tag{15.2}$$

we can determine the stress state for an isotropic material by applying Eqs. 1.120. Substituting Lamé's constants by elastic constants E, G, and ν, we can determine the maximum shear stress:

$$\tau_{\max} = \frac{\sigma_1 - \sigma_3}{2} = \frac{E\epsilon_1}{2(1+\nu)} = \frac{3(1-2\nu)P}{2(1+\nu)} \tag{15.3}$$

where $\sigma_1 > \sigma_2 > \sigma_3$ are the principal stresses and ϵ_1 is the uniform axial strain. Taking $\nu = 0.305$ for Ni, we have

$$\tau_{\max} = 0.45P \tag{15.4}$$

Thus a pressure of 20 GPa, common for such experiments, will give shear stresses of the order of 9 GPa. Such stresses are required for homogeneous nucleation of dislocations. Details about the dislocation nucleation mechanism are given in Section 6.8.1 and in [51], [56], and [57].

The study of the metallurgical effects of shock waves was initiated in the 1950s by Cyril Stanley Smith;[51] the understanding of the effects of shock waves on the structure and properties has increased steadily in this period. It is now recognized that there are two classes of important parameters: (1) shock-wave parameters: pressure, pulse duration, rarefaction rate, attenuation rate, planarity, and normality of wave; and (2) material parameters: grain size,[52,57] stacking-fault energy, existing substructure,[60] and so on.

Shock waves can produce diverse structural and substructural changes in metals, such as point-defect formation,[49] dislocations,[51] twins,[50] and phase transformations.[52,53] Consequently, the properties of metals are altered. Chapter 6 (Section 6.8.1) presents the generation of dislocations by shock-wave deformation.[57,59] The high density of dislocations and their uniform distribution is responsible for a unique response in nickel: work softening. This is described

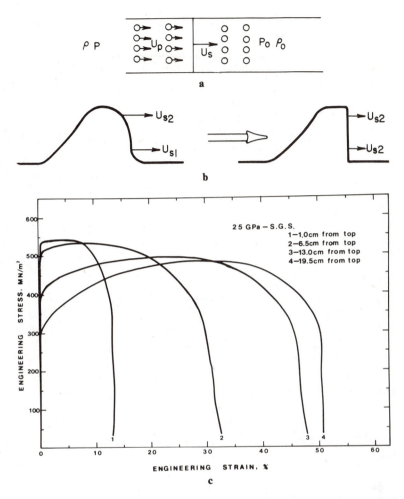

Figure 15.20 (a) Change in state variables induced by shock wave; pressure, temperature, and density are changed; (b) wave velocity increases with pressure producing a shock front; (c) change in tensile response of shock-hardened nickel as a function of distance from impact surface. Initial pressure of 25 GPa and pulse duration of 2 μs.

in greater detail in Chapter 9. The shock wave attenuates itself as it travels through a solid, loosing its energy; consequently, the residual strengthening effect slowly decreases. Figure 15.20(c) shows how the tensile properties of shock-loaded nickel decrease with distance from the impact interface. The study of effects of shock waves is of importance because:

1. There is a certain technological potential in the use of processing by shock with a view to increasing the strength of metals; railroad frogs and crushers of Hadfield steel are shock-hardened.
2. Shock waves are present in explosive welding and compaction processes and it is important to know their effect on residual properties.

3. The impact of a projectile against an armor generates in it a shock wave, which is partially responsible for the damage. The spalling of the internal surface of the armor is produced by the interaction of direct (compressive) and reflected (tensile) waves. From this one can see the importance of an understanding of shock waves in ballistic impact.
4. Consideration of shock-wave propagation through metals is very important in the design of nuclear warheads.
5. Pulsed lasers are being experimentally used as a means of shock hardening the metal surface.
6. Shock waves are used as a metallurgical tool. One can produce a homogeneous structure with a high dislocation density without altering the grain shape.

Shock waves can be introduced in solids by impact of one object with another, by pressure of gas produced by detonation of an explosive, or by large depositions of energy. Examples of the latter are the impacts caused by pulsed laser beams and nuclear explosions. This creates a shock wave in the air, which impacts the objects. For a systematic study of shock waves, we use the simplest wave geometries and configurations, avoiding reflection waves as much as possible. The impact of a plane projectile with a surface parallel to it provides an easily tractable situation. The plane projectile can be accelerated by various means.[54] The technique that requires the least capital investment is that of acceleration of a plate by explosives. The experimental technique that is used for hardening of Hadfield steel employs the detonation of explosive directly in contact with the metal. The description of experimental techniques and procedures is given by Orava and Wittman[55] and deCarli and Meyers.[58] It is possible to monitor the wave pressure and duration by, among other means, manganin piezoresistive gages.[60]

15.5 ORDER HARDENING

An alloy is said to be disordered when the atoms of species, say, A and B, are arranged in a random manner. As pointed out in Chapter 7, there is hardly any alloy system that is truly random; however, there are many that come close to it. On the other hand, there are a fairly large number of alloys in which it is energetically favorable for atoms A and B to segregate to preferred lattice sites below a critical temperature T_c, and generally in certain well-defined atomic proportions (i.e., AB_3, AB, etc.). Sometimes when the bonding is not totally metallic but partly ionic in nature, we call the structure an intermetallic compound. Westbrook[61] defines intermetallics as all compound phases of two or more normal metals (ordered or disordered). These intermetallic compounds constitute an interesting class of materials of great practical importance. The reader is referred to works listed in "Suggested Reading" for delving in greater depth in this area.

Mechanical properties of alloys are altered when they have an ordered structure. We define the degree of long-range order (LRO) by means of a parameter,

$$S = \frac{r - f_A}{1 - f_A} \qquad (15.5)$$

where r is the fraction of A sites occupied by A atoms, and f_A is the fraction of A atoms in the alloy. Thus, S goes from 0 (completely disordered) to 1 (perfectly ordered). The various dislocation morphologies observed in ordered alloys have been summarized by Marcinkowski[62] and are presented in Table 15.3.

A superdislocation in a perfectly ordered crystal and a single dislocation in a completely disordered crystal will both experience less friction stress than either of them will experience at an intermediate degree of order S. Thus, qualitatively, one would expect a yield stress maximum at an intermediate degree of order (i.e., the change in yield stress is not directly related to the degree of ordering). For example, Cu_3Au crystals show a lower yield stress when fully ordered than when only partially ordered. Experiments showed that this results from the fact that the maximum in strength is associated with a critical domain size. Short-range order (SRO) results in a distribution of neighboring atoms which is not random. Thus, the passage of a dislocation will destroy the SRO between the atoms across the slip plane. The stress to do this is large. A crystal of Cu_3Au in the quenched state (SRO) has nearly double the yield stress of the annealed state (LRO). The maximum in strength is exhibited by a partially ordered alloy with a critical domain size of about 6 nm. The transition from deformation by unit dislocations in the disordered state to deformation by superdislocations in the ordered state gives rise to a peak in the flow stress versus degree of order curve.

The presence of atomic order leads to a marked change in the flow curve of the alloy. Figure 15.21 shows the flow curves of a fully ordered FeCo alloy at low temperatures where the order is not affected.[63] Stage I in Fig. 15.21 is associated with a well-defined yield point. This is followed by a high linear work-hardening stage, II. Finally there occurs stage III, with nearly zero work hardening. The stress–strain curves of the same alloy in the disordered state are shown in Fig. 15.22. The curves in Fig. 15.21 (ordered) are markedly different from the ones in Fig. 15.22 (disordered). The sharp yield point and stage II are absent in the disordered alloy. The disordered alloy goes straight into stage III after gradual yielding. Fully ordered alloys deform by means of the movement of superlattice dislocations at rather low stresses. However, the superdislocations (i.e., closely spaced pairs of unit dislocations bound together by an antiphase boundary) must move as a group in order to maintain the ordered arrangement of atoms. This makes cross-slip difficult. Long-range order thus leads to high strain-hardening rates and frequently to brittle fracture.[64]

Figure 15.23 shows this effect of ordering on uniform elongation of FeCo–2%V at room temperature.[65] The ductility decreases with the degree of increasing LRO. Mg_3Cd is the only known exception to this tendency for brittleness because of a restricted number of slip systems or less easy cross-slip.

TABLE 15.3   Dislocation Morphologies in Some Ordered Alloys[a]

Superlattice Type (*Strukturbericht* Designation)	Chemical Designation	Unit Cell Dimensions	Alloy Types	Superlattice Dislocation Type	Burgers Vector of Each Dislocation	Antiphase Boundary Type[b]
B2	CsCl	a_0	NiAl, AgMg AuZn	⊥	$a_0\langle100\rangle$	None
			CuZn, FeCo FeAl, FeRh NiAl, AgMg AuZn		$\frac{1}{2}a_0\langle111\rangle$	NN
DO$_3$	Fe$_3$Al	a_0	Fe$_3$Al, Fe$_3$Si Fe$_3$B		$\frac{1}{4}a_0\langle111\rangle$	NN NNN
L1$_2$	Cu$_3$Au	a_0	Cu$_3$Au, Ni$_3$Mn Ni$_3$Al, Ni$_3$Fe Cu$_3$Pd, Ni$_3$Ti Ag$_3$Mg, Ni$_3$Ta Ni$_3$Si, Cu$_3$Pt Ni$_3$Ga		$\frac{1}{6}a_0\langle112\rangle$	NN NN + SF
DO$_{19}$	Mg$_3$Cd	a_0 c_0	Mg$_3$Cd		$\frac{1}{6}a_0\langle10\bar{1}0\rangle$ $\frac{1}{2}a_0\langle2\bar{1}\bar{1}0\rangle$	NN SF NNN
L1$_0$	CuAu	a_0 c_0	CuAu, CoPt FePt		$\frac{1}{6}a_0\langle112\rangle$	NN NN + SF SF

[a] Adapted with permission from [62], p. 189.

[b] NN, nearest neighbor; NNN, next-nearest neighbor; SF, stacking fault.

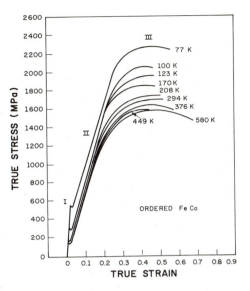

Figure 15.21 Stress–strain curves of ordered FeCo alloys at different temperatures. (Adapted with permission from [63], p. 1240.)

Ordered alloys such as FeCo[64] and Ni₃Mn[66] obey the Hall–Petch relationship between flow stress and grain size:

$$\sigma = \sigma_0 + kD^{-1/2}$$

where σ is the flow stress at a given strain, σ_0 and k are constants for that strain, and D is the grain diameter. In these alloys, long-range order increases k, as shown in Fig. 15.24 for Ni₃Mn. This increase in k with long-range order can be explained by the change in number of slip systems with order since the ease of spreading of slip across boundaries is controlled by the degree of order.

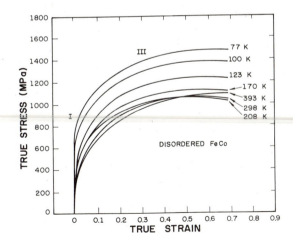

Figure 15.22 Stress–strain curves of fully disordered FeCo alloys at different temperatures. (Adapted with permission from [63], p. 1240.)

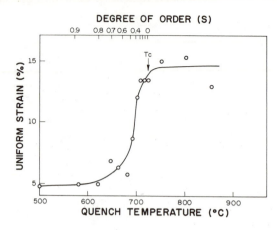

Figure 15.23 Effect of atomic order on uniform strain (ductility) of FeCo–2V at 25°C. (Adapted with permission from [65], p. 475.)

The effect of atomic ordering on fatigue behavior has been studied by Boettner et al.[67] Figure 15.25 shows the S–N curves for ordered and disordered Ni₃Mn. The improved fatigue performance in the ordered state is explained by less ease of cross-slip and a decrease in slip-band formation in the ordered state. These slip bands lead to the formation of extrusions and intrusions on the sample surface, which lead to fatigue crack nucleation.

Gamma-prime strengthened superalloys are an example of the effect of ordering on the strength. The Ni₃Al precipitate produces very low coherency stresses and is coherent with the austenitic matrix. The strengthening effect is clearly evident from Fig. 15.26(b), which shows the strength of the austenitic matrix and Ni₃Al separately, and the strength of MarM-200, composed of 65 to 85% gamma prime. In ordered structures it is energetically favorable for dislocations to move in groups, forming antiphase boundaries between them, as seen in the preceding paragraphs. Gleiter and Hornbogen[69] developed a

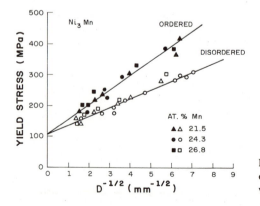

Figure 15.24 Hall–Petch relationship for ordered and disordered alloys. (Adapted with permission from [66], p. 308.)

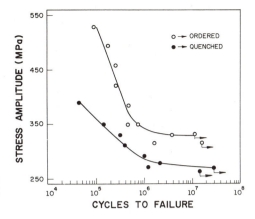

Figure 15.25 Effect of atomic order on fatigue behavior of Ni₃Mn. (Adapted with permission from [67], p. 132.)

theory for the strengthening due to coherent, stress-free, ordered particles, based on the hypothesis that dislocations glide in pairs. The equilibrium distance between the pairs and their form was found to depend on the particle size, particle distribution, energy of the antiphase boundary, elastic constants, and the external shear stress. The parameters above are part of the equations derived by Gleiter and Hornbogen[68,69] for the increase in the critical resolved shear stress, $\Delta\tau$. The results of calculations are compared with observed results for a Ni-Cr-Al alloy in Fig. 15.26(a). δ is the atomic percent aluminum. The experimental results are marked by dots and triangles; they refer to 0.5 and 1.8% aluminum, respectively. The correlation is good and maximum strengthening is obtained for particles having a diameter of 100 Å.

Another outstanding property of Ni₃Al is the increase in yield stress with temperature. This is clearly evident in Fig. 15.26(b). The yield stress increases by a factor of 5 when the temperature is raised from ambient temperature to 800°C. The inverse temperature dependence is unique and contrary to what would be expected based on thermally activated motion of dislocations. Thus, in spite of the normal temperature dependence of the austenite [also shown in Fig. 15.26(b)], the alloy MarM-200 exhibits a constant yield stress up to 800°C; the decrease in the flow stress of γ is compensated by the increase of γ'. Copley and Kear[70,71] discuss this unique behavior. It is interesting to notice that other ordered alloys, such as Cu₃Au and Ir₃Cr do not exhibit this unique behavior, while Ni₃Ge, Ni₃Si, Co₃Ti, Ni₃Ga do. The explanation provided by Copley and Kear[70,71] is based on the temperature dependence of the separation between partial dislocations in Ni₃Al. The dislocation doublets responsible for deformation of Ni₃Al are composed of two a/2 ⟨110⟩ perfect dislocations. At ambient temperature, they are decomposed into partial dislocations and their separation should be small: 0.3 to 0.4 nm. Hence, their cores should overlap. As the temperature increases, the changes in elastic anisotropy of the lattice would lead to a further constriction of the partials, forming a perfect dislocation. This would induce an increase in flow stress. At higher temperatures (>800°C) the partials

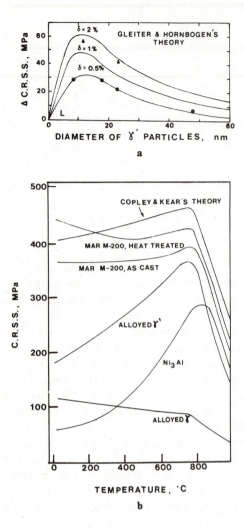

Figure 15.26 (a) Calculated and observed increase of the critical resolved shear stress (CRSS) in an Ni-Cr-Al alloy as a function of the diameter of the precipitate; full lines represent calculations. (● δ = 0.5% Al; ▲ δ = 1.8% Al) δ → atomic percent aluminum. (Adapted with permission from [68], p. 250.) (b) Effect of temperature on CRSS for Ni₃Al, γ, and Mar M-200 superalloy (γ + γ'). (Adapted with permission from [71], p. 989.)

dissociate themselves again as disordering sets in, and the yield stress decreases. High-voltage electron microscopy (2 MV) work conducted in Japan[72] disclosed that at higher temperatures (where the yield stress drops) slip changes from the normal {111} ⟨110⟩ to {100} ⟨110⟩. For Ni₃Ge, it was found[73] that the substructure at −196°C consisted roughly of an equal number of edge and screw dislocations, while at 27°C it consisted mostly of screw dislocations aligned along [1̄01]; this is shown in Fig. 15.27. This led Saburi, Nenno, Pak, and Hamana[72,73] to conclude that the decreased mobility of screw dislocations with increase in temperature was responsible for the strengthening effect. At temperatures above the one providing maximum strength, change in slip plane from {111} to {100} would be responsible for the strength decrease. This explanation is different from the one provided by Copley and Kear.[70,71] However, there are additional complications, as the work of Mulford and Pope[74] can attest.

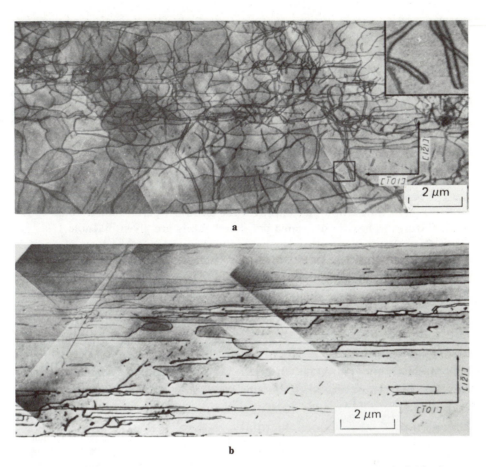

Figure 15.27 Effect of deformation temperature on the dislocation arrangement in the (111) primary slip plane of ordered Ni$_3$Ge. (a) $T = -196°C$; $\epsilon_p = 2.4\%$; (b) $T = 27°C$; $\epsilon_p = 1.8\%$. (Courtesy of H.-r. Pak, Michigan State University.)

15.6 TEXTURE STRENGTHENING

A single crystal rotates when it deforms plastically on a particular slip system (see Section 8.6). When a polycrystal is deformed in rolling, forging, drawing, and so on, the randomly oriented grains will slip on their appropriate glide systems and rotate from their initial conditions, but this time under a constraint from the neighboring grains. Consequently, a strong preferred orientation or texture develops after large strains; that is, certain slip planes tend to align parallel to the rolling plane, while certain slip directions tend to align in the direction of rolling or wire drawing. Annealing treatment of metals can also result in a texture generally different from that obtained by mechanical working, but dependent on the history of mechanical work.

A strongly textured material can exhibit highly anisotropic properties. This is not intrinsically bad. In fact, controlled anisotropy in sheet metals can be exploited to obtain an improved final product. The Young's modulus, E, of steel can, theoretically, have a value between the extreme values of the iron monocrystal [i.e., Fe[111] and Fe[100]], as shown in Fig. 15.28. The Young's modulus cannot be changed much by alloying, but texture can, theoretically, have some influence. The reader should be cautioned that the effect on E for all practical purposes is rather small (Fig. 15.28). This is not the case, however, for many other properties. Figure 15.29, for example, shows the rather marked orientation dependence of yield strength σ_y and strain to fracture ϵ_f of a rolled copper sheet. Clearly, cups made out of this material by deep drawing would show "earing" at 90° intervals due to this texture (see Section 18.1). Use is made of such texture development in Fe-3% Si laminations that make transformer cores, wherein thermomechanical treatments are given to develop a desirable magnetic anisotropy that improves the electrical performance. Some common textures in heavily deformed wires and sheets are given in Table 15.4.

Crystallographic texture is commonly represented in the form of normal pole or inverse pole figures. A normal pole figure is a stereographic projection showing the intensity of normals to a specific plane in all directions, while an inverse pole figure is a stereographic projection showing the intensities of all planes in a specific direction. The experimental procedure involves measuring relative intensities of X-ray reflections from the polycrystalline material at different angular settings. For details of the experimental determination of pole figures, the reader is referred to standard texts on the subject (e.g., that of Barrett and Massalski[75]). Figure 15.30 shows the [111] pole figure of a heavily deformed α-brass (70% Cu–30% Zn) sheet. This texture, called brass-type texture, is a

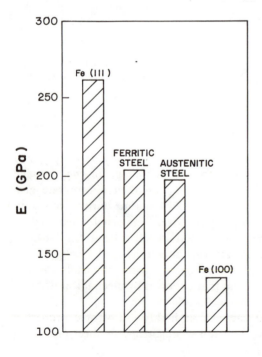

Figure 15.28 Theoretical bounds on the Young's modulus E of steel.

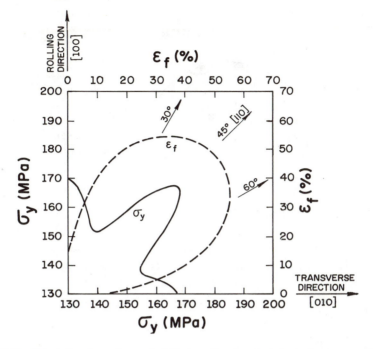

Figure 15.29 Orientation dependence of yield strength σ_y and strain to fracture ϵ_f of a rolled copper sheet.

(110)[1$\bar{1}$2] texture i.e., (110) planes parallel to the rolling plane and [1$\bar{1}$2] directions parallel to the rolling direction. The double texture indicated for FCC structures in Table 15.4 is not obtained in α-brass, but a single (110)[1$\bar{1}$2] texture develops due to its low stacking-fault energy or (probably) to mechanical twinning.[76,77]

A fuller description of the anisotropy of textured metals is obtained by using the crystalline orientation distribution function (CODF).[78–80] The CODF gives the probability of crystallite having an orientation described by the Euler angles ψ, θ, and ϕ and a series of generalized spherical harmonics. The analyses in terms of CODF are generally presented by plotting the probability contours of a crystallite having a given orientation in the Eulerian space (ψ, θ, and ϕ) at constant ϕ sections. A random polycrystalline aggregate will have a value of unity. Charts[80] are available for indexing which allow one to describe a texture in terms of ideal orientations. As an example, Fig. 15.31 shows a CODF

TABLE 15.4 Some Common Wire and Sheet Textures

	Wire (Fiber Texture)	Sheet (Rolling Texture)
FCC	[111] + [100]	(110)[1$\bar{1}$2] + (112)[11$\bar{1}$]
BCC	[110]	(100)[011]
HCP	[10$\bar{1}$0]	(0001)[11$\bar{2}$0]

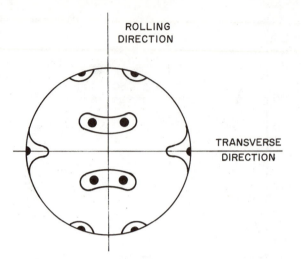

Figure 15.30 [111] pole figure of a rolled brass sheet.

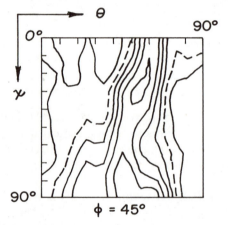

Figure 15.31 Crystallite orientation distribution function (CODF), section at $\phi = 45°$, of a 50% cold-rolled and annealed rimming steel. Unit contour is dashed line. (Courtesy of C. S. da C. Viana, Instituto Militar de Engenharia.)

section for $\phi = 45°$ of a 50% cold-rolled and annealed rimming steel. Davies and coworkers[78-80] have described methods for calculating elastic and plastic properties, determination of the yield loci, prediction of earing behavior, and formability limits of textured polycrystalline metals using the CODF. Calculations of CODF are rather laborious and use is made of computers.

EXERCISES

Thermomechanical Processing

15.1. (a) Describe all the strengthening mechanisms operating in the beta III–titanium alloy (Ti–11.5% Mo–5.5% Zr–4.55Sn) for which D. Kalish and H. Rack [*Met. Trans.*, *3* (1972) 1885] developed a TMT schedule.

(b) To what do you attribute the excellent fracture toughness after the 1175°C aging treatment?

15.2. (a) B. H. Kear, J. M. Oblak, and W. A. Owczarski [*J. Metals*, **24**, No. 4 (1972) 25] comment on the influence of deformation temperature on slip dispersal. What is the effect, and why is slip dispersal a positive feature?

(b) Would you expect a greater or smaller dispersal of slip in shock loading (versus conventional deformation)?

Dual-Phase Steels

15.3. Consider a 0.01%C steel containing about 1.5% Mn. This steel was subjected to intercritical annealing to produce a dual-phase microstructure. The intercritical annealing was carried out by holding samples for 15 min in salt baths at three different temperatures: 740°C, 800°C, and 840°C. Estimate the expected martensite volume fractions and their respective carbon contents for these three temperatures.

15.4. One of the limiting factors in specifying the exact conditions (temperature, time, and the quenching medium) of intercritical treatment for producing dual-phase steels is the fact that the proper equilibrium phase diagram is generally not available. The use of Fe-C diagram can lead to unsatisfactory results, as the alloy elements normally included in dual-phase steel compositions can affect the equilibrium conditions and the austenite decomposition mechanism. Explain how the addition of a third element to Fe-C steel can affect the critical temperatures. Explain how a change in the slope of A_3 line in the Fe-C diagram can be useful in terms of a greater flexibility in the heat treatment. (*Hint*: Compare, for example, Fe-C and Fe-Si-C systems.)

15.5. A fiber composite model involving the assumption of isostrain in the components has been frequently used to explain the strengthening in and stress–strain behavior of dual-phase steel. Comment. [See P. R. Rios, J. R. C. Guimarães, and K. K. Chawla, *Scripta Met.*, **15** (1981) 899.]

Radiation Damage

15.6. Voids can form in metals after irradiation in a reactor for doses $>10^{21}$ n/cm^2 and at about $0.3 T_m$. These voids grow and agglomerate with increasing irradiation. But irradiation produces interstitials and vacancies in equal numbers. How and why, then, do only vacancies agglomerate? [See, for example, R. Bullough and A. B. Lidiard, *Comments Solid State Phys.*, **4** (1972) 69.]

15.7. What is the effect of neutron irradiation on the parameters σ_o and k in the Hall–Petch relation? Why?

15.8. In low-alloy steels one can represent the change in yield strength on irradiation due to changes in the friction stress σ_i. Thus

$$\Delta\sigma_y = \Delta\sigma_i = A(\phi t)^{1/2}$$

where A is a constant that depends on irradiation temperature and the type of steel and ϕt is the integrated neutron dose in n/cm^2. It has been observed that for a given steel, A decreases as the irradiation temperature increases and that for steels irradiated above 350°C, there is little change in tensile and impact properties. Why?

Shock Hardening

15.9 Knowing that the shock wave produces a state of uniaxial strain (Eq. 15.2), derive the relationship between the maximum shear and the pressure (Eq. 15.4), assuming that the pressure is the hydrostatic component of the stress state. Consider the material response to be elastic.

15.10. The shock wave, upon passing through a material, compresses the lattice according to the plot in Fig. E15.10. Dislocations are generated at the front according to the model depicted in Chapter 6.

(a) Determine the total dislocation length per unit area of shock front if the

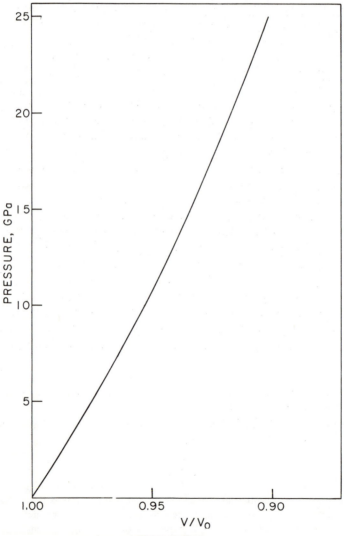

Figure E15.10

peak pressure is 10 GPa, the material has a simple cubic structure, and the lattice parameter is 4 Å. Use Fig. 6.28 for your computations.

(b) Assuming that the dislocation spacing along the front and perpendicular to it is the same, calculate the resultant dislocation density (assuming no dislocation motion).

Order Hardening

15.11. Describe some techniques of observing antiphase boundaries.

15.12. Discuss the phenomenon of strain hardening in ordered FCC superlattices.

15.13. Discuss the strain aging effect in superlattice alloys (FCC as well as BCC).

15.14. M. A. Meyers, C. O. Ruud, and C. S. Barrett [*J. Appl. Cryst.*, 6 (1973) 39] determined the type of long-range ordering existing in the Cu_2MnSn structure (see Fig. 4.1). If all the copper atoms are in their right positions, but 25% of the Mn atoms occupy Sn positions and vice versa, what is the degree of long-range ordering for the three elements?

REFERENCES

[1] E.B. Kula and M. Azrin, in *Advances in Deformation Processing*, Sagamore Army Materials Research Conf. Proc., Vol. 21, J.J. Burke and V. Weiss (eds.), Plenum Press, New York, 1978, p. 245.

[2] R.J. McElroy and Z.C. Szkopiak, *Int. Met. Rev.*, *17* (1972) 175.

[3] V.F. Zackay, *Mater. Sci. Eng.*, *25* (1976) 247.

[4] H.J. Henning, *Applications and Potential of Thermomechanical Treatment*, Battelle Memorial Institute, DMIC (Defense Materials Information Center) Rep. No. 251, Nov. 1970.

[5] S.V. Radcliffe and E.B. Kula, in *Fundamentals of Deformation Processing*, Sagamore Army Materials Research Conf. Proc., Vol. 9, W.A. Backofen, J.J. Burke, L.F. Coffin, Jr., N.L. Reed and V. Weiss (eds.), Syracuse University Press, Syracuse, N.Y., 1964, p. 321.

[6] J.C. Dunleavy and J.W. Spretnak, *Soviet Technology on Thermal–Mechanical Treatment of Metals*, Battelle Memorial Institute, DMIC (Defense Materials Information Center) Memo 244, Nov. 1969.

[7] T. Araki, in *Toward Improved Ductility and Toughness*, Climax Molybdenum Co., Ann Arbor, Mich., 1971, p. 358.

[8] V.F. Zackay, E.R. Parker, D. Fahr, and R. Bush, *ASM Trans. Quart.*, *60* (1967) 252.

[9] V.F. Zackay, E.R. Parker, J.W. Morris, Jr., and G. Thomas, *Mater. Sci. Eng. 16* (1974) 201.

[10] B.H. Kear, J.M. Oblak, and W.A. Owczarski, *J. Metals*, *24*, No. 6 (1972) 25.

[11] J.M. Oblak and W.A. Owczarski, *Met. Trans.*, *3* (1972) 617.

[12] M.A. Meyers, Ph.D. thesis, University of Denver, Denver, Colo., 1974.

[13] M.A. Meyers and R.N. Orava, *Met. Trans.*, *7A* (1976) 179.

[14] R.N. Orava, in *Advances in Deformation Processing*, Sagamore Army Materials Research Conf. Proc., Vol. 21, J.J. Burke and V. Weiss (eds.), Plenum Press, New York, 1978, p. 485.

[15] M. Conserva, M. Buratti, E. de Russo, and F. Gatto, *Mater. Sci. Eng.*, *11* (1973) 103.

[16] D. Kalish and H.J. Rack, *Met. Trans.*, *3* (1972) 1885.

[17] R.G. Davies and C.L. Magee, in *Dual Phase and Cold Pressing Vanadium Steels in the Automobile Industry*, Vanitec, London, 1979, p. 25.

[18] G.R. Speich and R.L. Miller, in *Structure and Properties of Dual Phase Steels*, R.A. Kot and J.W. Morris (eds.), TMS-AIME, Warrendale, Pa., 1979, p. 145.

[19] R.G. Davies, *Met. Trans.*, *9A* (1978) 41.

[20] W.C. Leslie and R.J. Sober, *Trans. ASM*, *60* (1967) 459.

[21] C.L. Magee and R.G. Davies, *Acta Met.*, *19* (1971) 345.

[22] P.M. Kelly and J. Nutting, *Proc. Roy. Soc. (London)*, *259A* (1960) 45.

[23] P.M. Kelly and J. Nutting, *J. Iron Steel Inst.*, *197* (1961) 109.

[24] M. Nishida, K. Hashiguchi, I. Takahashi, T. Kato, and T. Tanaka, *Proc. 10th IDDRG Biennial Congr.*, Warwick, England, 1978, p. 311.

[25] J.M. Rigsbee and P.J. Van der Arend, in *Formable HSLA and Dual Phase Steels*, A.T. Davenport (ed.), TMS-AIME, Warrendale, Pa., 1979, p. 56.

[26] R.G. Davies, *Met. Trans.*, *9A* (1978) 671.

[27] T. Furukawa, *Metal Prog.*, *116* (Dec. 1979) 36.

[28] A.E. Cornfield, J.R. Hiam and R.M. Hobbs, SAE Tech. Paper No. 790007, 1979.

[29] J.A. Brinkman, *J. Appl. Phys.*, *25* (1954) 961.

[30] A. Seeger, in *Proc. Symp. Radiat. Damage Solids React.*, Vol. 1, IAFA, Vienna, 1962, p. 101.

[31] A.C. Damask, *Proc. AIME Symposium on Radiation Effects*, TMS-AIME, New York, 1967, p. 77.

[32] J. Diehl, *Vacancies and Interstitials in Metals*, North-Holland, Amsterdam, 1970, p. 739.

[33] D. Hull and J.L. Mogford, *Phil. Mag.*, *3* (1958) 1213.

[34] R.W. Nichols and D.R. Harries, in *Radiation Effects on Metals and Neutron Dosimetry*, ASTM STP 341, ASTM, Philadelphia, 1963, p. 162.

[35] M.J. Makin and F.J. Minter, *Acta Met.*, *8* (1960) 691.

[36] A.L. Bement, *Proc. AIME Symposium on Radiation Effects*, TMS-AIME, New York, 1967, p. 671.

[37] H.C. van Elst, *Bull. Soc. Belge Phys.*, *4* (1964) 238.

[38] C. Cawthorne and F.J. Fulton, *Nature*, *216* (1967) 575.

[39] J.T.A. Roberts, *IEEE Trans. Nucl. Sci.*, *NS-22* (1975) 2219.

[40] P.G. Shewmon, *Science*, *173* (1971) 987.

[41] W.B. Hillig, *Science*, *191* (1976) 733.

[42] G. Carter, J.S. Colligon, and W.A. Grant (eds.), *Applications of Ion Beams to Metals*, Institute of Physics, London, 1975.

[43] G. Dearnaley, *Mater. Eng. Appl.*, *1* (Sept. 1978) 28.

[44] G. Dearnaley and N.E.W. Hartley, *Thin Solid Films*, *54* (1978) 215.

[45] G. Dearnaley, in *Ion Implantation Metallurgy*, C.M. Preece and J.K. Hirvonen (eds.), TMS-AIME, Warrendale, Pa., 1980, p. 1.

[46] G. Dearnaley and N.E.W. Hartley, in *Scientific and Industrial Applications of Small Accelerators*, J.L. Duggan and J.A. Martin (eds.), IEEE, New York, 1976, p. 20.

[47] J.A. Borders, *Annu. Rev. Mater. Sci.*, *9* (1979) 313.

[48] R. Andrew, *Phil. Mag.*, *35* (1977) 1153.

[49] H. Kressel and N. Brown, *J. Appl. Phys.*, *33* (1967) 33.

[50] D.C. Brillhart, R.J. De Angelis, A.G. Preban, J.B. Cohen, and P. Gordon, *Trans. AIME*, *239* (1967) 836.

[51] C.S. Smith, *Trans. AIME*, *212* (1958) 574.

[52] H.-J. Kestenbach and M.A. Meyers, *Met. Trans. 7A* (1976) 1943.

[53] M.A. Meyers and J.R.C. Guimarães, *Mater. Sci. Eng.*, *24* (1976) 286.

[54] A.J. Cable, in *High-Velocity Impact Phenomena*, Academic Press, New York, 1970, p. 1.

[55] R.N. Orava and R.H. Wittman, *Proc. 5th Conf. High Energy Rate Fabrication*, University of Denver, Denver, Colo., 1975, p. 1.1.1.

[56] M.A. Meyers, *Scripta Met.*, *12* (1978) 21.

[57] M.A. Meyers and M.S. Carvalho, *Mater. Sci. Eng.*, *24* (1976) 131.

[58] P.S. deCarli and M.A. Meyers, in *Shock Waves and High-Strain-Rate Phenomena in Metals*, M.A. Meyers and L.E. Murr (eds.), Plenum Press, New York, 1981, p. 341.

[59] M.A. Meyers and L.E. Murr, cited in [58], p. 487.

[60] C.Y. Hsu, K.C. Hsu, L.E. Murr, and M.A. Meyers, cited in [58], p. 433.

[61] J.H. Westbrook, *Met. Trans.*, *8A* (1977) 1327.

[62] M.J. Marcinkowski, in *Treatise on Materials Science and Technology*, Vol. 5, H. Herman (ed.), Academic Press, New York, 1974, p. 181.

[63] S.T. Fong, K. Sadananda, and M.J. Marcinkowski, *Met. Trans.*, *5* (1974) 1239.

[64] M.J. Marcinkowski and R.M. Fisher, *Trans. AIME*, *233* (1965) 293.

[65] N.S. Stoloff and R.G. Davies, *Acta Met.*, *12* (1964) 473.

[66] T.L. Johnston, R.G. Davies, and N.S. Stoloff, *Phil. Mag.*, *12* (1965) 305.

[67] R.C. Boettner, N.S. Stoloff, and R.G. Davies, *Trans. AIME*, *236* (1968) 131.

[68] H. Gleiter and E. Hornbogen, *Phys. Status Solidi*, *12* (1965) 235.

[69] H. Gleiter and E. Hornbogen, *Phys. Status Solidi*, *12* (1965) 251.

[70] S.M. Copley and B.H. Kear, *Trans. TMS-AIME*, *239* (1967) 977.

[71] S.M. Copley and B.H. Kear, *Trans. TMS-AIME*, *239* (1967) 984.

[72] T. Saburi, T. Hamana, S. Nenno, and H.-r. Pak, *Jap. J. Appl. Phys.*, *16* (1977) 267.

[73] H.-r. Pak, T. Saburi, and S. Nenno, *Trans, Jap. Inst. Met.*, *18* (1977) 617.

[74] R.A. Mulford and D.P. Pope, *Acta Met.*, *21* (1973) 1375.

[75] C.S. Barrett and T.B. Massalski, *Structure of Metals*, 3rd ed., McGraw-Hill, New York, 1966, p. 193.

[76] G. Wassermann, *Z. Metallk.*, *54* (1963) 61.

[77] J.S. Kallend and G.J. Davies, *Texture*, *1* (1972) 51.

[78] R.J. Roe, *J. Appl. Phys.*, *36* (1965) 2024; *37* (1966) 2069.

[79] H.J. Bunge, *Z. Metallk.*, *56* (1965) 872.

[80] G.J. Davies, D.J. Goodwill, and J.S. Kallend, *J. Appl. Cryst.*, *4* (1971) 67.

[81] G.J. Davies, D.J. Goodwill, and J.S. Kallend, *Met. Trans.*, *3* (1972) 1627.

[82] C.S. da C. Viana, J.S. Kallend, and G.J. Davies, *Int. J. Mech. Sci.*, *21* (1979) 355.

[83] C.S. da C. Viana, G.J. Davies, and J.S. Kallend, in *Proc. 3rd Int. Conf. Mech. Behaviour Materials*, Vol. 2, Pergamon Press, Elmsford, N.Y., 1979, p. 569.

SUGGESTED READING

Thermomechanical Processing

BACKOFEN, W.A., J.J. BURKE, L.F. COFFIN, JR., N.L. REED, and V. WEISS (eds.), *Fundamentals of Deformation Processing*, Syracuse University Press, Syracuse, N.Y., 1964.

BURKE, J.J., and V. WEISS, *Advances in Deformation Processing,* Sagamore Army Materials Research Conf. Proc., Vol. 21, J.J. Burke and V. Weiss (eds.), Plenum Press, New York, 1978.

Dual Phase Steels

DAVENPORT, A.T. (ed.), *Formable and Dual-Phase Steels*, TMS-AIME, Warrendale, Pa., 1979.

Dual Phase and Cold Pressing Vanadium Steels in the Automobile Industry, VANITEC, London, 1978.

KOT, R.A., and J.W. MORRIS (eds.), *Structure and Properties of Dual Phase Steels*, TMS-AIME, Warrendale, Pa., 1979.

Radiation Damage

DIEHL, J., *Vacancies and Interstitials in Metals*, North-Holland, Amsterdam, 1970.

Proc. Radiation Induced Voids in Metals, TMS-AIME, New York, 1971.

Radiation Damage in Reactor Materials, Vol. II, IAEA, Vienna, 1969.

ROSENBAUM, H.S., *Microstructures of Irradiated Materials*, Vol. 7, in series *Treatise on Materials Science and Technology*, H. Herman (ed.), Academic Press, New York, 1975.

Shock Hardening

BLAZYNSKI, T.Z. (ed.), *Explosive Welding, Forming and Compaction*, Applied Science Publishers, Essex, England, 1983.

BURKE, J.J., and V. WEISS (eds.), *Shock Waves and the Mechanical Properties of Solids*, Syracuse University Press, Syracuse, N.Y., 1971.

KINSLOW, R. (ed.), *High Velocity Impact Phenomena*, Academic Press, New York, 1973.

Kolsky, H., *Stress Waves in Solids*, Dover, New York, 1963.

Meyers, M.A., and L.E. Murr (eds.), *Shock Waves and High-Strain-Rate Phenomena in Metals: Concepts and Applications*, Plenum Press, New York, 1981.

Rinehart, J.H., and J. Pearson, *Behavior of Metals Under Impulsive Loads*, ASM, Metals Park, Ohio, 1954.

Rohde, R.W., B.M. Butcher, J.R. Holland, and C.H. Karnes (eds.), *Metallurgical Effects at High Strain Rates*, Plenum Press, New York, 1973.

Shewmon, P.G., and V.F. Zackay (eds.), *Response of Metals to High Velocity Deformations*, Interscience, New York, 1961.

Wasley, R.J., *Stress Wave Propagation in Solids*, Marcel Dekker, New York, 1973.

Order Hardening

Marcinkowski, M.J., in *Treatise on Materials Science and Technology*, Vol. 5, H. Herman (ed.), Academic Press, New York, 1974, p. 181.

Stoloff, N.S., and R.G. Davies, *Prog. Mater. Sci.*, *13* (1966) 1.

Westbrook, J.H. (ed.), *Intermetallic Compounds*, Wiley, New York, 1967.

Westbrook, J.H. (ed.), *Mechanical Properties of Intermetallic Compounds*, Wiley, New York, 1960.

Westbrook, J.H., *Met. Trans.*, *8A* (1977) 1327.

Texture Strengthening

Barrett, C.S., and T.B. Massalski, *Structure of Metals*, 3rd ed., McGraw-Hill, New York, 1966.

Dillamore, I.L., and W.T. Roberts, *Met. Rev.*, *10* (1965) 271.

Hu, H., R.S. Cline and S.R. Goodman, in *Recrystallization, Grain Growth and Textures*, ASM, Metals Park, Ohio, 1966, p. 295.

Wasserman, G. and J. Grewen, *Texturen metallischer Werkstoffe*, 2nd ed., Springer-Verlag, Berlin, 1962.

MECHANICAL TESTING

Part IV

The linkage between mechanical properties and performance is provided by mechanical testing. A range of different mechanical tests has evolved to characterize the different mechanical responses of metals and to predict their performance as realistically as possible. The most important mechanical tests are discussed in Part IV. Tensile testing is described in Chapter 16. The interactions between machine and specimen may lead to misinterpretation of the results and are discussed in detail. Stress-relaxation testing is also described. Metal deformation modeling, or the attempts to obtain a constitutive equation representing metal deformation, is discussed, as is the electroplastic effect. Hardness testing, the most common mechanical test, is presented in Chapter 17. The prediction and improvement of the formability of sheet metals has been an area of great concern and importance. There are three important material parameters: the work-hardening rate, the strain rate sensitivity, and the through-thickness anisotropy (R factor). They are discussed in Chapter 18, together with forming-limit curves. Fracture is, under most circumstances, an undesirable feature and the resistance to it is an important material parameter. Both the traditional Charpy and drop-weight tests and the more modern plane-strain fracture toughness test are described in Chapter 19. Creep and fatigue resistance are important material properties. The substructural processes responsible for creep and fatigue are described in Chapters 20 and 21, respectively, followed by the most common testing techniques.

Chapter 16

TENSILE TESTING

‡16.1 GENERAL DESCRIPTION

‡16.1.1 Introduction

The mechanical response of a metal can be studied under a variety of regimes; they are described in the various chapters in Part IV of this book. The mechanical strength under a steadily increasing load is determined in uniaxial tensile tests, compression (upsetting) tests, bend tests, shear tests, plane-strain tensile tests, plane-strain compression (Ford) tests, torsion tests, biaxial tests. The uniaxial tensile test consists of extending a specimen whose longitudinal dimension is substantially larger than the two lateral dimensions [Fig. 16.1(a)]. The upsetting test consists of compressing a cylinder between parallel platens; the height/diameter ratio has to be lower than a critical value in order to eliminate the possibility of instability (buckling) [Fig. 16.1(b)]. After a certain amount of strain, "barreling" takes place, destroying the state of uniaxial compression. The stress analysis is presented in Section 2.5.7. The three-point bend test is one of the most common bending tests. A specimen is simply placed between two supports; a wedge advances and bends it through its middle point [Fig. 16.1(c)]. Plane-strain tests simulate the conditions encountered by a metal in, for instance, rolling. Loading is imparted in such a way as to result in zero strain along one direction. The two most common geometries are shown in Fig. 16.1(d) and (e). In the tensile mode, two grooves are made parallel to each other, on opposite sides of a plate. The width of the plate is much greater than its thickness in the reduced thickness region; hence, flow is restricted in

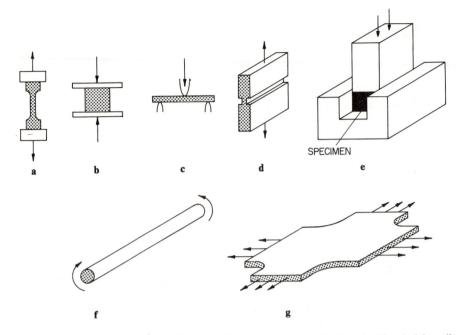

Figure 16.1 Common tests used to determine the monotonic strength of metals: (a) uniaxial tensile test; (b) upsetting test; (c) three-point bending test; (d) plane-strain tensile test; (e) plain-strain compression (Ford) test; (f) torsion test; (g) biaxial test.

the width direction. In the compressive mode (Ford test) a parallelepiped of metal is machined and inserted between the groove-and-punch setup of Fig. 16.1(e). As the top punch is lowered, the specimen is plastically deformed. Strain is restricted in one direction. In the torsion tests [Fig. 16.1(f)] the cylindrical (or tubular) specimen is subjected to a torque and undergoes an attendant angular displacement. One of the problems in the analysis of the torsion test is that the stress varies as a distance from the central axis of the specimen. The biaxial test is usually applied to thin sheets and one of the configurations is shown in Fig. 16.1(g). Other configurations are testing a tubular specimen in tension with an internal pressure or in tension with torsion. Chapter 18 (Section 18.4) provides a more detailed description of the biaxial test. The results of tests described above can be expressed graphically as stress versus strain curves. They can be compared directly by using effective stresses and effective strains. These are defined in Sections 1.10.7 and 1.11.4.

This chapter deals only with the tensile tests, since they are by far the most common monotonic strength tests for metals. In the tensile test, one applies an external load P so that the specimen is, macroscopically, in a state of uniaxial stress. Due to the anisotropy of the individual grains, the state of stress is not uniaxial at the microscopic level; stress and strain inhomogeneities establish themselves inside the individual grains. However, in the treatment given here, these localized variations are not considered. The state of stress is described by

$$\begin{pmatrix} \sigma_{11} & 0 & 0 \\ 0 & 0 & 0 \\ 0 & 0 & 0 \end{pmatrix} \qquad (16.1)$$

Applying Eqs. 1.11, one obtains the corresponding strains, in the elastic regime, for an isotropic material:

$$\begin{pmatrix} \dfrac{\sigma_{11}}{E} & 0 & 0 \\ 0 & -\dfrac{\nu\sigma_{11}}{E} & 0 \\ 0 & 0 & -\dfrac{\nu\sigma_{11}}{E} \end{pmatrix} \qquad (16.2)$$

There are no shear stresses or strains along the system of reference defined by the tensile axis. However, along any other directions these are shear stresses and strains, which can be found by transforming the reference system or applying Schmid's law.

The specific details on specimen dimensions, strain rates, gripping systems, and so on, are given in a number of ASTM standards.[1-9] This chapter restricts itself to a description of the fundamentals. Prior to conducting a tensile test, the ASTM standards should be consulted. Results of tests conducted under nonstandard conditions should not be directly compared with those of standard tests. When reporting the strength of an alloy, standard specimens and conditions should be used whenever possible.

There are essentially two types of machines used in tensile testing: (1) screw-driven machines, in which the velocity of the cross-head is constant (Fig. 1.1), and (2) servohydraulic machines, which are the most modern ones. One sometimes finds (especially in the "mining schools") hydraulic machines in which the movement of the cross head is produced by a hydraulic cylinder; they do not possess servocontrols and should be avoided if more modern machines are available. A modern servohydraulic testing machine is shown in Fig. 16.2, coupled with a computer.

There is always some degree of interaction between the machine and the specimen. Therefore, we cannot directly convert the cross-head motion velocity into deformation of the specimen without appropriate corrections. The reason for this is that the machine is also elastically deformed when the specimen is being tested. Hence, the rate at which the strain is applied to the specimen is lower than the velocity of the cross-head motion when no load is being applied. It is possible to determine the "stiffness" of the machine if we know the slope of the load-extension curve (linear portion) and the stiffness of the specimen. Figure 16.3 shows a direct load-extension curve. The linear portion has slope K_2. However, if we compute the stiffness of the specimen from its dimensions and elastic modulus, we obtain

$$K_{\text{spec}} = K_1 = \frac{EA}{L} \qquad (16.3)$$

Figure 16.2 Servohydraulic universal testing machine linked to computer. (Courtesy of MTS Systems Corp.)

where A and L are the cross-sectional area and length of the reduced gage section. The remainder of the elastic deformation is due to the machine. We have

$$\frac{1}{K_2} = \frac{1}{K_{\text{mach}}} + \frac{1}{K_{\text{spec}}} \tag{16.4}$$

If we want to subtract away the effect of the machine, we have to correct the whole curve. The arrows in Fig. 16.3 show how this correction is made. The

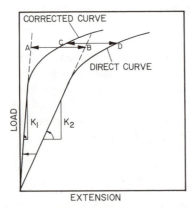

Figure 16.3 Load-elongation curves; the direct curve has to be corrected for machine stiffness.

curve is shifted to the left by an amount given by the distance between the two elastic lines K_1 and K_2 at that specified value of load. In Fig. 16.3 we have $AB = CD$. The strain rate is also affected by the machine stiffness. This correction can be very substantial. Often, the slope K_2 is only one-third of the slope K_1; this means that the amount of elastic deformation undergone by the machine is twice that of the specimen. This value might seem very high if we consider that the cross-sectional area of the specimen is much smaller than that of the working parts of the machine. However, we have to consider that the length of the machine also greatly exceeds that of the specimen. Tensile machines with slender members usually have low stiffness; they are called "soft." Robust machines, with high capacity, are "hard."

If an extensometer or strain gage is connected directly to the specimen to read the strain and provides the input for the abscissa axis in Fig. 16.3, this correction is not necessary, because the machine effects are bypassed. This effect is discussed further in Section 16.2.

‡16.1.2 Nominal (or Engineering) Stresses and Strains

The determination of the true stress, during a test, requires determination of the instantaneous cross-sectional area, because

$$\sigma_{11} = \sigma = \frac{P}{A} \tag{16.5}$$

If the deformation in the gage section of the specimen is uniform, assuming that plastic deformation takes place at constant volume (a reasonable assumption, as can be seen from Chapter 2),

$$V = V_0 = A_0 L_0 = AL \tag{16.6}$$

where L and A are the instantaneous values of the length and cross-sectional area and L_0 and A_0 are the initial values. Hence,

$$A = \frac{A_0 L_0}{L} \tag{16.7}$$

On the other hand, the longitudinal strain is defined as

$$d\epsilon = \frac{dL}{L} \tag{16.8}$$

For extended deformations, integration is required:

$$\epsilon = \int_{L_0}^{L} \frac{dL}{L} = \ln \frac{L}{L_0} \tag{16.9}$$

$$e^\epsilon = \frac{L}{L_0} \tag{16.10}$$

Substituting Eq. 16.10 into 16.7 and 16.7 into 16.5, we get

$$\sigma = \frac{P}{A_0} e^\epsilon \qquad (16.11)$$

The engineering (or nominal) stresses and strains are commonly used in tensile tests with the double objective of (1) avoiding complications in the computation of σ and ϵ, and (2) obtaining values that are more significant from an engineering point of view. Indeed, the load-bearing ability of a beam is better described by the engineering stress, referred to the initial area A_0. Hence, we have

$$\overline{\sigma} = \frac{P}{A_0} \qquad (16.12)$$

$$\overline{\epsilon} = \frac{\Delta L}{L_0} = \frac{L - L_0}{L_0} \qquad (16.13)$$

The $(-)$ is used to distinguish the true from the engineering values. It is possible to correlate engineering and true values. Hence,

$$\overline{\epsilon} = \frac{L}{L_0} - 1 \qquad \text{or} \qquad 1 + \overline{\epsilon} = \frac{L}{L_0} \qquad (16.14)$$

Substituting 16.14 and 16.12 into Eq. 16.11, we get

$$\sigma = \overline{\sigma}(1 + \overline{\epsilon}) \qquad (16.15)$$

Figure 16.4 shows engineering and true stress versus strain curves for the same hot-rolled AISI 4140 steel. In the elastic regime the coincidence is exact because strains are very small ($\sim 0.5\%$). From Eq. 16.15 we can see that we would have $\overline{\sigma} \simeq \sigma$. As plastic deformation increases, ϵ and $\overline{\epsilon}$ become progressively different. For $\epsilon = 0.20$ (a common value for metals) we have $\overline{\epsilon} = 0.221$. For this deformation the true stress is 22.1% higher than the nominal one. It can be seen that these differences become greater with increasing plastic deformation. Another *basic* difference between the two curves is the decrease in the engineering stress beyond a certain value of strain (~ 0.14 in Fig. 16.4). This phenomenon is described in detail in Section 16.1.4.

‡16.1.3 Parameters on the Tensile Curve

Figure 16.5 shows two types of stress–strain curves. The first does not exhibit a yield point, while the second does. A number of parameters are used to describe the various features of these curves. First, the elastic limit; since it is difficult to determine the maximum stress for which there is no residual deformation, the 0.2% offset yield stress (point A in Fig. 16.5) is commonly used instead; it corresponds to a residual strain of 0.2% after unloading. Actually, there is evidence of dislocation activity in a specimen at stress levels as low as 25% of the yield stress. This region (between 25 and 100% of the yield stress) is called the microyield region and has been the object of careful

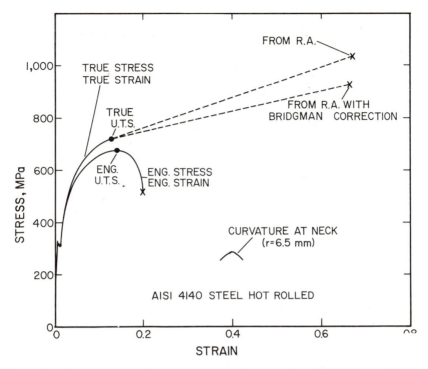

Figure 16.4 True and engineering stress versus strain curves for AISI 4140 hot-rolled steel.

investigations.[10] In case there is a yield drop, an *upper* (*B*) and a *lower* (*C*) *yield point* are defined. It should be emphasized that the lower yield point depends on the machine stiffness. This point is discussed in Section 16.2. A *proportionality limit* is also sometimes defined (*D*); it corresponds to the stress at which the curve deviates from linearity. The maximum engineering stress

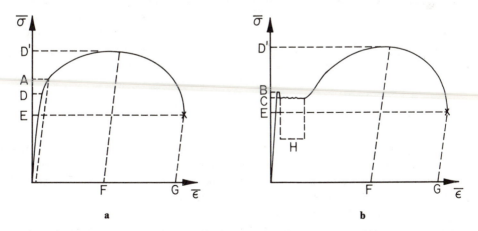

| a | b |

Figure 16.5 Engineering (or nominal) stress versus strain curves (a) without and (b) with yield point.

is called *ultimate tensile stress* (UTS); it corresponds to point D' in Fig. 16.5. Beyond the UTS, the engineering stress drops until the *rupture stress* (E) is reached. The *uniform elongation* (F) corresponds to the plastic strain that takes place uniformly in the specimen. Beyond that point, necking occurs. Necking is treated in detail in Section 16.1.4. G is the *total elongation*. Additional parameters can be obtained from the stress–strain curve: (1) the elastic energy absorbed by the specimen (area under the elastic portion of the curve) is called *resilience*; (2) the total energy absorbed by the specimen during deformation, up to fracture, the area under the whole curve, is called *toughness*. The strain rate undergone by the specimen, $\dot{\bar{\epsilon}} = d\bar{\epsilon}/dt$, is equal to the cross-head velocity divided by the initial specimen length.

The *reduction of area* is defined as

$$q = \frac{A_0 - A_f}{A_0} \tag{16.16}$$

A_0 and A_f are the initial area and cross-sectional area in the fracture region, respectively. The true strain at the fracture is defined as

$$\epsilon_f = \ln \frac{A_0}{A_f} \tag{16.17}$$

The true uniform strain is

$$\epsilon_u = \ln \frac{A_0}{A_u} \tag{16.18}$$

A_u is the cross-sectional area corresponding to the onset of necking (when the stress is equal to the UTS).

The stress–strain curves are not always smooth. They sometimes present irregularities, such as those shown in Fig. 16.6, observed in a nickel-iron base superalloy Inconel 718 tested in tension at 650°C after different processing schedules.[11] These different processing schedules affect the extent of the irregularities. They are known as the Portevin–Le Châtelier effect and are due to interactions between solute atoms and dislocations. When the dislocations move at a velocity approximately equal to the migration rate of the solute atoms, the dislocations periodically free themselves from their atmospheres, resulting in load drops. Cetlin et al.[12] studied this effect in detail.

‡16.1.4 Necking

Necking corresponds to the part of the tensile test in which instability exists. The neck is a localized region in the reduced section of the specimen in which the greatest portion of strain concentrates. The specimen "necks" down in this region. Several criteria for necking have been developed. The oldest one, due to Considère,[13] is described below. Hart's criterion is considered in Section 16.4. According to Considère, necking starts at the maximum stress (UTS), when the increase in strength of the material due to work hardening

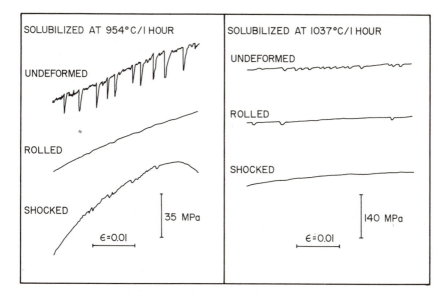

Figure 16.6 Serrated flow observed in tensile test performed at 650°C in Inconel 718 (nickel–iron-base superalloy) solubilized at two temperatures. Undeformed, cold-rolled (19.1% reduction) and shock-loaded (51 GPa peak pressure) conditions shown. (Reprinted with permission from [11], p. 206.)

is less than the decrease in the load-bearing ability due to the decrease in cross-sectional area. In other words, necking starts when the increase in stress due to the reduction in cross-sectional area starts to exceed the increase in load-bearing ability due to work hardening. We have, at the onset of necking,

$$\frac{d\overline{\sigma}}{d\overline{\epsilon}} = 0 \tag{16.19}$$

Substituting Eqs. 16.14 and 16.15 into 16.19 yields

$$\frac{d\sigma}{d\epsilon} = \sigma \tag{16.20}$$

Hence, when the slope of the true stress–true strain curve is numerically equal to the stress, necking is initiated. The Considère criterion can also be expressed in a different way, by using Hollomon's equation for the curve:

$$\sigma = K\epsilon^n \tag{16.21}$$

Substituting Eq. 16.21 into Eq. 16.20, we arrive at

$$\epsilon_u = n \tag{16.22}$$

The work-hardening coefficient is numerically equal to the true uniform strain and can be easily obtained in this way.

It is sometimes useful to present results of tensile tests under the form of $d\sigma/d\epsilon$ versus σ or $d\sigma/d\epsilon$ versus ϵ plots. In particular, Guimarães and de Angelis[14] developed a method for the analysis of stress–strain curves; they plot $(d\sigma/d\epsilon - 1)$ versus σ. An example of a log $(d\sigma/d\epsilon)$ versus log ϵ curve for AISI 302[15] stainless steel is given in Fig. 16.7. It can be seen that $d\sigma/d\epsilon$ decreases with ϵ, indicating that the necking tendency steadily increases. For metals that do not exhibit any work-hardening ability, necking should start immediately at the onset of plastic flow. Under certain conditions (predeformation at very low temperature or very high strain rate) some metals can exhibit this response, called work softening. This behavior is described in Section 9.8; Figure 9.11 shows a typical example.

The formation of the neck results in an accelerated and localized decrease in the cross-sectional area. Figure 16.4 shows how the true stress–true strain curve continues to rise after the onset of necking. It can also be seen that the true elongation at fracture is much higher than the "total elongation." The correct plotting of the true stress–true strain curve beyond UTS requires determination of the cross-sectional area in the neck region continuously after necking. This is difficult to do, and the simplest way is to obtain one single point on the plot, joining it to the point corresponding to the maximum load. For this reason a hatched line is used in Fig. 16.4. The deformation in the neck region is much higher than the one uniformly distributed in the specimen. It can be said that the neck acts as a second tensile specimen. Since its length is smaller than that of the specimen and the cross-head velocity is constant, the strain rate is necessarily higher.

The onset of necking is accompanied by the establishment of a triaxial state of stress in the neck; the uniaxial stress state is destroyed by the geometrical irregularity. Remembering the flow criteria (Chapter 2), it can be seen that the flow stress of a material is strongly dependent on the state of stress. Hence, a correction has to be introduced to convert the triaxial flow stress into an uniaxial one. If we imagine an elemental cube aligned with the tensile axis and situated in the neck region, it can be seen that it is subjected to tensile

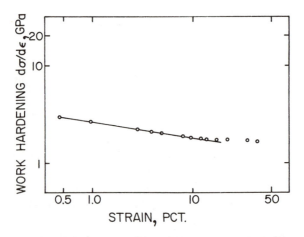

Figure 16.7 Log $d\sigma/d\epsilon$ versus log ϵ for stainless steel AISI 302. (Adapted with permission from [15], p. 420.)

stresses along three directions (the external boundaries of the neck generate the tensile components perpendicular to the axis of the specimen). The magnitude of the transverse tensile stresses depends on the geometry of the neck on the material, the shape of the specimen, the strain-rate sensitivity of the material, the temperature, the pressure, and so on. Bridgman[16] introduced a correction from a stress analysis in the neck. His analysis applies to cylindrical specimens. The following equation expresses the corrected stress:

$$\sigma = \frac{\sigma_{av}}{(1 + 2R/r_n)\, \ln\,(1 + r_n/2R)}$$

where R is the radius of curvature of the neck and r_n is the radius of the cross section in the thinnest part of the neck. Thus, one has to continuously monitor the changes in R and r_n during the test to perform the correction. McGregor Tegart[17] presents a plot in which the corrections have already been computed as a function of strain beyond necking. There are three curves, for copper, steel, and aluminum. The correction factor can be read directly from the plot shown in Fig. 16.8; ϵ_u is the true uniform strain (strain at onset of neck). In Fig. 16.4, the true stress–true strain curve that was corrected for necking by the Bridgman technique lies slightly below the one determined strictly from the reduction of area at fracture and the load at the breaking point. This is consistent with Fig. 16.8; σ is always lower than σ_{av}.

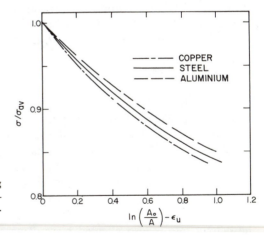

Figure 16.8 Correction factor for necking as a function of (strain in neck, $\ln(A_0/A)$— strain at necking, ϵ_u). (Adapted with permission from [17], p. 22.)

For flat specimens (sheets) the stress analysis was conducted by Aronofsky.[18] The ratio between the width and thickness is an important parameter.

It should be noted that necking is a characteristic of tensile stresses. Compressive stresses are not characterized by necking. Barreling is the corresponding deviation from uniaxiality in compressive tests. Hence, metals will only exhibit necking during deformation processing if the state of stress is conducive to it (tensile). Figure 16.9 shows eloquently how the work-hardening capacity of a metal exceeds greatly the one in an individual tensile test.[19] Wire was drawn

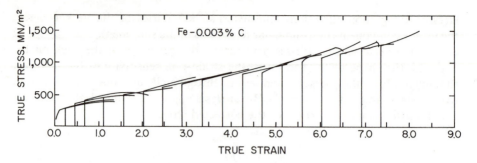

Figure 16.9 Stress–strain curves for Fe–0.003% C alloy wire deformed to increasing strains by drawing; each curve is started at the strain corresponding to the prior wire-drawing reduction. (Courtesy of H. J. Rack, Exxon Corp.)

to different strains (see Chapter 2): wire drawing consists of pulling it through a conical die; at each pass there is a reduction in cross section. Tensile tests were conducted after different degrees of straining by (0 to 7.4) wire drawing; it can be seen that the wire work hardens at each step. However, the individual tensile tests are interrupted by necking and fracture. In wire drawing, necking and fracture are inhibited by the state of stress in the deformation zone (compressive). The individual true stress–true strain curves were corrected for necking by Bridgman's technique; in each case, the individual curve fits fairly well into the overall work-hardening curve. It can be concluded that the individual tensile test gives only a very limited picture of the overall work-hardening response of a metal; for the wire in Fig. 16.9, the total strain can exceed 7.4.

‡16.1.5 Strain Rate

The stress–strain curves are, for the majority of metals, sensitive to the strain rate $\dot{\epsilon}$. Figure 16.10 shows the range of strain rates[20] for the different types of tests. The lowest range of strain rates corresponds to creep and stress-relaxation tests. The tensile tests are usually conducted in the range 10^{-4} s^{-1} < $\dot{\epsilon}$ < 10^{-3} s^{-1}. At strain rates of the order of 10^2 s^{-1} inertial and wave-propagation effects start to become important. The highest range of strain rates corresponds to the passage of a shock wave through the metal.

More often than not, the flow stress increases with strain rate; the work-hardening rate is also affected by it. Two parameters are defined to describe these effects:

$$m = \frac{\partial \ln \sigma}{\partial \ln \dot{\epsilon}} \bigg|_{\epsilon, T} \tag{16.23}$$

$$s = \frac{\partial \ln w}{\partial \ln \dot{\epsilon}} \bigg|_{\epsilon, T} \qquad \text{where } w = \frac{d\sigma}{d\epsilon} \bigg|_{\epsilon, T} \tag{16.24}$$

m is known as the strain rate sensitivity. Equations 16.23 and 16.24 can also be expressed as

STRAIN RATE	COMMON TESTING METHODS	DYNAMIC CONSIDERATIONS	
10^7	HIGH VELOCITY IMPACT —Explosives	SHOCK-WAVE PROPAGATION	INERTIAL FORCES IMPORTANT
10^6	—Pulsed laser	(See Section 15.3)	
10^5	—Projectile impact		
10^4	IMPACT Hopkinson Bar (plane stress)	PLASTIC WAVE PROPAGATION SHEAR WAVE PROPAGATION	
10^3	Inclined parallel impact (shear)		
10^2	DYNAMIC High-velocity hydraulic or pneumatic	MECHANICAL RESONANCE IN SPECIMEN AND MACHINE IS IMPORTANT	
10^1	machines; cam plastometer		
10^0	QUASI-STATIC Hydraulic, servo-hydraulic or screw-	TESTS WITH CONSTANT CROSS-HEAD VELOCITY STRESS THE SAME THROUGH-	INERTIAL FORCES NEGLIGIBLE
10^{-1}	driven testing machines	OUT LENGTH OF SPECIMEN	
10^{-2}			
10^{-3}			
10^{-4}			
10^{-5}			
10^{-6}	CREEP AND STRESS-RELAXATION —Conventional testing machines	VISCO-PLASTIC RESPONSE OF METALS	
10^{-7}	—Creep testers		
10^{-8}			
10^{-9}			

Figure 16.10 Range of strain rates in different tests of metals, experimental techniques, and phenomena involved.

$$\sigma = K\dot{\epsilon}^m$$

$$\frac{d\sigma}{d\epsilon} = K'\dot{\epsilon}^s \tag{16.25}$$

Materials can be tested in a wide range of strain rates; however, standardized tensile tests require well-characterized strain rates that do not exceed a critical value. High strain-rate tests are often used to obtain information on the performance of material under dynamic impact conditions. The cam plastometer is one of the instruments used. In certain industrial applications metals are also

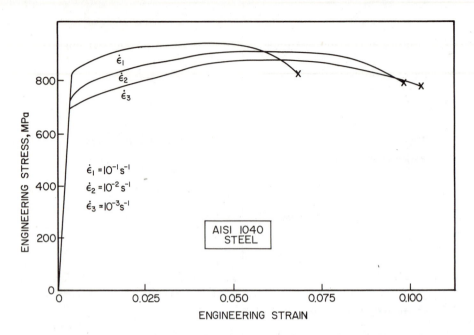

Figure 16.11 Effect of strain rate on the stress–strain curves for AISI 1040 steel.

deformed at high strain rates. Modern rolling mills, such as the Morgan mills, generate bar velocities of 180 km/h; the attendant strain rates are extremely high. In wire drawing the situation is similar.

Figure 16.11 shows the effect of different strain rates on the tensile response of AISI 1040 steel. The yield stress and flow stresses at different values of strain increase with strain rate. The work-hardening rate, on the other hand, is not as sensitive to strain rate. This illustrates the importance of correctly specifying the strain rate when giving the yield stress of a metal. Not all metals exhibit a high strain rate sensitivity. Aluminum and some of its alloys have either zero or negative m. In general, m varies between 0.02 and 0.2, for homologous temperatures between 0 and 0.9 (90% of melting point in K). Hence one would have, at the most, an increase of 15% in the yield stress by doubling the strain rate.

It is possible to determine m from tensile tests by changing the strain rate suddenly and by measuring the instantaneous change in stress. This technique is illustrated in Fig. 16.12(a). Applying Eq. 16.25 to two strain rates and eliminating K, we have

$$m = \frac{\ln (\sigma_2/\sigma_1)}{\ln (\dot{\epsilon}_2/\dot{\epsilon}_1)} \tag{16.26}$$

The reader can easily obtain m from the strain-rate changes in Fig. 16.12(a).

Some alloys have a peculiar plastic behavior and are called *superplastic*. When necking starts, the deformation concentrates itself at the neck. Since the velocity of deformation is constant and the effective length of the specimen

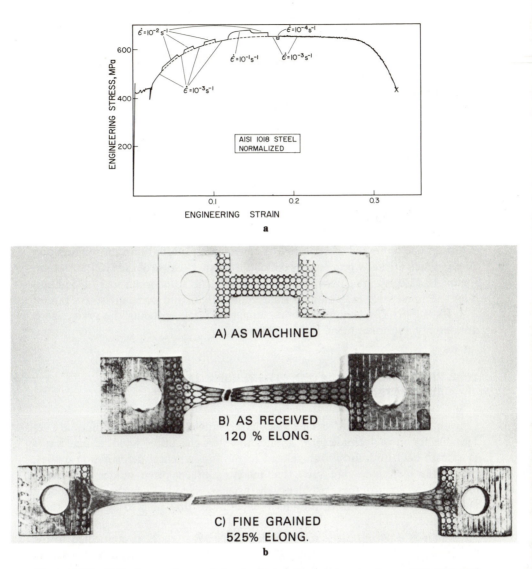

Figure 16.12 (a) Strain-rate changes during tensile test. Four strain rates shown: 10^{-1}, 10^{-2}, 10^{-3}, and 10^{-4} s^{-1}. (b) Superplastic behavior in 7475 Al alloy tensile test specimen (A: prior to test; B: large-grained alloy; C: fine-grained (D- 10 μm alloy). (Reprinted with permission from [23], p. 25.)

is reduced during necking, the strain rate increases ($\dot{\epsilon} = v/L$). If a material exhibits a positive strain-rate sensitivity, the flow stress in the neck region will increase due to the increased strain rate; hence, necking is inhibited. This topic is treated in greater detail in Section 16.4. This is what takes place in superplastic alloys. They can undergo plastic strains of up to 1000%. A 78% Zn–22% Al alloy was initially tested by Backofen et al.[21] in 1964; it was shown to be superplastic. Formability tests demonstrated the outstanding potential of this alloy. In the 1970s superplastic forming evolved in the United States from a laboratory curiosity to a practical and economical[22] production technique. Superplastic alloys owe their ductility to a very fine grain size; a great portion of the deformation takes place by sliding along grain boundaries. Additional information on the deformation mechanisms is given in Chapter 20. An extreme case of superplasticity is that exhibited by molten glass. A Newtonian viscous fluid, obeying the law $F = Bv$ or $\sigma = B\dot{\epsilon}$, has a coefficient $m = 1$ (see Eq. 16.25). Viscous glass can be stretched until very fine fibers are formed (fiberglass) without necking.

Superplasticity has been obtained in a number of alloy systems: titanium alloys (Ti–6% Al–4% V), iron-base alloys, aluminum alloys. High-strength nickel-base superalloys have been found to exhibit superplastic behavior and the process of "GATORIZING" (supposedly named after an alligator living in the lake in front of the research institute) has been developed by Pratt and Whitney. The potential of superplastic forming is especially bright for titanium alloys, which are known to be very difficult to form, because of the HCP structure. Figure 16.12(b) shows a conventional grain-size and a fine-grain size 7475 aluminum specimen after extension to fracture at 516°C and constant strain rate of $2 \times 10^{-4} \text{s}^{-1}$.[23] The change in total elongation upon reducing the grain size is dramatic; it increases from 120 to 525 pct.

‡16.2 PROBLEMS ASSOCIATED WITH TENSILE TESTING; AUTOMATED TENSILE TESTING

Conventional tensile tests using screw-driven universal testing machines are acceptable for most technological applications; on the other hand, if one wants to extract delicate information related to the fundamental deformation modes, the testing techniques are such that machine effects often render the results meaningless. Some of these problems are discussed below. Then the more modern machines will be discussed; they have capabilities that help to overcome these problems.

This is especially true if no strain-gage extensometer is connected to the specimen; in this case, one uses the velocity of the cross-head and assumes that it remains constant during testing. In effect, the elastic deformation of the machine during the loading of the specimen reduces the velocity of the cross-head. If a tensile specimen with dimensions A and L and Young's modulus E is being pulled by a tensile testing machine with stiffness K_{mach}, the velocity of the cross-head during the test, v_2, is related to the initial cross-head velocity v_1, by

$$\frac{v_1}{v_2} = 1 + \frac{K_2}{K_{mach}}$$

AE/L is the stiffness of the specimen. The initial engineering strain rate being equal to the ratio v_2/L, one can see that one may overestimate the strain rate very substantially if no correction is made for machine stiffness. If a strain-gage extensometer is used, on the other hand, this problem would not arise.

A second problem with machines having a constant cross-head velocity is that the true strain rate is not constant throughout the test. Since

$$v = \frac{dL}{dt}$$

and

$$\dot{\epsilon} = \frac{d\epsilon}{dt}$$

one has

$$\dot{\epsilon} = \frac{v}{L}$$

$\dot{\epsilon}_0$ and L_0 are the initial strain rate and specimen length, respectively. Figure 16.13 presents a plot of the $\dot{\epsilon}/\dot{\epsilon}_0$ ratio as a function of engineering strain. At small strains the deviation is not large, but at a strain of 0.8, the strain rate is reduced to half of its initial value.

A third problem is associated with the interpretation of the load drops observed during deformation. These load drops occur when the rate of the factor that caused them is much higher than the rate of extension of the specimen at conventional strain rates ($\sim 10^{-4}$ s^{-1}). Examples are yield-point drops due to Lüders band formation, serrations on the curve due to deformation twinning (see Section 7.4.1), stress-assisted martensitic transformations (see Section 13.5), and the Portevin–Le Châtelier effect. The analysis below shows how the load drop can be converted into a more fundamental parameter, an "instantaneous strain." During the loading cycle, the slope of the load-extension curve reflects the deformation of *both* specimen and machine. If we consider the machine–

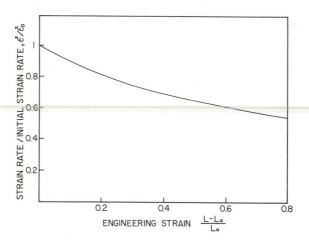

Figure 16.13 Change in true strain rate as specimen is deformed in tension at constant cross-head velocity.

specimen system as two springs associated in series, the slope of the load-extension curve in the elastic range can be expressed as

$$\frac{1}{S} = \frac{1}{K_{mach}} + \frac{L}{AE} \tag{16.27}$$

where K_{mach} is the machine stiffness and L, A, and E are the specimen length, cross-sectional area, and elastic modulus. We can express the slope in terms of changes in load ΔP and length ΔL:

$$\frac{\Delta L}{\Delta P} = \frac{1}{K_{mach}} + \frac{L}{AE} \tag{16.28}$$

If at a certain load level the specimen is unloaded, the unloading proceeds elastically with a slope equal in magnitude to the elastic loading line. If there is an instantaneous change in length of the specimen, the drop in load will be expressed by Eq. 16.28, with the appropriate change in sign. Upon unloading, the machine stiffness is unchanged:

$$\Delta L = -\Delta P \left(\frac{1}{K_{mach}} + \frac{L}{AE} \right) \tag{16.29}$$

The "instantaneous strain" can be expressed as

$$\epsilon_i = \frac{\Delta L}{L} = -\frac{\Delta P}{L} \left(\frac{1}{K_{mach}} + \frac{L}{AE} \right) \tag{16.30}$$

Let us assume that one Lüders band formed in an Fe-C specimen after the yield drop. This Lüders band generates a well-defined strain ϵ_i. A "soft" machine (K low) will show a small load drop ΔP, whereas a "hard" machine will show a larger drop ΔP. This effect is illustrated schematically in Fig. 16.14.

Gillis and Medrano[24] reported that the stiffness K of conventional screw-driven testing machines varied erratically. However, experiments conducted by Fortes and Proença[25] and Guimarães and Chawla[26] do not corroborate this hypothesis. Meyers et al.[27] showed that the technique used by Gillis and Medrano[24] to determine the machine stiffness was faulty. Figure 16.15 shows results obtained by Guimarães and Chawla to determine the stiffness of an Instron TT-DM machine. AISI 1020 steel and aluminum specimens were used and one parameter was systematily varied in each of them: the gage length L_0 for aluminum and the cross-sectional area for steel. The stiffness K of the machine is obtained by the equation

$$(\tan \theta_0)^{-1} = \frac{1}{K_{mach}} + \frac{1}{E} \frac{L_0}{A_0}$$

where θ_0 is the slope of the elastic range of the load-displacement plot. The intersection of the two lines with the ordinate axis should provide the stiffness. The fact that both lines are straight is an indication that the stiffness of the machine is reasonably constant. The difference between the stiffnesses obtained

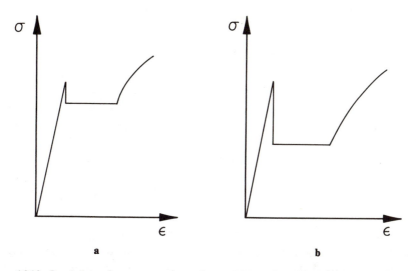

Figure 16.14 Load drops for same specimen due to Lüders band formation in (a) soft and (b) hard machines.

for the two metals is probably due to the slightly different gripping arrangements. The stiffness is around 35 MN/m. This value is consistent with the one obtained by Rohde and Nordstrom[28] for a machine of the same type by a completely different technique: 17.3 MN/m. One simpler way of obtaining the stiffness of a machine is to use a material with known elastic modulus and high yield stress and apply Eq. 16.4. It is possible to make a "hard" machine "soft" by attaching a low-stiffness spring in series with the specimen. One of the ways of making a "soft" machine "hard" is to use the experimental setup shown in Fig. 16.16. The jacket around the specimen has a higher stiffness than the specimen and should be completely characterized before the test.

Another source of problems in tensile testing is the alignment of the specimen with the grips. If the alignment is not good, the stress distribution along

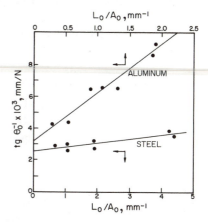

Figure 16.15 Determination of machine stiffness using steel specimens with different cross-sectional areas (but same length) and aluminum specimens with different lengths (but same cross-sectional areas.) (Adapted with permission from [26], p. 550.)

Figure 16.16 Specimen design for producing a hard machine test on a nominally soft machine. If the load on the outer cylinder is large compared with that on the specimen, the machine stiffness K_{tm} will be essentially that of the specimen itself. (Reprinted with permission from [29], p. 106.)

the cross section will not be uniform. This is very critical in establishing the yield stress; the plastic range of the curve is not as sensitive. Avery and Findley[29] present equations that give the stress variation as a function of misalignment for cylindrical and square cross sections. For precision yield stress determinations, self-aligning grips are recommended. For high-temperature tests a furnace is usually used, surrounding the grip system. In the temperature range 77–500 K, a simpler setup can be constructed, and it is a common practice to use fluids, such as liquid nitrogen, dry ice–acetone mixtures, brine, and heated oils. Figure 16.17 shows an inexpensive setup (for screw-driven Instron) in which the specimen is surrounded by the fluid.

The newer generation of servohydraulic testing machines presents significant advantages over the older screw-driven machines.[30] They lend themselves much better to interfacing with a computer. Figure 16.2 shows a modern automated testing system. It consists of three components: the testing machine,

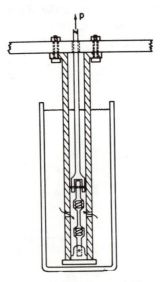

Figure 16.17 Coaxial tube test fixture for environment and thermal testing. (Reprinted with permission from [29], p. 120.)

the interface, and the computer. The testing machine is of the servohydraulic type. The moving piston is driven by a double-action hydraulic cylinder, so it can operate under tension and compression. The fluid pressure in the chamber is controlled by a servovalve. This servovalve responds to the difference between the measured signal and the desired signal. The signal is amplified to drive the valve so as to remove the error. There are three main modes of operating the machine, commonly referred to as stroke, strain, and load. Under the stroke mode, the movement of the piston is the controlling variable; under the load mode, it is the load acting on the specimen; and under the strain mode, it is the strain, as read from a strain-gage extensometer. For each of these modes, different time functions can be established. For a monotonic tensile test, one may choose the constant velocity by setting a ramp function under the strain or stroke mode. For fatigue, one may choose a sine or haversine function. The corrections between command and feedback are conducted in an analog way. The computer, shown on the left-hand side of Fig. 16.18, operates digitally. Hence, there is a need for a analog–digital converter. Usually a mini- or micro-computer is sufficient for the processor. There is a very large range of possible terminals available commercially to fit various needs. The programming capability gives a great degree of control of the "calculated" variable. Additionally, the results can be processed in the computer and displayed graphically as desired. The number of tests that use the calculated-variable control technique has been steadily expanding. As the control capability of the testing machines improves, new tests are developed. Table 16.1 shows some common calculated-variable tests. With the objective of better and better predicting the performance of

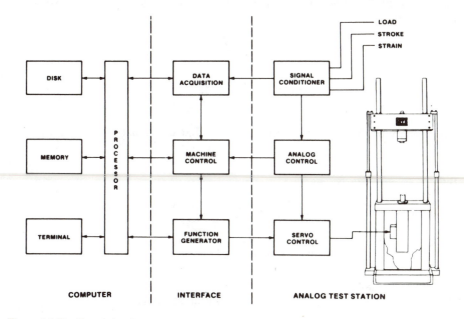

Figure 16.18 Conventional automated test system consisting of an analog machine, an interface, and a digital computer. (Reprinted with permission from [30], p. 18.)

TABLE 16.1   Common Calculated-Variable Tests[a]

Type	Test	Control Variable
Monotonic	Constant true strain rate	$\epsilon_{true} = \ln(\epsilon_{eng} + 1)$
	Yield surface probe	Yield condition $= f$(stress, strain)
Fracture mechanics	Constant crack-growth rate	$\Delta K = f$(stress, crack length) $=$ constant
	Crack-growth threshold test	$\Delta K = f$(stress, crack length), ΔK decreasing
Fatigue	Plastic strain limit control	$\epsilon_{plastic} = \epsilon_{total} - \sigma/E$
	Transition fatigue life	$\epsilon_{plastic} = \epsilon_{elastic}$
	Thermomechanical fatigue	$\epsilon_{mechanical} = \epsilon_{total} - \epsilon_{thermal}$

[a] Reprinted with permission from [30], p. 19.

material, increasingly sophisticated tests are conducted. One of them is discussed here as an example. A constant true strain-rate test is very difficult to perform with the older tensile testing machines; this was discussed earlier in this section. Hartley and Jenkins[31] describe a machine built at the University of Florida in which the cross-sectional area is continuously monitored after the onset of necking. This is accomplished by an optical scanning device; the specimen is illuminated from one side and its silhouette is projected into a flat surface. A photocell device determines the minimum cross-sectional area. Before the onset of necking the deformation is homogeneous and there is no need to monitor the cross-sectional area. The true plastic strain and strain rate can be expressed as

$$\epsilon = \ln \frac{L}{L_0}$$

$$\dot{\epsilon} = \frac{\dot{L}}{L} = -\frac{\dot{A}}{A}$$

The velocity of the cross-head must be such that $\dot{\epsilon}$ is constant.

$$v = \dot{L} = \dot{\epsilon} L$$

But

$$L = L_0 \exp(\epsilon) = L_0 \exp(\dot{\epsilon} t)$$

Hence,

$$v = \dot{\epsilon} L_0 \exp(\dot{\epsilon} t)$$

But

$$\dot{\epsilon} L_0 = v_0 \qquad \text{(the initial velocity)}$$

$$v = v_0 \exp(\dot{\epsilon} t) \tag{16.31}$$

If Eq. 16.31 provides the command for the machine, the velocity will continuously increase with time in order to provide a constant strain rate. Beyond necking, the cross-sectional area is continuously monitored, and the area has to vary as a function of time in such a way that

$$A = A_0 \exp(-\dot{\epsilon}t)$$

16.3 METAL DEFORMATION MODELING
(MECHANICAL EQUATION OF STATE)

The concept of metal deformation modeling—or mechanical equation of state—is an important one. Investigators have been attempting to establish an equation of state for mechanical deformation ever since Zener and Hollomon[32] suggested its possible existence. In a thermodynamic system one has, in equilibrium,

$$f(P, V, T) = 0 \qquad (16.32)$$

P, V, and T are state variables and define a thermodynamic state. Hence, if we know P, V, and T, we can determine all other state variables. We can also predict the effect of the change of one variable on the other ones, and we have

$$dP = \left(\frac{\partial P}{\partial V}\right)_T dV + \left(\frac{\partial P}{\partial T}\right)_V dT \qquad (16.33)$$

It is an attractive idea to establish that a certain number of parameters determine deformation and to correlate them in a similar manner, creating a "mechanical equation of state." There have been several attempts, some of which are described by Argon.[33] Essentially, they are based on this or similar functions:

$$f(\sigma, \epsilon, \dot{\epsilon}, T, \ldots) = 0 \qquad (16.34)$$

Other terms can be added to the function. One has to recognize, at the outset, that plastic deformation is an *irreversible* process and that no state is an equilibrium state. The variables ϵ and $\dot{\epsilon}$ do not describe a state such as P, V, T describe for ideal gases; they are part of what is called "metallurgical history." Hence, thermodynamic rigorousness does not exist, and the different models are attempts to correlate the foregoing parameters. The name "mechanical equation of state" is therefore not appropriate; "metal deformation model" is better. We have, from Eq. 16.34,

$$d\sigma = \left(\frac{\partial \sigma}{\partial \epsilon}\right)_{\dot{\epsilon}, T} d\epsilon + \left(\frac{\partial \sigma}{\partial \dot{\epsilon}}\right)_{\epsilon, T} d\dot{\epsilon} + \left(\frac{\partial \sigma}{\partial T}\right)_{\epsilon, \dot{\epsilon}} dT. \qquad (16.35)$$

The "metallurgical history" of the metal, so important in determining its deformation response, is not included directly in the expression above, because of the obvious complexity. Different deformation modes (e.g., twinning versus dislocation climb versus slip) operate at different temperatures and strain rates. Similarly, the substructure configuration changes with plastic strain. Hence, it is very difficult to assume a constant substructure and substructure behavior, as implied in Eq. 16.35, or to describe it by a few simple parameters. Nevertheless, metal deformation modeling has the virtue of attempting to rationalize the plastic deformation of metals. In Eq. 16.35, $(\partial\sigma/d\epsilon)_{\dot{\epsilon},T}$ is obtained from a constant true strain-rate test. The difficulty of conducting these tests with conventional testing machines is described in Section 16.2. The second term, $(d\sigma/d\dot{\epsilon})_{\epsilon,T}$, is easily obtained by means of stress-relaxation tests, as described in Section 16.5. The tensile test is stopped at a certain strain and the drop of stress is monitored. By plotting the resultant stress decrease one can find this parameter. The third term, $(d\sigma/dT)_{\epsilon,\dot{\epsilon}}$, is obtained by means of "instantaneous" temperature changes during a tensile test. Hart and coworkers[34-36] have proposed the following form of an equation of state:

$$f(\ln\sigma,\ln\dot{\epsilon}_p,\epsilon_p)=0$$

$$d\ln\sigma=\left(\frac{\partial\ln\sigma}{\partial\ln\dot{\epsilon}_p}\right)_{\epsilon_p}d\ln\dot{\epsilon}_p+\left(\frac{\partial\ln\sigma}{\partial\epsilon_p}\right)_{\dot{\epsilon}_p}d\epsilon_p$$

A term describing, in an approximate way, the metallurgical history of the metal was named "hardness." This "hardness state" should describe the amount of deformation previously received by the metal and the conditions under which this state was achieved. It should not be confused with the common metallurgical hardness test. Rohde and co-workers[37] have proposed another equation of state using another evolutionary variable that describes the "metallurgical history" of the metal; they considered Hart's hardness to lack generality. Both Hart's[34,36] and Rohde's[37] models rely heavily on stress-relaxation tests, in which the substructure is assumed to be constant; Meyers et al.[27] have shown that this hypothesis must be made only after careful experimentation. The presence of solute atoms often results in changes in the mobile dislocation density with time. Section 16.5 describes this matter in greater detail. Yet another attempt at metal deformation modeling is proposed by Ghosh;[38] he does not arrive at a "single metallurgical equation of state" per se, but at a number of constitutive relations that describe reasonably well the constant true strain-rate tensile test, the strain-rate change test, stress-relaxation test, creep and load-dip tests, unloading–holding–reloading test, and cyclic loading test.

In summary, it can be said that the ultimate goal of metal deformation modeling is to *predict* the response of a metal; data obtained under one set of deformation conditions, such as monotonic tensile test, could be used to predict the mechanical response under another set, such as cyclic deformation response (fatigue). The great difficulty stems from the complex nature of deformation modes and their interdependency. It is doubtful that a "universal equation of state" will be ever devised; rather, constitutive equations with limited applications such as the Larson–Miller and Manson–Haferd equations for creep (see Chapter 20) and Hart's and Rohde's formulations will be developed.

16.4 THEORETICAL CONSIDERATIONS

The theoretical treatment of the tensile test involves essentially three regions: the elastic regime, the uniform plastic regime, and the unstable deformation after the onset of necking. The former two are treated in different parts of the book. The elastic limit (micro- and macroyield stress) is determined by the generation and movement of dislocations, in general; Chapters 1 and 2 treat this topic from a mechanistic point of view, while Chapter 9 treats it from a dislocation angle. The shape of the plastic curve has been described by several empirical equations, such as Ludwik's and Hollomon's. The underlying deformation modes (work hardening, etc.) have been correlated to the actual shape. Only the onset of instability will be discussed here. It is of great importance, because plastic forming of sheets is highly dependent on their uniform elongation. The understanding of necking in biaxial loading rests, in its turn, on its understanding under uniaxial loading; formability testing is discussed in Chapter 18. The theories that have been developed on plastic instability seek to establish its onset and progression from the characteristics of the material.

The first attempt was due to Considère, and is discussed in Section 16.1. He observed that necking began at the point of maximum load and derived his criterion ($d\sigma/d\epsilon = \sigma$) and geometrical construction therefrom. Considère's criterion does not take into account the strain-rate sensitivity of the material. The latter parameter is very important—as important as the strain-hardening exponent, n, which determines $d\sigma/d\epsilon$ as a function of strain. In 1967, Hart[39] proposed a theory taking into account the strain-rate sensitivity; and, in rapid succession, a number of other criteria followed.[40-46] Hart's development will be presented here, in detail, while the other criteria will be discussed more briefly. The starting point is the equation of state (Eq. 16.35). Hart did not assume that the partial derivatives were integrable, and designated them simply by material parameters. However, they will be considered as partial derivatives here, for the sake of simplifying the interpretation of the results. Hart's postulate was that instability started when the difference in the areas between two cross sections increased with time as the test progressed. Figure 16.19 shows the portion of a tensile specimen being deformed. Hence, the difference $\delta A = A_2 - A_1$ (δ refers to fluctuations between sections) should decrease or stay constant with time. In other words, the variation in the area increment rate $\dot{A}$ as a function of the variation of cross-sectional area A should be

$$\frac{\delta \dot{A}}{\delta A} \leq 0$$

if deformation is stable. If necking starts, on the other hand, the differences in cross-sectional area increase with time.

The true stress is given by

$$\sigma = \frac{P}{A} \qquad \text{or} \qquad P = \sigma A$$

$$\delta P = 0 = \sigma \delta A + A \delta \sigma$$

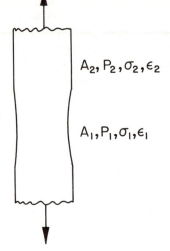

Figure 16.19 Section of specimen being deformed in tension.

This is so because the load P is the same throughout the whole specimen; in Fig. 16.19, $P_2 = P_1$. Applying the equation of state to the differences in stress between two sections (Eq. 16.35), we have

$$\delta\sigma = \left(\frac{\partial\sigma}{\partial\epsilon}\right)_{\dot\epsilon} \delta\epsilon + \left(\frac{\partial\sigma}{\partial\dot\epsilon}\right)_{\epsilon} \delta\dot\epsilon \qquad (16.36)$$

The strain is

$$\delta\epsilon = \frac{\delta L}{L} = -\frac{\delta A}{A} \qquad (16.37)$$

because of the constancy-of-volume condition. Taking the time derivative of Eq. 16.37, we arrive at

$$\delta\dot\epsilon = \frac{\dot A\,\delta A}{A^2} - \frac{\delta\dot A}{A} \qquad (16.38)$$

Substituting Eqs. 16.37 and 16.38 into Eq. 16.36 yields

$$\delta\sigma = \left(\frac{\partial\sigma}{\partial\epsilon}\right)_{\dot\epsilon}\left(-\frac{\delta A}{A}\right) + \left(\frac{\partial\sigma}{\partial\dot\epsilon}\right)_{\epsilon}\left(\frac{\dot A}{A}\frac{\delta A}{A} - \frac{\delta\dot A}{A}\right)$$

$$\frac{\delta\sigma}{\sigma} = \frac{1}{\sigma}\left(\frac{\partial\sigma}{\partial\epsilon}\right)_{\dot\epsilon}(-\delta \ln A) + \frac{1}{\sigma}\left(\frac{\partial\sigma}{\partial\dot\epsilon}\right)_{\epsilon}\left(\frac{\dot A}{A}\,\delta \ln A - \frac{\dot A}{A}\,\delta \ln \dot A\right)$$

But

$$-\frac{\dot A}{A} = \frac{\dot L}{L} = \dot\epsilon$$

Hence,

$$\frac{\delta\sigma}{\sigma} = \frac{1}{\sigma}\left(\frac{\partial\sigma}{\partial\epsilon}\right)_{\dot\epsilon}(-\delta \ln A) + \frac{\dot\epsilon}{\sigma}\left(\frac{\partial\sigma}{\partial\dot\epsilon}\right)_{\epsilon}(-\delta \ln A + \delta \ln \dot A) \qquad (16.39)$$

Defining, we have

$$\frac{1}{\sigma}\left(\frac{\partial\sigma}{\partial\epsilon}\right)_{\dot\epsilon} \equiv \gamma$$

$$\qquad (16.40)$$

$$\frac{\dot\epsilon}{\sigma}\left(\frac{\partial\sigma}{\partial\dot\epsilon}\right)_{\epsilon} \equiv m = \frac{\partial \ln \sigma}{\partial \ln \dot\epsilon}$$

where m is actually the strain-rate sensitivity, defined earlier (Eq. 16.23). From Eq. 16.39,

$$\frac{\delta\sigma}{\sigma} = -\frac{\delta A}{A} = -\delta \ln A = -\gamma\delta \ln A + m(-\delta \ln A + \delta \ln \dot A)$$

$$\delta \ln A(1 - \gamma - m) = -m\,\delta \ln \dot A \qquad (16.41)$$

$$\frac{\delta \ln \dot A}{\delta \ln A} = -\frac{1 - \gamma - m}{m}$$

The criterion for stability defined by Hart can be expressed in terms of the logarithm:

$$\frac{\delta \ln \dot A}{\delta \ln A} \geqslant 0$$

Hence,

$$-(1 - \gamma - m) \geqslant 0$$

or

$$\gamma + m \geqslant 1 \qquad (16.42)$$

This criterion is consistent with Considère's criterion if the material does not exhibit any strain-rate sensitivity ($m = 0$). In that case,

$$\gamma = 1$$

or

$$\frac{d\sigma}{d\epsilon} = \sigma$$

establishes the onset of necking; this equation is identical to Eq. 16.20. On the other end of the spectrum, Newtonian viscous flow is characterized by an absence of work hardening ($\gamma = 0$) and by a force proportional to velocity. This translates itself into a stress proportional to the strain rate, or

$$\frac{d\sigma}{d\dot\epsilon} = \frac{\sigma}{\dot\epsilon}$$

From Eq. 16.42, we have

$$m = 1$$

Hence, Hart's criterion is consistent with Considère's criterion, on the one side, and with the behavior of Newtonian viscous materials, on the other. For hot glass this criterion is obeyed, and $m = 1$; hence, it can be pulled up to very large strains. The physical significance of Hart's criterion is the following. As a material exhibits a more and more positive strain-rate sensitivity, the slope $d\sigma/d\epsilon$ at which necking is initiated is decreased from its lower bound σ to $\sigma(1 - m)$. This corresponds to a retardation of necking, as expected. The strain rate in the necking region is higher than the overall strain rate, and a positive strain-rate sensitivity (exhibited in a marked way by superplastic alloys) retards the onset of necking.

The additional criteria developed by other investigators are briefly described next. Campbell[40] based his criterion on the development of strain gradients in the tensile specimen. Argon[41] extended Hart's treatment, incorporating the effects of adiabatic heating (produced by plastic deformation) into the development of a neck. Jonas et al.[42] treated the instability as a nonuniformity in strain rather than in the cross-sectional area. They arrived at the following expression for the onset of necking:

$$\gamma = 1$$

This is identical to Considère's criterion. Demeri and Conrad[43] assumed that plastic flow became unstable when a local nonuniformity grew in length with deformation. It can be shown that it is expressed by

$$\gamma - m = 1$$

Ghosh[44] proposed the following criterion for stable deformation:

$$\gamma \geq \frac{1 - m}{1 - \dfrac{\delta \ln A_0}{\delta \ln A}}$$

He considers two necks: one "diffuse" or long-wavelength neck, and one localized or short-wavelength neck. The first one does not significantly disturb the state of uniaxial stress, while the second one introduces a triaxial state. Figure 16.20 shows schematically these two types of neck. Other criteria were proposed by Jones and Gillis[45] and Jaliner et al.[46]

16.5 STRESS-RELAXATION TESTING

Stress relaxation can be defined as the decrease in stress in a solid under a constant (or close to) strain, at a constant temperature. This decrease in stress is the result of dislocations moving, overcoming localized barriers.

Stress-relaxation tests have been and are widely used to obtain fundamental parameters that describe microdeformation modes in metals: internal and effec-

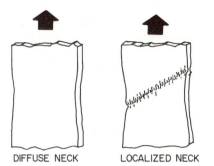

Figure 16.20 Diffuse and localized neck in sheet metals.

DIFFUSE NECK   LOCALIZED NECK

tive stresses[47-50] (treated in Section 8.5), activation volume[50] (treated in Section 8.5), and the dislocation-velocity stress exponent[47-50] (Section 8.4.1). Of more practical relevance is the ability to predict how stresses will relax, in structures in which the length of the component is kept constant. In prestressed concrete, an initial tension is applied to the steel tendons, resulting in compressive stresses in the aggregate. If the tensile stresses in the steel relax with time, the prestressed concrete will lose its unique properties. It is well known that the aggregate has a low tensile strength and that the function of the steel is to put it under compression.

Another application in which a low rate of relaxation is desired is in bolts. If the load relaxes, the bolt will loosen itself from the nut. Springs, rivets, pressure vessels in nuclear power stations, and tightness of gaskets are other applications in which stress-relaxation properties are of importance. The need for low-relaxation steels in prestressed concrete has led to the development of special compositions and heat treatments. The heat treatments are based on the pinning of dislocations by carbon atmospheres, impeding their motion. A practical application of this process is stabilized wire. The stresses in stabilized wire relaxed only one-fifth of conventionally treated wire. The "stabilizing" process consists of prestressing the wire at a specified temperature.

A number of experimental setups have been designed for the measurement of stress relaxation. ASTM Standard E328[51] describes some of them, as well as the experimental procedures. These tests have mostly practical applications as a goal; on the other hand, the tests used in the obtention of the fundamental micromechanical deformation parameters described above have mostly been conducted in tensile testing machines. The results have been plagued by extraneous effects due to the machines, and it can be said that a great number of the stress-relaxation results obtained on screw-driven tensile testing machines are worthless. On the other hand, the use of servohydraulic testing machines can eliminate some of these problems, and meaningful information is possible. The use of the tensile testing machine to obtain stress-relaxation results will be described next. The big problem with screw-driven machines is that they have a definite stiffness (see Section 16.3) and that they also relax during the test. These two effects are important. One can use the spring–dashpot analog to describe the response of the machine–specimen system to stresses. The elastic (time-independent) portion of deformation is represented by the spring, and

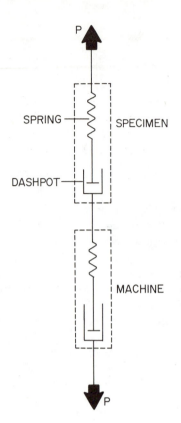

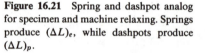

Figure 16.21 Spring and dashpot analog for specimen and machine relaxing. Springs produce $(\Delta L)_e$, while dashpots produce $(\Delta L)_p$.

the plastic or anelastic (time-dependent) portion is represented by the dashpots. This is shown in Fig. 16.21. The changes in length of machine and specimen after the initiation of a stress-relaxation test are interdependent; the change in length of specimen has to be equal to the change in length of the machine.

$$[(\Delta L)_e + (\Delta L)_p]_{\text{spec}} + [(\Delta L)_e + (\Delta L)_p]_{\text{mach}} = 0 \qquad (16.43)$$

The complications are great, as we will see below, because one has to determine the changes in length of the machine using a standard specimen; the best thing to use is a zero-length specimen, so that the left-hand side of the equation is zero. This is accomplished by having the grips attached to each other by means of a specimen that has only the two gripping regions. Actually, Meyers et al.[27] found that the grips could be a great contributor to the total relaxation. Another source of relaxation in the machine are the oil layers between the screws; as the oil is driven out of the region by the load, the machine relaxes. Figure 16.22 shows the relaxation of a screw-driven universal testing machine where the wedge-action grips were hand-tightened. This amount is clearly significant and was of the same order of magnitude as the specimens that were tested later. This can obviously lead to a great amount of confusion and misinterpretation. It is virtually impossible to make a meaningful correction for machine

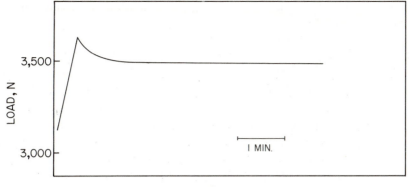

Figure 16.22 Relaxation of Instron 10,000-lb-capacity universal testing machine using hand-tightened wedge action grips (no specimen in system).

relaxation. This problem is further discussed by Meyers et al.[27] The other problem, machine stiffness, is shown in Fig. 16.23. The rate of relaxation is strongly affected by the machine stiffness when machine and specimen stiffnesses are of the same order of magnitude.

A machine with infinite stiffness and no relaxation, on the other hand, eliminates the above-mentioned problems, and Eq. 16.43 becomes

$$[(\Delta L)_e + (\Delta L)_p]_{\mathrm{spec}} = 0$$

or

$$(d\epsilon_e)_{\mathrm{spec}} = -(d\epsilon_p)_{\mathrm{spec}}$$

Servohydraulic systems operating under the strain-control mode provide this type of information. An extensometer provides the input to the machine;

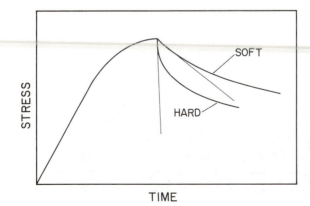

Figure 16.23 Relaxation from soft and hard machines.

the change in length is set at zero during the test, while the change in stress is monitored with time.

Although stress relaxation has existed for a long time,[52,53] the theoretical framework on which the analysis rests is due to Sargent[54,55] and Guiu and Pratt.[56] It assumes that there is no relaxation in the machine. Guiu and Pratt's[56] analysis is given below. Neglecting the machine relaxation term in Eq. 16.43, we have

$$[(\Delta L)_e + (\Delta L)_p]_{\text{spec}} + [(\Delta L)_e]_{\text{mach}} = 0$$

or

$$\frac{(dL)_{\text{es}}}{dt} + \frac{(dL)_{\text{ps}}}{dt} + \frac{(dL)_{\text{em}}}{dt} = 0$$

The stiffness of the machine K_{mach} (Section 16.1.1) is represented as

$$K_{\text{mach}} = \frac{dP}{(dL)_{\text{em}}}$$

$$\dot{L}_{\text{es}} + \dot{L}_{\text{ps}} + \frac{\dot{P}}{K_{\text{mach}}} = 0$$

Dividing this by the length of the specimen at the onset of relaxation, we have

$$\frac{\dot{L}_{\text{es}}}{L_0} + \frac{\dot{L}_{\text{ps}}}{L_0} + \frac{\dot{P}}{K_{\text{mach}}L_0} = 0$$

The change in length divided by the length of the specimen is equal to the strain:

$$\dot{\varepsilon}_{\text{es}} + \dot{\varepsilon}_{\text{ps}} = -\frac{A_0}{K_{\text{mach}}L_0} \dot{\sigma} \qquad (16.44)$$

The elastic strain in the specimen is related to the stress by Young's modulus, and

$$\dot{\varepsilon}_{\text{es}} = \frac{\sigma}{E} \qquad (16.45)$$

Substituting Eq. 16.45 into Eq. 16.44, we obtain

$$\dot{\varepsilon}_{\text{ps}} = -\left(\frac{A_0}{K_{\text{mach}}L_0} + \frac{1}{E}\right) \dot{\sigma} \qquad (16.46)$$

The stress-relaxation test provides $\dot{\sigma}$ as a function of stress (slope of the stress versus time curve) at various stress levels. From there, it is a simple matter to calculate the plastic strain rate of the specimen, $\dot{\varepsilon}_{\text{ps}}$, as a function of stress, since A_0, K_{mach}, L_0, and E are constants. Hence, one can write

$$\dot{\varepsilon}_{\text{ps}} = -M\dot{\sigma} \qquad (16.47)$$

One can see, from Eq. 16.46, how different stress rates are obtained for machines having different stiffnesses. For the same initial strain rate, a soft machine (low K_{mach}) will require a smaller value of $\dot{\sigma}$; this is shown graphically in Fig. 16.23.

Up to this point no assumption is made regarding the micromechanical deformation modes in the specimen. There are essentially two approaches that can be used in introducing the parameters defining the deformation processes. Perhaps the more fundamental one is to use the theory of thermal activation processes; it is described in detail by Krausz and Eyring.[50] This theory considers barriers to deformation, and the activation energy necessary to transpose these barriers determines the kinetics of deformation. It is an extension of the theory of chemical kinetics. When more than one barrier is present, the interrelationship between them is studied, just as in the case of chemical reactions; a rate-controlling mechanism is found. The second approach is to assume a preestablished relationship between dislocation velocity and applied stress and to obtain therefrom an equation for the relaxation of stress that matches the experimental results. This semiempirical approach is called an "engineering" approach by Krausz and Eyring. One has to carefully distinguish between the two approaches when analyzing the results.[57-61] The most simple equation relating stress and dislocation velocity is the Gilman equation (see Section 8.4.1). Li[47,48] uses it in the calculations performed, which are given briefly below. The Gilman equation can be expressed as

$$v \propto (\sigma - \sigma_i)^m \tag{16.48}$$

where σ_i is the internal stress (see Section 8.5) and m is the dislocation velocity stress exponent. Applying Orowan's equation (see Section 8.3), we have

$$\dot{\epsilon} = k\rho bv \tag{16.49}$$

Substituting Eq. 16.48 into Eq. 16.49 yields

$$\dot{\epsilon} = k'\rho b(\sigma - \sigma_i)^m$$

Applying Eq. 16.47 and assuming engineering and true stresses to be the same, we obtain

$$-M\dot{\sigma} = k'\rho b(\sigma - \sigma_i)^m$$

Assuming that the dislocation density and internal stresses are constant during stress relaxation, we have

$$\frac{d\sigma}{(\sigma - \sigma_i)^m} = k'' \, dt$$

Integration yields

$$\sigma - \sigma_i = k'''(t + a)^{\frac{1}{(1-m)}} \tag{16.50}$$

where

$$k''' = [k''(1 - m)]^{\frac{1}{(1-m)}}$$

By appropriate plotting of the stress-relaxation results it is possible to obtain σ_i and m. Figure 16.24 shows a curve obtained for zone-refined niobium by Gupta and Li.[48] The dislocation-velocity stress exponent is found from the slope; it is equal to 12. The internal stress σ_i is then calculated by applying Eq. 16.50 to two points on the curve and eliminating k''':

$$\frac{\sigma_1 - \sigma_i}{\sigma_2 - \sigma_i} = \frac{(t_1 + a)^{\frac{1}{(1-m)}}}{(t_2 + a)^{\left[\frac{1}{(1-m)}\right]}} \qquad (16.51)$$

a is the time corresponding to the nonlinear portion of the plot. In Fig. 16.24 it is approximately 20 s.

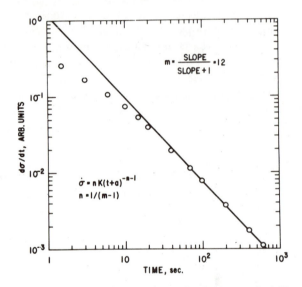

Figure 16.24 Rate of stress relaxation in one-zone-pass single-crystal niobium. (Adapted with permission from [48], p. 2325.)

The analysis presented above is applicable only if the dislocation density ρ is assumed to be constant. Rohde and Nordstrom[62] showed that such is not the case for many systems, invalidating many of the experimental results. Meyers et al.[27] proposed a method that takes into account the change in mobile dislocations with time. It is based on the simple concept that successive relaxations from the same stress will reflect the change in ρ and that it can be calculated therefrom. Pinning of dislocations by solute atoms is an effective mechanism of decreasing the mobile dislocation density. Figure 16.25 shows schematically the exhaustion in relaxation for a 6061-T6 aluminum alloy.[5] The sequential relaxations were conducted at three plastic strain levels, 0.0085,

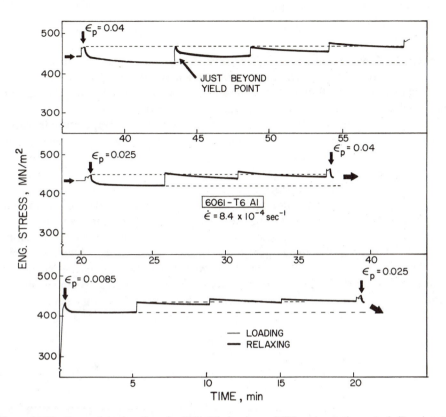

Figure 16.25 Sequential relaxations for 6061-T6 aluminum. MTS under strain control. (Reprinted with permission from [26], p. 35.)

0.025, and 0.04. The second relaxation cycle, after 5 min, is definitely characterized by a lower stress drop, at the three strain levels. This type of experiment can establish conclusively whether the hypothesis of constant mobile dislocation density is correct.

~~The activation volume for the deformation mode can also be determined~~ and it is given by both Eq. 8 in Conrad's[49] review article or Eq. 2.4.11 of Krausz and Eyring.[50] It is given, at constant temperature and substructure, by

$$v^* = kT \left(\frac{\partial \ln \dot{\gamma}}{\partial \tau} \right)_{T,\,\text{subs}}$$

$\dot{\gamma}$ and τ, the shear strain rate and stress, can be converted into $\dot{\epsilon}$ and σ, and we have, applying Eq. 16.47,

$$v^* = 2kT \left(\frac{\partial \ln - \dot{\sigma}}{d\,\sigma} \right)_{T,\,\text{subs}}$$

Hence, a plot of the stress versus the log of the stress rate yields the activation volume.

16.6 ELECTROPLASTIC EFFECT

Since 1969, Soviet scientists have been investigating an interesting phenomenon: that the application of a high-current pulse to a metal during deformation results in a decrease of the flow stress. Conrad and coworkers[63-66] and Varma and Cornwell[67] have continued these studies in the United States. It seems to be well established that the drop in flow stress is not due to a temperature increase; the nature of the phenomenon is not yet well understood, but seems to be related to interactions between the moving dislocations and electrons. The electroplastic effect has potential technological applications; it could be used in deformation processing of metals. Figure 16.26 shows the drops in stress observed in a tensile curve of titanium upon applying current densities of increasing amplitude. The drops in stress are the result of almost instantaneous increases in length of the specimen. These stress drops are dependent on the machine stiffness: the stiffer the machine, the larger the stress drop. This stress drop can be converted into an electroplastic strain[68] that represents the effect in a normalized way (independent of machine characteristics and specimen dimensions). The analysis performed in Section 16.2 shows how this parameter is obtained (see Eq. 16.30):

$$\epsilon_{ep} = \left(-\frac{\Delta P}{L}\right)\left(\frac{1}{K} + \frac{L}{AE}\right) \tag{16.52}$$

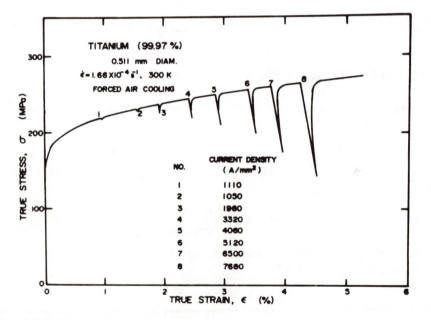

Figure 16.26 Electroplastic effect in titanium: effect of current pulses on the true stress–true strain curve. (Reprinted with permission from [64], p. 1066.)

For a machine with infinite stiffness, Eq. 16.52 reduces itself, as expected, to

$$\epsilon_{ep} = -\frac{\Delta P}{AE} = -\frac{\Delta \sigma}{E}$$

The reader should, as an exercise, calculate the electroplastic strains for the stress drops in Fig. 16.26, knowing that the tensile specimens have a cross-sectional diameter of 1.3 mm and a length of 50 mm; assume machine stiffnesses of 10,000 and 100,000 N/m.

EXERCISES

16.1. A tensile test on a steel specimen having a cross-sectional area of 2 cm² and length of 10 cm is conducted in an Instron universal testing machine with stiffness of 20 MN/m. If the initial strain rate is 10^{-3} s^{-1}, determine the slope of the load-extension curve in the elastic range ($E = 210$ GN/m²).

16.2. Determine all the parameters that can be obtained from a stress–strain curve from the load-extension curve below (cylindrical specimen), knowing that the initial cross-sectional area is 0.04 cm², the Young's modulus is 210 GN/m², the cross-head velocity is 3 mm/s, the gage length is 10 cm, the final cross-sectional area is 0.02 cm², and the radius of curvature of the neck is 1 cm.

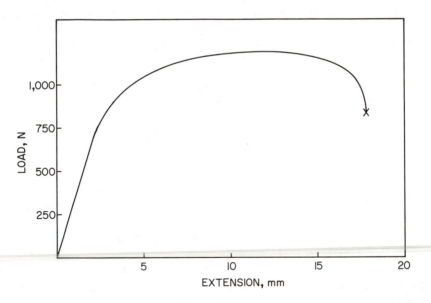

Figure E16.2

16.3. Draw the engineering stress–engineering strain and true stress–true strain (with and without Bridgman correction) curves for the curve in Exercise 16.2.

16.4. What is the strain-rate sensitivity of AISI 1040 steel at a strain of 0.01 and 0.04? (Obtain data from Fig. 16.11.)

16.5. Obtain the work-hardening exponent n using Considère's criterion for the curve of Exercise 16.2.

16.6. Determine the strain-rate sensitivity for the normalized AISI steel of Fig. 16.12.

16.7. The stress–strain curve of a 70–30 brass is described by the equation

$$\sigma = 600\epsilon_p^{0.35} \qquad \text{MPa}$$

until the onset of plastic instability.
(a) Find the 0.2% offset yield stress.
(b) Applying Considère's criterion, find the real and engineering stress at the onset of necking.

16.8. The onset of plastic flow in an annealed AISI 1018 steel specimen is marked by a load drop and the formation of a Lüders band (see Chapter 9). The initial strain rate is 10^{-4}s^{-1}; the length of the specimen is 5 cm; the Lüders plateau extends itself for a strain equal to 0.1. Knowing that each Lüders band is capable of producing a strain of 0.02 after its full motion, determine:
(a) The number of Lüders bands that traverse the specimen.
(b) The velocity of each Lüders band, assuming that only one band exists at each time.

16.9. A 10-cm-long tensile specimen (Al; $E = 70$ GPa) is being tested in a soft machine ($K = 1$ MN/m). To increase the stiffness of the system, a steel jacket ($E = 210$ GN/m^2) is constructed around it (as shown in Fig. 16.16). What is the required cross-sectional area of the jacket (length between bolts, 40 cm) if the resultant stiffness is to be equal to 20 MN/m? The cross-sectional area of the specimen is 2 cm^2.

16.10. When the cross-head is suddenly stopped for the specimen in Fig. 16.23 at a plastic strain level of 0.04 (mobile dislocation density of 10^{10} cm^{-2}), what is the velocity at which these dislocations will move at the first moment of the first relaxation cycle? The strain rate during loading is 8.4×10^{-4} s^{-1} and the initial specimen length is 10 cm. $E = 70$ GPa.

REFERENCES

[1] ASTM Standard E4–72, *Testing Machines*, ASTM, Philadelphia, 1978.

[2] ASTM Standard E9–77, *Compression Testing of Metallic Materials at Room Temperature*, ASTM, Philadelphia, 1977.

[3] ASTM Standard E290–77, *Semi-guided Bend Test for Ductility of Metallic Materials*, ASTM, Philadelphia, 1977.

[4] ASTM Standard E558–75, *Torsion Testing of Wire*, ASTM, Philadelphia, 1975.

[5] ASTM Standard A370–77, *Mechanical Testing of Steel Products*, ASTM, Philadelphia, 1977.

[6] ASTM Standard E345–69, *Tension Testing of Metallic Foil*, ASTM, Philadelphia, 1976.

[7] ASTM Standard E8–78, *Tension Testing of Metallic Materials*, ASTM, Philadelphia, 1978.

[8] ASTM Standard B557–74, *Tension Testing Wrought and Cast Aluminum—and Magnesium*, ASTM, Philadelphia, 1974.

[9] ASTM Standard B557M-76, *Tension Testing Wrought and Cast Aluminum—and Magnesium*, ASTM, Philadelphia, 1976.

[10] R.D. Carnahan, *J. Metals*, Dec. 1964, 990.

[11] M.A. Meyers, Ph.D. thesis, University of Denver, Denver, Colo., 1974, p. 206.

[12] P.R. Cetlin, A.S. Gülec, and R.E. Reed-Hill, *Met. Trans.*, *4* (1973) 513.

[13] A. Considère, *Ann. Ponts Chaussées*, *9*, Ser. 6 (1885) 574.

[14] J.R.C. Guimarães and R.J. de Angelis, *Mater. Sci. Eng.*, *13* (1974) 109.

[15] A.S. de S. e Silva and S.N. Monteiro, *Metalurgia-ABM*, *33* (1977) 417.

[16] P.W. Bridgman, *Trans. ASM*, *32* (1974) 553.

[17] W.J. McGregor Tegart, *Elements of Mechanical Metallurgy*, Macmillan, New York, 1964.

[18] J. Aronofsky, *J. Appl. Mech.*, *18* (1951) 75.

[19] H.J. Rack and M. Cohen, *Mater. Sci. Eng.*, *9* (1972) 175.

[20] U.S. Lindholm, in *Techniques of Metals Research*, Vol. 5, Pt. 1, R.F. Bunshah (ed.), Wiley-Interscience, New York, 1971, p. 199.

[21] W.A. Backofen, I.R. Turner, and D.H. Avery, *ASM Trans. Quart.*, *57*, 980 (1964).

[22] D.S. Fields, Jr., and J.S. Hubert, in *Advances in Deformation Processing*, J.J. Burke and V. Weiss (eds.), Plenum Press, New York, 1978, p. 441.

[23] N.E. Paton, C.H. Hamilton, J. Wert, and M. Mahoney, *J. Metals*, *34*, No. 8 (1981) 21.

[24] P.P. Gillis and R.E. Medrano, *J. Mater.*, *6* (1971) 514.

[25] M.A. Fortes and J.G. Proença, *J. Test. Eval.*, *4* (1976) 248.

[26] J.R.C. Guimarães and K.K. Chawla, *Metalurgia-A.B.M.*, *34* (1978) 549.

[27] M.A. Meyers, J.R.C. Guimarães, and R.R. Avillez, *Met. Trans.*, *10A* (1979) 33.

[28] R.W. Rohde and T.V. Nordstrom, *Scripta Met.*, *7* (1973) 317.

[29] D.H. Avery and W.N. Findley, "Quasistatic Mechanical Testing," in *Techniques of Metals Research*, Vol. 5, Pt. 1, R.F. Bunshah (ed.), Wiley-Interscience, New York, 1971, p. 104.

[30] W.D. Cooper and R.B. Zweigoron, *J. Metals*, *32*, No. 7 (1980) 17.

[31] C.S. Hartley and D.A. Jenkins, *J. Metals*, *32*, No. 7 (1980) 23.

[32] C. Zener and J.H. Hollomon, *J. Appl. Phys.*, *17* (1946) 69.

[33] A.S. Argon (ed.), *Constitutive Equations in Plasticity*, MIT Press, Cambridge, Mass., 1975.

[34] E.W. Hart, *Trans. ASME*, *98* (1976) 193.

[35] G.L. Wire, F.V. Ellis, and C.Y. Li, *Acta Met.*, *24* (1976) 677.

[36] E.W. Hart and A.D. Solomon, *Acta Met.*, *21* (1973) 195.

[37] J.C. Swearengen, R.W. Rohde, and D.L. Hicks, *Acta Met.*, *24* (1976) 969.

[38] A.K. Ghosh, *Acta Met.*, *28* (1980) 1443.

[39] E.W. Hart, *Acta Met.*, *15* (1967) 351.

[40] J. Campbell, *J. Mech. Phys. Solids*, *15* (1967) 359.

[41] A. Argon, in *The Inhomogeneity of Plastic Deformation*, ASM, Metals Park, Ohio, 1973, p. 161.

[42] J.J. Jonas, R. Holt, and C. Coleman, *Acta Met.*, *24* (1977) 43.

[43] M.Y. Demeri and H. Conrad, *Scripta Met.*, *12* (1978) 389.

[44] A.K. Ghosh, in *Formability: Analysis, Modeling, and Experimentation*, S.S. Hecker, A.K. Ghosh, and H.L. Gegel (eds.), TMS-AIME, New York, 1978, p. 14.

[45] S.E. Jones and P.P. Gillis, cited in [44], p. 46.

[46] J.M. Jaliner, M. Christodoulou, B. Baudelet, and J.J. Jonas, cited in [44], p. 46.

[47] J.C.M. Li, *Can. J. Phys.*, *45* (1970) 493.

[48] I. Gupta and J.C.M. Li, *Met. Trans.*, *1* (1970) 2323.

[49] H. Conrad, *Mater. Sci. Eng.*, *2* (1970) 265.

[50] A.S. Krausz and H. Eyring, *Deformation Kinetics*, Wiley-Interscience, New York, 1975, p. 226.

[51] ASTM Standard E328, ASTM, Philadelphia, 1978.

[52] P. Trouton and R.O. Rankine, *Phil. Mag.*, *8* (1904) 538.

[53] P. Feltham, *Phil. Mag.*, *6*, (1961) 259.

[54] G. Sargent, Ph.D. thesis, Imperial College, London, 1965.

[55] G. Sargent, *Acta Met.*, *13*, 663 (1965).

[56] F. Guiu and P.L. Pratt, *Phys. Status Solidi*, *6* (1964) 111.

[57] R.W. Rohde and T.V. Nordstrom, *Mater. Sci. Eng.*, *12* (1973) 179.

[58] W.R. Thorpe and I.O. Smith, *Mater. Sci. Eng.*, *18* (1975) 167.

[59] A.S. Krausz, *Mater. Sci. Eng.*, *22* (1976) 91.

[60] W.R. Thorpe and I.O. Smith, *Mater. Sci. Eng.*, *22* (1976) 94.

[61] R.W. Rohde and T.V. Nordstrom, *Mater. Sci. Eng.*, *22* (1976) 99.

[62] R.W. Rohde and T.V. Nordstrom, *Scripta Met.*, *7* (1973) 317.

[63] H. Conrad, *J. Metals*, *16* (1964) 582.

[64] K. Okazaki, M. Kagawa, and H. Conrad, *Scripta Met.*, *12* (1978) 1063.

[65] K. Okazaki, M. Kagawa, and H. Conrad, *Scripta Met.*, *13* (1979) 277.

[66] K. Okazaki, M. Kagawa, and H. Conrad, *Scripta Met.*, *13* (1979) 473.

[67] S.K. Varma and L.R. Cornwell, *Scripta Met.*, *13* (1979) 733.

[68] M.A. Meyers, *Scripta Met.*, *14* (1980) 1033.

SUGGESTED READING

Tensile Testing

AVERY, D.H., and W.N. FINDLEY, "Quasistatic Mechanical Testing," in *Techniques of Metals Research*, Vol. 5, Pt. 1, R.F. Bunshah (ed.), Wiley-Interscience, New York, 1971.

Deformation Modeling

ARGON, A.S. (ed.), *Constitutive Equations in Plasticity*, MIT Press, Cambridge, Mass., 1975.

KOCKS, U.F., A.S. ARGON, and M.F. ASHBY, *Thermodynamics and Kinetics of Slip*, Progress in Materials Science, Vol. 19, B. Chalmers, J.W. Christian, and T.B. Massalski (eds.), Pergamon Press, Oxford, 1975, p. 10.

Stress Relaxation

Fox, A. (ed.), *Stress Relaxation Testing*, ASTM STP No. 676, ASTM, Philadelphia, 1979.

Krausz, A.S., and H. Eyring, *Deformation Kinetics*, Wiley-Interscience, New York, 1975, p. 226.

Necking

Hecker, S.S., A.K. Ghosh, and H.L. Gegel (eds.), *Formability: Analysis Modeling and Experimentation*, TMS-AIME, New York, 1978.

Superplasticity

Paton, N.E. and C.H. Hamilton (eds.), *Superplastic Forming of Structural Alloys*, AIME, Warrendale, PA, 1982.

Chapter 17

HARDNESS TESTING

‡17.1 INTRODUCTION

The term "hardness" is rather vague, but represents the resistance of a body to any permanent change due to an external mechanical stimulus. In different technological areas, the term "hardness" has been associated with different specific tests establishing this resistance. Thus, a variety of testing methods has evolved; Shaw[1] classified them as follows:

1. *Static indentation tests*. In these tests, an indenter is forced against a surface perpendicularly to it; the dimension of the deformation zone is used to obtain a parameter. Examples are the diamond pyramid (DPH) or Vickers (HV), Rockwell (HR), Brinell (HB), Knoop (KHN), Monotron, and Meyer tests. These are by far the most popular tests in metals and are treated separately in Section 17.2.

2. *Scratch tests*. In these tests, the ability of materials to produce scratches in each other is measured. The most famous scratch test is the Mohs'; it dates from 1882 and is commonly used in mineralogy. The Mohs' scale is the following: talc (1), gypsum (2), calcite (3), fluorite (4), apatite (5), orthoclase (6), quartz (7), topaz (8), corundum (9), and diamond (10). If an unknown material is scratched by quartz but scratches orthoclase, its Mohs' hardness will be somewhere between 6 and 7. It should be noticed that the Mohs' scale is not linear, and does not increase monotoni-

cally as the hardness of the material increases. There is a much larger difference in the resistance to being scratched between 9 and 10 than between 1 and 2. So, small increases in Mohs' hardness in the upper range can be very significant. Tabor[2] investigated the characteristics of the Mohs' scale and concluded that the hardness increases exponentially with Mohs' scale. The hardness of each number is 60% higher than that of the preceding. The recent development of ultrahard materials for abrasives has led to the need of a better distinction between hard metals. This has led to an extension of the Mohs' scale in the following fashion: topaz (9), garnet (10), fused zirconia (11), fused alumina (12), silicon carbide (13), boron carbide (14), and diamond (15).

3. *Plowing tests*. In these tests, a blunt tool (diamond) is moved across a surface under controlled conditions. The width of the groove is a measure of hardness. The Bierbaum test[3] is an example.

4. *Rebound tests*. An object of standard shape and mass is dropped from a predetermined height onto the metal. The height of the rebound is measured in diameter. The Shore scleroscope uses this principle. A small pointed hammer ($\frac{1}{4}$ in. in diameter with a diamond tip) falls through a glass tube from a height of 10 in. (25.4 cm) and hits the surface. The resilience of the material being impacted is an important parameter. If no plastic deformation takes place, the rebound will, of course, be high. On the other hand, the more plastic deformation is generated in the metal, the more kinetic energy of the hammer is absorbed. The classic text by Tabor[4] describes this test in greater detail.

5. *Damping tests*. A change in amplitude or in frequency of a pendulum having a hard pivot resting on the test surface is the measure of hardness. The Herbert pendulum uses this principle.

6. *Cutting tests*. A sharp tool removes a chip of predetermined dimensions at low speed. The force required, divided by the cross-sectional area of the chip, provides the hardness.

7. *Abrasion tests*. A specimen is loaded against a rotating disk and the rate of wear is measured. Another type of test is the Deval test, in which 50 stone specimens are loaded into a rotating cylinder and the weight loss is measured after a predetermined time.

8. *Erosion tests*. Sand or another abrasive material is caused to impinge on the material; surface loss of material is measured.

Additional classes of hardness tests can be found. Petty[5] describes a "mutual indentation test" in which two specimens are rubbed against each other. Actually, the first hardness test ever devised (Reaumur, 1722) was of this type. Yet another type is a sharp specimen (material whose hardness one wants to measure) against a flat plate. The difficulty of preparing specimens made this "Polish" test very unpopular.

‡17.2 MICRO AND MACROINDENTATION TESTS

Indentation tests constitute the vast majority of metallurgical tests. They are essentially divided into two classes, commonly called microindentation and macroindentation tests, but improperly referred to as microhardness and macrohardness tests. The division between the two occurs for a load of approximately 200 gf (~2 N). The indentation tests in metals measure the resistance to plastic deformation; both the yield stress and the work-hardening characteristics of the metal are important in determining the hardness. In spite of the theoretical studies done on hardness, some of which are described in Section 17.3, hardness cannot be considered as a fundamental property of a metal. Rather, it represents a quantity measured on an arbitrary scale. Weiler[6] systematically studied hardness parameters and could find no fundamental correlation between them. His basic conclusion is that: "The hardness values calculated with a specified method represent a scale by themselves, allowing the comparison of materials. A comparison of the scales with each other and with other qualities of the material must be carried out empirically." Hence, hardness measurements should not be taken to mean more than what they are: an empirical, comparative test of the resistance of the metal to plastic deformation. Any correlation with a more fundamental parameter, such as the yield stress, is valid only in the range experimentally determined. Similarly, comparisons between different hardness scales are meaningful only through experimental verification. For steels, Table 17.1 gives a fair conversion of hardnesses and the UTS equivalents.

The most important macro- and microindentation tests for metallurgists are described next.

‡17.2.1 Macroindentation Tests

The impressions caused by these tests are shown in Fig. 17.1. The Brinell test produces by far the largest indentation. The Vickers test may produce very small indentations, depending on the load used. These tests are described in the following sections.

‡17.2.1.1 Brinell hardness test.

A steel sphere is pressed against a metal surface for a specified period of time (10 to 15 s, according to the ASTM), and the surface of the indentation is measured. The load (in kgf) divided by the area (in mm²) of the curved surface gives the hardness, HB.

$$\text{HB} = \frac{P}{\pi D \times \text{depth}} \tag{17.1}$$

$$= \frac{2P}{\pi D(D - \sqrt{D^2 - d^2})} \tag{17.2}$$

where D and d are the sphere and impression diameters, respectively. The parameters are indicated in Fig. 17.2. Since $d = D \sin \phi$, we have

Table 17.1 Approximate Hardness Conversions for Steels

Vickers	HB (Brinell) (10-mm ball, 30,000 kgf)		Rockwell				Rockwell Superficial		Shore	Approx. Tensile Strength
HV	Stand.	WC	A (Brale 60 kgf)	B (1/16-in. ball, Brale 100 kgf)	C (Brale 150 kgf)	D (Brale 100 kgf)	15N	45N	Scleroscope	(MPa)
940			85.6		68.0	76.9	93.2	75.4	97	
900			85.0		67.0	76.1	92.9	74.2	95	
860		757	84.4		65.9	75.3	92.5	73.1	92	
820		733	83.8		64.7	74.3	92.1	71.8	90	
780		710	83.0		63.3	73.3	91.5	70.2	87	
740		684	82.2		61.8	72.1	91.0	68.6	84	
700		656	81.3		60.1	70.8	90.3	66.7	81	2199
660		620	80.3		58.3	69.4	89.5	64.7	79	2061
620		582	79.2		56.3	67.9	88.5	62.4	75	1923
580		545	78.0		54.1	66.2	87.5	59.9	72	1792
540	496	507	76.7		51.7	64.4	86.3	57.0	69	1655
500	465	471	75.3		49.1	62.2	85.0	53.9	66	1517
460	433	433	73.6		46.1	60.1	83.6	50.4	62	1379
420	397	397	71.8		42.7	57.5	81.8	46.4	57	1241
380	360	360	69.8		38.8	54.4	79.8	41.7	52	1110
340	322	322	67.6		34.4	51.1	77.4	36.5	47	972
300	284	284	65.2		29.8	47.5	74.9	31.1	42	834
260	247	247	62.4		24.0	43.1	71.6	24.3	37	696
220	209	209		95.0					32	634
200	190	190		91.5					29	579
180	171	171		87.1					26	517
160	152	152		81.7					24	455
140	133	133		75.0					21	
120	114	114		66.7						
100	95	95		62.3						
85	81	81		41						

a Adopted with permission from [5], p. 180.

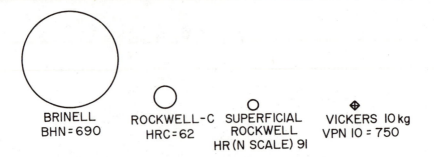

BRINELL
BHN = 690

ROCKWELL-C
HRC = 62

SUPERFICIAL
ROCKWELL
HR (N SCALE) 91

VICKERS 10 kg
VPN 10 = 750

Figure 17.1 Comparison of the impression sizes produced by various hardness tests on material of 750 HV. (Adapted with permission from [5], p. 174.)

$$HB = \frac{2P}{\pi D^2 (1 - \cos \phi)} \tag{17.3}$$

Different spheres produce different impressions and, if we want to maintain the same HB independent of the size of the sphere, the load has to be varied according to the relationship

$$\frac{P}{D^2} = \text{constant} \tag{17.4}$$

This assures the same geometrical configuration (the same ϕ). The diameter of the impressions between 0.25 and $0.5D$ produces good, reproducible results. The target sought is $d = 0.375D$. If the same d/D ratio is maintained (constant ϕ), the Brinell test is reliable. Spheres with diameters of 1, 2, 5, and 10 mm have been used, and some of the ratios P/D^2 that provide good d/D ratios for different metals are: steels and cast irons (30); Cu and Al (5); Cu and Al alloys (10); Pb and Sn alloys (1). The softer the material, the lower the P/D^2 ratio required to produce $d/D = 0.375$.

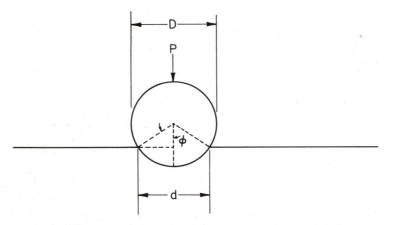

Figure 17.2 Impression caused by spherical indenter on metal plate.

One of the problems of the Brinell test is that HB is dependent on the load P for the same sphere. In general, HB decreases as the load is increased. ASTM standard E10–78 provides details and specifications for Brinell hardness tests. It states that the standard Brinell test is conducted under the following conditions:

Ball diameter: 10 mm

Load: 3000 kgf

Duration of loading: 10 to 15 s

In this case, 360 HB indicates a Brinell hardness of 360 under the testing conditions above. For different conditions, the parameters have to be specified. For example, 63 HB 10/500/30 indicates a Brinell hardness of 63 measured with a ball of 10 mm diameter and a load of 500 kgf applied for 30 s. Brinell tables and additional instructions are provided in ASTM E10–78. Meyer[7] was aware of this problem and proposed a modification of the Brinell formula (Eq. 17.1). He found out that the load divided by the *projected* area of the indentation $(\pi d^2/4)$ was constant. Hence, he proposed the following:

$$\text{Meyer} = \frac{4P}{\pi d^2} \tag{17.5}$$

where P is expressed in kilograms-force and d in millimeters. The Meyer hardness never gained wide acceptance, in spite of being more reliable than the Brinell hardness. For work-hardened metals it seems to be independent of P. Figure 17.3 shows the results of an experiment conducted by Bunshah and Armstrong.[8] They attached an indenter to a tensile testing machine and monitored the increase in the projected indentation as a function of load. The proportionality between P and A show that Meyer's precept is correct; the material tested was highly work-hardened copper. For annealed copper, on the other hand, the curve was not as linear and a slight change in slope could be observed.

‡**17.2.1.2 Rockwell**[9–13] **hardness test.** The most popular hardness test is also the most convenient, since there is no need to measure the depth or width of the indentation optically. This testing procedure is described basically in Fig. 17.4. A preload is applied prior to the application of the main load. The dial of the machine provides a number that is related to the depth of the indentation produced by the main load. Several Rockwell scales are used and the numbers refer to arbitrary scales and are not directly related to any fundamental parameter of the material. Two different types of indenters are used. The A, C, D, and N scales use the Brale indenter, which is a diamond cone with a cone angle of 120°. The other scales use either $\frac{1}{8}$-in. (3.175-mm) or $\frac{1}{16}$-in. (1.587-mm)-diameter steel spheres. The loads also vary, depending on the scale. Table 17.2 shows the various loads and typical applications. Usually, the C scale is used for harder steels and the B scale for softer steels; the A scale covers a wider range of hardness. Because of the nature of the measurement, any sagging of the test piece will produce changes in hardness. Therefore, it

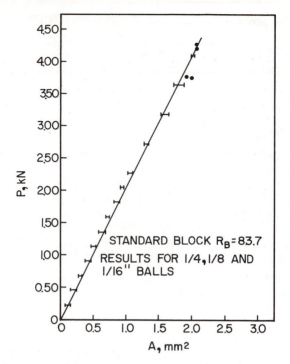

Figure 17.3 Load versus area for standard block HRB = 83.7, tested with $\frac{1}{4}$-, $\frac{1}{8}$-, and $\frac{1}{16}$-in balls. (Adapted with permission from [8], p. 325.)

is of utmost importance to have the sample well supported; specimens indented in Bakelite cannot be tested. The Brinell and Vickers tests, on the other hand, which are based on optical measurements, are not affected by the support.

For very thin samples, there is a special superficial Rockwell test. The testing procedure is described in detail in the ASTM Standard E18–74,[9] and conversion tables for a number of alloys is given in ASTM Standard E140–78.[12] The symbol used to designate this hardness is, according to the ASTM, HR; 64HRC corresponds to Rockwell hardness number 64 on the C scale.

Dieter[13] recommends the following precautions for reproducible results in Rockwell testing:

1. The indenter and anvil should be clean and well seated.
2. The surface to be tested should be clean and dry, smooth, and free from oxide. A rough ground surface is usually adequate for the Rockwell test.
3. The surface should be flat and perpendicular to the indenter.
4. Tests on cylindrical surfaces will give low readings, the error depending on the curvature, load, indenter, and hardness of the material. Corrections are given in ASTM E140–78.[12]
5. The thickness of the specimen should be such that a mark or bulge is not produced on the reverse side of the piece. It is recommended that

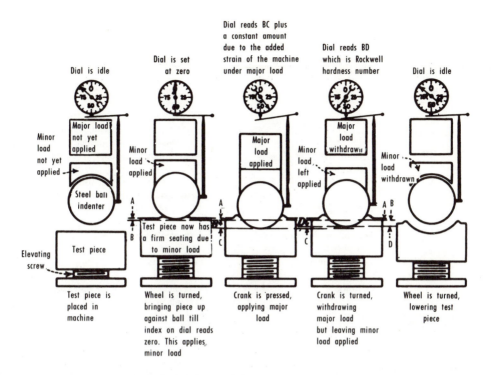

Figure 17.4 Procedure in using Rockwell hardness tester. (Reprinted with permission from [11], p. 149.)

the thickness be at least 10 times the depth of the indentation. Tests should be made on only a single thickness of material.

6. The spacing between indentations should be three to five times the diameter of the indentation.

7. The speed of application of the load should be standardized. This is done by adjusting the dashpot on the Rockwell tester. Variations in hardness can be appreciable in very soft materials unless the rate of load application is carefully controlled. For such materials the operating handle of the Rockwell tester should be brought back as soon as the major load has been fully applied.

‡17.2.1.3 Vickers (or diamond pyramid) hardness test. This test uses a pyramidal indenter with a square base, made of diamond. The angle between the faces is 136°. This test was introduced because of the problems encountered with the Brinell test. One of the known advantages of the Vickers

Scale Designation	Type of Indenter	Major Load (kgf)	Typical Field of Application
A	Brale	60	The only continuous scale from annealed brass to cemented carbide, but is usually used for harder materials
B	1/16-in.-diameter steel ball	100	Medium-hardness range (e.g., annealed steels)
C	Brale	150	Hardened steel > HRB100
D	Brale	100	Case-hardened steels
E	1/8-in.-diameter steel ball	100	Al and Mg alloys
F	1/16-in.-diameter steel ball	60	Annealed Cu and brass
L	1/4-in.-diameter steel ball	60	Pb or plastics
N	N Brale	15, 30, or 45	Superficial Rockwell for thin samples or small impressions

[a] Adapted with permission from [5], p. 173.

test is that one indenter covers all the materials, from the softest to the hardest. The load is increased with hardness, and there is a continuity in scale. The angle of 136° was chosen based on results with spherical indenters. For these, the best results were obtained (see Section 17.2.1.1) when $d/D = 0.375$. If we take the points at which the sphere touches the surface of the metal and the point of highest deformation (center of depression in metal) and calculate this angle, we obtain 136°. This exercise is left to the student. The description of the procedures used in testing is given in ASTM Standard E92–72,[14] which can be purchased at the address given in the reference. The Vickers hardness (HV) is computed from the following equation:

$$HV = \frac{2P \sin (\alpha/2)}{d^2} = \frac{1.8544P}{d^2} \qquad (17.6)$$

where P is the applied load (in kgf), d the average length of the diagonals (in mm), and α the angle between the opposite faces of the indenter (136°). The Vickers test described by ASTM E92–72 uses loads varying from 1 to 120 kgf. For example, 440HV30 represents a Vickers hardness number of 440 measured with a load of 30 kgf. Vickers testing requires a much better preparation of the material surface than Rockwell testing; hence, it is more time consuming. The surface has to be ground and polished, care being taken not to work-harden it. After the indentation, both diagonals of impression are measured and their average taken. If the surface is cylindrical or spherical, a correction factor has to be introduced. ASTM Standard E92 (Tables 4 through 6) provide correction factors. As with other hardness tests, the distance between the indentations has to be greater than two and one-half times the length of the indentation diagonal, to avoid interaction between the work-hardening regions.

The manner in which the material flows and work hardens (or work softens) beneath the indenter affects the shape of the impression. The sides of the square impression can be deformed into concave or convex curves, depending on the nature of the deformation process, and this results in reading errors. These effects are described in greater detail in Section 17.3; Fig. 17.15 shows the effect of deformation irregularities on the impression.

‡17.2.2 Microindentation Hardness Tests

Microindentation hardness tests—or microhardness tests—utilize a load lighter than 200 gf, and very minute impressions are thus formed; a load of 200 gf produces an indentation of about 50 μm for a medium-hardness metal. These tests are ideally suited to investigate changes in hardness at the microscopic scale. One can measure the hardness of a second-phase particle and identify regions within a grain where differences in hardness occur. Microhardness tests are also used to perform routine tests on very small precision components such as watch parts. These tests and their limitations are described in detail by Mott,[15] Bückle,[16] and Petty.[5]

The results shown in Fig. 17.5 illustrate well an application of microindentation testing. When a metal is alloyed, the distribution of the solute is not even throughout the grain, due to the stress fields produced by the solute atom (see Chapter 10). The solute atoms often tend to segregate at the grain boundaries. Figure 17.5(a) shows how the addition of aluminum to zinc is reflected by an increase in the hardness in the grain-boundary region, and the addition of gold results in a lowering of the grain-boundary hardness. This effect can be noted at extremely low concentrations of solute (a few parts per million). Figure 17.5(b) shows how this "excess" hardening increases with concentration of aluminum. Gold and copper do not seem to contribute significantly to grain-boundary hardening.

In spite of the attempts made, several problems have risen in the standardization of microindentation testing and its extrapolation to macroindentation results. There are several reasons for this. Almost invariably, the microhardness of any material is higher than its standard macrohardness. Additionally, the microhardness varies with load, as shown in Fig. 17.6. There is first a tendency to increase (up to a few grams); then the hardness value drops with load. At very low loads, one is essentially measuring the hardness of a single grain; the indenter "sees" a single crystal, and the plastic deformation produced by the indentation is contained in this grain. As the load is increased, plastic deformation of adjoining grains is involved, and a truly polycrystalline deformation regime is achieved. As we know well (see Chapter 14), the grain size has a marked effect on the yield strength and work-hardening characteristics of metals. Another reason for hardness variations is shown in Fig. 17.5. Yet another source of error is the work hardening introduced in the surface by polishing. The effect of crystallographic orientation, when the impression is restricted to a single

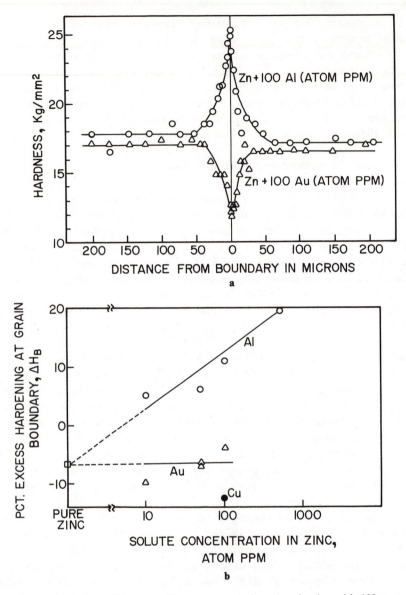

Figure 17.5 (a) Hardness–distance profiles near a grain boundary in zinc with 100 at. ppm of Al and zinc with 100 at. ppm of Au (1-gf load); (b) solute concentration dependence of percent excess boundary hardening in zinc containing Al, Au, or Cu. (3-gf load). (Adapted with permission from [17], pp. 293, 294.)

grain, is of utmost importance. It is well known that both the yield stress and the work hardening are dependent on crystallographic orientation. The Schmid law relates the applied stress to the shear stress "seen" by the various slip systems. Daniels and Dunn[18] considered the Knoop hardness test as a constrained tensile test and found that the Knoop hardness was determined by

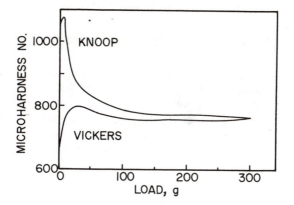

Figure 17.6 Typical variation of microhardness results with load. (Adapted with permission from [5], p. 190.)

an "effective resolved shear stress." Armstrong and Raghuram[19] extended this treatment, taking into account the work hardening of the different crystallographic orientations. There have been attempts to obtain corrected microhardness numbers that would be independent of load by introducing corrections for elastic recovery, increased surface area, and for the maximum load that would not give any permanent impression. These corrected values work, but only for materials in which the correction parameters were established.

The two most common microindentation tests are the Knoop[20] and Vickers tests. The Knoop indenter is an elongated pyramid, shown in Fig. 17.7. The hardness is obtained from the surface area of the impression:

$$\text{KHN} = \frac{14.228P}{d^2}$$

where P is the load of kgf and d is the length of major diagonal, in mm. The ratio between the dimensions of the impression is

$$h/w/l = 1:4.29:30.53$$

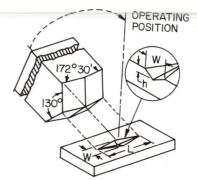

Figure 17.7 Some of the details of the Knoop indenter, together with its impression. (Adapted with permission from [21], p. 149.)

This results in an especially shallow impression, making this technique very helpful for testing brittle materials. Indeed, this was the purpose of introducing this test. The ratio between the major and minor diagonal of the impression is approximately 7:1, resulting in a state of strain in the material that can be considered as plane strain; the strain in the l direction can be neglected. This subject is treated again in Section 17.3. The very shallow Knoop impression is also helpful in testing thin components, such as electrodeposits or hardened layers. The Vickers microhardness test uses the same 136° pyramid with loads of a few grams. Both Knoop and Vickers indenters require prepolishing of the surface to a microscopic grade. Petty[5] provides (Table 5) a list of instruments and manufacturers.

17.3 STRESSES AND STRAINS IN HARDNESS TESTS

There have been several attempts to characterize the stresses and strains in the region under the indenter. The state of stress is, of course, tridimensional, and it is virtually impossible to find a direct equivalence between the uniaxial stress tensile test and the hardness test. However, these calculations have led to an improved understanding of metals under indenters and of the reasons why metals under different conditions act differently. The mathematical treatment used is rather involved and will not be discussed here.

The problem of a rigid sphere impinging on an elastic, semi-infinite plate was treated by Lee[21] and Hertz.[22] Muskhelishvili[23] provides an advanced treatment. The stresses generated when a rigid sphere impinges on a rigid plate are infinite at the point of contact; these stresses are known as Hertzian stresses. However, when the plate is elastic, the contact region increases from a point to a defined region as the load P increases. Figure 17.8(a) shows the isostress lines. These are maximum shear stress lines for each point, and plastic deformation should follow the pattern of maximum shear stress. The units are M', equal to the ratio between the maximum shear stress (τ_{max}) and the load divided by the maximum contact area ($\bar{p} = P/\pi a^2$). The radius of the contact region is given by

$$a = \left(\frac{4k}{3E} Pr\right)^{1/3} \tag{17.7}$$

where P is the load, r the sphere radius, E the Young's modulus of the plate, and k a parameter that can be taken to be equal to 1. The maximum shear stress occurs at a certain depth below the indentation. It corresponds to $M' = 0.468$, or to $\tau_{max} = 0.468\bar{p} = 0.468(P/\pi a^2)$. Substituting the value for a, we have

$$\tau_{max} = \frac{M'P}{\pi(4Pr/3E)^{2/3}} \tag{17.8}$$

But it can be assumed that the yield stress in uniaxial tension, σ_y, is equal to $2\tau_{max}$, at the onset of plastic flow. Hence,

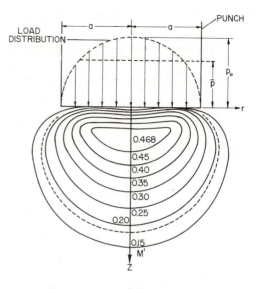

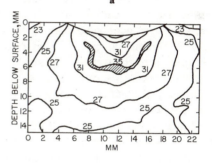

Figure 17.8 (a) Hertz lines of constant maximum shear stress for frictionless spherical indenter, $M' = \sigma_{max}/\bar{p}$. (Reprinted with permission from [1], p. 7.) (b) Hardness contours (isosclers) obtained with a pyramidal indenter at 5-kgf load on a section through an impression in aluminum made with 28.55-mm ball. (Reprinted with permission from [5], p. 164.)

$$\frac{\sigma_y}{2} = \frac{0.468\,P}{\pi(4Pr/3E)^{2/3}} \qquad (17.9)$$

and the load at which plastic deformation in the plate is initiated is

$$P = \frac{r^2(\pi\sigma_y)^3}{0.227\,E^2} \qquad (17.10)$$

One can compute the loads at which the plastic deformation zone expands to the various isostress lines in Fig. 17.8(a) by using the various values of M' in Eq. 17.8 and scaling Fig. 17.8(a) down to keep a (which changes as deformation takes place) in scale with its value calculated in Eq. 17.8. Figure 17.8(b) shows the hardness pattern in aluminum after extensive plastic deformation by a 28.55-mm sphere. The isohardness lines represent, to a certain extent, regions that have undergone the same plastic deformation; the higher the hardness, the higher the amount of plastic strain. There is a reasonable correspondence between the isohardness-line pattern and the pattern of isoshear. The region of maximum hardness occurs, as expected, at a certain depth below the surface.

For conical and pyramidal indenters a very simplified solution is presented by Shaw.[24] He considers the hardness, expressed as the load divided by the projected area, to be equal to three times the yield stress:

$$\frac{P}{A} = 3\sigma_y \tag{17.11}$$

This solution is the one obtained for a flat cylindrical punch using the maximum shear stress yield criterion.

The Knoop hardness indenter, on the other hand, generates a state of plane strain due to the ratio between the length of the major and minor diagonals: 7:1. The theory developed by Hill et al.[25] establishes the pattern of maximum shear stresses beneath a wedge penetrating in a rigid, plastically non-work-hardening material with a constant volume. The simplifying assumptions above, especially the non-work-hardening restriction, have a strong bearing on the resultant deformation pattern, as will be seen later. The non-work-hardening assumption is valid for a metal that has been heavily prestrained, so that the work-hardening capability has already saturated, or when it exhibits a sharp yield point with wide yield-point extension. The slip-line field theory, described briefly in Chapter 2, can be used only in plane-strain configurations; Hill et al.[25] apply it to the wedge. The slip lines constitute a field of lines along which the shear stresses are maximum; hence, they determine the pattern of plastic deformation. Hill et al.[25] assume, for simplicity, that there is no friction between the wedge and the specimen; additionally, the slip lines intersect the free surface at 45° angles and each other at 90°. The configuration is geometrically similar for all the stages of penetration of the wedge; hence it is only necessary to calculate the parameters at one position. Figure 17.9 shows the appearance of the deformation zone, divided into three distinct regions, I, II, and III. In I and III the slip lines are orthogonal, making 90° angles with AB and AC.

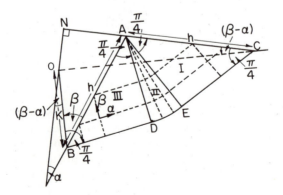

Figure 17.9 Wedge with semiangle β penetrating metal. Constancy of volume requires material to be pushed upward. (Adapted with permission from [25], p. 278.)

The region AC is extruded upward by the downward motion of the wedge. The development of slip-line field theory shows that AC is a straight line. Applying the constancy of volume to the deformation zone in Fig. 17.9, it is possible to obtain a relationship between h and α as a function of the wedge semiangle β and the depth of indentation.

$$h \cos \beta - k = h \sin (\beta - \alpha) \tag{17.12}$$

Proof of this equation is left to the student; it stems from purely geometrical considerations. The angle $(\beta - \alpha)$ should be independent of the depth of penetration if the configuration is to remain geometrically equivalent throughout penetration. This implies the following: An increase in penetration Δk will be such that the ratio $\Delta k \sin \beta / \Delta k$ is equal to the ratio ON/k. Hence,

$$ON = k \sin \beta \tag{17.13}$$

But it can be shown by geometrical manipulation that

$$ON = h \cos \alpha - k \cos (\beta - \alpha) \tag{17.14}$$

Hence,

$$h \cos \alpha = k[\sin \beta + \cos (\beta - \alpha)]$$

or

$$\frac{h}{k} = \frac{\sin \beta + \cos (\beta - \alpha)}{\cos \alpha} \tag{17.15}$$

Expressing the ratio h/k in Eq. 17.12 and making it equal to Eq. 17.15:

$$\frac{1}{\cos \beta - \sin (\beta - \alpha)} = \frac{\sin \beta + \cos (\beta - \alpha)}{\cos \alpha} \tag{17.16}$$

or

$$\cos (2\beta - \alpha) = \frac{\cos \alpha}{1 + \sin \alpha} \tag{17.17}$$

Equation 17.17 allows the determination of α as a function of β. Figure 17.10 presents a plot of α as a function of β. As β increases, so does α. The vertical load divided by the area of the indentation H is also plotted. Here, it is divided by the maximum shear stress. If we apply the von Mises yield criterion to a plane-strain configuration, we obtain $\tau_{max} = \sigma_y/\sqrt{3}$ and find that the hardness will vary in the following range: $(\beta = 0°)$ $1.15\sigma_y < H < 3\sigma_y$ $(\beta = 90°)$.

In real life there is always a certain degree of friction; it will have a definite effect on the indentation parameters. Intuitively, one can see that the angle $(\beta - \alpha)$ will be decreased because the material will be pushed inward as the wedge moves down. The slip lines no longer intersect the sides of the wedge at 45°. Figure 17.11 shows the slip lines in the presence of friction. On the right-hand side one has sticking conditions, and $(\beta - \alpha)$ is reduced to 3°. Applying Eq. 17.17 one can calculate $(\beta - \alpha)$ in the absence of friction. One

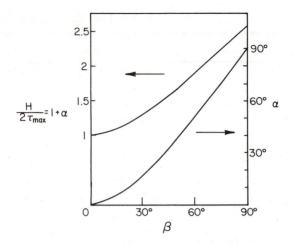

Figure 17.10 Variation of pressure P on wedge with semiangle β. (Adapted with permission from [25], p. 286.)

finds a value close to 7°; as the friction coefficient increases, the slip-line field reaches a limiting condition whereupon it becomes inclined at 45° to the vertical (symmetry axis). In this case, a 90° wedge-shaped cap of dead metal forms under the indenter. This shows that friction can play a significant role in indentation, changing the mode of deformation under the indenter. This dead metal does not flow and acts like a plug, advancing with the wedge. In Figure 17.11 the region *OAC* represents such a dead zone on the right side. The friction between the indenter and the material can affect the load. The hardness (mean pressure) would be given by $2.81\sigma_y$ for the right side, compared to $2.78\ \sigma_y$ on the left side.[24] All indenters whose fronts create a 45° dead zone present essentially the same solution.

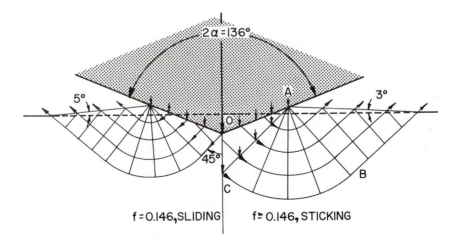

Figure 17.11 Slip-line fields for wedge indentation at the transition between sliding and sticking. (Adapted with permission from [24], p. 454.)

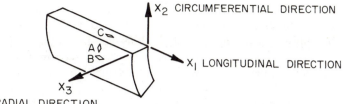

Material condition	Knoop hardness numbers (500-g load) for indenter orientations		
	a	*b*	*c*
Cold worked	254 ± 8	293 ± 7	269 ± 5
Held at 510 C for 4 hr	244 ± 5	257 ± 5	256 ± 6
Held at 750 C for 4 hr	209 ± 8	237 ± 6	218 ± 6

Figure 17.12 Plastic anisotropy as measured by Knoop hardness number. (Adapted with permission from [21], p. 163.)

A potentially interesting application of hardness tests is the determination of the anisotropy of metals. The analysis is conducted by Lee.[21] Knoop hardness tests are used to determine the anisotropy of sheet material (orthotropic) and to predict its response under biaxial loading. The Knoop test was used because it generates a state of plane strain and the stress analysis is somewhat simplified and because it essentially measures the deformation in one plane. Figure 17.12

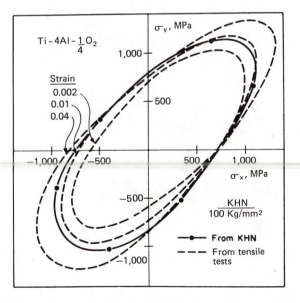

Figure 17.13 Yield locus diagram for Ti–4% Al–1/4% O_2 at 0.2, 1, and 4% strain compared with that from [21], p. 160.

shows how the Knoop hardness of a pipe varies with orientation; as the material is annealed, the texture decreases and the Knoop hardness becomes more uniform. By applying a plasticity analysis described in his paper, Lee[21] was able to obtain yield loci using hardness tests; as can be seen from Fig. 17.13, there is a reasonable agreement between the flow locus for a titanium–aluminum–oxygen alloy obtained by tensile testing (dashed curves for strains of 0.002, 0.01, and 0.04) and the curve for hardness (full line).

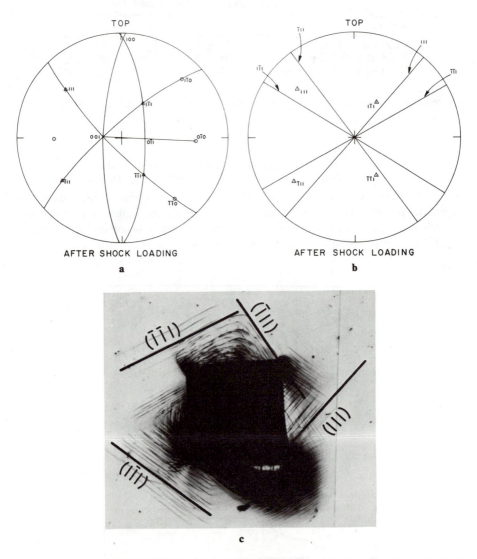

Figure 17.14 Fe–34.5% Ni after shocking: (a) crystallographic orientation of platform on which hardness measurements were made; (b) traces of slip planes on plane of platform; (c) typical hardness indentation showing "bulging out" of material in dark.

17.4 ADDITIONAL METALLURGICAL CONSIDERATIONS

Hardness tests contain a considerable amount of information that is usually neglected by the user. For example, the slip markings at the surface of a carefully polished monocrystalline specimen are consistent with the crystallographic orientation. This is illustrated in Fig. 17.14, for a shock-loaded Fe–34.5% Ni monocrystal. The crystallographic orientation is shown in Fig. 17.14(a); Fig. 17.14(b) shows the traces of the {111} on the plane corresponding to the surface of the crystal. The hardness indentation of Fig. 17.14(c) produces a clear pattern of slip markings which are parallel to the respective traces in Fig. 17.14(b). This actually serves to establish the slip planes for this austenitic (FCC) alloy. As expected, they are {111}.

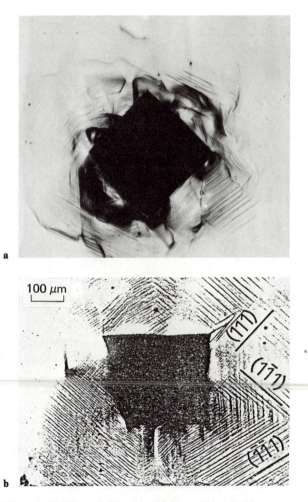

Figure 17.15 (a) Hardness indentation in Fe–34.5% Ni polycrystal in *postexplosionem* condition; (b) Fe–34.5% Ni monocrystal prior to shocking. Typical hardness indentation with slip markings covering a greater extension than in Fig. 17.14(c) and without inducing substantial "bulging out."

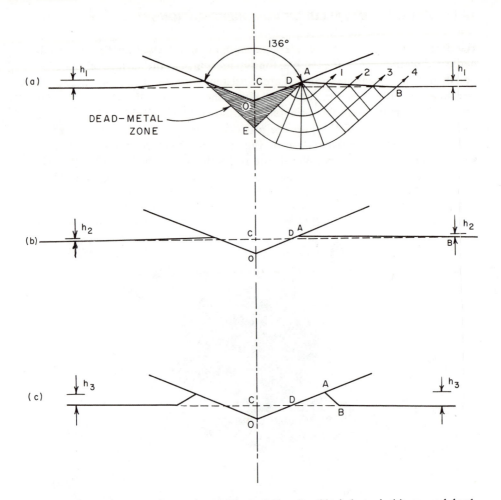

Figure 17.16 Schematic representation of deformation produced by indenter in (a) non-work-hardening metal; (b) work-hardening metal; (c) work-softening metal.

Figure 17.15(a) shows a Vickers indentation for the same material under a polycrystalline condition. The irregular bulging of the surface is clearly shown; the compatibility requirements between the deformation of the various grains results in the irregular deformation. The grain boundaries are made visible by the intersection of different slip systems.

In contrast with Fig. 17.14, Fig. 17.15(b) shows the crystal with almost the same orientation, but in the annealed condition (prior to shock hardening). The slip markings extend to regions much farther away from the indentation. The bulging at the regions adjoining the indentation is minimum compared with Fig. 17.14(c). In the latter, the photograph is partially darkened because of the shade created by the bulge. This darkening renders the measurement of hardness very difficult. The explanation for this phenomenon is provided by Fig. 17.16. The tensile responses of annealed and shock-hardened material are given at the top of the figure. Whereas annealed material work hardens when subjected to a tensile stress, Meyers et al.[26] showed that the response of the shocked alloy was characterized by work softening. The arguments given below are not restricted to work-softening materials; they are also applicable to materials exhibiting no work hardening or a low work-hardening rate.

In the material exhibiting a high work-hardening rate, the region $\tau_{max} > \tau_y$ (where τ_y is the shear yield stress) flows plastically, becoming intrinsically stronger. Hence, its ability to withstand higher elastic stresses increases, and the elastic stresses are not so easily accommodated. A rather large plastic deformation region is generated under the indenter. This is shown in the sequence in the right-hand side of Fig. 17.16. On the other hand, in the non-work-hardening material, the resistance to plastic deformation does not increase in the plastically deformed region, and deformation is concentrated there; a much more localized plastic deformation zone results. The requirement of constant volume (volume displaced by indenter = volume of bulge) establishes that the bulge will have a much greater height. In Fig. 17.16 we see that $h_1 > h_2$ because of the smaller extent of the plastic deformation region. The extent of the plastic deformation region along the surface was found to be over six times the diameter of the indentation for annealed nickel (which has a high initial work-hardening rate). We can calculate the corresponding angle $\beta - \alpha$ (see Fig. 17.13) and will find it to be so small that the bulging will barely be perceptible; consequently, h is very low.

Böklen[27] did a careful analysis of the "bulge" profiles with a special apparatus, and the results confirm the ones presented in Figs. 17.14 to 17.16. Figure 17.17 shows these profiles for annealed and work-hardened steel. Notice that the vertical and horizontal scales are different. A 90° conical indenter was used and it can be seen that the work-hardened mild steel exhibited a bulge twice as high as the annealed one (60 μm versus 30 μm). These results clearly show that the analysis conducted by Shaw and de Salvo[28] should be

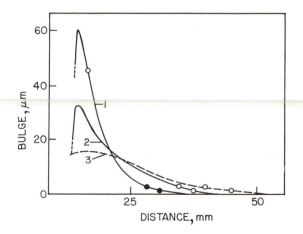

Figure 17.17 Mean wall curves for rolled (curve 1) and annealed (curve 2) mild steel and for annealed 18–8 stainless steel (curve 3). (Adapted with permission from [27], p. 112.)

revised. They assumed that in annealed metals there was *no* bulging, but that the volume was actually decreased by the elastic residual stresses. A more feasible explanation is that the constancy of volume is maintained but that the region of plastic deformation is greatly increased.

EXERCISES

17.1. The following hardness data were obtained on nickel using a 10-mm spherical indenter:

Load (kgf)	Indentation Diameter (mm)
500	3.3
1000	3.9
1500	4.8
2500	5.9
3000	6.4

(a) Which is the best load to obtain the HB?

(b) Find the Brinell and Meyer hardness for all the loads.

(c) If a load of 2000 kgf were applied, what indentation diameter would you expect?

17.2. There is one value of Brinell hardness that should correspond numerically with the Vickers hardness V. Determine it.

17.3. Obtain a relationship between the Rockwell B hardness and the Brinell hardness. The Rockwell hardness is given by

$$HR = 130 - 500 \, \Delta t$$

where t is the depth of the impression in millimeters. Some assumptions are needed in the derivation.

17.4. Prove Eq. 17.12.

17.5. Schematically show, drawing their stress–strain curves in uniaxial tension, how two metals having widely different yield stresses can have the same hardness.

REFERENCES

[1] M.C. Shaw, in *The Science of Hardness Testing and Its Research Applications*, J.H. Westbrook and H. Conrad (eds.), ASM, Metals Park, Ohio, 1973, p. 1.

[2] D. Tabor, *Brit. J. Appl. Phys.*, 7 (1956) 159.

[3] C.H. Bierbaum, *Trans. Am. Soc. Steel Treat.*, 80 (1930) 1009.

[4] D. Tabor, *The Hardness of Metals*, Oxford University Press, London, 1951, p. 115.

[5] E.R. Petty, in *Techniques of Metals Research*, Vol. 5, Pt. 2, R.F. Bunshah (ed.), Wiley-Interscience, New York, 1971, p. 157.

[6] W. Weiler, cited in [1], p. 16.

[7] E. Meyer, *Z. Ver., Dtsch. Ing.*, *52* (1980) 645, 740, and 835.

[8] R.F. Bunshah and R.W. Armstrong, cited in [1], p. 318.

[9] *Standard Test Methods for Rockwell Hardness and Rockwell Superficial of Metallic Materials*, ASTM 18–74, ASTM, Philadelphia.

[10] S.P. Rockwell, *Trans. Am. Soc. Steel Treat.*, *2* (1922) 1013.

[11] H.E. Davis, G.E. Troxel, and C.T. Wiskocil, *The Testing and Inspection of Engineering Materials*, McGraw-Hill, New York, 1941, p. 149.

[12] *Standard Hardness Conversion Tables for Metals*, ASTM 140–78, ASTM, Philadelphia.

[13] G.E. Dieter, *Mechanical Metallurgy*, 2nd ed., McGraw-Hill, New York, 1976, p. 398.

[14] *Standard Methods for Vickers Hardness of Metallic Materials*, ASTM E92–72, ASTM, 1916 Race St., Philadelphia, 19103.

[15] B.W. Mott, *Micro-indentation Hardness Testing*, Butterworth, London, 1956.

[16] H. Bückle, *Met. Rev.*, *4* (1959) 49.

[17] K.T. Aust, R.E. Hanemann, P. Niessen, and J.H. Westbrook, *Acta Met.*, *16* (1968) 291.

[18] F.W. Daniels and C.G. Dunn, *Trans. ASM*, *41* (1949) 419.

[19] R.W. Armstrong and A.C. Raghuram, cited in [1], p. 174.

[20] F. Knoop, G. Peters, and W.B. Emerson, *J. Res. Natl. Bur. Stand.*, *23* (1939) 39.

[21] D. Lee, cited in [1], p. 147.

[22] H. Hertz, *Gesammelte Werke* (Collected Papers), Leipzig, 1894; also in *Hertz's Miscellaneous Papers*, Macmillan, London, 1896.

[23] N.E. Muskhelishvili, *Some Basic Problems of the Mathematical Theory of Elasticity*, Noordhoff, Leyden, The Netherlands, 1975, p. 510.

[24] M.C. Shaw, in *Mechanical Behavior of Materials*, F.A. McClintock and A.S. Argon (eds.), Addison-Wesley, Reading, Mass., 1966, p. 443.

[25] R. Hill, E.H. Lee, and S.J. Tupper, *Proc. Roy. Soc.* (*London*), *188*, (1947) 273.

[26] M.A. Meyers, L.E. Murr, C.Y. Hsu, and G.A. Stone, *Mater. Sci. Eng.*, *57* (1983) 113.

[27] R. Böklen, cited in [1], p. 109.

[28] M.C. Shaw and C.J. de Salvo, *Metals Eng. Quart.*, *12* (1972) 1.

SUGGESTED READING

DAVIS, H.E., G.E. TROXELL, and C.T. WISKOCIL, *The Testing and Inspection of Engineering Materials*, McGraw-Hill, New York, 1941, p. 137.

DIETER, G.E., *Mechanical Metallurgy*, 2nd ed., 1976, p. 383.

O'NIEL, H., *Hardness Measurements of Metals and Alloys*, 2nd ed., Chapman & Hall, London, 1967.

PETTY, E.R., "Hardness Testing," in *Techniques of Metals Research*, Vol. 5, Pt. 2, R.F. Bunshah (ed.), Wiley-Interscience, New York, 1971., p. 157.

SHAW, M.C., in *Mechanical Behavior of Materials*, F.A. McClintock and A.S. Argon (eds.), Addison-Wesley, Reading, Mass., 1966, p. 443.

TABOR, B., *The Hardness of Metals*, Oxford University Press, London, 1951.

WESTBROOK, J.H., and H. CONRAD (eds.), *The Science of Hardness Testing and Its Research Applications*, ASM, Metals Park, Ohio, 1973.

WILLIAMS, S.R., *Hardness and Hardness Measurements*, ASM, Cleveland, Ohio, 1942.

1978 Annual Book of ASTM Standards, Pt. 10, ASTM, Philadelphia, 1978.

Chapter 18

FORMABILITY TESTING

‡18.1 IMPORTANT PARAMETERS

The essential features of sheet-metal forming are described in Section 2.5.8 and the reader is referred to it. An excellent introductory overview is provided by Hecker and Ghosh [1]. *Deep drawing* and *stretching* are the two main processes involved in most sheet-metal forming operations. In a stamping operation one part of the blank might be subjected to a deformation process similar to deep drawing (thickness increasing with time) while a neighboring region is being stretched (thickness decreasing with time). In deep drawing the material is required to contract circumferentially, while in stretching the stresses applied on the sheet are tensile in all directions. Sheet-metal forming is evolving from an art into a science, and important material parameters have been identified. These material properties are obtained in special tests and allow a reasonable prediction of the blank in the actual sheet-forming operation.

The work-hardening rate n is important, because it determines the onset of necking (tensile instability), an undesired feature. According to Considère's criterion (see Chapter 16), n is equal to ϵ_u, the uniform strain. Hence, the higher the n, the higher the ϵ_u. The strain-rate sensitivity m is also an important parameter, because it also helps to avoid necking. If m is positive, the material becomes stronger at incipient necks because the strain rate in the neck region is higher (see Section 16.1.5). The parameter R (through-thickness plastic anisotropy) is also important; it is defined and discussed in Section 18.2. The greater the resistance to "thinning" in stretching, the better is the formability. This resistance to thinning corresponds to a R value greater than 1: The strength in the thickness direction is higher than the strength in the plane of the sheet. The three parameters n, m, and R are readily obtained in a tensile test (Section 18.3).

Additional important information on workability of sheets is provided by the yield and flow loci. Chapter 2 gives a preliminary description of yield criteria and how they are graphically presented in a plane-stress situation. The experimental determination of the yield locus and its expansion as plastic deformation takes place is conducted in biaxial tests (Section 18.4).

Punch–stretch tests have been conducted for many years to determine the ability of sheets to withstand stretching. Two of these tests are the Olsen and Erichsen tests, which consist simply of stretching a sheet over an advancing punch with a rounded head and determining the fracture point. The modern counterpart of these older but reliable tests is the forming-limit curve; they are described in Section 18.5. The circle-grid analysis, which consists of applying a circle grid to the blank and of measuring the strains in the critical regions of the stamped part, is also described in this section.

18.2 PLASTIC ANISOTROPY

Elastic deformation under anisotropic conditions is described by elastic constants, whose number can vary from 21 for the most anisotropic solid to 3 for one exhibiting cubic symmetry (for isotropic solids, the number of independent elastic constants is two). In a similar way, plasticity increases in complexity as the anisotropy of the solid increases. Chapter 2 covers only the isotropic case, and in a very superficial way. In polycrystals, anisotropy in plasticity is more the rule than the exception. Essentially, there are two sources of anisotropy. First, texture, in which the grains are not randomly oriented, but have one or more preferred orientations. Texturing is often introduced by deformation processing. Well-known and well-characterized textures accompany cold rolling, wire drawing, and extrusion. This type of anisotropy is also called "crystallographic anisotropy." Second, anisotropy is produced by the alignment of inclusions or second-phase particles along specific directions. When steel is produced, the inclusions existing in the ingot take the shape and orientation of the deformation process (rolling). These inclusions, such as MnS, produce mechanical effects called "fibering." This type of anisotropy is also known as "mechanical anisotropy." Whereas crystallographic anisotropy can strongly affect the yield stress, mechanical anisotropy manifests itself (usually) only in the later stages of deformation, influencing fracture.

Figure 18.1 shows the effect of texture on a deep-drawn cup. This effect is known as "earing." The sheet exhibited, prior to drawing, different yield stresses along different directions. The orientation in which the sheet is softer is "drawn in" faster than the "harder" direction, resulting in "ears." The number of these ears (four) actually shows the type of texture. Figure 18.2, on the other hand, illustrates the effect of inclusions on the formability of an alloy.

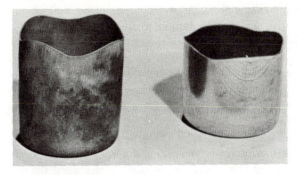

Figure 18.1 "Ears" formed in deep-drawn cups due to in-plane anisotropy. (Reprinted with permission from [2], p. 189.)

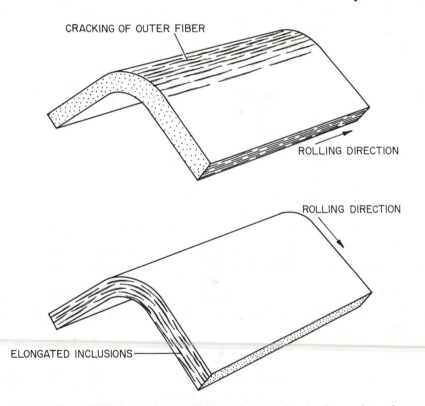

CRACKING OF OUTER FIBER

ROLLING DIRECTION

ROLLING DIRECTION

ELONGATED INCLUSIONS

Figure 18.2 Effect of "fibering" on formability. Bending operation is often an integral part of sheet-metal forming, particularly in making flanges so that the part can be attached to another part. During bending the fibers of the sheet on the outside of the bend are under tension and the inside ones are under compression. Impurities introduced in the metal as it was made become elongated into "stringers" when the metal is rolled into sheet form. During bending the stringers can cause the sheet to fail by cracking if they are oriented perpendicularly to the direction of bending (top). If they are oriented in the direction of the bend (bottom), the ductility of the metal remains normal. (Adapted with permission from [1], p. 107.)

Fracture is much more probable if the sheet is bent along the second-phase strings than if it is bent perpendicular to them.

In the paragraphs that follow, the equation that describes the yield locus for anisotropic materials is derived; this equation is an ellipse essentially identical to that one described by the von Mises yield criterion in plane stress (Section 2.2.4). The ellipse is distorted, however, as will be seen later.

While the most anisotropic crystal would render the plasticity treatment prohibitively complex, there is one type of anisotropy that can be studied without excessive complications. The type of response displayed by wood is a good illustration of this anisotropy. Wood has different yield stresses along the three directions defined by the wood fibers and by the normals to it. Similarly, a rolled sheet or slab of metal will exhibit orthotropic plastic properties; the rolling direction, transverse direction, and thickness direction define the three axes. The condition of orthotropy is that there are three mutually perpendicular symmetry planes whose intersection defines the axes of orthotropy. Hill[3] developed an analysis that begins with the general equation

$$F(\sigma_{22} - \sigma_{33})^2 + G(\sigma_{33} - \sigma_{11})^2 + H(\sigma_{11} - \sigma_{22})^2$$

$$+ 2L\sigma_{23}^2 + 2M\sigma_{13}^2 + 2N\sigma_{12}^2 = 1 \tag{18.1}$$

This equation reduces itself to Eq. 18.2 for the case of isotropy when

$$F = G = H = L = M = N = \frac{1}{2\sigma_y^2}$$

where σ_y is the yield stress (in any direction):

$$\sigma_y = \frac{\sqrt{2}}{2} [(\sigma_{11} - \sigma_{22})^2 + (\sigma_{11} - \sigma_{33})^2 + (\sigma_{22} - \sigma_{33})^2$$

$$+ 2\sigma_{12}^2 + 2\sigma_{13}^2 + 2\sigma_{23}^2]^{1/2} \tag{18.2}$$

Equation 18.2 is the von Mises yield criterion when it is referred to a general system of axes. It reduces itself to Eq. 2.5 when $\sigma_{12} = \sigma_{13} = \sigma_{23} = 0$.

Differentiating Eq. 18.1 with respect to σ_{13} and following the procedure described by Hill[3] and Backofen,[4] we obtain equations similar to the Levy–von Mises equations (Eqs. 2.19):

$$d\epsilon_{11} = ds[H(\sigma_{11} - \sigma_{22}) + G(\sigma_{11} - \sigma_{33})]$$

$$d\epsilon_{22} = ds[F(\sigma_{22} - \sigma_{33}) + H(\sigma_{22} - \sigma_{11})] \tag{18.3}$$

$$d\epsilon_{33} = ds[G(\sigma_{33} - \sigma_{11}) + F(\sigma_{33} - \sigma_{22})]$$

Here, ds has the same significance as $d\epsilon_{eff}/\sigma_{eff}$ has in the Levy–von Mises equations. For the case of isotropy, ds can be calculated by comparing the two sets of equations.

Two particular cases of orthotropic anisotropy are usually discussed where applications in thin sheets are concerned. The first may be called in-plane isotropy. If one defines the plane of the sheet as $x_1 x_2$, the transverse direction is

ox_3. In-plane isotropy exists when the yield stress in all directions within the sheet plane are the same (i.e., $\sigma_{y1} = \sigma_{y2}$). However, σ_{y3} is not necessarily the same. In-plane anisotropy exists when $\sigma_{y1} \neq \sigma_{y2}$, usually $\sigma_{y1} > \sigma_{y2}$ in rolling (σ_{y1} is the rolling and σ_{y2} is the transverse direction). In deep drawing, this causes the well-known phenomenon of earing. The material is drawn more easily along the soft directions, and the resultant cup has "ears" along the stronger directions. Figure 18.1 shows a typical example.

It is assumed that the axes Ox_1, Ox_2, and Ox_3 (thickness direction) are the principal axes of stress which coincides with the principal axes of strain. It is assumed that three tensile tests are conducted along these directions, and that yield stresses of σ_{y1}, σ_{y2}, and σ_{y3} are obtained.

Parameters R and P are defined as the tensile transverse strain ratio:

$$R = \frac{d\epsilon_2}{d\epsilon_3} \tag{18.4}$$

$$P = \frac{d\epsilon_1}{d\epsilon_3} \tag{18.5}$$

If the material exhibits in-plane isotropy,

$$R = P$$

If we could experimentally determine the yield stresses σ_{y1}, σ_{y2}, and σ_{y3} of uniaxial tensile specimens along the principal directions Ox_1, Ox_2, and Ox_3, we could obtain relationships between the coefficients F, G, and H. Since these are the principal directions, the terms having shear stresses in Eq. 18.1 fall off:

$$F(\sigma_2 - \sigma_3)^2 + G(\sigma_3 - \sigma_1)^2 + H(\sigma_1 - \sigma_2)^2 = 1 \tag{18.6}$$

For a tensile test along direction Ox_1, yielding is reached at

$$G + H = \frac{1}{\sigma_{y1}^2} \tag{18.7}$$

For a tensile test conducted along Ox_2, yielding is reached at

$$F + H = \frac{1}{\sigma_{y2}^2} \tag{18.8}$$

Finally, for a test conducted along Ox_3,

$$F + G = \frac{1}{\sigma_{y3}^2} \tag{18.9}$$

For the in-plane isotropy case,

$$\sigma_{y1} = \sigma_{y2} = \sigma_{y(1,2)}$$

Equations 18.7 and 18.8 yield

$$G = F$$

Hence,

$$G = F = \frac{1}{2\sigma_{y3}^2}$$

and

$$H = \frac{1}{\sigma_{y(1,2)}^2} - \frac{1}{2\sigma_{y3}^2}$$

Equation 18.6 becomes

$$(\sigma_2 - \sigma_3)^2 + (\sigma_3 - \sigma_1)^2 + \left(\frac{2\sigma_{y3}^2}{\sigma_{y(1,2)}^2} - 1\right)(\sigma_1 - \sigma_2)^2 = 2\sigma_{y3}^2 \qquad (18.10)$$

The next step in obtaining a relationship between the stresses in the deformation of thin sheets is to apply the plane stress condition to Eq. 18.10. This corresponds to making $\sigma_3 = 0$, because the normal stresses along the thickness directions are zero due to the reduced thickness of the sheet.

$$\sigma_1^2 + \sigma_2^2 + 2\sigma_1\sigma_2 \left(\frac{\sigma_{y(1,2)}^2}{2\sigma_{y3}^2} - 1\right) = \sigma_{y(1,2)}^2 \qquad (18.11)$$

When $\sigma_{y(1,2)} = \sigma_{y3}$, Eq. 18.11 reduces to

$$\sigma_1^2 + \sigma_2^2 - \sigma_1\sigma_2 = \sigma_y^2$$

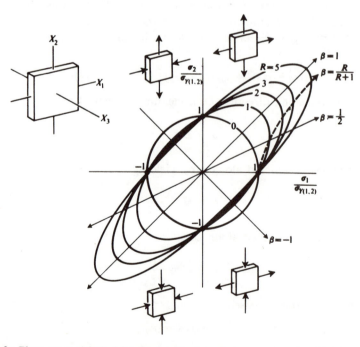

Figure 18.3 Plane–stress yield loci for sheets with planar isotropy or textures that are rotationally symmetric about the thickness direction, x_3. (Values of R indicate the degree of anisotropy $= \sigma_2/\sigma_1$.) (Adapted with permission from [4], p. 49.)

This is the equation of an ellipse that describes the yield locus in plane stress; as we recall from Chapter 2, it corresponds to the von Mises yield (or flow) criterion. In the case of $\sigma_{y(1,2)} \neq \sigma_{y3}$, important conclusions can be drawn from Eq. 18.11, or its equivalent form below, obtained by applying Eqs. 18.3 and 18.4 to Eq. 18.11:

$$\frac{\sigma_{y3}}{\sigma_{y(1,2)}} = \sqrt{\frac{1+R}{2}} \tag{18.12}$$

When $R < 1$, or $\sigma_{y3} < \sigma_{y(1,2)}$, the ellipse is shortened along its major axis. When $R > 1$, or $\sigma_{y3} > \sigma_{y(1,2)}$, the ellipse is stretched along its major axis. This has an important effect on formability, and can be seen in Fig. 18.3. Values of R varying from 0 to 5 are represented. When a sheet is stretched uniformly, we can assume that $\sigma_y = \sigma_2$; hence, we are stationed along the major axis of the ellipse. The stress at which the sheet will plastically flow is strongly dependent on the $\sigma_{y3}/\sigma_{y(1,2)}$ ratio. Amazingly, we can say that the flow in the Ox_1Ox_2 plane is determined by the flow stress in the Ox_3 direction. This is an important conclusion. The particular case in which $\sigma_{y3} > \sigma_{y(1,2)}$ is called "texture hardening"; this corresponds to values of the parameter R greater than 1.

"Texture strengthening" has been taken advantage of in some alloy systems; HCP alloys are especially amenable to "texture hardening" because they twin readily, and twinning is an inherently anisotropic phenomenon. Figure 18.4 shows how a Ti–4% Al–0.25% O_2 alloy is affected by in-plane anisotropy. High values of R lead to desirable formability, because the alloy resists thinning during deep drawing. FCC metals, on the other hand, tend to have values of R[6,7] below 1. Favorable R ratios are readily developed in low-carbon steels; for AK (aluminum killed) and IF (interstitial free) steels they vary typically between 1.5 and 2.[8]

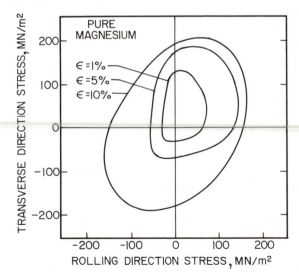

Figure 18.4 Yield loci for pure magnesium. (Adapted with permission from [5], p. 659.)

18.3 TENSILE TESTS

The tensile test (described in detail in Chapter 16) can provide values of the parameters m, n, and R. Strain-rate changes readily provide values of m as described in Section 16.1.5 and shown in Fig. 16.12 (a $\sigma = k\dot{\epsilon}^m$ is assumed). The work-hardening rate n can be found from Considère's criterion ($n = \epsilon_u$). A $\sigma = k\epsilon^n$ equation is usually assumed. Monitoring the change in thickness of a sheet tensile specimen as a function of longitudinal strain, one obtains R. Table 18.1 shows the mechanical properties of a number of sheet alloys.

18.4 BIAXIAL TESTS

Multiaxial loading experiments are, obviously, much more complex than uniaxial tests. Hecker[7] describes these tests, pointing out their limitations. His conclusions are summarized here. The objective of a biaxial test is to provide a flow (or yield) locus, similar to the ones shown in Fig. 18.4. Several types of experimental setups have been used over the years for sheets:

1. Combined axial load–torsion in a thin-walled tube.
2. Combined axial load–internal pressure in a thin-walled tube.
3. In-plane biaxial tension of a cross-shaped sheet specimen.
4. Hydraulic bulging of thin sheets.
5. Microhardness indentations on three orthogonal planes in a sheet; this technique is described in Section 17.3.

A plane strain biaxial test on sheet was developed by Lee and Backofen.[9]

Of all these techniques, the ones preferred by Hecker are the tubular biaxial tests (types 1 and 2). The introduction of servohydraulic biaxial testing machines has been a significant development in biaxial testing. Some of the difficulties are the precision machining and elaborate fixturing required. It can be easily seen that the axial-tension biaxial tubular test easily generates instability (buckling). An additional problem is that the metallurgical structure of the tubular material may not be identical to that of the sheet.

One of the applications of biaxial testing is the study of the response of fuel-cladding materials such as stainless steel and Zircaloy.

‡18.5 PUNCH–STRETCH TESTS AND FORMING-LIMIT CURVES (OR KEELER–GOODWIN DIAGRAMS)

The ideal test is the one that predicts exactly the performance of a material. The m, n, and R values are insufficient to predict the formability, and tests more closely resembling the actual plastic-forming operations have been used for a long time. The main parameter that they can provide is the strain to fracture. These tests are called punch-stretch tests or simply "cupping" tests.

The punch–stretch (or cupping) test consists simply of clamping a blank firmly on its edges between two rings or dies; the next step is to force a plunger or punch through the center area of the specimen enclosed by the area of the ring until the blank fractures. Figure 18.5 shows the dimensions recommended by the ASTM for the test; the testing procedure is described in ASTM Standard E643. Several punch-stretch tests have been developed throughout the years; the Olsen, Erichsen, Guillery, Wazau, and so on. In the Olsen test, the depth of the cup is indicated by a special gage. The pressure applied is measured during the test. The point at which it starts to drop indicates the onset of instability (necking), closely followed by fracture. The Erichsen test uses a conical punch with a smooth spherical end; as in the Olsen test, the depth of the cup is measured by means of a mirror device. Unfortunately, these simple "cupping" tests do not predict satisfactorily the formability of a sheet; only gross differences in formability can be predicted. This has led to the development of improved tests, described in the next paragraphs. Nevertheless, "cupping" tests are routinely used for inspection purposes since they provide a quick indication of ductility; they also show the change in surface appearance of the sheet upon forming. Two important defects appear in stamping.

1. The orange-peel effect (surface rugosity), due to the large grain size of the blank. The anisotropy of plastic deformation of the individual grains results in an irregular surface, perfectly visible to the naked eye, when the grain size is large.
2. Stretcher strains are produced when Lüders bands are formed in the forming process. The interface between the Lüders band and undeformed materials exhibits a step perfectly visible to the naked eye. This is an undesirable feature that can be eliminated by either prestraining the sheet prior to forming (beyond the Lüders band region) or by alloying the material in such a way as to eliminate the yield drop and plateau from the stress–strain curve. In low-carbon steels, Lüders bands are formed by the interactions of carbon and nitrogen atoms with dislocations. After a process called temper rolling, the susceptibility is eliminated; however, it can return after aging. This problem is easily solved by flexing the sheet by effective roller leveling just prior to forming.[11]

The poor correlation between the common "cupping" test and the actual performance of the metal led investigators to look at some more fundamental parameters. The first breakthrough came in 1963, when Keeler and Backofen[12] found that the localized necking required a critical combination of major and minor strains (along two perpendicular directions in the sheet plane). This concept was extended by Goodwin to the negative strain region, and the resulting diagram is known as the Keeler–Goodwin[13] or forming-limit curve (FLC). The FLC is an important addition to the arsenal of testing techniques in formability and is described below after the testing technique presented by Hecker.[14]

Hecker[14] developed a punch–stretch apparatus and technique well suited for the determination of FLC; it is shown schematically in Fig. 18.6(b). It

TABLE 18.1 Formability Properties of Some Sheet Alloys[a,b]

Material	Yield Strength (MN/m²)	Ultimate Tensile Strength (MN/m²)	n	R	Elongation(%)		m^d	Thickness (mm)
					Uniform	Total[c]		
Steel 1A[e]	174.3	292.1	0.240	1.61	26.1	44.3	—	0.84
1B	217.4	299.4	0.250	1.2.	26.9	44.3	0.010	0.86
2A	156.0	296.4	0.232	1.74	27.2	46.7	—	0.84
2B	167.4	290.6	0.244	1.80	27.6	45.3	—	0.99
P2A	178.0	311.1	0.235	1.22	26.6	44.1	0.011	0.86
5	174.5	297.5	0.231	1.68	26.2	44.8	—	0.89
E5	170.9	294.6	0.208	1.79	25.1	43.7	0.012	0.86
E1	210.4	301.6	0.207	1.26	25.0	45.6	—	0.84
E2	213.4	306.4	0.235	1.23	26.7	44.2	—	0.81
10	318.8	363.6	0.160	1.19	18.4	28.0	—	0.89
12	246.3	313.9	0.210	1.05	23.6	40.0	—	0.89
14	225.1	293.0	0.217	0.89	25.5	40.8	—	0.97
E28	255.2	336.0	0.180	0.88	20.0	32.3	—	1.08

E32	272.5	338.4	0.181	0.94	20.8	35.3	—	0.81
31	261.8	381.8	0.183	0.96	21.4	35.9	0.010	0.81
IF	159.2	313.8	0.245	2.00	25.4	43.4	0.012	0.97
GIF	151.9	311.9	0.243	1.98	24.6	45.0	—	1.22
409 Stainless	282.7	452.8	0.210	0.95	21.0	32.2	—	1.27
304 Stainless	354.6	717.8	—	0.96	41.0	45.0	0.010	0.18
70-30 Brass	122.0	336.4	0.540	0.70	55.0	60.0	0	0.94
61-39 Brass	148.3	350.5	0.460	0.60	46.4	51.0	—	0.88
Titanaloy	161.0	183.6	0.054	0.37	4.5	29.0	0.052	0.42
2036-T4 Al	189.3	319.6	0.245	0.78	21.6	23.8	-0.006	1.02
X5020-T4 Al	210.3	348.2	0.250	0.65	22.0	25.0	—	0.89
6151-T4 Al	162.0	270.3	0.235	0.48	20.0	23.0	—	0.81

[a] Reprinted with permission of [8], p. 178.

[b] All properties are averages of tests taken at 0°, 45°, and 90° to the rolling direction. $x_{ave} = (x_0 + x_{90} + 2x_{45})/4$. Hence, the n and R values shown here and used in the text are typically denoted as $\bar{n}$ and $\bar{R}$.

[c] In 50.8-mm gage length.

[d] Strain-rate hardening exponent from $\sigma = K\dot{\epsilon}^m$ determined from a strain-rate change test.

[e] Compositions and heat treatments given by [8], p. 177.

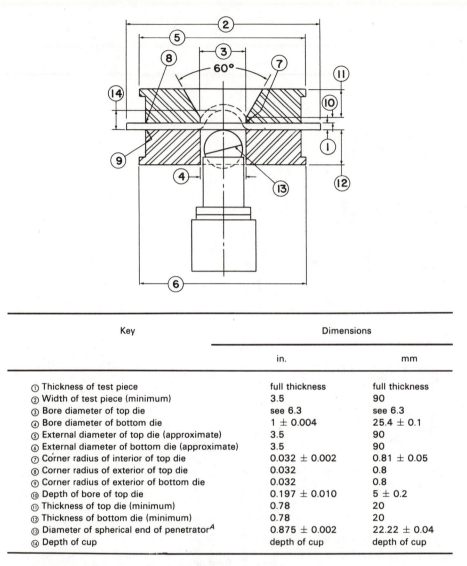

Key	Dimensions	
	in.	mm
① Thickness of test piece	full thickness	full thickness
② Width of test piece (minimum)	3.5	90
③ Bore diameter of top die	see 6.3	see 6.3
④ Bore diameter of bottom die	1 ± 0.004	25.4 ± 0.1
⑤ External diameter of top die (approximate)	3.5	90
⑥ External diameter of bottom die (approximate)	3.5	90
⑦ Corner radius of interior of top die	0.032 ± 0.002	0.81 ± 0.05
⑧ Corner radius of exterior of top die	0.032	0.8
⑨ Corner radius of exterior of bottom die	0.032	0.8
⑩ Depth of bore of top die	0.197 ± 0.010	5 ± 0.2
⑪ Thickness of top die (minimum)	0.78	20
⑫ Thickness of bottom die (minimum)	0.78	20
⑬ Diameter of spherical end of penetrator[A]	0.875 ± 0.002	22.22 ± 0.04
⑭ Depth of cup	depth of cup	depth of cup

[A] "Olsen" Ball, 22.22 mm (⅞ in.); "Erichsen" Ball, 20 mm.

Figure 18.5 Ball punch deformation test tooling. (Reprinted with permission from [10], p. 4.)

consists of a punch with hemispherical head with a 101.6 mm (4 in.) diameter. The die plates are mounted in a servohydraulic testing machine with the punch mounted on the actuator. The hold-down pressure on the die plates (rings) is provided by three hydraulic jacks (the hold-down load is 133 kN). The bead-and-groove arrangement in the rings eliminates any possible drawing in. The specimens are all gridded with 2.54-mm circles by a photoprinting technique. The load versus displacement is measured and recorded during the test, and the maximum load is essentially coincident with localized instability and the

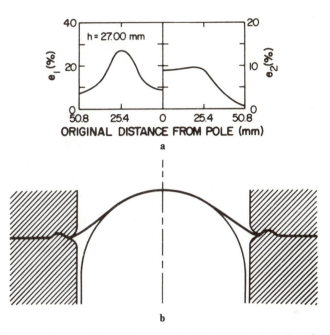

Figure 18.6 Schematic of sheet deformed by punch stretching. (a) Representation of strain distribution: ϵ_1, meridional strains; ϵ_2, circumferential strains; h, cup height. (b) Geometry of deformed sheet.

onset of fracture. A gridded specimen after failure is shown in Fig. 18.7. The circles become distorted into ellipses. The clear circumferential mark is due to necking. The strains ϵ_1 and ϵ_2 are called meridian and circumferential strains, respectively, and are measured at various points when the test is interrupted. Figure 18.6(a) shows how these strains vary with distance from the axis of symmetry of the punch, at the point where the punch has advanced a total distance of $h = 27$ mm. ϵ_1, the meridional strain, is highest at about 25 mm. from the center ($\epsilon_1 \simeq 0.25$); ϵ_2, the circumferential strain, shows a definite plateau. By using sheets with different widths and varying lubricants between the sheet and the punch, different strain patterns are obtained. These tests are conducted to obtain different combinations of minor–major strains leading to failure. Figure 18.8 shows how the FLC curve is obtained. The minor strain (circumferential) is plotted on the abscissa and the major strain (meridional) is plotted on the ordinate axis. Four different specimen geometries are shown. The V-shaped curve (FLC) marks the boundary of the safe–fail zone. The region above the line corresponds to failure; the region below is safe. In order to have both major and minor strains positive, we use a full-sized specimen. By increasing lubrication the major strain is increased; a polyurethane spacer is used to decrease friction. The drawings on the lower left- and right-hand corners of Fig. 18.8 show the deformation undergone by a circle of the grid. When both strains are positive, there is a net increase in area. Consequently, the thickness of the sheet has to decrease proportionately. On the left-hand side of the plot, negative strains are made possible by reducing the lateral dimensions of

Figure 18.7 Sheet specimen subjected to punch–stretch test until necking; necking can be seen by clear line. (Courtesy of S. S. Hecker, Los Alamos National Laboratory.)

the blank. This allows free contraction in this dimension. The strains in a FLC diagram are obtained by carefully measuring the dimensions of the ellipses adjacent to the neck-failure region. It is interesting to notice that diffuse necking (thinning) starts immediately after deformation, whereas localized necking occurs only after substantial forming. Ghosh and Hecker[15,16] develop a semiempirical criterion for localized necking that agrees well with experimental results. This diffuse versus localized neck dicotomy is also analyzed by Ghosh[17] and is discussed in Chapter 16. Ghosh and Hecker[15,16] also compare FLCs determined using a hemispherical punch with an in-plane stretching technique.

FLCs are presently available for a large number of alloys and have proven to be technologically helpful. Figure 18.9 shows the experimental band obtained from various heats for a number of AISI 1008 steels. Hecker obtained his band from the FLCs of about 20 steels with gage thickness 20; rimmed (both annealed-last and temper-rolled) and aluminum-killed steels (in both coated and uncoated conditions) were included. The results show that both Hecker's test and the Keeler–Goodwin press-shop results agree reasonably well. Hence, one can conclude that the band in Fig. 18.9 represents the formability limit of low-carbon steel. Whereas the right-hand portion of the curve represents the strains in stretching, the left-hand portion simulates the strains in deep-drawing ($\epsilon_2 < 0$).

Figure 18.10 shows the FLCs for α- and β-brasses and for three aluminum alloys. The curves have essentially the same shape as the one for low-carbon steels. The poor formability exhibited by aluminum, in comparison with low-

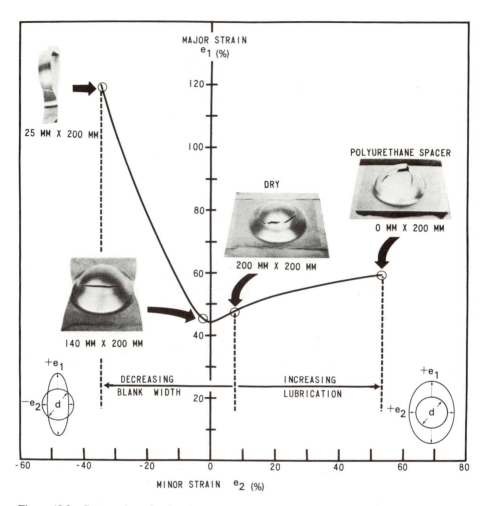

Figure 18.8 Construction of a forming-limit curve (or Keeler–Goodwin diagram). (Courtesy of S. S. Hecker, Los Alamos National Laboratory.)

carbon steel, comes from the strain-rate insensitivity or strain-rate softening exhibited by most alloys.

These FLCs provide helpful guidelines for press-shop formability. Coupled with circle-grid analysis, they can serve as a guide in modifying the shape of stampings. Circle-grid analysis consists of photoprinting a circle pattern on a blank and stamping it, determining the major and minor strains in its critical areas. This is then compared with the FLC to verify the available safety margin. The strain pattern can be monitored with changes in lubrication, hold-down pressure, and size and shape of drawbeads and the blank; this can lead to changes in experimental procedure. Circle-grid analysis also serves, in conjunction with the FLC, to indicate whether a certain alloy might be replaced by another one, possibly cheaper or lighter. During production, the use of occasional

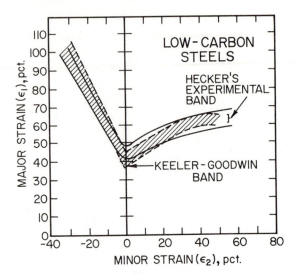

Figure 18.9 Forming-limit curve for low-carbon steels.

circle-grid stampings provides a valuable help with respect to wear, faulty lubrication, and changes in hold-down pressure. Hecker and Ghosh[1] claim that the circle-grid analysis has replaced the craftsman's "feel" for the proper flow of the metal.

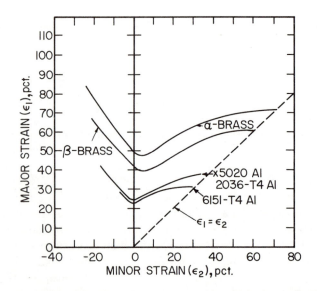

Figure 18.10 Forming-limit curves for α (70–30)-brass, β (61–39)-brass, $\times$5020-T4 Al, 2036-T4 Al, and 6151-T4 Al. (Courtesy of S. S. Hecker, Los Alamos National Laboratory.)

EXERCISES

18.1. From Equation 18.10, obtain Equation 18.11. Then prove that Equation 18.11 represents an ellipse rotated 45° from its principal axis.

18.2. From the data of Table 18.1 would you expect 70–30 or 69–31 brass to exhibit a better formability? Explain in terms of three parameters.

18.3. An annealed sheet of AISI 1040 steel (0.85 mm thick; in-plane isotropy) was tested in uniaxial tension until the onset of necking to determine its formability. The initial specimen length and width were 20 and 2 cm, respectively. At the onset of necking, the length and width were 25 and 1.7 cm, respectively.
  (a) Determine the ratio between the through-thickness and the in-plane yield stress, assuming that R does not vary with strain.
  (b) Draw the flow locus of this sheet, assuming that $\sigma_{y(1,2)} = 180 \text{MN/m}^2$.

18.4. Repeat Exercise 18.3 if the final width of the specimen were 1.9 cm and explain the differences. Which case has a better formability?

18.5. Imagine that you want to perform a circle-grid analysis but you do not have the facilities for photoprinting. Hence, you decide to make a grid of perpendicular and equidistant lines. Can you still, after plastic deformation, determine the major and minor strains from the distorted grid? (Hint: Use the method for determining principal strains.)

REFERENCES

[1] S.S. Hecker and A.K. Ghosh, *Sci. Am.*, Nov. (1976), p. 100.

[2] R.L. Whiteley, W.A. Backofen, J.J. Burke, L.F, Coffin, Jr., N.L. Reed, V. Weiss (eds.), *Fundamentals of Deformation Processing,* Sagamore Army Materials Research Conf. Proc., Vol. 9. New York: Syracuse University Press, 1964, 150, 644.

[3] R. Hill, *Proc. Roy. Soc. (London), A193* (1948) 281.

[4] W.A. Backofen, *Deformation Processing*, Addison-Wesley, Reading, Mass., 1972, p. 46.

[5] E.W. Kelly and W.F. Hosford, Jr., *Trans. TMS-AIME, 242* (1968) 654.

[6] S.S. Hecker, in *The Office of Naval Research Plasticity Workshop*, RPT-75-51, Texas A&M, College Station, Tex., June 1975, p. 190.

[7] S.S. Hecker, in *Constitutive Equations in Viscoplasticity: Conventional and Engineering Aspects*, ASME, New York, 1976, p. 1.

[8] S.S. Hecker, in *Formability: Analysis, Modeling, and Experimentation*, S.S. Hecker, A.K. Ghosh, and H.L. Gegel (eds.), TMS-AIME, New York, 1977, p. 150.

[9] D. Lee and W.A. Backofen, *Trans. AIME, 236* (1966) 1077.

[10] ASTM E643, ASTM, Philadelphia.

[11] H.E. McGannon (ed.), *The Making, Shaping, and Treating of Steel*, U.S. Steel, 9th ed., Pittsburgh, Pa., 1971, pp. 1126, 1260.

[12] S.P. Keeler and W.A. Backofen, *Trans. ASM, 56* (1963) 25.

[13] G.M. Goodwin, "Application of Strain Analysis to Sheet Metal Forming Problems in the Press Shop," *SAE Automotive Eng. Congr.*, Detroit, Jan. 1968, SAE Paper No. 680093.

[14] S.S. Hecker, *Metals Eng. Quart.*, *14* (1974) 30.

[15] A.K. Ghosh and S.S. Hecker, *Met. Trans.*, *6A* (1975) 1065.

[16] A.K. Ghosh and S.S. Hecker, *Met. Trans.*, *5* (1974) 2161.

[17] A.K. Ghosh, in *Formability*: *Analysis, Modeling, and Experimentation*, S.S. Hecker, A.K. Ghosh, and H.L. Gegel (eds.), TMS-AIME, New York, 1978, p. 15.

SUGGESTED READING

HECKER, S.S., Experimental Studies of Yield Phenomena in Biaxially Loaded Metals," in *Constitutive Equations in Viscoplasticity*: *Computational and Engineering Aspects*, ASME, New York, 1976, p. 1.

HECKER, S.S., and A.K. GHOSH, *Sci. Am.*, Nov. 1976, p. 100.

HECKER, S.S., A.K. GHOSH, and H.L. GEGEL (eds.), *Formability*: *Analysis, Modeling, and Experimentation*, TMS-AIME, New York, 1978.

KEELER, S.P., "Forming Limit Criteria-Sheets," in *Advances in Deformation Processing*, J.J. Burke and V. Weiss (eds.), Plenum Press, New York, 1978, p. 127.

KOISTINEN, D.P., and N.M. WANG (eds.), *Mechanics of Sheet-Metal Forming*, Plenum Press, New York, 1978.

Chapter 19

FRACTURE TESTING

‡19.1 WHY FRACTURE TESTING?

There are several good reasons why we should be interested in fracture behavior and fracture testing of metals. Fracture of any material (be that a recently acquired child's toy or a nuclear pressure vessel) is generally an undesirable happening, resulting in economic loss, an interruption in the availability of a desired service, and possibly damage to human beings. Besides, one has good technical reasons to do fracture testing. These include:[1] to compare and to select from candidate materials the toughest (and most economic) one for given service conditions; to compare a given material's fracture characteristics against a specified standard; to be able to predict the effects of service conditions (e.g., corrosion, fatigue, stress corrosion, etc., on the material toughness); and to study the effects of metallurgical changes on material toughness. One or more of these reasons for fracture testing may be involved during the design, material selection, construction, and/or operation of metallic structures. There are two broad categories of fracture tests, qualitative and quantitative. The Charpy impact test exemplifies the former, and the plane-strain fracture toughness (K_{Ic}) test illustrates the latter. We describe briefly important tests in both these categories.

‡19.2 IMPACT TESTING

We saw in Chapter 3 that stress concentrations, like cracks and notches, are sites where failure of a material starts. It has been long appreciated that the failure of a given material in the presence of a notch is controlled by its fracture toughness. A number of tests have been developed and standardized to measure this "notch toughness" of a material. Almost all of these are qualitative and comparative in nature. As pointed out in Chapter 3, triaxial stress state, high strain rate, and low temperature all contribute to a brittle failure of the material. Thus, in order to simulate the most service conditions, almost all of these tests involve a notched sample to be broken by impact over a range of temperatures.

‡19.3 CHARPY IMPACT TEST

The Charpy V-notch impact test is an ASTM standard.[2] The notch is located in the center of test sample. The test sample, supported horizontally at two points, receives an impact from a pendulum of a specific weight on the side opposite that of the notch (Fig. 19.1). The specimen fails in flexure under impact ($\dot{\varepsilon} \sim 10^3$ s^{-1}).

In the region around the notch in the test piece, there exists a triaxial stress state due to plastic yielding constraint there. This triaxial stress state and the high strain rates used propitiate the tendency for a brittle failure. Generally, we present the results of a Charpy test as the energy absorbed in fracturing the test piece. An indication of the tenacity of the material can be obtained by an examination of the fracture surface. Ductile materials show a fibrous aspect, whereas brittle materials show a flat fracture.

A Charpy test at only one temperature is not sufficient, however, because the energy absorbed in fracture drops with decreasing test temperature. Figure 19.2 shows this variation of energy absorbed as a function of temperature for a steel in the annealed and in the quenched and tempered state.[3] The temperature

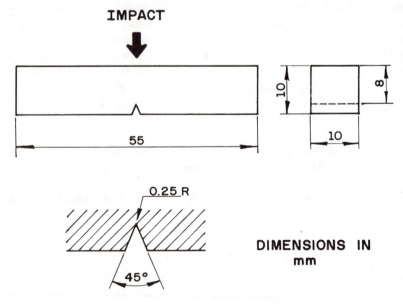

Figure 19.1 Charpy impact test specimen.

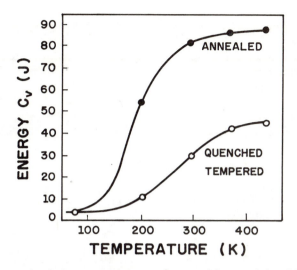

Figure 19.2 Energy absorbed versus temperature for a steel in annealed and in quenched and tempered states. (Adapted with permission from [3], p. 825–829.)

at which there occurs a change from a high-energy fracture to a low-energy one is called the ductile–brittle transition temperature (DBTT). However, as in practice there does not occur a sharp change in energy but instead there occurs a transition zone, it becomes difficult to obtain this DBTT with precision. Figure 19.3 shows how the morphology of the fracture surface changes in the transition region. The greater the fraction of fibrous fracture, the greater the energy that is absorbed by the specimen. The brittle fracture has a typical cleavage appearance and does not require as much energy as the fibrous fracture. BCC and HCP metals or alloys show a ductile-brittle transition, whereas FCC structures do not. Thus, generally, a series of tests at different temperatures is conducted which permits us to determine a transition temperature. This ductile–brittle transition temperature, however arbitrary, is an important parameter in material selection from the point of view of tenacity or the tendency of occurrence of brittle fracture. As this transition temperature is, generally, not very well defined, there exist a number of empirical ways of determining it, based on a certain absorbed energy (e.g., 15 J), change in the fracture aspect (e.g., the temperature corresponding to 50% fibrous fracture), or lateral contraction (e.g., 1%) that occurs at the notch root. The transition temperature depends on the chemical composition, heat treatment, processing, and microstructure of the material. Among these variables, grain refinement is the only method that results in an increase in strength of the material in accordance with the Hall–Petch relation and at the same time reduction in transition temperature. Heslop and Petch[4] showed that the transition temperature T_c depended on the grain size, D, in the following way:

$$\frac{dT_c}{d \ln D^{1/2}} = -\frac{1}{\beta}$$

a

b

c

d

Figure 19.3 Effect of temperature on the morphology of fracture surface of Charpy steel specimen. Test temperatures $T_a < T_b < T_c < T_d$. (a) Fully brittle fracture; (b, c) mixed-mode fractures; (d) fully ductile (fibrous) fracture.

where β is a constant. Thus, a graph of T_c against D will be a straight line with slope $-1/\beta$.

19.4 DROP-WEIGHT TEST

This test is used to determine a reproducible and well-defined ductile–brittle transition in steels. The specimen consists of the steel plate containing a brittle weld on one surface. A saw cut is made in the weld to localize the fracture (Fig. 19.4). The specimen is treated as "simple edge-supported beam" with a stop placed below the center to limit the deformation to a small amount (3%)

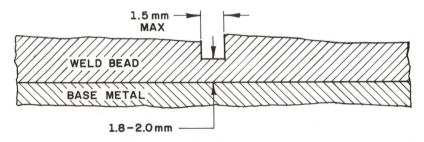

1.5 mm
MAX

WELD BEAD

BASE METAL

1.8–2.0 mm

Figure 19.4 Drop-weight test specimen.

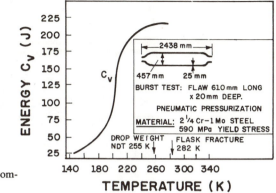

Figure 19.5 Charpy V-notch curve compared with drop-weight NDT.

and prevent general yielding in different steels. The load is applied by means of a freely falling weight striking the specimen side opposite to the crack starter. Tests are conducted at 5-K intervals and a break/no break temperature, called the nil ductility transition (NDT) temperature, is determined. NDT temperature is thus the temperature below which a fast unstable fracture (i.e., brittle fracture) is highly probable. Above this temperature, the toughness increases rapidly with temperature. This transition temperature is more precise than one of the Charpy-based transition temperatures. The drop-weight test uses a sharp crack that moves rapidly from a notch in a brittle weld material, and thus the NDT temperature correlates better with the information from a K_{Ic} test, described in Section 19.6. This test provides a useful link between the qualitative "transition temperature" approach and the quantitative "K_{Ic}" approach to fracture.[1]

The drop-weight test provides a simple means of quality control through the NDT temperature. It (the NDT temperature) can be used to group and classify various steels. For some steels, identification of the NDT temperature can be used to indicate safe minimum operating temperatures for a given stress. That this drop-weight NDT test is more reliable than a Charpy V-notch value of transition temperature is illustrated by Fig. 19.5 for a pressure-vessel steel.[1] The vessel fractured in an almost brittle manner near its NDT temperature, although according to the Charpy curve it was still very tough.

The drop-weight test is applicable primarily to steels in the thickness range 18 to 50 mm. NDT temperature is unaffected by section sizes above about 12 mm; because of the small notch and the limited deformation due to brittle weld bead material, sufficient notch-tip restraint is ensured.

19.5 INSTRUMENTED CHARPY IMPACT TEST

The Charpy impact test described above is one of the most common tests for characterizing the mechanical behavior of materials. The principal advantages of the Charpy test are the ease of specimen preparation and the execution of the test proper, speed, and low cost. However, one must recognize that the common Charpy test basically furnishes information of only a comparative char-

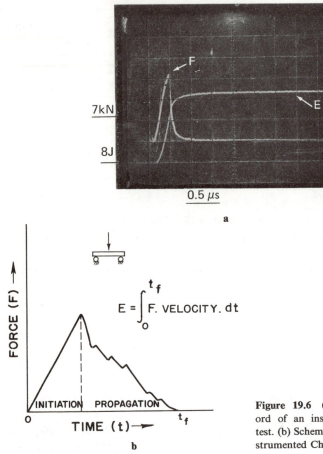

$$E = \int_0^{t_f} F \cdot VELOCITY \cdot dt$$

INITIATION   PROPAGATION

FORCE (F) →

TIME (t) →

7kN

8J

0.5 μs

a

b

Figure 19.6 (a) Typical oscilloscope record of an instrumented Charpy impact test. (b) Schematic representation of an instrumented Charpy impact test.

acter. The transition temperature, for example, depends on the specimen thickness (hence, the need to use standard samples); that is, this transition temperature can be used to compare, say, two steels, but it is not an absolute material property. Besides, the common Charpy test measures the total energy absorbed (E_T), which is the sum of energies spent in initiation (E_i) and in propagation (E_p) of crack (i.e., $E_T = E_i + E_p$). In view of this problem, a test has been developed called the instrumented Charpy impact test.[5] This instrumented impact test furnishes, besides the absorbed energy, the variation of applied load with time. The instrumentation involves the recording of the signal from a load cell on the pendulum by means of an oscilloscope in the form of a load–time curve of the test sample. Figure 19.6(a) shows a typical oscilloscope record and Fig. 19.6(b) shows a schematic representation. This type of curve can provide information about the load at general yield, maximum load, load at fracture, and so on. The energy spent in impact can also be obtained by integration of the load–time curve. From the load–time curve, one can obtain the energy of fracture if the pendulum velocity is known. Assuming this velocity to be constant during the test, we can write the energy of fracture as

$$E' = V_0 \int_0^t P \, dt \tag{19.1}$$

where E' is the total fracture energy based on the constant pendulum velocity, V_0 the initial pendulum velocity, P the instantaneous load, and t the time.

In fact, the assumption of a constant pendulum velocity is not a valid one. According to Augland,[7]

$$E_t = E'(1 - \alpha) \tag{19.2}$$

where E_t is the total fracture energy, $E' = V_0 \int_0^t P \, dt$, $\alpha = E'/4E_0$, and E_0 is the initial pendulum energy. The values of total energy absorbed in fracture computed this way from the load–time curves show a one-to-one correspondence with the energy-absorbed values determined in a conventional Charpy test.[8]

Based on this, we can use Eq. 19.2 for computing, at a given temperature, the initiation and propagation energies. This information, together with the load at yielding, maximum load, and load at fracture, gives the capacity to identify the various stages of the fracture process.

It is well known (see Section 3.4) that the plane-strain fracture toughness (K_{Ic}) test gives a much better and precise idea of the material tenacity and is a material property. However, this test, as will be seen below, possesses certain disadvantages: equipment and specimen preparation is rather expensive, the test is relatively slow and not simple to execute, and so on. Consequently, there have been attempts at developing empirical correlations between the energy absorbed in a conventional Charpy test (C_v) and the plane-strain fracture toughness (K_{Ic}).[9] The reader is warned that such correlations are completely empirical and are valid only for the specific metals tested.[9] The instrumented Charpy test, with samples precracked and containing side grooves in order to assure a plane-strain condition, can be used to determine the dynamic fracture toughness K_{ID}. For ultra-high-strength metals (σ_y very large), $K_{ID} \simeq K_{Ic}$. Thus, we may use instrumented Charpy test to determine K_{Ic} or K_{ID} for very high strength steels. But there is still the need to verify this better and we must check the results obtained from the instrumented Charpy impact test with those obtained from standard ASTM K_{Ic} test, as described in the next section.

19.6 PLANE-STRAIN FRACTURE TOUGHNESS TEST

The fracture toughness K_{Ic} may be determined according to the following standards: ASTM E399/79 or BS 5447/77.[10] The essential steps in the fracture toughness tests involve measurement of crack extension and load at the sudden failure of sample. As it is difficult to measure crack extension directly, one measures the relative displacement of two points on the opposite sides of the crack plane. This displacement can be calibrated and related to real crack front extension.

The typical test samples of tension and bending that are used in the fracture toughness tests according to the ASTM standard are shown in Fig. 19.7. Figure 19.7(c) shows how the size of the specimens can be varied. Tensile and Charpy specimens are also shown for comparison. The relation between the applied load and the crack opening displacement depends on the size of the crack and

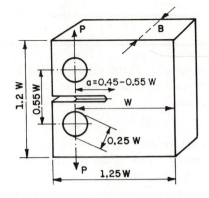

a

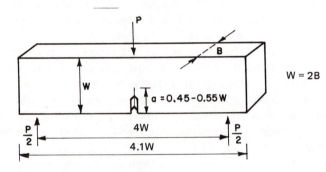

$W = 2B$

b

c

Figure 19.7 Typical ASTM standard plane-strain fracture toughness test specimens: (a) compact tension; (b) bending; (c) photograph of specimens of various sizes showing Charpy and tensile specimens for comparison purposes. (Courtesy of MPA, Stuttgart)

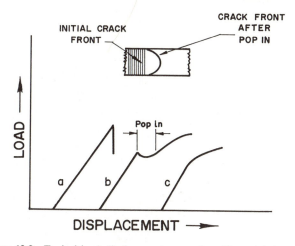

Figure 19.8 Typical load–displacement curves in a K_{Ic} test (schematic).

thickness of the sample in relation to the extent of plastic zones. When the crack length and the sample thickness are very large in relation to the quantity $(K_{Ic}/\sigma_y)^2$, the load–displacement curve is of the type shown in Fig. 19.8(a). The load at the brittle fracture that corresponds to K_{Ic} is then well defined. When the specimen is of reduced thickness, a step called "pop-in" occurs in the curve, indicating an increase in the crack opening displacement without an increase in the load Fig. 19.8(b). This phenomenon is attributed to the fact that the crack front advances only in the center of the plate thickness, where the material is constrained under plane-strain condition. However, near the free surface, plastic deformation is much more pronounced than that at the center, and it approaches the conditions of plane stress. Consequently, the plane-strain fracture advances much more in the central portion of the plate thickness, and in regions of material near the specimen surfaces the failure eventually is by shear.

Figure 19.9 shows the plastic zone at the crack front in a plate of finite thickness. At the edges of the plate ($x_3 \rightarrow \pm B/2$) the stress state approaches that of plane stress. At the center of a sufficiently thick plate, the stress state approaches that of plane strain. This is due to the fact that the ϵ_{33} component of strain is equal to zero at the center as the material there in this direction is constrained, whereas near the edges the material can yield in the x_3 direction and ϵ_{33} is different from zero.

When the test piece becomes even thinner, the plane-stress condition prevails and the load–displacement curve becomes as shown in Fig. 19.8(c). To make valid fracture toughness measurements in plane strain, the influence of the free surface, which relaxes the constraint, must be maintained small. This enables the plastic zone to be constrained completely by elastic material. The crack length must also be maintained greater than a certain lower limit.

Up to this point, the conditions of sample size and the crack length have been discussed in a qualitative way. The lower limits on width, thickness, and

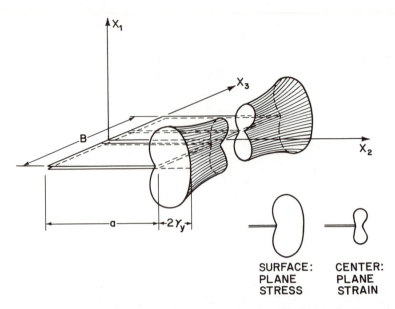

Figure 19.9   Plastic zone at the crack tip in a plate of finite thickness.

crack length all depend on the extent of plastic deformation through the $(K_{Ic}/\sigma_y)^2$ factor. In view of the lack of current knowledge about the exact plastic zone size for the crack in mode I (crack opening mode), it is not yet possible to determine the lower limits of dimension of the test piece theoretically. These lower limits above which K_{Ic} remains constant are determined by means of trial tests. Samples of dimensions smaller than these limits tend to overestimate the K_{Ic} limit.

In the fracture toughness tests, the crack is preferably introduced by fatigue from a starter notch in the sample. The fatigue crack length should be long enough to avoid interference in the crack-tip stress field by the shape of notch.

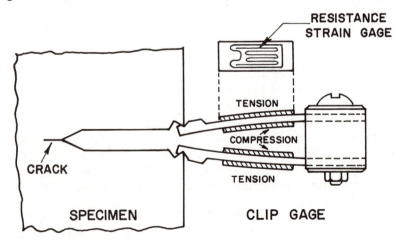

Figure 19.10   Assembly for measuring displacement in a notched specimen.

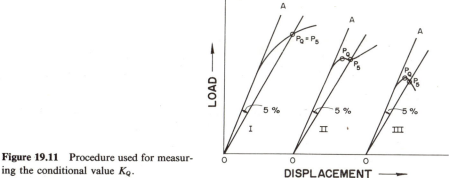

Figure 19.11 Procedure used for measuring the conditional value K_Q.

Under an applied load, the crack opening displacement can be measured between two points on the notch surfaces by various types of transducers. Figure 19.10 shows an assembly for measuring displacement in a notched sample. Electrical resistance measurements have also been used to detect crack propagation. Calibration curves are used for converting displacement measurements and resistance measurements into crack extension.

The load–displacement curves generally show a gradual deviation from linearity and the "pop-in" step is very small (Fig. 19.11). The procedure used in the analysis of load–displacement records of this type can be explained by using the Fig. 19.11. Let us designate the linear slope part as OA. A secant line, OP_5, is then drawn at a slope 5% less than that of line OA. The point of intersection of the secant with the load–displacement record is called P_5. We define the load P_Q, for computing a conditional value of K_{Ic}, called K_Q, as follows: If the load on every point of curve before P_5 is less than P_5, then $P_5 = P_Q$ (case 1, Fig. 19.11). If there is a load more than P_5 and before P_5, this load is considered to be P_Q (cases II and III in Fig. 19.11). In these cases if $P_{max}/P_Q > 1.1$, the test is not a valid one; K_Q does not represent the K_{Ic} value and a new test needs to be done. After determining the point P_Q, K_Q is calculated according to the known equation for the geometry of the test piece used. A checklist of points is given in Table 19.1, and Fig. 19.12 shows schematically the variation of K_c with flaw size, specimen thickness, and specimen width.

TABLE 19.1   Checklist for the K_{Ic} Test

---

1. Dimensions of test piece
   a. Thickness, $B \geq 2.5(K_{Ic}/\sigma_y)^2$
   b. Crack length, $a \geq 2.5(K_{Ic}/\sigma_y)^2$
2. Fatigue pre-cracking
   a. $K_{max}/K_{Ic} \leq 0.6$
   b. Crack front curvature $\leq 5\%$ of crack length
   c. Inclination $\leq 10°$
   d. Length between $0.45W$ and $0.55W$, where W is the width of the test sample
3. Characteristics of load–displacement curve. This is effectively to limit the plasticity during the test and determines if the gradual curvature in the load–displacement curve is due to plastic deformation or crack growth.
   a. $P_{max}/P_Q \leq 1.1$

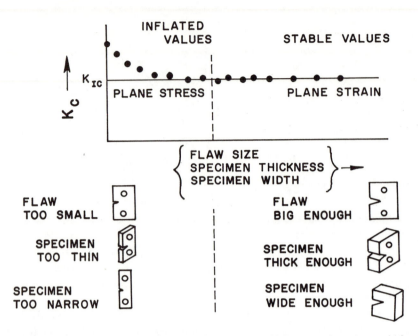

Figure 19.12 Variation of K_c with flaw size, specimen thickness, and specimen width.

19.7 CRACK OPENING DISPLACEMENT TESTING

For crack opening displacement (COD) testing, there exists a British Standards Institution BS 5762:1979 method for crack opening displacement (COD) testing." Basically, the test sample for determining δ_c is a slow bend test specimen similar to the one used for K_{Ic} testing. The proposed BSI method is very similar to the ASTM E399 method for K_{Ic}.

A clip gage is used to obtain the crack opening displacement. During the test, one obtains a continuous record of load, P, versus opening displacement, Δ (Fig. 19.13). In the case of a smooth P–Δ curve, the critical value, Δ_c, is the total value (elastic + plastic) corresponding to the load maximum [Fig. 19.13(a)]. In case the P–Δ curve shows a region of increase in displacement at a constant or decreasing load, followed by an increase in load before fracture, one needs to make auxiliary measurements to determine that this is associated with crack propagation. Should this be so, Δ_c will correspond to the first instability in the curve. If the P–Δ curve shows a maximum and Δ increases with a reduction in P, then either a stable crack propagation is occurring or a "plastic hinge" is being formed. The "Δ_c" in this case [Fig. 19.13(b)], according to the British Standards Institution, is the value corresponding to the point at which a certain specified crack growth has started. If it is not possible to determine this onset of crack propagation, one cannot measure the COD at the start of crack propagation. However, we can measure, for comparative purposes, an opening displacement δ_m, computed from the clip gage output Δ_m, corresponding to the first load maximum. The results in this case will depend on the specimen geometry.

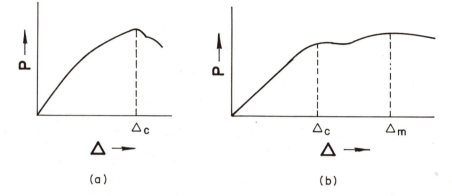

Figure 19.13   Load P versus crack opening displacement Δ records (schematic).

19.7.1 Computation of δ_c

Experimentally, we obtain Δ_c, the critical displacement of the clip gage. We need to obtain δ_c, the critical CTOD (crack-tip opening displacement). The British (COD) Standard contains various methods, all based on the hypothesis that the deformation occurs by a "hinge" mechanism around a center of rotation at a depth of $r(w-a)$ below the crack tip (Fig. 19.14). Experimental calibrations,

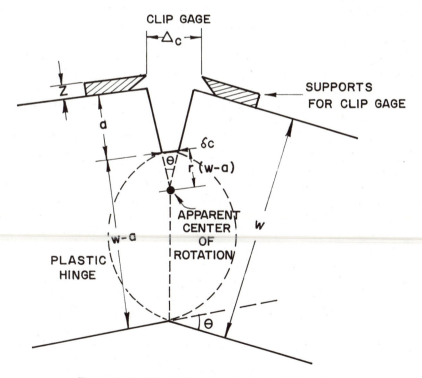

Figure 19.14   "Plastic hinge" mechanism of deformation.

using specimens of up to 50 mm thickness, have shown that for COD in the range 0.0625 to 0.625 mm, to a very good approximation δ_c can be obtained from the relation

$$\delta_c = \frac{(w - a)\Delta_c}{w + 2a + 3z}$$

This relation is derived on the assumption that the deformation occurs by a hinge mechanism about a center of rotation at a depth of $(w - a)/3$ below the crack tip (i.e., $r = \frac{1}{3}$). However, r can be smaller for smaller values of Δ_c. $r \simeq 0$ in the elastic case (very limited plastic deformation at the crack tip) and $r \simeq \frac{1}{3}$ for a totally plastic ligament.

19.8 *J*-INTEGRAL TESTING

The *J*-integral test was standardized by ASTM in 1981. The "Standard Test for J_{Ic}, A Measure of Fracture Toughness" bears the designation E813-81 and covers the determination of J_{Ic}, the critical value of *J*. As pointed out in Section 3.8, the physical interpretation of the *J*-integral is related to the area under the curve load versus the load-point displacement for a cracked sample. Both compact tension

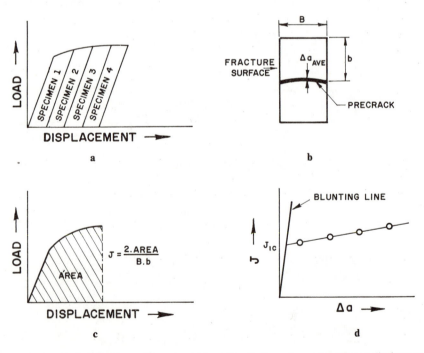

Figure 19.15 Method for determining J_{Ic}: (a) load identical specimens to different displacements; (b) measure average crack extension by heat tinting; (c) calculate *J* for each specimen; (d) plot *J* versus Δa to find J_{Ic}.

and bend specimens can be used. The ASTM recommended procedure requires at least four specimens to be tested. Each specimen is loaded to different amounts of crack extensions (Fig. 19.15). One calculates J for each specimen from the expression

$$J = \frac{2A}{Bb}$$

where A is the area under load versus the load-point displacement curve, B the specimen thickness, and b the uncracked ligament. The value of J so derived is plotted against Δa, the crack extension of each specimen. One way of obtaining Δa is to heat-tint the specimen after test and then break it open. A "best line" through the J points and a "blunting line" from the origin are then drawn. This blunting line (indicating onset of crack blunting due to plastic deformation) is obtained as follows:

$$J = 2\sigma_{\text{flow}} \Delta a$$

where

$\sigma_{\text{flow}} = \dfrac{\sigma_y + \sigma_{\text{UTS}}}{2}$, σ_y is the yield stress and σ_{UTS} is the ultimate tensile stress.

The intercept of the J line and the blunting line gives J_{Ic}. J is related to K:

$$J_{Ic} = \frac{K_{Ic}^2}{E}$$

J_{Ic} defines the onset of crack propagation in a material where large-scale plastic yielding makes direct measurement almost impossible.

Thus, one can use J-integral testing to find K_{Ic} value of a very ductile material from a specimen of dimensions too small to satisfy the requirements of a proper K_{Ic} test.

EXERCISES

19.1. The unique feature of grain refining is that it is the only strengthening mechanism that increases strength as well as toughness (as measured by the Charpy impact test). However, the grain size, generally, does not show any effect on uniform deformation. Why?

19.2. How does the Charpy impact transition temperature of a steel vary with:
(a) Pearlite content?
(b) Free-nitrogen content?
(c) Substitutional solutes?

19.3. Why are impact tests commonly used for steels but not so commonly for nonferrous metals and alloys?

19.4. Describe a few methods of monitoring crack length, pointing out the advantages and disadvantages associated with each one.

19.5. The fracture behavior of metals is affected by temperature, strain rate, and thickness. Show schematically the effect of these three variables on the K_c behavior of a metal.

19.6. Plot schematically the Charpy curves, energy absorbed versus temperature, for three different thicknesses $t_1 > t_2 > t_3$.

19.7. A "stiff" machine is one that imposes a given external displacement irrespective of the load level, while a "soft" machine is one that imposes a given loading rate. What do you think is the effect of machine stiffness on the determination of K_{Ic}?

19.8. "Acoustic emission" is a term applied to low-amplitude stress waves emitted by a solid when it is deformed. Extremely sensitive transducers can be used to study the acoustic emission from structures containing cracklike flaws. Write a short description of the application of acoustic emission to fracture mechanics. (*Ref.*: *Acoustic Emission*, ASTM STP 505, ASTM, Philadelphia, 1972.)

19.9. Describe the use of photoelasticity for determining stress-intensity factors in some fracture mechanics problems.

19.10. Comment on the use of K_{Iscc} as an engineering design parameter.

REFERENCES

[1] W.J. Langford, *Can. Met. Quart.*, *19* (1980) 13.

[2] *Standard Methods and Definitions for Mechanical Testing of Steel Products*, ASTM Standard Method A370, ASTM Annual Standards, Part 10, ASTM, Philadelphia.

[3] J.C. Miguez Suarez and K.K. Chawla, *Metalurgia-ABM*, *34* (1978) 825.

[4] J. Heslop and N.J. Petch, *Phil. Mag.*, *3* (1958) 1128.

[5] *Instrumented Impact Testing*, ASTM STP 563, ASTM, Philadelphia, 1974.

[6] K.K. Chawla and M.R. Krishnadev, unpublished research.

[7] B. Augland, *Brit. Weld. J.*, *9* (1962) 434.

[8] G.D. Fearnehough and C.J. Hoy, *J. Iron Steel Inst.*, *202* (1964) 912.

[9] S.T. Rolfe and J.M. Barsom, *Fracture and Fatigue Control in Structures*, Prentice-Hall, Englewood Cliffs, N.J., 1977, p. 140.

[10] *Standard Test Method for Plane Strain Fracture Toughness of Metallic Materials*, ASTM Standard Method E399-78, ASTM Annual Standards, Part 10, ASTM, Philadelphia.

SUGGESTED READING

ASTM STP 466, ASTM, Philadelphia, 1970.

ASTM STP 514, ASTM, Philadelphia, 1972.

ASTM STP 536, ASTM, Philadelphia, 1973.

ASTM STP 563, ASTM, Philadelphia, 1974.

Rolfe, S.T., and J.M. Barsom, *Fracture and Fatigue Control in Structures*, Prentice-Hall, Englewood Cliffs, N.J., 1977.

Chapter 20

CREEP AND CREEP TESTING

‡20.1 INTRODUCTION

The technological developments since the 1930's have required materials that resist higher and higher temperatures. The applications lie mainly in these three areas:

1. Turbines
2. Chemical and petrochemical industries
3. Nuclear reactors

The degradation undergone by materials under these extreme environments can be classified into two groups:

1. *Mechanical degradation*. The material, in spite of initially resisting the applied loads, undergoes anelastic deformation; its dimensions change with time.
2. *Chemical degradation*. This is due to the reaction of the material with the chemical environment and to the diffusion of external elements into it. Chlorination (which negatively affects the properties of superalloys used in jet turbines) and internal oxidation are examples of chemical degradation.

This chapter deals exclusively with mechanical degradation. The anelastic response of the material is known as "creep"; a great number of high-temperature failures are attributable either to creep or to a combination of creep and fatigue. Creep is characterized by a slow flow of the material; it behaves as if it were viscous. If a mechanical component of a structure is subjected to a constant tensile load, the decrease in cross-sectional area (due to the increase in length resulting from creep) generates an increase in stress; when the stress reaches the stress at which failure occurs statically (ultimate tensile stress), failure occurs. The temperature regime for which creep is important is $0.5T_m < T < T_m$; T_m is the melting point in Kelvin. This is the temperature range in which diffusion is a significant factor. Diffusion, being a thermally activated process, shows an exponential temperature dependence. Below $0.5T_m$ the diffusion coefficient is so low that any deformation mode exclusively dependent on it can effectively be neglected. Section 20.3 presents the various mechanisms responsible for creep. Hence, the critical temperature for creep varies from metal to metal; lead creeps at ambient temperature, whereas for iron, creep becomes important above 600°C.

In spite of the fact that creep has been known since 1834, when Vicat[1] conducted the first experiments assessing the phenomenon, it is only in this century that systematic investigations have been conducted. The creep test is rather simple and consists of subjecting a specimen to a constant load (or stress) and measuring its length as a function of time, at a constant temperature. Figure 20.1 shows the characteristic curve; the ordinate axis shows the strain and the abscissa axis shows time. Three tests are represented in Fig. 20.1; three constant loads corresponding to three stresses, σ_a, σ_b, and σ_c, were used. The creep curves are usually divided into three stages: I, primary or transient; II, secondary, constant rate, or quasi-viscous; and III, tertiary. These stages are shown in Fig. 20.1. This division into stages was made by Andrade,[2] one of the pioneers in its study. Creep stage II in which the creep rate $\dot{\epsilon}$ is constant, is the most important. It can be seen $\dot{\epsilon}_a > \dot{\epsilon}_b > \dot{\epsilon}_c$ as a consequence of $\sigma_a > \sigma_b > \sigma_c$. This creep rate is also known as minimum creep rate, because it corresponds to the inflection point of the curve. In stage III there is an acceleration in the creep rate, leading to eventual rupture of the specimen. Since the strain is a function of time, temperature, and stress, it is natural

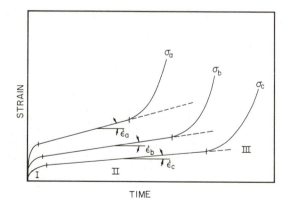

Figure 20.1 Creep tests at constant load and for different engineering stress levels ($\sigma_a > \sigma_b > \sigma_c$).

that these variables have been mathematically correlated. Le May[3] summarizes these equations; the best known are

$$\epsilon = a(1 + \beta t^{1/3}) \exp{(kt)} \qquad \text{Andrade's}[4]$$

$$\epsilon = a + bt^n \qquad \text{Bailey's}[5]$$

Garofalo[6] developed an equation consisting of an instantaneous strain ϵ_0 (the strain at the instant of application of load), a linear time function $\dot{\epsilon}_s t$ (depicting stage II), and an exponential term representing stage I:

$$\epsilon_t = \epsilon_0 + \epsilon_p[1 - \exp{(-mt)}] + \dot{\epsilon}_s t$$

The creep theories attempt to calculate the minimum creep rate from the micromechanical deformation modes. If the temperature is changed and the load is kept constant, we would have a sequence similar to the one in Fig. 20.1; as the temperature increases, so does $\dot{\epsilon}$.

From a fundamental point of view there are significant differences between the constant-load and constant-stress creep tests. The first investigator who was aware of this was Andrade;[2] he built a constant true stress creep machine. For this type of machine, the load has to decrease with an increase in length in such a way that the true stress remains constant. Harrigan[7] describes such a machine. Fullman et al.[8] developed a rather simple system, based on a system of levers.

Cunha and Kestenbach[9] compared the creep rate for stage II in commercially pure aluminum under constant stress and load. As shown in Fig. 20.2, they found a significant difference. The slope of the log ($\dot{\epsilon}_s/D$) versus log σ curve is greater for the constant-load condition [D is the self-diffusion coefficient for aluminum at the temperature of test (554 and 704 K)]. This shows that the creep rate is higher for constant-load tests. Indeed, this should be the case, because the true stress increases continuously due to the reduction in cross-sectional area. The volume can be assumed to be constant and the rate of decrease of the cross-sectional area is equal to the rate of increase in length.

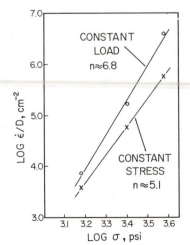

Figure 20.2 Comparison between creep tests at constant stress and load. (Adapted with permission from [9], p. 258.)

Cunha and Kestenbach[9] concentrated on the intermediate stress range, in which the response of the material can be satisfactorily described by a power law ("power law creep"). The following equation applies:

$$\dot{\epsilon}_s = A \sigma^n e^{-Q_c/kT} \tag{20.1}$$

where A and n are constants, $\dot{\epsilon}_s$ the creep rate for stage II, and Q_c the activation energy for creep. This equation will be analyzed in Section 20.3. It is experimentally observed that n is equal to 5 for a number of pure metals. Cunha and Kestenbach[9] concluded that results obtained in constant-load tests cannot be accepted without reticence as representative of constant true stress tests.

Another important difference between the two tests is that the onset of stage III is greatly retarded at constant stress. The dashed lines in Fig. 20.1 show the trajectory that a constant true stress test would follow. From an engineering point of view the creep test at constant load is more important than the one at constant stress; this is because it is the *load*, not the stress, that is maintained constant in engineering applications. On the other hand, fundamental studies with the objective of elucidating the underlying mechanisms should be conducted at constant stress; the study of the substructure evolution of an alloy under an increasing stress condition would be excessively complex.

Another test commonly used in substitution of the creep test is the stress-rupture (or creep-rupture) test. It consists of an accelerated creep test leading to rupture. The important parameter obtained from this test is the time to rupture, while in the regular creep test the minimum creep rate is the experimental parameter sought.

The sections that follow deal with several important aspects. Section 20.2 describes the extrapolation methods used to obtain the response at very large times after conducting more accelerated tests. Some of the creep theories are described in Section 20.3. The very helpful deformation-mechanism maps (Weertman–Ashby maps) are presented in Section 20.4. Some important heat-resisting alloys are described in Section 20.5.

‡20.2 CORRELATION AND EXTRAPOLATION METHODS

The central theme of physical metallurgy is the structure–property–performance triangle. In creep the correlation between properties and performance is very critical, because in certain applications we want to know the performance during an extended period (20 or more years) while the properties (secondary creep rate or stress-rupture life) are only known for a shorter period. In general, industrial equipment operating at a high temperature is designed admitting a certain life. For jet turbines, 10,000 h (about 1 year) is a reasonable value. For stationary turbines, the weight of the components is not so critical; a life of 100,000 h (about 11 years) is the goal. For nuclear reactions, for obvious reasons, we use the criterion of 350,000 h[10] (40 years). A great number of advanced alloys are used in these projects, and the metallurgist does not have on hand results of such lengthy tests. Hence, several extrapolation methods have been developed; they seek to predict the performance of alloys based on tests performed over a shorter period. Silveira and Monteiro[10] state that there is a certain consensus in considering safe extrapolation amplitudes below 3, independent of the extrapolation method. Hence, a test that lasts 33,000 h should be the minimum duration for prediction of 100,000 h of life. The number of parametric methods developed exceeds 30; they are described in a series of articles published in honor of Goldhoff in the *Journal of Engineering Materials and Technology*.[11-14]

Larson and Miller[15] proposed in 1952 a method that is well known to this day; this method correlates the temperature T (in degrees Rankine) with the time for failure t_r, at a *constant* engineering stress σ. The Larson–Miller equation has the form

$$T(\log t_r + C) = m \tag{20.2}$$

In this equation, C is a constant that depends on the alloy and m is a parameter that depends on stress. Hence, if C is known for an alloy, one can find m in a single test. From this result, one can find the rupture times at any temperature, *as long as the same engineering stress is applied*. Hence, the following procedure is adopted. If we want to know the rupture time at a certain stress level σ_a and temperature T_a, we conduct the test at $T_b > T_a$ and stress level σ_a. Substituting these values into Eq. 20.2, we find m. The latter test has a short duration, because the rupture time decreases with temperature at a constant stress. Figure 20.3 shows schematically the family of lines for different levels of applied stress. This figure is the graphic representation of Eq. 20.2. It can be seen that C does not depend on the stress; it is the intersection between the various lines. On the other hand, each line has a different slope, m. Since

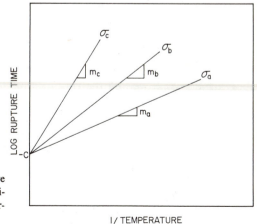

Figure 20.3 Relationship between rupture time and temperature at three levels of engineering stress σ_a, σ_b, and σ_c using Larson–Miller equation ($\sigma_a > \sigma_b > \sigma_c$).

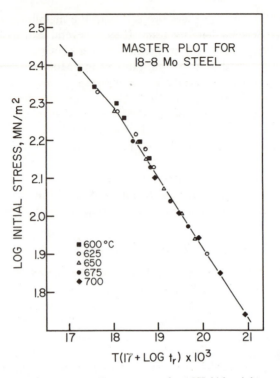

Figure 20.4 Master plot for Larson–Miller parameter for AISI 316 stainless steel. (Adapted with permission from [10], p. 258.)

the value of C is constant for an alloy, we can build a "master plot" that represents the creep response of an alloy over a range of stresses and temperatures. Figure 20.4 shows, as an example, the master plot determined by Silveira and Monteiro[10] for stainless steel AISI 316. Silveira and Monteiro also present typical C values (when t_r is expressed in hours and T in degrees Rankine = degrees Fahrenheit + 460): low-carbon steel ($C = 18$); 18–8 stainless steel ($C = 18$); 25–20 stainless steel ($C = 15$); D − 9 titanium ($C = 20$); and Haynes Stellite No. 34 superalloy ($C = 20$). It can be seen that C fluctuates around 20. Hence, from the master plot we readily obtain m at a certain stress level; applying Eq. 20.2, we find the rupture time for the derived temperature.

Soon after Larson and Miller proposed their parameter, Manson and Haferd[16] presented the results of their experiments, which disagreed with Eq. 20.2 on the following points:

1. The family of lines does not intersect in the ordinate axis ($1/T = 0$) but at a specific point (t_a, T_a).
2. A better linearization is obtained if the results are plotted as log t_r versus T instead of log t_r versus $1/T$. This led Manson and Haferd[16] to propose the following equation:

$$\frac{\log t_r - \log t_a}{T - T_a} = m \tag{20.3}$$

This equation is represented graphically in Fig. 20.5. We use the same extrapolation procedure as that of Larson and Miller to obtain rupture times at different times and temperatures.

Later, in 1968, Manson[17] observed that the Larson–Miller parameter tended to predict exceedingly optimistic values for the rupture times, whereas the Manson–Haferd parameter tended to predict pessimistic values; based on this, he proposed a mixed parameter, of the form

$$\log t_r + \frac{\log^2 t_r}{40} - \frac{40,000}{T} = m \tag{20.4}$$

where T is expressed in degrees Rankine. Apparently, this parameter predicts more realistic values.

Another method that has found considerable success is the Sherby–Dorn[18] method. It is based on fundamental studies conducted by Sherby, Dorn, and coworkers with the objective of better understanding creep. The result of these investigations is shown in Section 20.3; in spite of the fact that the equations were derived for pure elements, they have been successfully applied to alloys. The Sherby–Dorn method is based on the fundamental result found by them— that the activation energy for diffusion is equal to the activation energy for creep. This leads to the equation

$$\log t_r - \frac{Q}{kT} = m \tag{20.5}$$

where Q is the activation energy for diffusion (or creep), m is a parameter, and t_r is the rupture time. Figure 20.6 shows the graphic representation of the Sherby–Dorn parameter. It differs from the Larson–Miller parameter in that the isostress lines are parallel. Equation 20.5 has a certain fundamental justification. Monkman and Grant[19] and others observed that for a great number

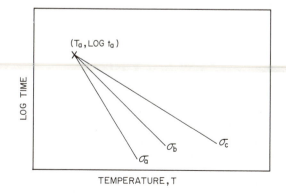

Figure 20.5 Relationship between rupture time and temperature at three levels of stress σ_a, σ_b, σ_c using Manson–Haferd parameter ($\sigma_a > \sigma_b > \sigma_c$).

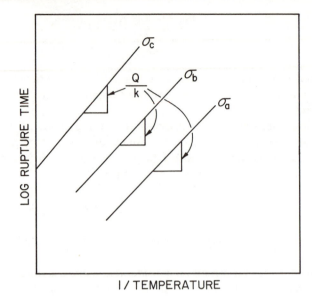

Figure 20.6 Relationship between rupture time and temperature at three levels of stress $\sigma_a >$ $\sigma_b > \sigma_c$ using Sherby–Dorn parameter.

of alloys, the minimum creep rate $\dot{\epsilon}_s$ was inversely proportional to the rupture time t_r:

$$\dot{\epsilon}_s t_r = k' \qquad (20.6)$$

Applying Eq. 20.1, which states that creep is a thermally activated mechanism and that the minimum creep rate increases exponentially with temperature at the same value of stress, and combining the preexponential terms, we have

$$\dot{\epsilon}_s = A' e^{-Q_c/kT} \qquad (20.7)$$

Substituting Eq. 20.6 into Eq. 20.7 yields

$$t_r = \frac{k'}{A'}\, e^{Q_c/kT}$$

or

$$\mathrm{T}\left(\log t_r - \log \frac{k'}{A'}\right) = \frac{Q_c}{k}$$

The slope of the lines in Fig. 20.6 is Q_c/k, which is equal to Q_D/k. Hence, if we know the activation energy for diffusion and one point of the line, we have all the other points. The activation energy for self-diffusion is obtained from the diffusion coefficients at two different temperatures. The diffusion, being a thermally activated process, obeys the equation

$$D = D_0 e^{-Q_D/kT} \qquad (20.7a)$$

where D is the diffusion coefficient at T.

Mechanical Testing   Part IV

A proliferation of parameters has taken place in the last years: Goldhoff–Sherby, White–LeMay, Manson–Brown, Grounes, and Barrett–Ardell–Sherby, among others. In an attempt to reduce the confusion, the American Society for Metals organized a symposium in 1967 under the chairmanship of Goldhoff. The most important result of this symposium was the proposal of a new method, the "minimum commitment method" (MCM). This method recommends analysis of the results without a preestablished bias; then the parametric method providing the best fit with the experimental results is shown. If none of the existing methods is satisfactory, a new parametric equation is shown. Essentially, one has a curve-fitting technique. The equation proposed for the MCM is

$$P(T) + Q(\log t) = Z(\sigma)$$

where P, Q, and Z are general functions. Initially, they are not expressed analytically, and different values of t and σ are determined at various temperatures T. The values of P, Q, and Z are determined for different nodal points. From these values the analytical expressions for P, Q, and Z are obtained. A problem that has been a great cause of concern in the high-temperature-alloy industry is the variation in properties between different heats of the same alloy. Manson and Ensign[20] extended the MCM to take into account the variability in different heats. The heat having the highest UTS might very well have a reduced rupture time.

20.3 FUNDAMENTAL MECHANISMS RESPONSIBLE FOR CREEP

According to Sherby and Miller,[13] the history of progress in the understanding of creep can be divided into two periods: before and after 1954. In that year Orr et al.[18,21] introduced the important concept that the activation energy for creep and diffusion are the same for an appreciable number of metals (more than 25).

Figure 20.7 shows their results. The activation energy for diffusion is connected to the diffusion coefficient by Eq. 20.7a. It should be emphasized that several mechanisms can be responsible for creep; the rate-controlling mechanism depends both on the stress level and on the temperature, as will be seen in Section 20.4. For temperatures below $0.5 T_m$ (T_m is the melting point in Kelvin), the activation energy for creep tends to be lower than the one for self-diffusion. Sherby and Miller[13] interpret this effect as being due to diffusion taking place preferentially along dislocations (pipe diffusion), instead of bulk diffusion. Figure 20.8 shows the variation of Q_C/Q_D for some metals. The activation energy for diffusion through dislocations is considerably lower than the one for bulk diffusion; this lends support to Sherby and Miller's[13] argument.

For the temperature range $T > 0.5 T_m$ the mechanisms responsible for creep can be conveniently described as a function of the applied stress. The mechanisms can be divided into four major groups.

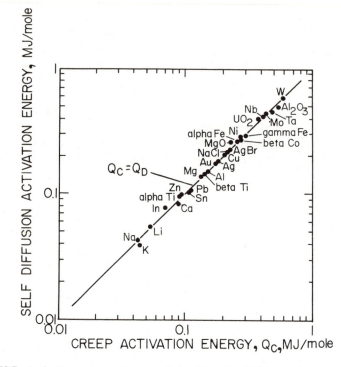

Figure 20.7 Activation energies for creep (stage II) and self-diffusion for a number of metals. (Adapted with permission from [13], p. 388.)

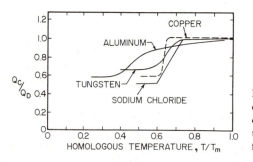

Figure 20.8 Ratio between activation energy for secondary creep and the activation energy for bulk diffusion as a function of temperature. (Adapted with permission from [13], p. 389.)

20.3.1 Diffusion Creep

Diffusion creep tends to occur for $\sigma/G \leq 10^{-4}$ (this value depends, to a certain extent, on the metal). Two mechanisms are considered important in this region. Nabarro[22] and Herring[23] proposed the mechanism shown schematically in Fig. 20.9. It involves the flux of vacancies inside the grain. The vacancies move in such a way as to produce an increase in length of the grain along the direction of applied (tensile) stress. Hence, the vacancies move from the top and bottom regions in Fig. 20.9 to the lateral regions of the grain. Coble[24] proposed the other mechanism. It is based on diffusion in the grain boundaries

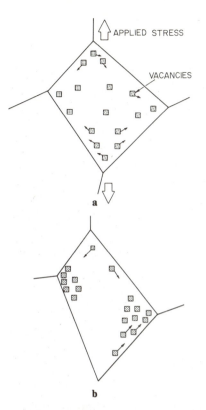

Figure 20.9 Flow of vacancies according to Nabarro–Herring mechanism, resulting in increase in length of grain.

a

b

instead of in the bulk. This diffusion results in sliding of the grain boundaries. Hence, if a fiducial scratch is made on the surface of the specimen prior to creep testing, the scratch will show a series of discontinuities (at the grain boundaries) after testing if Coble creep is operative. A practical way of having an alloy with high resistance to Nabarro–Herring or Coble creep is to increase the size of the grains. This method is used in superalloys; a fabricating technique called directional solidification has been developed to eliminate virtually all grain boundaries perpendicular and inclined to the tensile axis. Ashby[25] expressed the combined Nabarro–Herring and Coble diffusional creep by a Newtonian viscous equation (strain rate proportional to stress) of the form

$$\dot{\epsilon} = 14 \frac{\sigma \Omega}{kT} \frac{1}{D^2} D_v \left(1 + \frac{\pi \delta D_B}{D D_v}\right) \tag{20.8}$$

where D_B and D_v are the boundary and bulk diffusion coefficients, respectively; D is the grain size; δ is the effective cross section of a grain boundary for diffusional creep; Ω is the atomic volume; and k, T, and σ have their usual values.

Harper and Dorn[26] observed another type of diffusional creep in aluminum; it occurred at high temperatures and low stresses and the creep rates were over 1000 times greater than the one predicted by Nabarro–Herring (and little Coble creep was observed). They concluded that creep occurred exclusively by dislocation climb. Le May[3] extended their concept and concludes that the mechanism is vacancy diffusion between sources and sinks, which are climbing dislocations. One can readily see that the diffusion path is much smaller than the one in Nabarro–Herring creep.

20.3.2 Dislocation Creep

In the stress range $10^{-4} < \sigma/G < 10^{-2}$, creep tends to occur by dislocation glide aided by vacancy diffusion (when an obstacle is to be overcome). This mechanism should not be confused with Harper–Dorn creep, which relies exclusively on dislocation climb. Orowan[27] proposed the concept of creep being a balance between work hardening (due to the plastic strain) and recovery (due to the high-temperature exposure). Hence, the increase in stress is, at a constant temperature,

$$d\sigma = \left(\frac{\partial \sigma}{\partial \epsilon}\right)_{t,\sigma} d\epsilon + \left(\frac{\partial \sigma}{\partial t}\right)_{\sigma,\epsilon} dt \qquad (20.9)$$

where

$$\left(\frac{\partial \sigma}{\partial \epsilon}\right)_{t,\sigma} \rightarrow \text{rate of hardening}$$

$$\left(\frac{\partial \sigma}{\partial t}\right)_{\epsilon,\sigma} \rightarrow \text{rate of recovery}$$

The strain rate $\dot{\epsilon}$ can be expressed as a ratio between the rate of recovery and the rate of hardening. This is the Bailey–Orowan equation. It has been successfully applied by Gittus[28] to primary creep considering immobilization of dislocations and their remobilization by a thermally activated process.

The treatment given by Weertman[29,30] constitutes the fundamental theory for dislocation creep. He developed a theory for the minimum creep rate based on dislocation climb as the rate-controlling step. In his first theory, Weertman[29] presented Cottrell–Lomer locks as barriers to plastic deformation; his second theory[30] applies to HCP metals in which these barriers do not exist. Hence, different barriers were assumed. Figure 20.10 shows schematically how the mechanism based on Cottrell–Lomer locks operates. Dislocations are pinned by obstacles; they overcome them by climb, aided by either interstitial or vacancy generation or destruction. The obstacles are assumed to be Cottrell–Lomer locks. These Cottrell–Lomer locks are formed by dislocations that intersect and react (see Chapter 6). They are shown by circles in Fig. 20.10(c). Hence, dislocations are pinned between the locks and have to climb to overcome them [Fig. 20.10(d)]. However, dislocations are continuously generated by the Frank–Read source in the horizontal plane in Fig. 20.10, and the ones overcoming the obstacle are replaced by others. To calculate the creep rate, we have to find the rate of escape of the dislocations from the locks. The height a dislocation has to climb in order to pass through a lock, h, is the position at which the applied stress on the dislocation, due to the other dislocations in the pile-up, is equal

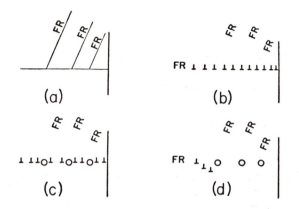

Figure 20.10 Dislocation overcoming obstacles by climb according to Weertman theory. (Reprinted with permission from [29], p. 1213.)

to the repulsive force due to the stress field of the lock. The stress exerted by the dislocation due to the pile-up effect is given in Section 6.1 and is

$$\sigma^* = n'\sigma \qquad (20.10)$$

where n'^* is the number of the dislocations in the pile-up and σ is the stress applied on one dislocation. Now, taking the stress field around a dislocation as a function of distance (Section 6.4.2) and equating it to Eq. 20.10, Weertman arrives at

$$h = \frac{2Gb^2/[12\pi(1-\nu)]}{n'\sigma b} \qquad (20.11)$$

The rate of climb is determined by the rate at which the vacancies (Weertman did the derivation for vacancies and not interstitials) arrive at or leave the dislocation. The concentration gradient of vacancies was calculated by Weertman and he obtained a rate of climb r:

$$r = \frac{N_0\, Dn'\sigma b^5}{kT} \qquad (20.12)$$

when $n\sigma b^3/kT < 1$.

With a known climb height and rate of climb it is possible to calculate the rate of creep. If M is the number of active Frank–Read sources per unit volume, L the distance the edge portion of a dislocation loop moves after break away from a barrier, and L' the distance the screw portion moves, then the creep rate $\dot{\epsilon}$ is given by

$$\dot{\epsilon} = \frac{6\pi(1-\nu n'^2 b^5 N_0 LL'MD\sigma^2}{kGT} \qquad (20.13)$$

where N_0 is the equilibrium concentration of vacancies and D the diffusion coefficient at the test temperature T. The temperature dependence of Eq. 20.13 resides in these two terms. Further manipulation of Eq. 20.13 leads to

$$\dot{\epsilon}_s = \frac{C}{kT}\,\sigma^n \exp\left(-\frac{Q_D}{kT}\right) \qquad (20.14)$$

*n' is used in order to avoid confusion with n (the exponent in power-law creep).

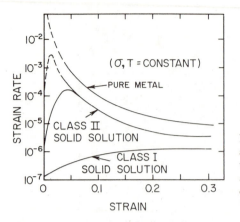

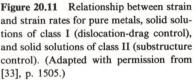

Figure 20.11 Relationship between strain and strain rates for pure metals, solid solutions of class I (dislocation-drag control), and solid solutions of class II (substructure control). (Adapted with permission from [33], p. 1505.)

Equation 20.14 is fairly close to Eq. 20.1 and represents "power law creep." The climb of jogs establishes the rate of movement of dislocations and, consequently, the creep rate. Barrett and Nix[3] proposed an alternative theory for this stress and temperature range based on the drag (and not climb) of jogs aided by diffusion.

Solid solutions have also been investigated for this range of stresses and two very distinct responses were found. Sherby and Burke[32,33] divided the solid solutions into two classes: class I alloys, in which the movement of dislocations is controlled by the drag of solutes; the parameter n is approximately 2 for these alloys. For class II solid solutions the creep is controlled by climb of dislocations, as in the case of pure metals; hence, n is approximately 5. Figure 20.11 shows the differences in response between the different conditions. In a pure metal the initial creep rate is very high because the existing dislocations can freely move. As the metal work hardens the density of dislocations increases and the strain rate decreases. The decrease is due to an increased density of barriers. For class I solid solutions, since the initial density of dislocations is low, there are a great number of solute atoms pinning each dislocation. As the deformation proceeds, the dislocation density increases and the number of solute atoms per unit length of dislocation decreases. Hence, the creep rate increases with deformation. For class II solid solutions there is initially a control of the dislocation drag by solutes; after some deformation, dislocation climb is the rate-controlling step. In addition to the classes noted above, Monteiro and Silveira[34] consider dispersion-strengthened alloys as a separate group. These are characterized by an exponent higher than 7.

20.3.3 Dislocation Glide

Dislocation glide occurs for $\sigma/G > 10^{-2}$. At a certain stress level the power law breaks down. Both Monteiro and Silveira[34] and Kestenbach et al.[35] analyzed this situation for AISI 316 stainless steel. Kestenbach et al.[35] identified this alloy as being of class II according to Sherby and Burke's classification. Figure 20.12 presents clearly the region in which the power law ($n = 4$) breaks

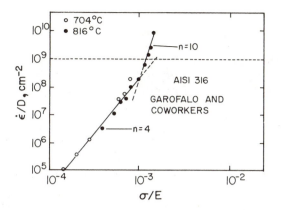

Figure 20.12 Power relationship between $\dot{\epsilon}$ and σ for AISI 316 stainless steel. (Adapted with permission from [34], p. 327.)

down. For $\dot{\epsilon}_s/D > 10^9$ the exponent n increases to 10. Kestenbach et al.[35] conducted an analysis of the deformation substructure by transmission electron microscopy and concluded that at high stresses dislocation climb was replaced by dislocation glide, which does not depend on diffusion. Hence, when $\dot{\epsilon}_s/D > 10^9$, thermally activated dislocation glide is the rate-controlling step; this is the same deformation mode as the one in conventional deformation at ambient temperature. Kestenbach et al.[35] observed that the substructure changes from equiaxed subgrains to dislocation tangles and elongated subgrains when the stress reaches a critical level. A similar effect is observed when the temperature is decreased and the stress maintained constant. Figure 20.13 shows the substructures at various values of stress and temperatures for secondary creep. Another important observation made by Kestenbach et al. is that the substructure varies with the orientation of the grain with respect to the tensile axis.

20.3.4 Grain-Boundary Sliding

Grain-boundary sliding usually does not play an important role during primary and secondary creep. However, in tertiary creep it does contribute to the initiation and propagation of intercrystalline cracks.[3] Another deformation process to which it contributes significantly is superplasticity; it is thought that most of the deformation in superplastic forming takes place by grain-boundary sliding.

Raj and Ashby[36] developed a theory for grain-boundary sliding of polycrystalline aggregates. They found that the sliding rate was controlled by the accommodating processes where the sliding surface deviated from a perfect plane. One can readily see that we cannot have a perfect plane defined by the boundaries between different grains; we cannot look separately at the sliding between two grains having a common interface. The requirements of strain compatibility are such that we have to model the interface as having the sinusoidal nature depicted in Fig. 20.14(a). The applied stress τ_a can produce sliding

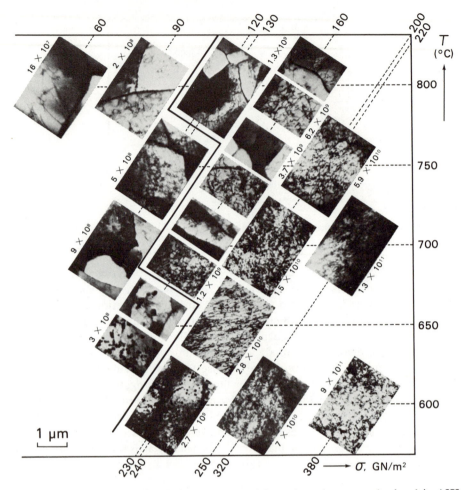

Figure 20.13 Effect of stress and temperature on deformation substructure developed in AISI 316 stainless steel in middle of stage II. (Reprinted with permission from [35], p. 667.)

only if it is coupled with diffusional flow that transports material (or vacancies) over a maximum distance of λ, the wavelength of the irregularities. Figure 20.14(b) shows the same effect in a polycrystalline aggregate. The individual grain boundaries are translated by a combination of sliding and diffusional flow under the influence of the applied stress. Raj and Ashby[37] applied these concepts to the generation of fractures at grain boundaries by nucleation and growth of voids.

The manner in which the individual grains move and change their relative positions by sliding and diffusional accommodation led Ashby and Verrall[38] to propose a mechanism for diffusion. Its essential features are shown in Fig. 20.15. The sliding of grains under the influence of σ, coupled with minor changes in shape, makes possible the sequence a-b-c, which results in a strain of 0.55; the unique feature of this mechanism is that this is accomplished with relatively little strain *within* the grains. Figure 20.16 compares the diffusional flow required

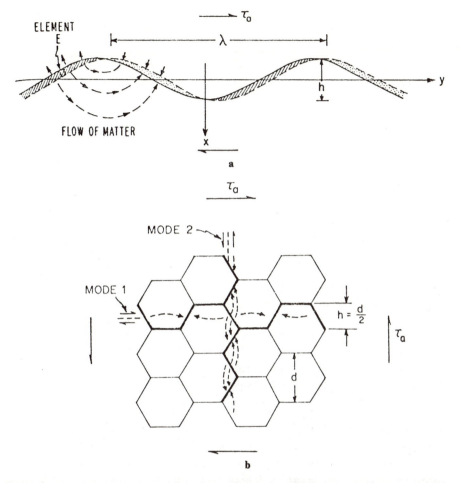

Figure 20.14 (a) Steady-state grain-boundary sliding with diffusional accommodation; (b) same process as in (a), in idealized polycrystal; dashed lines show flow of vacancies. (Reprinted with permission from [36], p. 1120.)

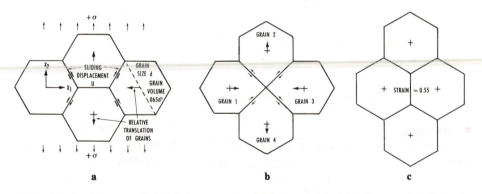

Figure 20.15 Grain-boundary sliding assisted by diffusion in Ashby–Verrall's model. (Reprinted with permission from [38], p. 159.)

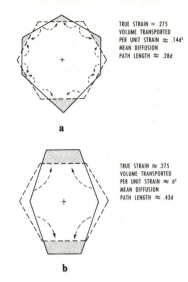

TRUE STRAIN = .275
VOLUME TRANSPORTED
PER UNIT STRAIN ≈ .14d³
MEAN DIFFUSION
PATH LENGTH ≈ .28d

a

TRUE STRAIN ≈ .275
VOLUME TRANSPORTED
PER UNIT STRAIN ≈ d³
MEAN DIFFUSION
PATH LENGTH ≈ .43d

b

Figure 20.16 Transport of matter required to produce a strain of 0.275 according to: (a) Ashby–Verrall's mechanism; (b) Nabarro–Herring or Coble's mechanism. (Reprinted with permission from [38], p. 155.)

to produce half of the strain of Fig. 20.15 (0.275) for Ashby–Verrall's mechanism [Fig. 20.16(a)] and Nabarro–Herring's or Coble's mechanisms [Fig. 20.16(b)]. For the latter two mechanisms each grain undergoes the same strain as the specimen. The volume transported per unit strain is equal to 0.14 d^3 for Ashby–Verrall's model, while it is equal to d^3 for the other mechanisms; additionally, the diffusion path is shorter for Ashby–Verrall's mechanism. These features were compared with experimental results obtained in alloys exhibiting superplastic behavior (Zn–0.2% Al); superplasticity is discussed in Section 16.1.5. Notable features of superplastic behavior are little change of grain shape, no important cell formation, destruction of texture, grain-boundary sliding and rotation. They are consistent with the Ashby–Verrall mechanism. Ashby–Verrall's mechanism also predicts a value of m (strain-rate sensitivity) close to 1; this agrees with experimental results on superplastic alloys. The following strain-rate is predicted from their mechanism:

$$\dot{\epsilon} = 98 \frac{\Omega D_v}{kTD^2} \left(\sigma - \frac{0.72\Gamma}{D} \right) \left(1 + \frac{\pi \delta D_B}{D D_v} \right)$$

where all the parameters have the same meaning as in Eq. 20.8. Γ is the surface free energy of the alloy.

‡20.4 DEFORMATION-MECHANISM (WEERTMAN–ASHBY) MAPS

These maps, introduced by Weertman[39] and Ashby,[25,40] are a graphical description of creep, representing the ranges in which the various deformation modes are rate-controlling steps in the stress versus temperature space. The Weertman–Ashby plots assume, for simplicity, that there are only six independent and distinguishable ways by which a polycrystal can be deformed, retaining its crystallinity.

1. Above the theoretical shear strength, plastic flow of the material can take place without dislocations, by simple glide of one atomic plane over another.
2. Movement of dislocations by glide.
3. Dislocation creep; this includes glide and climb, both being controlled by diffusion.
4. Nabarro–Herring creep.
5. Coble creep.
6. Twinning; Monteiro and Silveira[41] correctly added to this group stress-assisted and strain-induced martensitic transformations.

The theories developed for these different modes propose constitutive equations (Section 20.3) that are used in the establishment of the ranges. This procedure is described by Ashby.[25] Figure 20.17 shows a typical map for silver. The theoretical shear stress is approximately equal to $G/20$ (see Chapter 4) and is practically independent of temperature. A small temperature dependence is exhibited by G and is built into the ordinate of Fig. 20.17. For values of σ/G between 10^{-1} and 10^{-2}, slip by dislocation movement is the controlling mode at all temperatures. It can be seen that the grain size affects the extent of the fields. Three grain sizes are represented: 10, 32, and 100 μm. The fields also depend on strain rate. The map of Fig. 20.18 was made for a strain rate of 10^{-8} s^{-1}. The Coble and Nabarro–Herring mechanisms, especially, are affected by the grain size, because of their nature.

Deformation-mechanism maps have found technological applications. The

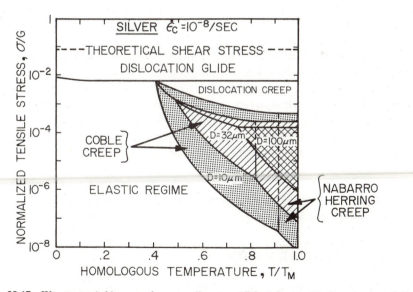

Figure 20.17 Weertman–Ashby map for pure silver, established for a critical strain rate of 10^{-8} s^{-1}; it can be seen how the deformation-mechanism fields are affected by the grain size. (Adapted with permission from [25], p. 891.)

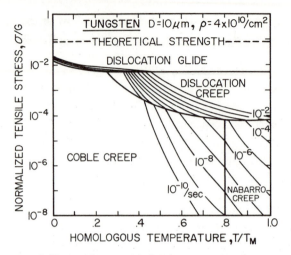

Figure 20.18 Weertman–Ashby map for tungsten showing constant strain-rate contours. (Reprinted with permission from [25], p. 890.)

example given by Ashby[25] illustrates an application. He assumed that a turbine blade operated in a temperature and stress range that are known. The specific stress-temperature profile can be plotted in the map in the form of a line. Different parts of the blade undergo different deformation modes. These different deformation modes, the rate of creep of each portion, and the respective constitutive equation can be read from the map. Multiaxial stress states can be resolved calculating the maximum shear stress or the effective stress. Ashby[25] emphasized that a strengthening mechanism is helpful only if it retards the creep rate in the correct portion of the map. For instance, dispersion hardening is effective in controlling dislocation glide and climb, but cannot effectively stop Nabarro–Herring and Coble creep.

From the deformation-mechanism map we can, in addition to determining the dominant mechanism for a certain combination of stress and temperature, find the strain rate (creep rate) that will result. For this we have to apply the appropriate constitutive equations and plot the constant strain-rate contours. This is shown in Fig. 20.18 for tungsten. The lines allow ready identification of the creep rate. The region in Fig. 20.17 consisting of the elastic regime is occupied by Coble creep in Fig. 20.18. The reason for this is that Fig. 20.17 applies to one constant strain rate (10^{-8} s^{-1}), whereas Fig. 20.18 is built for a whole range of strain rates. Hence, at a strain rate of 10^{-8} s^{-1} the metal might respond elastically, whereas at a strain rate orders of magnitude lower, Coble creep becomes significant.

‡20.5 HEAT-RESISTANT ALLOYS

High-temperature materials can be classified into two groups: metals and ceramics. High-temperature alloys are, in their turn, classified into superalloys and refractory alloys. The latter are alloys of elements with high melting points, such as tantalum, molybdenum, and tungsten. The superalloys can be defined as "alloys developed for elevated temperature service, usually based on group VIIIa elements, where relatively severe mechanical stressing is encountered and where high surface stability is frequently required."[42]

The development of superalloys was initiated in the 1930s and their first use was in turbosuperchargers of reciprocating airplane engines.[43] The introduction of the turbine in the 1940s was a strong motivator for subsequent developments. The technological importance of superalloys can be assessed by the sales volume, which reached $10 billion in 1972.[44] Superalloys encompass the nickel, iron, cobalt, and iron–nickel systems. The majority of authors do not include chromium-based alloys in this group. The maximum service temperature (temperature capability) has increased continuously in the past; it is presently around 1200°C. Figure 20.19 illustrates the increase in temperature capability of superalloys; it has averaged out to 9°C per year. The life of turbines has increased from 5000 to over 20,000 h.[16] The combined effects of high stresses, temperatures, and long times have required improvements in the following properties:

1. *Short-term mechanical properties*: yield stress, ductility.
2. *Long-term mechanical properties*: low- and high-cycle fatigue; creep; creep-fatigue.
3. *Hot corrosion resistance*: the principal deterioration processes are oxidation, chlorination, sulfidation, and carburization.

Nickel-base superalloys are the most important group; the majority of commercial alloys has more than 10 constituent elements and over 10 trace

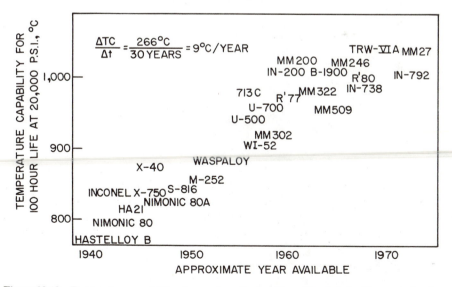

Figure 20.19 Temperature capability of a number of superalloys. (Reprinted with permission from [45], p. 8.)

elements. These can be divided into categories depending on their function and position in the periodic chart.

1. Elements that form substitutional solid solutions in the austenitic matrix: cobalt, iron, chromium, vanadium, molybdenum, tungsten.
2. Elements that form precipitates: aluminum, titanium, niobium, tantalum. Figure 20.20 shows the cuboidally shaped γ' precipitates [Ni_3Al or Ni_3Ti or $Ni_3(Al,Ti)$] that are aligned along specific planes of the austenitic matrix.
3. Carbide-forming elements: chromium, molybdenum, tungsten, vanadium, niobium, tantalum, titanium.
4. Elements that segregate along the grain boundaries: magnesium, boron, carbon, zirconium.
5. Elements forming protective and adherent oxides: chromium and aluminum.
6. Rare earths.

The microstructure of superalloys reflects the concern of using all possible strengthening mechanisms to retard creep. It is discussed in refs. [47–49], among

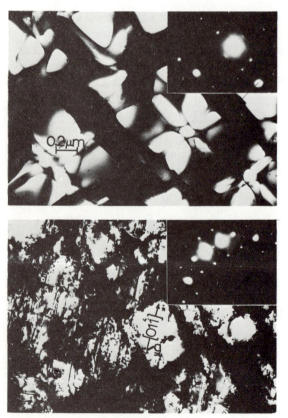

a

b

Figure 20.20 Transmission electron micrograph of Mar M-200; notice cuboidal γ' precipitates.

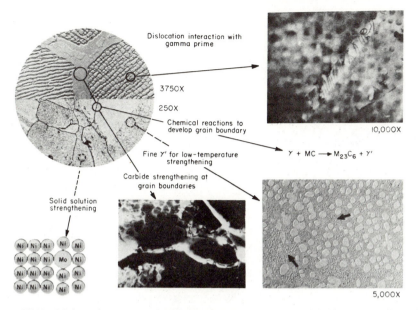

Figure 20.21   Major microstructural strengthening mechanisms in nickel-base superalloys. (Reprinted with permission from [49], p. 74.)

other sources. Figure 20.21 is a composite of these features. One has to retard the movement of dislocations. This is achieved by substitutional solid solution atoms and by a great volume percentage of the $Ni_3(Ti,Al)$ phase γ'. The grain boundaries are strengthened by precipitation of $M_{23}C_6$[50] carbides on them. Secondary γ', very fine, is precipitated between the primary γ', which is larger. One also wants to carefully avoid the TCP (topologically close-packed phases) R, mu, and sigma phases that occur accidentally and after long exposure to high temperatures, embrittling the alloy.[51]

Figure 20.22 shows the stress-rupture properties of a number of nickel-base superalloys. The stress required for rupture in 1000 h is plotted against the temperature. The load-bearing ability in the upper range of their use is only a fraction of the one at lower temperatures. The range 800 to 1000°C is a very critical one.

A technological development that seems to have great potential is rapid solidification processing. It is hoped that it will provide a quantum jump in the temperature capability of alloys. The principles of obtaining and processing the alloy are unique. First, a very fine powder is obtained by extremely rapid cooling. Pratt & Whitney has developed the RSR process, in which a liquid jet impinges on a turbine rotating at high velocity; the cooling rates are reported to be 10^4 K/s or higher. Then this powder with unique properties is compacted.

The opening lecture of Cohen et al.[52] at the Second International Conference on Rapid Solidification Processing emphasized the bright potential of rapid solidification processing (RSP) as a large-scale industrial method for superalloy component production. Unique microstructures can result from the rapid solidification of fine powders: metastable phases, metallic glasses, supersaturated solid

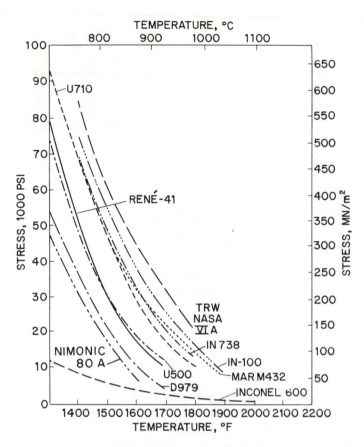

Figure 20.22 Stress versus temperature curves for rupture in 1000 h for selected nickel-base superalloys. (Reprinted with permission from [42], p. 591.)

solutions, particles covered with fine carbide coatings, dendritic structures, and so on. The refined and novel microstructures characteristic of RSP are obtained at cooling rates of 10^4 K/s or higher. Patterson et al.[53] recently reported results of transmission electron microscopy of RSP powders; unique microstructures were obtained in MAR-M200 and IN-100 alloys. The associated mechanical properties and resistance to degradation by oxidation compare favorably with alloys processed conventionally.

The consolidation of RSP powders into tridimensional shapes can be accomplished by several techniques.[52] The most promising seem to be hot extrusion, hot isostatic pressing,* incremental solidification, and dynamic compaction. In dynamic shock compaction, a shock wave is passed through the powder,

* Hot isostatic pressing ("HIP") consists of filling a container with dimensions slightly larger than the final shape with the RSP powder, sealing the container after evacuation and putting it into an autoclave where appropriate pressure and temperature are applied during a specified period of time. This will cause the container to contract, promoting sintering of the powder. More details can be found in F. V. Lenel and G. S. Ansell, in "Metallurgical Treatises," J. K. Tien and J. F. Elliott (eds.), AIME, Warrendale, PA., 1981, p. 345.

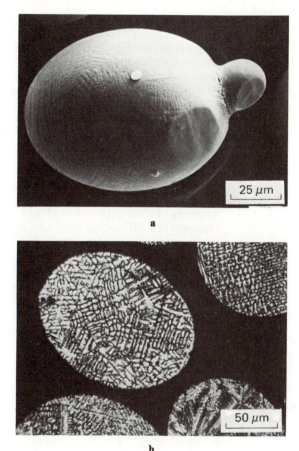

Figure 20.23 Mar M-200 particles prepared by rapid solidification processing: (a) scanning electron micrograph; (b) optical micrograph of section.

compacting it in the process. The shock wave may be induced by the detonation of an explosive in contact with the system or by impact with a high-velocity projectile. The essential problem is to produce a tridimensional shape from the powder without destroying its unique properties, and with a minimum of machining (net shape capability). Of the techniques cited above, all by dynamic compaction require high-temperature environments.

It has been demonstrated[54,55] that RSP powders can be consolidated by dynamic (shock) compaction. These somewhat preliminary results,[54] yielding specimens with relatively small dimensions, were obtained with the use of gas-gun arrangements consisting of a long gun tube through which a projectile is accelerated to impact the target assembly containing the fine powder configuration.

Figure 20.23 shows a particle of MAR-M 200 prepared by the Pratt & Whitney RSR process. The microdendrites can be seen both on the surface [Fig. 20.23(a)] and in the section of the particle, after etching [Fig. 20.23(b)]. The very fine spacing between dendrites (~5 μm) is one of the strengthening mechanisms.

EXERCISES

20.1. A cylindrical specimen creeps at a constant rate during 10,000 h when subjected to a constant load of 1000 N. The initial diameter and length are 10 and 200 mm, respectively, and the creep rate is 10^{-5} h^{-1}. Find:
   (a) The length of the specimen after 100, 1000, and 10,000 h.
   (b) The real and nominal strains after these periods.
   (c) The real and nominal stresses after these periods.

20.2. Give three reasons why the extrapolation of creep data obtained over a short period can be dangerous over long periods.

20.3. T. E. Howson, D. A. Mervyn, and J. K. Tien [*Met. Trans.*, *11A* (1980) 1609] studied the creep and stress-rupture response of oxide dispersion strengthened (ODS) superalloys produced by mechanical alloying. They determined the activation energy for creep $Q_c = 619$ kJ/mol by conducting tests at a constant applied stress of 558.7 MPa at the three temperatures of 746, 760, and 774°C.
   (a) The results shown below were found for experimental alloy MA 6000 E at 760°C. Estimate the value of n and discuss this value in terms of the microstructure exhibited by the alloy (dispersion strengthening by inert yttrium oxide dispersoids plus precipitation strengthening by gamma prime).
   (b) By applying Eq. 20.1, show how this activation energy can be found. Make the appropriate plot and find the minimum creep rate at these three temperatures. Note that the activation energy is given per mole.

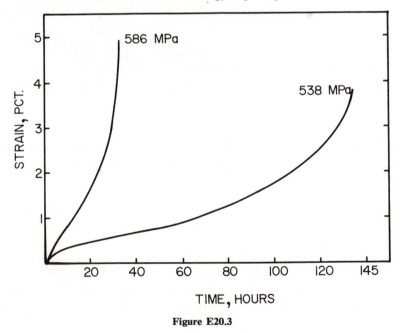

Figure E20.3

20.4. T. E. Howson, J. E. Stulga, and J. K. Tien [*Met. Trans.*, *11A* (1980) 1599] obtained the following stress-rupture results for the superalloy Inconel MA 754 (a dispersion-strengthened alloy).
   (a) Verify whether this alloy obeys a Larson–Miller relationship and find C. Then prepare a master plot.

(b) Determine the predicted stress-rupture life if the alloy is stressed at 1000°C and 50 MPa.

Temperature (°C)	Applied Stress (MPa)	Rupture Life (h)
760	189.7	—
760	206.9	83.9
760	206.9	111.2
760	224.2	38.6
760	224.2	29.0
760	241.4	6.9
760	258.7	1.8
746	206.9	320.8
774	206.9	65.0
788	206.9	33.2
982	110.4	195.1
982	113.8	136.6
982	113.8	106.9
982	116.5	27.6
982	117.3	106.3
982	120.7	13.0
982	120.7	39.0
996	110.4	52.6
996	110.4	41.3
1010	110.4	20.3
1010	110.4	41.7
1024	110.4	9.4

20.5. What is the predicted stress-rupture life of AISI 316 steel at 800°C and 160 MPa? (See Fig. 20.4.)

20.6. Verify whether the data of Exercise 20.4 obey the Manson–Haferd correlation.

20.7. Assuming that pure silver creeps according to the Dorn equation, estimate the rupture time at 400°C when subjected to a stress of 50 MPa, knowing that at 300°C and at the same stress level the rupture time is 2000 h.

20.8. Tungsten is being used at half its melting point ($T_m \sim 3400°C$) and a stress level of 160 MPa. An engineer suggests increasing the grain size by a factor of 4 as an effective means of reducing the creep rate.
 (a) Do you agree with her? Why? What if the stress level was equal to 1.6 MPa?
 (b) What is the predicted increase in length of the specimen after 10,000 h if the initial length is 10 cm?
 (*Hint*: Use a Weertman–Ashby map.)

20.9. In Exercise 20.4, verify how closely the Monkmon–Grant relationship is obeyed.

20.10. By means of plots, show how isochronal stress versus strain curves can be constructed from creep curves for various stresses at a certain temperature.

REFERENCES

[1] M. Vicat, *Ann. Chim. Phys.*, *54* (1833) 35.

[2] E.N. da C. Andrade, *Proc. Roy. Soc.* (*London*), *A84* (1911) 1.

[3] I. Le May, *Principles of Mechanical Metallurgy*, Elsevier, New York, 1981, p. 354.

[4] E.N. da C. Andrade, *Proc. Roy. Soc. (London)*, *A90* (1914) 329.

[5] R.W. Bailey, *Proc. Inst. Mech. Eng.*, *131* (1935) 131.

[6] F. Garofalo, *Fundamentals of Creep and Creep Rupture in Metals*, Macmillan, New York, 1965.

[7] W.C. Harrigan, Jr., "A Study of the Effects of Shock-Loading and Cold Rolling on High Temperature Creep Properties of Metals and Alloys," Ph.D. thesis, Stanford University, Stanford, Calif., 1971.

[8] R.L. Fullman, R.P. Carrecker, Jr., and J.C. Fisher, *Trans. AIME*, *197* (1953) 657.

[9] P.C.R. Cunha and Hans-Jurgen Kestenbach, *Metalurgia-ABM*, *35* (1979) 257.

[10] T.L. da Silveira and S.N. Monteiro, *Proc. 35th Annu. Meet. Brazilian Soc. Metals*, July 1980, Vol. 1., p. 415.

[11] S.S. Manson and C.R. Ensign, *J. Eng. Mater. Technol.*, *101* (1979) 317.

[12] I. Le May, *J. Eng. Mater. Technol.*, *101* (1979) 326.

[13] O.D. Sherby and A.K. Miller, *J. Eng. Mater. Technol.*, *101* (1979) 387.

[14] P.P. Pizzo, *J. Eng. Mater. Technol.*, *101* (1979) 387.

[15] F.R. Larson and J. Miller, *Trans. ASME*, *74* (1952) 765.

[16] S.S. Manson and A.M. Haferd, Rep. NACA TN 2890, March 1953.

[17] S.S. Manson, Publ. No. DB-100, ASM, Metals Park, Ohio, 1968.

[18] R.L. Orr, O.D. Sherby, and J.E. Dorn, *Trans. ASM*, *46* (1954) 113.

[19] F.C. Monkman and N.J. Grant, *Proc. ASTM*, *56* (1956) 593.

[20] S.S. Manson and C.R. Ensign, "Interpolation and Extrapolation of Creep Rupture Data by the Minimum Commitment Method—Part III: Analysis of Multiheats," ASME Publ. MPC-7, ASME, New York, 1978.

[21] O.D. Sherby, R.L. Orr, and J.E. Dorn, *Trans. AIME*, *200* (1954) 71.

[22] F.R.N. Nabarro, "Deformation of Crystals by the Motion of Single Ions," Report of a Conference on Strength of Solids, Physical Society, London, 1948, p. 75.

[23] C. Herring, *J. Appl. Phys.*, *21* (1950) 437.

[24] R.L. Coble, *J. Appl. Phys.*, *34* (1963) 1679.

[25] M. F. Ashby, *Acta Met.*, *20* (1972) 887.

[26] J. Harper and J.E. Dorn, *Acta Met.*, *5* (1957) 654.

[27] E. Orowan and J.W. Scott, *J. Iron Steel Inst.*, *54* (1946) 45.

[28] J. H. Gittus, *Phil. Mag. 21* (1970) 495.

[29] J. Weertman, *J. Appl. Phys.*, *26* (1955) 1213.

[30] J. Weertman, *J. Appl. Phys.*, *28* (1957) 362.

[31] C.R. Barrett and W.D. Nix, *Acta Met.*, *13* (1965) 1247.

[32] O.D. Sherby and P.M. Burke, *Proc. Mater. Sci.*, *13* (1968) 325.

[33] W.D. Nix and B. Ilschner, in *Strength of Metals and Alloys*, P. Haasen, V. Gerold, and G. Kostorz (eds.), Pergamon Press, Elmsford, N.Y., p. 1503, vol. 3.

[34] S.N. Monteiro and T.L. da Silveira, *Metalurgia-ABM*, *35* (1979) 327.

[35] H.-J. Kestenbach, W. Krause, and T.L. da Silveira, *Acta Met.*, *26* (1978) 661.

[36] R. Raj and M.F. Ashby, *Met. Trans.*, *2A* (1971) 1113.

[37] R. Raj and M.F. Ashby, *Acta Met.*, *23* (1975) 653.

[38] M.F. Ashby and R.A. Verrall, *Acta Met.*, *21* (1973) 149.

[39] J. Weertman, *Trans. AIME*, *227* (1963) 1475.

[40] M.F. Ashby and H.J. Frost, in *Frontiers in Materials Science*, L.E. Murr and C. Stein (eds.), Marcel Dekker, New York, 1976, p. 391.

[41] S.N. Monteiro and T.L. da Silveira, *Metalurgia-ABM*, *36* (1980) 17.

[42] C.T. Sims and W.C. Hagel (eds.), *The Superalloys*, Wiley, New York, 1972, p. vii.

[43] R.G. Dunn, D.L. Sponseller and J.M. Dahl, in *Toward Improved Ductility and Toughness*, Climax Molybdenum Co., Ann Arbor, Mich., 1971, p. 319.

[44] R.F. Decker and R.R. de Witt, *J. Metals*, *15* (1965) 139.

[45] R.W. Fawley, in source cited in [42], p. 8.

[46] J. Vaccari, *Mater. Eng.*, *69*, No. 5 (1969). 21.

[47] C.T. Sims, "Nickel Alloys—The Heart of Gas Turbine Engines," ASME 70-6T-24, ASME, New York, 1970.

[48] C.T. Sims, *J. Metals*, *18* (1966) 1119.

[49] R.F. Decker and C.T. Sims, cited in [42], p. 33.

[50] J. Marcantonio and J.B. Newkirk, *Electrochem. Technol.*, *6*, No. 11–12 (1968) 447.

[51] R.G. Barrows, Ph.D. thesis, University of Denver, Col., 1970, p. 15.

[52] M. Cohen, B.H. Kear, and R. Mehrabian, in "Rapid Solidification Processing, Principles, and Technologies," M. Cohen, B.H. Kear, and R. Mehrabian (eds.), Claitors' Publishing Div., Baton Rouge, LA 1980, p. 1.

[53] R.J. Patterson, A.R. Cox, and E.C. van Reuth, *J. Metals*, *32*, No. 9, 34 (1980).

[54] D. Raybould, in *Shock Waves and High-Strain-Rate Phenomena in Metals: Concepts and Applications*, M.A. Meyers and L.E. Murr (eds.), Plenum Press, New York, 1981, p. 895.

[55] M.A. Meyers, B.B. Gupta, and L.E. Murr, *J. Metals*, *33* (Oct. 1981) 21.

SUGGESTED READING

CONWAY, J.B., *Stress Rupture Parameters: Origin, Calculation and Use*, Gordon and Breach, New York, 1969.

GAROFALO, F., *Fundamentals of Creep and Creep Rupture in Metals*, Macmillan, New York, 1965.

GITTUS, J., *Creep, Viscoelasticity and Creep Fracture in Solids*, Halsted Press (Wiley), New York, 1975.

J. Eng. Mater. Technol., *101* (1979) 317ff.

Chapter 21

FATIGUE AND FATIGUE TESTING

This chapter is divided in two parts. Section 21.1 covers the fundamental aspects of fatigue, and Section 21.2 describes the most common testing techniques.

21.1 FATIGUE

Fatigue is the phenomenon of failure of a material under cyclic loading. It is a problem that affects any component or part that moves. Automobiles on roads, airplanes (principally the wings) in the air, ships on the high sea constantly battered by waves, nuclear reactors and turbines under cyclic temperature (i.e., cyclic thermal stresses) conditions, and many others are examples of situations where fatigue behavior of a material assumes a singular importance.

It is known that failure under cyclic stress or strain occurs at a stress level much lower than that under conditions of monotonic loading. It is estimated that 90% of service failures of components that undergo movement of one form or another can be attributed to the phenomenon of fatigue. This is the traditional aspect of the problem of fatigue. Today, the problem of structural fatigue has assumed an even greater importance. There are two main reasons for this:

1. An ever-increasing use of high-static-strength materials
2. Desire for an ever-higher structural performance from these materials

As it happens, these improvements in static strength are rarely accompanied by improvements in fatigue strength. This fundamental inequality in the material properties has resulted in high-performance structures that are not completely reliable in service.

Up until the 1960s, almost all fatigue failures, and consequently all the research in the field, was confined to moving mechanical components (e.g., axles, gears, etc.). Starting in the late 1950s, entire structures or very large structural elements (e.g., pressure vessels, rockets, airplane fuselages, etc.) have been studied and tested for fatigue. This can be attributed to the use of materials such as high-strength alloys, together with the advances in the fabrication technology, resulting in monolithic structures meant to undergo high cyclic stresses in service. It is this class of materials which has shown catastrophic failures in fatigue, and it is for this kind of material that fracture mechanics is being applied, with considerable success, to fatigue problems.

In this part we first present a traditional view of the fatigue phenomenon and then the application of fracture mechanics concepts to fatigue crack propagation in high-strength materials.

21.1.1 S–N Curves

Traditionally, the behavior of a material under conditions of fatigue has been studied by obtaining the S–N curves (Fig. 21.1), where S is the stress and N is the number of cycles to failure. For steels, in general, one observes

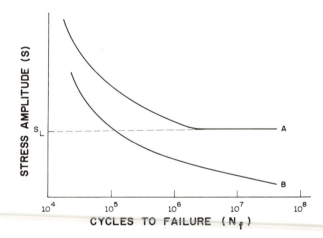

Figure 21.1 S (stress)–N (cycles to failure) curves. A, ferrous metals; B, nonferrous metals. S_L is the endurance limit.

a fatigue limit or endurance limit (curve A in Fig. 21.1) which represents a stress level below which the material does not fail and can be cycled infinitely. Such an endurance limit does not exist for nonferrous metals (curve B in Fig. 21.1). The relation between S and N, it must be pointed out, is not a single-value function but serves to indicate a statistical tendency.

Here, it would be in order to define formally some important parameters

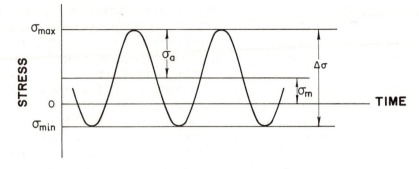

Figure 21.2 Fatigue parameters.

which will be useful in later discussion of the fatigue phenomenon. These parameters are (Fig. 21.2):

Cyclic stress range: $\quad \Delta\sigma = \sigma_{max} - \sigma_{min}$

Cyclic stress amplitude: $\quad \sigma_a = \dfrac{\sigma_{max} - \sigma_{min}}{2}$

Mean Stress: $\quad \sigma_m = \dfrac{\sigma_{max} + \sigma_{min}}{2}$

Stress ratio: $\quad R = \dfrac{\sigma_{min}}{\sigma_{max}}$

The statistical treatment of cyclic loading (whether rotating beam bending, pulsating tension, or axial tension–compression) data is described in Section 21.2.

21.1.2 Cyclic Loading

In the cyclic loading in elastic regime, stress and strain are related through the elastic modulus. In such a case it is enough to measure either stress or strain. Normally, one uses an X–Y recorder and a stress–strain curve is obtained

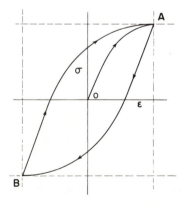

Figure 21.3 Hysteresis loop for cyclic loading of an elastic–plastic material.

directly. For any controlling function, σ or ϵ, the signal goes from 0 to (σ, ϵ) to ($-\sigma$, $-\epsilon$) and back to zero. The points 0; σ,ϵ; and $-\sigma$,$-\epsilon$ are located on the same cyclic stress–strain curve—a straight line through the origin.

For cyclic loading conditions that produce plastic strains, the responses are more complex. Figure 21.3 shows the result schematically. From point O to A there is tension. Unloading from A and entering in compression, we get to point B. Unloading from B and reversing the stress direction, we return to point A. A complete cycle thus gives a hysteresis loop which furnishes a means to describe the material behavior under cyclic loading.

A hysteresis loop serves to show the cyclic variation of stress, but more important, it gives a means to measure plastic strain per cycle. It can be said[1] that plastic strain is a measurable physical quantity that can be better correlated than any other parameter to damage caused by fatigue. Let us consider the notation indicated in Fig. 21.4 for a loop symmetrical with respect to the coordinates σ,ϵ. A complete cycle consists of start at any point of the loop, trace of the loop in the clockwise direction, and termination at the starting point. Independent of the location of the starting point or the control condition, points A and B represent the cyclic stress and strain limits. The total strain $\Delta\epsilon$ consists of elastic and plastic components. The elastic component is $\Delta\epsilon_e = \Delta\sigma/E$ and the plastic strain $\Delta\epsilon_p$ can be written as

$$\Delta\epsilon_p = \Delta\epsilon - \Delta\epsilon_e$$

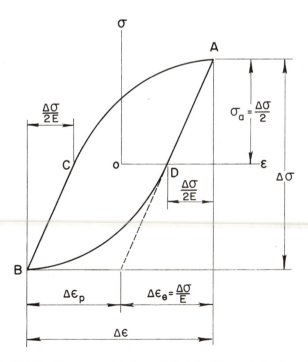

Figure 21.4 Notation for a symmetric loop. (Adapted with permission from [1], p. 22.)

where $\Delta\epsilon_p$ is equal to the width of the loop at its center (i.e., the distance *CD*). The area of the hysteresis loop is equal to the work done or energy loss per cycle.

21.1.2.1 Stress-controlled cycling.

In this case, the controlling function is the stress which oscillates between two extremes (Fig. 21.5); that is, the cyclic stress amplitude σ_a is a constant. The strain, however, does not have a constant amplitude. The resistance of the material to further deformation may increase with cycling; in such a case the cyclic strain becomes ever smaller under the same stress amplitude [Fig. 21.5(b)]. We call such behavior "cyclic hardening." The envelope of strain peaks is generally an exponential function. On the other hand, a material can show the phenomenon of increasing deformation with cycling, again under an exponential envelope. This is the case with cyclic softening [Fig. 21.5(c)].

We can better describe this cyclic response of a material in the form of a σ–ϵ hysteresis loop. Under stress control, the sample is cycled between horizontal limits of σ_a [Fig. 21.5(d) and (e)]. The material loaded from the origin in the tensile direction reaches the imposed tensile limit at point 1. Notice that this stretch, from 0 to 1, coincides with the monotonic stress–strain curve. At 1 the material suffers a change in the loading direction and reaches point 2, the limit in compression. The numbers 3, 4, 5, and so on, indicate the strains at which the stress limits are reached subsequently.

Strictly speaking, it is incorrect to speak of hysteresis loop, as the loop never returns to the initial point. But the differences are negligible and even an imperfect hysteresis loop can show the material behavior, principally the fact that the loop width represents the range of plastic deformation, $\Delta\epsilon_p$.[1] Figure 21.5(d) shows cyclic hardening, in the hysteresis form, under stress-

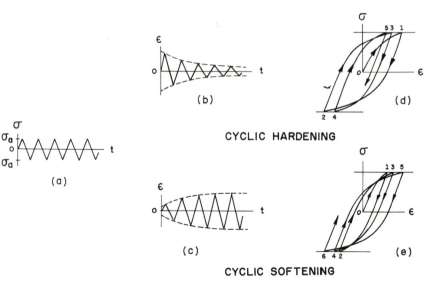

Figure 21.5 Cyclic behavior of a material under stress control.

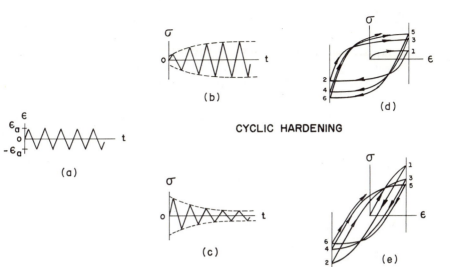

Figure 21.6 Cyclic behavior of a material under strain control.

control conditions. $\Delta\epsilon_p$ gradually decreases with cycling. Figure 21.5(e) shows cyclic softening under stress control, wherein one sees an increasing plastic strain with cycling.

21.1.2.2 Strain-controlled cycling. In this case the controlling function, strain, has a constant amplitude ϵ_a, and the stress changes under exponential envelopes at the two extremes [Fig. 21.6(a), (b), and (c)]. The cyclic hardening in this case implies that resistance of the material to deformation increases with cycling and that we need ever higher stresses to strain the material to the imposed constant strain limits. The cyclic softening implied that the material deforms easily, and the stress necessary to deform the material to the imposed strain limits decreases with continued cycling.

Figure 21.6(d) shows cyclic hardening, in hysteresis form, under strain control. One notes that the imposed limits in this case are vertical. The stress necessary to deform the material to a certain strain increases with cycling. Cyclic softening under strain control is shown in Fig. 21.6(e). In this case, the resistance of material against deformation decreases gradually, the stress necessary to impose the set limits of deformation decreases with cycling, and the plastic strain per cycle increases.

21.1.3 Cyclic Stress–Strain Curves

A given material shows cyclic hardening or softening depending, generally, on its initial condition and in some cases on the magnitude of imposed cyclic loads. For example, annealed oxygen-free high-conductivity (OFHC) copper cyclically hardens, whereas the same copper, cold worked, shows cyclic softening.

Another important fact is that the stress adjustments occur early in the fatigue life; that is, the stress amplitude is more or less constant during most of the fatigue life. In other words, the hysteresis loops become stabilized after the initial cycling (in general, fewer than 100 cycles). A stabilized hysteresis loop represents an equilibrium condition for the material under the imposed limits. The stress–strain curves of such a stabilized material are important in characterizing the cyclic behavior of the material. We can obtain such curves by connecting the tips of stable hysteresis loops for a series of "companion" samples (of the same material and with the same initial conditions) cycled to different strain amplitudes (Fig. 21.7). Landgraf et al.[2] have suggested other methods of obtaining the cyclic stress–strain curves. An important one of these is the incremental step test, in which a *single* specimen is subjected to blocks of gradually decreasing and then increasing strain amplitudes. A maximum strain amplitude of ±1.5 to 2.0% is enough to cyclically stabilize the material without necking or buckling of the specimen. The locus of the superimposed loop tips then determines the cyclic stress–strain curve. The cyclic stress–strain curve (or the cyclic flow curve, as it is sometimes called) consists of the saturation stress in the steady state versus $\Delta\epsilon_p/2$, where $\Delta\epsilon_p$ is the imposed plastic strain range. Figure 21.7 shows a comparison of monotonic (tensile) and cyclic curves.

The cyclic flow curves of a large number of materials follow, approximately, a power law:

$$\sigma_{\text{cyclic}} = \sigma_0 \left(\frac{\Delta\epsilon_p}{2}\right)^{n'}$$

where σ_0 is the cyclic strength coefficient and n' is the cyclic hardening exponent.

There are no quantitative rules for predicting when cyclic hardening or softening will occur under a given set of conditions. Manson and Hirschberg[3] have given a general rule of thumb, which predicts from the monotonic σ–ϵ curves which of the two will occur: cyclic hardening or softening. The rule states that if the ratio between the ultimate tensile strength and the 0.2% tensile strength (i.e., UTS/$\sigma_{0.2\%}$) is more than 1.4, cyclic hardening will occur. If (UTS/$\sigma_{0.2\%}$) < 1.2, cyclic softening will occur. For values of this ratio between

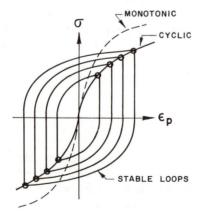

Figure 21.7 Monotonic and cyclic stress (σ)–strain (ϵ) curves.

1.2 and 1.4, the material behavior is uncertain. Feltner and Landgraf[4] tested this rule successfully for 35 different materials. Thus, we can say, in a general way, that initially hard and strong materials show cyclic softening, whereas initially soft materials show cyclic hardening. Microscopically, in an initially soft material dislocation density increases rapidly, thus producing hardening. At some point during cycling, the dislocations reach a stable configuration for that material and for that imposed cyclic strain. An originally work-hardened material has a high dislocation density to start with. On cycling, there occurs a rearrangement of dislocations into a configuration which offers small resistance to further strain, thus producing softening.

We know that the mobility of dislocations depends on the stacking-fault energy (SFE) of the material. The larger the SFE, the greater is the dislocation mobility and the material shows a predominantly wavy slip due to the greater ease of cross-slip. For example, initially work-hardened copper, a material that has a relatively high SFE, shows cyclic softening, whereas initially soft copper shows cyclic hardening.[5] In this case the cyclically stabilized state is unique, irrespective of the initial state [Fig. 21.8(a)]. In other words, for copper (high SFE, wavy slip) the material properties in the stabilized state are independent of the prior strain history.

On the other hand, a material of low SFE and planar slip, Cu–7.5% Al, which does not possess the ease of cross-slip, does not show this type of common stabilized cyclic state,[5] and a different cyclic stress–strain curve results for a different initial condition. In addition, a low-SFE material hardens or softens much more slowly. The material does cyclically harden or soften, but a unique stabilized state for two different starting conditions is never reached. The final cyclically stabilized state in such materials does depend on the prior strain history [Fig. 21.8(b)].

21.1.4 Fatigue Strength or Fatigue Life

Traditionally, fatigue life has been presented in the form of an S–N curve (Fig. 21.1). In this manner, fatigue strength refers to the capacity of a material to resist conditions of cyclic loading. But we just saw that in the presence of measurable plastic deformation, metals respond differently to strain cycling than to stress cycling. Thus, one would expect that the response of metals to fracture under cyclic conditions would also show a similar difference. In this section, we treat fatigue life in terms of strain versus number of cycles to failure N_f, or number of reversals to failure $2N_f$.

It is convenient to consider separately the elastic and the plastic components of strain. The elastic component can be readily described by means of a relation between the true stress amplitude and the number of reversals (i.e., twice the number of cycles),

$$\frac{\Delta \epsilon_e}{2} = \frac{\sigma_a}{E} = \left(\frac{\sigma_f'}{E}\right)(2N_f)^b \qquad (21.1)$$

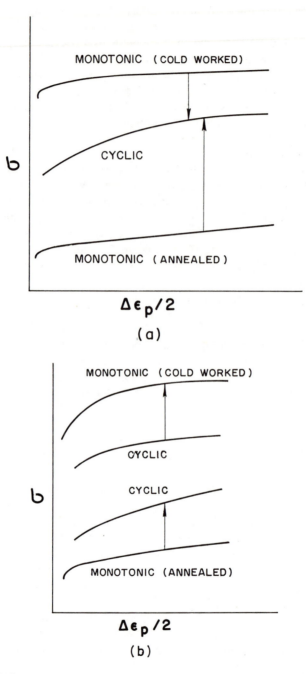

Figure 21.8 Cyclic response of (a) Cu–high SFE; (b) Cu–7.5% Al–low SFE. Note that the cyclic state is path independent for high-SFE Cu. (Adapted with permission from [5], p. 1624.)

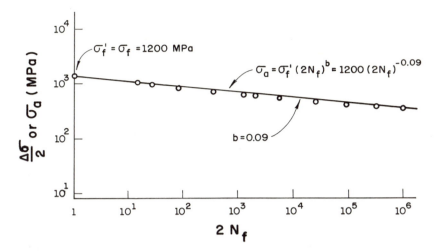

Figure 21.9 Stress amplitude ($\Delta\sigma/2$) versus number of reversals ($2N_f$) for AISI 4340 steel. (Adapted with permission from [6], p. 70.)

where $\Delta\epsilon_e/2$ is the elastic strain amplitude, σ_a the true stress amplitude, σ_f' the fatigue strength coefficient (equal to stress intercept at $2N_f = 1$), N_f the number of cycles to failure, and b the fatigue strength exponent.

This relation is an empirical representation of the S–N curve above the fatigue limit in Fig. 21.1. Figure 21.9 shows an application of this relation to SAE 4340 steel.[6] It was observed that fatigue life increased with decreasing b. Morrow,[6] based on energy considerations, showed that the fatigue strength exponent is given by

$$b = -\frac{n'}{1 + 5n'}$$

where n' is the cyclic hardening coefficient.

Thus, the fatigue life under elastic cyclic conditions (whether stress or strain-controlled) increases with a reduction in n'. Of course, the higher the material coefficient σ_f', the better it is for fatigue. There is evidence that σ_f' is approximately equal to σ_f, the monotonic fracture strength.

The plastic strain component is better described by the Manson–Coffin relation:[5,7]

$$\frac{\Delta\epsilon_p}{2} = \epsilon_f'(2N_f)^c \tag{21.2}$$

where $\Delta\epsilon_p/2$ is the plastic strain amplitude, ϵ_f' is the ductility coefficient in fatigue and is equal to strain intercept at $2N_f = 1$, $2N_f$ is the number of reversals to failure, and c is the ductility exponent in fatigue.

When plotted on log-log paper, the Manson–Coffin relation gives a straight line of slope c. Figure 21.10 shows the application of this relation to SAE 4340 steel. A smaller value of c, it has been observed, results in a longer fatigue

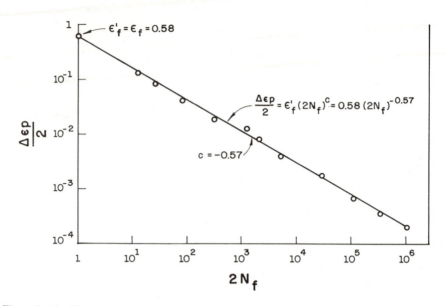

Figure 21.10 Plastic strain amplitude ($\Delta\epsilon_p/2$) versus number of reversals ($2N_f$) for a AISI 4340 steel. (Adapted with permission from [6], p. 70.)

life. In the regime of high-strain, low-cycle fatigue, the Manson–Coffin relation assumes great importance. Morrow[6] showed that in this case

$$c = -\frac{1}{1+5n'}$$

This relation indicates that fatigue life under plastic strain control conditions is superior in materials that have a large value of n'. The reader should note that according to the Manson–Coffin relation, the capacity for plastic deformation for a given value of ($2N_f$) also depends on ϵ'_f, the fatigue ductility coefficient. The estimates of ϵ'_f vary from $\epsilon'_f = 0.35\epsilon_f$ to $1.0\epsilon_f$, where ϵ_f is the true monotonic fracture strain.

Experimentally, it is frequently more convenient to control the total strain. In many structural components, the material in a critical place (say, at a notch root) may be subjected, essentially, to strain control conditions due to the elastic constraint of the surrounding material.

Manson and Hirschberg[3] have shown that for a material subjected to a total strain range of $\Delta\epsilon_t$ (elastic + plastic), we can determine the fatigue strength by a superposition of the elastic and plastic strain components.

In Morrow's notation, we have, from Eqs. 21.1 and 21.2,

$$\frac{\Delta\epsilon_t}{2} = \frac{\Delta\epsilon_e}{2} + \frac{\Delta\epsilon_p}{2} = \frac{\sigma'_f}{E}(2N_f)^b + \epsilon'_f(2N_f)^c \tag{21.3}$$

Thus, we expect that the fatigue life curve in terms of total strain will tend to the plastic curve at large total strain amplitudes, whereas it will tend to the elastic curve at low total strain amplitudes, as shown schematically in Fig.

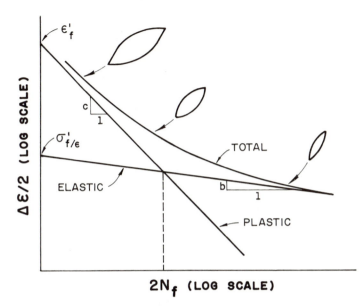

Figure 21.11 Superposition of elastic and plastic curves gives the fatigue life in terms of total strain. (Adapted with permission from [8], p. 24.)

21.11. An example from a real material (an 18% Ni maraging steel) is shown in Fig. 21.12.

21.1.5 Effect of Mean Stress on Fatigue Life

The mean stress σ_m can have an important effect on the fatigue strength of a material. A simple and crude way to demonstrate the effect of σ_m would be to present an S–N curve of a given material for various σ_m values, on the same graph. Such a graph would show that for a given stress level, fatigue life decreases with increasing mean stress.

A better way is to incorporate the effect of σ_m on S–N curves through some empirical mathematical relation. Sandor[1] has described the effect of σ_m in a very simple and elegant manner. We recognize that the limiting value of any combination of stresses is σ_f, the monotonic true fracture stress. We can think of other arbitrary limits, such as the ultimate tensile stress, σ_{UTS} or the yield stress, σ_y, but σ_f is the maximum allowable true stress. Figure 21.13 shows a schematic plot of alternating stress σ_a (or S) versus σ_m. One notes that for $\sigma_m = 0$, the alternating stress σ_a is a maximum and reaches σ_f. For $\sigma_a = \sigma_f$, the fatigue life is simply one-fourth of a cycle. For an ideal material, one would expect that the relation $\sigma_a + \sigma_m \leq \sigma_f$, the limiting value of any combination of stresses, to be always valid. Thus, one can expect a straight line to join points A and B in Fig. 21.13. Cyclic loading is not possible to the right of line AB. Then, in the presence of a mean stress of, say, $\sigma_f/3$, we shall have a maximum allowable stress equal to $\frac{2}{3}\sigma_f$. This type

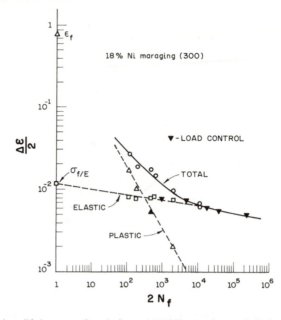

Figure 21.12 Fatigue life in terms of strain for an 18% Ni maraging steel. (Adapted with permission from [8], p. 25.)

of ideal behavior will be valid if the damages produced in each cycle by cyclic plastic strain were independent and noncumulative. Various empirical expressions have been proposed that take into account the effect of mean stress on fatigue life. Some of these are:

$$\text{Goodman's relation:} \quad \sigma_a = \sigma_0 \left(1 - \frac{\sigma_m}{\sigma_{\text{UTS}}} \right)$$

a linear effect of mean stress between $\sigma_m = 0$ and σ_{UTS};

$$\text{Gerber's relation:} \quad \sigma_a = \sigma_0 \left[1 - \left(\frac{\sigma_m}{\sigma_{\text{UTS}}} \right)^2 \right]$$

a parabolic effect of mean stress between $\sigma_m = 0$ and σ_{UTS}; and

$$\text{Soderberg's relation:} \quad \sigma_a = \sigma_0 \left(1 - \frac{\sigma_m}{\sigma_y} \right)$$

a linear effect of mean stress between $\sigma_m = 0$ and σ_y, where σ_m is the mean stress, σ_a is the fatigue strength in terms of stress amplitude when $\sigma_m \neq 0$, σ_0 is the fatigue strength in terms of stress amplitude when $\sigma_m = 0$, σ_{UTS} is the ultimate tensile stress, and σ_y is the tensile yield stress.

Figure 21.14 shows these three relations schematically. Experimentally, it has been observed that the great majority of data fall between Gerber and Goodman lines. Thus, the Goodman diagram represents a conservative estimate of the main stress effect.

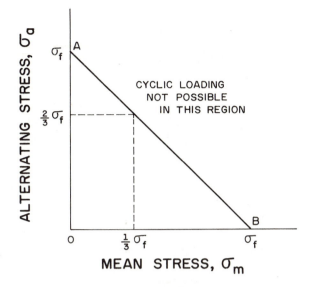

Figure 21.13 Effect of mean stress.

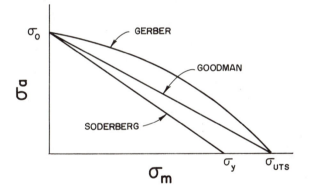

Figure 21.14 Gerber, Goodman, and Soderberg diagrams showing mean stress effect on fatigue.

21.1.6 Cumulative Damage and Life Exhaustion

The discussion in the preceding sections has been restricted to fatigue under simple condition of constant amplitude, constant frequency, and so on. In real life, the service conditions are rarely so simple. Many components and structures are subject to a range of fluctuating loads, mean stress levels, and variable frequencies. Thus, it is of great importance to be able to predict, starting from data obtained in simple constant-amplitude tests, the life of a component subjected to variable-amplitude conditions. The cumulative damage theories attempt to do this.

Basically, these theories consider the fatigue process to be a process of damage accumulation until a certain maximum tolerable damage. In other words, one can say that fatigue is an exhaustion process of a material's inherent life (or ductility).

Consider Fig. 21.15, showing schematically a fatigue life diagram. At a constant stress of σ_1, say, the life is of 150 cycles, while at σ_2 it is 300 cycles. According to the cumulative damage theory, in going from A to B or C to D, we gradually exhaust the fatigue life. That is, at points A and C, 100% of life at that level is available, while at points B and D, the respective lives are completely exhausted. If the fatigue damage does indeed accumulate in a linear manner, each cycle contributes the same amount of damage at a given stress level. For example, at σ_1, on cycling the material from A to E, we exhaust one-third of the fatigue life available at σ_1. If we now change the stress level to σ_2, then the percentage of life already exhausted at σ_1 is equivalent to the percentage of life exhausted at σ_2. That is, one-third of fatigue life at σ_2 is equivalent to one-third of fatigue life at σ_1. Thus, in descending from E to F, we get from 50 to 100 cycles, and as only one-third of fatigue life was exhausted at σ_1, two-thirds of fatigue life is still available at σ_2 (i.e., 200 cycles). The same kind of change can be described for a low to high stress traverse.

This model does not concern itself with the physical picture of the fatigue damage. It does, however, give an approximate empirical way of predicting the fatigue life after a complex loading sequence. The method is generally known as Palmgren–Miner rule or simply as linear cumulative damage theory.[9,10] The Palmgren–Miner rule says that the sum of all life fractions is unity:

$$\sum_{i=1}^{k} \frac{n_i}{N_i} = 1 \qquad \text{or} \qquad \frac{n_1}{N_1} + \frac{n_2}{N_2} + \frac{n_3}{N_3} + \cdots + \frac{n_k}{N_k} = 1$$

where k is the number of stress levels in the block spectrum loading; N_1, N_2, . . . , N_i are the fatigue lives corresponding to stress levels σ_1, σ_2, . . . ,

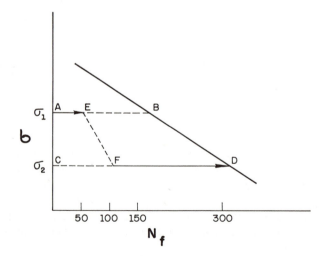

Figure 21.15 Damage accumulation in high-to-low loading sequence. (Adapted with permission from [1], p. 69.)

σ_i, respectively; and, n_1, n_2, . . . , n_i are the number of cycles carried out at specific levels. This rule is obeyed by a series of materials when the limiting assumptions are satisfied. The principal assumption is that the damage accumulation rate at any level must not depend on prior loading history; that is, the damage per cycle is the same in the beginning or in the end of fatigue life, at each level. This implies that the magnitude of amplitude change, as well as the direction of the change (low to high or high to low), do not have an effect on fatigue life. We also assume that in each block the loading is totally reversible (i.e., $\sigma_m = 0$). All these assumptions are potentially dangerous. For example, it is quite likely that for blocks identical in size and amplitude, a change in load from high to low would be much more dangerous than one from low to high. Cracks initiated at high loads can continue to grow at low loads, whereas in the reverse case, at low loads, the cracks would not have perhaps initiated.

21.1.7 Crack Nucleation in Fatigue

Fatigue cracks nucleate at singularities or discontinuities in metals. Such discontinuities may be on the surface or near the surface. The singularities can be structural (such as inclusions or second-phase particles) or geometrical (such as scratches). These singularities may be present from the beginning or may develop during cyclic deformation, as, for example, the formation of intrusions and extrusions at the persistent slip bands (PSBs). The persistent slip bands were first observed by Thompson et al. in Cu and Ni.[11] These bands appeared after cyclic deformation and "persisted" even after electropolishing. On retesting, slip bands appeared again in these places. The dislocation structure in the PSBs has been investigated extensively. Lukáš and Klesnil[12] showed that stacking fault energy and the attendant ease of cross-slip plays an important role in the development of the dislocation structure in the PSBs. Kuhlmann-Wilsdorf and Laird[13-15] have given an extensive discussion of models for forming the PSBs. The explanation of preferential nucleation of fatigue cracks at surfaces resides perhaps in the fact that plastic deformation is easier there and that slip steps form on the surface. Such steps alone can be responsible for crack initiation, or they can interact with existing structural or geometric defects to produce cracks.

Kuhlmann-Wilsdorf and Laird[13] compared the deformation substructures of unidirectional and cyclic (fatigue) deformation and interpreted them in terms of the differences between the two modes of deformation. The principal differences are:

1. Due to the much larger time spans of deformation, the dislocation structures formed in fatigue are much closer to the configurations having minimum energy than the ones generated by monotonic straining (more stable dislocation arrays for fatigue).
2. The oft-repeated to-and-fro motions in fatigue minimize the buildup of local Burgers vectors surpluses, which are fairly prevalent after unidirectional (monotonic) strain.

3. Much higher local dislocation densities are found in fatigued specimens.

Laird[15] depicts a scenario for the formation of the dislocation arrangements in FCC metals that will be briefly reviewed here. There is some difference in the development of the substructure between mono- and polycrystals. In monocrystals, we first have uniform fine slip, followed by the formation of veins; these veins are dense bundles of dipoles and other debris. After this, persistent slip bands (PSBs) are formed. The PSBs occur with the onset of saturation and are often associated with slight work softening. There also seems to be a threshold strain for PSB formation. It seems to be equal to 8×10^{-5} for copper monocrystals. Figure 21.16 shows two parallel PSBs (diagonally

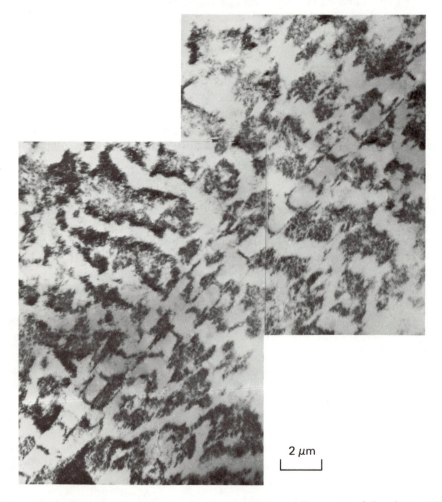

$2\,\mu m$

Figure 21.16 Persistent slip bands in vein structure. Polycrystalline copper fatigued at a total strain amplitude of 6.4×10^{-4} for 3×10^5 cycles. Fatiguing carried out in reverse bending at room temperature and at a frequency of 17 Hz. The thin foil was taken 73 μm below the surface. (Courtesy of J. R. Weertman and H. Shirai, Northwestern University.)

across micrograph) embedded in a veined structure in polycrystalline copper. The PSBs are clearly distinguished and consist of a series of parallel "hedges" (a ladder). These ladders are channels through which the dislocations move and produce intrusions and extrusions at the surface.

The crack nucleation, in fatigue, at slip bands has been observed in Fe, Cu, Ni, Al–4.5%, low-carbon steels, and other alloys. The model for this form of nucleation is shown in Fig. 21.17. During the loading part of the cycle, slip occurs on a favorably oriented plane, and during the unloading part of the cycle, reverse slip occurs on a parallel plane, as the slip on the original plane is inhibited due to hardening or perhaps due to the oxidation of the newly created free surface. The first cyclic slip may create an extrusion or an intrusion at the surface. An intrusion may grow and form a crack by continued plastic deformation during subsequent cycles. Even during cyclic stressing in the tension–tension mode, this mechanism can function as the plastic strain occurring at the peak load may lead to residual compressive stresses during the decreasing load part of the cycle.

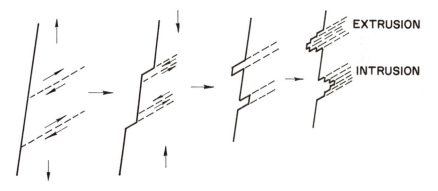

Figure 21.17 Fatigue crack nucleation at slip bands.

Twin boundaries are important crack nucleation sites in hexagonal close-packed materials such as titanium and its alloys. Inclusions and second-phase particles are the dominant nucleation sites in commercial-purity materials, for example, aluminum and high-strength steels.

Grain boundaries become important nucleation sites at large strain amplitudes and at temperatures greater than about $0.5 T_m$, where T_m is the melting point in Kelvin, or in the presence of impurities that produce grain boundary embrittlement (e.g., O_2 in iron). These mechanisms are illustrated schematically in Fig. 21.18.

Ignoring the details of the mechanisms of fatigue crack nucleation, one can suggest some general guidelines for delaying the initiation of cracking in fatigue:[17]

1. The slip (i.e., the plastic deformation) should be homogenized to prevent localized plastic strain accumulation. One way of realizing this is to use

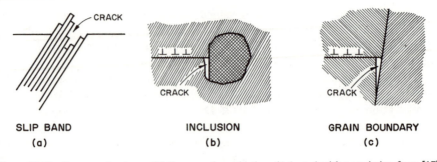

SLIP BAND
(a)

INCLUSION
(b)

GRAIN BOUNDARY
(c)

Figure 21.18 Some mechanisms of fatigue crack nucleation. (Adapted with permission from [17], p. 15.)

materials that exhibit planar slip instead of materials that show wavy slip. Examples of materials showing these two types of slip are:

   Planar slip: α-brass, Mg, Ti, Ni-base superalloys, stainless steel
   Wavy slip: Cu, Al, Fe, Ni, Ag, low-C steels

2. Increase the surface yield stress (i.e., prevent the initiation of plastic deformation there). The traditional way of doing this is the technique of "shot peening," which introduces compressive residual stresses. Another method, more complex, is to introduce an alloy in the surface layer in order to increase the yield stress there. An example of this is carburized steel.

3. Reduce the inclusion content. In this way we can eliminate these crack nucleation sites. There is a good deal of evidence that this will result not only in suppression of the fatigue crack initiation, but also in an inhibition of crack growth and improvement in fracture toughness.

4. Choose materials that resist cyclic softening.

The fraction of fatigue life spent in crack nucleation, N_i/N_f, increases with decreasing load amplitude (i.e., at high N's). Thus, we would expect that the guidelines and treatments suggested above will produce a great effect in large fatigue life regimes (i.e., under conditions where initiation of fatigue crack is more important than its propagation). Figure 21.19 shows this phenomenon in the case of pure aluminum and aluminum with a copper surface layer.

21.1.8 Fatigue Crack Propagation[19,20]

At large amplitudes, a very large fraction ($\sim$90%) of fatigue life is involved in the growth or propagation of crack. For a piece that contains a notch, this fraction becomes even larger. For real structures, with the inevitable presence of defects, crack propagation is generally the most important aspect of fatigue, irrespective of the load amplitude.

A brief description of the crack propagation process follows. Cracks start in a crystallographic shear mode (stage I), penetrate a few tenths of a millimeter,

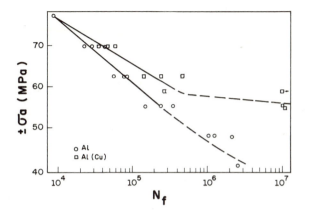

Figure 21.19 Stress amplitude σ_a versus number of cycles to failure N_f for Al and Al with a Cu layer. Note the pronounced improvement in the latter at large N_f. (Adapted with permission from [17], p. 17.)

and from there on propagate in a direction normal to the stress axis in the tensile mode (stage II, Fig. 21.20). The ratio of the extent of stage I to stage II decreases with an increase in tensile stress amplitude. The stress concentration at the crack tip causes local plastic deformation in a zone in front of the crack. With crack growth, this plastic zone increases in size until it becomes comparable to the specimen thickness. When this occurs, the plane-strain conditions at the crack front in stage II do not exist any more, the crack plane undergoes a rotation, and the final part of rupture occurs in plane stress (shear mode), as indicated in Fig. 21.21. Microscopic observations of fatigue fracture surfaces in stage II for various materials show that crack propagation frequently occurs in each load cycle. During stage I (cracking along the crystallographic slip

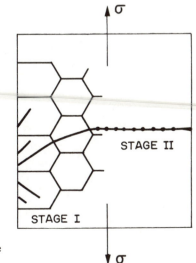

Figure 21.20 Stages I and II of fatigue crack propagation (schematic).

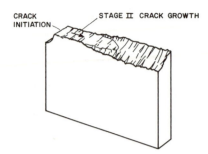

CRACK INITIATION

STAGE II CRACK GROWTH

Figure 21.21 Fatigue fracture surface (schematic). Near the crack initiation site, the crack propagates under plane strain conditions (90° to the stress axis). Later, the crack propagates under plane-stress conditions (45° to the stress axis). (Adapted with permission from [19], p. 132.)

planes) the crack growth is of the order of a few nanometers per cycle and little is known about crack propagation in this stage. Many consider this stage to be an extension of the nucleation process. Once a crack is initiated at a slip band, it continues along the slip band until it encounters a grain boundary. In stage II, propagation occurs in a direction perpendicular to the tensile stress, and in a large number of metals and alloys (principally of Al and Cu) at high amplitudes, the fracture surface shows the characteristic striations indicated in Fig. 21.22. Quite frequently, each striation is thought to represent one load cycle. Indeed, it has been observed by means of programmed amplitude fatigue tests that in many materials these striations do represent the crack front position in each cycle. However, the reader is warned that this is not invariably true. If this is true, one should be able to relate striation spacing to ΔK (see Section 21.1.7) and obtain a one-to-one correspondence with the macroscopic growth rate and ΔK relationship. This is not always true, indicating that the crack front may have advanced by a combination of striation formation and other

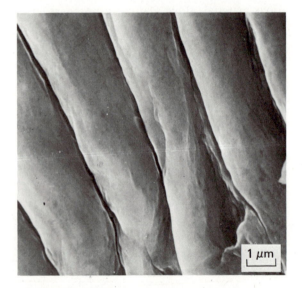

1 μm

Figure 21.22 Striations on fatigue fracture surface in AISI 304 stainless steel. (Reprinted with permission from [21], p. 2148.)

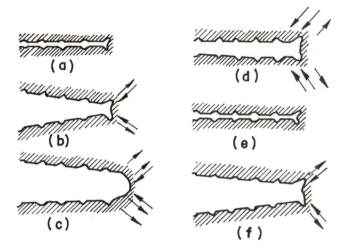

Figure 21.23 Fatigue crack growth by plastic blunting mechanism: (a) zero load; (b) small tensile load; (c) maximum tensile load; (d) small compressive load; (e) maximum compressive load; (f) small tensile load. The loading axis is vertical. (Adapted with permission from [19], p. 137.)

fracture mechanisms. At higher ΔK values, striations become less important in the overall crack propagation rate. According to Laird,[19] fatigue crack growth by striation mechanism occurs by repetitive blunting and sharpening of the crack front, as shown in Fig. 21.23. During the tensile part of the load cycle, plastic strains at the crack tip cause localized slip on planes of maximum shear. The reversal of loading direction forces the crack faces to join; however, the new surface created during tension is not completely rehealed, due to slip in the reverse direction. Depending on the material and the environment, a large part of slip during compression occurs on new slip planes and the crack tip assumes a bent form with "ears." At the end of the compression half of cycle, the crack tip is resharpened and the propagation sequence of the next cycle is restarted. There is evidence that the crack propagates in a similar manner in stage I, but with only a group of slip planes at 45° operating. However, one must bear in mind that although the presence of striations confirms a fatigue failure mechanism, an absence of striations does not necessarily preclude fatigue.

21.1.9 Linear Elastic Fracture Mechanics Applied to Fatigue Analysis

The principles of fail-safe and safe life form the basis of a large part of modern design philosophy in the aerospace industry. In recent times, as pointed out in the introduction, road-vehicle and nuclear industries which use monolithic structures have also shown great interest in the fracture mechanics discipline, in general, and crack propagation in fatigue, in particular, for the same reasons of safety and effective material utilization. Again, the basic assumption here is that the cracks preexist in a structural component and that they will grow in service. In terms of fatigue crack growth studies, it is also implied that the fatigue life of a component is determined by the crack growth rate under cyclic loading.

We can determine the K_{Ic} or K_c of a given material in the laboratory and can use this to obtain a failure locus in terms of a critical applied stress and a corresponding critical crack length, or vice versa (Fig. 21.24). For example, in Fig. 21.24 we can observe that for a given crack length a_1, there is a critical failure stress σ_1 of the material. Conversely, for a given design stress σ_2, there is a critical crack length a_2. In principle, then, the region to the left of the failure locus represents the safe region with respect to a catastrophic failure. Consider, for example, a component containing a crack length of a_1 at a stress of σ_2, where $\sigma_2 < \sigma_1$. The component under these conditions will be safe because a_1 is smaller than the critical defect size a_2 which corresponds to the applied stress σ_2. This security is based on the assumption that loading is static and that the crack does not grow in service. But we know very well that cracks in structures grow during service. An increase in crack length, in service, from a_1 to a_2 will eventually lead to structural failure. The fracture toughness establishes the failure condition and the residual strength of a structural component. But the component's service life or durability is mainly a function of its resistance to subcritical crack growth (i.e., its resistance to crack growth by fatigue, creep, stress corrosion, etc.).

21.1.9.1 Linear elastic fracture mechanics (LEFM) applied to fatigue. LEFM (Chapter 3) accepts the preexistence of cracks in a structural member. This being so, we focus our attention on the propagation of these

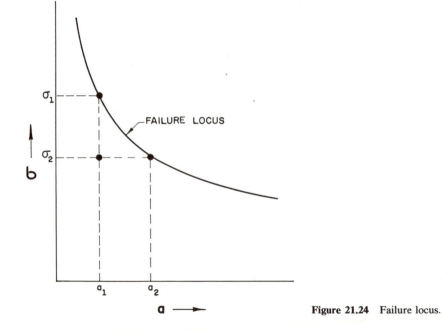

Figure 21.24 Failure locus.

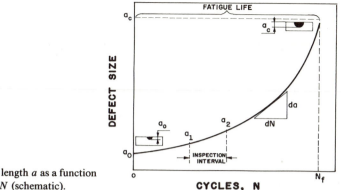

Figure 21.25 Crack length a as a function of number of cycles N (schematic).

cracks under conditions of fatigue. Once again, and it is worth repeating, we do not concern ourselves here with the crack nucleation problem under fatigue. The cracks grow from an initial size a_0 to a critical size a_c corresponding to failure as a function of the number of load cycles (Fig. 21.25). The basic problem is thus reduced to one of characterizing the crack growth kinetics in terms of an appropriate driving force. From there on one can estimate the service life and/or inspection intervals under designed loading conditions and service environments. As the crack growth starts from the most highly stressed region at the crack tip, we characterize the driving force in terms of the stress intensity factors at the crack tip; that is, the range of the stress intensity factor $\Delta K = K_{max} - K_{min}$.

The crack growth rate per cycle, da/dN, can be expressed as a function of the stress intensity factor at the crack tip. Hence, if a mathematical equation describing the crack growth process and the appropriate boundary conditions are available, we can, in principle, compute the fatigue life (i.e., number of cycles to failure). The model for the crack tip is the same as that described for nonfatigue regimes (see Chapter 3). The material containing a crack, under tension, has a small plastic zone at the crack tip, and this plastic zone is surrounded by a rather large elastic region.

Many variables affect the crack growth rate in fatigue. We can write for the crack growth rate,

$$\frac{da}{dN} \approx \frac{\Delta a}{\Delta N} = F(\Delta K, K_{max}, R, \text{frequency, temperature, } \ldots) \quad (21.4)$$

Clearly, one cannot obtain this ideal characterization. In practice, one obtains data under restrictive conditions but consistent with the applications in service. In principle, the rate equation 21.4 can be integrated to determine the service life N_f, or an appropriate inspection interval ΔN, for a structural component. Thus,

$$N_f = \int_{a_0}^{a_f} \frac{da}{F(\Delta K, \ldots)} \quad (21.5)$$

or

$$\Delta N = N_2 - N_1 = \int_{a_1}^{a_2} \frac{da}{F(\Delta K, \ldots)} \tag{21.6}$$

Rewriting, we have

$$N_f = \int_{K_{i\,\text{max}}}^{K_{f\text{max}}} \frac{dK}{(dK/da)F(\Delta K, \ldots)} \tag{21.7}$$

or

$$\Delta N = N_2 - N_1 = \int_{K_1}^{K_2} \frac{dK}{(dK/da)F(\Delta K, \ldots)} \tag{21.8}$$

The relation describing the propagation of a fatigue crack in metals is shown schematically in Fig. 21.26. The logarithm of crack growth rate

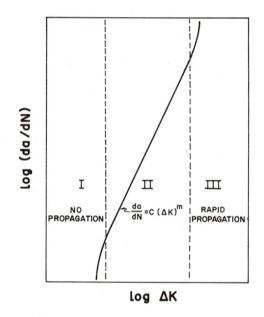

Figure 21.26 Crack propagation rate (da/dN) versus alternating stress intensity factor (ΔK) (schematic).

(da/dN) is plotted against the logarithm of alternating stress intensity factor ($\Delta K = K_{\text{max}} - K_{\text{min}}$) at the crack tip. The curve has a sigmoidal form with a power law that connects the upper and lower limiting regions. The power law (region II in Fig. 21.26), known as the Paris equation,[22] has the form

$$\frac{da}{dN} = C(\Delta K)^m \tag{21.9}$$

where parameters C and m are numerical constants that represent material constants. There is another empirical relation connecting the parameters C and m,[23]

$$C = \frac{A}{(\Delta K_0)^m}$$

where A and K_0 are some other material constants. The lower limit or threshold (region I, Fig. 21.26) indicates the fatigue conditions under which the crack does not propagate. The upper limit (region III, Fig. 21.26) indicates the conditions of accelerated crack growth rate associated with the start of final rupture. For a great majority of situations in engineering, the conditions of interest are described by the power law regime. The threshold conditions for the fatigue crack growth indicate that for steel, ΔK_{th}, the lower threshold alternating stress intensity factor, is between 6 and 9 MPa $\sqrt{m}$, for titanium between 2 and 7 MPa $\sqrt{m}$, and for aluminum between 1 and 3 MPa $\sqrt{m}$. Consider a steel with a $\Delta K_{th} = 8$ MPa $\sqrt{m}$. This value of ΔK_{th} represents a small semicircular crack (about 1.3 mm deep and 12.5 mm long) at a cyclic stress level of about 120 MPa. This level of stress is very small compared to the yield stress of high-strength alloys. That is, we can consider, for a large majority of cases, that the cracks that matter in fatigue are the ones that propagate, and under these conditions the situation is described fairly well by the power law (Eq. 21.9). The upper limit of the usefulness of this relation, called the Paris relation, varies and depends on the material and its thickness. However, a value of 80% of K_{Ic} would give a conservative estimate of the starting point of accelerated crack growth in fatigue. Above this point, the crack tip plasticity limits a generalized application of fracture mechanics data and one is well advised to obtain specific information for analysis and description of region III of the crack growth curve.

Ritchie[24] has summarized the primary mechanisms and the important variables in these three stages of fatigue crack propagation in metals:

Stage I: Crack propagation mechanisms characteristic of a noncontinuum medium. Large influence of microstructure, the stress ratio R, and environment.

Stage II: Crack propagation mechanisms characteristic of a continuum medium. Small influence of microstructure, R, environment, thickness, and so on.

Stage III: Crack propagation mechanisms similar to those in static mode (cleavage, intergranular, microcavities). Large influence of microstructure, R, and thickness, but small influence of environment.

The Paris power law (Eq. 21.9) which describes the crack propagation rate in stage II for a series of metals, is very useful because of its extreme simplicity. For example, it has been observed experimentally that data points in the form of log (da/dN) versus log ΔK for a given material (constant metallurgical structure) from three different samples—edge crack in a compact tension

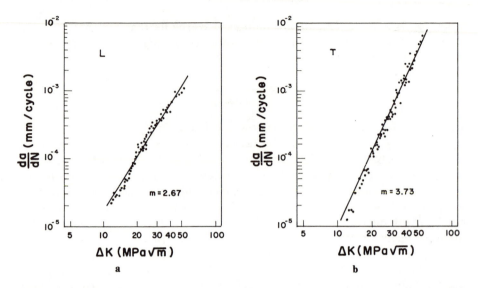

Figure 21.27 Fatigue crack propagation in an AISI 4140 steel: (a) longitudinal direction (parallel to rolling direction); (b) transverse direction (perpendicular to rolling direction). (Reprinted with permission from [25], p. 350.)

sample, through-thickness central crack in a plate, and plate containing a partially through thickness crack—all fall on the same line. Also, there is experimental evidence that shows that the stress level by itself does not influence the fatigue crack growth rate for stress levels below the general yielding. Thus, we can consider that the parameter ΔK describes uniquely the crack growth rates for many engineering applications. However, the structure of material can influence fatigue crack growth rates drastically; the value of m can change a lot. Figure 21.27 illustrates the directionality in the fatigue crack propagation rate in an AISI 4140 steel.[25] The exponent m has a much higher value in the transverse direction than in the longitudinal (rolling) direction, due to the presence of elongated inclusions.

In the literature there already exists a compilation of da/dN versus ΔK data for various structural materials. This information is also starting to appear in manuals and handbooks (see "Suggested Reading").

21.1.9.2 Effect of *R* ratio on crack propagation.

The crack propagation relations of the type $da/dN = C(\Delta K)^m$ apply primarily to constant-amplitude loading with $R = 0$ ($R = \sigma_{min}/\sigma_{max} = K_{min}/K_{max}$). Experimental evidence is conflicting on the R-ratio effect. In the case of $R > 0$ in a majority of cases, an increase in K_{max} results in an increase in da/dN (Fig. 21.28). In the literature, there exist various empirical relations that attempt to take this effect into account. For example, a relation due to Erdogan[26] gives

$$\frac{da}{dN} = C(\Delta K)^m (K_{max})^n$$

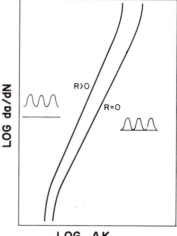

Figure 21.28 Effect of stress ratio R on fatigue crack propagation rate (schematic).

while Forman et al.[27] have proposed

$$\frac{da}{dN} = \frac{C(\Delta K)^m}{(1 - R)K_c - \Delta K}$$

Figure 21.29 shows that the data scatter is very much less when the rate da/dN is normalized according to the equation due to Forman et al. For $R < 0$, the validity of these equations has not been shown.

21.1.9.3 Overloads in tension and crack retardation.
Application of tensile overloads such as that shown in Fig. 21.30 results in a decrease in da/dN and consequently in a longer fatigue life than that predicted on the basis of the sum of the growth rate in each cycle using the crack growth rates under constant amplitude.

This delay in crack growth rate after an overload is related to crack closure phenomenon, originally pointed out by Elber.[28] He indicated that cyclic plastic deformation due to the tensile overload spectrum results in residual compressive stresses in the vicinity of the crack tip and that there occurs a crack closure for some positive load value on subsequent cycling. Accordingly, the crack growth kinetics is controlled by $\Delta K_{\text{effective}} = \Delta K_{\text{applied}} - \Delta K_{\text{residual}}$. Thus, the crack growth rate should be related to this $\Delta K_{\text{effective}}$

$$da/dN = C'(\Delta K_{\text{effective}})^m$$

or

$$da/dN = C'(U\Delta K)^m$$

where U, called the effective stress range factor, equals $(\sigma_{\text{max}} - \sigma_{\text{open}})/(\sigma_{\text{max}} - \sigma_{\text{min}})$. σ_{max}, σ_{min}, and σ_{open} are, respectively, the maximum, minimum, and crack opening stresses. The magnitude of U depends on R ratio, the ratio of stress range and yield stress ($\Delta\sigma/\sigma_y$), and the cyclic strain hardening exponent

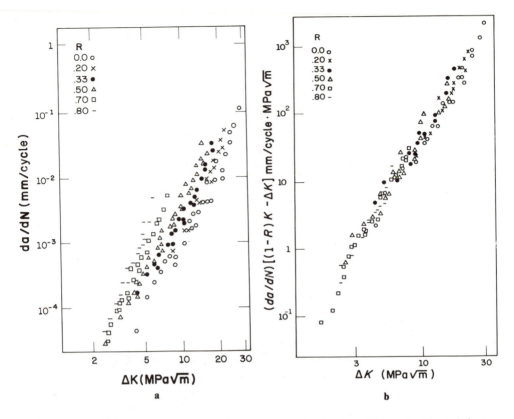

Figure 21.29 Fatigue crack propagation in Al 7075-T6 showing the effect of R ratio and the applicability of the Forman, Keraney, and Engle relation. The scatter in the data is much less in the latter. (Adapted with permission from [27], p. 969.)

n. The reader is warned not to confuse this beneficial effect of a tensile overload on crack propagation with the adverse effect of such an overload on crack initiation as discussed in Sec. 21.1.6.

In this type of complex loading, the important points are the load peaks. These peak loads control the events that follow. This is due to the fact that the residual stresses at crack tips at peak loads affect da/dN through local variations in R during a period of small decay after the application of the peak load. A tensile overload followed by low tensile loads (hi-lo sequence) leaves a residual compressive stress at the crack tip. This results in a decrease in the crack propagation rate during successive applications of low loads (Fig. 21.30). A compressive overload can produce the reverse effect.

21.1.9.4 Effects of aggressive environment on fatigue.
Aggressive environments, including humid environments and high temperatures, can have deleterious effect on fatigue crack growth (Fig. 21.31). In fact, all materials show accelerated propagation rate due to environmental action. A quantitative

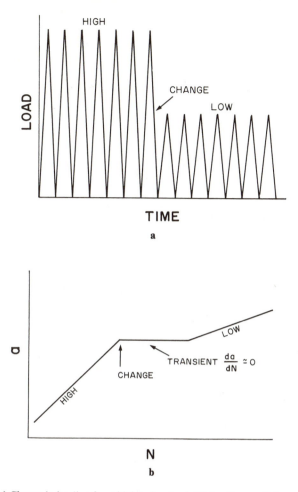

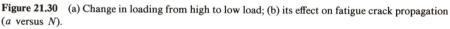

Figure 21.30 (a) Change in loading from high to low load; (b) its effect on fatigue crack propagation (a versus N).

analysis of fatigue–environment interaction is extremely difficult because fatigue is a cycle dependent phenomenon, whereas environmental attack is a time-dependent phenomenon.

In a great majority of structural fatigue cases, the frequency or the cyclic loading rate in large structures is small. Generally, therefore, data of crack growth kinetics obtained at frequencies of a few cycles per minute or less are applicable in structural design. Some systematic studies[29,30] of time-dependent effects on fatigue crack propagation in a corrosive environment show that the problem is more serious at low frequencies. Studies of the effect of high temperature on crack growth show that the problem becomes more serious with increasing temperature.[31,32]

The environmental effect, as mentioned, has its major effect at low frequencies and at low ΔK's (Fig. 21.31). This is the reason that environment has a

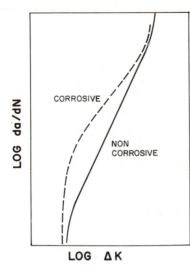

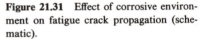

Figure 21.31 Effect of corrosive environment on fatigue crack propagation (schematic).

powerful effect on the finite fatigue life of a structure. The presence of an aggressive environment eliminates any chances of long, slow crack growth stages and, instead, transforms small cracks rather quickly into very large cracks. Thus, the structure generally ends up not having that tranquil period in which slow crack growth occurs. A reduction of 50% in fatigue life due to environmental effects is quite common, and reductions of up to 90% are not rare.

Another important point is that the crack growth data obtained in the presence of aggressive environments do not smoothly follow a power law function; instead, they show inflection points. Figure 21.32 shows the effect of water vapor partial pressure on fatigue of Al 7075-T6.[33] For two stress-intensity levels, a large increase in the crack propagation rate was observed for a rather small change in the amount of humidity—an order-of-magnitude increase in crack growth rate resulted due to a change from a dry to a wet environment.

Various material–environment systems have been studied under corrosion fatigue conditions. Important parameters are frequency, load profile, stress ratio R, and temperature. It has been verified that crack growth rate increases when more time (i.e., low frequency) is permitted for environmental attack during fatigue. It is worth mentioning that in an inert atmosphere, no effect of frequency is generally observed. It has also been found that the environment effect on crack growth rate is negligible at high rates (or high ΔK's), where the mechanical process of fatigue probably occurs much more quickly than chemical attack.

21.1.9.5 Linear elastic fracture mechanics applied to stress corrosion cracking (SCC).

The delayed fracture of a component subjected to aggressive environments can occur under a static stress well below its yield stress. This type of failure occurs due to the well-known phenomenon of stress corrosion cracking. Traditionally, the susceptibility of a material to stress corrosion cracking in a given environment was measured by the time to failure of smooth or not very sharply notched samples subjected to different stress levels

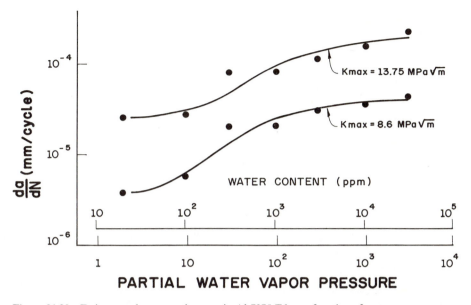

Figure 21.32 Fatigue crack propagation rate in Al 7075-T6 as a function of water vapor content. (Adapted with permission from [33], p. 637.)

in that environment. However, this method combines, as does the traditional *S–N* curve method of fatigue, the time required for crack initiation and the time required for propagation of the crack to a critical size. The need to separate stress corrosion cracking into the initiation and propagation stages has been verified experimentally. Application of the LEFM to stress corrosion cracking is obvious, as the crack growth is aided by the environment and the corrosion attack under stress would be expected to be very strong at the crack tip. Thus, we would expect that the stress intensity factor, K_I, will be the parameter characterizing the driving force for SCC. There is experimental evidence for this.

The use of stress intensity factor K_I in stress corrosion studies is based on the same hypotheses and is subject to the same limitations as those of fracture toughness studies. The primary requisite to be satisfied is the existence of a state of plane strain at the crack tip. The crack growth under SCC is due to interactions between the chemical and the mechanical processes occurring at the crack tip. The highest value of the stress intensity factor in plane strain at which the subcritical crack growth does not occur in a material loaded statically in an aggressive environment is called K_{ISCC} (Fig. 21.33). One uses precracked samples under different loads (i.e., with different initial K_I's) and one determines the limiting K_I (i.e., K_{ISCC}) below which, supposedly, failure due to SCC does not occur. The level of K_{ISCC} with respect to K_{Ic} of material gives a measure of its susceptibility to SCC (Fig. 21.33). We can also measure the rate of crack growth, da/dt, as a function of K_I, the mechanical driving force, under controlled conditions. This condition is more difficult and requires more sophisticated instrumentation. However, this method gives information more useful for quantita-

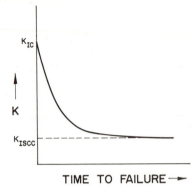

Figure 21.33 $K_{I\text{SCC}}$ is the limiting K_I below which there is no failure due to SCC (schematic).

TIME TO FAILURE →

tive design and life prediction. The reader should consult [34] and [35] for more details about this subject.

21.1.10 Creep–Fatigue Interaction

Many service components (e.g., aircraft gas turbines, nuclear pressure vessels, heat exchangers and fuel elements, steam turbines and power plant components) are subjected to conditions where both fatigue and creep processes are involved. It is easy to visualize the complexity of the problem. Some of the parameters, generally thought to be unimportant in ambient temperature fatigue, assume a very important role in the presence of creep conditions. Examples are: frequency, loading waveform, and atmospheric attack of the surface. If a mean stress is present durin̩ cyclic loading, the specimen may creep appreciably. In such a case creep fracture mechanisms may play an important part in both initiation and propagation of fracture.

There exists some controversy as to the mechanisms of high temperature–time dependent fatigue. According to Coffin,[36] it is the time-dependent localized oxidation, interacting with crack nucleation and growth process in the crack-tip proximity, that is responsible for decreasing fatigue life as test frequency decreases. Coffin[37] found practically no frequency effects in tests conducted on several alloys in high vacuum (i.e., under conditions that preclude localized oxidation). Solomon,[38] however, found that the plastic crack propagation rate in vacuum did depend on frequency, but at low frequencies where creep deformation would be important. This leads to the other viewpoint—that it is the creep damage, not the environment, that leads to a reduction in high-temperature fatigue life. According to Wareing,[39,40] these points of view are not entirely incompatible. Environment would be quite important in case the crack growth is occurring at or near the surface. In such a case, one would not expect the environment to affect the creep damage in the form of cavitation in the interior of the material. Thus, environment would affect the results of tests conducted with no "hold time" or rather short hold times, whereas tests conducted with long "dwell periods" (i.e., large creep components) would be unaffected by environment changes.

Coffin[41] has reviewed the role of environment and waveform in high-temperature and low-cycle fatigue. Multiple crack initiation processes occur due to the interactions between chemical and mechanical factors. In vacuum, the low-cycle fatigue life increases, particularly at low frequencies. Environmental damage is predominant when the hysteresis loop is balanced, whereas creep damage is a waveform effect that results from unbalanced loop cycling.

Under conditions involving fatigue–creep interactions, cracks can grow into a cavitated, but unfailed material, with a resultant acceleration of final failure. Such "creep–fatigue" failures do not involve a new failure process but an interaction of cracks with cavitated material. It is easy to imagine the great difficulty involved in formulating design codes to cope with these interactive failures. The traditional design philosophy, based on the attainment of a failure stage (be it ductility, endurance, or time to rupture) is all right as long as one failure process operates. Then, we can apply a Palmgren–Miner type of "damage criterion" to sum fractions leading to the failure state. When a linear summation like that of Palmgren–Miner's (Section 21.1.6) is applied, path independence is assumed. Interactive failures such as creep–fatigue failures are, however, path dependent. Robinson[42] proposed a linear summation of creep and fatigue life fractions to give failure when the damage sum is unity, a simple extension of Miner's linear damage rule. Thus,

$$\sum_1^n \frac{N_i}{N_{fi}} + \sum_1^n \frac{t_i}{t_{ri}} = 1$$

where N_{fi} and t_{ri} are the fatigue and creep failure states, respectively, at strain ranges and stress levels $\Delta\epsilon_i$ and σ_i, respectively. Tomkins and Wareing[43] outline some improved design approaches.

‡21.2 FATIGUE TESTING

Fatigue testing is one of the most important material and structural characterization techniques. Among the reasons for fatigue testing we may include the need to develop a better understanding (fundamental or empirical) of the fatigue behavior of a given material or to obtain more practical information on the fatigue response of a component or a structure. The fatigue test samples may thus range from tiny samples tested within, say, the specimen chamber of a scanning electron microscope to complete aircraft wings weighing many tons. It would be futile to try to include everything known about fatigue testing within the scope of this book. Instead, we present below some of the common techniques currently employed in fatigue testing and point out some of their salient points, with particular emphasis on the modern fatigue crack propagation techniques.

‡21.2.1 Conventional Fatigue Tests

Conventionally fatigue testing has been done by cycling a given material through ranges of stress amplitude and recording the number of cycles of failure. The results are reported in the form of *S–N* curves (Fig. 21.1).

There are two main types of loading: rotating bending tests and direct stress tests (Fig. 21.34). In direct stress machines the stress distribution over any cross section of the specimen is uniform, and we can easily apply a static mean tensile or compressive load (i.e., the *R* ratio can be varied). However, the more common and popular type has been the rotating bending beam test, described in the next section. The direct loading machines are included in the discussion of nonconventional fatigue testing (Section 21.2.2).

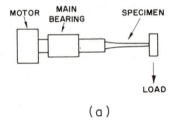

(a)

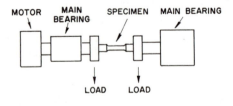

(b)

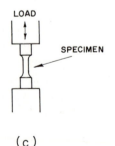

(c)

Figure 21.34 Various loading configurations used in fatigue testing: (a) cantilever loading, bending moment increases toward the fixed end; (b) two-point beam loading, bending moment constant; (c) pulsating tension or tension–compression axial loading.

‡21.2.1.1 Rotating bending machine. This is perhaps one of the simplest and oldest fatigue tests. It provides a simple method of determining fatigue properties at zero mean load by applying known bending moments to rotating round specimens. Commercially, a number of versions are available, the main difference being in the application of load: at a single point, as in a cantilever loading machine, or by some kind of two- or four-point loading (Fig. 21.34). In the latter case, the bending moment is constant over the entire test section of the specimen, and thus we use a specimen of constant diameter. In the cantilever type of loading machine, the specimen is either waisted (hourglass type) so that the maximum bending stress occurs at the smallest diameter, or the specimen has a tapered cross section such that the maximum bending stresses are constant at all cross sections. The stress at a point on the surface of a rotating bending specimen varies sinusoidally between numerically equal maximum tensile and compressive values in every cycle. Assuming the specimen to be elastic, we have

$$\pm S = \frac{32M}{\pi d^3}$$

where S is the maximum surface stress, M the bending moment at the cross section under consideration, and d the specimen diameter. In such a test we obtain the number of cycles to failure at a given stress level. The stress level S is continually reduced and the cycles to failure N_f increase, a logarithmic scale being used for N. Thus, we obtain an S–N curve. In the case of ferrous materials, we generally attain a fatigue limit or endurance limit S_L (Fig. 21.1). Cycling below this stress level S_L can be done indefinitely without resulting in material failure. Such an endurance limit is not encountered in nonferrous metals.

‡21.2.1.2 Statistical analysis of S–N curves. If a sufficiently large number of identical specimens are fatigue tested at the same stress amplitude, it has been observed that a Gaussian or normal distribution describes the logarithm of the fatigue life distribution. Figure 21.35 shows a schematic S–N diagram with log-normal distribution of lives at various stress levels. There is a greater spread in the lives of a group of specimens tested at a stress level greater than their fatigue limit than in the stress levels necessary to cause failure at a given life.

The data from cyclic loading tests (whether rotating bending beam, pulsat-

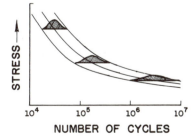

Figure 21.35 *S–N* curve showing log-normal distribution of lives at various stress levels.

TABLE 21.1 q-values for S–N Data Assuming a Normal Distribution[a]

$P(\%)$	C = 0.50					C = 0.75				
n	75	90	95	99	99.9	75	90	95	99	99.9
4	0.739	1.419	1.830	2.601	3.464	1.256	2.134	2.680	3.726	4.910
6	0.712	1.360	1.750	2.483	3.304	1.087	1.860	2.336	3.243	4.273
8	0.701	1.337	1.719	2.436	3.239	1.010	1.740	2.190	3.042	4.008
10	0.694	1.324	1.702	2.411	3.205	0.964	1.671	2.103	2.927	3.858
12	0.691	1.316	1.691	2.395	3.183	0.933	1.624	2.048	2.851	3.760
15	0.688	1.308	1.680	2.379	3.163	0.899	1.577	1.991	2.776	3.661
18	0.685	1.303	1.674	2.370	3.150	0.876	1.544	1.951	2.723	3.595
20	0.684	1.301	1.671	2.366	3.143	0.865	1.528	1.933	2.697	3.561
25	0.682	1.297	1.666	2.357	3.132	0.842	1.496	1.895	2.647	3.497

$P(\%)$	C = 0.90					C = 0.95				
n	75	90	95	99	99.9	75	90	95	99	99.9
4	1.972	3.187	3.957	5.437	7.128	2.619	4.163	5.145	7.042	9.215
6	1.540	2.494	3.091	4.242	5.556	1.895	3.006	3.707	5.062	6.612
8	1.360	2.219	2.755	3.783	4.955	1.617	2.582	3.188	4.353	5.686
10	1.257	2.065	2.568	3.532	4.629	1.465	2.355	2.911	3.981	5.203
12	1.188	1.966	2.448	3.371	4.420	1.366	2.210	2.736	3.747	4.900
15	1.119	1.866	2.329	3.212	4.215	1.268	2.068	2.566	3.520	4.607
18	1.071	1.800	2.249	3.106	4.078	1.200	1.974	2.453	3.370	4.415
20	1.046	1.765	2.208	3.052	4.009	1.167	1.926	2.396	3.295	4.319
25	0.999	1.702	2.132	2.952	3.882	1.103	1.838	2.292	3.158	4.143

[a] Reprinted with permission from [44], p. 67.

ing tension, or axial tension–compression) must be analyzed statistically. The mean value $\bar{x}$ and the standard deviation σ for a given set of data are given by

$$\bar{x} = \frac{\Sigma\, x}{n}$$

and

$$\sigma = \left[\frac{\Sigma\, (x - \bar{x})^2}{n - 1} \right]^{1/2}$$

where x is the cyclic life at a given stress (test value) and n is the number of test values (i.e., the number of samples tested to failure at a given stress). With these statistical parameters one can obtain the confidence limits for the survival probability. The anticipated fatigue life with a desired level of confidence ($C\%$) that at least $P\%$ of the samples will not fail may be written as

$$\text{anticipated life } (C, P) = \bar{x} - q\sigma \tag{21.10}$$

where q is a function of $C\%$, $P\%$, and the number of test samples used to determine $\bar{x}$ and s. The selection of a particular confidence limit ($C\%$) depends on the importance of the component for structural integrity. The more important the component, the higher should be the confidence limit, and the lower the stress. The q-values assuming a distribution are available in tabulated form in the literature. Table 21.1 presents the q-values assuming a normal distribution. With the Eq. 21.10 of anticipated life and the q tables, we can develop a family of curves showing the survival probability or failure of a component (Fig. 21.36).

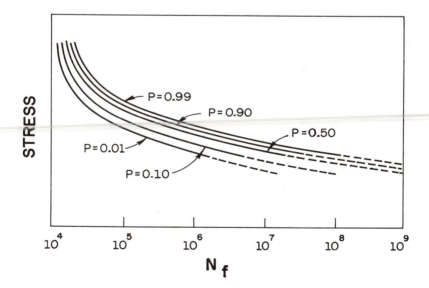

Figure 21.36 Family of curves showing the survival probability or failure of a component.

‡21.2.2 Nonconventional Fatigue Testing

In this category we include practically all modern fatigue testing other than that involving *S–N* curve determination. The machines used are direct loading machines. The drive system of the load train receives a time-dependent signal from the controls, converts it into a force or displacement—time excitation, and transfers this excitation to the fatigue specimen. The three common control parameters are (1) force, (2) deflection or displacement, and (3) strain. For most constant-amplitude fatigue tests we simply program a simple harmonic motion into the drive system. Electronic function generators are quite common these days. They generate an electrical signal that varies with time in the way that the fatigue control parameter is desired to vary with time. A variety of signals can be programmed: for example, constant amplitude, constant frequency, and zero mean stress; constant amplitude, constant frequency, with a non-zero-mean stress level; random loading; and so on.

‡21.2.2.1 Servohydraulic machines.

Servohydraulically operated fatigue machines have shown a marked increase in recent years. Figure 21.37

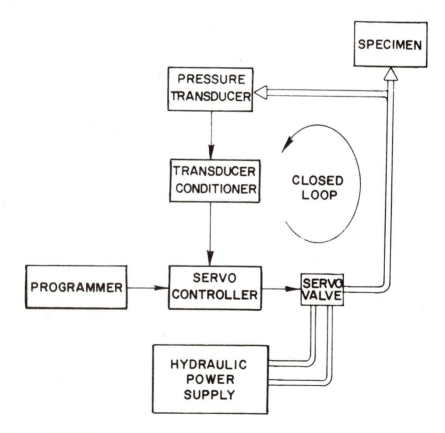

Figure 21.37 Line diagram of a hydraulically operated closed-loop system.

shows a line diagram of a servohydraulically operated closed-loop system. The load applied through a hydraulic actuator is measured by a load cell in series with the specimen. The amplified signal from the load cell is compared in a differential amplifier with the desired signal obtained from say, a function generator. Thus, this system forms a closed-loop load control system. We can also have a displacement or strain control from a transducer or a strain gage on the specimen instead of the load cell. The actual value measured by a load cell, displacement transducer, or strain gage is compared continuously with the desired value, and continuously corrected by the high-response electromagnetic servo valve. The energy is provided by a hydraulic power supply.

The main advantage of such machines is that they afford a greater degree of flexibility. Larger specimen deflections are possible than are possible in electromechanical machines. Thus, we can test components involving large deflections as well as conventional stiff specimens. Another big advantage has to do with the versatility of the system in regard to the input signal that can be used. Virtually any analog signal from a function generator, magnetic tape, random-noise generator, or punched paper tape reader is acceptable. This enables us to use not only constant-amplitude waveforms or block program spectrum loading, but also random waveforms, such as those obtained from actual service conditions. This means that the materials or components can be subjected to more realistic fatigue testing. The main disadvantage of servohydraulic machines is, of course, that of much higher power consumption. Haas and Kreiskorte have compared servohydraulic and conventional fatigue machines.[45]

‡21.2.2.2 Low-cycle fatigue tests.

Under conditions of high nominal stresses (i.e., short lifetimes), the constant stress amplitude test gives only limited information. Large plastic strain components become involved. In such cases we are more interested in cyclic stress–strain curves obtained under strain control (see Section 21.1). Servohydraulic machines are generally used in a closed-loop mode. Figure 21.38 shows schematically a cyclic straining facility.[46] Axial tension–compression is generally employed. We measure stress as a function of the number of strain reversals. Stress and strain signals are generally fed to an X-Y recorder and the complete hysteresis loop is obtained. The frequencies used are generally low enough (1 to 10 cycles per minute) for plotting on an X-Y recorder. The area of the hysteresis loop is the plastic strain energy per cycle.

Cyclic Stress–Strain Curves. We can obtain cyclic stress–strain curves by linking the tips of a series of hysteresis loops obtained from equivalent specimens tested at different plastic strain amplitudes ($\Delta\epsilon_p$). There are methods of obtaining cyclic stress–strain curves from one single specimen. The hysteresis loop adjusts rather quickly following a sudden change in $\Delta\epsilon_p$. Thus, we can obtain a cyclic stress–strain curve from only one specimen tested at several strain amplitudes. This is called a multiple-step test. Another method is the incremental step test with one specimen. This method consists of gradually increasing the cyclic strain range until a cyclic strain of about ±1% is attained.

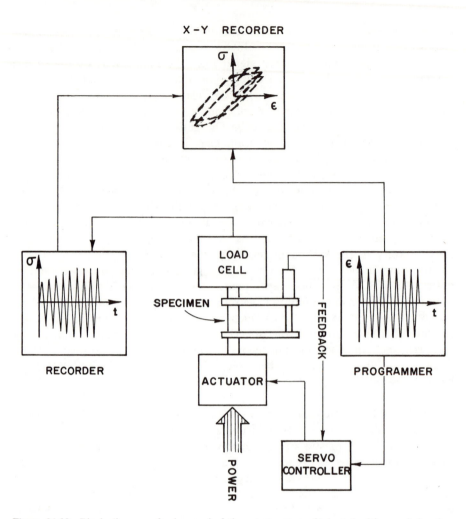

X-Y RECORDER

σ

ε

σ
t

RECORDER

LOAD
CELL

SPECIMEN

FEEDBACK

ACTUATOR

POWER

SERVO
CONTROLLER

ε
t

PROGRAMMER

Figure 21.38 Block diagram of a low-cycle fatigue test system. (Adapted with permission from [46].)

The strain range is then slowly reduced, and the procedure is repeated until the metal is stabilized.[44]

21.2.3 Fatigue Crack Propagation Testing

As pointed out in Section 21.1, we can consider the process of fatigue to be made up of the following stages:

1. A certain number of cycles N_i in which a small crack is initiated. Some

people include in this stage early growth of this microcrack to a somewhat larger crack.

2. Propagation of the crack. Generally, this occurs in such a way that we are able to describe the propagation behavior by some kind of standard rule, say, the Paris rule (Section 21.1.9.1). There is a substage of this propagation stage wherein the final rupture occurs. This occurs when the crack has reached a certain critical length for the material, the applied stress, and the test piece or structural component.

With the advent of fracture mechanics, a good deal of attention has been paid to the crack propagation behavior of materials in fatigue. Fatigue crack growth rates under service conditions can be of great importance, especially in determining inspection intervals. For example, wheels on large aircraft may have an ample safe lifetime after the appearance of detectable cracks. What we want to be sure of is that the detectable cracks will not grow to a size that is critical for the part during the time available before the next periodic inspection.

Flat sheet specimens are commonly chosen for crack propagation studies. The starter notch can be a side edge notch, a central through-the-thickness hole, or some other form appropriate to the form of defects obtained in service. These notches can be made by a mechanical saw, electrical discharge machining, and so on. Usually, crack growth measurements are made after a small initial propagation where there is an atomically sharp fatigue crack. We measure crack length as a function of number of cycles and subsequent analysis is made in terms of fracture mechanics concepts.

Synchronized strobe lighting can be used to illuminate the sample surface in order to provide a stable, vibration-free crack-length reading capability or, in more sophisticated cases, a movie record can be obtained of crack length increase.

Traveling or stereo zoom microscopes are used in manual monitoring of crack length. Such devices typically can read up to 0.01 mm. We may have scale markings photographically prepared on the sample or have a scale inserted in the microscope ocular piece. We can also use crack propagation gages, consisting of a series of 20 or 25 parallel, equally spaced wires in the form of a grid. Crack length is measured by monitoring the overall change in resistance.

In the electric potential drop method, a constant direct current is passed through the specimen containing a crack. The resistance of the specimen changes as the crack grows. This is detected by measuring the potential drop across the mouth of the starter notch. Figure 21.39 shows the setup for bend and compact tension specimen.[47]

As a crack is observed propagating, the number of cycles required for each increment is recorded, and a crack growth rate, da/dN, is computed from the curve showing crack length a versus the number of cycles N. The cyclic stress intensity factor at the crack tip (ΔK) can be computed from the crack

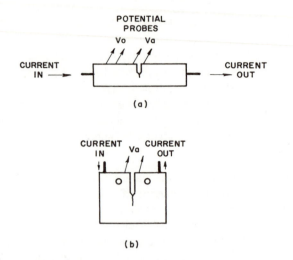

POTENTIAL
PROBES

Figure 21.39 Electric potential drop method for crack growth measurements: (a) bend specimen; (b) compact tension specimen. (Adapted with permission from [47], p. 467.)

length and load. By plotting da/dN versus ΔK, the fatigue crack growth characteristics of a material can be obtained.

EXERCISES

21.1. The fatigue properties of ordinary cast iron are about the same in a notched or unnotched state. Why?

21.2. A steel has the following properties:

Young's modulus $E = 210$ GPa
Monotonic fracture stress $\sigma_f = 2.0$ GPa
Monotonic strain at fracture $\epsilon_f = 0.6$
Exponent n' (cyclic) $= 0.15$

Compute the total strain that a bar of this steel will experience under cyclic straining before failing at 1500 cycles.

21.3. What is the effect of stacking fault energy on the fatigue life of FCC metals?

21.4. What is the effect of the size of a material on its fatigue properties?

21.5. Explain the following terms and describe where and under which conditions they will be of importance.
   (a) Thermal fatigue.
   (b) Sonic fatigue.

21.6. A microalloyed steel was subjected to two fatigue tests at ± 400 MPa and ± 250 MPa. Failure occurred after 2×10^4 and 1.2×10^6 cycles, respectively, at these two stress levels. Making appropriate assumptions, estimate the fatigue life at ± 300 MPa of a part made from this steel that has already suffered 2.5×10^4 cycles at ± 350 MPa.

21.7. Steel A514 has the following properties:

Yield stress $\sigma_y = 700$ MPa
Fracture toughness $K_{Ic} = 165$ MPa $\sqrt{m}$

A plate of this steel containing a single edge crack was tested in fatigue under $\Delta\sigma = 140$ MPa, $R = 0.5$, and $a_0 = 2$ mm. It was observed experimentally that fatigue crack propagation in this steel could be described by the following Paris-type relationship:

$$\frac{da}{dN} \text{ (m/cycle)} = 0.66 \times 10^{-8}(\Delta K)^{2.25}$$

where ΔK is measured in MPa $\sqrt{m}$.
(a) What is the critical crack size a_c at σ_{max}?
(b) Compute the fatigue life of this steel.

21.8. During a fatigue test of a steel with $\sigma_{min} = 138$ MPa, $\sigma_{max} = 992$ MPa, it was observed that the crack grew 0.16 mm in the last two cycles when ΔK was 82.5 MPa $\sqrt{m}$. Compute K_{Ic} or K_c for this steel.

21.9. L. P. Pook (in *Stress Analysis and Growth of Cracks*, ASTM STP 513, 1972, p. 106) showed how one can obtain the fatigue crack growth rate, indirectly, by a consideration of the total fatigue life of a precracked specimen. Assume that the fatigue life of the precracked specimen is totally occupied by crack propagation. If in a certain case, the initial crack growth rate is given by

$$\frac{da}{dN} = C(\Delta K)^2$$

and $\Delta K = 2\alpha\Delta\sigma\sqrt{\pi a}$, show that

(a)
$$C = \frac{1}{4N(\Delta\sigma)^2\alpha^2\pi} \ln\frac{a_f}{a_0}$$

where N is the fatigue life, a_0 the initial crack length, and a_f the final crack length.
(b) The initial crack growth rate can be expressed as

$$\frac{da}{dN} = \frac{a_0}{N} \ln\frac{a_f}{a_0}$$

21.10. It has been observed experimentally that in cold-worked brass under stress-corrosion conditions, the crack propagation is adequately described by

$$\frac{da}{dt} = AK^2$$

where A is a constant and the other symbols have their normal significance. Derive an expression for the time to failure of the material t_f, in terms of A, the applied stress σ, the initial crack length a_0, and the critical stress intensity corresponding to a_f (i.e., K_{Ic}).

21.11. Assuming that fatigue failures are initiated at the "weakest link," we may use the Weibull frequency distribution function to represent the fatigue lives of a group of specimens tested under identical conditions:

$$f(N) = \frac{b}{N_a - N_0}\left(\frac{N - N_0}{N_a - N_0}\right)^{b-1} \exp\left[-\left(\frac{N - N_0}{N_a - N_0}\right)^b\right]$$

where N is the specimen fatigue life, N_0 the minimum life ≥ 0, N_a the characteristic life at 36.8% survival of the population (36.8% = $1/e$, e = 2.718), and b is the shape parameter of the Weibull distribution curve. Letting $x = (N - N_0)/(N_a - N_0)$, plot frequency curves $f(N)$ versus x for $b = 1, 2,$ and 3.

21.12. Fatigue data are, generally, analyzed cumulatively to determine the survival percentage. The Weibull cumulative function for the fraction of population failing at N is an integration of the expression for $f(N)$ in Exercise 21.11. Show that it is

$$F(N) = 1 - \exp\left[-\left(\frac{N - N_0}{N_a - N_0}\right)^b\right]$$

Transform this function into a straight-line relationship by logarithm of the logarithm of the equation. Show how this relationship can be used on log-log paper for graphical fitting of the Weibull cumulative distribution and for graphically estimating the parameters b, N_0, and N_a [see, e.g., C. S. Yen, in *Metal Fatigue: Theory and Design*, A. F. Madayag (ed.), Wiley, New York, 1969, p. 140.]

21.13 (For the discriminating student) Calculate the life of a 15 cm-thick panel between 0 and $0.2\sigma_y$ ($\sigma_y = 345$ MPa). The photomicrograph of Fig. 21.22 shows the fatigue striations at 4cm of the initial flaw (that had a diameter of 200 μm) which gave rise to failure. The frequency of cycling is 10Hz and the plane strain fracture toughness is 44 MPa $\sqrt{m}$. The separation between fatigue striations at 4mm from the flaw is 100 times smaller than the ones of Fig. 21.22.

REFERENCES

[1] B.I. Sandor, *Fundamentals of Cyclic Stress and Strain*, University of Wisconsin Press, Madison, Wis., 1972.

[2] R.W. Landgraf, J.D. Morrow, and T. Endo, *J. Mater.*, *4* (1969) 176.

[3] S.S. Manson and M.H. Hirschberg, in *Fatigue: An Interdisciplinary Approach*, Syracuse University Press, Syracuse, N.Y., 1964, p. 133.

[4] C.E. Feltner and R.W. Landgraf, "Selecting Materials to Resist Low Cycle Fatigue," ASME Paper No. 69-DE-59, ASME, New York, 1969.

[5] C.E. Feltner and C. Laird, *Acta Met.*, *15* (1967) 1621.

[6] J.D. Morrow, in *Internal Friction, Damping and Cyclic Plasticity*, ASTM STP No. 378, ASTM, Philadelphia, 1965, p. 72.

[7] L.F. Coffin, *Trans. ASME*, *76* (1954) 931.

[8] R.W. Landgraf, ASTM STP 467, ASTM, Philadelphia, 1970, p. 3.

[9] A. Palmgren, *Z. Ver. Dtsch. Ing.*, *53* (1924) 339.

[10] M.A. Miner, *J. Appl. Mech.*, *12* (1945) 159.

[11] N. Thompson, N.J. Wadsworth, and N. Lovat, *Phil. Mag.*, *1* (1956) 113.

[12] P. Lukáš and M. Klesnil, *Phys. Status Solidi*, *37* (1970) 833.

[13] D. Kuhlmann-Wilsdorf and C. Laird, *Mater. Sci. Eng.*, *27* (1977) 137.

[14] D. Kuhlmann-Wilsdorf and C. Laird, *Mater. Sci. Eng.*, *37* (1979) 111 and 127.

[15] C. Laird, in *Work Hardening in Tension and Fatigue*, A.W. Thompson (ed.), TMS-AIME, New York, 1977, p. 150.

[16] W.A. Wood, ASTM STP 237, 1958, ASTM, Philadelphia, p. 110.

[17] J.C. Grosskreutz, Tech. Rep. AFML-TR-70–55, Air Force Materials Laboratory, Wright-Patterson AFB, Ohio, 1970.

[18] J.C. Grosskreutz and D.K. Benson, in *Surfaces and Interfaces II*, Syracuse University Press, Syracuse, N.Y., 1968, p. 61.

[19] C. Laird, in *Fatigue Crack Propagation*, ASTM STP 415, ASTM, Philadelphia, 1967, p. 131.

[20] J. Schijve, in *Fatigue Crack Propagation*, ASTM STP 415, 1967, ASTM, Philadelphia, p. 415.

[21] K.K. Chawla and P.K. Liaw, *J. Mater. Sci.*, *14* (1979) 2143.

[22] P.C. Paris and F. Erdogan, *J. Basic Eng.*, *Trans. ASME*, *85* (1963) 528.

[23] K. Tanaka, C. Masuda, and S. Nishijama, *Scripta Met.*, *15* (1981) 259.

[24] R.O. Ritchie, *Int. Metals Rev.*, *245* (1979) 205.

[25] E.G.T. De Simone, K.K. Chawla, and J.C. Miguez Suarez, *Proc. 4th CBECIMAT*, Florianópolis, Brasil, 1980, p. 345.

[26] F. Erdogan, *Crack Propagation Theories*, NASA-CR-901, 1967.

[27] R.G. Forman, V.E. Kearney, and R.M. Engle, *J. Basic Eng.*, *Trans. ASME*, *89* (1967) 459.

[28] W. Elber, *Eng. Fract. Mech.*, *2* (1970) 37; see also ASTM STP 486, 1971, p. 230, and ASTM STP 559, 1974, p. 45.

[29] J.A. Feeney, J.C. McMillan, and R.P. Wei, *Met. Trans.*, *1* (1970) 741.

[30] D.A. Meyn, *Met. Trans.*, *2* (1971) 853.

[31] W.G. Clark and H.E. Trout, *Eng. Fract. Mech.*, *2* (1970) 107.

[32] L.A. James and E.B. Schwenk, *Met. Trans.*, *2* (1971) 491.

[33] R.P. Wei, *Eng. Fract. Mech.*, *2* (1970) 633.

[34] R.P. Wei, S.R. Novak, and D.P. Williams, *Mater. Res. Stand.*, *12* (Sept. 1972) 25.

[35] R.P. Wei, in *Stress Corrosion*, H. Arup and R.N. Parkins (eds.), Sijthoff & Noordhoff, Alphen aan den Rijn, The Netherlands, 1979, p. 65.

[36] L.F. Coffin, *Met. Trans.*, *3* (1972) 1777.

[37] L.F. Coffin, in *Fatigue at Elevated Temperatures*, ASTM STP 520, ASTM, Philadelphia, 1973, p. 1.

[38] H.D. Solomon, in *Fatigue at Elevated Temperatures*, ASTM STP 520, ASTM, Philadelphia, 1973, p. 112.

[39] J. Wareing, *Met. Trans.*, *6A* (1975) 1367.

[40] J. Wareing, *Met. Trans.*, *8A* (1977) 711.

[41] L.F. Coffin, in Fracture 1977—*Advances in Research on the Strength and Fracture of Materials*, D.M.R. Taplin (ed.), Pergamon Press, New York, 1978, Vol. 1, p. 263.

[42] E.L. Robinson, *Trans. ASME*, *74* (1952) 777.

[43] B. Tomkins and J. Wareing, *Metal Sci.*, *11* (1977) 414.

[44] ASTM STP No. 91, 1963, p. 67.

[45] T. Haas and H. Kreiskorte, *Symp. Developments in Material Testing and Machine Design*, *Proc. Inst. Mech. Eng.*, *180(3A)* (1965–6) 155.

[46] J.D. Morrow, ASTM STP 378, ASTM, Philadelphia, 1965, p. 45.

[47] R.O. Ritchie, G.G. Garrett, and J.F. Knott, *Int. J. Fract. Mech.*, *7* (1971) 462.

SUGGESTED READING

Fatigue

Damage Tolerant Design Handbook: *A Compilation of Fracture and Crack Growth Data for High Strength Alloys*, Metals and Ceramics Information Center, Batelle Laboratory, Columbus, Ohio.

Fatigue Crack Growth Under Spectrum Loads, ASTM STP 595, ASTM, Philadelphia, 1976.

Fatigue Crack Propagation, ASTM STP 415, ASTM, Philadelphia, 1967.

Fatigue and Microstructure, ASM Symposium, St. Louis, 1979.

FINE, M.E., *Met. Trans.*, *11A* (1980) 365.

FROST, N.E., K.J. MARSH, and L.O. POOK, *Metal Fatigue*, Oxford University Press, London, 1974.

HERTZBERG, R.W., *Deformation and Fracture Mechanics of Engineering Materials*, Wiley, New York, 1976.

KLESNIL, M., and LUKÁŠ, P., *Fatigue of Metallic Materials*, Elsevier, New York, 1980.

LAIRD, C., in *Treatise on Materials Science and Technology*, Vol. 6, H. Herman (ed.) Academic Press, New York, 1975, p. 101.

MANSON, S.S., *Thermal Stress and Low-Cycle Fatigue*, McGraw-Hill, New York, 1966.

PELLOUX, R.M. and N.S. STOLOFF (eds.), *Creep–Fatigue–Environment Interactions*, TMS-AIME, Warrendale, Pa., 1980.

ROLFE, S.T., and J.M. BARSOM, *Fracture and Fatigue Control in Structures*, Prentice-Hall, Englewood Cliffs, N.J., 1977.

SANDOR, B.I., *Fundamentals of Cyclic Stress and Strain*, University of Wisconsin Press, Madison, Wis., 1972.

THOMPSON, A.W. (ed.), *Work Hardening in Tension and Fatigue*, TMS-AIME, New York, 1977.

Fatigue Testing

Cyclic Stress-Strain Behavior—Analysis, Experimentation and Fatigue Prediction, ASTM STP 519, ASTM, Philadelphia, 1973.

Flaw Growth and Fracture, ASTM STP 631, ASTM, Philadelphia, 1974.

A Guide for Fatigue Testing and the Statistical Analysis of Fatigue Data, ASTM STP 91, ASTM, Philadelphia, 1963.

JOHNSON, J.G., *The Statistical Treatment of Fatigue Experiments*, Elsevier, New York, 1964.

MADAYAG, A.F. (ed.), *Metal Fatigue: Theory and Design*, Wiley, New York, 1969.

Manual on Low Cycle Fatigue Testing, ASTM STP 465, ASTM, Philadelphia, 1970.

OSGOOD, C.C., *Fatigue Design*, Wiley, New York, 1970.

WEIBULL, W., *Fatigue Testing and Analysis of Results*, Pergamon Press, Oxford, 1961.

AUTHOR INDEX

Numbers in italics refer to page where complete reference is given.

SUBJECT INDEX

High-strength low alloy (HSLA) steels (*see* Steels)
Hot isostatic pressing (HIP), 95, 682
Hot working, 96, 97
Hydrodynamic lubrication, 114
Hydrogen embrittlement, 154
  hydride formation, 155
  methods of eliminating, 155

Impact:
  dynamic, 571
Impact testing (*see* Fracture testing)
Impact transition temperature (*see* Ductile-brittle transition temperature)
Inclusions, 147, 148
  effect on fatigue, 705, 706, 714
  MnS, 626
Incompatibility strains, 340, 341
Inconel 718, 84, 521–23
  stress-strain curve, 566, 567
Indentation (*see* Hardness testing)
Indentation tests (*see* Hardness testing)
Indicial notation, 18, 19
  dummy suffix, 18, 19
  Einstein's summation rule, 19
Instability (*see also* Necking and Tensile test):
  plastic, 583
Interaction:
  chemical, 392
  dislocation-precipitation, 405
  elastic, 387
  electrical, 392
  energy, 383–92
    dilatation misfit, 384
    due to modulus difference, 392
    elastic misfit, 384, 388, 389
    variation with lattice parameter, 192
  fatigue-creep, 720
  force, 192, 395, 396
    variation with lattice parameter, 192
  local order, 393
  modulus difference, 391
  solute atoms with dislocations, 386
  Suzuki, 392
Interfacial defects, 271 *passim* (*see also* Defects)
Intergranular fracture, 153 (*see also* Fracture)
Internal friction, 221
  vacancy-dislocation interaction, 221, 222
  Zener relaxation, 221
Interstices:
  in BCC structure, 214
    octahedral void, 215
    tetrahedral void, 215
  in FCC structure, 214
    octahedral void, 214
    tetrahedral void, 214
  in HCP structure, 214
    octahedral void, 215
    tetrahedral void, 215
Interstitial(s), 215 *passim*
  self-interstitials, 216, 217
    concentrations, 216, 217
Interstitial atom(s) (*see also* Solid solution):
  tetragonal strain field, 389
    interaction with dislocations, 390
Interstitial position(s), 387, 388
Inverse pole figures (*see* Texture)
Ionic crystals, 205, 206
  electrostatic forces, 206

Ion implantation, 534
  applications, 534–35
  modification of composition and structure, 534
  surface chemistry, 534
  technique, 534
    advantages, 534
Iron-3% silicon:
  yield stress-grain size relationship, 344
Irradiation:
  damage, 530
    Seeger model, 530
  electron, 528, 529
  gamma-rays, 528, 529
  neutron, 528, 529
  protons, 528, 529
  of solids (*see* Radiation damage)
Iron:
  cleavage planes, 150
Isotropic materials:
  stress-strain relations, 55
  Young's modulus, 52

J-integral, 172, 175, 656, 657 (*see also* **Fracture toughness**)
  blunting line, 657
  determination, 656, 657
  • line integral, 175
    path independence, 176
      deformation theory of plasticity, 176
    physical interpretation, 176
J-integral testing, 656 (*see also* Fracture testing)
J_{Ic}, 657
Jogs, 297, 298 (*see also* Dislocation)

Keeler-Goodwin diagrams, 632 (*see also* **Formability testing**)
K_{Ic}, 166, 167, 173
  relation with COD, 173
K_{Ic} test, 649 (*see also* Fracture testing)
Kinks, 297, 298 (*see also* Dislocation)
Knoop hardness number:
  measure of anisotropy, 617

Lame's constant, 55
Laplace equation, 306, 308
Larson-Miller equation, 582, 663–65 (*see also* Creep testing)
Larson-Miller parameter, 663 (*see also* Creep testing)
  for AISI 316 stainless steel, 664
Lath martensite (*see* Martensite)
Laue back reflection, 423
Lenticular martensite (*see* Martensite)
Linear elastic fracture mechanics, (LEFM), 133, 155
  (*see also* Fracture toughness)
Linear hardening, 320
Liquid metal embrittlement, 154
  reduction in specific surface energy, 154
Lorentz transformation, 307
Lüders band(s), 197, 396, 632
  propagation, 397

Machine stiffness (*see* **Tensile test**)
Macroplasticity, 72
Macroyielding, 505, 506
Magnesium:
  yield loci, 631
Magnetic bubble memories, 200
Manson-Coffin relation, 697 (*see also* Fatigue)
Manson-Haferd equation, 582